2017 47th European Solid-State Device Research Conference (ESSDERC 2017)

Leuven, Belgium
11-14 September 2017

IEEE Catalog Number: CFP17543-POD
ISBN: 978-1-5090-5979-9

**Copyright © 2017 by the Institute of Electrical and Electronics Engineers, Inc.
All Rights Reserved**

Copyright and Reprint Permissions: Abstracting is permitted with credit to the source. Libraries are permitted to photocopy beyond the limit of U.S. copyright law for private use of patrons those articles in this volume that carry a code at the bottom of the first page, provided the per-copy fee indicated in the code is paid through Copyright Clearance Center, 222 Rosewood Drive, Danvers, MA 01923.

For other copying, reprint or republication permission, write to IEEE Copyrights Manager, IEEE Service Center, 445 Hoes Lane, Piscataway, NJ 08854. All rights reserved.

*** This is a print representation of what appears in the IEEE Digital Library. Some format issues inherent in the e-media version may also appear in this print version.

IEEE Catalog Number:	CFP17543-POD
ISBN (Print-On-Demand):	978-1-5090-5979-9
ISBN (Online):	978-1-5090-5978-2
ISSN:	1930-8876

Additional Copies of This Publication Are Available From:

Curran Associates, Inc
57 Morehouse Lane
Red Hook, NY 12571 USA
Phone: (845) 758-0400
Fax: (845) 758-2633
E-mail: curran@proceedings.com
Web: www.proceedings.com

ESSDERC 2017 Table of Contents

A1L-A   **Joint Plenary 1: Françoise Chombar, Melexis - Engineering a Safe, Clean & Comfortable Future**

Date: Tuesday, September 12, 2017
Time: 09:00 - 09:40
Room: Vesalius
Chair(s):   Georges Gielen; *KU Leuven*

Engineering a Safe, Clean & Comfortable Future ....**No paper**
Françoise Chombar
Melexis, Belgium

A2L-A   **Joint Plenary 2: Peter Real, Analog Devices - Navigating without a Moore's Law Compass**

Date: Tuesday, September 12, 2017
Time: 09:40 - 10:20
Room: Vesalius
Chair(s):   Georges Gielen; *KU Leuven*

Navigating Without a Moore's Law Compass ....**No paper**
Peter Real
Analog Devices Inc, Ireland

B1L-A   **Joint Plenary 3: Hans Stork, ON Semiconductor - Smart Power for Automotives**
Date: Wednesday, September 13, 2017
Time: 08:45 - 09:25
Room: Vesalius
Chair(s):   Jo De Boeck; *imec*

Smart Power for Automotives ....**No paper**
Johannes Stork
ON Semiconductor, United States

C1L-A   **Joint Plenary 4: Tetsuo Endoh, Tohoku University – Spintronics Applications: STT-MRAM & Non Volatile Logic**
Date: Thursday, September 14, 2017
Time: 09:-00 - 09:40
Room: Vesalius
Chair(s):   Jo De Boeck; *imec*

Spintronics Applications: STT-MRAM and Non Volatile Logic ....**No paper**
Tetsuo Endoh
Tohoku University, Japan

A5L-F   **ESSDERC Keynote 1: Ian Young - Principles & Trends in Quantum Nano-Electronics & Nano-Magnetics for Beyond-CMOS Computing**

Date: Tuesday, September 12, 2017
Time: 15:50 - 16:30
Room: AP01.30
Chair(s):   Marc Heyns; *imec*

Principles and Trends in Quantum Nano-Electronics and Nano-Magnetics for Beyond-CMOS Computing 1
Ian A. Young, Dmitri E. Nikonov
Intel Corp., United States

B3L-F   **ESSDERC Keynote 2: Siva Sivaram - Storage Class Memories: Desire Meets Reality**

Date: Wednesday, September 13, 2017
Time: 13:30 - 14:10
Room: AP01.30
Chair(s):   Bogdan Govoreanu; *imec*

Storage Class Memories: Desire Meets Reality ....**No paper**
Siva Sivaram
Western Digital Corporation, United States

C3L-F   **ESSDERC Keynote 3: David DiVincenzo - Control Systems for Quantum Computers**

Date: Thursday, September 14, 2017
Time: 13:30 - 14:10
Room: AP01.30
Chair(s):   Guido Groeseneken; *imec*

Control Systems for Quantum Computers ....**No paper**
David DiVincenzo
RWT Aachen, Germany

A3L-F	Resistive RAM
Date:	Tuesday, September 12, 2017
Time:	11:00 - 12:20
Room:	AP01.30
Chair(s):	Thomas Mikolajick; *NaMLab gGmbH*
	Paolo Pavan; *Università di Modena e Reggio Emilia*

22% Higher Performance, 2x SCM Write Endurance Heterogeneous Storage with Dual Storage Class Memory and NAND Flash 6
Chihiro Matsui, Ken Takeuchi
Chuo University, Japan

Study of Error Repeatability and Recovery in 40nm TaOx ReRAM 10
Takashi Inose[1], Seiichi Aritome[1], Ryutaro Yasuhara[2], Satoshi Mishima[2], Ken Takeuchi[1]
[1]Department of Electrical, Electronic, and Communication Engineering, Chuo University, Tokyo, Japan, Japan; [2]Panasonic Semiconductor Solutions Co., Ltd., 1 Kotari-yakemachi, Nagaokakyo, Kyoto, Japan, Japan

Optimization of Writing Scheme on 1T1R RRAM to Achieve Both High Speed and Good Uniformity 14
Shan Wang, Huaqiang Wu, Bin Gao, Ning Deng, Dong Wu, He Qian
Tsinghua University, China

Analyzing Inference Robustness of RRAM Synaptic Array in Low-Precision Neural Network 18
Rui Liu[1], Heng-Yuan Lee[2], Shimeng Yu[1]
[1]Arizona State University, United States; [2]Electronics and Optoelectronics Research Laboratory, Industrial Technology Research Institute, Taiwan

A3L-G   Cross-Domain Compact Modelling

Date: Tuesday, September 12, 2017
Time: 11:00 - 12:20
Room: AV00.17
Chair(s):   Wladek Grabinski; 0
            Cristell Maneux; *Laboratoire de l'Intégration du Matériau au Système*

INVITED: SPICE Modeling in Verilog-A: Successes and Challenges 22
Colin McAndrew
NXP Semiconductors, United States

SPICE Modeling of Light Induced Current in Silicon with 'Generalized' Lumped Devices 26
Chiara Rossi[2], Pietro Buccella[2], Camillo Stefanucci[1], Jean-Michel Sallese[2]
[1]AMS, AG, Schloss Premstaetten, Austria; [2]École Polytechnique Fédérale de Lausanne (EPFL), Switzerland

Total Ionizing Dose Effects on Analog Performance of 28 nm Bulk MOSFETs 30
Chun-Min Zhang[3], Farzan Jazaeri[3], Alessandro Pezzotta[3], Claudio Bruschini[3], Gulio Borghello[2], Serena Mattiazzo[1], Andrea Baschirotto[4], Christian Enz[3]
[1]Department of Information Engineering, University of Padova, Italy; [2]EP-ESE Group, CERN and University of Udine, Switzerland; [3]Integrated Circuits Laboratory, École Polytechnique Fédérale de Lausanne, Switzerland; [4]Microelectronic Group, INFN Milano-Bicocca and University of Milano-Bicocca, Italy

A3L-H	Focus Session: Beyond CMOS Devices I
Date:	Tuesday, September 12, 2017
Time:	11:00 - 12:20
Room:	AV02.17
Chair(s):	Joachim Knoch; *Rheinisch-Westfälische Technische Hochschule Aachen*
	Thomas Zimmer; *Laboratoire de l'Intégration du Matériau au Système*

1/f Noise in 3D Vertical Gate-All-Around Junction-Less Silicon Nanowire Transistors 34

Chhandak Mukherjee[1], Julien Pezard[2], Guilhem Larrieu[2], Cristell Maneux[1]
[1]IMS Laboratory, University of Bordeaux, France; [2]LAAS-CNRS, Université de Toulouse CNRS, INP, Toulouse, France

Random Telegraph Signal Noise in Tunneling Field-Effect Transistors with S Below 60 mV/decade 38

Markus Hellenbrand, Elvedin Memišević, Johannes Svensson, Erik Lind, Lars-Erik Wernersson
Lund University, Sweden

Experimental Characterization of the Static Noise Margins of Strained Silicon Complementary Tunnel-FET SRAM 42

Gia Vinh Luong[1], Sebastiano Strangio[2], Andreas Tiedemann[1], Patrick Bernardy[1], Stefan Trellenkamp[1], Pierpaolo Palestri[2], Siegfried Mantl[1], Qing-Tai Zhao[1]
[1]Forschungszentrum Juelich, Germany; [2]University of Udine, Italy

A4L-F   Alternative Device Simulations

Date: Tuesday, September 12, 2017
Time: 14:00 - 15:20
Room: AP01.30
Chair(s):   Francois Triozon; *Commissariat à l'Energie Atomique et aux Energies Alternatives*
            Fabian Bufler; *Synopsys, Inc.*

INVITED: Advances in the Understanding of Microscopic Switching Mechanisms in ReRAM Devices 46

Benoît Sklénard[1], Philippe Blaise[1], Boubacar Traoré[2], Alberto Dragoni[1], Cécile Nail[1], Elisa Vianello[1]
[1]CEA LETI, France; [2]Institut des Sciences Chimiques de Rennes, France

Modeling the Effect of Surface Roughness on the Performance of Line Tunnel Fets 50

Saurabh Sant, Andreas Schenk
Integrated Systems Laboratory, ETH Zurich, Switzerland

Material Selection and Device Design Guidelines for Two-Dimensional Material Based TFETs 54

Tarun Agarwal[3], Bart Soree[3], Iuliana Radu[1], Praveen Raghavan[1], Gianluca Fiori[4], Marc Heyns[3], Wim Dehaene[2]
[1]imec, Belgium; [2]KU Leuven, Belgium; [3]KU Leuven/imec, Belgium; [4]Univ of Pisa, Italy

A4L-G	Parameter Extraction
Date:	Tuesday, September 12, 2017
Time:	14:00 - 15:20
Room:	AV00.17
Chair(s):	Thierry Poiroux; *Commissariat à l'Energie Atomique et aux Energies Alternatives*
	Marco Bellini; *ABB Group*

Nanometer CMOS Characterization and Compact Modeling at Deep-Cryogenic Temperatures 58

Rosario Marco Incandela[1], Lin Song[2], Harald Homulle[1], Fabio Sebastiano[1], Edoardo Charbon[1], Andrei Vladimirescu[1]

[1]*Delft University of Technology, Netherlands;* [2]*Tsinghua University, China*

Cryogenic Characterization of 28 nm Bulk CMOS Technology for Quantum Computing 62

Arnout Beckers[3], Farzan Jazaeri[3], Andrea Ruffino[1], Claudio Bruschini[2], Andrea Baschirotto[4], Christian Enz[3]

[1]*EPFL AQUA, Switzerland;* [2]*EPFL AQUA-ICLAB, Switzerland;* [3]*EPFL ICLAB, Switzerland;* [4]*INFN Milano Bicocca, Italy*

A New Method for Junctionless Transistors Parameters Extraction 66

Renan Trevisoli[2], Rodrigo Doria[2], Michelly de Souza[2], Sylvain Barraud[1], Marcelo Pavanello[2]

[1]*CEA, LETI, Minatec Campus, France;* [2]*Centro Universitário FEI, Brazil*

Avalanche Compact Model Featuring SiGe HBTs Characteristics Up to BVCBO 70

Mathieu Jaoul[5], Didier Céli[4], Cristell Maneux[3], Michael Schröter[2], Andreas Pawlak[1]

[1]*Dresden University of Technology, Germany;* [2]*Dresden University of Technology, University of California, Germany;* [3]*IMS Bordeaux, France;* [4]*STMicroelectronics, France;* [5]*STMicroelectronics, IMS, France*

A4L-H Focus Session: Beyond CMOS Devices II

Date: Tuesday, September 12, 2017
Time: 14:00 - 15:20
Room: AV02.17
Chair(s): Joachim Knoch; *Rheinisch-Westfälische Technische Hochschule Aachen*
Thomas Zimmer; *Laboratoire de l'Intégration du Matériau au Système*

Utilizing I-V Non-Linearity and Analog State Variations in ReRAM-Based Security Primitives 74
Gina Cristina Adam[1], Hussein Nili[3], Jeeson Kim[2], Brian Douglas Hoskins[3], Omid Kavehei[2], Dmitri Strukov[3]
[1]National Institute for R&D in Microtechnologies, Bucharest, Romania; [2]RMIT University, Australia; [3]University of California Santa Barbara, United States

Negative Capacitance Field Effect Transistors: Capacitance Matching and Non-Hysteretic Operation 78
Ali Saeidi, Farzan Jazaeri, Francesco Bellando, Igor Stolichnov, Christian Enz, Adrian Ionescu
Ecole Polytechnique Fédérale de Lausanne (EPFL), Switzerland

Buried Multi-Gate InAs-Nanowire FETs 82
Thomas Grap, Felix Riederer, Charu Gupta, Joachim Knoch
RWTH Aachen, Germany

A6L-G	Modelling of Emerging Devices
Date:	Tuesday, September 12, 2017
Time:	16:40 - 18:00
Room:	AV00.17
Chair(s):	Jean-Michel Sallese; *École polytechnique fédérale de Lausanne*
	Daniel Tomaszewski; *Institute of Electron Technology*

Equivalent Circuit Model for the Electron Transport in 2D Resistive Switching Material Systems 86

Enrique Miranda[2], Chengbin Pan[1], Marco Villena[1], Na Xiao[1], Jordi Suñe[2], Mario Lanza[1]
[1]Soochow University, China; [2]Universitat Autonoma de Barcelona, Spain

Analytical Drain Current Model for Non-Ballistic Schottky-Barrier CNTFETs 90

Igor Bejenari, Michael Schroter, Martin Claus
Technische Universitat Dresden, Germany

A General Circuit Model for Spintronic Devices Under Electric and Magnetic Fields 94

Meshal Alawein, Hossein Fariborzi
King Abdullah University of Science and Technology, Saudi Arabia

Compact Physical Model of a-IGZO TFTs for Circuit Simulation 98

Matteo Ghittorelli[3], Fabrizio Torricelli[3], Carmine Garripoli[2], Jan-Laurens van der Steen[1], Gerwin Gelinck[1], Sahel Abdinia[2], Eugenio Cantatore[2], Zsolt Kovàcs-Vajna[3]
[1]TNO, Netherlands; [2]Tu/e, Netherlands; [3]University of Brescia, Italy

A6L-H	2D Material Devices
Date:	Tuesday, September 12, 2017
Time:	16:40 - 18:00
Room:	AV02.17
Chair(s):	Cees de Groot; *University of Southampton*
	Gianluca Fiori; *Università di Pisa*

Complementary Black Phosphorous FETs by Workfunction Engineering of Pre-Patterned Au and Ag Embedded Electrodes.. 102
Nicolo Oliva, Emanuele Andrea Casu, Wolfgang Amadeus Vitale, Igor Stolichnov, Adrian Mihai Ionescu
EPFL, Switzerland

Tunneling Transistors Based on MoS2/MoTe2 Van der Waals Heterostructures.................. 106
Yashwanth Balaji, Quentin Smets, Cesar J Lockhart de la Rosa, Anh Khoa Augustin Lu, Daniele Chiappe, Tarun Agarwal, Dennis Lin, Cedric Huyghebaert, Iuliana Radu, Dan Mocuta, Guido Groeseneken
imec, Belgium

Temperature Dependence of Contact Resistance for Gold-Graphene Contacts................... 110
Amit Gahoi, Satender Kataria, Max Christian Lemme
University of Siegen, Germany

Radical Oxidation Process for Hybrid SAM/HfOx Gate Dielectrics in MoS2 FETs 114
Takamasa Kawanago, Ryo Ikoma, Tomoaki Oba, Hiroyuki Takagi
Tokyo Institute of Technology, Japan

B2L-F	**Widebandgap Power Devices**
Date:	Wednesday, September 13, 2017
Time:	10:10 - 11:50
Room:	AP01.30
Chair(s):	Mikael Ostling; *KTH Royal Institute of Technology*
	Susanna Reggiani; *Università di Bologna*

INVITED: CoolSiC(TM) and Major Trends in SiC Power Device Development 118
Roland Rupp
Infineon Technologies AG, Germany

Gated Base Structure for Improved Current Gain in SiC Bipolar Technology 122
Gunnar Malm, Hossein Elahipanah, Arash Salemi, Mikael östling
KTH, Sweden

On the Understanding of Cathode Related Trapping Effects in GaN-on-Si Schottky Diodes 126
William Vandendaele, Thomas Lorin, Romain Gwoziecki, Yannick Baines, Jérome Biscarrat, Marie-Anne Jaud, Charlotte Gillot, Matthew Charles, Marc Plissonnier, Gilles Reimbold
CEA-LETI, France

Temperature Dependent Substrate Trapping in AlGaN/GaN Power Devices and the Impact on Dynamic Ron 130
Arno Stockman[2], Michael Uren[4], Alaleh Tajalli[5], Matteo Meneghini[5], Benoit Bakeroot[1], Peter Moens[3]
[1]*CMST imec/Ghent University, Belgium;* [2]*Ghent University, Belgium;* [3]*ON Semiconductor, Belgium;* [4]*University of Bristol, United Kingdom;* [5]*University of Padova, Italy*

B2L-G Advanced CMOS Characterization and Reliability

Date: Wednesday, September 13, 2017
Time: 10:10 - 12:10
Room: AV00.17
Chair(s): Erik Bury; *imec*
Maryline Bawedin; *Minatec*

INVITED: Material and Device Innovation Impact on Reliability for Scaled CMOS Technologies 134
Tanya Nigam, Andreas Kerber, Tian Shen, Rakesh Ranjan, Linjun Cao
GLOBALFOUNDRIES, United States

Carrier Lifetime Evaluation in FD-SOI Layers 140
Kyung Hwa Lee, Maryline Bawedin, Hyungjin Park, Mukta Singh Parihar, Sorin Cristoloveanu
IMEP-LAHC, France

Precise EOT Regrowth Extraction Enabling Performance Analysis of Low Temperature Extension First Devices 144
Jessy Micout[2], Quentin Rafhay[3], Xavier Garros[1], Mikael Casse[1], Jean Coignus[1], Luca Pasini[1], Cao-Minh Vincent Lu[1], Nils Rambal[1], Claire Fenouillet-Beranger[1], Laurent Brunet[1], Giovanii Romana[4], Rémy Gassilloud[1], Perrine Batude[1], Maud Vinet[1], Gérard Ghibaudo[3]
[1]*CEA-LETI, France;* [2]*CEA-LETI, IMEP-LAHC, France;* [3]*IMEP-LAHC, France;* [4]*STMicroelectronics, France*

Back-Gate Bias Effect on UTBB-FDSOI Non-Linearity Performance 148
Babak Kazemi Esfeh[3], Valeria Kilchytska[3], Bertrand Parvais[1], N. Planes[2], M. Haond[2], Denis Flandre[3], Jean-Pierre Raskin[3]
[1]*imec, Belgium;* [2]*ST-Microelectronics, France;* [3]*UCL, Belgium*

Evolution of Oxygen Vacancies Under Electrical Characterization for HfOx-Based ReRAMs 152
Behnoush Attarimashalkoubeh, Jury Sandrini, Elmira Shahrabi, Yusuf Leblebici
Microelectronic Systems Laboratory (LSM),EPFL, Switzerland

B2L-H   **Emerging Memory Technologies**

Date: Wednesday, September 13, 2017
Time: 10:10 - 12:10
Room: AV02.17
Chair(s):   Jerome Dubois; *NXP*
           Dirk Wouters; *RWTH Aachen*

INVITED: Emerging Memory Technologies for High Density Applications............................156
Giorgio Servalli
Micron Technology, Italy

Anti-Ferroelectric ZrO2: An Enabler for Low Power Non-Volatile 1T-1C and 1T Random Access Memories....................160
Milan Pesic[2], Michael Hoffmann[2], Claudia Richter[2], Stefan Slesazeck[2], Thomas Kämpfe[1], Lukas Eng[1], Thomas Mikolajick[3], Uwe Schroeder[2]
[1]IAPP/TU Dresden, Germany; [2]NaMLab gGmbH, Germany; [3]NaMLab gGmbH/TU Dresden, Germany

From Planar to Vertical Capacitors : a Step Towards Ferroelectric V-FeFET Integration........................164
Karine Florent[2], Simone Lavizzari[1], Luca Di Piazza[1], Mihaela Popovici[1], Goedele Potoms[1], Tom Raymaekers[1], Guido Groeseneken[2], Jan Van Houdt[2]
[1]imec, Belgium; [2]imec/KU Leuven, Belgium

Doped GeSe Materials for Selector Applications.........................168
Naga Sruti Avasarala[3], Bogdan Govoreanu[2], Karl Opsomer[2], Wouter Devulder[2], Sergiu Clima[2], Christophe Detavernier[1], Marleen van der Veen[2], Jan Van Houdt[3], Marc Heyns[3], Ludovic Goux[2], Gouri Sankar Kar[2]
[1]Ghent University, Belgium; [2]imec, Belgium; [3]imec, KU Leuven, Belgium

Multilevel SOT-MRAM Cell with a Novel Sensing Scheme for High-Density Memory Applications........................172
Behzad Zeinali, Mahsa Esmaeili, Jens Kargaard Madsen, Farshad Moradi
Aarhus University, Denmark

B4L-F   **FinFET and Nanowire Simulations**

Date: Wednesday, September 13, 2017
Time: 14:20 - 15:40
Room: AP01.30
Chair(s):   Bernd Meinerzhagen; *TU Braunschweig*
            Mathieu Luisier; *ETHZ*

On the Ballistic Ratio in 14nm-Node FinFETs 176
Fabian Bufler, Kenichi Miyaguchi, Thomas Chiarella, Naoto Horiguchi, Anda Mocuta
imec, Belgium

Three-Dimensional Multi-Subband Simulation of Scaled FinFETs 180
Luca Donetti, Carlos Sampedro, Francisco García Ruiz, Andrés Godoy, Francisco
Gámiz
Universidad de Granada, Spain

Study of Strained Effects in Nanoscale GAA Nanowire FETs Using 3D Monte Carlo
Simulations 184
Muhammad Elmessary[1], Daniel Nagy[2], Manuel Aldegunde[3], Antonio Garcia-Loureiro[2],
Karol Kalna[1]
[1]*Swansea University, United Kingdom;* [2]*Universidade de Santiago de Compostela,
Spain;* [3]*University of Notre Dame, United States*

Investigation of Electrically Gate-All-Around Hexagonal Nanowire FET (HexFET)
for 5nm Node Logic and SRAM Applications 188
Jeffrey Smith[2], Kai Ni[2], Ram Krishna Ghosh[2], Jeff Xu[1], Mustafa Badaroglu[1], Pr Chidi
Chidambaram[1], Suman Datta[2]
[1]*Qualcomm, United States;* [2]*University of Notre Dame, United States*

B4L-G   Traps and Noise

Date: Wednesday, September 13, 2017
Time: 14:20 - 15:40
Room: AV00.17
Chair(s): Benjamin Iniguez; *Universitat Rovira i Virgili*
          Sadayuki Yoshitomi; *Toshiba*

Modeling of Dynamic Trap Density Increase for Aging Simulation of Any MOSFET Circuits..192
Mitiko Miura-Mattausch[3], Hidenori Miyamoto[3], Hideyuki Kikuchihara[3], Dondee Navarro[3], Tapas K. Maiti[3], Nezam Rohbani[4], Chenyue Ma[2], Hans Juergen Mattausch[3], Alexander Schiffmann[1], Alexander Steinmair[1], Ehrenfried Seebacher[1]
[1]*ams, Austria;* [2]*ASTRI, Hong Kong;* [3]*Hiroshima University, Japan;* [4]*Sharif University, Iran*

Comprehensive Compact Electro-Thermal GaN HEMT Model................................196
Muhammad Alshahed, Mina Dakran, Lars Heuken, Mohammed Alomari, Joachim Burghartz
IMS CHIPS, Germany

Trap-Assisted Carrier Transport Through the Multi-Stack Gate Dielectrics of HKMG nMOS Transistors: a Compact Model 200
Apoorva Ojha, Nihar Ranjan Mohapatra
IIT Gandhinagar, India

A New Verilog-A Compact Model of Random Telegraph Noise in Oxide-Based RRAM for Advanced Circuit Design..204
Francesco Maria Puglisi, Nicolò Zagni, Luca Larcher, Paolo Pavan
Università di Modena e Reggio Emilia, Italy

B4L-H **2D Material Integration**

Date: Wednesday, September 13, 2017
Time: 14:20 - 15:40
Room: AV02.17
Chair(s): Thomas Ernst; *Commissariat à l'Energie Atomique et aux Energies Alternatives*
          Max Lemme; *Universität Siegen*

INVITED: Ink-Jet Printed 2D Crystal Heterostructures 208
Francesco Bonaccorso
Istituto Italiano di Tecnologia, Italy

WS2 Transistors on 300 mm Wafers with BEOL Compatibility 212
Tom Schram[2], Quentin Smets[2], Benjamin Groven[2], Markus Heyne[2], Eddy Kunnen[2], Arame Thiam[2], Katia Devriendt[2], Annelies Delabie[2], Dennis Lin[2], Marcel Lux[2], Daniele Chiappe[2], Inge Asselberghs[2], Stephan Brus[2], Cedric Huyghebaert[2], Safak Sayan[2], A. Juncker[1], Matty Caymax[2], Iuliana Radu[2]
[1]COVENTOR, France; [2]imec, Belgium

200 mm Wafer Level Graphene Transfer by Wafer Bonding Technique 216
Mesut Inac[1], Grzegorz Lupina[2], Matthias Wietstruck[2], Marco Lisker[2], Mirko Fraschke[2], Andreas Mai[2], Fabio Coccetti[3], Mehmet Kaynak[2]
[1]Berlin Technical University, Germany; [2]IHP, Germany; [3]RF Microtech, Italy

C2L-E **Focus Session: Neuromorphic Computing**
Date: Thursday, September 14, 2017
Time: 10:00 - 12:00
Room: AV03.12
Chair(s): Cor Claeys; *imec*

INVITED: Neuromorphic Computing: Architectures and Applications..........................No paper
Karlheinz Meier
University of Heidelberg, Germany

INVITED: RRAM: Does It Have the Potential for Neuromorphic Computation..............No paper
Praveen Raghavan
imec, Belgium

C2L-F   Photonics / Microwave / Harsh Enviroment

Date: Thursday, September 14, 2017
Time: 10:20 - 12:00
Room: AP01.30
Chair(s):   Denis Flandre; *Université catholique de Louvain*
         Dana Cristea; *IMT Bucharest*

Epitaxial Growth and Diffusion Characteristics Analysis of Vertical Thin Poly-Si Channel Transfer Gate Structured Pixels for 3D CMOS Image Sensor 220

Sung-Kun Park[2], Donghyun Woo[2], Min-Ki Na[2], Pyong-Su Kwag[2], Ho-Ryeong Lee[2], Kyoung-Wook Ro[2], Kyung-Hwan Kim[2], Dong-Kyu Lee[2], Chris Hong[2], In-Wook Cho[2], Kyung-Dong Yoo[1]
[1]Hanyang University, Korea; [2]SK hynix, Korea

Modelling, Design and Characterization of Schottky Diodes in 28nm Bulk CMOS for 850/1310/1550nm Fully Integrated Optical Receivers 224

Wouter Diels, Michiel Steyaert, Filip Tavernier
KULeuven, Belgium

Importance of Buffer Configuration in GaN HEMTs for High Microwave Performance and Robustness.. 228

Romain Pecheux, Riad Kabouche, Ezgi Dogmus, Astrid Linge, Etienne Okada, Malek Zegaoui, Farid Medjdoub
IEMN-CNRS, France

Shunt Capacitive Switches Based on VO2 Metal Insulator Transition for RF Phase Shifter Applications 232

Emanuele Andrea Casu, Wolfgang Amadeus Vitale, Michele Tamagnone, Mariazel Maqueda Lopez, Nicolò Oliva, Anna Krammer, Andreas Schüler, Montserrat Fernandez-Bolaños, Adrian Mihai Ionescu
EPFL, Switzerland

Single Event Effects and Total Ionising Dose in 600V Si-on-SiC LDMOS Transistors for Rad-Hard Space Applications 236

Khaled Ben Ali[3], Peter Gammon[4], Chunwa Chan[4], Fan Li[4], Vasantha Pathirana[1], Tanya Trajkovic[1], Farzan Gity[2], Denis Flandre[3], Valeriya Kilchytska[3]
[1]Cambridge Microelectronics Limited, United Kingdom; [2]Tyndall National Institute at University of Ireland, Ireland; [3]Université catholique de Louvain, Belgium; [4]University of Warwick, United Kingdom

C2L-G Advanced CMOS Technology

Date: Thursday, September 14, 2017
Time: 10:00 - 12:00
Room: AV00.17
Chair(s): Francois Andrieu; *Commissariat à l'Energie Atomique et aux Energies Alternatives*
          Blandine Duriez; *TSMC*

PPAC Scaling Enablement for 5nm Mobile SoC Technology...240
Mustafa Badaroglu[4], Jeff Xu[4], John Zhu[4], Da Yang[4], Jerry Bao[4], Sc Song[4], Peijie Feng[4],
Romain Ritzenthaler[3], Hans Mertens[3], Geert Eneman[3], Naoto Horiguchi[3], Jeffrey Smith[5],
Suman Datta[5], David Kohen[2], Po-Wen Chan[1], Keagan Chen[1], Chidi Chidambaram[4]
[1]*Applied Materials, United States;* [2]*ASM International, Belgium;* [3]*imec, Belgium;*
[4]*Qualcomm Technologies Inc., Belgium;* [4]*Qualcomm Technologies Inc., United States;*
[5]*University of Notre Dame, United States*

**INVITED: Hybrid InGaAs/SiGe CMOS Circuits with 2D and 3D Monolithic
Integration** 244
Veeresh Deshpande, Herwig Hahn, Vladimir Djara, Eamon O'Connor, Daniele Caimi,
Marilyne Sousa, Jean Fompeyrine, Lukas Czornomaz
IBM Research GmbH, Switzerland

**Tunable ESD Clamp for High-Voltage Power I/O Pins of a Battery Charge Circuit in
Mobile Applications** 248
Mirko Scholz, Geert Hellings, Shih-Hung Chen, Dimitri Linten
imec, Belgium

Guidelines for Intermediate Back End of Line (BEOL) for 3D Sequential Integration...........252
Claire Fenouillet-Beranger[1], Sylvain Beaurepaire[1], Fabien Deprat[1], Alexandre Ayeres de
Sousa[1], Laurent Brunet[1], Perrine Batude[1], Paul Besombes[1], Marie-Pierre Samson[2],
Fabrice Nemouchi[1], François Andrieu[1], Romain Famulok[1], Nils Rambal[1], Viorel Balan[1],
Vincent Jousseaume[1], Vincent Delaye[1], Xavier Federspiel[2], Maud Vinet[1], O. Rozeau[1], B.
Previtali[1], G. Rodriguez[1], P. Rodriguez[1], Z. Saghi[1], C. Guerin[1], F. Ibars[2], F. Proud[2], D.
Nouguier[2], D. Ney[2], H. Dansas[1]
[1]*CEA/LETI, France;* [2]*STMicroelectronics, France*

**Device Circuit and Technology Co-Optimisation for FinFET Based 6T SRAM Cells
Beyond N7** 256
Mohit Gupta[2], Pieter Weckx[1], Stefan Cosemans[1], Pieter Schuddinck[1], Rogier Baert[1],
Dmitry Yakimets[1], Doyoung Jang[1], Yasser Sherazi[1], Praveen Raghavan[1], Alessio
Spessot[1], Anda Mocuta[1], Wim Dehaene[2]
[1]*imec, Belgium;* [2]*KU Leuven/imec, Belgium*

C2L-H Microfluidics and TFTs

Date: Thursday, September 14, 2017
Time: 10:00 - 12:00
Room: AV02.17
Chair(s): Montserrat Fernandez-Bolanos; *École Polytechnique Fédérale de Lausanne*
Joachim Burghartz; *IMS Chips*

INVITED: Microfluidic Technology: New Opportunities to Develop Physiologically Relevant in vitro Models 260
Séverine Le Gac
University of Twente, Netherlands

Development of Ultrasensitive Extended-Gate Ion-Sensitive-Field-Effect-Transistor Based on Industrial UTBB FDSOI Transistor 264
Getenet Tesega Ayele[3], Stephane Monfray[4], Frederic Boeuf[4], Jean-Pierre Cloarec[1], Serge Ecoffey[3], Dominique Drouin[3], Etienne Puyoo[2], Abdelkader Souifi[2]
[1]Ecole Central de Lyon, France; [2]INSA Lyon, France; [3]Sherbrooke University, Canada; [4]STMicroelectronics, France

Ultrathin Lateral Unidirectional Bipolar-Type Insulated-Gate Transistor as pH Sensor 268
Qinghua Han[1], Anran Gao[1], Keyvan Narimani[1], Yuelin Wang[2], Tie Li[2], Siegfried Mantl[1], Qing-Tai Zhao[1]
[1]Forschungszentrum Jülich, Germany; [2]Shanghai Institute of Microsystem and Information Technology, China

A Novel Approach for Scalable Sensor Arrays Using Cantilever Field-Effect Transistors 272
Andreas Hessel, Stefan Scholz, Alexander Pelger, Albert Pfander, Joachim Knoch
RWTH Aachen University, Germany

ESD Characterization of a-IGZO TFTs on Si and Foil Substrates 276
Nian Wang[3], Shih-Hung Chen[1], Geert Hellings[1], Kris Myny[1], Soeren Steudel[1], Mirko Scholz[1], Roman Boschke[2], Dimitri Linten[1], Guido Groeseneken[2]
[1]imec, Belgium; [2]imec, KULeuven, Belgium; [3]KULeuven, imec, Belgium

C4L-F	**Modelling and Measurement of Alternative Material Devices**
Date:	Thursday, September 14, 2017
Time:	14:20 - 15:40
Room:	AP01.30
Chair(s):	Denis Rideau; *STMicroelectronics*
	Massimo Rudan; *Università di Bologna*

INVITED: Dopant Diffusion and Segregation, Si-Ge Interdiffusion and Defect Engineering in SiGe Devices 280
Guangrui Xia
University of British Columbia, Canada

Physical Modeling of the Hysteresis in MoS2 Transistors 284
Theresia Knobloch[2], Gerhard Rzepa[2], Yury Yuryevich Illarionov[2], Michael Waltl[2], Franz Schanovsky[1], Markus Jech[2], Bernhard Stampfer[2], Marco Mercurio Furchi[3], Thomas Müller[3], Tibor Grasser[2]
[1]Global TCAD Solutions, Bösendorferstraße 1/12, 1010 Wien, Austria; [2]Institute for Microelectronics, TU Wien, Gußhausstraße 27–29/E360, 1040 Wien, Austria; [3]Institute for Photonics, TU Wien, Gußhausstraße 27–29/E387, 1040 Wien, Austria

Impact of Impurities, Interface Traps and Contacts on MoS2 MOSFETs: Modelling and Experiments 288
Gioele Mirabelli, Farzan Gity, Scott Monaghan, Paul Hurley, Ray Duffy
Tyndall National Institute, University College Cork, Ireland

C4L-G　　**High Mobility Materials and Nanowires**

Date:　　　Thursday, September 14, 2017
Time:　　　14:20 - 15:40
Room:　　　AV00.17
Chair(s):　Lukas Czornomaz; *IBM Zurich*
　　　　　　Nadine Collaert; *imec*

Electron Mobility in Thin In0.53Ga0.47As Channel.. 292
Eduard Cartier, Amlan Majumdar, Ko-Tao Lee, Takashi Ando, Martin M Frank, John Rozen, Keith A Jenkins, Cheng-Wei Cheng, John Bruley, Marinus Hopstaken, Pranita Kerber, Jeng-Bang Yau, Xiao Sun, Renee T Mo, Chun-Chen Yeh, Effendi Leobandung, Vijay Narayanan, C. Liang
IBM Corporation, United States

Understanding of Slow Traps Generation in Plasma Oxidation GeOx/Ge MOS Interfaces with ALD High-K Layers.. 296
Mengnan Ke, Mitsuru Takenaka, Shinichi Takagi
the University of Tokyo, Japan

Isolation of Nanowires Made on Bulk Wafers by Ground Plane Doping.......................... 300
Romain Ritzenthaler, Hans Mertens, An de Keersgieter, Jerome Mitard, Dan Mocuta, Naoto Horiguchi
imec, Belgium

In-Depth Electrical Characterization of Carrier Transport in Ambipolar Si-NW Schottky-Barrier FETs 304
Dae-Young Jeon[2], Tim Baldauf[5], So Jeong Park[3], Sebastian Pregl[4], Larysa Baraban[1], Gianaurelio Cuniberti[1], Thomas Mikolajick[4], Walter M. Weber[4]
[1]Institute for Materials Science and Max Bergmann Center of Biomaterials, Germany; [2]KIST, Jeonbuk, Korea; [3]Korea University, Korea; [4]Namlab gGmbH, Germany; [5]University of Applied Sciences Dresden, Germany

C4L-H   **Focus Session: Quantum Computing**
Date: Thursday, September 14, 2017
Time: 14:20 - 15:40
Room: AV02.17
Chair(s):   Iuliana Radu; *imec*

INVITED: Superconducting Qubit Research at Chalmers ....No paper
Jonas Bylander
Chalmers University, Sweden

INVITED: A 'Spins Inside' Quantum Processor ....No paper
Juan Pablo Dehollain
TU Delft, Netherlands

Principles and Trends in Quantum Nano-Electronics and Nano-Magnetics for Beyond-CMOS Computing

Ian A. Young and Dmitri E. Nikonov
Components Research, Technology & Manufacturing Group
Intel Corp.
Hillsboro, Oregon, USA
ian.young@intel.com

Abstract—An analysis of research in quantum nanoelectronics and nanomagnetics for beyond CMOS devices is presented. Some device proposals and demonstrations are reviewed. Based on that, trends in this field are identified. Principles for development of competitive computing technologies are formulated. Results of beyond-CMOS circuit benchmarking are reviewed.

Keywords— beyond-CMOS, ferroelectric, magnetoelectric switching, spin-orbit coupling, interconnects, majority gate, neuromorphic

I. Introduction

Unprecedented success of the information technology in the past 50 years was based on Moore's law [1] scaling and mostly one underlying technology – complementary metal-oxide-semiconductor (CMOS) field effect transistors (FET). Present day research effort on logic technologies going beyond-CMOS [2] has an objective to complement CMOS (rather than replacing it). It is also planned as the continuation of Moore's law, which persists [3] contrary to multiple claims of the opposite [4].

Beyond-CMOS research effort has been underway for 10 years, being funded in the USA in a large part via the Semiconductor Research Corporation (SRC) [5]. Research in Europe and Asia resulted (among other) in breakthroughs in spin-orbit effects [6] and tackled problems such as non-volatile devices and their impact on architecture [7].

The expectation in the research community 10 years ago was that this field will produce a computing technology which is better than CMOS in the majority of its characteristics. Reality showed that, among many impressive proposals and demonstrations, none of them beats CMOS. However they possess many valuable features, such as low energy operation and non-volatility. Thus the current vision is that beyond-CMOS circuits will replace CMOS in some types of computing or some information processing applications. They would be monolithically integrated with CMOS on the same chip or packaged together in a multi-chip module.

Another expectation was that beyond-CMOS circuits will not require any MOSFETs as part of them and maybe even get rid of any charge currents in the quest for energy efficiency. This did not come true: a thorough circuit analysis reveals that a MOSFET transistor is needed to supply power, and to clock and control the logic circuit operation.

Principle 1: Beyond-CMOS circuits require CMOS as an integral part. They will work alongside and augment CMOS computing blocks.

Beyond-CMOS benchmarking [8,9] was helpful in evaluating the potential of various computing technologies. It enabled setting expectations for performance and showed pathways for improvement. Experimental demonstrations do not yet achieve theoretical projections put forward in the benchmarking. One of the reasons is that each computing technology requires solving numerous fabrication challenges, see e.g. [10].

A historic similarity is fitting – the disruption of bipolar transistors by CMOS [11]. The latter had the advantage of smaller static current, but was much more difficult to manufacture. We believe that the same drive toward lower power computing will compel technologists to solve implementation problems for beyond-CMOS computing. One should understand that a 100x improvement in characteristics (e.g. the energy-delay product) is not required from an emerging device. Even 4x improvement will correspond to a couple of generations of historic scaling and will justify its integration on the computer system.

II. From Computation Variables to Materials

We classify beyond-CMOS devices by the computation variable – a physical quantity which encodes information. According to this we consider 4 classes: electronic; ferroelectric; spintronic; orbitronic devices, see Table I. The computation variables dictate the choice of materials that were the focus of development in the last few years.

(1) In electronic devices, 2D layered materials such as graphene and transition-metal dichalcogenides (TMDs) were prominent. They enable a wide range in the choice of the band

978-1-5090-5979-9/17 $31.00 © 2017 IEEE

structure even for channels of conventional transistors. Some devices exploit their unique transport properties such electron reflection [12]. Multiple tunnel device [13] proposals promise better transistor energy efficiency and rely on tunneling along the layers [14] or between the layers [15].

(2) The interest in beyond-CMOS logic led to a revival of research on ferroelectrics, which also present an attractive non-volatile option. This material class significantly overlaps with piezo-electric devices. They can be used in their direct capacity – to induce and utilize strain, as well as for the magnetostrictive type of magneto-electric (ME) switching.

(3) Multiferroic materials (i.e. ones simultaneously having electric dipole and spin ordering) were in focus for spintronic devices as means of magneto-electric switching. Spin-orbital effects gained prominence as an efficient spin to charge and vice versa transduction [6]. Non-volatility of not only ferromagnetic, but also anti-ferromagnetic (AFM) orders are now envisioned as computation variables.

(4) Among orbitronic devices, metal insulator transition FETs made progress [16]. Also strides were made in generating and controlling transport of excitons [17]. Strong interaction with light makes exciton devices an attractive option for optical fiber transducers.

Principle 2: Some devices utilize collective states; this confers advantages of non-volatility or more energy efficient operation.

TABLE I. VOLTAGE LIMITATIONS FOR COMPUTATION VARIABLES

Comp. variable	Collective state	Vola-tile?	Write limit	Read limit
Charge	None		Capacitor, none	FET, 600mV
	None		Capacitor, none	Tunnel, 300mV
Electric Dipole	Ferroelectric	NV	FE, 50mV	Negative Cap., 300mV
Spin	Ferromagnetic	NV	ME, 50mV	Inverse Spin-orbit, 50mV
		NV	ME, 50mV	Inverse ME, 50mV
	Anti-Ferromagnetic	NV	ME, 50mV	??
Orbital State	Metal-insulator transition		MIT, 1000mV	FET, 600mV
	Excitons		Binding, 300mV	Recombination, 300mV

III. FROM MATERIALS TO DEVICES

The minimum signal voltage in electrical interconnects is limited by the switch. An interconnect can operate above the noise limit even at millivolt levels [18]. A major portion of the switching energy in a circuit is determined by CV^2, the product of the square of the operating (signal swing) voltage

and the capacitance of a dielectric barrier or an interconnect wire (with an increasing proportion of the later). In hindsight, a lot of beyond-CMOS research evolved into the search for a lower-voltage device. Meanwhile, spintronic devices only make sense if they are efficiently interconnected, see Section IV.

Principle 3: The choice for an optimal beyond-CMOS device will be determined by compatibility with an efficient and effective interconnect.

The nature of the phenomena to write and read the computation variables in devices is tied to voltages of operation. Material benchmarking [19] is very helpful in cataloging materials, structures, and parameters for these phenomena. In Table I we present our educated projection of the practical limit of voltage for various mechanisms.

(1), and (2) In the write regime of a MOSFET, the gate capacitor can be charged to an arbitrarily small voltage. Its state is read off by the value of the source-to-drain current. The supply voltage is limited by the on-current I_{on}, off-current I_{off}, and their ratio. (The off-current is connected to standby power dissipation.) They depend on the energy barrier of the channel for its "off state" vs. "on state". Negative capacitance FET [20] is based on an idea of switching this energy barrier with a smaller voltage. Tunnel FETs aim to shape the energy distribution of electrons to increase the I_{on} and decrease the I_{off} achieved by a certain switching voltage.

(3) In spintronic devices, the departure from spin transfer torque switching towards more efficient magnetoelectric switching [21] led to a significant decrease of switching energy [22]. The advantage of ferroelectric and magnetoelectric layers is that they can potentially be switched by a voltage of tens of millivolts provided that they can be made sufficiently thin. Eliminating leakage through thin layers would be a challenge for material science. Low I_{on}/I_{off} is in general still a problem for many spintronic devices. However the off-current problem is solved by eliminating low-resistance current paths from supply to ground voltages and relying on the capacitive nature of magnetoelectric switching [22].

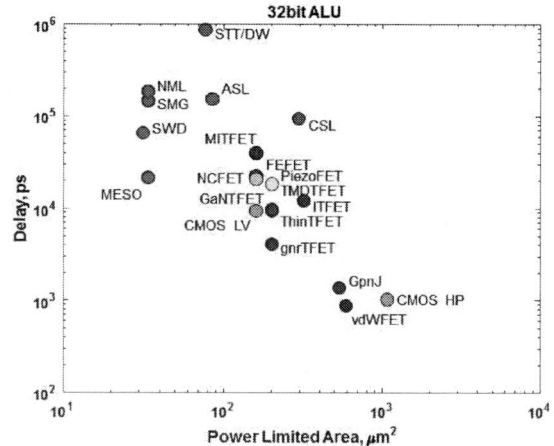

Fig. 1. Delay vs. power limited area in an ALU implemented with beyond-CMOS devices.

Fig. 2. Energy vs. delay in an ALU implemented with beyond-CMOS devices.

(4) Orbitronic devices have made good progress in decreasing the operating voltage. However there are presently no indications that their switching or reading voltages would reach record low enough values.

Principle 4: Low voltage devices – most direct way to low energy operation.

IV. FROM DEVICES TO CIRCUITS

Beyond CMOS benchmarking [8,9] was a theoretical exercise in estimating the future performance of devices, provided they could be successfully implemented. It starts from considering the physical design, estimates of area of elementary circuits and the associated parasitics. Estimates for performance of devices and interconnects are combined to sequentially build up more complicated logic circuits such as an arithmetic logic unit (ALU). Benchmarking is still actively continuing [23-26]. Even though estimates change quantitatively, the overall picture of relative performance of various classes of devices remains similar.

Principle 5: Start benchmarking with bottom up modeling of devices, build up from simple to more complicated circuits.

A result of such benchmarking is shown in Figs. 1-3. For details of the method and the explanation of acronyms see [8,9]. Fig. 1 shows area and delay for an ALU based on various beyond CMOS devices. One can see that spintronic circuits are more compact. They proved to be not as amenable to realizing complementary logic gates (such as NAND, NOR). However the alternative logic function – the majority gate – could be implemented more naturally as a single logic element. Besides, far fewer majority gates than transistors were required to implement certain logic functions. The synthesis of various circuits from majority gates [27] shows that:

Principle 6: Majority gates (if easily implemented in a certain technology) enable more efficient circuits, especially for more complex computation functions.

Also Fig. 2 shows that spintronic circuits are slower. This switching speed problem is caused by a slower time constant for magnetization precession than for an electron traversing a logic gate. A potential solution is replacing ferromagnetic magnetic with antiferromagnetic (AFM) ones [28]. They have intrinsically shorter time constants and thus enable faster switching. Also they still can be switched with the magnetoelectric effect. However an efficient read mechanism for the AFM state (other than coupling it to a ferromagnet) is still lacking.

Fig. 2 shows the corresponding switching energy vs. delay for the ALUs based on various beyond-CMOS devices. Spintronic circuits have lower switching energy (i) due to lower operating voltage (as discussed above) and (ii) more efficient circuits based on majority gates. This energy advantage exists despite the fact that CMOS transistors are used in every logic stage of spintronic circuits. They are needed to turn on and off the current to power on/off spintronic devices. However complementary CMOS which turns off static current, does not carry the logical signal, rather the interconnect on the CMOS gate output carries the signal. Also spintronic devices are non-volatile, i.e. present a latch in every gate output. The CMOS transistors are used to clock them and thus advance the logic pipeline.

Until a few years ago, not much thought was given to how spintronic devices would be interconnected. Since then, schemes of interconnection were explored which were based on spin polarized diffusion [29], domain wall motion [30], and spin waves [31]. Also a practical and robust way to generate and detect spin wave packets and to pipeline them was worked out [32].

It was originally believed that spin-based interconnects were necessary for spintronic devices. The thinking behind this was that conversion of a signal from spin to electrical and vice versa would be prohibitively energy expensive. However the realization that spin interconnects have a limited propagation span and require frequent repeaters [33], changed the calculation. On the other hand, the advent of magneto-electric and spin-orbit effects made the spin to electrical conversions energy affordable. This is demonstrated by outstanding predicted performance of MESO circuits (Fig. 2) which are based on electrical interconnects.

Principle 7: Use electrical interconnects for longer propagation spans.

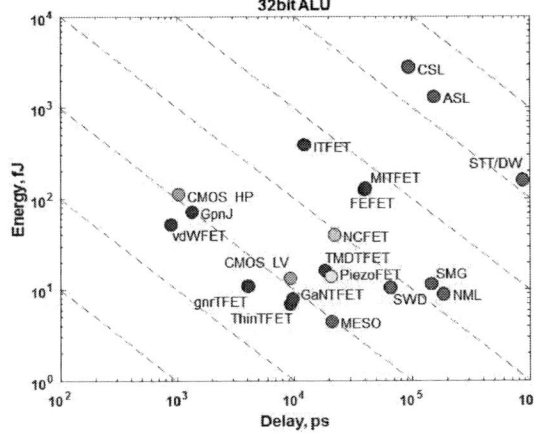

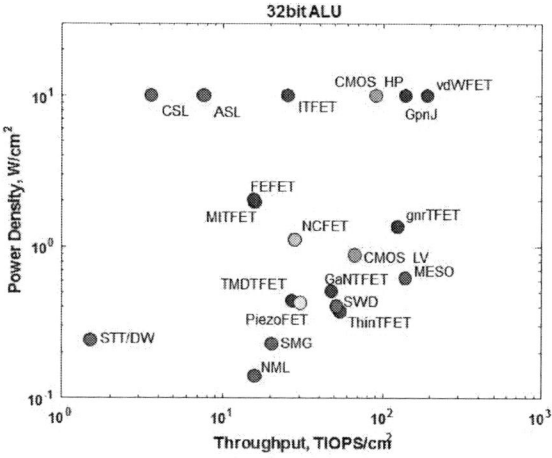

Fig. 3. Capped power density vs. computation throughput.

V. FROM CIRCUITS TO ARCHITECURES

Beyond CMOS circuits may result in various architectures. One possibility for general purpose computing would be a "drop-in" replacement of CMOS circuits. A good opportunity for this exists with MOSFETs based on emerging materials. Also tunnel FETs will require only modest modifications of existing circuit blocks [34]. Benchmarking of ALU throughput vs. power density in Fig. 3 assumes an approach where circuits are replaced on the lowest level, but the high level architecture remains the same. An important feature in this calculation is that a "limit" or "cap" is applied on the power dissipation – resulting from limitations on the heat removal from a chip or the energy available to a mobile or an IoT chip. To meet this "cap", we choose to increase the area allocated per circuit. It is this adjusted area which is plotted in Fig. 1. The optimistic take-away from this figure of merit is that even slower beyond CMOS circuits can beat CMOS in computation throughput while maintaining lower dissipated power.

Non-volatility of beyond-CMOS logic elements can be effectively utilized in "compute-in-memory" architectures, which are already being developed independently based on CMOS or other types of novel memory. However one can expect a more dramatic benefit from special architectures designed to make use of non-volatile elements. Modern processors are frequently put to "sleep states", in which power is turned off to blocks of logic or memory. Necessary data from the processor needs to be saved in another memory to be retained and then fetched to be processed again when it is returned to the working state. An integrated non-volatile memory will save time and energy on these operations. Even though multiple publications advocated this idea [7,35-38], we do not think a convincing value proposition has been made for this yet.

Principle 8: To convince the wider community, a non-volatile computing paradigm needs to be general enough to prove that it is valid for more than one architecture; while

it needs to be specific enough to dispel claims that an essential aspect is missed.

More importantly, the wide architecture community has not risen to the challenge of innovating and implementing such specially designed new architecture in a practical chip.

A promising direction is stochastic computing [39]. It attracted attention because beyond-CMOS devices (e.g. magnetics) are more affected by thermal fluctuations than present-day transistors. This thrust focuses on methods of processing and exploiting fluctuations to still achieve a computing performance goal. Another twist on it is utilizing the stochastic property to perform certain computing functions [40].

Neuromorphic computing became popular since it is related to recent information processing breakthroughs such machine learning [41]. It crudely can be pictured as combinations of inference (finding the distance between patterns) and classification (separating patterns to groups with lowest mutual distance). Even though most of the real world machine learning is performed with traditional CMOS ICs, it is attractive to attempt to build special purpose ICs with beyond CMOS devices which might perform machine learning functions better.

In a recent intriguing study, it was argued that beyond-CMOS devices can perform significantly better than CMOS in a type of neural circuits, CNN, used for inference of black-and-white patterns [25]. Many neuromorphic specific circuits can be represented as neural gates performing the operation of a "dot product" of vectors of analog inputs and stored weights (synapses), followed by a non-linear thresholding function (a neuron). Such analog products and the non-linearity can be implemented more efficiently with beyond-CMOS devices. In this case a single device can be used for a neuron instead of tens of transistors [42].

Also neuromorphic circuits have much larger numbers of fan-in and fan-out wires than typical CMOS architectures. This stronger connectivity may be implemented more efficiently with low-voltage interconnects in emerging computing technologies.

Principle 9: Neuromorphic computing can be done more efficiently with beyond-CMOS circuits.

VI. CONCLUSION

In summary, we have presented our view on the recent trends in quantum nanoelectronics and nanomagnetics for beyond CMOS devices, and outlined a few principles to make them realize practical computing technologies.

We can pose a question for ourselves: What are the most promising directions of research? Where to double down on the effort? Among many equally important thrusts, our subjective preference is for magnetoelectric switching and neuromorphic beyond-CMOS circuits.

REFERENCES

[1] G. E. Moore, ```Cramming more components onto integrated circuits,'' *Electronics*, vol. 38, no. 8, pp. 114-117, 1965.

[2] W. M. Holt, "1.1 Moore's law: A path going forward," *2016 IEEE International Solid-State Circuits Conference (ISSCC)*, San Francisco, CA, 2016, pp. 8-13.

[3] M. Bohr, available onine: https://newsroom.intel.com/news/intel-presents-technology-manufacturing-day-live-video-updates/, Mar. 2017.

[4] T. N. Theis and H. S. P. Wong, "The End of Moore's Law: A New Beginning for Information Technology," *Computing in Science & Engineering*, vol. 19, no. 2, pp. 41-50, Mar.-Apr. 2017.

[5] I. A. Young. "Technology Options for Beyond-CMOS", *Proceedings of the 2017 ACM on International Symposium on Physical Design*, Portland, Oregon, USA, p. 1-1, 2017.

[6] A. Soumyanarayanan, N. Reyren, A. Fert, C. Panagopoulos, "Emergent phenomena induced by spin–orbit coupling at surfaces and interfaces", - *Nature*, vol. 539, pp. 509-517, 2016.

[7] T. Hanyu et al., "Standby-Power-Free Integrated Circuits Using MTJ-Based VLSI Computing," *Proceedings of the IEEE*, vol. 104, no. 10, pp. 1844-1863, Oct. 2016.

[8] D. E. Nikonov and I. A. Young, ``Overview of beyond-CMOS devices and a uniform methodology for their benchmarking," *Proceedings of the IEEE*, vol. 101, no. 12, pp. 2498-2533, Dec. 2013.

[9] D. E. Nikonov and I. A. Young, "Benchmarking of Beyond-CMOS Exploratory Devices for Logic Integrated Circuits," *IEEE Journal on Exploratory Solid-State Computational Devices and Circuits*, vol. 1, pp. 3-11, Dec. 2015.

[10] I. P. Radu *et al.*, "Spintronic majority gates," *IEEE International Electron Devices Meeting (IEDM)*, Washington, pp. 32.5.1-32.5.4, 2015.

[11] S. Borkar, "Electronics beyond nano-scale CMOS," *43rd ACM/IEEE Design Automation Conference*, San Francisco, CA, pp. 807-808, 2006.

[12] S. Chen et al,. "Electron optics with p-n junctions in ballistic graphene", *Science*, vol. 353, no. 6307, pp. 1522-1525, Sep. 2016.

[13] D. J. Fitzgerald and A. S. Grove, "Mechanisms of Channel Current Formation in Silicon P-N Junctions," *Fourth Annual Symposium on the Physics of Failure in Electronics*, Chicago, IL, USA, 1965, pp. 315-332.

[14] M. O. Li, D. Esseni, G. Snider. D. Jena, and H. G. Xing, "Single particle transport in two-dimensional heterojunction interlayer tunneling field effect transistor", *J. of Appl. Phys.*, vol. 115, p. 074508, 2014.

[15] E. Tutuc *et al.*, "Gate tunable resonant tunneling in graphene-based heterostructures and device applications," *73rd Annual Device Research Conference (DRC)*, Columbus, OH, pp. 269-270, 2015.

[16] K. Ahadi, O. F. Shoron, P. B. Marshall, E. Mikheev, and S. Stemmer, "Electric field effect near the metal-insulator transition of a two-dimensional electron system in SrTiO3", *Appl. Phys. Lett.*, vol. 110, p. 062104, 2017.

[17] A. A. High, J. R. Leonard. A. T. Hammack, M. M. Fogler, L. V. Butov, A. V. Kavokin, K. L. Campman, and A. C. Gossard, "Spontaneous coherence in a cold exciton gas", *Nature*, vol. 483, pp. 584–588, 2012.

[18] E. Yablonovitch, "Will a New Milli-Volt Switch Replace the Transistor for Digital Applications?" *APS March Meeting*, #S2.001, Mar. 2010.

[19] K. Galatsis *et al.*, "A Material Framework for Beyond-CMOS Devices," *IEEE Journal on Exploratory Solid-State Computational Devices and Circuits*, vol. 1, pp. 19-27, Dec. 2015.

[20] S. Salahuddin and S. Datta, ``Use of negative capacitance to provide voltage amplification for low power nanoscale devices," *Nano Lett.*, vol. 8, no. 2, pp. 405-410, 2008.

[21] F. Matsukura, Y. Tokura, and H. Ohno, "Control of magnetism by electric fields", *Nature Nano.*, vol. 10, p. 209, 2015.

[22] S. Manipatruni, D. E. Nikonov, and I. A. Young, "Spin-Orbit Logic with Magnetoelectric Nodes: A Scalable Charge Mediated Nonvolatile Spintronic Logic", available online https://arxiv.org/abs/1512.05428, 2015.

[23] C. Pan and A. Naeemi, "Interconnect Design and Benchmarking for Charge-Based Beyond-CMOS Device Proposals," *IEEE Electron Device Letters*, vol. 37, no. 4, pp. 508-511, April 2016.

[24] C. Pan, S.-C. Chang and A. Naeemi, "Performance analyses and benchmarking for spintronic devices and interconnects," *IEEE International Interconnect Technology Conference / Advanced Metallization Conference (IITC/AMC)*, San Jose, CA, pp. 56-58, 2016.

[25] C. Pan and A. Naeemi, "Non-Boolean Computing Benchmarking for Beyond-CMOS Devices Based on Cellular Neural Network," *IEEE Journal on Exploratory Solid-State Computational Devices and Circuits*, vol. 2, no. , pp. 36-43, Dec. 2016.

[26] C. Pan and A. Naeemi, "Nonvolatile Spintronic Memory Array Performance Benchmarking Based on Three-Terminal Memory Cell," *IEEE Journal on Exploratory Solid-State Computational Devices and Circuits*, vol. 3, no. , pp. 10-17, Dec. 2017.

[27] O. Zografos *et al.*, "Wave pipelining for majority-based beyond-CMOS technologies," *Design, Automation & Test in Europe Conference & Exhibition (DATE)*, Lausanne, Switzerland, pp. 1306-1311, 2017.

[28] V. Baltz, A. Manchon, M. Tsoi, T. Moriyama, T. Ono, and Y. Tserkovnyak, "Antiferromagnetic spintronics", available online https://arxiv.org/abs/1606.04284, 2016.

[29] S. C. Chang, R. M. Iraei, S. Manipatruni, D. E. Nikonov, I. A. Young and A. Naeemi, "Design and Analysis of Copper and Aluminum Interconnects for All-Spin Logic," *IEEE Transactions on Electron Devices*, vol. 61, no. 8, pp. 2905-2911, Aug. 2014.

[30] S. C. Chang, S. Dutta, S. Manipatruni, D. E. Nikonov, I. A. Young and A. Naeemi, "Interconnects for All-Spin Logic Using Automotion of Domain Walls," *IEEE Journal on Exploratory Solid-State Computational Devices and Circuits*, vol. 1, no. , pp. 49-57, Dec. 2015.

[31] S. Dutta, D. E. Nikonov, S. Manipatruni, I. A. Young and A. Naeemi, "SPICE Circuit Modeling of PMA Spin Wave Bus Excited Using Magnetoelectric Effect," *IEEE Transactions on Magnetics*, vol. 50, no. 9, pp. 1-11, Sept. 2014.

[32] S. Dutta, S.-C. Chang, N. Kani, D. E. Nikonov, S. Manipatruni, I. A. Young, and A. Naeemi, "Non-volatile Clocked Spin Wave Interconnect for Beyond-CMOS Nanomagnet Pipelines", *Scientific Reports*, vol. 5, p. 9861, 2015.

[33] R. M. Iraei, S. Manipatruni, D. E. Nikonov, I. A. Young, and A. Naeemi, "Electrical-Spin Transduction for CMOS-Spintronic Interface and Long-Range Interconnects", *IEEE Journal on Exploratory Solid-State Computational Devices and Circuits*, vol. 3, 2017.

[34] D. H. Morris, K. Vaidyanathan, U. E. Avci, H. Liu, T. Karnik and I. A. Young, "Enabling high-performance heterogeneous TFET/CMOS logic with novel circuits using TFET unidirectionality and low-VDD operation," *2016 IEEE Symposium on VLSI Technology*, Honolulu, HI, 2016, pp. 1-2.

[35] H. Ohno, T. Endoh, T. Hanyu, N. Kasai and S. Ikeda, "Magnetic tunnel junction for nonvolatile CMOS logic," *2010 International Electron Devices Meeting*, San Francisco, CA, 2010, pp. 9.4.1-9.4.4.

[36] K. Ando *et al.*, "Spin-RAM for Normally-Off Computer," *2011 11th Annual Non-Volatile Memory Technology Symposium Proceeding*, Shanghai, 2011, pp. 1-6.

[37] E. Kitagawa *et al.*, "Impact of ultra low power and fast write operation of advanced perpendicular MTJ on power reduction for high-performance mobile CPU," *2012 International Electron Devices Meeting*, San Francisco, CA, 2012, pp. 29.4.1-29.4.4.

[38] K. Nomura, K. Abe, H. Yoda, and S. Fujita, "Ultra low power processor using perpendicular-STT-MRAM/SRAM based hybrid cache toward next generation normally-off computers", *J. Appl. Phys.*, vol. 111, p. 07E330 2012.

[39] N. Shanbhag, "Statistical information processing: Computing for the nanoscale era," *2015 IEEE/ACM International Symposium on Low Power Electronics and Design (ISLPED)*, Rome, Italy, 2015, pp. 1-1.

[40] B. Sutton, K. Y. Camsari, B, Behin-Aein, S. Datta, "Intrinsic optimization using stochastic nanomagnets", *Scientific Reports*, vol. 7, p. 44370, 2017.

[41] Y. LeCun, Y. Bengio, and G. Hinton, "Deep learning", *Nature*, vol. 521, no. 7553, pp. 436–444, 2015.

[42] G. Srinivasan, A. Sengupta, and K. Roy, "Magnetic Tunnel Junction Based Long-Term Short-Term Stochastic Synapse for a Spiking Neural Network with On-Chip STDP Learning", *Scientific Reports*, vol. 6, p. 29545, 2016.

22% Higher Performance, 2x SCM Write Endurance Heterogeneous Storage with Dual Storage Class Memory and NAND Flash

Chihiro Matsui and Ken Takeuchi

Dept. of Electrical, Electronic and Communication Engineering, Chuo University, Tokyo, Japan
E-mail: matsui@takeuchi-lab.org

Abstract—Storage class memories (SCMs); for instance, (STT-)MRAM, ReRAM, PRAM, and 3D XPoint, have much attention from storage systems. Each SCM has different characteristics, such as read/write latency, endurance, and bit cost. For example, MRAM has short latency and high endurance, but its cost is high. In contrast, ReRAM, PRAM, and 3D XPoint have lower endurance, but their cost is lower than MRAM. From these SCM characteristics, MRAM is classified into memory-type SCM (M-SCM), and ReRAM, PRAM, and 3D XPoint are called storage-type SCM (S-SCM). Previous studies show that SCM improves NAND flash based storage performance. However, cost of M-SCM is too high for storage usage. To bring out the best of M-SCM and S-SCM characteristics, this paper proposes a heterogeneous storage with two types of SCM, called dual SCM, and NAND flash memory. In the proposed storage, M-SCM stores very frequently accessed super-hot data and S-SCM stores hot data. By considering the storage performance and total storage cost, only 1% of M-SCM improves the storage performance by 22% and write endurance of S-SCM by 2 times compared with conventional S-SCM and NAND flash hybrid storage. With this storage configuration, M-SCM and S-SCM should have write cycles 3.2×10^2 and 4.5×10^1 times higher than MLC NAND flash.

Keywords— storage class memory (SCM); memory-type SCM; storage-type SCM; NAND flash memory; heterogeneous storage; dual SCM

I. INTRODUCTION

Because of spread of Internet of Things (IoT) technology in every field, storage in data centers is required to have higher speed to process wide variety of data. NAND flash is faster than hard disk drives. However, NAND flash's performance is limited because of the unique characteristics: read/write and erase unit asymmetry, slower write operation than read, and *erase-before-write* requirement [1]. To obtain the higher speed,

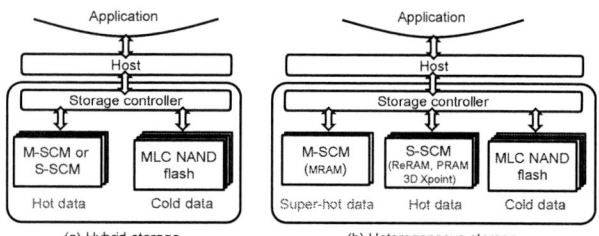

Fig. 1 (a) Conventional hybrid storage with SCM and NAND flash [3]. (b) Proposed heterogeneous storage with dual SCM and NAND flash.

a storage class memory (SCM) and NAND flash hybrid storage (Fig. 1(a)) is previously proposed [2, 3]. In the conventional SCM and NAND flash hybrid storage, frequently accessed (hot) data are stored in SCM. As shown in Table I [4], SCMs [5, 6] are classified into two types: memory-type SCM (M-SCM) and storage-type SCM (S-SCM) [7]. In general, M-SCM has high speed and high endurance, and S-SCM has large capacity and the cost is low. Thus, using M-SCM as storage is not cost effective because M-SCM has 10 times higher cost than multi-level cell (MLC) NAND flash [4]. To bring out the best of M-SCM and S-SCM characteristics, this paper proposes a heterogeneous storage with dual SCM and NAND flash memory as shown in Fig. 1(b). With the proposed heterogeneous storage, small capacity of M-SCM improves write endurance of S-SCM, and leverages the performance of conventional S-SCM and NAND flash hybrid storage [8].

II. MEMORY-TYPE AND STORAGE-TYPE STORAGE CLASS MEMORY (SCM)

Fig. 2 shows SCM candidates: (a) spin-transfer torque magnetoresistive RAM (STT-MRAM) [9], (b) resistive RAM

Table I Characteristics of storage class memory (SCM) and NAND flash [4]

Memory device	Read latency	Write latency	Erase latency	I/O frequency	V_{DD} (Core, I/O)	Access unit	Acceptable endurance [5]	Bit cost [4]
M-SCM (MRAM)	0.1us/sector	0.1us/sector	Not required	1066 MHz	1.8V, 1.2V	Sector (512B)	10^{12}	10
S-SCM (ReRAM, PRAM, 3D XPoint)	10us/sector	10us/sector					10^8	4
MLC NAND flash	52us/page (U) 36us/page (L)	2000us/page (U) 370us/page (L)	3300us /block	400 MHz	3.3V, 1.8V	Page (16KB)	10^4	1

U: Upper page, L: Lower page of MLC NAND flash

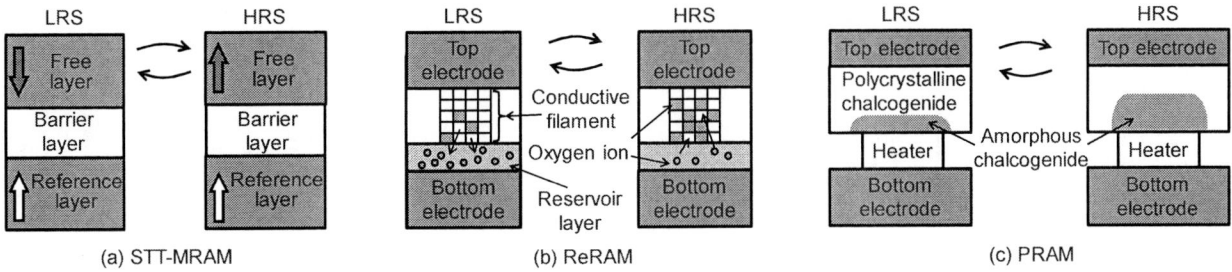

Fig. 2 State switching mechanisms of SCMs between low-resistance state (LRS, "1") and high-resistance state (HRS, "0"). (a) STT-MRAM [9], (b) ReRAM [10], and (c) PRAM [11].

(ReRAM) [10], and (c) phase-change RAM (PRAM) [11, 12]. Each SCM has different switching mechanism between low-resistance state (LRS, "1") and high-resistance state (HRS, "0"). By spin torque effect, state of STT-MRAM [9] is changed to anti-parallel (LRS) or parallel (HRS) between the free and reference layers. ReRAM's switching between LRS and HRS is caused by diffusion of oxygen vacancies or oxygen ion between conductive filament and metal oxide layer [10]. In PRAM [11], chalcogenide material is changed to crystal (LRS) or amorphous (HRS) by controlling the heating temperature.

SCMs have different characteristics depending on their switching mechanisms. SCMs can be divided into memory-type SCM (M-SCM) and storage-type SCM (S-SCM) [7]. Using the SCM classification, MRAM will be M-SCM because of its high cost and high endurance. Instead, ReRAM, PRAM, and 3D XPoint, which have lower cost but lower endurance than MRAM, are classified as S-SCM. As illustrated in Table I, the specific differences between M-SCM and S-SCM are: 1) M-SCM has 10^2 times shorter latency for one sector read/write than S-SCM, 2) M-SCM has 10^4 times higher acceptable endurance than S-SCM [5], and 3) M-SCM is 2.5 times higher cost than S-SCM [4].

III. HETEROGENEOUS STORAGE WITH DUAL SCM AND NAND FLASH

By using M-SCM, S-SCM, and MLC NAND flash, a heterogeneous storage is proposed (Fig. 1(b)). The proposed storage brings the best of M-SCM and S-SCM characteristics, and compensate for the SCMs' shortcomings. Capacity of each memory in the proposed storage is flexibly changed depending on application characteristic in order to obtain higher performance. As shown in Fig. 3(a), the proposed storage uses M-SCM and S-SCM as non-volatile cache. In contrast to SRAM cache and DRAM in main memory system [13], all of memory used in the proposed heterogeneous storage are non-volatile. Therefore, the proposed storage does not require periodical data flush from M-SCM/S-SCM to the lower-level memory. In the heterogeneous storage, the fastest M-SCM stores super-hot data, which are very frequently accessed data. In addition, the second fastest S-SCM stores hot data, and the slowest but large-capacity NAND flash stores rarely-accessed cold data.

Detailed data management algorithm used in the proposed heterogeneous storage is explained in Fig. 3(b). The two least recently used (LRU) tables manage access order and clean/dirty flag of data stored in M-SCM and S-SCM. The clean data are stored in both M-SCM (S-SCM) and S-SCM (NAND flash); instead, the dirty data are not. All data of new write request is stored in M-SCM. When M-SCM or S-SCM becomes almost full of data, evict operation is triggered. Eviction data is chosen from the tail of the LRU list. If the dirty data are chosen for eviction, the data are moved from M-SCM (S-SCM) to S-SCM (NAND flash); however, when the clean data are chosen to evict, data movement is not required. If read data or overwrite data is stored in S-SCM or NAND flash (cache miss), only dirty data are copied from S-SCM (NAND flash) to M-SCM (S-SCM).

IV. PERFORMANCE EVALUATION

Performance of the proposed heterogeneous storage with

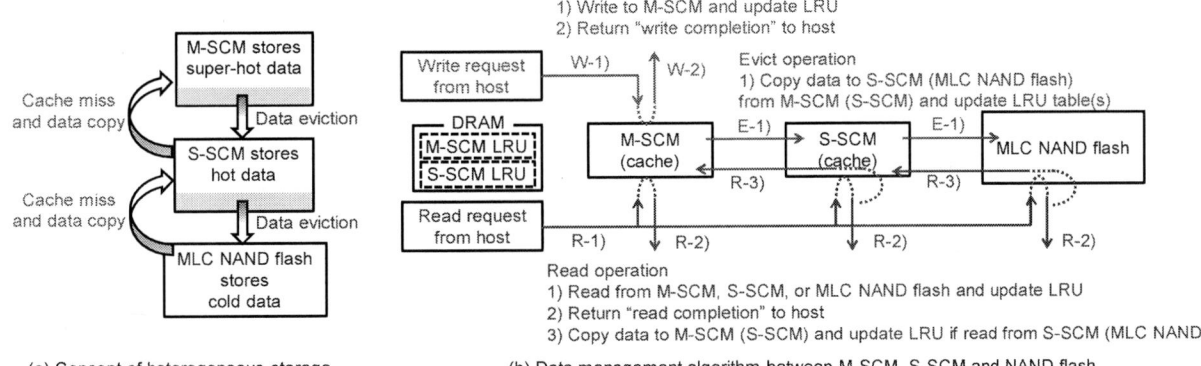

(a) Concept of heterogeneous storage   (b) Data management algorithm between M-SCM, S-SCM and NAND flash

Fig. 3 (a) Concept and (b) data management algorithm of heterogeneous storage with dual SCM and NAND flash.

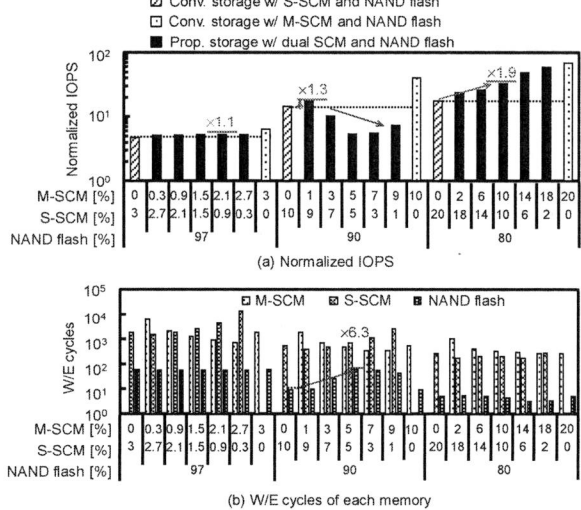

Fig. 4 Performance of heterogeneous storage with prxy_0. (a) IOPS performance normalized by NAND flash only storage and (b) W/E cycles of each memory.

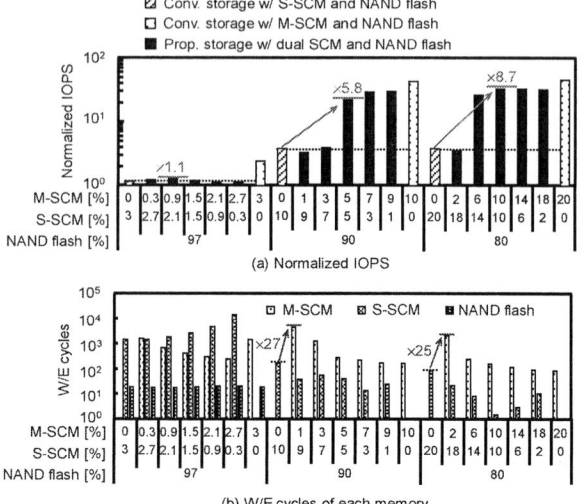

Fig. 5 Performance of heterogeneous storage with prxy_1. (a) IOPS performance normalized by NAND flash only storage and (b) W/E cycles of each memory.

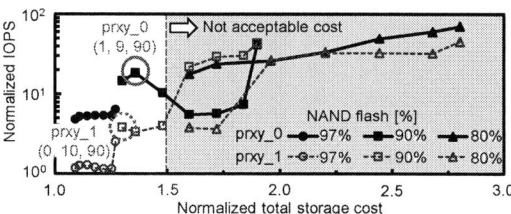

Fig. 6 Relationship between total storage cost and IOPS performance of heterogeneous storage with different memory capacity ratios. Cost of NAND flash only storage being 1.

dual SCM and NAND flash is evaluated with a transaction-level modeling-based emulator [2]. Two applications are input: super-hot (prxy_0) and hot (prxy_1) [14]. The definition of applications comes from the previous study [3], in which 1% and 5% of SCM boosts the NAND flash based storage for prxy_0 and prxy_1, respectively. Memory characteristics in Table I are used in the evaluation. In the following evaluations, storage IOPS performance and each memory capacity ratio are calculated as relative values with MLC NAND flash only storage being 1 (= 100%). Note that write/erase (W/E) cycles of NAND flash are average erase count per NAND flash block, while W/E cycles of SCM are average writes per SCM sector.

A. Storage performance with constant storage capacity

First, performance of the proposed heterogeneous storage is evaluated with constant total storage capacity, which is same as the capacity of NAND flash only storage (100% storage capacity). The storage performance and W/E cycles show different tendency per memory capacity ratio and application.

Fig. 4 shows (a) normalized IOPS performance and (b) W/E cycles of prxy_0 application. When NAND flash capacity is 97%, there is 1.1 times performance improvement by adding M-SCM. When NAND flash capacity is 90%, 1% of M-SCM improves the performance by 1.3 times compared with conventional S-SCM and NAND flash hybrid storage; however, more than 3% of M-SCM degrades the performance because large number of data are evicted from S-SCM to NAND flash, and the W/E cycles of NAND flash is increased (Fig. 4(b)). In addition, when NAND flash is 80%, more than 10% of S-SCM prevents frequent data eviction to NAND flash and improves performance of S-SCM and NAND flash hybrid storage by upto 1.9 times.

With prxy_1 application (Fig. 5(a)), by adding M-SCM, there is 1.1 times performance improvement with 97% of NAND flash capacity. When NAND flash capacity is 90% and 80%, the storage performance improves as M-SCM capacity increases, compared to conventional S-SCM and NAND flash hybrid storage. However, small M-SCM capacity degrades the

performance because a large number of data write is occurred in M-SCM as illustrated as W/E cycles in Fig. 5(b).

The storage cost is also important to design a heterogeneous storage because M-SCM and S-SCM have 10 and 4-times higher cost than MLC NAND flash [4] as shown in Table I. The total storage cost is calculated by Equation (1).

$$\text{Total storage cost} = \Sigma(\text{bit cost} \times \text{memory capacity ratio}) \quad (1)$$

Assuming that 1.5-times higher cost is acceptable compared to NAND flash only storage, some of memory capacity combinations exceeds the threshold as shown in Fig. 6. If the total storage cost is capped by 1.5, dual SCM and NAND flash storage with 1% M-SCM, 9% S-SCM, and 90% NAND flash is the best for prxy_0. For prxy_1, however, 10% S-SCM and 90% NAND flash hybrid storage shows the best performance.

B. SCM write endurance with constant storage cost

Next, constant total storage cost is assumed, which is capped by 1.5, same as the previous subsection IV-A. Note that the total storage capacity becomes smaller or larger than the capacity of 100% NAND flash storage. With prxy_0 application (Fig. 7(a)), by adding 1% M-SCM, the storage performance is improved by over 1.2 times compared to conventional S-SCM and NAND flash hybrid storage. As intended in the design of the heterogeneous storage with dual

978-1-5090-5979-9/17 $31.00 © 2017 IEEE

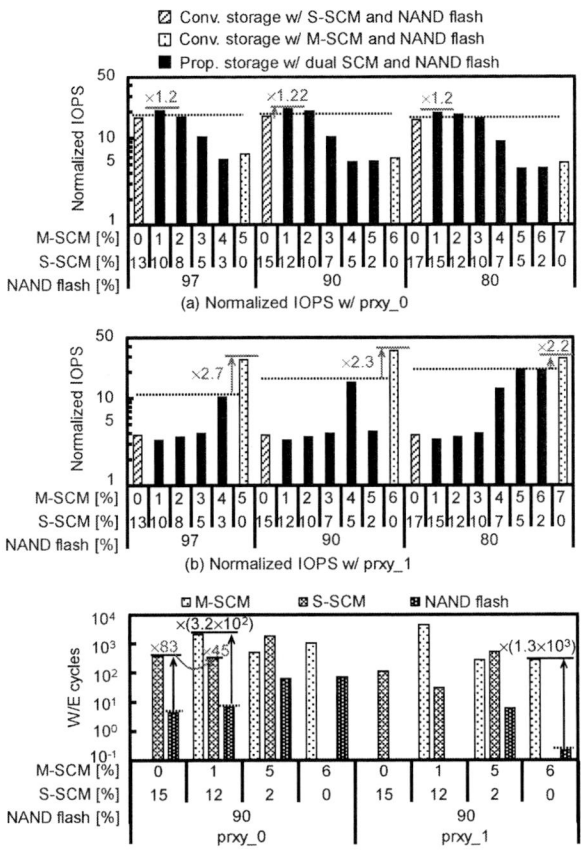

Fig. 7 Performance comparison with constant storage cost. (a) IOPS performance with prxy_0, (b) IOPS performance with prxy_1, and (c) required W/E cycles of M-SCM and S-SCM with the storage configuration which obtains the best performance.

Table II Summary of this work

Storage architecture for prxy_0	Capacity ratio [1] M-SCM, S-SCM, NAND flash	IOPS performance [1]	Required Write cycles [2]	
			M-SCM	S-SCM
Dual SCM and NAND flash (this work)	1%, 12%, 90%	×22 ↑ ×1.22 (+22%)	3.2×10^6	4.5×10^5 ⎫ ×1/2
S-SCM and NAND flash [8]	0%, 15%, 90%	×18	-	8.3×10^5 ⎭

[1] Capacity and IOPS performance of MLC NAND flash only storage being 1 (= 100%).
[2] Required W/E cycles of MLC NAND flash being 10^4.

and 90% NAND flash. Compared to the conventional storage with S-SCM and NAND flash [8], only 1% M-SCM boosts the performance by 22%, and reduces write cycles of S-SCM by half. The latter means that 1% M-SCM extends write endurance of S-SCM by 2 times. With NAND flash W/E cycles being 10^4, the required write endurance of M-SCM and S-SCM is 3.2×10^6 and 4.5×10^5, respectively.

ACKNOWLEDGMENT

The authors thank Y. Tsukamoto for his support and discussion. This work is partly supported by NEDO.

REFERENCES

[1] K. Takeuchi, "Novel co-design of NAND flash memory and NAND flash controller circuits for sub-30 nm low-power high-speed solid-state drives (SSD)," *IEEE J. Solid-State Circuits (JSSC)*, vol. 44, no. 4, pp. 1227-1234, Apr. 2009.

[2] H. Fujii et al., "x11 performacne increase, x6.9 endurance enhancement, 93% energy reduction of 3D TSV-integrated hybrid ReRAM/MLC NAND SSDs by data fragmentation suppression," *in Proc. IEEE Symp. VLSI Circuits*, Jun. 2012, pp. 134-135.

[3] S. Okamoto et al., "Application driven SCM and NAND flash hynrid SSD desing for data-centric computation system," *in Proc. IEEE Int. Memory Workshop (IMW)*, May 2015, pp. 157-160.

[4] C. Matsui et al., "Optimal memory configuration analysis in tri-hybrid solid-state drives with storae class memory and multi-level cell/triple-level cell NAND flash memory," *Jpn. J. Applied Physics (JJAP)*, vol. 56, no. 4S, pp. 04CE02-1-04CE02-9, Apr. 2017.

[5] R. F. Freitas and W. W. Wilcke, "Storage-class memory: The next storage system technology," *IBM J. Research and Development*, vol. 52, no. 4/5, pp. 439-447, Jul. 2008.

[6] G. W. Burr et al., "Overview of candidate device technologies for storage-class memory," *IBM J. Research and Development*, vol. 52, no. 4/5, pp. 449-464, Jul. 2008.

[7] http://researcher.watson.ibm.com/researcher/files/us-gwburr/ Almaden_SCM_overview_Jan2013.pdf.

[8] Intel Optane Technology, http://www.intel.com/content/www/us/en/ architecture-and-technology/intel-optane-technology.html.

[9] S.-W. Chung et al., "4Gbit density STT-MRAM using perpendicular MTJ realized with compact cell structure," *in Proc. IEEE Int. Electron Devices Meeting (IEDM)*, Dec. 2016, pp. 27.1.1-27.1.4.

[10] K. Kawai et al., "Highly-reliable TaOx ReRAM technology using automatic forming circuit," *in Proc. IEEE Int. Conf. IC Design & Technology (ICICDT)*, May 2014, pp. 100-103.

[11] Y. Choi et al., "A 20nm 1.8V 8Gb PRAM with 40MB/s program bandwidth," *in Proc. IEEE Int. Solid-State Circuits Conf. (ISSCC)*, Feb. 2012, pp. 46-48.

[12] Micron 3D XPoint Technology, https://www.micron.com/about/our-innovation/3d-xpoint-technology.

[13] J. L. Hennessy and D. A. Patterson, Computer Architecture: A Quantitative Approach 5th ed., Morgan Kaufmann, 2012.

[14] MSR Cambridge Traces, http://iotta.snia.org/traces/388.

SCM and NAND flash (Fig. 1(b)), M-SCM and S-SCM stores super-hot and hot data, respectively. With prxy_1 application (Fig. 7(b)), however, more than 5% M-SCM and NAND flash hybrid storages, in which NO S-SCM is used, show the best performance as discussed in IV-A. Thus, large capacity of M-SCM is required to improve the storage performance of prxy_1.

Fig. 7(c) shows W/E cycles of each memory when the storage obtains the best performance with prxy_0 and prxy_1 application. For prxy_0, W/E cycles of M-SCM and S-SCM are sufficient if they are 3.2×10^2 and 4.5×10^1 times higher than NAND flash W/E cycles. The S-SCM write cycles becomes half of conventional hybrid storage. In contrast, for prxy_1, M-SCM requires to have 1.3×10^3 times higher W/E cycles than NAND flash.

V. SUMMARY

This paper proposes a heterogeneous storage with dual SCM (M-SCM and S-SCM) and NAND flash memory. If the total storage cost is capped by 1.5, the IOPS performance of prxy_0 obtains the maximum with 1% M-SCM, 12% S-SCM,

Study of Error Repeatability and Recovery in 40nm TaOx ReRAM

Takashi Inose[1], Seiichi Aritome[1], Ryutaro Yasuhara[2], Satoshi Mishima[2], and Ken Takeuchi[1]

[1]Department of Electrical, Electronic, and Communication Engineering, Chuo University, Tokyo, Japan
[2]Panasonic Semiconductor Solutions Co., Ltd., 1 Kotari-yakemachi, Nagaokakyo, Kyoto, Japan
E-mail: inose@takeuchi-lab.org

Abstract— The repeatability of set/reset errors has been investigated in 40nm TaOx based ReRAM cells. Errors of the Low Resistance State (LRS) in specific cells are observed repeatedly, and such cells are recovered by DC read operation. When error cells are recovered, the LRS cell current of the recovered cells shows a sudden jump up to large cell current in certain set cycles. Then, the High Resistance State (HRS) current is also recovered from almost no current to normal HRS current. The phenomena of repeating error can be explained as lack of oxygen vacancy in specific cells which have a wide conductive filament. Moreover, the error repeated cells have larger LRS cell current than the normal LRS cell current after recovery. Thus, the potential error repeated cells can be detected by cell current check sequences, which detect to repeat the larger LRS cell current. The failure rate can be reduced by replacing the error repeated cells with redundant cells.

Index terms—ReRAM, LRS, redundant cell.

I. INTRODUCTION

Recently, resistive random access memories (ReRAMs) are considered as promising candidates for the next generation non-volatile memory devices due to its properties of fast and low current switching, simple structure and process, scalability, good retention and endurance [1]-[7]. However, it has been reported that BER (Bit Error Rate) is high due to random bit errors [8][9]. In this paper, the properties of error cells are studied. Errors are repeated in set/reset cycling in specific error cells, and are recovered by DC read operation. An operation sequence to detect the error repeated cells is proposed. BER can be reduced by replacing these error repeated cells with redundant cells.

II. DECIVE AND MEASUREMENT METHOD

Fig. 1 shows the structure of a ReRAM cell, which is constructed by one-transistor and one-resistor (1T1R). The resister is composed of $Ta_2O_{5-\delta}$ and TaO_x layers. Table 1 describes the operation conditions. The resistance switching between the high resistance state (HRS) and the low resistance state (LRS) are performed by a diffusion of oxygen vacancy (Vo) or oxygen ion (O^{2-}) between a conductive filament (CF) and a $TaOx$ layer [10]. In set operation, LRS is made by removing O^{2-} from the CF region to the $TaOx$ layer by applying set voltage to the source-line (SL). On the other hand, in reset operation, HRS is made by moving O^{2-} from the $TaOx$ layer to the CF region by applying a reset voltage to the bit-line (BL). Fig. 2 shows the measurement sequence. The

measurement of 10^3 set/reset cycles with the DC read operation follows 10^6 set/reset cycles without the DC read, and the mesurement are repeated three times (1st to 3rd mesurement). The DC read is a cell current measurment test, and has a longer read time than the normal read operation. The DC read operation is applied during the 10^3 set/reset cycles, although it is not applied during the 10^6 set/reset cycles as shown in Fig. 2. Fig. 3 shows BER on LRS and HRS during the 10^3 cycles with the DC read operation. High BER of LRS is observed after the 10^6 set/reset cycles, and decreases during the 10^3 set/reset cycles by the DC read operation. It is considered that error cells are recovered to normal LRS by DC read effects [11]. For convenience of analysis, the test device is operated in a severe condition and errors are intentionally caused.

III. ERROR CLASIFICATION AND ERROR CHARACTERISITICS

Table 2 shows classification of LRS errors and recovery cells. Group A is the good cell group, which has no errors in the 1st ~ 3rd measurements. Group B~E are the 1-time error cells groups. Group F~L are the 2- or 3-times error repeated cells groups. Fig. 4 shows the example of cell current of LRS and HRS in a single cell, (a) and (b) are for the good cells, (c) and (d) are error recovered cells, and (e) and (f) are permanent error cells. As shown in (c) and (d) in Fig. 4, the cell current in LRS suddenly jumps up and is recovered to normal cell current. At same time, cell current in HRS is raised up from almost zero current to normal HRS current, as shown (d) in Fig. 4. In Fig. 4, (e) and (f) show the permanent error cells. Their cell current is almost zero both in LRS and HRS.

IV. ERROR RECOVERY

Fig. 5 shows the average cell current in LRS and HRS of the good cells (Group A) and the 1-time error cells (Group B ~ E) during the 10^3 set/reset cycles with the DC read operation. In Group B, the average current of LRS is gradually recovered in the 0 ~ 600 set/reset cycles in the 1st measurement, as shown in (a) of Fig. 5. This means the cell current of each error cell is sequentially recovered in the 0 ~ 600 set/reset cycles. In Group D and E, the average cell currents of LRS are also recovered in the 0 ~ 100 set/reset cycles in the 2nd and 3rd measurements.

Fig. 6 shows the average cell current of Group F ~ L, compared with Group A. In the 2-time error cells (Group F ~ H), the average current of LRS is recovered in the 0~800 set/reset cycles, as well as the current recovery in the 1-time error cells shown in Fig. 5. It is newly observed that, after cell current is recovered, the average cell currents of the 3-time

978-1-5090-5979-9/17 $31.00 © 2017 IEEE

error cells in Group I and L exceeds the LRS cell current of the good cells, as shown in (a),(c),(e) and (k) in Fig. 6. Fig. 7 shows the average cell current of the 2-time and the 3-time error cells, which is highlighted in (e) of Fig.6. The LRS cell current of Group I (3-time error) is clearly larger than that of Group A (the good cells) after recovery.

Fig. 8 shows the cell current distributions of LRS, (a) is for the good cells (Group A) and (b) is for the 1- ~ 3-times error cells (Group B~L). Error cells (Fig. 8(b)) show high population at cell current around 5.0. Such cells are mainly caused after recovering LRS cell current, as shown in Figs. 6 and 7. Fig. 9 shows the cell current distribution of HRS, (a) is for the good cells and (b) is for the 1- ~ 3-time error cells (Group B~L). The cell current of HRS error cells (Fig. 9(b)) cannot be made to low current (the normal HRS current), and has also larger cell current than the good cells (over 5.0). Therefore, the error repeated cells in LRS have a tendency to have the larger cell current than the good cells in LRS.

V. Mechanism

Fig. 10 shows the resistive switching and the larger cell current mechanism in ReRAM cells. The resistive switching has been explained by providing and removing oxygen vacancy in the conductive filament (CF) area [12,14], as shown in Pattern 1 of Fig. 10 marked "Good cell". Error cells are caused by lack of oxygen vacancy in LRS, as shown in "Failure cell" in Pattern 1 of Fig. 10. The recovery is considered as compensation of oxygen vacancy by the DC read operation [11]. After a large number (10^6) of set/reset cycles (Pattern 2 in Fig. 10), the area of CF expands due to heating by set/reset current during the cycles, as shown in Pattern 2 (wide filament) in Fig. 10[4,13]. A concentration of oxygen vacancy is low in wide CF, then error cells are easily caused by lack of oxygen vacancy. In the error cells with wide CF, when they are recovered by the DC read, cell current become larger than the normal cells (with narrow CF) because larger cell current flows through wide CF area. Therefore, it is considered that the cells with wide CF suffer repeated errors and have the larger LRS cell current, because of lower density of oxygen vacancy and the wide conductive filament (CF) area.

VI. Detection of error repeated cell

Fig. 11 shows a method to check the potential error repeated cells in the following set/reset cycles. Read cell current levels are changed to "Low level", "Middle level" and "High level". Fig. 12 shows the read success (pass) rate in the three read level, (a) for "Low level", (b) for "Middle level" and (c) for "High level". The read success (pass) rate means the probability that cell current is less than the read level. By using the "High level" read (Fig. 12(c)), the read success (pass) rate of the 3-times error cells can be kept high, while that of the good cells and the 1- and 2-times error cells are kept to low enough. Therefore, the 3-times error repeated cells can be detected in a higher rate. Therefore, it is possible to reduce the error rate by replacing the detected cells (the error repeated cells) to redundant cells.

VII. Method of redundandacy replacement

After a certain number of set/reset cycles (e.g. 10^6 cycles), the cells with high read current in LRS are detected by "High level read" operation, and their addresses are stored in a controller, as shown in Fig.13. After a certain number of set/reset cycling, "High level read" is applied again. If the number of times for detecting high current exceeds 3 times, the detected cells are replaced with the redundant cells, as show in Fig.13. A peripheral circuit for realizing the proposed method is also shown in Fig. 13. By using a sense amplifier, it is judged whether the cell current exceeds a predetermined value or not. If it exceeds the predetermined value, a command is sent to the controller and such cells are defined as cells that need replacement.

VIII. Conclusiton

The repeatability of set/reset errors has been investigated in 40nm TaOx based ReRAM cells. It is found that the error on LRS are repeated during set/reset cycles, and recovered by the DC read operation. The potential error repeated cells can be detected by the cell current check sequences, which detect the higher cell current in LRS. The failure rate can be reduced by replacing the error repeated cells with the redundant cells.

Acknowledgement

This work is partially supported by CREST, JST.

Reference

[1] Z. Wei et al., "Switching and reliability mechanisms for ReRAM", *IITC*, pp. 349-352, 2014.

[2] Z. Wei et al., "Innovation culture, transformational capability and transformational performance:An empirical study of SMEs in China", *IMW*, pp.1-4 , 2012.

[3] A. Kawahara et al., "Filament Scaling Forming Technique and Level-Verify-Write Scheme with Endurance Over 107 Cycles in ReRAM", *ISSCC*, pp.220-223, 2013.

[4] Y. Hayakawa et al., "Highly reliable TaOx ReRAM with centralized filament for 28-nm embedded application", *VLSI Tech, Dig.*, pp. 2015.

[5] F. Nardi et al., " Sub-10 μA reset in NiO-based resistive switching memory (RRAM) cells", *IMW*, pp.1-4, 2010.

[6] Z. Wei et al., "Demonstration of High-density ReRAM Ensuring 10-year Retention at 85oC Based on a Newly Developed Reliability Model", *IEDM*, pp.729-732, 2011.

[7] T. Cabout et al., "Effect of SET temperature on data retention performances of HfO2-based RRAM cells", *IMW*, pp1-4, 2014.

[8] A. Alvarez et al., "14.3 15fJ/b static physically unclonable functions for secure chip identification with <2% native bit instability and 140× Inter/Intra PUF hamming distance separation in 65nm", *ISSCC Tech. Dig.*, pp. 256-257, 2015.

[9] Y. Yoshimoto et al., "A ReRAM-based physically unclonable function with bit error rate < 0.5% after 10 years at 125°C for 40nm embedded application", *VLSI Tech. Dig.*, pp., 2016.

[10] K. Kawai et al., "Highly-reliable TaOx reram technology using automatic forming circuit", *ICICDT*, pp.1-4, 2014.

[11] K. Maeda et al., "Error Recovery of Low Resistance State in 40nm TaOx-based ReRAM", *IRPS*, pp.1-6, 2017.

[12] M.Liu et al., "Formation and Annihilation of Cu Conductive Filament in the Nonpolar Resistive Switching Cu/ZrO2:Cu/Pt ReRAM", *ISCAS*, pp.1-4, 2010.

[13] D. Ielmini, *et al.*, "Modeling the Universal Set/Reset Characteristics of Bipolar RRAM by Field- and Temperature-Driven Filament Growth", *TED*, vol.58, pp3246-3253,2011.

[14] A. Hayakawa, et al, "Resolving Endurance and Program Time Trade-off of 40nm TaOx-based ReRAM by Co-optimizing Verify Cycles, Reset Voltage and ECC Strength", IMW, 2017

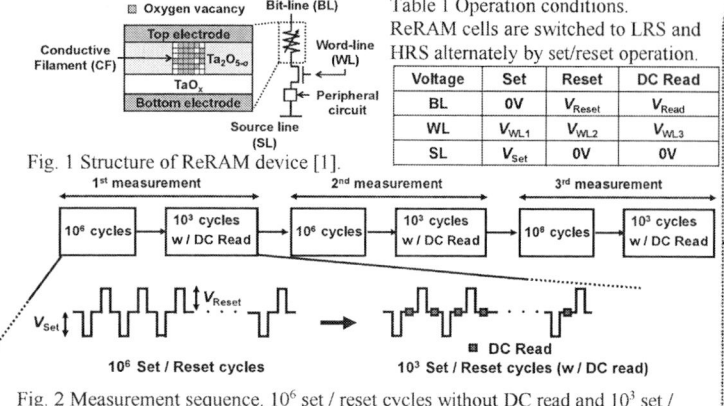

Fig. 1 Structure of ReRAM device [1].

Table 1 Operation conditions. ReRAM cells are switched to LRS and HRS alternately by set/reset operation.

Voltage	Set	Reset	DC Read
BL	0V	V_{Reset}	V_{Read}
WL	V_{WL1}	V_{WL2}	V_{WL3}
SL	V_{Set}	0V	0V

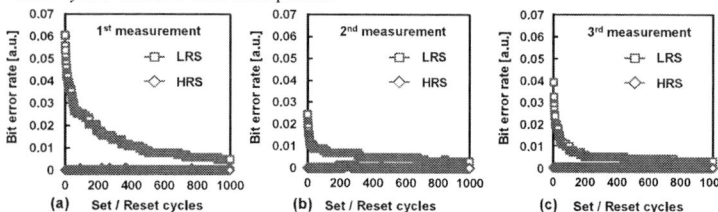

Fig. 2 Measurement sequence. 10^6 set / reset cycles without DC read and 10^3 set / reset cycles with DC read are repeated.

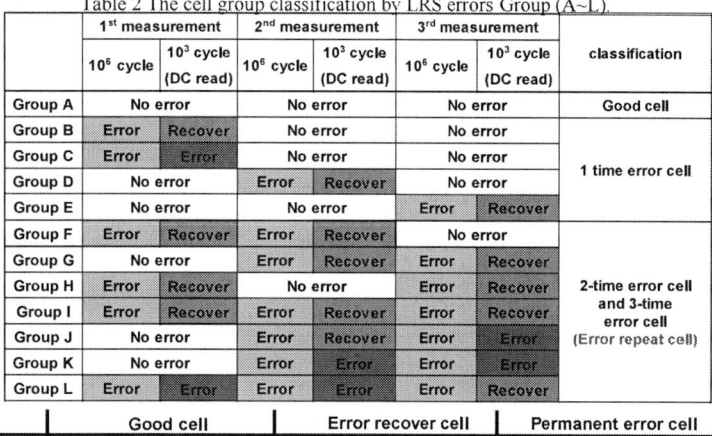

Fig. 3 Bit Error Rate (BER) of LRS and HRS during 10^3 cycles with DC read. High BER of LRS is caused by 10^6 set / reset cycles, and decreases during 10^3 set / reset cycles with DC read.

Table 2 The cell group classification by LRS errors Group (A~L).

	1st measurement		2nd measurement		3rd measurement		classification
	10^6 cycle	10^3 cycle (DC read)	10^6 cycle	10^3 cycle (DC read)	10^6 cycle	10^3 cycle (DC read)	
Group A	No error	No error	No error	No error	No error	No error	Good cell
Group B	Error	Recover	No error	No error	No error	No error	1 time error cell
Group C	Error	Error	No error	No error	No error	No error	
Group D	No error	No error	Error	Recover	No error	No error	
Group E	No error	No error	No error	No error	Error	Recover	
Group F	Error	Recover	Error	Recover	No error	No error	2-time error cell and 3-time error cell (Error repeat cell)
Group G	No error	No error	Error	Recover	Error	Recover	
Group H	Error	Recover	No error	No error	Error	Recover	
Group I	Error	Recover	Error	Recover	Error	Recover	
Group J	No error	No error	Error	Recover	Error	Error	
Group K	No error	No error	Error	Error	Error	Error	
Group L	Error	Error	Error	Error	Error	Recover	

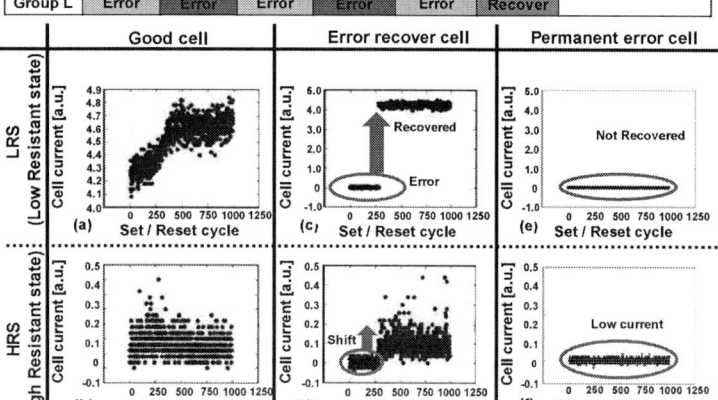

Fig. 4 Example of cell current of LRS and HRS in a single cell, (a) and (b) for good cells, (c) and (d) for error recovered cells, (e) and (f) for permanent error cells. In (c) and (d), cell current in LRS suddenly increases and recovers to the normal cell current. At the same time, the cell current in HRS increases.

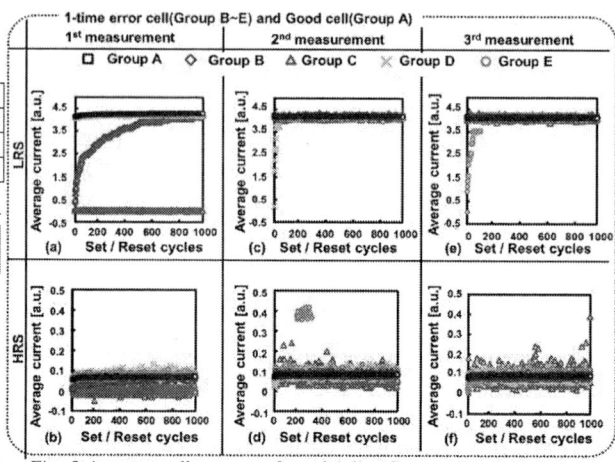

Fig. 5 Average cell current of good cells (Group A) and the 1-time error cells (Group B~E). The average cell current of the 1-time error cells is gradually recovered from the 0~800 set/reset cycles with the DC read.

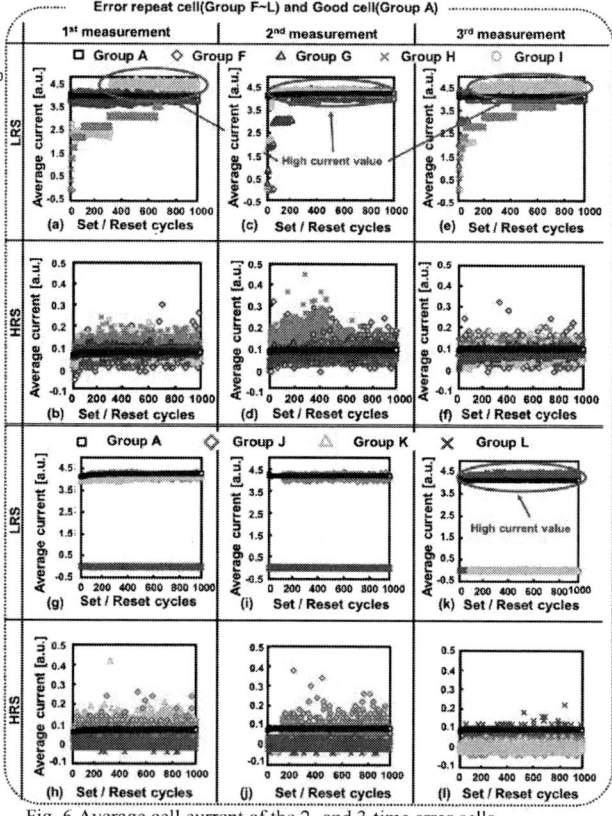

Fig. 6 Average cell current of the 2- and 3-time error cells. These cells show higher LRS cell current than the good cells of (a),(c),(e) and (k).

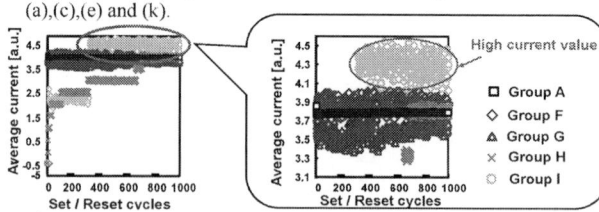

Fig. 7 Average cell current of 2- and 3-time error cells (Group F,G,H,I), which is highlighted in Fig. 6(a). The LRS cell current of Group I is clearly larger than that of Group A.

978-1-5090-5979-9/17 $31.00 © 2017 IEEE

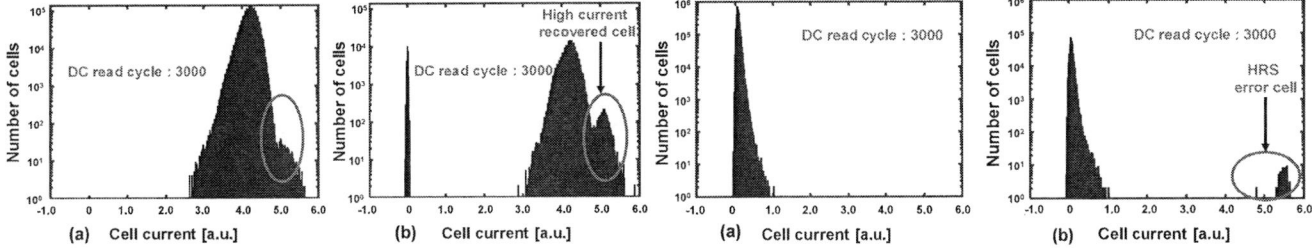

Fig. 8 Cell current distribution of LRS in (a) good cell and (b) 1~3 time error cell.   Fig. 9 Cell current distribution of HRS in (a) good cell and (b) 1~3 time error cell.

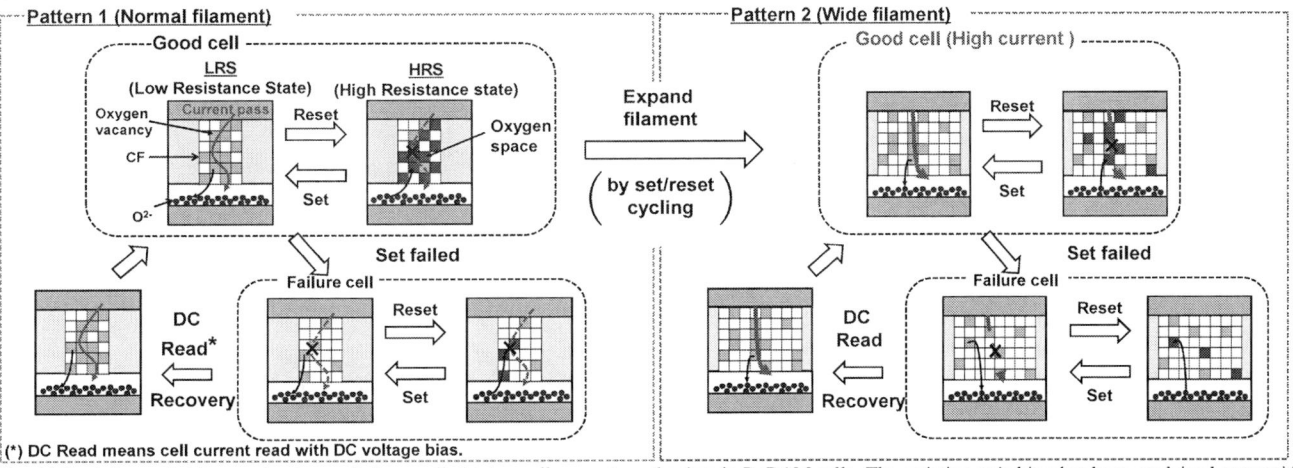

Fig. 10 Resistive switching mechanism and abnormally higher cell current mechanism in ReRAM cells. The resistive switching has been explained as providing and removing oxygen vacancy in the conduction filament, as shown in "Good cell" in Pattern 1 (Normal filament). Error cells are caused by lack of oxygen vacancy in LRS, as shown in "Failure cell" in Pattern 1. Recovery is considered as compensation of oxygen vacancy by DC read operation. After the large number of set/reset cycles (Pattern 2), the area of a filament expands due to heating by set/reset current, as shown in Pattern 2 (Wide filament). The cells with wide CF have larger cell current than the normal cells because of larger cell current paths in a wide CF area. It is considered that the cells with wide CF suffer from errors again, and have abnormally larger LRS cell current due to extended current paths of wide CF.

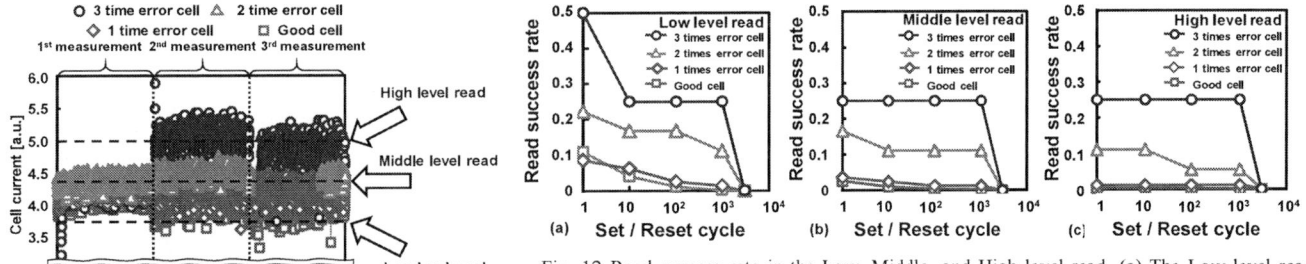

Fig. 11 Proposed method to detect abnormally higher LRS cells after high endurance cycles by using the multi-level read.

Fig. 12 Read success rate in the Low, Middle, and High level read. (a) The Low level read cannot be used to judge whether cells are good or will cause multiple errors in the following endurance. (c) The High level read can efficiently detects future multiple error cells by detecting the abnormally higher LRS current.

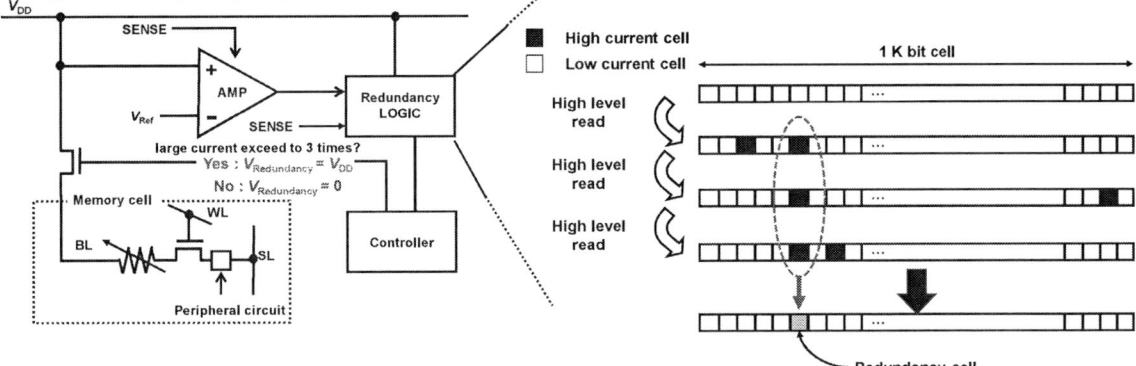

Fig. 13 Proposed circuit of redundant logic and the peripheral circuits. If the large current is detected over 3 times, the cells are regarded as bad and will be replaced by the redundancy cells.

978-1-5090-5979-9/17 $31.00 © 2017 IEEE   13

Optimization of Writing Scheme on 1T1R RRAM to Achieve both High Speed and Good Uniformity

Shan Wang, Huaqiang Wu*, Bin Gao, Ning Deng, Dong Wu, and He Qian

Institute of Microelectronics, Tsinghua University, Beijing 100084, China

*wuhq@tsinghua.edu.cn

Abstract—**This paper systematically analyzed the tradeoff between writing operation time and tail bit of LRS, and provided the optimal writing operation time for 1T1R RRAM with the target LRS 500kΩ and HRS 10MΩ. Under three different cases of pulse width, the experiment results all show that the optimal voltage amplitude and step could achieve a good tradeoff between writing operation time and tail bits of LRS. Based on the analysis of three kinds of pulse width, we find that the verification operation strategy using the smallest pulse width (10ns) shows the best performances. The RESET process needs three pulses and SET process needs four pulses at most. And the tail bit rate is lower than 7%. The related voltage overshoot effect on the pulse rising edge was discussed to explain the origin that the short pulse width is superior to long pulse width.**

Keywords—RRAM, writing time, tail bit, tradeoff, overshoot effect.

I. INTRODUCTION

Resistive random access memory (RRAM) has attracted great interests as one of the most promising technologies for the next generation non-volatile memory due to its outstanding performances [1], [2]. Variability is a key issue which blocks the application of RRAM [3]-[4]. Various verify writing schemes are used to suppress the variations of parameters in a RRAM array. Using this method, a series of variable pulses are applied on a memory cell. A read process follows each write pulse to decide whether the resistance reaches the target value. This method requires a relative long time to write a cell. Therefore, although it is well-known that the write speed of RRAM cell can be as short as several nano-seconds, the average total time to write a cell might be much longer in a RRAM array. Therefore, reducing the verify process is very crucial. Actually, increasing operation voltage and step is beneficial to reduce the required writing pulse number, but the tail bit of LRS will be largely increased, which can cause large power consumption and circuit instability. So the pulse conditions during a verify process should be carefully designed to make a good tradeoff between writing operation time and uniformity.

In this work, we successfully demonstrated a good tradeoff between writing operation time and tail bit of LRS by adjusting an appropriate pulse height and step. Meanwhile, it was also experimentally demonstrated that the verification operation scheme with short pulse width was superior to that with long pulse width and the related voltage overshoot effect on the pulse rising edge was discussed. Based on the above analysis, the optimal performance on the HfO$_x$/TaO$_x$ bilayer 1T1R (one-transistor-one-resistor) RRAM devices fabricated with 130nm CMOS process was achieved.

II. DEVICE INTRODUCTION

The diagrammatic sketch of the fabricated 1T1R device is shown in Fig. 1(a). In the 1T1R architecture, the transistor is used to control the current through the RRAM cell by applying various voltages on the gate terminal. Fig. 1(b) is the architecture of the HfOx/TaOx based RRAM stack. For the fabrication of 1T1R RRAM devices, the MOSFETs were fabricated with 130nm CMOS process in foundry and the RRAM devices were fabricated using BEOL (back-end-of-the-line) process in the lab. HfOx/TaOx bilayer structure is used as the resistive switching layers and TiN layers are used for both bottom and top electrode. Meanwhile, the TaOx layer serves as an oxygen reservoir and act as an in-built current compliance layer, which helps the transistor reduce the surge current generated by the parasitic capacitance and the Vth shift of transistor [5]. The HfOx layer was deposited by atomic layer deposition (ALD) and the TaOx layer was deposited by physical vapor deposition (PVD). The device exhibits excellent resistance switching performance with RESET current less than 20µA and ON/OFF ratio larger than 20 times.

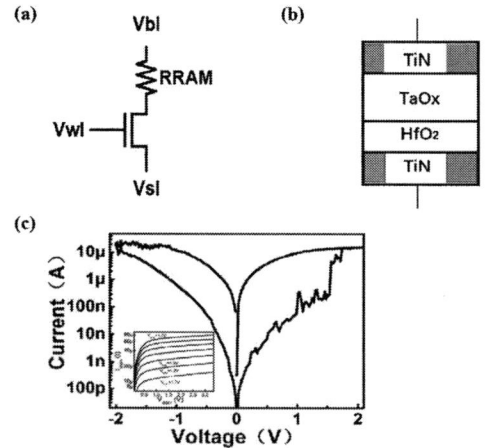

Fig. 1. (a) Diagrammatic sketch of 1T1R device. (b) Structure of RRAM stack. (c) DC I-V of the 1T1R RRAM, inset is the transistor I-V curves.

The typical DC sweep I-V curve of the 1T1R RRAM device measured by Agilent B1500A semiconductor parameter analyzer is shown in Fig. 1(c), and inset is the I-V curve of the integrated transistor. The HfOx/TaOx bilayer RRAM shows bipolar switching property, which means that the switching

978-1-5090-5979-9/17 $31.00 © 2017 IEEE

direction depends on the polarity of the applied voltage [6]. During SET process, voltage pulse is applied on the top electrode of RRAM and the conductive filaments (CFs) are formed, which switches the RRAM device from high resistance state (HRS) to low resistance state (LRS). And during RESET process, voltage pulse is applied on the source of transistor and the exerting CF's break down, which leads to the RRAM device switching from LRS to HRS [7]. Gradual RESET switching and sharp SET switching are observed.

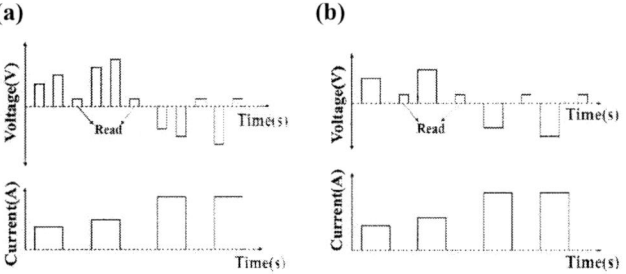

Fig. 2. (a) Sketch of operation scheme with pulse width of 10ns. (b) Sketch of operation scheme with pulse width of 20ns and 50ns.

Fig. 2 shows the verification operation scheme of three kinds of pulse width separately. During RESET process, the RRAM device adopted incremental step pulse programming (ISPP) strategy. If the high resistance cannot reach to the target HRS after a RESET pulse, another writing pulse with increased amplitude and the same pulse width was applied. And during SET process, traditional ISPP was used to adjust the SET pulse voltage amplitude first, and then current compliance was gradually increased by transistor gate controlling. For the three kinds of pulse width, the difference is that the scheme using the pulse width of 10ns adopted the measurement that reading once after applying two write pulses to reduce reading time during SET process, and during RESET process, the measurement condition was that reading once after applying two write pulses first, and then reading once after applying one pulse and so on, which is shown in Fig. 2(a). The other two schemes using the pulse width of 20ns and 50ns and both used the condition that reading once after one write pulse during both SET and RESET processes, which are shown in Fig. 2(b).

III. RESULTS AND ANALYSIS

Taking the verification operation scheme with pulse width of 10ns as example, Fig. 3(a) shows the correlation between writing error rate and operation time of the 1T1R RRAM device with the target LRS 500kΩ during SET process. The lines with various colors express different operation voltage amplitudes applied on the gate terminal of the transistor. The operation voltage amplitude of black line, red line and blue line are in ascending order. The related resistance distribution of three kinds of operation voltage are shown in Fig. 3(b). Obviously, the Fig. 3(a) shows that the larger the operating voltage applied on the device, the less the number of pulses is required. However, the Fig. 3(b) shows that the distribution of tail bits in LRS degrades as the operation voltage increases. Even though the scheme with large operation voltage only needs 2 pulses at most, the large tail bits (lower than 100kΩ) rate leads to higher power consumption. Therefore, to strike a

balance between the writing time and power consumption, the appropriate operation voltage makes good tradeoff, which needs 4 pulses and the tail bit rate of LRS is about 7%.

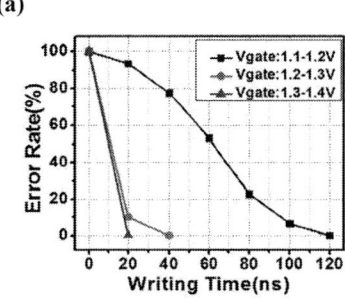

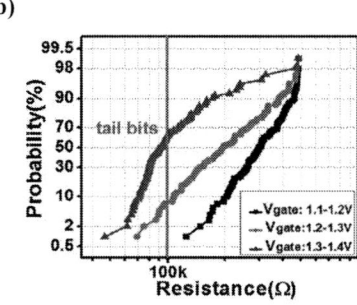

Fig. 3. (a) The correlation between writing error rate and operation time. The target LRS is lower than 500kΩ during SET process. (b) The resistance distribution of three kinds of operation voltage during SET process.

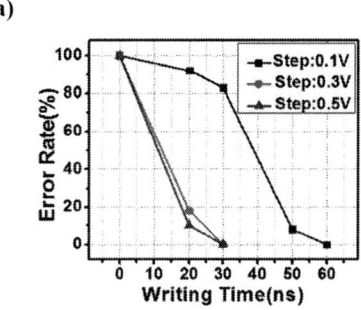

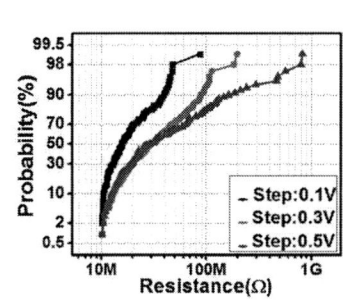

Fig. 4. (a) The correlation between writing error rate and operation time. The target HRS is higher than 10MΩ during RESET process. (b) The resistance distribution of three kinds of operation voltage during RESET process.

The correlation between writing error rate and operation time with the target HRS 10MΩ during RESET process is shown in Fig. 4(a), and Fig. 4(b) shows the related resistance

978-1-5090-5979-9/17 $31.00 © 2017 IEEE   15

distribution. The line with various colors express different pulse voltage amplitude and step applied on the source terminal of the transistor. The pulse voltage amplitude and step of black line, red line and blue line are in ascending order as well. According to the black and red line in Fig. 4(a), increasing pulse voltage amplitude and step appropriately can reduce pulse number effectively to a large extent. But if the pulse voltage amplitude and step continue to increase, as shown in the blue line, the number of pulses will no longer be reduced obviously. However, it can be seen from Fig. 4(b) that the high resistance state can be very high when the pulse voltage amplitude and step is large. Under this circumstance, the uniformity of HRS is poor and the RRAM devices are easily broken down. Therefore, the appropriate voltage amplitude and step is essential as well in the RESET process.

In addition to testing on a single cell, the experiments were also performed on 1k-bit array during SET process. Taking the verification operation scheme with pulse width of 50ns as example, the statistical experiment results of device-to-device are shown in Fig. 5. Fig. 5(a) shows the correction between writing error rate and operation time on 1k-bit RRAM array with the target LRS 100kΩ during SET process. The related resistance distribution of three kinds of operation voltage are shown in Fig. 5(b). From the statistical results point of view, similar conclusions can be drawn.

(a)

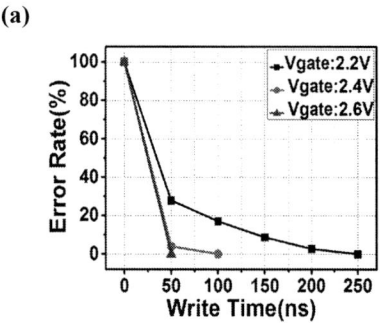

(b)

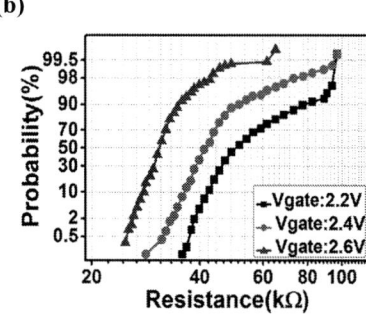

Fig. 5. (a) The correlation between writing error rate and operation time. The target LRS is lower than 100kΩ during SET process. (b) The resistance distribution of three kinds of operation voltage during SET process.

Furthermore, pulse width is also a critical ingredient for optimizing the writing operation time. As the above method, the optimal tradeoff can be obtained separately by applying on pulse width of 20ns and 50ns as well. During SET process, the correlation between writing error rate and operation time for three kinds of pulses width and the related resistance

distributions are exhibited in Fig. 6(a) and Fig. 6(b), and during RESET process they are shown in Fig. 7(a) and Fig. 7(b) separately.

(a)

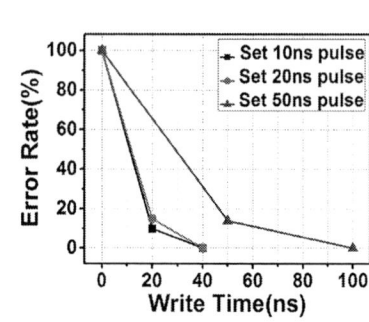

(b)

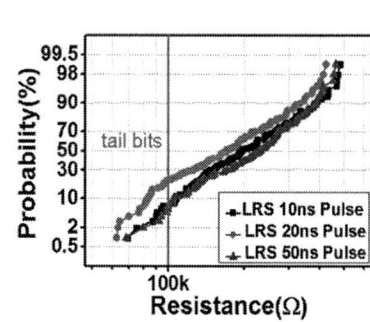

Fig. 6. (a) The correlation between writing error rate and operation time for three kinds of pulses width during SET process. (b) The resistance distribution of during SET process.

(a)

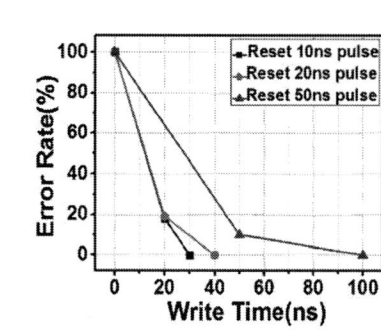

(b)

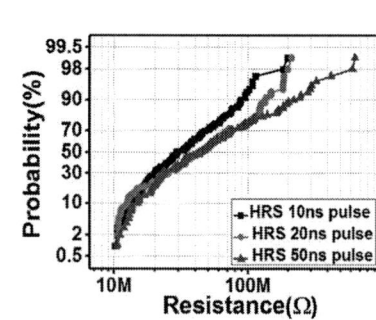

Fig. 7. (a) The correlation between writing error rate and operation time during RESET process. (b) The resistance distribution during RESET process.

978-1-5090-5979-9/17 $31.00 © 2017 IEEE

Table I lists the writing operation time and tail bit rate of the LRS lower than 100kΩ for pulse width of 10ns, 20ns, and 50ns separately. By comparison, the operation scheme with pulse width of 10ns shows the best performances. The RESET process and SET process need three pulses and four pulses respectively, and 7% tail bit rate is reached. The experimental results reveal that several short pulses are better than one long pulse both in terms of writing operation time and power consumption.

Table I. Statistics from 1T1R RRAM device for different pulse widths.

Operation Mode / Pulse Width	SET			RESET	
	Writing Time/Error Rate		Tail Bit Rate	Writing Time/Error Rate	
10ns	20ns/10%	40ns/0	7%	20ns/18%	30ns/0
20ns	20ns/15%	40ns/0	21%	20ns/19%	40ns/0
50ns	50ns/14%	100ns/0	7%	50ns/10%	100ns/0

We propose a model to explain this mechanism. In the model, a simplified equivalent circuit is used to describe the RRAM, and the sketch is shown in Fig. 8 [8].

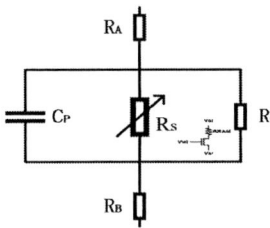

Fig. 8. Simplified model sketch of RRAM cell composed resistors and parasitic capacitor, inset is the diagrammatic sketch of 1T1R cell.

The related physical mechanisms can be expressed as follow: Due to the parasitic capacitance of RRAM cell and the Vth sift of transistor caused by voltage drop on the RRAM, there is voltage overshoot effect on the pulse rising edge as shown in Fig. 9.

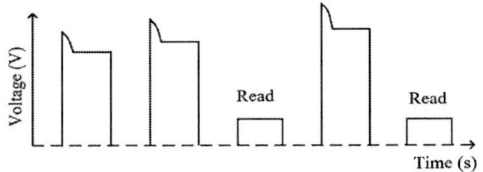

Fig. 9. Diagrammatic sketch of real voltage pulse applied on the RRAM due to the voltage overshoot effect.

The migration rate of oxygen ion is exponentially dependent on the voltage amplitude while linearly dependent on pulse width [6], which can be describes as:

$$V = a\, t\, \exp[(qlE-\varepsilon)/KT] \qquad (1)$$

where t is operation time, which is determined by the pulse width and the number of pulses. E is electric field in the gap region, which is determined by the voltage amplitude.

Therefore, even a small overshoot voltage can cause rapid oxygen ion migration significantly. Under this circumstance, increasing the number of pulses is more effective than increasing the pulse width to optimize the writing time.

The parasitic capacitance is approximately dozens of fF per square microns, so it is estimated that the overshoot voltage is about 0.3V. After a simple calculation, the effect of two or three 10ns pulses and one 50ns pulse is similar. So the calculated results are in agreement with the experimental results.

IV. CONCLUSION

By comparing various operation voltage amplitudes and steps, an optimal voltage amplitude and step solve the tradeoff between writing operation time and uniformity. Due to the voltage overshoot effect on the pulse rising edge, increasing the number of pulses is more effective than increasing the pulse width to reduce the writing time. This work provides valuable guidelines for optimizing writing operation time and non-volatile storage technology.

ACKNOWLEDGMENT

The authors would like to thank ICAC@IMECAS for the device fabrication. This work was supported in part by: China key research and development program (2016YFA0201801), National Hi-tech (R&D) project of China (2014AA032901), National Natural Science Foundation of China (61076115, 61674089, 61674087), and 863 program (2015AA016501).

REFERENCE

[1] Rainer Waser, Regina Dittmann, Georgi Staikov and Kristof Szot, "Redox-Based Resistive Switching Memories – Nanoionic Mechanisms, Prospects, and Challenges," in Advanced Materials, vol. 21, pp. 2632-2663, 2009.

[2] Byung Joon Choi, et al., "High-speed and low-energy Nitride memristors," in Adv. Funct. Mater., may 2016.

[3] Z. Fang, H. Y. Yu, X. Li, N. Singh, G. Q. Lo, and D. L. Kwong. "HfOx/TiOx/HfOx/TiOx Multilayer Based Forming Free RRAM Devices With Excellent Uniformity," in IEEE Electron Device Letters, April, 2011, vol. 32, no. 4, pp.566-568.

[4] B. Govoreanu, G. S. Kar, Y-Y. Chen, V. Paraschiv, S. Kubicek and A. Fantini et al., "10x10nm² Hf/HfOx Crossbar Resistive RAM with Excellent Performance, Reliability and Low-Energy Operation," in 2011 IEEE International Electron Devices Meeting (IEDM), 2011, pp. 31.6.1-31.6.4.

[5] Xueyao Huang, et al., "HfO2/Al2O3 multilayer for RRAM arrays: a technology technique to improve tail-bit retention," Nanotechnology, vol. 27, 395201, June 2016.

[6] H.-S. Philip Wong, Heng-Yuan Lee, Shimeng Yu, Yu-Sheng Chen, Yi Wu and Pang-Shiu Chen et al., "Metal–Oxide RRAM," Proceedings of the IEEE, vol. 100, No. 6, June 2012.

[7] H. Y. Lee, P. S. Chen, T. Y. Wu, Y. S. Chen, C. C. Wang and P. J. Tzeng et al., "Low Power and High Speed Bipolar Switching with A Thin Reactive Ti Buffer Layer in Robust HfO2 Based RRAM," in 2008 IEEE International Electron Devices Meeting (IEDM), 2008, pp.297-300.

[8] Haitong Li, Peng Huang, Bin Gao, Bing Chen, Xiaoyan Liu, Jinfeng Kang, "A Spice Model of Resistive Random Access Memory for Large-Scale Memory Array Simulation," in IEEE Electron Device Letters, vol. 35, NO .2, February 2014.ch. 08.

Analyzing Inference Robustness of RRAM Synaptic Array in Low-Precision Neural Network

Rui Liu[1], Heng-Yuan Lee[2], and Shimeng Yu[1]

[1]School of Electrical, Computer, and Energy Engineering, Arizona State University, Tempe, AZ 85287, USA
[2]Electronics and Optoelectronics Research Laboratory, Industrial Technology Research Institute (ITRI), Hsinchu 31040, Taiwan
Email: shimengy@asu.edu

Abstract—In this work, we investigate the robustness of 1-transistor-1-resistor (1T1R) synaptic array to implement a low-precision neural network. The experimental results on 1 kb HfOx-based RRAM array show a large on/off ratio (i.e. $> 10^5 \times$) and 5 stable resistance states can be reliably achieved with $10\times$ window between adjacent two states. As the RRAM has the resistance drift over time under read voltage stress, the impact of read disturbance occurred in 1T1R synaptic array on the neural network classification accuracy is analyzed with the RRAM compact model fitted with experimental data. The simulation results of a single-layer perceptron with compressed MNIST dataset indicate that 1) more stable multi-level states are desired to have higher mapping capability of weights, thus achieving a higher initial classification accuracy; 2) good mapping strategies that avoid the read disturbance-induced sign change on the most significant weight levels are very important to mitigate the classification accuracy loss.

Keywords—Neural network, RRAM, 1T1R, multi-level states, read disturbance, inference, classification accuracy.

I. INTRODUCTION

Neuro-inspired machine learning is now state-of-the-art computing paradigm in several application domains, e.g. visual object recognition and objection detection [1]. A typical neuromorphic computing system consists of two parts, neurons and synapses. The number of synapses is much larger than the number of neurons. For example, AlexNet has 60 million synapses and 650 thousand neurons [2], which need 230 MB of on-chip memory to store these parameters in a high precision format. Therefore, it is a grand challenge to implement those learning algorithms on mobile devices where the power and area are constrained. In order to reduce the hardware overhead, network pruning and parameter compression is desired in the learning algorithms [3, 4]. Furthermore, it is also attractive to use much more compact emerging non-volatile memory devices (i.e. PCM [5] or RRAM [6] based synapses) to overcome the large-area and leakage power of SRAM based synapses [7]. Recently, a low-precision neural network (with ternary level weights) has been experimentally demonstrated with binary RRAM arrays (using two cells representing -1, 0, +1 weights) [8]. However, the impact of the reliability issues such as read disturbance on the neural network performance has not been discussed in the prior works, which will be systematically analyzed in this work.

In this work, 1kb HfOx-based 1T1R array is used to implement the low precision synaptic weights in a single-layer perceptron to recognize the compressed MNIST dataset. By controlling the gate voltage of the selection transistor in the SET operation, the RRAM cell can be programmed to multi-level states, which makes it possible to map the synaptic weights to RRAM conductance. HfOx-based RRAM can be programmed up to 5 stable states thus can be used to map 1-bit with sign (-1, 0, +1) or 2-bit with sign (-1, -0.5, 0, +0.5, +1) weight levels in different ways. Here the relationship between the precision of weights with sign (denoted as n-bit) and the number of weight levels (denoted as L) is $L = 2^n + 1$. In the offline trained neural network, the weights are pre-trained in software and programmed into the synaptic devices for the subsequent inference tasks in a read-only operation mode. Because RRAM has the read disturbance issue (i.e. resistance drift over time under read voltage stress), the shit of weight from one level to another may adversely affect the neural network performance (i.e. the classification accuracy). We model the lifetime of each state by a RRAM compact model fitted with our experimental data, and perform the neural network simulation to investigate the impact of the read disturbance on the proposed low-precision neural network performance.

II. ELECTRICAL CHARACTERIZATION

In this work, ITRI's 1kb HfOx-based 1T1R array was used to characterize the electrical properties of the synaptic weights. The RRAM devices consisting of TiN/Ti/HfOx/TiN structure were integrated in 0.18 µm CMOS technology. More detail processes can be found in [9]. Transistors are used in this design to allow for individual access of each memory cell with minimized sneak path current during cell programming. This is critical for accurate tuning of conductance states for implementing synaptic weights.

In our experiments, NI PXI system with analog output module PXIe-6738 and source measurement unit (SMU) PXIe-4140 module were used to perform all the electrical measurements. Labview was used to communicate with the instruments and implement the programming operations. For the DC characterization, the SMUs from PXIe-4140 were used for performing voltage sweeps and current measurements. All the address signals and gate voltage (WL) of selection transistor were applied through analog outputs from PXIe-6738. To program all the cells in the 1T1R array, only one cell is selected at any given time by corresponding address signals. In the Forming/SET/Read operations, a positive voltage ($V_{SET}/V_{Forming}/V_{Read}$) is applied to the top electrode of RRAM (BL) while the bottom electrode is connected to ground by grounding the source (SL) of the transistor. In the RESET operation, a negative voltage (V_{RESET}) is applied to BL and SL is grounded. In all the cases, WL is used to select the cell to be programmed by applying a high enough voltage, however, during the SET operation the transistor gate voltage (V_{WL}) is an analog voltage that sets the current compliance.

Fig. 1(a) shows the DC sweeps of SET and RESET operations for different WL voltages. The increasing WL

978-1-5090-5979-9/17 $31.00 © 2017 IEEE

voltage results in increasing current compliance. The read-out current of RRAM is obtained by a read operation with biasing the WL to a high voltage (i.e. 1.5 V) and applying a 0.2 V read voltage at BL. Fig. 1(b) shows the read current distribution of 5 different 1T1R cells after the DC SET operations. This experimental result indicates the current limiting capability of the selection transistor to program the RRAM cell to multi-level states. This property enables the 1T1R array to implement the tunable synaptic weight.

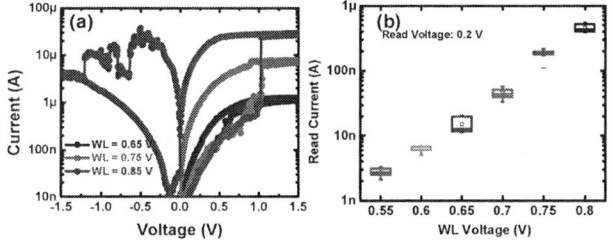

Fig. 1. (a) DC switching I-V curves for a 1T1R cell with increasing WL voltage from 0.65 to 0.75 V and 0.85 V. (b) The distribution of read currents with a low read voltage (i.e. 0.2 V) of the 1T1R cell after the DC SET operation as a fucntion of WL voltage.

III. LOW-PRECISION NEURAL NETWORK

In this work, a single-layer perceptron neural network is employed as a case study considering the available array size (32×32) we have in the 1kb 1T1R array. In the single-layer perceptron, the neurons are arranged in two layers, i.e. an input layer and an output layer. The input layer is fully connected to the output layer with a weight matrix. Also because of the relatively small array size, the MNIST handwritten digits [10] are employed as training and testing dataset but with scaling to 8×8 pixels, namely compressed MNIST dataset. First, the images in compressed MNIST dataset was converted to binary image (black and white), as shown in Fig. 2(a). Some of the digits after scaling become difficult to distinguish even by our human naked eyes, for example, 7 and 9 shown in Fig. 2(a). This could result in a lower classification accuracy baseline to start with as compared to the original MNIST dataset.

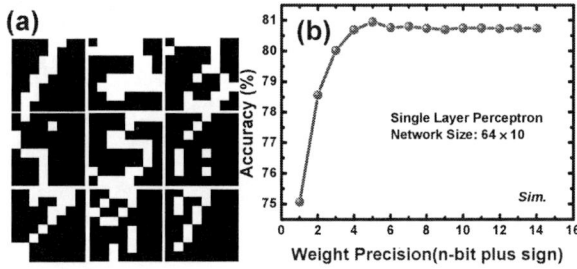

Fig. 2. (a) Samples of compressed MNIST handwritten digit dataset binarized to black and white. The compressed samples are 8×8-pixel by cropping edges and resizing. (b) Classification accuracy as a fucntion of synpatic weight precision. The accuracy slightly decreases as the weight precision is truncated to below 4-bit.

A. Offline Training Neural Network in Software

As a case of study, we trained a small single layer network structure 64-10 with the compressed MNIST dataset. The input layer has 64 neurons corresponding to the 8×8-pixel images and output layer of 10 neurons corresponding to the 10 classes of digits. For offline training, we first performed software training with floating point (32-bit) precision for a certain number of epochs with 60,000 training images in each epoch.

The classification accuracy (on 10,000 testing images) saturates ~ 81% after tens of training epochs. Finally, we chose a weight matrix that achieves classification accuracy of 80.7% for the network 64-10 as the ideal weight pattern to truncate the precision and being loaded in the 1T1R array experimentally.

B. Truncation of Weight Precision for Classification

It is unpractical to store the software trained weighs with high precision (e.g. 32-bit) on-chip, as it will contribute substantially to the area overhead and power consumption of the neuromorphic computing system. Therefore, it is necessary to compress the weights to lower precision at the expense of slight accuracy loss. Fig. 2(b) shows the classification accuracy as a function of the weight precision. The classification accuracy presents negligible deterioration when the weight precision is larger than 4-bit. However, the classification accuracy decreases gradually when the weight precision is smaller than 4-bit. The simulation result shows the accuracy is 75.07% with ternary weight levels (-1, 0, +1) and 78.57% with 5 weight levels (-1, -0.5, 0, +0.5, +1). Such low precision weight matrix can be implemented by a 1T1R array based on our HfOx RRAM with different conductance states. In general, the more states available, the higher mapping capability of weights can be achieved with a single RRAM cell. Fig. 3 shows an example of using a 1T1R array with 3-level states to implement a 64×10 ternary weight matrix. The RRAM cells in the array were SET to ~100nA to implement weight 1, ~10nA to implement -1 and ~100nA to implement 0, respectively. The current is measured at 0.2 V. Fig. 3 (a) shows the digitalized the weight pattern and Fig. 3(b) presents the actual measured current pattern of the 1T1R synaptic array. It should be noted that there are different ways to map weight values to RRAM conductance states. We will discuss it in the following section.

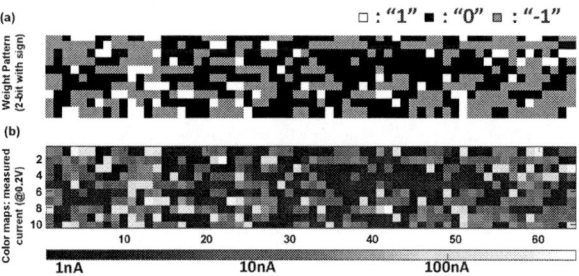

Fig. 3. (a) The weight pattern of a 64×10 single-layer peceptron neural network after software training. The weights are truncated to (+1, 0, -1). (b) The meausred weight pattern of 1T1R synaptic array when RRAM cells are programmed to with 3 levels to implement the ternary weights.

IV. IMPACT OF READ DISTURBANCE ON CLASSIFICATION ACCURACY

A. Read Disturbance Experiments and Modeling

When the RRAM is used as the synaptic device for the classification applications (read-intensive tasks), it is required that RRAM should have a good data retention even under read voltage stress and elevated temperature. The HfOx based RRAM device used in this work has a large on/off ratio (i.e. > 10^5) between the lowest conductance state and highest conductance state, and multi-level states can be achieved by controlling the gate voltage in the SET operation. However, to maintain enough window between two adjacent states, only 5 states (i.e. Level 1 to Level 5, as shown in Fig. 4) with 10× window for two adjacent states, could be reliably used against

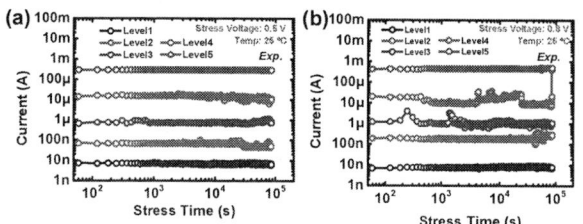

Fig. 4. Measured retention degradation of the RRAM cells with difference read-out current levels at 25 °C under stress voltage of (a) 0.5 V and (b) 0.8 V, respectively.

any temporal fluctuations of the read-out current in this work. To evaluate the RRAM's read robustness over time, high stress voltage is used to accelerate the failure. There is no error happened for any conductance states within 24 hours when the stress voltage is 0.5 V at room temperature, as shown in Fig. 4(a). When the stress voltage is increased to 0.8 V, error occurs only for resistance state Level 4 at ~23.5 hours. To capture the read robustness degradation behavior, a well-accepted conductive filament (CF) growth/dissolution based compact model [11] is employed to simulate the data retention under different voltage stress and temperature. The I-V relationship of the RRAM model is expressed as

$$I = I_0 \exp(-\frac{g}{g_0}) \sinh(\frac{V}{V_0}) \quad (1)$$

where I_0, g_0 and V_0 are the fitting parameter, i.e. 76.4 μA, 0.275 nm and 0.35 V, respectively, in this work. Under the voltage stimulus, the CF will grow or dissolute due to oxygen vacancies generation or recombination with oxygen ions at the tip of CF, which can be expressed by

$$\frac{dg}{dt} = -v_0 [\exp(-\frac{qE_{ag}}{kT}) \exp(\frac{\gamma a_0}{L} \frac{qV}{kT})$$
$$- \exp(-\frac{qE_{ar}}{kT}) \exp(-\frac{\gamma a_0}{L} \frac{qV}{kT})] \quad (2)$$

$$g(t + dt) = g(t) + dg \quad (3)$$

where dg/dt is the gap growth/dissolution velocity. E_{ag} and E_{ar} are the activation energy for oxygen vacancy generation and recombination, respectively, which are 1.5 eV and 1.501 eV, respectively. L is the oxide thickness (5 nm); a_0 is the atomic hopping distance (0.35 nm); γ is the g-dependent local field enhancement factor (7.7). The simulation results in Fig. 5(a) present the same failure time for Level 4 as experimental data as shows in Fig. 4(b). We assume the read voltage is 0.2 V in the inference application and the neuromorphic computing system works at 85 °C in the worst case. The simulation results

of data retention under stress voltage 0.2 V at 85 °C are shown in Fig. 5(b). The cell tends to drift towards higher conductance levels under positive read voltage stress as the CF tends to grow, and this process could be accelerated at elevated temperature. In practice, a current reference will be set between each two adjacent states to differentiate them by sensing circuit (e.g. sense amplifier). Error occurs when the read-out current drifts across the reference. However, error will never happen to L5 because it is the highest conductance state. Fig. 5(c) is the extracted lifetime for each level before the error occurs under continuous reading but with 50% read duty cycle assumed.

B. Weight Mapping Strategies

For a given weight precision of the weight levels, we can have different strategies to map them in the synaptic array. Since the device can reliably achieve 5 levels, the weight matrix 2-bit with sign (-1, -0.5, 0, +0.5, +1) can be implemented. In order to save power, the lowest 3 conductance states (i.e. Level 1 to Level 3) can also be employed to implement the weight matrix 1-bit with sign (-1, 0, +1). Table I and II present some examples of mapping strategies for 3-level and 5-level weight matrices, respectively.

C. Impact of Read Disturbance on Classification Accuracy

In this section, we incorporate the read disturbance-induced error model into the neural network simulation to study the

TABLE I.   MAPPING THE TRUNCATED 3-LEVEL WEIGHTS TO 3 DIFFERENT RRAM LEVELS WITH 3 DIFFERENT STRATEGIES

Weight	RRAM Read Current @ 0.2 V		
	S1	S2	S3
1	Level 1	Level 2	Level 3
0	Level 2	Level 3	Level 1
-1	Level 3	Level 1	Level 2

TABLE II.   MAPPING THE TRUNCATED 5-LEVEL WEIGHTS TO 5 DIFFERENT RRAM LEVELS WITH 5 DIFFERENT STRATEGIES

Weight	RRAM Read Current @ 0.2 V				
	S1	S2	S3	S4	S5
1	Level 1	Level 5	Level 4	Level 3	Level 2
0.5	Level 2	Level 1	Level 5	Level 4	Level 3
0	Level 3	Level 2	Level 1	Level 5	Level 4
-0.5	Level 4	Level 3	Level 2	Level 1	Level 5
-1	Level 5	Level 4	Level 3	Level 2	Level 1

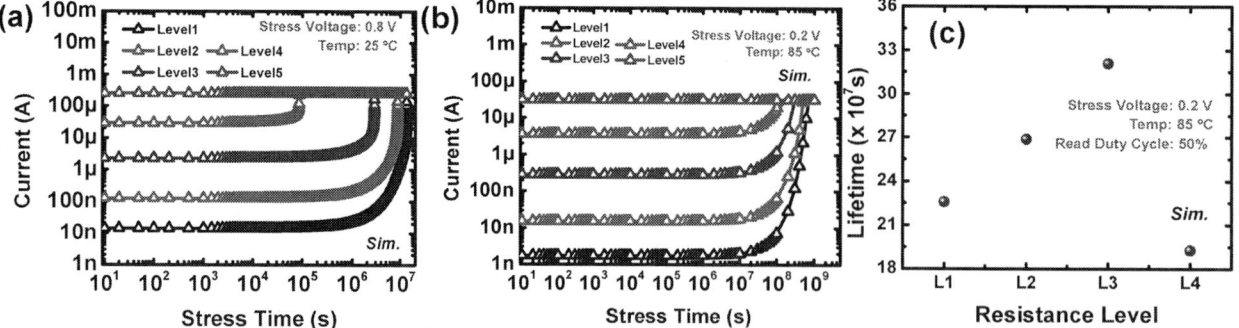

Fig. 5 Simulated read robustness degradation of the RRAM cells with difference condutance levels under stress voltage of (a) 0.8 V at 25 °C and (b) 0.2 V at 85 °C, respectively. (c) Extracted lifetime for each conductance level under stress voltage of 0.2 V at 85 °C with assumption of 50% read duty cylce.

978-1-5090-5979-9/17 $31.00 © 2017 IEEE

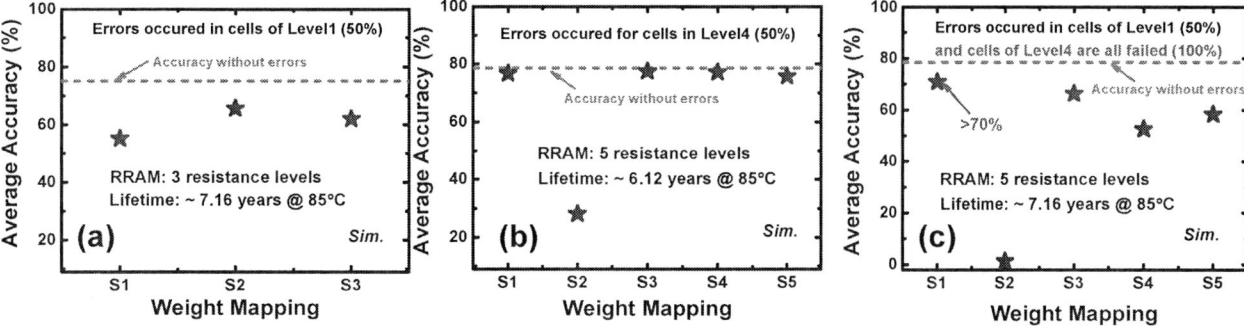

Fig. 6 (a) The accuracy loss of (-1, 0, +1) weight matrix over different mapping strategies when 50% of synaptic cells in Level 1 failed. The accuracy degration of (-1, -0.5, 0, +0.5, +1) weight matrix over different mapping strategies (b) when 50% of synaptic cells in Level 4 failed and (c) when all the synaptic cells in Level 4 failed and 50% of synaptic cells in Level 1 failed.

classification accuracy loss. Fig. 6 shows the classification accuracy for different mapping strategies as listed in Table I and II at different lifetimes. To map the (-1, 0, +1) weight matrix, states from Level 1 to Level 3 are employed. The synaptic 1T1R array can maintain no errors up to ~7.16 years under continuous reading (50% duty cycle). After that, errors start to happen in RRAM cells at Level 1. In the simulation, we assumed the error rate is 50% after ~3.58 years and error locations are random distribution in the array. In addition, since the cells of Level 1 drifts above the reference current between Level 1 and Level 2, we assume the error output of the sensing circuit is Level 2. When the weight is aggressively truncated to (-1, 0, +1), the classification accuracy loss caused by error in synaptic 1T1R array becomes remarkable (>10% compared to the ideal value) no matter what mapping strategy is used, as shown in Fig. 6(a). This is probably because each weigh level is significant, thus any error in (-1, 0, +1) weights will change the results of the weighted sum results remarkably.

To map the (-1, -0.5, 0, +0.5, +1) weight matrix, states from Level 1 to Level 5 are employed. While the synaptic 1T1R array can only maintain no errors for ~6.12 years under continuous reading (50% duty cycle). Later, error begins to occur in Level 4 and the corresponding error output is L5. However, the simulation results in Fig. 6(b) suggest that there is a negligible accuracy loss compared to the ideal value for all the mapping strategies except for S2 when Level 4 has 50% error rate. The large accuracy loss of S2 is probably because the error in Level 4 (shifting to Level 5) changed the sign of the most significant weight level from -1 to 1. We also investigated the classification accuracy loss when the error also happens in the second state (Level 1) at ~7.16 years. In the simulation, we assume all the cells in Level 4 and 50% of cells of Level 1 are failed. With a better mapping strategies (S1, S3, S4), the classification accuracy loss (< 5%) is still less than that of the (-1, 0, +1) weight matrix at the same lifetime, as shown in Fig. 6(c). This is because there is no sign change involved for the most significant weight levels 1 or -1 in theses mapping strategies.

V. CONCLUSION

In summary, we experimentally demonstrated the multilevel capability of HfOx-based 1T1R array for implementing the low-precision neural network, where the offline trained weights can be reduced to a few bits for inference in hardware. The simulation of neural network with RRAM read disturbance model evaluated the classification accuracy loss for different weight precision and mapping

strategies. Still more stable multi-level states are desired since it could achieve a slightly higher initial classification accuracy and shows better error resiliency. A smart weight mapping strategy that avoids the sign change for the most significant weight levels when error happens is also important to mitigate the impact of read disturbance on the neural network.

ACKNOWLEDGMENT

This work is supported by NSF-CCF-1552687. The authors thank Pai-Yu Chen and Zhiwei Li for the help in the algorithm.

REFERENCES

[1] Y. LeCun, Y. Bengio, and G. Hinton, "Deep learning," in *Nature*, vol. 521, pp. 436-444, 2015.

[2] A. Krizhevsky, I. Sutskever and G. Hinton, "Imagenet classification with deep convolutional neural networks," in *Advances in Neural Information Processing Systems* (*NIPS*), 2012.

[3] M. Courbariaux, Y. Bengio, J. P. David, "BinaryConnect: Training Deep Neural Networks with binary weights during propagations," in *Advances in Neural Information Processing Systems (NIPS)*, 2015.

[4] M. Rastegari, V. Ordonez, J. Redmon, A. Farhadi, "XNOR-Net: ImageNet Classification Using Binary Convolutional Neural Networks," in *European Conference on Computer Vision (ECCV)*, 2016.

[5] S. Kim, M. Ishii, S. Lewis, T. Perri, M. BrightSky, W. Kim, R. Jordan, G.W. Burr, N. Sosa, A. Ray, J.-P. Han, C. Miller, K. Hosokawa, and C. Lam, "NVM Neuromorphic Core with 64k-cell (256-by-256) Phase Change Memory Synaptic Array with On-Chip Neuron Circuits for Continuous In-Situ Learning," in *IEEE International Electron Devices Meeting (IEDM)*, 2015

[6] D Garbin, E Vianello, O Bichler, Q Rafhay, C Gamrat, G Ghibaudo, B. DeSalvo and L. Perniola, "HfO2-based OxRAM devices as synapses for convolutional neural networks," in *IEEE Transactions on Electron Devices*, vol. 62, no. 8, pp. 2494-2501, Aug. 2015.

[7] S. Yu, P.-Y. Chen, Y. Cao, L. Xia, Y. Wang, H. Wu, "Scaling-up resistive synaptic arrays for neuro-inspired architecture: challenges and prospect," in *IEEE International Electron Devices Meeting (IEDM)*, 2015.

[8] S. Yu, Z. Li, P.-Y. Chen, H. Wu, B. Gao, D. Wang, W. Wu, H. Qian, "Binary neural network with 16 Mb RRAM macro chip for classification and online training," in *IEEE International Electron Devices Meeting (IEDM)*, 2016

[9] H. Y. Lee, P. S. Chen, T. Y. Wu, Y. S. Chen, C. C. Wang, P. J. Tzeng, C. H. Lin, F. Chen, C. H. Lien, and M.-J. Tsai, "Low power and high speed bipolar switching with a thin reactive Ti buffer layer in robust HfO2 based RRAM," in *IEEE International Electron Devices Meeting (IEDM)*, 2008.

[10] Y. LeCun, C. Cortes, and C.J. Burges. MNIST handwritten digit database. AT&T Labs. [Online]. Available: http://yann.lecun.com/exdb/mnist/

[11] P.-Y. Chen and S. Yu, "Compact modeling of RRAM devices and its applications in 1T1R and 1S1R array design, " *IEEE Transactions on Electron Devices*, vol. 62, pp. 4022-4028, Dec. 2015.

978-1-5090-5979-9/17 $31.00 © 2017 IEEE

SPICE Modeling in Verilog-A: Successes and Challenges

Invited Paper

Colin C. McAndrew

NXP Semiconductors, Tempe, AZ 85284

Abstract—**Compact modeling has evolved considerably since SPICE was announced to the world in 1973. Many challenging model formulation problems have been solved, and model code itself has changed from being tightly integrated within simulators to being defined in a stand-alone manner. Decades of research led to the former, Verilog-A enabled the widespread adoption of the latter. This paper reviews key steps along both paths, with emphasis on the role of Verilog-A. There are still outstanding challenges, and these are outlined.**

I. Introduction

The design of modern integrated circuits (ICs) is based on a flow of an ever increasing assortment of computer-aided design tools. Circuit simulation is the foundation on which this design flow is built. Since the advent of SPICE almost 45 years ago [1], [2] there have been significant additions and improvements to the analysis algorithms used in circuit simulators[1]. However, even the most sophisticated and accurate simulation algorithms are useless without accurate models of the components ("devices") that are used to make the IC. One of the core strengths of SPICE was that it placed as much importance on models as on algorithms; to quote from the abstract of [1]: "particular emphasis is placed upon the circuit models for the BJT and the FET" (today "circuit models" are commonly referred to "compact models" or "SPICE models").

In the decades since SPICE spread throughout the semiconductor industry the growing complexity of technology[2] and circuits, increasingly stringent IC performance specifications, and an industry-wide push to reduce IC design cycle time have all focused attention on model accuracy and computational efficiency. This has led to significant improvements in the basic capabilities of models, i.e. in the physics that underlies them, and in how models are implemented in simulators.

This paper presents my perspective, and opinions, on the model improvement path the industry has followed, often reluctantly and always slowly, over the 30 years I have been involved in compact modeling.

II. Models in (Original) SPICE

If you have ever looked at the code (FORTRAN) of SPICE2G6 one thing is immediately apparent: it is no hacked-together, get-the-job-done-to-graduate program. The quality

```
action(inst, devType):
  if   devType == 'MOS1':
    codeForActionForMos1
  elif devType == 'MOS2':
    codeForActionForMos2
  elif devType == 'SGP':
    codeForActionForSgp
  elif devType == 'JFET':
    codeForActionForJfet
  ...
```

Fig. 1. Simplified representation of model implementation in SPICE. action is one specific operation, such as temperature mapping, dc evaluation, small-signal evaluation, noise evaluation, etc. inst is an integer index of the device instance in the netlist. devType is the type of device model.

of the coding is simply outstanding[3], and this is all the more remarkable because (speaking from my own personal experience) high-level coding skills, and knowledge of good software practices, were uncommon in electrical engineering students in the early 1970's.

One feature of the code was partitioning into subroutines that performed common operations, see Fig. 1. This simplified writing the high-level simulation flow control algorithms, and initially it enabled reasonably efficient model implementation[4]. However, as model formulation complexity increased it became a significant drawback.

Although simulation control could be done at a high level, some numerical method specific calculations (e.g. time integration for large-signal analysis) were embedded in the model code. The models and algorithms were thus intertwined, which unnecessarily added to the burden of model implementation: Model developers had to know in detail how the simulation algorithms worked. This also erected a barrier to improving simulation algorithms: To implement a new algorithm the code for *every* model had to be updated. In SPICE there were around 40 types of "action" subroutines like in Fig. 1. There were specific data structures to pass information between each of these action codes, so if the data structures for one device model were updated the code had to be updated in up to 40 places. As the computational complexity of models increased, to improve their accuracy, the size of code per model crept into the tens of thousands of lines. And a large part of this involved hand-coding of derivatives.

---

[1]Although the work-horse analyses (dc, large-signal, small-signal, noise), and the core algorithms that underlie them, have changed remarkably little.

[2]Here the term "technology" encompasses IC manufacturing technologies, devices, device structures, and layout dependent effects.

[3]It has been ventured that SPICE stalled circuit simulator development for over a decade because it was *so* good [3].

[4]The first MOS transistor model in SPICE, the "level 1" model [4], was implemented over a single weekend [5].

978-1-5090-5979-9/17 $31.00 © 2017 IEEE

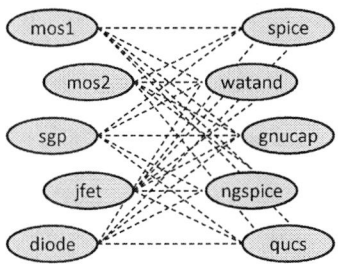

Fig. 2. Key issue with integrating model code within simulator code, and *vice versa*.

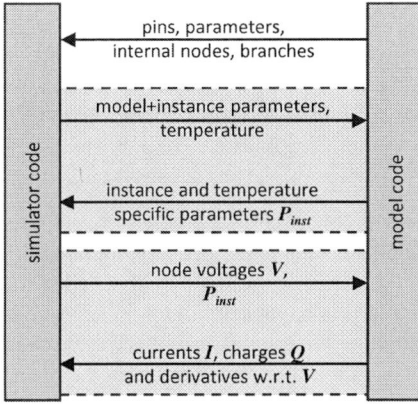

Fig. 3. Representation (simplified) of hand-shaking when model code is separated from simulator code.

The net result was that model implementation time expanded to be months instead of days, and as simulators other than SPICE emerged, with different data structures and algorithms, models had to be hand-implemented multiple times, see Fig. 2. With hindsight, a positive aspect of this situation was that it erected a barrier to "amateurs" dabbling in compact model development; anyone developing a model had to spend significant time learning the details of the model/simulator interface, and in hand-coding, which unintentionally selected for "professional" model developers.

III. SEPARATING MODELS AND SIMULATION ALGORITHMS

The obvious solution to the "implement every model in every simulator" quandary was to completely separate model code from simulator code. Conceptually, this can be done in two ways: either define a standard model/simulator interface, or define a standard language for model definition and develop compilers to generate simulator specific code. These are not mutually exclusive; standardized model/simulator interfaces reduce the number of model compilers needed.

Fig. 3 shows, at a high level of abstraction, how model code can be separated from simulator code. The first reports I am aware of that describe separation of model and simulator code in this manner are [6], [7].

Once code for a model is separated from simulator code, the need for every model to implement interaction with simulator-specific data structures and numerical methods is averted.

However, small-signal analysis, and numerical methods for many other circuit analyses, require derivatives of $I(V)$ and $Q(V)$ of the large-signal model with respect to the node voltages. Historically, coding of these derivatives was the most tedious, time-consuming, non-value-added engineering time-wise, and error-prone part of implementing a compact model.

The solution to this bottleneck is to use automated (symbolic, not numerical) differentiation. Although never published, I believe that the first practical, industrial-strength, high-level model description language (without explicit coding of derivatives) plus associated model compiler was developed by Tektronix in the late 1980's [8]. The driving force behind this was the need for accurate models of components for the design of ICs for high-speed test equipment. An alternative, that ended up being the first high-level compact modeling language implemented in a commercial circuit simulator, was also under development around the same time [9].

In the early 1990's my colleagues at AT&T Bell Laboratories and I developed several model compilers, and associated high-level model definition languages, that automatically generated code consistent with the model interface of [6]. The compilers used the symbolic differentiation capability of Maple [10]. Direct generation of final derivatives turns out not to work well, but by keeping track of bias dependencies for every line of code and every intermediate variable it is simple to apply the chain rule to generate concise and regular code.

So, by the early- to mid-1990's the technology and tools for dragging model definition and implementation into the 21st century were available.

But what about the capabilities of the basic models?

IV. MODEL PHYSICAL BENCHMARKING

The asymmetry between textbooks on circuit design and textbooks on modeling is striking. The former invariably include introductory chapters on device fabrication, operation, and modeling. The latter almost always include introductory chapters on basic device physics, but never include material on core design requirements[5]. This lack of awareness of design needs has lead to many model publications showing essentially perfect fits to measured device data while in reality the model is unfit for real design use, especially for analog and RF.

Besides the implementation issues with models, for over 35 years flags have been raised about fundamental problems with models for IC design. That a model can predict digital gate delays reasonably well in no way means it is accurate for analog circuits. The first exposé of the inadequacies of MOS transistor models for analog CMOS IC design was published in 1982 [14]. The impact that paper had on MOS transistor model development was unexpected: Precisely zero. Fig. 4 shows $g_m(V_{gs})$ characteristics from several state-of-the-art MOS transistor models from about a decade after [14] was published. Clearly, the message about needing smooth behavior in derivatives, which is critical for analog design,

---

[5]I have observed that model developers, in the past myself included, are primarily wanna-be physicists, although this often transforms over time into being wanna-be designers.

978-1-5090-5979-9/17 $31.00 © 2017 IEEE

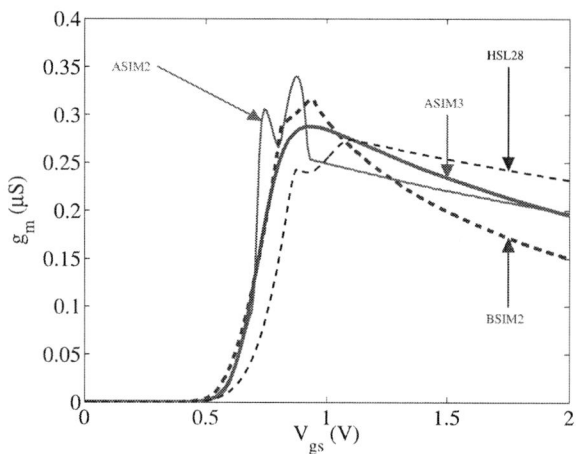

Fig. 4. Transconductance modeling from leading-edge MOS transistor models of the early 1990's. ASIM2 is from [11], BSIM2 is from [12], HSL28 is the HSPICE level 28 model, ASIM3 is from [13]. $V_{ds} = 0.1$ V, $V_{sb} = 0$ V.

had failed to get through[6]. In part, I believe, the lack of improvement was because model developers were focused on device physics and not on design needs. But how come designers were not holding model developers' feet to the fire?

Prof. Tsividis again attempted to enlighten the modeling world about a decade later [15], [16]. At the presentation of that work, the authors exhorted designers in their audience to go back to their companies or institutions and challenge their modeling engineers to apply the benchmarks of [15] to their models[7] (note that at the time there was no industry standard advanced MOS transistor model, it would be three years before BSIM3 was standardized, and most semiconductor companies had their own proprietary models). The alarm bells rung in [15], [16] were heard a little more widely than when [14] appeared more than a decade before, and although they did not immediately trigger fundamental improvements in many models, they did lead to significant effort to expand MOS transistor model benchmarks. These are summarized in [17] and Appendix K of [18] and the references therein.

V. Taking the Big Step Forward

By the mid 1990's the need for standardized models was apparent. The use of foundries for manufacturing was growing, and (analogous to Fig. 2) supporting every proprietary model for every foundry and manufacturing technology was not feasible. To address this need a "Compact Models Workshop" was convened in March 1995, at SEMATECH in Austin, Texas. The meeting was co-organized by Reddy Manukonda

[6]The ASIM2 model was mathematically *guaranteed* to have continuity of I_d and its derivatives with respect to all biases; clearly this is not sufficient. The BSIM2 model used to generate Fig. 4 used a cubic spline to connect weak and strong inversion operation, and again that is clearly not adequate for accurate modeling of g_m in moderate inversion (the cubic spline approach was abandoned in later versions of BSIM). Although the HSPICE level 28 model did a better job of improving BSIM1, on which it was based, it still has a slight "wobble" through moderate inversions.

[7]I was at AT&T Bell Laboratories at the time, and one of the expert analog designers there, Dave Rich, who had attended the presentation, gave me a copy of [15] when he returned and (politely) asked if my model, ASIM3 in Fig. 4, passed the tests. As I recall, it did fairly well, but was not perfect.

of SEMATECH and me, then at AT&T Bell Laboratories; Fig. 5 lists the presentations at the workshop. The ostensible reason for the workshop was to assemble a group of modeling experts, from industry, academia, and CAD companies, to make recommendations on the path forward for modeling. In this the group was successful: we agreed to continue to meet, which led to the establishment of the CMC (then the "Compact Model Council," now the "Compact Model Coalition"), the industry-wide body that oversees model standardization.

I also had two "hidden agenda" items for the workshop. First, to make public some benchmark tests developed by Hermann Gummel at AT&T Bell Laboratories, to try to "force the physics" back into MOS transistor models. Second, to enlighten the modeling world that the only logical way forward was to move to a high-level-language-plus-model-compilers approach to model definition and implementation. In these, I was spectacularly unsuccessful, at least in the short term.

The Gummel tests added to the suite of benchmarks of Prof. Tsividis, helped spur the continued development of such tests, and fostered the expectation that models needed to pass those tests. Eventually this happened, although it was disheartening both that it took so long and that on several occasions I saw a model developer proudly present test results without realizing that their model *failed* the benchmark.

What transpired for model definition and implementation was more interesting. The discussion split into two, one focused on language/compilers and another focused on a standard model interface (see Fig. 3). By far the more emphasis was placed on the latter, not the former as I wanted.

I believe this was because most modeling people at the time were not software engineers: The concept of a program that took code as an input and generated a different sort of code as output was foreign (programs generate numbers, right?). There were some enlightened individuals who immediately saw the benefit, but very few. Not to exaggerate too much, people spent more time telling me why it wouldn't work than it had taken me to write model compilers to prove that it did work: "you can't get the model into a parameter extractor"–yes you can because your extractor calls a simulator, and once a model is in that simulator it works fine; "you'll get different results in different simulators"–yes because different model compilers can order computations differently, but the same thing happens if you use different C or FORTRAN compilers, different –O optimization flags for those compilers, or different operating systems; "it runs too slow"–yes, initially, but later compilers generated code that was as efficient as, sometimes faster than [19], hand-coded models.

The advent of Verilog-A, the provision of a public domain Verilog-A compiler [20], experimental dabbling (once you have written a model in Verilog-A you will *never* go back to hand-coding in C), and persistence eventually led to Verilog-A becoming the *de facto* standard for model implementation. But it was a long, slow journey.

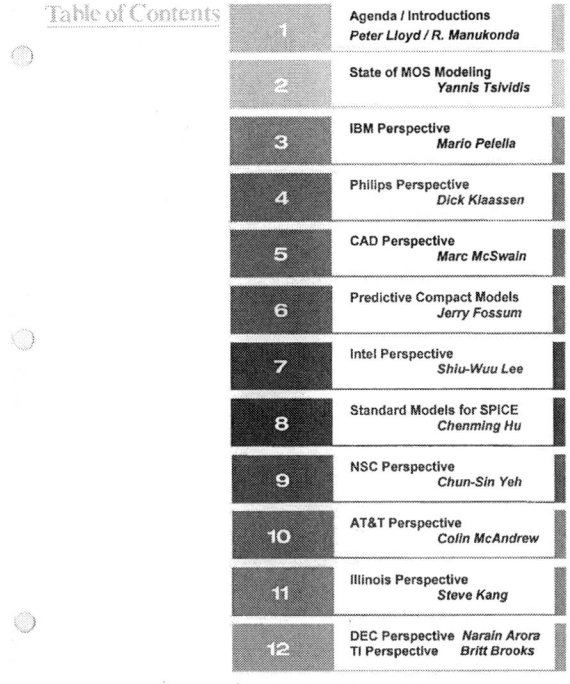

1	Agenda / Introductions	*Peter Lloyd / R. Manukonda*
2	State of MOS Modeling	*Yannis Tsividis*
3	IBM Perspective	*Mario Pelella*
4	Philips Perspective	*Dick Klaassen*
5	CAD Perspective	*Marc McSwain*
6	Predictive Compact Models	*Jerry Fossum*
7	Intel Perspective	*Shiu-Wuu Lee*
8	Standard Models for SPICE	*Chenming Hu*
9	NSC Perspective	*Chun-Sin Yeh*
10	AT&T Perspective	*Colin McAndrew*
11	Illinois Perspective	*Steve Kang*
12	DEC Perspective *Narain Arora* TI Perspective *Britt Brooks*	

Fig. 5. 1995 SEMATECH compact models workshop contents.

VI. FINAL REMARKS

The inertia to changing models, both their formulation and the language in which they are defined, is large. It took more than 30 years of constant nagging to force "compliance" of the most widely used MOS transistor models with benchmark tests. And it took more than 10 years of behind-the-scenes wrangling to enlighten the industry that Verilog-A, and not hand-coded C, was the way to define models.

Once a minimum acceptable quality level became the expected standard the worst problems in models disappeared[8]. And it only took a few years after Verilog-A was accepted as an industry standard language for the question ''Why is your model not available in Verilog-A?'' to be asked if a model was still coded in C.

Perhaps the biggest benefit of the high-level-language-plus-model-compilers approach to modeling is how it enables efficient implementation of new developments. If a new model/simulator interface is developed that significantly speeds up simulation, you only have to implement it once in a model compiler and then it *automatically* gets applied to *every* model. Similarly, if new code generation techniques are developed that minimize cache misses (a not well known but common source of model evaluation inefficiencies), not only does this benefit all models, but it is essentially impossible to duplicate in hand-crafted code for each model.

The lowering of the entry barrier to model development has had the undesired effect of lowering the quality of many experimental models. The development of "best practices" [21]

hopefully can help raise the bar of model coding standards. Per "Доверяй, но проверяй,"[9] automated checking for adherence to the coding standards also helps [22].

Many long standing problems in modeling have been solved over the past decades. In addition, tomorrow's transistors must "look" like today's transistors (in their electrical behavior), otherwise the accumulated world-wide circuit design know-how will become useless. The pace of core model development has therefore slowed down. But layout dependent and proximity effects, parasitics, and variability have become much more important, and are a key focus of modeling today.

Besides quality code, models should be based on quality physics; an incomplete list of requirements is given in [21], including the need for passivity, C^∞ continuity, and symmetry. There have been some attempts to implement algorithms to automatically detect "physics problems" with models, but to my knowledge none have been successful. This is a hard problem. Hopefully, one that some smart graduate student or professor will find to be attractive.

REFERENCES

[1] L. W. Nagel and D. O. Pederson, "Simulation program with integrated circuit emphasis (SPICE)," in *16th Midwest Symp. Circuit Theory*, 1973.

[2] L. W. Nagel, *SPICE2, A Computer Program to Simulate Semiconductor Circuits,*. PhD thesis, EECS Department, UC Berkeley, May 1975.

[3] I. E. Getreu, "Who wants to be a modelaire?," in *IEEE BMAS*, 2001.

[4] H. Shichman and D. A. Hodges, "Modeling and simulation of insulated-gate field-effect transistor switching current," *IEEE J. Solid-State Circuits*, vol. 3, pp. 285–289, Sep. 1968.

[5] L. W. Nagel. personal communication.

[6] S. Liu, K. C. Hsu, and P. Subramanian, "ADMIT-ADVICE modeling interface tool," in *Proc. IEEE CICC*, pp. 6.6.1–6.6.4, 1988.

[7] A. T. Yang and S. M. Kang, "iSMILE: a novel circuit simulation program with emphasis on new device model development," in *Proc. ACM/IEEE DAC*, pp. 630–633, 1989.

[8] E. McReynolds. personal communication.

[9] M. Vlach, "Modeling and simulation with saber," in *Proc. 3rd Annu. ASIC Sem. Exhibit*, pp. T11.1–T11.9, 1990.

[10] MapleSoft, "Maple." http://www.maplesoft.com/products/maple/. Accessed: Apr. 2017.

[11] S.-W. Lee and R. C. Rennick, "A compact IGFET model–ASIM," *IEEE Trans. CAD*, vol. 7, pp. 952–1113, Sep. 1988.

[12] M.-C. Jeng *et al.*, "A deep-submicrometer MOSFET model for analog/digital circuit simulations," in *IEDM Tech. Dig.*, pp. 114–117, 1988.

[13] C. C. McAndrew *et al.*, "A single piece C_∞-continuous MOSFET model including subthreshold conduction," *IEEE EDL*, vol. 12, Oct. 1991.

[14] Y. Tsividis, "Problems with precision modeling of analog MOS LSI," in *IEEE IEDM Tech. Dig.*, pp. 274–277, 1982.

[15] K. Suyama and Y. Tsividis, "MOSFET modeling for analog circuit CAD: problems and prospects," in *Proc. IEEE CICC*, pp. 14.1.1–14.1.6, 1993.

[16] Y. Tsividis and K. Suyama, "MOSFET modeling for analog circuit CAD: problems and prospects," *IEEE JSSC*, vol. 29, pp. 210–216, Mar. 1994.

[17] X. Li *et al.*, "Benchmark tests for MOSFET compact models," in *Compact Modeling: Principles, Techniques and Applications* (G. Gildenblat, ed.), ch. 3, pp. 75–104, Springer, 2010.

[18] Y. Tsividis and C. McAndrew, *Operation and Modeling of the MOS Transistor*. Oxford University Press, third ed., 2011.

[19] B. Wan *et al.*, "MCAST: an abstract-syntax-tree based model compiler for circuit simulation," in *Proc. IEEE CICC*, pp. 11–4–1–11–4–3, 2003.

[20] L. Lemaitre, C. C. McAndrew, and S. Hamm, "ADMS-automated device model synthesizer," in *Proc. IEEE CICC*, pp. 27–30, 2002.

[21] C. C. McAndrew *et al.*, "Best practices for compact modeling in Verilog-A," *IEEE J. Electron Dev. Society*, vol. 3, pp. 383–396, Aug. 2015.

[22] "VALint: the NEEDS Verilog-A checker." https://nanohub.org/tools/vachecker. Accessed: Apr. 2017.

---

[8]And will not reappear if the industry maintains its "institutional memory."

[9]Trust, but verify.

SPICE Modeling of Light Induced Current in Silicon with 'Generalized' Lumped Devices

Chiara Rossi[1], Pietro Buccella[1], Camillo Stefanucci[2] and Jean-Michel Sallese[1]

[1]École Polytechnique Fédérale de Lausanne (EPFL), 1015 Lausanne, Switzerland
[2]AMS AG, Schloss Premstaetten, Austria

Abstract— SPICE-compatible modeling with generalized lumped devices is used to simulate the spatial and time dependence of photogenerated carriers with standard circuit simulators. Equivalent voltages and currents are used in place of minority carrier excess concentrations and minority carrier currents respectively. The initial light-induced excess carrier concentration in silicon is accounted by means of distributed external voltage sources. Generation, propagation and collection of these minority carriers is analyzed for three pertinent structures and compared with TCAD numerical simulations.

Keywords: Modeling, SPICE, photocurrent, photogeneration, photosensor, recombination, minority carriers, pn junction.

I. INTRODUCTION

A SPICE-compatible model has been previously developed in [1-2] to simulate injection and propagation of minority carriers in the substrate of Smart Power ICs through the definition of generalized models for resistances and diodes. Next, an automated meshing tool was developed [3] which could simulate complex 3D minority carriers dynamic inside the substrate of an IC with inherent related latch-up phenomena. In an attempt to account for free carrier generation upon light absorption in semiconductors, we propose to generalize this approach to simulate photogenerated carriers in three relevant 1D semiconductor structures.

II. PRINCIPLE OF GENERALIZED LUMPED DEVICES

According to [2], simulation of minority carriers in a piece of semiconductor involves three lumped elements: a generalized Diode, a generalized Homojunction and a generalized Resistor as shown in Fig. 1. These devices are characterized by an additional node which 'propagates' the information on the minority carriers through equivalent voltages and currents. For instance, the diode can inject and collect carriers, the resistor is used to propagate them in the semiconductor, and the homojunction is used to account for doping discontinuities implemented at the contact-semiconductor boundaries. The Finite Difference Method (FDM) is used to convert the set of drift-diffusion partial differential equations in a system of equations whose number depends on the discretization step, i.e. the 'meshing' [2]. This FDM gives rise to generalized lumped devices where input and output variables are electrical quantities which can be solved by standard SPICE-like softwares. Therefore, solving the system of equations in the electrical domain gives back the local

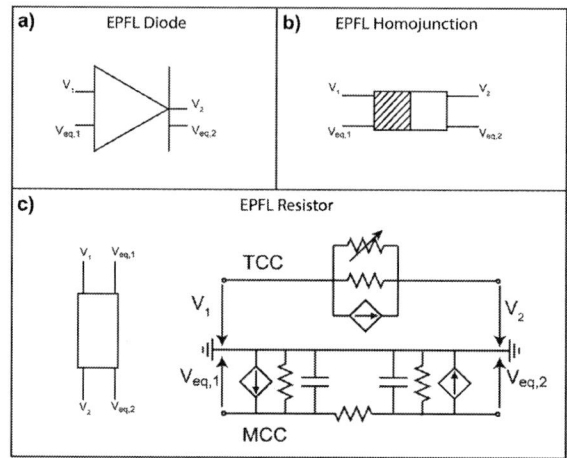

Fig. 1. Generalized lumped devices: a) EPFL Diode, b) EPFL Homojunction, c) EPFL Resistor with schematic representation of Total Current Circuit (TCC) and Minority Carrier Circuit (MCC) for the p-doped case.

distribution of charges and currents arising from majority and minority carriers.

Each of the three lumped devices embeds two equivalent circuits: the Total Current Circuit (TCC) which accounts for currents and voltages, and the Minority Carrier Circuit (MCC) which accounts for minority carriers through the definition of an equivalent voltage V_{eq} proportional to the excess minority carriers concentration. Similarly, an equivalent current I_{eq} is proportional to the current of these minority carriers. Combining TCC and MCC, four terminal 'generalized' devices are obtained as reported in Fig. 1. The two circuits are coupled together, especially in high injection conditions [2] where the electric field adds a drift component to the minority carriers transport, and where the injected carriers back modulate the conductivity of the substrate.

The TCC and MCC subcircuits are represented for a p-type resistor element in Fig. 1c. The TCC includes a resistance which takes into account the modulation of the conductivity due to minority carriers, and a current source whose value considers both the current proportional to the applied voltage (drift) and the difference between majority and minority diffusion currents. In addition, the MCC includes a conductance depending on the diffusivity which regulates the diffusion current, and a conductance depending on the minority lifetime (SHR model) which regulates the fraction of minority carriers that recombine in the volume. Finally, drift is also included in the MCC through current sources depending on the potential drop. Importantly, the model also defines diffusion capacitance to take into account transient effects of minority carriers, see [4] for a more in depth analysis.

III. Modeling Charge and Photo-Current under Pulsed Light Excitation

Since the generalized elements are able to cope with minority carriers, we propose to investigate how these latter can be used to simulate light absorption and induced photocurrent in simple 1D semiconductor structures based on SPICE-like simulations. To this purpose, generation and propagation of carriers upon a pulse of light is studied in a p-doped ($N_a = 10^{16}$ cm^{-3}) silicon resistor of 20 µm length, 5 µm wide and 1 µm thick (default value in 2D Sentaurus simulations). The resistor is illuminated uniformly on one side with a 1 ns light pulse and a wavelength of 600 nm. The current is extracted at the two side contacts. The intensity of light is set to 10^3 W/cm^2. Light absorption in silicon is assumed to follow the Beer-Lambert law ($I = I_0 \exp(-\alpha x)$) where α is the absorption coefficient equal to $4.14 \cdot 10^3$ cm^{-1} at $\lambda = 600$ nm for silicon. This kind of experiment has been used to measure carriers mobility in semiconductors [5, 6]. Three different structures have been considered, each of which introduces different categories of generalized elements:

S1: Electrodes are in contact with the p-type silicon (Ohmic contacts have been imposed).

S2: Electrodes are contacting two highly-doped layers ($N_a = 2 \cdot 10^{19}$ cm^{-3}) of 0.1 µm thickness, which constitute two homojunctions.

S3: A reverse-biased diode terminates the resistor (note that two homojunctions are introduced as well, $N_{a/d} = 2 \cdot 10^{19}$ cm^{-3}).

Fig. 2 is the layout for the structure S3 with different zones highlighted. Generalized homojunctions (n and p), resistors and pn junction schematics are superimposed. Since illumination is uniform on the side, a simple 1D discretization scheme can be used. A similar decomposition is done for S1 and S2.

The initial x-dependent excess minority carrier concentration is obtained from the amount of the light absorbed, and its value is assigned to each elements by imposing the corresponding equivalent voltage V_{eq} on the MCC node through a switch. During the short pulse, the switch closes to set the equivalent voltage, then remains open after the light pulse. The number of lumped elements introduced to model the p-type silicon slab is essentially imposed by the number of points needed to account for the optical generation. In order to minimize the number of components, still keeping the mismatch with numerical simulations below 2%, 8 lumped elements (and so 8 voltage sources) are distributed at x = 0.5 µm, 1 µm, 2 µm, 3 µm, 4 µm, 5 µm, 6 µm and 8 µm. As soon as the generation of minority carriers becomes negligible (around x > $4\alpha^{-1} \sim 8$ µm), a single generalized resistance would be needed since it accounts for drift, diffusion and recombination 'internally'. However, in order to extract the minority carrier density as a function of space and time, additional lumped components have been introduced.

TCAD simulations have been run for each structure imposing a doping-dependent mobility and Shockley-Read-Hall recombination, and light excitation was defined as a time dependent monochromatic source.

Importantly, the same physical parameters have been used for generalized devices, i.e. no fitting parameter was introduced in the models for S1, S2 and S3.

IV. Simulations

In the following sections, simulations performed using the lumped devices model and the TCAD software are presented for structures S1, S2 and S3.

A. Uniformly p-doped silicon layer

The device S1 consists in a uniformly p-doped silicon layer with a doping density of 10^{16} cm^{-3}. Ideal contacts are defined at x = 0 µm and x = 20 µm, meaning that full recombination of excess minority carriers is imposed at these interfaces. The resistor is biased with a DC voltage of 0.5 V applied on the illuminated side. The densities of electrons (minority carriers) at different coordinates obtained from SPICE and TCAD simulations are plotted in Fig. 3. Continuous lines hold for the lumped elements model and symbols for TCAD. The origin of time is set after the pulse of light, i.e. after 1 ns.

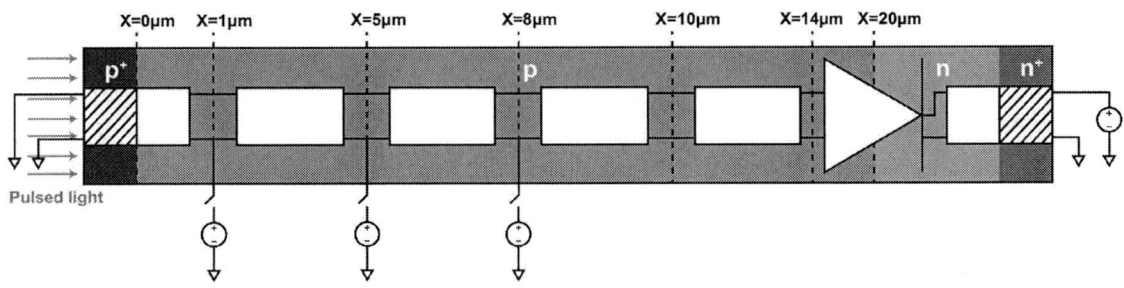

Fig. 2. Simulated structure (case S3) with equivalent circuit obtained with lumped devices; the external voltage sources are connected to the MCC node and simulate the excess minority carrier concentration generated by the impinging light.

978-1-5090-5979-9/17 $31.00 © 2017 IEEE

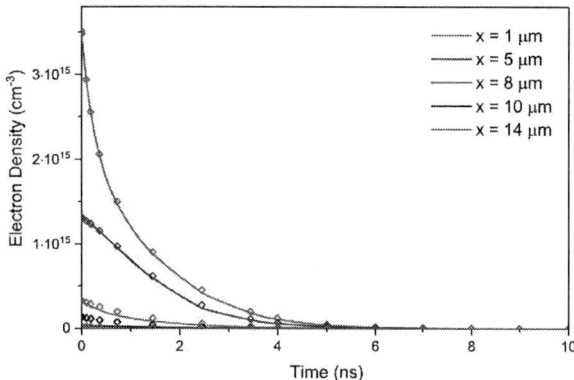

Fig. 3. Simulated electron density for structure S1 (continuous line for model and symbols for TCAD).

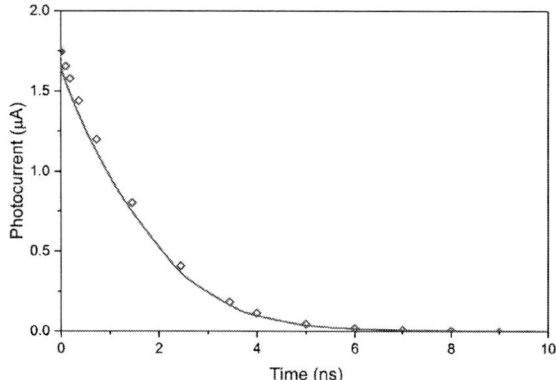

Fig. 4. Simulated photocurrent, i.e. total current minus dark current, for structure S1 (continuous line for model and symbols for TCAD).

Fig. 3 shows that the excess minority carrier concentrations decay rapidly, about 8 ns after the light pulse, as electrons are accelerated and collected by the applied electric field before they diffuse inside the silicon. In fact, for positions beyond 10 μm, the electron density becomes negligible. As evidenced, the equivalent SPICE model is in good agreement with numerical simulations. The lumped modeling scheme is able to predict the time dependence of the minority carrier concentration all along the piece of silicon. Concerning the photocurrent (total current minus the current at dark) at the output (x = 20μm), again SPICE simulations fit well TCAD results, see Fig. 4. Note that the carrier density is about one order of magnitude lower than the doping, meaning that low injection conditions are fulfilled.

B. P-type doped silicon with highly p-type doped homojunctions

A slightly different structure S2 is investigated where highly doped p-type regions ($2 \cdot 10^{19}$ cm^{-3}) are implemented at x = 0 μm and at x = 20 μm to create Ohmic contacts. These highly-doped layers have a negligible extension of 0.1 μm. Fig. 5 plots the electron density versus time at different coordinates with the generalized devices approach and TCAD simulations. Again, a good agreement is obtained with the generalized lumped element network. Even though the pulse of light is the same for

S2 and S1, the density of minority carriers is about one order of magnitude higher than before, i.e. 10^{16} instead of 10^{15} cm^{-3}, and lasts μs instead of some tens of ns. Similarly, the minority carrier concentration remains quite high even beyond 10 μm, and the transient current at x = 20 μm decays in the μs range (see Fig. 6) indicating that drift is no longer the dominant mechanism in regard to diffusion.

In fact, the additional homojunctions exert a 'repelling' effect on minority carriers due to the barrier created by the doping gradient. Electrons cannot be efficiently collected by the contact and remain longer inside the silicon. Excess of holes could be collected at the counter electrode as they encounter no barrier, but, since propagation of photogenerated e/h pairs is driven by ambipolar transport, holes will move with electrons as a whole. Note that these fine points of electrostatic confinement and transport are accurately depicted by the equivalent SPICE 'generalized' circuit.

C. Resistance with reverse-biased pn junction

The last structure investigated is S3. It consists of a diode in series with a uniformly doped silicon layer as represented in Fig. 2. Highly doped p and n homojunctions are also included

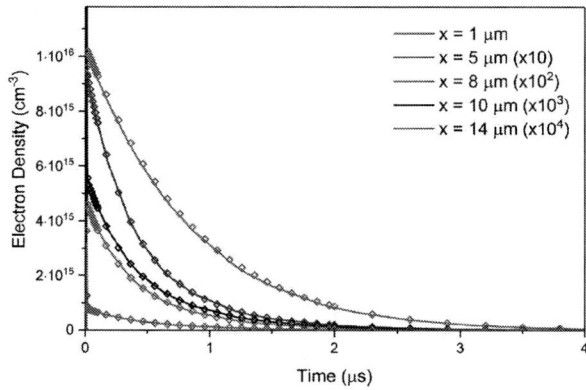

Fig. 5. Simulated electron density for structure S2 (continuous line for model and symbols for TCAD).

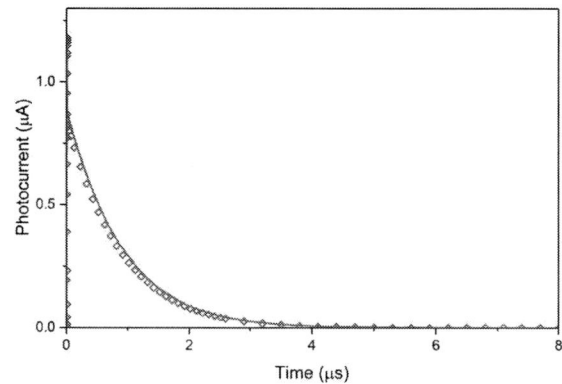

Fig. 6. Simulated photocurrent, i.e. total current minus dark current, for structure S2 (continuous line for model and symbols for TCAD).

978-1-5090-5979-9/17 $31.00 © 2017 IEEE

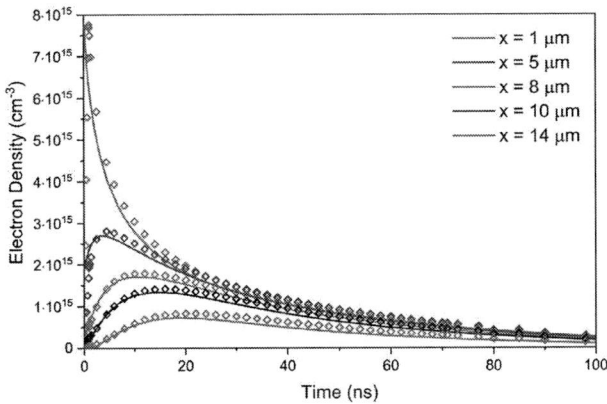

Fig. 7. Simulated electron density for structure S3 (continuous line for model and symbols for TCAD).

for Ohmic contacts. In this case a DC voltage of 0.5 V is applied on the n-doped (non-illuminated) side to set the diode in reverse mode. Again, the electron density in transient mode at different coordinates is well predicted by the lumped device approach in Fig. 7. Note that the time range is mid-way between that of S1 and S2. In S3, minority carriers are collected mainly by diffusion since the potential drops almost entirely across the pn junction, and not in the silicon layer as for S1. On the other hand, unlike for S2 there is no homojunctions repelling them at x = 20 μm. Here the maximum value is reached after some tens of ns, and this value increases when moving towards the pn junction while it also spreads with time. These travelling-widening bell shapes are consistent with a drift-diffusion process towards the junction. Again, the lumped modeling approach captures these processes accurately.

The transient current simulated at x = 20 μm is plotted in Fig. 8. As for minority carrier concentrations, the correspondence between numerical TCAD simulations and the predictions of SPICE simulations is fair.

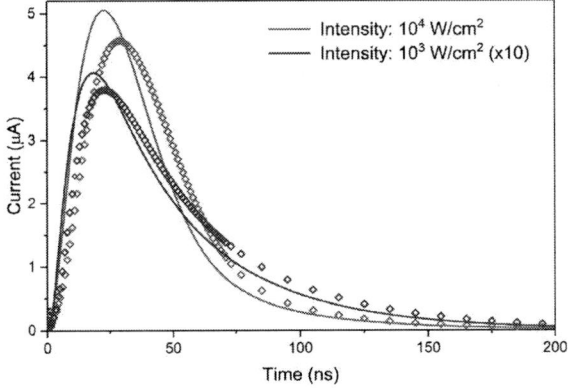

Fig. 8. Photocurrent collected by the reverse-biased diode in S3 for two light pulse power densities (continuous line for model and symbols for TCAD).

D. High Injection limitations

It is interesting to increase the intensity of light from 10^3 W/cm^2 to 10^4 W/cm^2 to investigate high injection regime. Indeed, simulations confirm that for this intensity the density of the photogenerated carriers becomes higher than the doping concentration (the electron density reaches ~ 10^{17} cm^{-3} at the left homojunction interface). Fig. 8 shows the photocurrent for low (10^3 W/cm^2) and high (10^4 W/cm^2) injection conditions for S3. The effect of high injection degrades slightly the matching with TCAD, but the agreement is still acceptable. We also noticed (not reported here) that the lumped model is still accurate for structure S1, while it degrades significantly for S2 where the relative error is about 15%. In fact, the model is able to cope with the high injection for cases S1 and S3 because the minority carrier concentration goes back to steady state in a quite short time, while for case S2 the concentration remains longer above the doping concentration, leading to a mismatch which amplifies with time.

V. CONCLUSION

A SPICE-compatible model developed for substrate coupling in Smart Power ICs has been adapted to simulate photogenerated carriers propagation and collection in a silicon slab. Good agreement has been obtained between TCAD and SPICE-based simulations. It comes out that the generalized elements arranged in a lumped topology are able to predict the behavior of excess carriers concentrations and photocurrents without the need for any fitting parameters. This work is a first step towards optoelectronics simulations based on generalized elements with standard circuits simulation tools.

ACKNOWLEDGMENT

The authors would like to thank Jacopo Bronuzzi, Nicolò Oliva and Mathieu Coustans for their support. This work was funded by the Swiss National Science Foundation - grant 200021_165773.

REFERENCES

[1] F. Lo Conte, J.M. Sallese, M. Pastre, F. Krummenacher, M. Kayal, "Global Modeling Strategy of Parasitic Coupled Currents Induced by Minority-Carrier Propagation in Semiconductor Substrates", IEEE Trans. Elect. Dev. Vol. 57 Issue 1, pp. 263-272 (2010).

[2] C. Stefanucci , P. Buccella, M. Kayal, J.M. Sallese, "Spice-compatible modeling of high injection and propagation of minority carriers in the substrate of Smart Power ICs", Solid State Elect. vol. 105, p.21 (2015).

[3] P. Buccella, C.Stefanucci, H. Zou et al., "Methodology for 3-D Substrate Network Extraction for SPICE Simulation of Parasitic Currents in Smart Power ICs", IEEE Trans. Comp. Aided Design, Vol 35, Iss. 9, pp. 1489-1502, (2016).

[4] C. Stefanucci , P. Buccella, M. Kayal, J.M. Sallese, "Modeling Minority Carriers Related Capacitive Effects for Transient Substrate Currents in Smart Power ICs", IEEE Trans. Elect. Dev. Vol. 62 Issue 4 p. 1215 (2015).

[5] J. R. Haynes, W. Shockley, "The Mobility and Life of Injected Holes and Electrons in Germanium", Phys. Rev., Vol. 81, Number 5, p. 835 (1951).

[6] S. M. Sze, "Physics of Semiconductor Devices", 2nd Edition, John Wiley & Sons (1981).

Total Ionizing Dose Effects on Analog Performance of 28 nm Bulk MOSFETs

C.-M. Zhang*, F. Jazaeri*, A. Pezzotta*, C. Bruschini*, G. Borghello[†], S. Mattiazzo[‡], A. Baschirotto[§], and C. Enz*

*ICLAB, EPFL, Neuchatel, Switzerland
[†]EP-ESE Group, CERN, Geneva, Switzerland & University of Udine, Udine, Italy
[‡]Department of Information Engineering, University of Padova, Padova, Italy
[§]Microelectronic Group, INFN Milano-Bicocca & University of Milano-Bicocca, Milano, Italy

Abstract—This paper uses the simplified charge-based EKV MOSFET model for studying the effects of total ionizing dose (TID) on analog parameters and figures-of-merit (FoMs) of 28 nm bulk MOSFETs. These effects are demonstrated to be fully captured by the five key parameters of the simplified EKV model. The latter are extracted from the measured transfer characteristics at each TID. Despite the very few parameters, both the simplified large- and small-signal models present an excellent match with measurements at all levels of TID. The impacts of TID on essential parameters, including the drain leakage current, the threshold voltage, the slope factor, and the specific current, are then evaluated. Finally, TID effects on the transconductance G_m, the output conductance G_{ds}, the intrinsic gain G_m/G_{ds} and the transconductance efficiency G_m/I_D are investigated.

Index Terms—analog parameters, transconductance efficiency, EKV, intrinsic gain, inversion coefficient, TID, 28 nm bulk MOS-FETs

I. INTRODUCTION

The forthcoming high-luminosity Large Hadron Collider (HL-LHC) at CERN is expected to experience an unprecedented radiation level up to 1 Grad of total ionizing dose (TID) over ten years of operation. This will be a challenge for the electronic components of the innermost detectors, including analog amplifiers, filters, analog-to-digital converters and application-specific integrated circuits. For a long-term reliable operation, highly upgraded detecting systems are needed with a higher bandwidth and more radiation-tolerant front-end electronics. Aggressively downscaled CMOS technologies bring a high operation speed and demonstrate the capability of mitigating TID effects [1], [2]. Therefore, we have started investigating the radiation tolerance of a commercial 28 nm bulk CMOS process up to 1 Grad with the perspective of using it for the future radiation-tolerant detectors [3], [4].

Our previous studies have shown that the dominant TID effects on 28 nm bulk MOSFETs include a significant drain leakage current increase, a subthreshold swing degradation, a substantial drive current loss, a threshold voltage shift, and a mobility reduction. Prior investigations have attributed these parameter variations to radiation-induced charge generation

This work is part of the GigaRadMOST project funded by the Swiss National Science Foundation (SNSF) under grant number 200021_160185, in collaboration with the ScalTech28 project funded by the Istituto Nazionale di Fisica Nucleare (INFN).

and trapping related to dielectrics [5]. Although significant progress has been made in understanding the underlying physics of TID effects, not much work has been undertaken about the effects on analog parameters and figures-of-merit (FoMs). As a first step towards the characterization of these effects, this work proposes to use the simplified charge-based EKV MOSFET model which requires only five parameters to fully describe the large- and small-signal characteristics over a wide range of bias from weak to strong inversion and even for advanced technology nodes [6], [7]. Additionally, the inversion coefficient offers an efficient way to normalize the MOSFET characteristics and to evaluate the transconductance efficiency G_m/I_D over a wide range of operating points.

This paper first describes the simplified EKV model, the parameter extraction methodology and the model validation with measurements in Section II. TID effects on extracted model parameters are discussed in Section III, while the impacts of TID on the intrinsic gain and the transconductance efficiency are illustrated in Section IV and Section V. The paper ends up with a brief conclusion.

II. THE SIMPLIFIED EKV MODEL AND THE RELATED PARAMETERS

A. Model Description

The simplified charge-based EKV MOSFET model introduces *inversion coefficient IC* to replace the overdrive voltage $V_G - V_{T0}$ as an essential design parameter that spans the entire range of operating points from weak via moderate to strong inversion. IC identifies the channel inversion level of a MOSFET and is defined as the normalized drain current [8]:

$$IC \triangleq I_D|_{saturation}/I_{spec}, \qquad (1)$$

where I_{spec} is the *specific current* defined as [8]

$$I_{spec} = I_{spec\square} \cdot W/L \quad \text{with} \quad I_{spec\square} \triangleq 2n \cdot \mu_0 \cdot C_{ox} \cdot U_T^2, \quad (2)$$

where $I_{spec\square}$ is the *specific current per square*, n is the *slope factor*, μ_0 is the *low-field channel mobility*, C_{ox} is the *gate oxide capacitance per unit area*, and $U_T \triangleq kT/q$ is the *thermal voltage*. Based on IC, operation regions of a MOSFET can be classified as *weak inversion* (WI) for $IC \leqslant 0.1$, *moderate inversion* (MI) for $0.1 < IC \leqslant 10$, and *strong inversion* (SI) for $IC > 10$.

978-1-5090-5979-9/17 $31.00 © 2017 IEEE

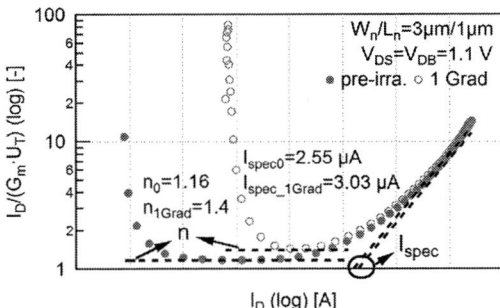

Fig. 1. Extraction of the slope factor n and the specific current I_{spec} with an example of a wide/long-channel nMOSFET.

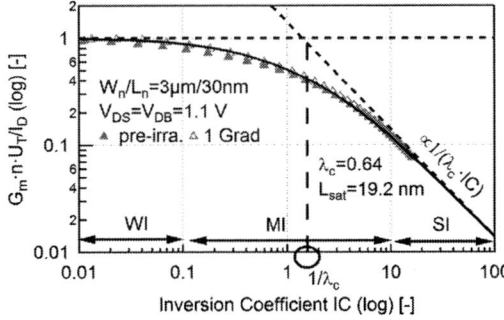

Fig. 2. Extraction of the velocity saturation parameter λ_c with an example of a wide/short-channel MOSFET.

To model the impact of *velocity saturation* (VS) on short-channel devices, the parameter λ_c is introduced as [6]

$$\lambda_c = L_{sat}/L, \qquad (3)$$

with L_{sat} being the portion of the channel over which the carrier drift velocity saturates. Including the effect of VS, the normalized saturation voltage is then expressed as

$$
\begin{aligned}
(V_P - V_S)/U_T &= (V_G - V_{T0} - n \cdot V_S)/(n \cdot U_T) \\
&= \ln\left[f(IC)/2\right] + f(IC), \qquad (4) \\
f(IC) &= \sqrt{(\lambda_c \cdot IC + 1)^2 + 4 \cdot IC} - 1,
\end{aligned}
$$

where V_{T0} is the *threshold voltage*.

Since TID has a strong impact on the *drain leakage current*, the latter is approximated by a constant current adding to the drain current. It constitutes an additional parameter I_{leak} to the simplified EKV model which is straightforward to extract.

Note that the above normalization makes the simplified EKV model independent of technology, which is entirely captured by five parameters (V_{T0}, n, I_{spec}, L_{sat} and I_{leak}). This is the main reason for which we use it to investigate the effects of TID on MOSFET characteristics.

B. Parameter Extraction

The parameter extraction starts with extracting the *slope factor* n from the plateau of $I_D/(G_m \cdot U_T)$ versus I_D curve in WI for a long-channel device, as illustrated in Fig. 1 for

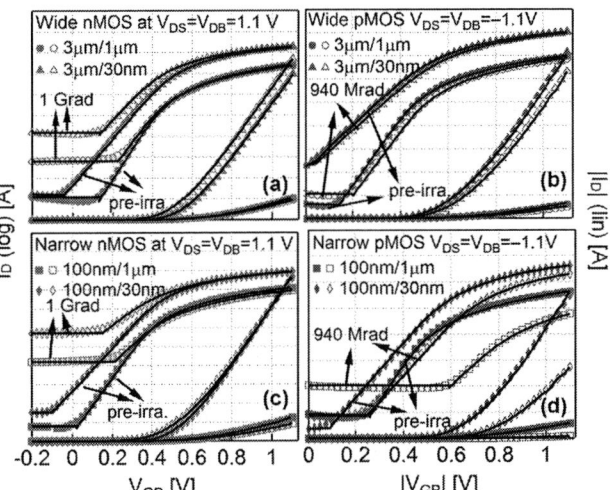

Fig. 3. Modeled and measured transfer characteristics of n (a,c) and p (b,d) types of MOSFETs with respect to pre-irradiation (solid markers) and a high TID (open markers).

measurements with respect to pre-irradiation and a high TID. It is followed by the extraction of the *specific current* I_{spec} from the same plot as the current at the intersection of the SI asymptote and the unity horizontal line. Having n and I_{spec}, and therefore $I_{spec\square}$, we can then plot the normalized transconductance efficiency $G_m \cdot n \cdot U_T/I_D$ versus IC for a short-channel device, as shown in Fig. 2. $1/\lambda_c$ is extracted at the intersection of the SI asymptote $1/(\lambda_c \cdot IC)$ and the unity horizontal line. Having now extracted n, I_{spec}, and λ_c, we can use (4) to calculate the overdrive voltage. The *threshold voltage* V_{T0} is then extracted to fit the model with the measured I_D versus V_{GB} curve. The last parameter is the *drain leakage current* I_{leak} that is easily extracted from the plateau of the I_D versus V_{GB} curve in WI, when available.

C. Large-signal Transfer Characteristics Validation

Fig. 3 compares the model to the measured transfer characteristics for all investigated devices with respect to pre-irradiation and a high TID. The simplified model matches pre-irradiation measurements very well in all regions of operation, demonstrating the capability of fully capturing the technology with only five parameters. This excellent agreement extends to all levels of TID, as plotted for the highest one in Fig. 3, which confirms the promising use of this simplified model in radiation-tolerant circuit design in nanoscale CMOS processes.

The comparison of measurements at pre-irradiation and a high TID indicates a significant *drain leakage current* increase for nMOSFETs (Fig. 3ac) and a substantial drive current loss for pMOSFETs (Fig. 3bd) as well as a *threshold voltage* shift and a *slope factor* degradation. Relevant parameters extracted with the simplified model are discussed in Section III.

III. TID EFFECTS ON ANALOG PARAMETERS

The five model parameters are extracted for each TID by following the procedures in Section II. Fig. 4 presents the

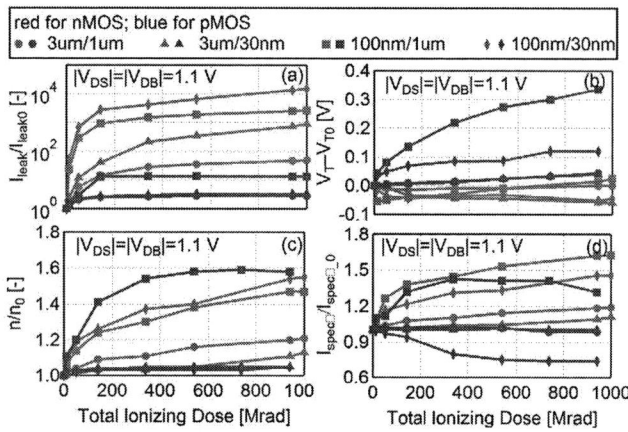

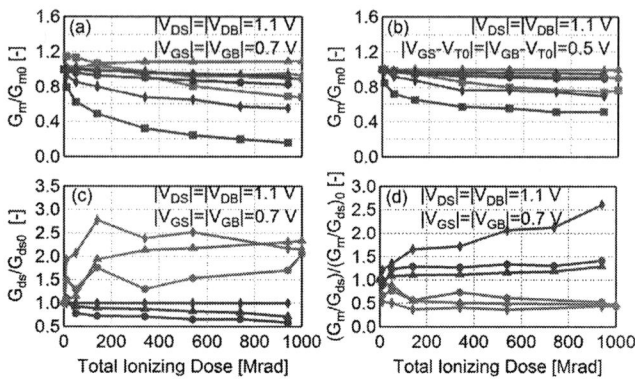

Fig. 5. TID effects on the transconductance (ab), the output conductance (c), and the intrinsic gain (d) of n- and pMOSFETs. This plot shares the legend with Fig. 4.

Fig. 4. TID effects on the drain leakage current (a), the threshold voltage (b), the slope factor (c), and the specific current per square (d) of n- and pMOSFETs.

evolution of extracted parameters with TID. Fig. 4a illustrates the significant *drain leakage current* increase for nMOSFETs, as also seen in Fig. 3ac. This increase is closely related to the radiation-induced parasitic leakage current along the sidewalls of the shallow-trench isolation (STI) [9].

Short-channel nMOSFETs present a slight *threshold voltage* decrease in Fig. 4b. Due to the counterbalance of interface over oxide charged traps, the *threshold voltage* of long-channel nMOSFETs first decreases and then increases [10]. In contrast, the superposed effects of oxide and interface charged traps result in a *threshold voltage* increment for all pMOSFETs. The increase reaches almost 350 mV for the narrow/long-channel pMOSFET, which corresponds to the significant drive current loss observed in Fig. 3d.

Except the narrow/long-channel pMOSFET, all other pMOSFETs show a negligible *slope factor* increase in Fig. 4c. The *slope factor* of nMOSFETs, especially two narrow-channel devices, is more sensitive to TID. The *slope factor* increase in narrow-channel MOSFETs is mostly due to the charge trapping at border traps or interface traps related to STI.

For most of MOSFETs, the *specific current* (Fig. 4d) follows the trend of the *slope factor* (Fig. 4c). However, the *specific current* of pMOSFETs, particularly two narrow-channel devices, starts to decrease at a certain TID. This demonstrates the radiation-induced mobility reduction, as predicted by (2).

As a first conclusion, the model and measurements demonstrate the high radiation tolerance of most of studied MOSFETs except some narrow-channel devices. Designers should therefore be careful when using narrow-channel MOSFETs for radiation-tolerant circuit design. Note that the *velocity saturation* parameter λ_c remains almost constant with TID, which is not shown here.

IV. TID EFFECTS ON THE INTRINSIC GAIN

The correct operation of many analog circuits relies on a sufficient *intrinsic gain* of the transistors. Therefore, it is

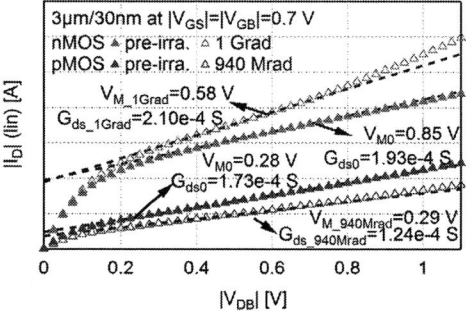

Fig. 6. Output characteristics of wide/short-channel (3 μm/30 nm) n- and pMOSFETs.

crucial to investigate TID effects on this FoM. Fig. 5acd show the *transconductance* G_m, the *output conductance* G_{ds} and the *intrinsic gain* G_m/G_{ds} evaluated at the same operating point $|V_{DB}| = 1.1$ V and $|V_{GB}| = 0.7$ V as a function of TID. The *transconductance* plotted in Fig. 5a includes the effects of both *threshold voltage* shift and *mobility reduction*. It is basically following the *threshold voltage* shift shown in Fig. 4b. The *transconductance* of long-channel and narrow/short-channel nMOSFETs first increases and then decreases, while that of the narrow/short-channel device continuously increases until the highest TID. Due to the significant *threshold voltage* increment, pMOSFETs present a substantial *transconductance* loss that is higher than the same size of nMOSFETs. To eliminate the effect of *threshold voltage* shift, the *transconductance* at a constant overdrive voltage ($|V_{GB} - V_{T0}| = 0.5$ V) is shown in Fig. 5b. All studied MOSFETs are still having a *transconductance* loss but less compared with Fig. 5a. The remaining loss in Fig. 5b is from the radiation-induced *mobility reduction*. Two narrow-channel pMOSFETs present the most serious *mobility reduction* that is consistent with their $I_{spec\square}$ decrease shown in Fig. 4d.

Fig. 5c presents the evolution of *output conductance* with TID. It increases for nMOSFETs and decreases for pMOSFETs. This behavior can be clearly seen from the out-

978-1-5090-5979-9/17 $31.00 © 2017 IEEE

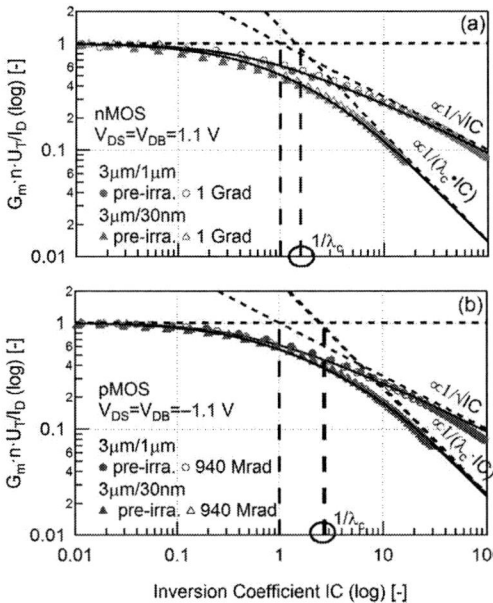

Fig. 7. Normalized transconductance efficiency of n (a) and p (b) types of wide/long- and wide/short-channel MOSFETs with respect to pre-irradiation and a high level of TID.

put characteristics of wide/short-channel ($3\,\mu m/30\,nm$) n- and pMOSFETs plotted in Fig. 6. After a high TID, the wide/short-channel nMOSFET shows a lower Early voltage and hence a higher *output conductance*, whereas the wide/short-channel pMOSFET presents a higher Early voltage and a lower *output conductance*. The observed modification in *output conductance* demonstrates the radiation-enhanced drain-induced barrier lowering (DIBL) for nMOSFETs and the radiation-suppressed DIBL for pMOSFETs. The *transconductance* loss and the *output conductance* increase result in a degraded *intrinsic gain* for nMOSFETs, as shown in Fig. 5d. Owning to a lower *output conductance* after irradiation, pMOSFETs obtain a slightly improved *intrinsic gain*.

V. TID EFFECTS ON THE TRANSCONDUCTANCE EFFICIENCY

The *transconductance efficiency* G_m/I_D is an important analog FoM for low-power analog circuit design. The normalized *transconductance efficiency* of wide/long- and wide/short-channel MOSFETs is illustrated in Fig. 7 which highlights the VS-induced degradation at a high IC. Moreover, after a proper normalization with extracted parameters, all the measured points with respect to pre-irradiation and a high TID are seen to nicely fall on the curves of the simplified EKV model. This demonstrates the negligible effects of TID on the *normalized* G_m/I_D. Although the ionizing radiation affects the analog parameters, the normalization strips off the TID effects from the normalized *transconductance efficiency*. This inversion coefficient based simplified model can therefore be promising for radiation-tolerant low-power analog circuit design.

VI. CONCLUSION

The effects of high total ionizing dose on analog performance of a 28 nm bulk CMOS process are investigated using the simplified charge-based EKV MOSFET model that requires only five parameters to fully describe the dc behavior of MOSFETs. The parameters are first extracted for non-irradiated n- and pMOSFETs of various sizes, hence validating the model. The same parameters are then extracted at each TID up to 1 Grad. The *drain leakage current* is the most seriously affected parameter, particularly for narrow-channel nMOSFETs. The *threshold voltage* demonstrates a slight change for nMOSFETs and a large increase for narrow-channel pMOSFETs. The impact of TID on the *intrinsic gain* is also investigated. It shows a reduction by half for nMOSFETs, whereas it increases by up to 2.5x for pMOSFETs. Finally, TID effects on the *transconductance efficiency* G_m/I_d is evaluated. After the proper parameter extraction and normalization, TID shows a negligible influence on the normalized *transconductance efficiency*. This study shows that overall, the 28 nm bulk CMOS process is rather radiation tolerant except for narrow/long-channel MOSFETs. It also highlights the advantage of using the simplified charge-based EKV MOSFET model for analyzing TID effects on analog performance.

ACKNOWLEDGMENT

The authors would like to thank Dr. A. Marchioro from the CERN EP-ESE Group for the fruitful collaboration and Dr. F. Faccio from the CERN EP-ESE Group for the constructive discussions and the help with the measurements.

REFERENCES

[1] F. Ellinger, M. Claus *et al.*, "Review of advanced and beyond CMOS FET technologies for radio frequency circuit design," in *MTT-S Int. Microwave & Optoelectronics Conference (IMOC)*, 2011, pp. 347–351.

[2] J. M. Benedetto, H. E. Boesch *et al.*, "Hole removal in thin-gate MOSFETs by tunneling," *IEEE Trans. on Nuclear Science*, vol. 32, no. 6, pp. 3916–3920, 1985.

[3] A. Pezzotta, C.-M. Zhang *et al.*, "Impact of GigaRad Ionizing Dose on 28nm Bulk MOSFETs for Future HL-LHC," in *European Solid-State Device Research Conference (ESSDERC)*, 2016, pp. 146–149.

[4] C.-M. Zhang, F. Jazaeri *et al.*, "GigaRad Total Ionizing Dose and Post-Irradiation Effects on 28 nm Bulk MOSFETs," in *IEEE Nuclear Science Symposium*, no. EPFL-CONF-221644, 2016.

[5] D. M. Fleetwood, "Total ionizing dose effects in MOS and low-dose-rate-sensitive linear-bipolar devices," *IEEE Trans. on Nuclear Science*, vol. 60, no. 3, pp. 1706–1730, 2013.

[6] C. Enz, M.-A. Chalkiadaki, and A. Mangla, "Low-power analog/RF circuit design based on the inversion coefficient," in *European Solid-State Circuits Conference (ESSCIRC), ESSCIRC 2015-41st*. IEEE, 2015, pp. 202–208.

[7] C. Enz and A. Pezzotta, "Nanoscale MOSFET modeling for the design of low-power analog and RF circuits," in *Mixed Design of Integrated Circuits and Systems Conf. (MIXDES)*. IEEE, 2016, pp. 21–26.

[8] C. C. Enz and E. A. Vittoz, *Charge-based MOS transistor modeling: the EKV model for low-power and RF IC design*. John Wiley, 2006.

[9] F. Faccio and G. Cervelli, "Radiation-induced edge effects in deep submicron CMOS transistors," *IEEE Transactions on Nuclear Science*, vol. 52, no. 6, pp. 2413–2420, 2005.

[10] F. Faccio, S. Michelis *et al.*, "Radiation-induced short channel (RISCE) and narrow channel (RINCE) effects in 65 and 130 nm MOSFETs," *IEEE Trans. on Nuclear Science*, vol. 62, no. 6, pp. 2933–2940, 2015.

$1/f$ Noise in 3D Vertical Gate-all-around Junction-less Silicon Nanowire Transistors

Chhandak Mukherjee and Cristell Maneux

IMS Laboratory
University of Bordeaux
351, Cours de la Libération – 33405, Talence, France
chhandak.mukherjee@ims-bordeaux.fr

Julien Pezard and Guilhem Larrieu

LAAS-CNRS
Université de Toulouse
CNRS, INP, Toulouse, France

Abstract— **Low-frequency noise characteristics have been investigated in arrays of 14 nm gate-all-around vertical silicon junction-less nanowire transistors. Extensive measurements have been performed to study the evolution of the $1/f$ noise as a function of bias for nanowire arrays with different nanowire diameters and several numbers of nanowires in parallel. Measured drain current noise can be explained well by correlated mobility fluctuation noise theory. Although the conduction is mainly limited by the bulk, *i.e.*, the core of the nanowire, additional trapping/release of charge carriers is observed due to an accumulation channel formed at higher gate bias. Additionally, for the first time in junction-less transistors, evidence of significant noise contribution from access regions at higher bias is observed that provides insight into $1/f$ noise origin.**

Keywords— *Vertical junction-less nanowire transistors, nanowire arrays, 1/f noise, carrier number fluctuation, mobility fluctuation, correlated mobility fluctuation noise model.*

I. INTRODUCTION

Nanowire transistors are being considered as crucial building blocks for the ultimate scaling of MOS transistors, capable of sustaining extreme miniaturization and complex architectures compatible for 3D vertical integration, owing to their physical and geometrical properties. In particular, nanowires' suitability for forming a gate-all-around (GAA) configuration offers an optimum electrostatic control of the gate over the conduction channel and a better immunity against the short channel effects (SCE). Junction-less nanowire transistors (JLNT) [1-5] have been proposed to further circumvent the problems due to SCE. JLNTs are accumulation mode transistors that are fully depleted below threshold. Above threshold the current flows through the bulk while an accumulation channel forms for sufficiently large gate voltages [1-2, 4]. However, many issues still remain unaddressed regarding the impact of the transition from 1D scaling to transistors scaled in all dimensions. Classical characterization techniques, such as mobility extraction, are insufficient to provide information on the devices quality at ultimate scaling, since mobility can collapse at such small gate lengths [5]. Low-frequency noise characterization can be a very precise technique for analyzing electronic transport in such nanometer sized devices. Previous studies in junction-less transistors with nanowire (NW) channels have addressed the potential of this architecture for extremely low noise level [3-5] due to bulk conduction. Even though the $1/f$ noise in junction-less NWTs is expected to be originated from bulk mobility fluctuations

(Hooge mobility fluctuations, HMF) due to carrier scattering in the channel, the junction-less FETs also have an additional surface conduction when the gate voltage is larger than the flat-band voltage [4]. This indicates that the noise in JLNTs can be explained by the widely accepted noise theory of correlated mobility fluctuations (CMF) coupled with number fluctuation (CNF) [6-7]. Our previous work [5] presented preliminary noise characteristics of 14 nm GAA JLNTs, demonstrating extremely low noise in JLNTs. In this work, extensive $1/f$ noise characterization is performed on vertical 14 nm GAA JLNTs for wide range of bias and geometries (several diameters and several numbers of NWs in parallel), demonstrating the advantage of higher number of nanowire parallelization for efficient noise reduction, followed by validation of CNF+CMF noise model. Moreover, for the first time in JLNTs, individual noise contributions from the gate-enclosed channel and the NW access regions are deeply analyzed and a significant noise contribution from the access regions is observed.

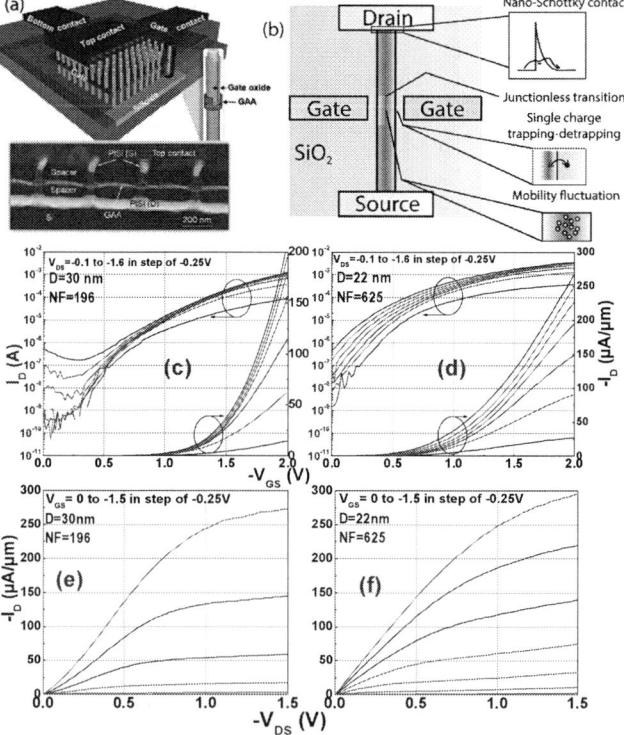

Fig. 1: (a) 3D areal Schematic and TEM image of NW arrays, (b) Schematic view of the device, (c), (d) I_D-V_{GS} and (e), (f) I_D-V_{DS} of two p-FET devices with NW diameter of 30 nm (196 NWs) and 22 nm (625 NWs), respectively.

This work is partially supported by LAAS-CNRS micro and nanotechnologies platform, a member of the French RENATECH network.

978-1-5090-5979-9/17 $31.00 © 2017 IEEE

II. Device Fabrication and DC Characteristics

The nanowire transistors have a junction-less architecture composed of a homogenous highly doped nanowire channel (channel length of 14 nm), fabricated on boron doped (2×10^{19} /cm^3) Si Substrate, with the current flowing between the silicided source/drain contacts being controlled by a gate-all-around structure (Fig.1a). More details on the fabrication steps can be found in [8]. Fig. 1b shows a schematic view of the device illustrating the different sources of noise. It includes noise in the access regions of the NW (beyond the gate-enclosed NW channel, *i.e.*, after the junction-less transition to the access), bulk mobility fluctuation due to defects within the crystal, *i.e.* NW core, and single-charge trapping-detrapping at gate oxide interface. In this work, extensive noise characterization has been performed on 16 geometries including 4 NW diameters (D = 22, 30, 42 and 80 nm), and for each, 4 different combinations of number of nanowires in parallel (NF = 36, 81, 196 and 625) with effective gate width being, $W_{eff} = \pi D \times NF$. The threshold voltage (V_T) of the JLNTs vary from (-0.1 to -0.7 V), roughly, between the largest and the smallest NW diameter). Examples of static characteristics (I_D-V_{GS} and I_D-V_{DS}) of two p-FET devices with NW diameter of 30 nm (196 NWs) and 22 nm (625 NWs) are presented in Fig. 1 (c), (d) and (e), (f), respectively.

III. 1/f Noise Characteristics

A. Bias-dependence of 1/f noise

Figure 2(a) and (b) show the drain current noise spectral density, S_{ID}, as a function of frequency for a JLNT with NW diameter of 30 nm and 196 nanowires in parallel, at different V_{GS} and I_D, respectively. The noise spectra reveal a $1/f^\gamma$ noise behavior and the frequency exponent, γ, is observed close to 1 in all geometries. The value of γ close to 1 signifies a uniform trap distribution within the gate oxide. This is consistent with the absence of generation-recombination (G-R) or RTS noises which indicate non-uniform distribution of traps. Although evidence of RTS has been reported from the same technology [5] for single nanowires, it has been demonstrated that the RTS disappears in transistors with higher number of nanowires in parallel. This is consistent with the observation of this work. In Fig. 2 (c), S_{ID} is shown as a function of V_{GS} at 10 Hz for three different values of V_{DS}. The results show an increasing trend of the noise till it reaches a maximum value and starts to roll off. This can be explained by the carrier number fluctuation theory [6]. Fig. 2(d) shows the input-referred gate voltage noise spectral density extracted using S_{VG} = S_{ID}/g_m^2 [6], with $S_{VG}^{1/2}$, as a function of I_D/g_m. Even though Hooge's mobility fluctuation (HMF) is expected to be the major source of noise in the JLNTs due to the fact that major current conduction is through the bulk [4], the dependence between $S_{VG}^{1/2}$ and I_D/g_m illustrates that correlated mobility fluctutation (CNF+CMF) model (dashed red line in Fig.2d) seems to describe the results seemingly better. S_{VG} can be written using the CNF+CMF model as [6-7],

$$\sqrt{S_{V_G}} = \sqrt{S_{V_{fb}}}\left(1+\frac{\alpha_{sc}\mu_{eff}C_{ox}I_D}{g_m}\right) = \sqrt{S_{V_{fb}}}\left(1+\Omega\frac{I_D}{g_m}\right). \quad (1)$$

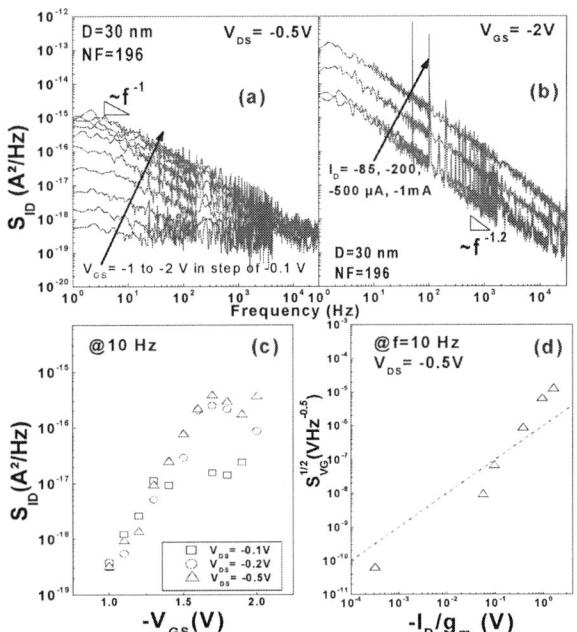

Fig. 2: Measured S_{ID} as a function of frequency under fixed (a) V_{GS}, and (b) I_D, (c) S_{ID} as a function of V_{GS} at f=10 Hz, (d) S_{VG} as a function of I_D/g_m.

With S_{Vfb} given by [6], $\quad S_{V_{fb}} = \dfrac{q^2kT\lambda N_T}{f^\gamma W_{eff}L_G C_{ox}^2} \quad (2)$

Here, S_{Vfb} is the flat band voltage power spectral density, α_{sc} is the coupling parameter and $\Omega = \alpha_{sc}\mu_{eff}C_{ox}$ is the Coulomb scattering parameter; λ is the tunneling constant in the dielectric (0.1 nm) [6], and N_T is the oxide trap density. In our results, Ω is observed to be constant (without any V_{DS} dependence) when the JLNT channel is in accumulation mode ($V_{GS} > V_T$). This implies that α_{sc} and μ_{eff} compensate each other. In surface channel MOS devices, μ_{eff} decreases at strong inversion due to surface roughness scattering [9], whereas α_{sc} also reduces at strong inversion due to screening effects [10]. In contrast, in JLNTs, the product $\alpha_{sc}\mu_{eff}$ remains constant owing to its operation in the accumulation mode.

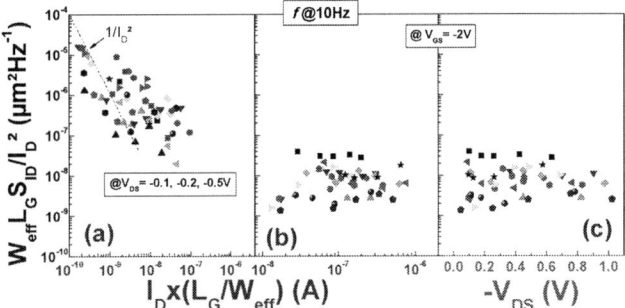

Fig. 3: Normalized drain current noise (S_{ID}/I_D^2) $\times W_{eff}L_G$ with normalized drain current $I_D \times L_G/W_{eff}$ at different (a) V_{GS} (constant V_{DS} = -0.1, -0.2, -0.5V) and (b) I_D (constant V_{GS}= -2V), (c) (S_{ID}/I_D^2) $\times W_{eff}L_G$ as a function of V_{DS}.

B. Geometry-dependence of 1/f noise

Fig. 3 (a) and (b) shows normalized S_{ID} as a function of I_D. The S_{ID}/I_D^2-I_D characteristics are normalized by channel area parameters (W_{eff} and L_G) for all device structures. Noise for

NWTs having 36 to 625 nanowires in parallel with NW diameters ranging from 22 nm to 90 nm are observed to almost merge, indicating that LFN behavior has a simple W_{eff} dependence. The dispersion observed can be attributed to sample-to-sample LFN variability reinforced in ultra-scaled devices. V_{DS} dependent LFN characteristics of JLNTs operating in accumulation mode are shown in Fig. 3(c). S_{ID}/I_D^2 normalized by channel area is practically independent of V_{DS} for values of V_{DS} up to -1.0 V, when the channel is in accumulation mode ($V_{GS} > V_T$) and even when the channel is partially depleted at the drain-end [4] ($V_{DS} > V_{GS}-V_T$).

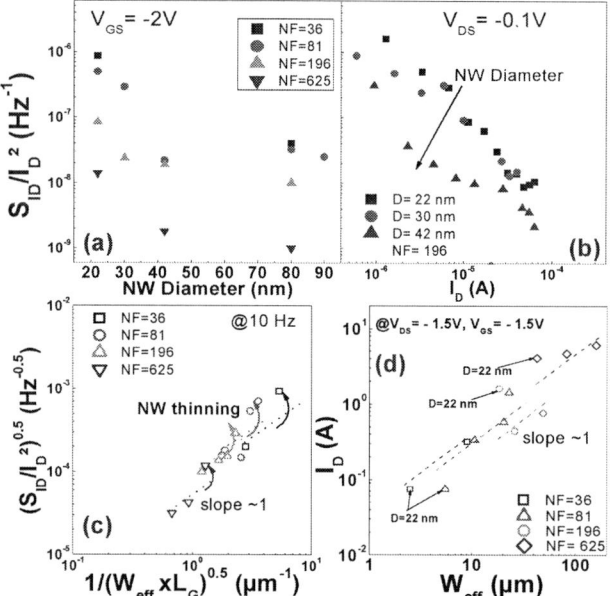

Fig. 4: Normalized drain current noise (S_{ID}/I_D^2) as a function of (a) NW diameter for different number of NWs in parallel, (b) I_D for different NW diameters, (c) inverse of the effective gate area ($W_{eff}L_G$) and (d) I_D as a function of effective NW width showing effects of NW narrowing.

Fig. 4 (a) shows S_{ID}/I_D^2 as a function of nanowire diameter depicting an inversely proportional behavior which is consistent with the CNF+CMF noise theory that describes an inverse dependence of the noise magnitude on the effective channel area. An additional observation is that increasing the number of nanowires in parallel reduces the noise magnitude as well. The reduction of the noise magnitude with nanowire diameter is explicitly observed from Fig. 4 (b) which shows S_{ID}/I_D^2 as a function of I_D for NWTs with 196 nanowires in parallel. Interestingly, when $\sqrt{(S_{ID}/I_D^2)}$ is plotted as a function of the inverse of the effective nanowire gate area, $W_{eff}L_G$, a linear dependence is observed (Fig. 4(c)), as expected from the conventional noise theory. However, in smaller geometries S_{ID}/I_D^2 rapidly increases, thus deviating from linearity. This means that the rapid increase in the noise at narrow W_{eff} cannot be explained only by the reduction in channel area as in the conventional noise theory. Firstly, we observe a small deviation of I_D from linearity in case of the smallest NW diameter with different number of NWs in parallel (Fig. 4(d)). So effectively, this could indeed impact the deviation of S_{ID} from the linearity. However, the deviation is observed in case

of normalized S_{ID} which should diminish the effect due to drain current deviation. Besides the deviation is significant for the case of smaller number of NWs in parallel (36, 81) and therefore, there must be an additional effect in play. According to [11], the NW structure itself could be responsible for this effect. In a narrow NW channel, an electron in a single trap on the NW channel increases the potential of the entire channel, and the current flow is strongly blocked. In this case, the Coulomb potential caused by an electron in an oxide trap becomes compatible with the thermal energy. As a result, large drain current noise is observed because of electron trapping/de-trapping (bottleneck effects) [11]. This leads to a rapid increase in noise for narrow nanowires (especially for NWTs with diameters of 22 nm).

IV. NOISE PARAMETER EXTRACTION

Fig. 5(a) shows $(S_{ID}/I_D^2) \times$frequency as a function of the effective NW area (A_{eff}) of the different transistor geometries, depicting the well-known inverse dependence of the SPICE noise parameter K_F (represented by $S_{ID}/I_D^2 = K_F/f$). Also, the oxide trap density, N_T, which acts as an interface/oxide quality indicator, has been extracted using (2) (Fig. 5(b)) for different NW geometries with different diameters. An average value of N_T is found to be around 10^{18} eV^{-1}cm^{-3}, which is close to values reported in GAA Si/SiO$_2$ nanowire FETs [12]. A slight decrease of N_T for higher number of NWs in parallel and smaller NW diameter has been observed possibly due to quantum confinement effects [12] since GAA architectures are mostly immune to short channel effects.

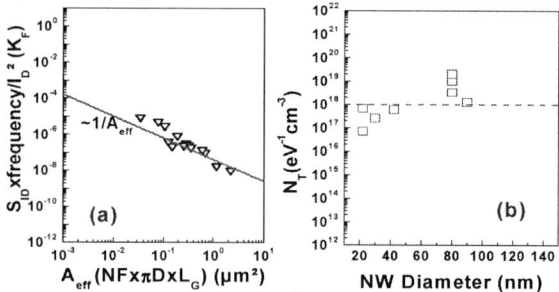

Fig. 5: (a) $(S_{ID}/I_D^2) \times f$ as a function of effective NW area showing the SPICE noise parameter, K_F, and (b) Oxide trap density for different NW diameters.

V. NOISE ANALYSIS

A. Validity of CNF+CMF Noise Model

The 1/f noise in the JLTNs under study can be explained using correlated mobility fluctuation model (CNF + CMF) [4, 6-7]:

$$ S_{I_D} = S_{V_{fb}} \left(1 + \frac{\alpha_{sc}\mu_{eff}C_{ox}I_D}{g_m} \right)^2 g_m^2 = S_{V_{fb}} \left(1 + \Omega \frac{I_D}{g_m} \right)^2 g_m^2 \quad (3) $$

S_{ID}/I_D^2 is plotted as a function of I_D in Fig. 6 for 3 geometries (D = 22, 30 and 42 nm) with 196 nanowires in parallel. S_{ID}/I_D^2 rolls-off gradually when V_{GS} is well above the flat band voltage and additional to bulk mobility fluctuation, surface effects can be observed in the accumulation channel. When accumulation regime starts, the noise mechanisms can be attributed principally to carrier number fluctuation, where the carrier trapping/de-trapping in to the gate oxide traps is independent of the small gate bias, which explains a slow variation of the S_{ID}/I_D^2. When mobility fluctuation starts to dominate, S_{ID}/I_D^2 rolls off. To offer a better perspective, in Fig. 6, the HMF model is also shown qualitatively along the CNF+CMF model, in comparison with measurement. A very good agreement between S_{ID}/I_D^2 and $(g_m/I_D)^2$ (here β is a constant) can be observed, in accordance with (3). Clearly, this shows that CNF+CMF model explains the noise behavior better than HMF, in the entire bias range.

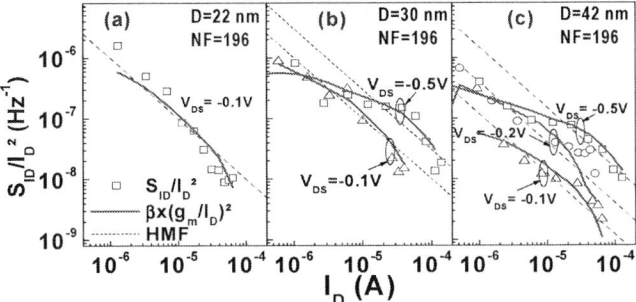

Fig. 6: Normalized drain–current spectral density S_{ID}/I_D^2 (Symbols) compared with $(g_m/I_D)^2$ and HMF (lines) for NW diameter (a) 22, (b) 30 and (c) 42 nm.

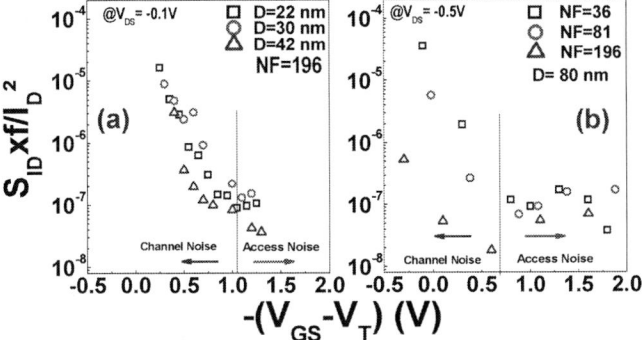

Fig. 7: S_{ID}/I_D^2 as a function of $(V_{GS}-V_T)$ for (a) different NW diameters and (b) different number of NW parallelization, indicating noise contributions from the nanowire channel and access regions.

B. Channel and Access Noise Components

Contrary to the classical MOSFETs, in nanowire devices, at high drain bias, the current and noise characteristics are heavily influenced by the source and drain access regions. The dominance of access region noise contribution over the channel noise at higher bias has been demonstrated by [13] for InAs NW transistors. In this work, for the first time, we demonstrate similar noise contributions of the total measured noise in junction-less NWs with several different diameters and number of NWs in parallel. At high bias, the dominant noise contribution from the access regions (refer to Fig. 1b that shows the access regions outside the gate region) over the gate-enclosed channel region of the JLNTs can be observed as shown in Fig. 7. Fig. 7 (a) shows S_{ID}/I_D^2 versus $(V_{GS}-V_T)$ for transistors with 196 nanowires in parallel having diameters of 22, 30 and 42 nm at a V_{DS} of -0.1V. We observe an inverse $(V_{GS}-V_T)$ dependence of S_{ID}/I_D^2 ($\sim (V_{GS}-V_T)^{-p}$) for lower bias, due to the channel contribution, whereas at higher bias, the total noise becomes almost constant and majorly dominated by

the access noise contributions [13]. The effect of the access noise can be seen to start at lower V_{GS} for V_{DS} of -0.5V (as shown in Fig. 7(b) for JLNTs having diameter of 80 nm with different number of nanowires in parallel) than at V_{DS} of -0.1V. In fact, the access noise contribution ($S_{ID}\times f/I_D^2 \approx 10^{-7}$) to the total noise remains the same regardless of the NW diameter or number of NWs in parallel. Additionally, increasing the nanowire parallelization reduces the channel noise significantly (see Fig. 7(b) at lower V_{GS}), while the access noise still dominates the total noise at higher bias.

VI.   CONCLUSION

Low-frequency noise has been investigated in arrays of 14 nm gate-all-around vertical junction-less nanowire Si transistors. Very low noise level has been achieved using large number of nanowires in parallel. At moderate bias, the total drain current noise is mainly dominated by the gate-enclosed channel region, corresponding to correlated mobility fluctuation (CNF+CMF) noise model. At higher bias, significant noise contributions from the access regions to the total noise has been demonstrated for all available diameters and number of NWs in parallel. To the best of our knowledge, such noise contributions have been analyzed for the first time in junction-less nanowire transistors.

REFERENCES

[1] J.-P. Colinge and J. C. Greer, Nanowire Transistors: Physics of Devices and Materials in One Dimension. Cambridge University Press, 2016.

[2] J. P. Colinge, C. W. Lee, A. Afzalian, et al., "Nanowire transistors without junctions", Nat. Nanotechnol. vol. 5, pp. 225-229, 2010.

[3] D.-Y. Jeon, S. J. Park, M. Mouis, et al., "Low-frequency noise behavior of junctionless transistors compared to inversion-mode transistors", Solid-State Electronics, vol. 81, pp. 101-104, March 2013.

[4] D. Jang, J. W. Lee, C.-W. Lee, et al., "Low-frequency noise in junctionless multigate transistors", Appl. Phys. Lett., vol. 98, pp. 133502-1-3, 2011.

[5] N. Clément, X. L. Han and G. Larrieu, "Electronic transport mechanisms in scaled gate-all-around silicon nanowire transistor arrays", Appl. Phys. Lett., vol. 103, pp. 263504-1-5, 2013.

[6] M. von Haartman and M. Östling, Low-Frequency Noise in Advanced MOS Devices, 1st ed. Springer Netherlands, 2007.

[7] G. Ghibaudo, et al., "Improved analysis of low frequency noise in field-effect MOS transistors," Phys. Stat. Sol.(a), vol. 124, pp. 571-581, 1991.

[8] G. Larrieu and X.–L. Han, "Vertical nanowire array-based field effect transistors for ultimate scaling", Nanoscale, vol. 5, pp. 2437-2441, 2013.

[9] M. Koyama, M. Cassé, R. Coquand, et al., "Study of carrier transport in strained and unstrained SOI tri-gate and omega-gate Si-nanowire MOSFETs," Proc. ESSDERC, pp. 73-76, Sep. 2012.

[10] K. K. Hung, et al., "Random telegraph noise of deep-submicrometer MOSFET's," IEEE Electron Dev. Lett., vol. 11, pp. 90-92, Feb. 1990.

[11] M. Saitoh, K. Ota, C. Tanaka, and T. Numata "Channel size dependence of low-frequency noise in tri-gate silicon nanowire transistors", Jpn. J. Appl. Phys., vol. 54, pp. 044201(1-3), 2015.

[12] R.-H. Baek, C.-K. Baek, H.-S. Choi, et al., "Characterization and modeling of 1/f noise in Si-nanowire FETs: effects of cylindrical geometry and different processing of oxides," IEEE Trans. Nanotechnol., vol. 10, pp. 417-423, May 2011.

[13] C. J. Delker, Y. Zi, C. Yang and D. B. Janes, "Low-Frequency Noise Contributions From Channel and Contacts in InAs Nanowire Transistors," IEEE Trans Electron Dev., vol. 60, pp. 2900-2905, 2013.

Random Telegraph Signal Noise in Tunneling Field-Effect Transistors with S below 60 mV/decade

Markus Hellenbrand, Elvedin Memišević, Johannes Svensson, Erik Lind, and Lars-Erik Wernersson

Department of Electrical and Information Technology

Lund University, Box 118

S-22100 Lund, Sweden

Abstract—Single gate oxide defects in strongly scaled Tunneling Field-Effect Transistors with an inverse subthreshold slope well below 60 mV/decade are investigated by Random Telegraph Signal (RTS) noise measurements. The cause for RTS noise are electrons being captured in and released from individual defects in the gate oxide. Under the assumption that elastic tunneling is the underlying capture and emission mechanism, the measured RTS time constants vary with the relative position of the channel Fermi level and the defect energy level while the amplitudes – independent of the capture and release mechanism – follow the inverse of the inverse subthreshold slope.

Keywords—Tunneling Field-Effect Transistors, Nanowires, Below 60 mV/decade, Random Telegraph Signal Noise, Elastic Tunneling

I. INTRODUCTION

Transistor dimensions are an important scaling parameter to continue Moore's law. Also for devices beyond Moore's law, certain transistor dimensions remain important metrics for optimized device performance. For vertical nanowire transistors, one of the key dimensions is the diameter of the nanowires, strongly affecting the electrostatic control of the gate over the channel [1]. With excellent electrostatic control and a carefully designed III-V heterojunction, we demonstrated Tunneling Field-Effect Transistors (TFETs) with an inverse subthreshold slope (S) well below 60 mV/decade [2]. At the same time, while improving the electrostatic gate control, scaling down transistor dimensions to a few nanometers limits the number of gate oxide defects to a few individual locations. Capturing and releasing individual electrons in and from these defects changes the channel potential energy by a discrete amount and thus – specific for TFETs with a dominant defect close to the junction – also the reverse bias of the tunneling junction. This leads to discrete steps in the device current, so-called Random Telegraph Signal (RTS) noise, and results in degradation of transistor and circuit performance [3]. In TFETs, this effect has been simulated thoroughly [3, 4], but rarely been observed experimentally [5, 6] and explained in detail even less. Here, we study RTS noise in TFETs with S well below 60 mV/decade, which allows estimations of both the spatial as well as energetical position of the dominant defect and its electrostatic effect on the channel potential.

II. DEVICE AND MEASUREMENT SPECIFICATIONS

A. Device Fabrication

To clearly separate and identify all effects, only transistors made of one single nanowire were studied for this work. A schematic illustration is shown in Fig. 1. All nanowires were grown by metalorganic vapor phase epitaxy (MOVPE) from an Au seed particle defined by electron beam lithography (EBL). The substrate consisted of a MOVPE-grown InAs buffer layer on top of Si. The InAs bottom segment of the nanowire was n-doped with Sn to reduce series resistance. In order to achieve a gate control as good as possible, the channel segment on top was not intentionally doped. The tunneling junction was created

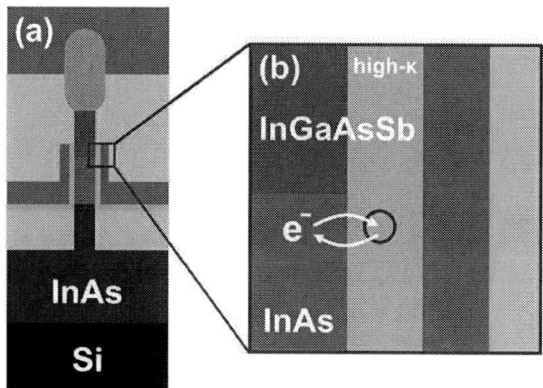

Fig. 1: (a) Schematic illustration of the studied devices. Dark blue indicates InAs, doped in the bottom, intrinsic in the channel region, purple indicates the InGaAsSb segment, yellow the GaSb segment, light blue indicates the gate oxide, grey are metals, green spacers. (b) Schematic of the tunneling process causing RTS noise. The red circle indicates a single defect in the gate oxide.

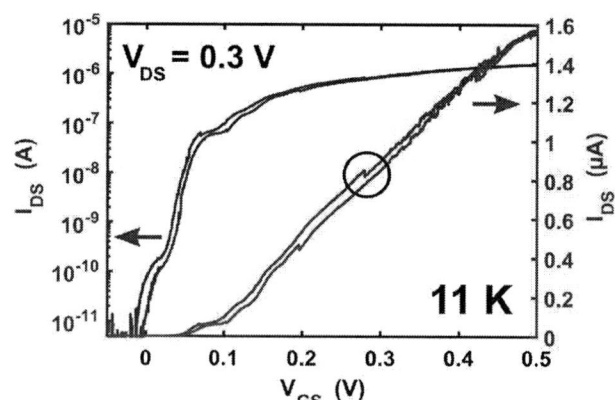

Fig. 2: Transfer characteristics for an example device at 11 K with $V_{DS} = 0.3$ V. Blue shows a logarithmic scale (left), red a linear scale (right). The circle indicates the RTS step more closely examined in this article. For the highest currents in the linear scale there are many more RTS steps.

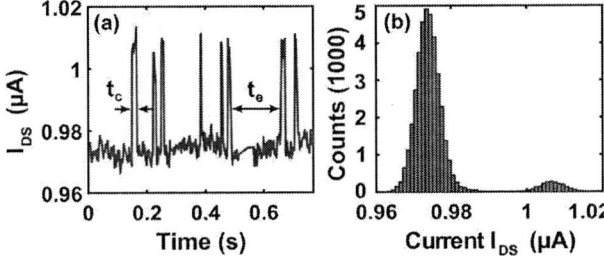

Fig. 3: (a) Excerpt from an RTS measurement close to the bias point indicated in Fig. 2. t_c and t_e denote the capture and the emission times, respectively. (b) Current histogram for the complete measurement in (a), clearly indicating two current levels.

by changing from the un-doped InAs channel segment to p-doped (using Zn) InGaAsSb. For good source contacts, the InGaAsSb segment was followed by a p-doped GaSb top segment. After growth, the InAs parts of the nanowire were thinned down to a final diameter of 20 nm by first oxidizing the surface with ozone plasma and then etching this oxide with citric acid. Immediately afterwards, a high-κ gate oxide consisting of approximately 1 nm of Al_2O_3 and 3.5 nm of HfO_2 was applied by atomic layer deposition (ALD). The estimated equivalent oxide thickness (EOT) is 1.4 nm. A SiO_x bottom spacer was deposited to separate the drain from the gate contact and a 60-nm-thick W gate metal was sputtered over the whole sample. To define the gate length from this coverage, another (temporary) spacer was applied and etched back so that all W beyond the intended gate contact could be removed from the nanowire. The gate pad in the lateral direction was defined by UV-lithography. After exchanging the temporary spacer for the final photoresist top spacer, the high-κ oxide was removed from the top of the nanowire and a 10 nm Ni/150 nm Au source metal contact was deposited. As a last step, drain and gate vias were etched and the top metal source, drain, and gate pads were defined, all by UV-lithography. The transfer characteristics of a representative device is shown in Fig. 2. For consistency, all data presented here are from this same device and the analysis is focused on the RTS step marked in Fig. 2. Further details on fabrication can be found in [2].

B. Measurement Setup and Technique

The measurement setup consisted of a Lake Shore Cryotronics CRX-4K probe station and an Agilent B2912A

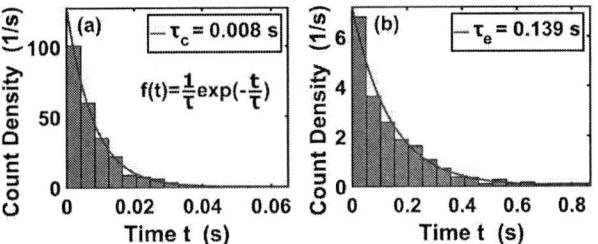

Fig. 4: Histograms of the capture (a) and emission (b) times for the measurement from Fig. 3. The red lines indicate the exponential distribution (equation in (a)) fitted to the data.

Source/Measure Unit. RTS noise measurements were carried out at room temperature, at 150 K, and at 11 K. However, due to the high sensitivity of the noise signal to even small changes in the device behavior and due to threshold voltage shifts at different temperatures, it was not possible to identify and measure with certainty the same defect at different temperatures. In each measurement like the one presented in Fig. 3, the drain current (InAs side) of a transistor was measured in sections of up to 200 seconds at constant gate and drain bias, while the source (GaSb top contact) was grounded. For each defect and each bias point, at least a hundred and up to 2200 transitions between current levels were recorded with measurement resolutions between 0.2 ms and 2 ms. The capture and emission time constants were determined by fitting an exponential distribution to the histograms of the times spent in the high current states (capture times) and the low current states (emission times), respectively (Fig. 4). The data presented here are from a measurement at 11 K, but are representative for measurements at all temperatures.

III. RESULTS AND DISCUSSION

Both the measured capture and emission time constants τ_c and τ_e vary with varying gate bias (Fig. 5(a)), which can be qualitatively explained with the help of the process illustrated in Fig. 5(b) under the assumption that elastic tunneling is the capture and emission mechanism. If the channel Fermi level is increased towards the defect energy level from below, electron tunneling into the oxide defect exhibits a rather long time constant due to the energy difference between the two states. The emission time constant in this configuration is rather short since the electron can always tunnel out of the defect and relax to the energy corresponding to the channel Fermi level. Increasing the channel Fermi level above the trap energy level reverses this behavior. Since the effect of the defect considered here shows up in the on-state, the defect energy level must be located quite far below the channel conduction band in the unbiased state.

Besides the energy difference explained above, the tunneling time constant is determined by the depth z of the defect from the channel into the oxide. According to standard quantum-mechanical tunneling the probability of a state in a forbidden area – and thus the tunneling time constant τ – decays exponentially as

$$\tau_{c/e} = \tau_0 \exp(z/\lambda) \qquad (1)$$

where τ_0 is a constant, and λ the tunneling attenuation constant according to the Wentzel-Kramers-Brillouin approximation [7] given by

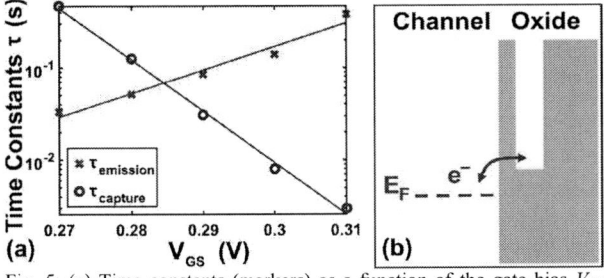

Fig. 5: (a) Time constants (markers) as a function of the gate bias V_{GS} together with exponential fits (lines). A decreasing capture time constant for an increasing emission time constant was observed for all TFETs. (b) Schematic illustration of the capture and emission process for a single oxide defect. The trends in (a) can be explained by the difference between the channel Fermi level and the defect energy level.

$$\lambda = \left(\frac{4\pi}{h}\sqrt{2m^*\phi_B}\right)^{-1} = 0.13 \text{ nm}, \qquad (2)$$

where h is the Planck constant, $m^* = 0.23 \cdot m_0$ the effective mass in the gate oxide (m_0 is the electron rest mass), and $\phi_B = 2.3$ eV the barrier height from the channel to the gate oxide [8]. With the assumptions made above, z can be determined from the biasing point where $\tau_c = \tau_e$ since in this configuration the channel Fermi level and the defect energy level are aligned and there is no energy difference, which distorts the time constants. With $\tau_{e/c} \approx 6\cdot10^{-2}$ s from Fig. 5(a), $\tau_0 = 4\cdot10^{-11}$ s from [9], and (1) rearranged to $z = \lambda \ln(\tau/\tau_0)$, the depth z amounts to approximately 2.7 nm from the channel interface into the gate oxide and thus to approximately halfway between the channel and the gate metal. From this calculation it is obvious that z depends on τ_0, for which different values can be found in literature, ranging from 10^{-10} s [7] to $6.6\cdot10^{-14}$ s [10]. Between these two extreme values, the position of the defect in the oxide would change by up to 1 nm, which does, however, not influence the underlying tunneling model. The value of $\tau_0 = 4\cdot10^{-11}$ s chosen here is taken from our own high-frequency characterizations of Metal Oxide Semiconductor Field-Effect Transistors (MOSFETs) similar to the presented TFET structures and corresponds to tunneling into defects which are very close to the channel/gate oxide interface.

Independent of the exact capture and emission mechanism, the RTS amplitude can be used to estimate the change in the channel potential energy that results from a single electron captured in a defect in the oxide. With the simplified expression

$$I_{1D} = a\,(V_R - V_T)\exp\left(-\frac{b}{\xi}\right) \qquad (3)$$

for a 1D TFET [11], it can be readily shown that the change in the TFET current from one single charge q_{ox} in the gate oxide amounts to

$$\frac{\partial I_{1D}}{\partial q_{ox}} = \frac{\partial V_R}{\partial q_{ox}}g_m, \qquad (4)$$

which also holds for a 3D expression of the current. In (3) and (4), a and b are constants summarizing elementary constants and material parameters, V_R is the reverse bias applied to the tunneling junction, V_T is the thermal voltage kT/q, ξ denotes the

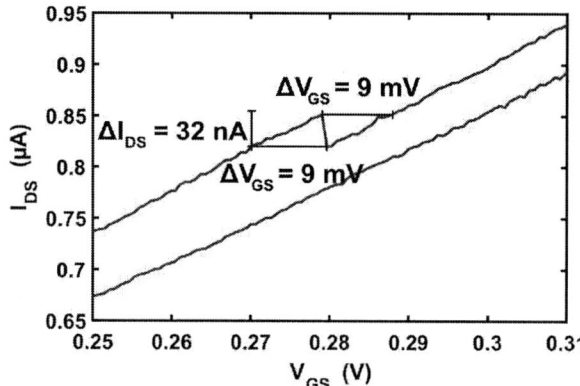

Fig. 6: Detailed view of the RTS step indicated in Fig. 2. As shown in (4), the change of the channel potential energy can be directly determined from the step.

electrical field across the tunneling junction and g_m is the transconductance. From (4) it is clear that the change in the channel potential energy can be determined from the transfer characteristics. For this purpose, a more detailed illustration of the encircled RTS step in Fig. 2 is provided in Fig. 6.

According to a simple model of the surface potential Ψ_S, it depends on the gate voltage V_{GS} as

$$\Delta\Psi_S = \Delta V_{GS}\frac{C_{ox}}{C_{ox} + C_q + C_{it}}, \qquad (5)$$

where C_{ox} is the gate oxide capacitance, C_q is the semiconductor capacitance resulting from a finite density of states in the channel, and C_{it} represents interface defects between channel and gate oxide. The centroid capacitance C_c is neglected for the sake of simplicity, and furthermore it is masked by the small C_q. Since V_R denotes the change of Ψ_S due to the change in V_{GS}, (5) can be applied to V_R in exactly the same way as to Ψ_S.

The step ΔI_{DS} in the current indicated in Fig. 6 is caused by an abrupt change $\partial V_R/\partial q_{ox}$ in the tunnel junction reverse bias V_R due to a single captured charge q_{ox} in the gate oxide. To obtain the same ΔI_{DS} by changing the gate bias V_{GS}, a ΔV_{GS} of 9 mV would be required, as obtained from Fig. 6. To estimate the actual $\partial V_R/\partial q_{ox}$ from (5), we need to calculate C_{ox}, C_q, and C_{it}. C_{ox} can be simply calculated as a cylindrical capacitor, and C_q can be calculated as

$$C_q = \sqrt{\frac{2m^*}{E_F - E_1(0)}}\frac{q^2}{\pi\hbar} \qquad (6)$$

for the approximation of a 1D channel with a large C_{ox} [12], where $m^* = 0.023 \cdot m_0$ is the effective mass in the channel, $E_F - E_1(0)$ denotes the position of the Fermi level above the first subband, $\hbar$ the reduced Planck constant, and q the elementary charge. C_{it} can be calculated as $C_{it} = q^2 D_{it}$, where $D_{it} \approx 5\cdot10^{12}$ eV^{-1} cm^{-2} close to the conduction band edge [13]. With values of $C_{ox} \approx 0.033$ F/m^2, $C_q \approx 0.002$ F/m^2 ($E_F - E_1(0) \approx 0.15$ eV from Fig. 2 and (5) and C_q divided by the channel circumference for equal units), $C_{it} \approx 0.008$ F/m^2, and $\Delta V_{GS} = 9$ mV from Fig. 6, (5) results in $\partial V_R/\partial q_{ox} \approx 6.9$ mV.

With this value for the change in the channel potential energy, the plausibility of the calculated depth of the defect z can be validated further by electrostatic calculations. Approximating the cylindrical nanowire by a square nanowire with the same cross-sectional area and modelling a single charge at the calculated depth of 2.7 nm with the method of image charges results in a change in the channel potential energy of around 2.2 meV. Although this value is somewhat lower than the experimental one, it is still very close, given the uncertainties in some of the parameters used. Possible explanations are as follows. First of all, the assumption of a 1D density of states might be too optimistic and a larger 2D C_q would yield even more similar results for the calculated and the measured values ($\partial V_R/\partial q_{ox} \approx 5$ meV for C_q in 2D). Also, C_{it} might increase from the value from [13] when moving the bands further into the on-state, or the single dominant defect could be located closer to the channel interface than 2.7 nm. This would imply that τ_0 is larger than our choice of $4\cdot10^{-11}$ s. Furthermore, of course, the simulated model of image charges might be too simple. Again, however, with these sources of uncertainty taken into account,

978-1-5090-5979-9/17 $31.00 © 2017 IEEE

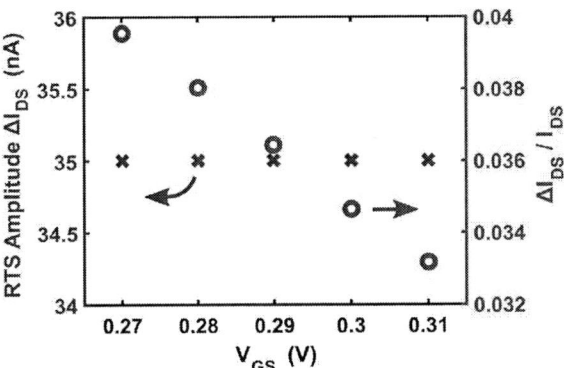

Fig. 7: Absolute (blue, left) and relative (red, right) RTS amplitude for the same RTS step as shown in the other figures. In accordance with (7), the relative RTS amplitude decreases with increasing gate voltage. Note that in accordance with (4), g_m is constant in this area.

both the experimental and the simulated values agree very well, which strongly supports the model of simple elastic tunneling as the mechanism for the observed RTS noise.

Finally, independent of the capture and release mechanism, the behavior of the relative RTS amplitude $\Delta I_{DS}/I_{DS}$ for the whole operation range can be described by extending (4) to

$$\frac{\Delta I_{DS}}{I_{DS}} = \frac{\partial V_R}{\partial q_{ox}} \frac{\ln(10)}{S}, \qquad (7)$$

where ΔI_{DS} corresponds to $\partial I_{ID}/\partial q_{ox}$, I_{DS} to I_{ID} from (3) and S denotes the subthreshold slope calculated from (3) as well. Just as with (4), (7) also holds for the 3D case. This expression suggests a peak of the relative RTS amplitude in the subthreshold region, where S is steepest. Before and after this region it suggests a decay. And indeed, for the measurement series presented here, $\Delta I_{DS}/I_{DS}$ shows a monotonous decrease with increasing gate voltage, depicted in Fig. 7. For other devices and bias points, we were able to measure the other regions described by (7) as well, both increasing relative RTS amplitudes far into the subthreshold region and values around the expected peak. In the on-state, where the measurement series presented here was carried out, the relative RTS amplitude is as low as $3-4$ %, but measurements in the subthreshold region showed values of up to 50 %. This large relative amplitudes strongly suggest a defect location close to the tunneling junction since defects far away from the junction will not affect the tunneling probability in a similarly strong way [4].

IV. CONCLUSIONS

We presented a detailed explanation of RTS noise in III-V nanowire TFETs with inverse subthreshold slopes well below 60 mV/decade. The origin of RTS noise are electrons captured

in and released from individual gate oxide defects close to the tunneling junction and the behavior of the RTS amplitude can be explained by the expression of the TFET tunneling current in a straight forward manner. In the case presented here, the energy level of the defect is located far below the channel conduction band. For the explanation of the RTS time constants we assume elastic tunneling as the capture and emission process and the resulting depth of the oxide defect, $z \approx 2.7$ nm from the channel interface, agrees well with electrostatic calculations.

ACKNOWLEDGMENTS

This work was supported in part by the Swedish Foundation for Strategic Research, Swedish Research Council, and the European Union Seventh Framework Program E2SWITCH under Grant 619509.

REFERENCES

[1] E. Memišević, J. Svensson, M. Hellenbrand, E. Lind, and L.-E. Wernersson, "Scaling of Vertical InAs-GaSb Nanowire Tunneling Field-Effect Transistors on Si," *IEEE Electron Device Letters*, vol. 37, no. 5, pp. 549-552, 5 2016.

[2] E. Memišević, J. Svensson, M. Hellenbrand, E. Lind, and L.-E. Wernersson, "Vertical InAs/GaAsSb/GaSb tunneling field-effect transistor on Si with S = 48 mV/decade and Ion = 10 µA/µm for Ioff = 1 nA/µm at Vds = 0.3 V," in *2016 IEEE International Electron Devices Meeting (IEDM)*, 2016, pp. 19.1.1-19.1.4.

[3] R. Pandey, B. Rajamohanan, H. Liu, V. Narayanan, and S. Datta, "Electrical Noise in Heterojunction Interband Tunnel FETs," *IEEE Transactions on Electron Devices*, vol. 61, no. 2, pp. 552-560, 2 2014.

[4] M. L. Fan, V. P. H. Hu, Y. N. Chen, P. Su, and C. T. Chuang, "Analysis of Single-Trap-Induced Random Telegraph Noise and its Interaction With Work Function Variation for Tunnel FET," *IEEE Transactions on Electron Devices*, vol. 60, no. 6, pp. 2038-2044, 2013.

[5] J. Wan, C. L. Royer, A. Zaslavsky, and S. Cristoloveanu, "Low-frequency noise behavior of tunneling field effect transistors," *Applied Physics Letters*, vol. 97, no. 24, p. 243503, 2010.

[6] Q. Huang *et al.*, "Deep insights into low frequency noise behavior of tunnel FETs with source junction engineering," in *2014 Symposium on VLSI Technology (VLSI-Technology): Digest of Technical Papers*, 2014.

[7] M. v. H. a. M. Östling, *Low-Frequency Noise in Advanced MOS Devices*. Springer, 2007.

[8] N. Li *et al.*, "Properties of InAs metal-oxide-semiconductor structures with atomic-layer-deposited Al2O3 Dielectric," *Applied Physics Letters*, vol. 92, no. 14, p. 143507, 2008.

[9] S. Johansson, M. Berg, K. M. Persson, and E. Lind, "A High-Frequency Transconductance Method for Characterization of High- Border Traps in III-V MOSFETs," *IEEE Transactions on Electron Devices*, vol. 60, no. 2, pp. 776-781, Feb 2013.

[10] I. Lundström and C. Svensson, "Tunneling to traps in insulators," *Journal of Applied Physics*, vol. 43, no. 12, pp. 5045-5047, 1972.

[11] A. C. Seabaugh and Q. Zhang, "Low-Voltage Tunnel Transistors for Beyond CMOS Logic," *Proceedings of the IEEE*, vol. 98, no. 12, pp. 2095-2110, December 2010.

[12] M. Lundstrom and J. Guo, *Nanoscale Transistors*. Springer, 2006.

[13] J. Wu *et al.*, "Low Trap Density in InAs/High-k Nanowire Gate Stacks with Optimized Growth and Doping Conditions," *Nano Letters*, vol. 16, no. 4, pp. 2418-2425, 2016.

Experimental characterization of the Static Noise Margins of strained Silicon complementary Tunnel-FET SRAM

G.V. Luong[1], S. Strangio[2,3], A.T. Tiedemann[1], P. Bernardy[1], S. Trellenkamp[*], P. Palestri[2], S. Mantl[1] and Q.T. Zhao[1]

[1]Peter Grünberg Institut 9, JARA-FIT, Forschungszentrum Jülich, 52425 Jülich, Germany
[2]DPIA, University of Udine, Via delle Scienze 206, 33100 Udine, Italy
[3]now with LFoundry S.r.l., a SMIC company, 6051 Avezzano, Italy.
Email: q.zhao@fz-juelich.de

Abstract—**Half SRAM cells with strained Si nanowire complementary Tunnel-FETs (CTFET) have been fabricated to explore the capability of TFETs for 6T-SRAM. Static measurements on cells with outward faced n-TFET access transistors have been performed to determine the SRAM butterfly curves, allowing the assessment of cell functionality and stability. The forward p-i-n leakage at certain bias configuration of the access transistor may lead to malfunctioning storage operation, even without the contribution of the ambipolar behavior. Lowering the bit-line bias is found to mitigate such effect resulting in functional hold, read and write operation.**

Keywords— nanowire, Silicon, SRAM, tunneling FET

I. INTRODUCTION

The technology trends towards more compact electronics increase the demand for supply voltage V_{dd} scaling to reduce the power dissipation in microprocessors. For this scenario Tunnel-FETs (TFETs) emerge as energy efficient switches featuring large I_{on}/I_{off} ratios at small voltage swings. This is enabled by the exploitation of band-to-band tunneling which allows subtreshold swing (SS) steeper than 60mV/dec at room temperature [1][2]. Over the last decade, many groups succeed in fabricating TFETs with different device architectures showing the potential of sub-60mV/dec switching operation [3][4]. Among them, experimental Si TFETs show competetive results in comparison to smaller band gap materials, as indicated by the benchmarks for I_d vs. SS [3][5] and I_{on}/I_{off} [6][7]. The succsess of Si TFETs is based on the access to the more mature CMOS technology. In addition, it enables the processing of p- and n-type devices on the same chip and thus the adoption of circuits based on complementary TFETs (CTFETs). Along this line, first fabricated logic circuits emplyoing TFETs were Si CTFET inverters [8][9]. The SRAM memory cell is one of the most important digital building block, widely used as data caches in processors. The most common design is the symmetric 6T-SRAM, displayed in Fig.1(a), with two cross-coupled inverters (to store a logical state) and two access transistors (AT) allowing read and write operations. So far, assessment of TFETs on the SRAM performance has been analyzed by means of mixed-mode device/circuit TCAD

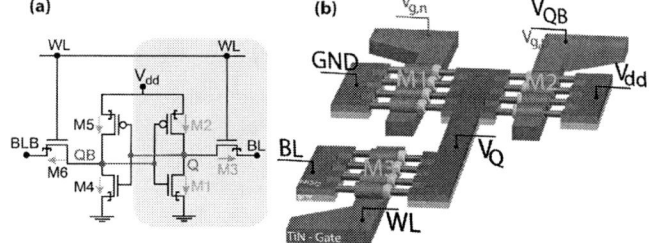

Fig. 1. (a) Symmetric 6T SRAM design composed of two cross-coupled inverters and two access transistors in outward configuration. (b) Schematic of the fabricated half SRAM layout composed of three strained Si NW TFETs (M1-M3) to investigate the 6T-SRAM static operation

simulations [10-15] and with Verilog-A models based on look-up tables for the TFETs [16][17]. Such early analysis relied mainly on idealized TFET templates, neglecting fundamental aspects of experimental devices, such as parasitic phenomena leading to ambipolarity and the forward p-i-n leakage. While the ambipolar behavior certainly needs to be suppressed to enhance SRAM stability, the unidirectional conduction of TFETs is the main hindrance for advanced circuits utilizing pass-transistors (as SRAMs). This limits the functionality of the 6T-SRAM since the access transistor is required to conduct current from source to drain and vice versa. Several modified SRAM topologies are proposed to circumvent this issue using outward (O-AT) or inward faced (I-AT) AT configurations and/or adding more transistors at cost of space to enhance the operation stability. Recent simulation results with ideal and calibrated TFET templates emphasize acceptable performances with the employment of the 6T-SRAM design with outward ATs. However, to our knowledge no experimental verification has been reported so far. In this regard, the purpose of this paper is to experimentally demonstrate the functionality of TFET 6T-SRAM by characterization of half SRAMs (HSRAM) based on strained Si NW TFETs as depicted in Fig.1(b). The half-cell allows to measure the inverter voltage transfer characteristics (VTC) as deformed by the AT influence, biased in various conditions, in order to reconstruct the butterfly curves for each SRAM operation. From the butterfly curves the static noise

978-1-5090-5979-9/17 $31.00 © 2017 IEEE

margins (SNM) can be extracted and used to understand and proof the cell stability.

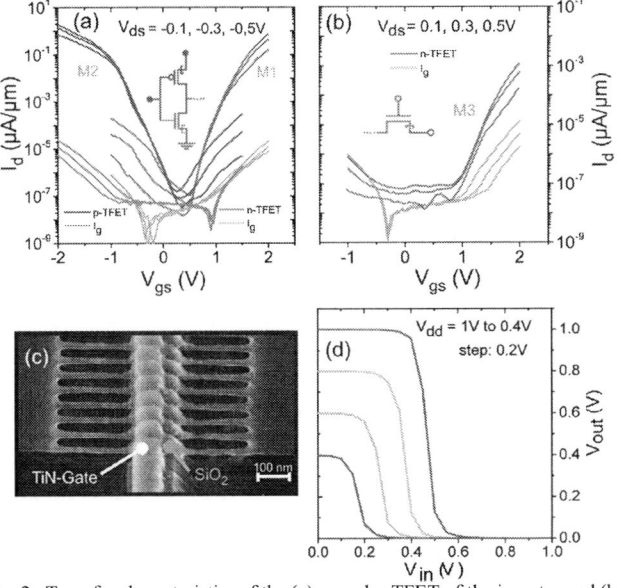

Fig. 2. Transfer characteristics of the (a) p- and n-TFET of the inverter and (b) the n-TFET used as AT. (c) The TEM image depicts the free standing NW-array of a single TFET including the TiN/HfO$_2$ wrap gate and the SiO$_2$ drain spacer. (d) The VTC of the corresponding inverter for different V$_{dd}$ after threshold voltage adjustment of the n-TFET.

II. TFET STRUCTURE AND ELECTRICAL CHARACTERIZATION

Fabricated strained Si nanowire CTFETs with HfO$_2$/TiN wrap gate are used for the HSRAM investigations. The gate-all-around architectures on thin NWs provides improved gate electrostatics. The tensile strain of about 0.8% along the NW in the [110] direction induces band splitting and reduces the Si bandgap and the effective masses. Both parameters affects the tunneling current so that for high tensile strain values, higher overall current is expected compared to unstrained Si [18]. Each device consists of an array of 60 NWs with a dimension of 30nm in width and 10nm in thickness (30x10nm^2 cross-section) and a gate length of about 100nm (Fig.2(c)). The detailed process flow for the CTFET fabrication is given in [19]. Fig.2(a) presents the measured I$_d$-V$_{gs}$ transfer characteristics of the p- and n-TFET (M1 & M2) of the HSRAM inverter while the n-TFET (M3), whose I$_d$-V$_{gs}$ characteristics are displayed in Fig.2(b), is employed as O-AT. The reported device characteristics indicate reduced ambipolar current in the OFF-state as a result of a selective SiO$_2$ spacer at the drain side (Fig.2(c)) [19]. Due to process variation the n-TFET in Fig.2(b) seems to have larger spacer length resulting in a fully suppression of the ambipolarity. However, in turn, the on-current I$_{on}$ is also decreased since the enlarged intrinsic Si region increases the total resistance. For the present devices the on-current I$_{on}$(V$_{gs\_min}$+1V) corresponds to 0.06 µA/µm, 0.01 µA/µm, 1.1 nA/µm at |V$_{ds}$|=0.5V, the off-current I$_{off}$(V$_{gs\_min}$) is 0.11 pA/µm, 0.2 pA/µm, 0.15 pA/µm and the minimum SS is 100, 160, 220 mV/dec for the devices M1, M2 and M3, respectively. The relative small on-current compared to the

devices shown in [19] is due to the larger NWs cross-section resulting in poor electrostatics.

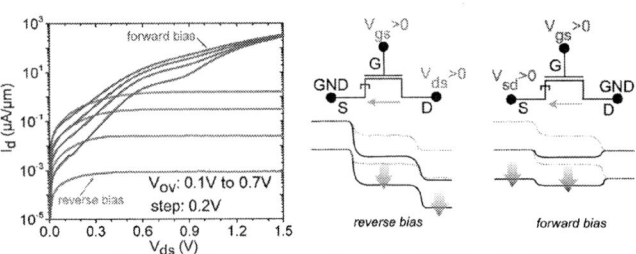

Fig. 3. Output characteristics of an experimental n-TFET with different source/drain biasing scheme. The blue curves show the standard reverse bias operation. In contrast, positive bias (red curve) at the source leads to forward biased p-i-n diode operation resulting in undesired high S/D current leakage almost independent of the applied gate bias.

Since our analysis is mainly concerned with static investigation, I$_{on}$ is not critical here. On the other hand, the relative shift in threshold voltage of the devices in Fig.2(a) and (b) as well as the degree of the ambipolar current are of great importance in the SRAM static behavior. Instead of a common gate for the pull-up and pull-down TFETs (M1 & M2), we have chosen to keep them independent in order to compensate for the threshold voltage mismatch due to the limitations of our fabrication process. This workaround allows us to electrically adjust the threshold voltage of each device, by adding an offset to the applied gate voltages. The threshold voltage offset for the n-TFET M1 is of 0.5V (see Fig.2(a)), whereas the p-TFET M2 does not need any correction. This "virtual" gate work-function tuning results in a shift of the transfer curves by 0.5V (M1) to the left. The inverter VTC of the electrically matched p- and n-TFETs is depicted in Fig.2(d) for V$_{dd}$ starting from 1V down to 0.4V, providing constant high and low voltage levels with a transition at V$_{dd}$/2, in contrast to previous reported inverter VTCs with ambipolar TFETs [8]. Hence, the small degree of ambipolarity for the TFET inverter (Fig2.(a)) does not significantly degrade the inverter VTC. Fig.3 illustrates the I$_d$-V$_{ds}$ output characteristics of an experimental n-TFET emphasizing the non-symmetric current transport in TFETs. A positive V$_{ds}$ voltage leads to reverse biasing (positive voltage at the n-region) of the p-i-n band structure and thus creates the desired output behavior with saturated currents (blue curves). In contrast, when a high potential is applied at the source terminal (i.e V$_{sd}$>0.5V), our TFETs exhibit large current flow due to the forward biased p-i-n structure (red curves). In comparison with the standard operation, this forward p-i-n leakage current is barely controlled by the gate voltage. Furthermore, the currents at different V$_{gs}$ converge for V$_{sd}$>1.2V, meaning that the gate loses control of the drain current. The current, hence, scales with V$_{sd}$. In order to prevent uncontrolled large parasitic currents, the forward biasing configuration needs to be avoided [15].

III. STATIC PERFORMANCE OF THE CTFET HSRAM

The fabricated HSRAM, as sketched in Fig.1(b), is composed of three TFETs. The inverter pull-down n-TFET

978-1-5090-5979-9/17 $31.00 © 2017 IEEE

(M1), the inverter pull-up p-TFET (M2) and the n-TFET

TABLE I
LOW/HIGH GATE VOLTAGE DEFINITION FOR EACH DEVICE AT V_{DD}=0.8V.

TFET devices		V_{low}	V_{high}
(M1) n-TFET, pull-down tranistor	$V_{g,n}$ =	0.5V	1.3V
(M2) p-TFET, pull-up transistor	$V_{in} = V_{g,p}$ =	0V	0.8V
(M3) n-TFET, access transistor	WL =	1V	1.8V

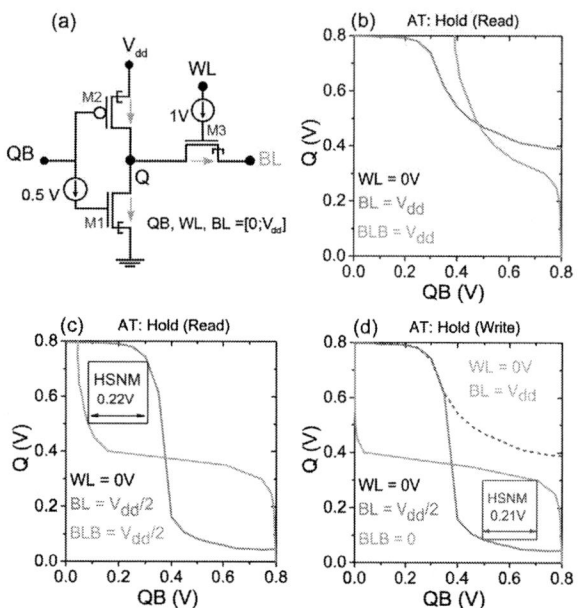

Fig. 4. (a) Circuit of the HSRAM including the voltage generator to drive the TFETs to the desired voltage range. (b)-(d) depicts the butterfly curves generated from the measured VTCs of the HSRAM in hold for various BLB/BL combinations with V_{dd} and V_{dd}/2. Only the butterfly curves with three crossing points confirms hold stability.

access-transistor (M3). In the SRAM terminology, the Q/QB node stores the differential voltage level needed to define the logical state in the SRAM cell. The stored potential can be accessed or manipulated with the bit-line (BL) and word-line (WL) of the AT during read or write operations, respectively. Otherwise, the voltage level in Q/QB is preserved during the hold state. For the investigation of the HSRAM static behavior, distinct gate voltage windows needs to be defined for each individual device to compensate the threshold voltage shift. The low and high gate potential definitions for all devices are summarized in Table I. In the following, the static performance of the HSRAM is presented for various bit- (BL) and word-line (WL) voltage levels in order to observe the potential behavior in node Q as a function of QB under different SRAM operations for V_{dd}=0.8 V. The hold ability is investigated with WL=V_{low} for the AT while keeping the BL(B)=V_{dd} to mimic a read operation of a neighboring cell within the same column (Fig.4(a)). In this case, the AT should be ideally off, screening the cell from the influence of BL(B), maintaining a VTC equal to the one in Fig.2(d) at V_{dd}=0.8V. However, the resulting HSRAM VTC is strongly distorted (orange curve in Fig.4(b),

where the green curve is just the mirrored version of the orange curve to account for the other half of the SRAM completing the full 6T cell). The reason for the VTC distortion in the hold case is caused by the forward p-i-n leakage in the AT, as explained in section II, since the high potential BL is applied at its source terminal. A significant current flow is established pulling up the potential of node Q and hence lift up the low level of the VTC. As result, the butterfly curve is deformed, a sign that data retention is not possible. Only reducing the BL(B) to V_{dd}/2 mitigates the parasitic current and maintains the VTC behavior (orange curve in Fig.4(c)). The butterfly plot in Fig.4(c) proves a successful hold operation with BL(B)=V_{dd}/2 retaining the VTC voltage levels of the HSRAM, opposite to the read operation with full V_{dd}. To test the hold condition under write of neighboring SRAM cells, we force the BL/BLB pair to differential voltage levels. The results in Fig.4(d) demonstrate cell stability for V_{dd}/2 as high potential for BL. However, when driving BL=V_{dd}, the butterfly curve (dashed-grey and green curve) leads to only one crossing-point resulting in an unintended write operation of the cell. For both hold-operations with BL_{high}=V_{dd}/2, the extracted HSNM (according to [20]) is slightly above 0.2V. As we found that the cell in hold can only preserve the logic state operating with BL(B) voltages lower than V_{dd}/2=0.4V, we continue the analysis by assuming the use V_{dd}/2 as high logic level at the bit-lines. Reading the cell is performed by applying high potential at both BL(B)=V_{dd}/2 (as proposed in [13]) and activating the AT with WL=V_{high}. Compared to the hold state in Fig.4(c) where the AT is switched OFF, the wings of the butterfly curve in Fig.5(a) is compressed but still preserve the three crossing points. The extracted noise margin for read (RSNM) is about 0.15V. With regard to the write operation, the butterfly curves shift apart due to the HSRAM VTC whose bit-line is connected to the low potential with active AT (green curve in Fig.5(b)). Such strong deformation of the corresponding HSRAM VTC results from the pull-down action of the n-TFET as AT forcing the storage node to follow the BLB=0V potential. Consequently, the write butterfly curve indicate only one crossing point between the VTCs, which corresponds to the logical value intended to be written. The write becomes more robust the larger the VTCs are separated [13]. In our case the extracted WSNM amounts to 0.11V.

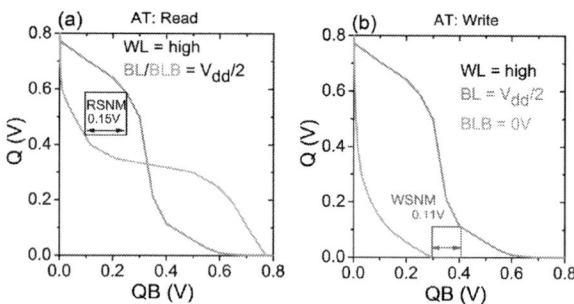

Fig. 5. The butterfly curves for (a) read operation forming a butterfly curve from which we extract the RSNM as the maximum square that can fit inside the two VTCs. In comparison, a successful (b) write operation requires a large

deformation of the VTCs with only one crossing point between the VTC curves. The WSNM is then extracted as the minimum square in between the two VTCs.

IV. CONCLUSION

In this work, for the first time, experimental HSRAM based on strained Si NW gate-all-around complementary TFETs with supressed ambipolar behavior have been investigated to understand the suitability of TFETs for the 6T-SRAM design. Although the considered TFET devices turned out to be somehow underperforming (SS>60mV/dec), they allow us to perform a systematical analysis on the static figures of merit of the cell. On the base of the butterfly curves, we have shown that with outward faced n-TFETs as ATs functional operation for hold, read and write can be achieved when the high voltage potential at the bitlines BL(B) is limited to $V_{dd}/2$. This requirement diminishes the impact of the forward p-i-n biasing of the O-AT by minimizing parasitic current leakage and thus reduces distortion of the resulting butterfly curves The present analysis is restricted to static figures of merit. Characterization of the full cell will be needed to evaluate the penalty in terms of write delay due to the use of $V_{dd}/2$ instead of V_{dd} at the bit-lines.

ACKNOWLEDGMENT

This work was supported in part by the German Federal Ministry of Education and Research under the project "UltraLowPower" (No. 16ES0060K) and in part by E2SWITCH from the European Community's Seventh Framework Program under grant agreement 619509

REFERENCES

[1] A. M. Ionescu and H. Riel, "Tunnel field-effect transistors as energy-efficient electronic switches," Nature, vol. 479, no. 7373, pp. 329–337, Nov. 2011. DOI: 10.1038/nature10679

[2] A. C. Seabaugh and Q. Zhang, "Low-Voltage Tunnel Transistors for Beyond CMOS Logic," Proc. IEEE, vol. 98, no. 12, pp. 2095–2110, Dec. 2010. DOI: 10.1109/JPROC.2010.2070470.

[3] H. Lu and A. Seabaugh. "Tunnel Field-Effect Transistors: State-of-the-Art," IEEE J. Electron Devices Soc., vol. 2, no. 4, pp. 44–49, Jul. 2014. DOI: 10.1109/JEDS.2014.2326622

[4] E. Memisevic, J. Svensson, M. Hellenbrand, E. Lind, and L.-E. Wernersson, "Vertical InAs/GaAsSb/GaSb Tunneling Field-Effect Transistor on Si with S = 48 mV/decade and Ion = 10 µA/µm for Ioff = 1 nA/µm at VDS = 0.3 V," in IEEE International Electron Devices Meeting, 2016, pp. 6–9. DOI: 10.1109/IEDM.2016.7838450

[5] D. Cutaia, K. E. Moselund, M. Borg, H. Schmid, L. Gignac, C. M. Breslin, S. Karg, E. Uccelli, and H. Riel, "Vertical InAs-Si Gate-All-Around Tunnel FETs Integrated on Si Using Selective Epitaxy in Nanotube Templates," IEEE J. Electron Devices Soc., vol. 3, no. 3, pp. 176–183, May 2015. DOI: 10.1109/JEDS.2015.2388793

[6] S. Datta, H. Liu, and V. Narayanan, "Tunnel FET technology: A reliability perspective," Microelectron. Reliab., vol. 54, no. 5, pp. 861–874, May 2014. DOI: 10.1016/j.microrel.2014.02.002

[7] H. Riel, L.-E. Wernersson, M. Hong, and J. a. del Alamo, "III–V compound semiconductor transistors—from planar to nanowire structures," MRS Bull., vol. 39, no. 08, pp. 668–677, Aug. 2014. DOI: 10.1557/mrs.2014.137

[8] L. Knoll, Q.-T. Zhao, A. Nichau, S. Trellenkamp, S. Richter, A. Schafer, D. Esseni, L. Selmi, K. K. Bourdelle, and S. Mantl, "Inverters With Strained Si Nanowire Complementary Tunnel Field-Effect Transistors,"

IEEE Electron Device Lett., vol. 34, no. 6, pp. 813–815, Jun. 2013. DOI: 10.1109/LED.2013.2258652

[9] G. V. Luong, S. Strangio, A. Tiedemann, S. Lenk, S. Trellenkamp, K. K. Bourdelle, Q. T. Zhao, and S. Mantl, "Experimental demonstration of strained Si nanowire GAA n-TFETs and inverter operation with complementary TFET logic at low supply voltages," Solid. State. Electron., vol. 115, pp. 152–159, Jan. 2016. DOI: 10.1016/j.sse.2015.08.020

[10] J. Singh, K. Ramakrishnan, S. Mookerjea, S. Datta, N. Vijaykrishnan, and D. Pradhan, "A novel Si-Tunnel FET based SRAM design for ultra low-power 0.3V VDD applications," in 2010 15th Asia and South Pacific Design Automation Conference (ASP-DAC), 2010, pp. 181–186. DOI: 10.1109/ASPDAC.2010.541989

[11] Xuebei Yang and K. Mohanram, "Robust 6T Si tunneling transistor SRAM design," in 2011 Design, Automation & Test in Europe, 2011, pp. 1–6. DOI: 10.1109/DATE.2011.5763126

[12] S. Strangio, P. Palestri, D. Esseni, L. Selmi, and F. Crupi, "Analysis of TFET based 6T SRAM cells implemented with state of the art silicon nanowires," in 2014 44th European Solid State Device Research Conference (ESSDERC), 2014, pp. 282–285. DOI: 10.1109/ESSDERC.2014.6948815

[13] S. Strangio, P. Palestri, D. Esseni, L. Selmi, F. Crupi, S. Richter, Q.-T. Zhao, and S. Mantl, "Impact of TFET Unidirectionality and Ambipolarity on the Performance of 6T SRAM Cells," IEEE J. Electron Devices Soc., vol. 3, no. 3, pp. 223–232, May 2015. DOI: 10.1109/JEDS.2015.2392793

[14] Y. Chen, M. Fan, V. P. Hu, P. Su, and C.-T. Chuang, "Design and Analysis of Robust Tunneling FET SRAM," IEEE Trans. Electron Devices, vol. 60, no. 3, pp. 1092–1098, Mar. 2013. DOI: 10.1109/TED.2013.2239297

[15] A. Makosiej, R. K. Kashyap, A. Vladimirescu, A. Amara, and C. Anghel, "A 32nm tunnel FET SRAM for ultra low leakage," in 2012 IEEE International Symposium on Circuits and Systems, 2012, pp. 2517–2520. DOI: 10.1109/ISCAS.2012.6271814

[16] Y. Lee, D. Kim, J. Cai, I. Lauer, L. Chang, S. J. Koester, D. Blaauw, and D. Sylvester, "Low-Power Circuit Analysis and Design Based on Heterojunction Tunneling Transistors (HETTs)," IEEE Trans. Very Large Scale Integr. Syst., vol. 21, no. 9, pp. 1632–1643, Sep. 2013. DOI: 10.1109/TVLSI.2012.2213103

[17] D. H. Morris, U. E. Avci, and I. A. Young, "Variation-tolerant dense TFET memory with low V_{MIN} matching low-voltage TFET logic," in 2015 Symposium on VLSI Technology (VLSI Technology), 2015, no. 4, pp. T24–T25. DOI: 10.1109/VLSIT.2015.7223688

[18] Peng-Fei Guo, Li-Tao Yang, Yue Yang, Lu Fan, Gen-Quan Han, G. S. Samudra, and Yee-Chia Yeo, "Tunneling Field-Effect Transistor: Effect of Strain and Temperature on Tunneling Current," IEEE Electron Device Lett., vol. 30, no. 9, pp. 981–983, Sep. 2009. DOI: 10.1109/LED.2009.2026296

[19] G. V. Luong, K. Narimani, A. T. Tiedemann, P. Bernardy, S. Trellenkamp, Q. T. Zhao, and S. Mantl, "Complementary Strained Si GAA Nanowire TFET Inverter With Suppressed Ambipolarity," IEEE Electron Device Lett., vol. 37, no. 8, pp. 950–953, Aug. 2016. DOI: 10.1109/LED.2016.2582041

[20] F. J. List, "The Static Noise Margin of SRAM cells," Solid-State Circuits Conf. 1986. ESSCIRC '86. Twelfth Eur., pp. 16–18, 1986

Advances in the understanding of microscopic switching mechanisms in ReRAM devices

(Invited Paper)

Benoît Sklénard*, Philippe Blaise*, Boubacar Traoré[†], Alberto Dragoni*[‡], Cécile Nail*[§] and Elisa Vianello*

*Univ. Grenoble Alpes, F-38000 Grenoble France
CEA, LETI, MINATEC Campus, F-38054 Grenoble, France — E-mail: benoit.sklenard@cea.fr
[†]Institut des Sciences Chimiques de Rennes (ISCR), Université de Rennes 1, CNRS, UMR 6226, 35042 Rennes, France
[‡]CNRS, Institut Néel, F-38042 Grenoble, France
[§]LTM CNRS, F-38054 Grenoble, France

Abstract—In this paper we present the recent advances in the understanding of microscopic mechanisms driving the resistive switching in ReRAM devices using *ab initio* theoretical methods. We highlight the complex interplay between interface reactions and charge injection in the generation of oxygen Frenkel pairs during the forming step. Energy barrier calculations suggest that the formation/destruction of the conductive filament can be due to movements of oxygen vacancies composing the filament or interaction with oxygen atoms released from the metal electrode.

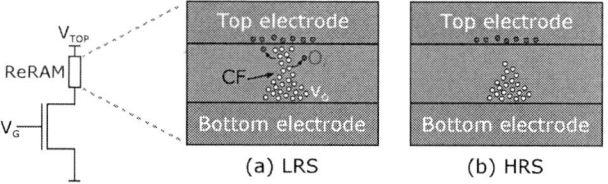

Fig. 1. Cartoon of a ReRAM structure. The formation of a conductive filament (CF) gives rise of the low resistive state (LRS) and while its partial destruction increases the resistivity to an high resistive state (HRS) respectively shown in (a) and (b).

I. Introduction

Resistive memories (ReRAM) are one of the most promising candidates for future generations of Non Volatile Memories (NVM) for both storage class applications [1] and embedded products [2]. They use the resistive–switching (RS) phenomenon of an insulating material driven by an applied electric field in a metal–insulator–metal (MIM) cell structure as shown in Fig. 1. RS relies on the formation and destruction of a conductive filament (CF) through the dielectric allowing to achieve low and high resistive states. This filamentary process has been observed in a wide variety of materials such as NiO [3,4], TiO_2 [5,6], SiO_x [7], HfO_2 [8] or Ta_2O_5 [9]. However the driving forces leading to the formation of the CF, its nature and composition remain unclear and direct experimental observations are challenging [10]. ReRAM are usually classified in two main categories: the conductive bridge RAM (CBRAM) — also referred to as electrochemical metallization memories (ECM) — and oxide based RAM (OxRAM) — also referred as valence change memories (VCM) — depending on whether the memory relies on the incorporation of metallic impurities into the dielectric or the movements of oxygen atoms. In this paper we will focus on OxRAM in which the CF is attributed to the creation of oxygen–poor conductive paths. We will present recent theoretical insights into the microscopic mechanisms taking place during RS by means of *ab initio* simulations. We will limit our review to HfO_2 and Ta_2O_5–based memories since they are CMOS compatible and exhibit remarkable switching properties including low operating voltages (< 2 V), nanosecond switching capability and good endurance [9, 11–13].

Experimentally, the dielectric material is deposited in thin films (< 10 nm) using atomic layer deposition (ALD) or physical vapor deposition (PVD) and is observed to be amorphous. For thicker layers, or after annealing, the material can crystallize and becomes polycristalline. The size and phase of the crystalline grains depend on many factors such as the nature of the substrate (in the case of a ReRAM a metallic electrode), the anneal temperature or the presence of impurities. In the case of HfO_2, the monoclinic structure (m-HfO_2, spacegroup $P2_1/c$) is the low–temperature stable phase [14] and has been observed in HfO_2–based memories [15]. In contrast, the low–temperature crystalline structure of Ta_2O_5 is orthorhombic but difficult to characterize experimentally. X-ray diffraction studies show that it is composed of disordered 2D layers of Ta_2O_3 connected by twofold coordinated O atoms along the c direction [16] and various high–symmetry models were proposed to describe its structure. Recently, Lee *et al.* proposed a new orthorhombic structural model (λ-Ta_2O_5, spacegroup $Pbam$) that is lower in energy than the other reported models. [17]. In this paper, we calculated defects properties in these two crystalline structures.

II. Computational Methodology

The calculations based on density functional theory (DFT) have been performed using the VASP package [18, 19] with the semilocal functional of Perdew–Burke–Ernzerhof (PBE) and the screened hybrid functional of Heyd–Scuseria–Ernzerhof with a range separation of 0.2 Å (HSE06) [20, 21]. We

978-1-5090-5979-9/17 $31.00 © 2017 IEEE

TABLE I
LATTICE PARAMETERS IN Å, ELECTRONIC BANDGAP (E_g) IN eV AND ENTHLAPY OF FORMATION ($\Delta_f H$) IN eV/f.u. OF m–HfO$_2$ AND λ–Ta$_2$O$_5$.

	HfO$_2$			Ta$_2$O$_5$		
	PBE	HSE	Expt.	PBE	HSE	Expt.
a	5.15	5.11	5.12[a]	6.62	6.21	6.20[d]
b	5.19	5.16	5.18[a]	3.71	3.67	3.66[d]
c	5.33	5.28	5.29[a]	3.83	3.79	3.89[d]
β	99.71	99.57	99.22[a]	—	—	—
E_g	4.02	5.67	5.7–5.9[b]	2.12	4.42	3.9–4.5[e]
$\Delta_f H$	−10.73	−11.25	−11.86[c]	−19.85	−25.92	−21.20[c]

[a]Ref. [14]
[b]PES+IPS from Ref. [22, 23]
[c]Ref. [24]
[d]Ref. [16]
[e]Optical bandgap from Ref. [25]

used projector augmented-wave (PAW) datasets with valence configurations $5p^6 6s^2 5d^2$, $5p^6 6s^2 5d^3$ and $2s^2 2p^4$ for Hf, Ta and O respectively. The electronic wave functions were expanded in a plane wave basis up to a kinetic energy cutoff of 500 eV. For conventional cells, the Brillouin zone was sampled with $4 \times 4 \times 4$ and $4 \times 4 \times 6$ Γ–centered grids for HfO$_2$ and Ta$_2$O$_5$, respectively. Structures were relaxed until the maximum residual forces were less than 0.01 eV/Å. Calculated material properties are summarized in Table I and compared with experimental data.

Defects were studied using 96 ($2 \times 2 \times 2$) and 378 ($3 \times 3 \times 3$) atoms supercells of HfO$_2$ and Ta$_2$O$_5$, respectively. The formation energy of an oxygen point defect X in charge state q is given by:

$$E_f[X^q] = E_{tot}[X^q] - E_{tot}[\text{bulk}] + n_O \mu_O + q(\epsilon_F + \epsilon_v + \Delta V) + E_{corr}(q), \quad (1)$$

where $E_{tot}[X^q]$ is the total energy of the supercell containing the defects, $E_{tot}[\text{bulk}]$ the total energy of the bulk supercell, n_O the number of removed ($n_O > 0$) or added ($n_O < 0$) oxygen atoms, μ_O the chemical potential of oxygen, ϵ_F the Fermi energy, ϵ_v the valence band maximum, and $E_{corr}(q) + q\Delta V$ a correcting term due to image charge interractions (in this paper we used a monopole chage correction).

In HfO$_2$, two positions of v_O were considered corresponding to the 3–fold (v_{O3}) and 4–fold (v_{O4}) coordinated sites. O$_i$ forms a strong bond with a 3–fold coordinated O (1.46 Å in the HSE calculation) due to the high ionicity of HfO$_2$.

For λ-Ta$_2$O$_5$, very large supercell sizes are required to achieve a good convergence. Such large supercells allow to capture complex distortions of O vacancies in the Ta$_2$O$_3$ layers as pointed out in [26]. Three different positions of v_O were considered corresponding to 3–fold (v_{O3}) and 2–fold (v_{O2a}) sites in a Ta$_2$O$_3$ layer and the 2–fold coordinated O atom connecting the Ta atoms in adjacent layers (v_{O2b}). Similarly to HfO$_2$, the most stable site for O$_i$ is a peroxide position forming a strong bond with a 2–fold O in a Ta$_2$O$_3$ plane.

III. THERMODYNAMIC PROPERTIES OF DEFECTS

Formation energies of oxygen vacancies (v_O) and interstitial (O$_i$) calculated using HSE06 are shown in Fig. 2 and 3 for HfO$_2$ and Ta$_2$O$_5$, respectively. Formation energies are reported for two limiting cases of the O chemical potential corresponding to O–rich and O–poor conditions [27]. In the

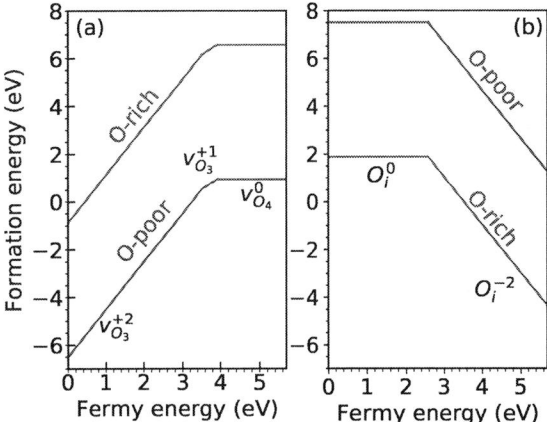

Fig. 2. Formation energies calculated with the HSE06 functional of (a) v_O and (b) O$_i$ for different charge states in m-HfO$_2$.

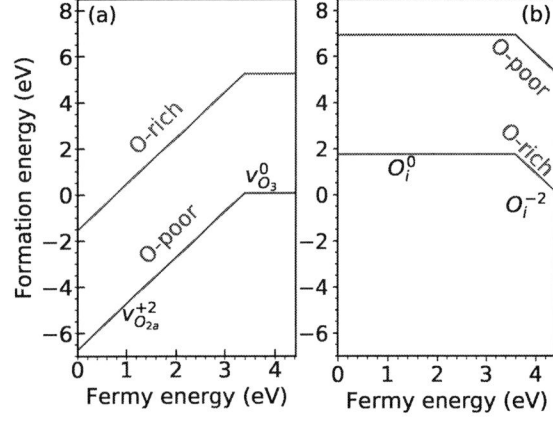

Fig. 3. Formation energies calculated with the HSE06 functional of (a) v_O and (b) O$_i$ for different charge states in λ-Ta$_2$O$_5$.

case of a ReRAM, the presence of a metal electrode will shift μ_O to O–poor conditions depending on its reactivity with O. Therefore, v_O are more likely present at the interface with the metal electrode than in the bulk. To highlight this effect, we calculated the formation enthalpy of a neutral v_O where the removed O is inserted in the metal (M) in an interstitial position, corresponding to the reactions:

$$M + HfO_2 \rightarrow HfO_2{:}v_O + M{:}O_i \quad (2)$$

and

$$M + Ta_2O_5 \rightarrow Ta_2O_5{:}v_O + M{:}O_i. \quad (3)$$

TABLE II
PBE CALCULATIONS OF THE v_O FORMATION ENERGIES (in eV) WHERE THE REMOVED O IS INSERTED IN INTERSTITIAL POSITION IN THE METAL.

	Spacegroup	HfO_2	Ta_2O_5
Ti	$P6_3/mmc$	0.77	−0.72
Hf	$P6_3/mmc$	0.80	−0.70
Zr	$P6_3/mmc$	0.81	−0.69
Ta	$Im\bar{3}m$	2.65	1.16
Pt	$Fm3m$	9.57	8.08

The results are summarized in Table II. For Ti, Zr and Hf, O can be easily inserted in octahedral interstitial sites. The reaction enthalpy is very low (< 1 eV), favoring the formation of an oxidized electrode and a v_O–rich dielectric after the deposition of the metal electrode. Such reaction was observed experimentally for Ti/HfO_2 memories [28]. In contrast, the process is much more endothermic for Ta or Pt metals. Deposition of O–reactive metal electrodes on top of the dielectric have been reported to significantly improve the performances of OxRAM devices [29, 30] likely because of the dissolution of O atoms in the layer allowing to limit their interaction with a v_O CF.

Under high electric fields, intrinsic defects start being generated inside the dielectric and inducing its resistance change. For HfO_2 or Ta_2O_5–based ReRAM, valence change of cation species have been experimentally measured during the *forming* step and can be attributed to the formation of O Frenkel Pairs (FP). In bulk HfO_2 and Ta_2O_5, the formation energy of an O FP is very high (repectively 6.16 eV and 7 eV from HSE calculations). We recently reported calculations of the activation energy for FP emission in HfO_2 along the $[00\bar{1}]$ direction [31]. The high activation barrier ($\sim E_f$) make this process highly improbable in the bulk material without any additional driving force. The application of an electric field E along a given direction, reduces this energy barrier by a factor $p_0(2 + \varepsilon)/3E$, where p_0 is the dipole moment corresponding to the FP defect with opposite charges localised on v_O and O_i and ε the permittivity of the dielectric [32]. We evaluated $p_0 \sim 6$ e.Å at the saddle point of the energy barrier during the emission of an O FP in HfO_2, and assuming $\varepsilon \sim 20$ we estimate a breakdown field of 14 MV/cm in good agreement with the value of 13 MV/cm measured experimentally [33]. However in HfO_2 OxRAMs, lower critical fields around 6 MV/cm have been reported and are further reduced to 4 MV/cm for sub–stoichimetric HfO_2 obtained by depositing an O reactive electrode [34]. Charge injection is likely a mechanism playing an important role in the degradation of the dielectric. It has been demonstrated that the activation energy barrier for the generation of O FP is lowered by the addition of electrons [29, 31, 35]. Extra electrons localize on the v_O (see Sec.V) and limit the electrostatic interaction between interstitial and vacancy defects, as shown in Fig. 4. Under the influence of an electric field, the emitted interstitials (O_i^{-2}) easily drift towards the electrode and leave O vacancies in the dielectric forming the CF.

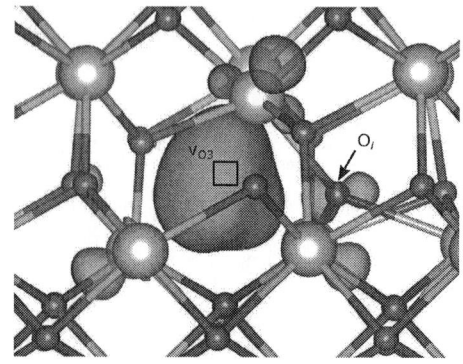

Fig. 4. Charge density isosurface showing the localization of two electrons on a 3–fold O vacancy during the emission of a FP. The O_i is located approximately at a 1st neighbor distance of the v_O. Hf and O atoms are shown in grey and red respectively.

IV. DEFECTS KINETICS

The evolution of generated point defects during the *forming* step (*i.e.* v_O and O_i) is driven by their migration barrier and is expected to control the RS. In addition, electric drift occurs when the defects are charged. For a point defect with charge q with a local electric field $E(\mathbf{r})$ along a given direction, the migration probabilities ν_+ and ν_- respectively in the positive and negative directions is:

$$\frac{\nu_+}{\nu_-} = \exp(\beta q E(\mathbf{r})\lambda), \tag{4}$$

where λ is the jump length and $\beta \equiv 1/k_B T$.

Migration energies of v_O and O_i calculated with DFT in m-HfO_2 and λ-Ta_2O_5, are summarized in Table III. From these values, v_O migration is easier when it is 2 times positively charged for both materials. Therefore, it cannot be excluded that v_O^{+2} play a role in the RS process, during *set* and *reset* steps. On the other hand, in HfO_2 the energy barrier for the migration of O_i is very small and could also explain an oxidation/reduction of the CF. In this case, the kinetics at the metal/oxide interface is important, in particular for an oxidized electrode.

TABLE III
MIGRATION BARRIERS (in eV) OF OXYGEN DEFECTS IN HfO_2 AND Ta_2O_5.

Defect	Charge	HfO_2		Ta_2O_5	
		E_m	Ref.	E_m	Ref.
v_O	0	2.4	[36]	1.26	[26]
v_O	+2	0.7	[36]	0.75	[26]
O_i	0	0.7	[37]	—	—
O_i	−1	0.1	[37]	—	—
O_i	−2	0.2	[37]	—	—

V. CHARGE INJECTION

In previous sections we pointed out that importance of charge injection in the generation of O FP. Electron–phonon coupling is usually strong in transition metal oxides giving rise to polaronic states. For example, in HfO_2, localized electron and hole polaron states were reported because of small distortions in the perfect crystal [38]. The presence of polaron–like bound states on v_O [39, 40] or more recently in amorphous HfO_2 [41] were also evidenced.

In ReRAM devices, the injection process results from the electron trapping on localized states into the dielectric from an electrode or another localized state occurring through non-radiative transitions. Electron trapping/detrapping at v_O sites are expected to control the current before the *forming* step [42].

VI. DOPING AND ALLOYING

Material properties can be modified by inserting impurities into the dielectric layer in order to optimize ReRAM performances. It can be achieved by incorporating dopants during the deposition process [43] or by ion implantation [44]. Experimentally, Si, Ti and Al–doped HfO_2 [43–45] and Si–doped Ta_2O_5 [46] ReRAM were investigated and has been observed to affect RS, endurance and retention characteristics. The critical field during the *forming* step is influenced by a modification of the dielectric response of the dielectric. For example, in $Hf_{1-x}Al_{2x}O_{2+x}$ alloys, the forming voltage increases with the Al mole fraction and can be explained by a lower dielectric constant which reduces the local electric field (see Sec. III) [43, 44]. However, charge injection can also be enhanced as in Si–doped HfO_2 likely due to the formation of dopant–defects aggregates and resulting in lower *forming* voltages [44].

VII. CONCLUSION

In this paper we studied using *ab initio* theoretical methods the microscopic mechanisms playing a role in the RS of ReRAM devices. O Frenkel pairs formation assisted by an electric field likely take place during the *forming* step and is affected by reactions at the insulator/metal interface and charge injection into the dielectric. After a CF has been formed, RS may result from O vacancies migration (dissolution of the CF in the dielectric) or interaction with O released from an electrode (local oxidation of the CF).

ACKNOWLEDGMENT

Part of the calculations were run on TGCC/Curie using allocations from GENCI.

REFERENCES

[1] S. Sills *et al.*, in *Proc. Symposium on VLSI Technology*, Jun. 2014, pp. 1–4.
[2] Z. Wei *et al.*, in *Proc. IEEE International Electron Devices Meeting (IEDM)*, Dec. 2015, pp. 7.7.1 – 7.7.4.
[3] S. Seo *et al.*, *Appl. Phys. Lett.*, vol. 85, no. 23, pp. 5655–5657, 2004.
[4] D. C. Kim *et al.*, *Appl. Phys. Lett.*, vol. 88, no. 20, p. 202102, 2006.
[5] D.-H. Kwon *et al.*, *Nat Nanotechnol*, vol. 5, no. 2, pp. 148–153, 2010.
[6] D. S. Jeong *et al.*, *Phys. Rev. B*, vol. 79, p. 195317, May 2009.
[7] A. Mehonic *et al.*, *J. Appl. Phys.*, vol. 111, no. 7, p. 074507, 2012.
[8] Y.-M. Kim and J.-S. Lee, *J. Appl. Phys.*, vol. 104, no. 11, p. 114115, 2008.
[9] M.-J. Lee *et al.*, *Nat. Mater.*, vol. 10, pp. 625–630, 2011.
[10] U. Celano *et al.*, *AIP Advances*, vol. 6, no. 8, p. 085009, 2016.
[11] H. Y. Lee *et al.*, *IEEE Electron Device Lett.*, vol. 31, no. 1, pp. 44–46, Jan 2010.

[12] A. Grossi *et al.*, in *Proc. IEEE International Electron Devices Meeting (IEDM)*, Dec 2016, pp. 4.7.1–4.7.4.
[13] M. Azzaz *et al.*, in *2016 IEEE 8th International Memory Workshop (IMW)*, May 2016.
[14] R. E. Hann *et al.*, *J. Am. Ceram. Soc.*, vol. 68, no. 10, pp. C–285–C–286, 1985.
[15] P. Calka *et al.*, *Nanotechnology*, vol. 24, no. 8, p. 085706, 2013.
[16] K. Lehovec, *J. Less Common Met.*, vol. 7, no. 6, pp. 397 – 410, 1964.
[17] S.-H. Lee *et al.*, *Phys. Rev. Lett.*, vol. 110, p. 235502, Jun 2013.
[18] G. Kresse and J. Furthmüller, *Phys. Rev. B*, vol. 54, pp. 11 169–11 186, Oct 1996.
[19] G. Kresse and D. Joubert, *Phys. Rev. B*, vol. 59, pp. 1758–1775, Jan 1999.
[20] J. Heyd *et al.*, *The Journal of Chemical Physics*, vol. 118, no. 18, pp. 8207–8215, 2003.
[21] J. Heyd and G. E. Scuseria, *The Journal of Chemical Physics*, vol. 121, no. 3, pp. 1187–1192, 2004.
[22] S. Sayan *et al.*, *J. Appl. Phys.*, vol. 96, no. 12, pp. 7485–7491, 2004.
[23] E. Bersch *et al.*, *Phys. Rev. B*, vol. 78, p. 085114, Aug 2008.
[24] W. M. Haynes, *CRC Handbook of Chemistry and Physics*, W. M. Haynes, Ed., 2017.
[25] C. Chaneliere *et al.*, *Materials Science and Engineering: R: Reports*, vol. 22, no. 6, pp. 269–322, 1998.
[26] H. Jiang and D. A. Stewart, *J. Appl. Phys.*, vol. 119, no. 13, p. 134502, 2016.
[27] C. Freysoldt *et al.*, *Rev. Mod. Phys.*, vol. 86, pp. 253–305, Mar 2014.
[28] P. Calka *et al.*, *ACS Applied Materials & Interfaces*, vol. 6, no. 7, pp. 5056–5060, 2014, pMID: 24625458.
[29] B. Traoré, P. Blaise, E. Vianello, L. Perniola, B. D. Salvo, and Y. Nishi, *IEEE Trans. Electron Devices*, vol. 63, no. 1, pp. 360–368, Jan 2016.
[30] M. Azzaz *et al.*, 2016.
[31] B. Traoré *et al.*, *J. Phys. Chem. C*, vol. 120, no. 43, pp. 25 023–25 029, 2016.
[32] J. McPherson *et al.*, *Appl. Phys. Lett.*, vol. 82, no. 13, pp. 2121–2123, 2003.
[33] C. Sire *et al.*, *Appl. Phys. Lett.*, vol. 91, no. 24, p. 242905, 2007.
[34] A. Padovani *et al.*, *IEEE Electron Device Lett.*, vol. 34, no. 5, pp. 680–682, May 2013.
[35] S. R. Bradley *et al.*, *Phys. Rev. Applied*, vol. 4, p. 064008, Dec 2015.
[36] N. Capron *et al.*, *Appl. Phys. Lett.*, vol. 91, no. 19, 2007.
[37] Z. F. Hou *et al.*, *Journal of Physics: Condensed Matter*, vol. 20, no. 13, p. 135206, 2008.
[38] D. Muñoz Ramo *et al.*, *Phys. Rev. Lett.*, vol. 99, p. 155504, Oct 2007.
[39] ——, *Phys. Rev. B*, vol. 75, p. 205336, May 2007.
[40] P. Broqvist and A. Pasquarello, *Appl. Phys. Lett.*, vol. 89, no. 26, p. 262904, 2006.
[41] M. Kaviani *et al.*, *Phys. Rev. B*, vol. 94, p. 020103, Jul 2016.
[42] L. Vandelli *et al.*, *IEEE Trans. Electron Devices*, vol. 58, no. 9, pp. 2878–2887, Sept 2011.
[43] B. Traoré *et al.*, in *Proc. IEEE International Electron Devices Meeting (IEDM)*, Dec 2014, pp. 21.5.1–21.5.4.
[44] M. Barlas *et al.*, in *Proc. European Solid-State Device Research Conference (ESSDERC)*, Sept 2016, pp. 168–171.
[45] Y. Y. Chen *et al.*, in *Proc. Symposium on VLSI Technology*, Jun. 2014, pp. 1–2.
[46] B. Y. Kim *et al.*, *Jpn. J. Appl. Phys.*, vol. 55, no. 4S, p. 04EE09, 2016.

978-1-5090-5979-9/17 $31.00 © 2017 IEEE

Modeling the Effect of Surface Roughness on the Performance of Line Tunnel FETs

Saurabh Sant and Andreas Schenk

Integrated Systems Laboratory,
ETH Zurich,
Zurich, Switzerland
sasant@iis.ee.ethz.ch

Abstract— Surface roughness causes random shifts in the lowest sub-band level around its ideal position. This gives rise to tail states of an otherwise step-like DOS of the 2D electron gas in the channel. These tail states cause a gradual onset of tunneling in a TFET with vertical tunnel paths and degrade the sub-threshold swing. The impact of roughness of the semiconductor/oxide interface on the transfer characteristics is analyzed in this paper. Quantum-mechanical calculations are performed on a (pseudo)-one-dimensional TFET to obtain the drain current in the presence of randomly shifted sub-band levels. It is found that, the larger the roughness amplitude, the stronger the degradation of the sub-threshold swing.

Keywords— *Surface roughness, Tunnel FETs, Line tunneling.*

I. INTRODUCTION

To minimize the frequency of charging batteries, it is necessary to reduce the energy footprint of integrated circuits used in mobile communication equipment. Supply voltage scaling is one way of reducing energy consumption in integrated circuits. However, due to the physical limit on the mechanism of operation of Metal Oxide Semiconductor Field Effect Transistors (MOSFETs), the supply voltage of MOSFET based integrated circuits cannot be further scaled at this stage. Therefore, alternative energy-efficient solid-state electronic switches are sought after to replace MOSFETs. The Tunnel Field Effect Transistor (TFET) is a switch which works on the principle of band-to-band tunneling (BTBT) in semiconductors. It offers the possibility of scaling down the supply voltage [1].

In a TFET, BTBT either takes place parallel to the gate (often referred to as point tunneling), perpendicular to the gate (often called line tunneling), or inclined. In a vertical TFET, called Line TFET (LTFET) in the following, vertical tunneling prevails. With the special device geometry shown in Fig. 1(a) point tunneling can be completely suppressed. Line tunneling takes place between two-dimensional electronic states in the triangular well at the oxide-semiconductor interface and the three-dimensional hole states in the bulk region. A sharp step-like onset of BTBT is a peculiar characteristic of the 2D-3D DOS matching. This sharp onset of BTBT results in a small sub-threshold swing (SS). Scaling of the gate area can scale up the on-state current in a LTFET. However, field-induced quantum confinement in the channel degrades the on-current [2,3]. Additionally, the rough oxide-semiconductor interface

Funding from the European Community's Seventh Frame-Work Program under Grant Agreement No. 619509 (Project E²SWITCH) is acknowledged.

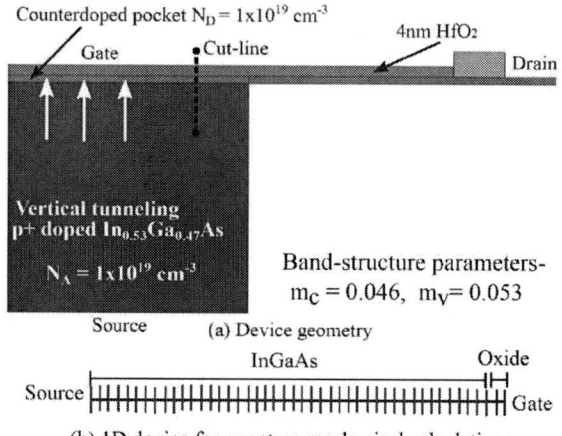

(a) Device geometry

(b) 1D device for quantum mechanical calculations

Figure 1: (a) InGaAs LTFET with counter-doped pocket. The special geometry favors vertical ("line") tunneling and is used to analyze the impact of channel quantization and surface roughness. (c) (Pseudo-) one-dimensional TFET used for quantum-mechanical calculations.

results in band tail states [4] which cause a more gradual onset of line tunneling and thereby degrades the SS.

In this work, the impact of surface roughness on the steepness of transfer characteristics of LTFETs is analyzed. Quantum-mechanical calculations are performed to obtain the tunnel current for sub-band edges at different energies adjacent to the energy of the first sub-band. The drain current for a LTFET with surface roughness is calculated by an ensemble average of these tunnel currents. The variation of the influence of surface roughness with source and channel doping is analyzed. Finally, the drain current resulting from the quantum-mechanical calculations is compared with that obtained from the semi-classical model from [3].

II. SIMULATION SET-UP

The special geometry of a LTFET which suppresses point tunneling is shown in Fig. 1(a). The wide gate overlapped on the source region provides a large area for vertical tunneling. The thin long undercut region between the source edge and drain ensures low off-state leakage. A 5 nm thick counter-doped pocket is introduced under the gate to advance the onset of the device. A two-dimensional quantum transport simulation of such a large device is computationally not feasible. Therefore, a cut-line normal to the interface was used for one-dimensional quantum-mechanical calculations. This uni-dimensional device is schematically shown in Fig. 1(b).

A. Quantum-mechanical Calculation of the Tunnel Current

The one-dimensional device in Fig. 1(b) is simulated using a self-consistent 1D k·p-Poisson solver available in the S-Band package [5]. For each gate bias, the energy levels and the quantized electron wave functions of the first sub-band are obtained from the solver. The Numerov method is used to calculate the valence band (VB) wave functions in the bulk region for all energies $E^\perp$ of the tunnel window. The tunnel probability is calculated for each $E^\perp$ using Fermi's golden rule. The total drain current per unit gate area is then calculated in the Landauer formalism. The entire procedure is described below.

The triangular potential well in the channel results in the quantization of the conduction band (CB) states. These states are obtained by solving the following envelope equation numerically using a 1D k·p-Poisson solver available in S-Band [5]:

$$\left(-\frac{\hbar^2}{2m_c}\frac{d^2}{dx^2} + U(z)\right)\chi_{c,n}(z) = E_n\,\chi_{c,n}(z). \quad (1)$$

For Eq. (1) it is assumed that the device is homogeneous in the xy-plane. Hence, the x- and y-dependent components of the envelope function are given by plane waves. This adds an additional energy term to the total energy. The total energy thus becomes $E_{n,\perp} = E_n + E^\perp$.

Since VB states are not quantized, an envelope function exists for each available value of $E^\perp$. The envelope functions for the VB states are obtained solving the following equation by the Numerov algorithm:

$$\left(-\frac{\hbar^2}{2m_v}\frac{d^2}{dx^2} + (U(z) - E - E')\right)\chi_v(z, E') = 0. \quad (2)$$

Hard-wall boundary conditions at the oxide-semiconductor interface ($\chi_v(z = 0, E^\perp) = 0$) are used as boundary condition for the wave function as required for the Numerov algorithm.

The envelope functions for the CB and VB states obtained by the above method are exact solutions of the Schrödinger envelope equation. The inter-band matrix element as a function of $E^\perp$ can be obtained by treating the field-dependent inter-band coupling term as perturbation,

$$M_{n,v}\left(E_{n,\perp}, E'\right) = \int_0^\infty \chi_{c,n}(x)\,H'(x)\,\chi_v(x, E')dx, \quad (3)$$

$$\text{where } H'(x) = \frac{\hbar \cdot P}{m_0 \cdot E_g}\nabla U(x).$$

Here, P is the momentum matrix element and E_g is the band gap. The tunnel probability can then be calculated using Fermi's golden rule,

$$T(E') = \frac{2\pi}{\hbar}\int_{E_n}^{E_{max}}\left|M_{n,v}\left(E_{n,\perp}, E'\right)\right|^2 \delta\left(E_{n,\perp} - E'\right)dE_{n,\perp}. \quad (4)$$

Note that although E_n is discrete, $E_{n,\perp}$ is continuous as it involves the transverse energies. The above integral gives a Heaviside function on integration, which implies that $E' > E_n$ for tunneling to take place. The total electron current can be calculated by integrating the tunnel probability over all available transverse energies,

$$J_T = \int_0^{E_{max}} T(E'^\perp)\rho(E'^\perp)dE'^\perp. \quad (5)$$

The current density obtained by this approach is in units of A/m^2. It can be multiplied by the gate area to obtain the drain current. In this way, the drain current is calculated per unit area for the given sub-band energy.

B. Introducing the Effect of Surface Roughness in the Calculation

The oxide-semiconductor interface (assumed to be in the xy-plane) is not atomistically smooth. Its roughness causes random shifts of the (assumed infinitely high) potential wall of the triangular well along the z-axis. This results in random shifts of the first sub-band level around its mean energy as

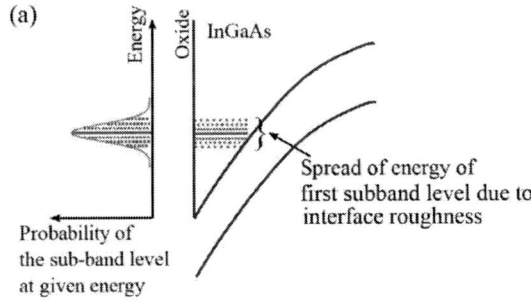

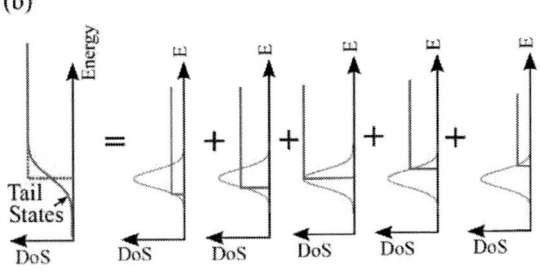

Figure 2: (a) Schematic showing the Gaussian spread of the energy of the first sub-band level. (b) Sketch of the weighted average of the DOS of all the randomly shifted first sub-band levels resulting in an error-function like DOS with tail states below the mean sub-band level.

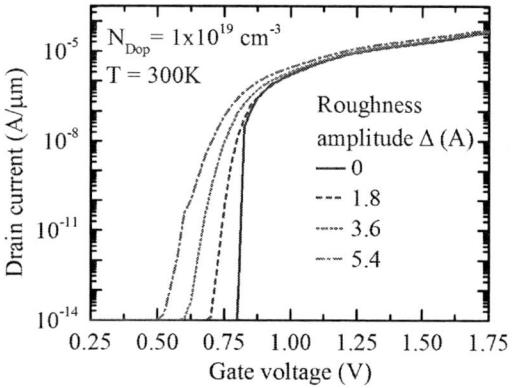

Figure 3: Comparison of transfer characteristics of the uni-dimensional LTFET for different values of the roughness amplitude.

shown in Fig. 2(a). This means that the value of the first sub-band energy level is different at different locations in the xy-plane. If the sub-system at each location in the xy-plane acts as an independent 2D sub-system, the DOS of the quantized electronic states in the sub-system is step-like. The average DOS of the total system is obtained by taking the weighted average of the DOS of each of such localized sub-systems at each location in the xy-plane. The weights are the probability of existence of a sub-band level at a given energy. For a rough interface with Gaussian correlation, the spread of the sub-band levels is a Gaussian with the center at the mean energy of the first sub-band as shown in Fig. 2(a). Averaging of the DOS as described above with a Gaussian as a weighing function results in the average DOS with the shape of an error function as depicted schematically in Fig. 2(b). The aggregate effect of this random shift of the sub-band level is the formation of band tail states below the mean energy.

The quantum-mechanical treatment of a 2D electronic system in the presence of an arbitrary random potential in 2D by Quang and Tung [6] yields an error-function-like DOS which is in agreement with our argument described above. The standard deviation (η) appearing in the Gaussian distribution and the error function is calculated in [8] by inserting Ando's roughness potential [7] as an arbitrary random potential. It is given by,

$$\eta = F_{\text{eff}} \Delta, \quad (6)$$

where η is the characteristic energy, F_{eff} is the effective electric field given by $F_{\text{eff}} = \frac{\int_0^T dz\, F(z) \cdot n(z)}{\int_0^T dz\, n(z)}$, and Δ is the roughness amplitude.

Assuming that the tunnel rate can be independently calculated for each localized sub-system, the total tunnel current in the presence of surface roughness is given by

$$I_{\text{DS}} = \frac{A}{\eta \sqrt{\pi}} \int_{-\infty}^{E_{\max}} J_{\text{T}}(E) \exp\left(-\frac{(E - E_0)^2}{2\eta^2}\right) dE. \quad (7)$$

Here, $J_{\text{T}}(E)$ is the drain current calculated by the quantum-mechanical treatment described in sub-section II-A (see Eqs. (1) - (5)) for a system with the sub-band lying at energy E. In this way, the *total* drain current is calculated by taking the ensemble average of the drain currents arising from the spread out levels of the first sub-band at each of the energies E ($-\infty < E < E_{max}$) around the mean sub-band level (E_0).

III. SIMULATION RESULTS AND DISCUSSION

The drain current in the presence of broadened sub-band energies due to surface roughness was calculated using the above described procedure. A roughness amplitude Δ of 1.8 Å and a correlation length L of 1.9 nm was experimentally extracted for a InAs/GaSb hetero-junction by Feenstra *et al.* [9] Also, roughness amplitudes ranging between 5 Å and 11 Å have been reported for thin $In_{0.75}Ga_{0.25}As$ pHEMT structures [10]. Taking these values as guidelines, the quantum-mechanical calculations were performed with $\Delta = 1.8$ Å, 3.6 Å, and 5.4 Å. The simulation results are presented in Fig. 3. As observed in the figure, an increasing roughness amplitude degrades the sub-threshold swing, as one would

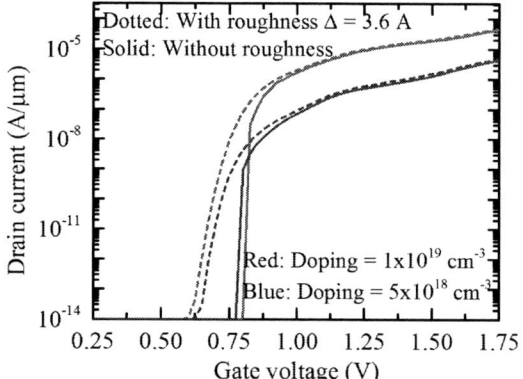

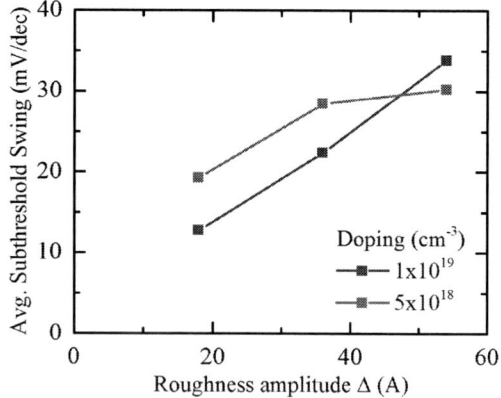

Figure 4: Comparison of (a) transfer characteristics and (b) sub-threshold swing of the LTFET with and without considering surface roughness for two different values of the bulk doping.

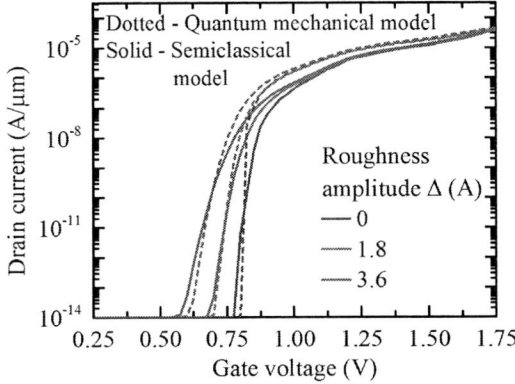

Figure 3: Comparison of transfer characteristics of the 1D LTFET calculated quantum-mechanically and by the semi-classical model for different values of the roughness amplitude.

expect. The sub-threshold swing averaged over four orders of magnitude of the drain current is plotted vs. roughness amplitude in Fig. 4(b) (blue line). The plot confirms that the steepness of the *IV*-plots at the onset degrades with increasing Δ.

The effect of the source doping level on the *IV*-plots with and without roughness was studied by calculating the transfer characteristics for two doping concentrations, 5×10^{18} cm^{-3} and 1×10^{19} cm^{-3}. The *IV*-curves are compared in Fig. 4(a). The average sub-threshold swing of the LTFET is plotted in Fig. 4(b) vs. roughness amplitude for the two doping values. The lower the doping, the smaller F_{eff} and the characteristic tail energy η. Hence, the reduction of doping is expected to result in a lowering of the sub-threshold swing. On the other hand, a smaller doping concentration increases the tunnel length which in turn degrades the sub-threshold swing. A degradation of the swing due to the reduced doping is observed for smaller Δ. For $\Delta = 5.4$ Å, the effect of a smaller η due to lower doping becomes dominant and the subthreshold swing is improved with decreasing doping concentration.

A. Comparison with Semi-classical Model

A semi-classical TCAD model was presented in [8] to account for the formation of band tails due to interface roughness and their impact on line tunneling. Here, this model is employed on the uni-dimensional LTFET for comparison. The band diagrams extracted from the k·p-Poisson solver at each bias point are processed to obtain the active tunnel paths. The semi-classical model is then used on these tunnel paths to calculate the tunnel current. In Fig. 5, the drain current obtained from the quantum-mechanical calculations is compared with the semi-classical drain current. The good agreement confirms the validity of the semi-classical TCAD model.

It must be noted that the above treatment is only applicable for vertical (line) tunneling. Furthermore, various non-idealities such as bulk band tails, interface and bulk traps, etc. affect the TFET performance in addition to interface roughness. It is observed that, among all the non-idealities, traps have the strongest impact [11]. However, the detrimental effect of traps is going to fade out with improving technology. If that happens, interface roughness may become an important non-ideality that degrades the sub-threshold swing.

IV. CONCLUSION

In summary, the impact of a rough semiconductor/oxide interface is assessed by quantum-mechanical calculations of a uni-dimensional LTFET. The simulations show that surface roughness degrades the sub-threshold swing and results in a gradual turn-on of the device instead of a sharp, step-like onset. The larger the roughness amplitude Δ, the more severe the degradation of the sub-threshold swing. Also, for larger values of Δ, reducing the source doping reduces the degradation of the swing. It must be noted, that the above results only apply to vertical (line) tunneling.

REFERENCES

[1] A. M. Ionescu, H. Riel,"Tunnel field-effect transistors as energy-efficient electronic switches, Nature vol. 479, pp. 329–337, 2011.

[2] S. Goodnick, R. Gann, D. Ferry, C. Wilmsen, and O. Krivanek, "Surface roughness induced scattering and band tailing, Surface Science vol. 113, pp. 233–238, 1982.

[3] W. G. Vandenberghe, B. Soree, W. Magnus, G. Groeseneken, A. Verhulst, and M. V. Fischetti, "Field induced quantum confinement in indirect semiconductors: Quantum mechanical and modified semiclassical model, SISPAD'11, pp. 271–274 (2011).

[4] S. Sant, A. Schenk, K. Moselund, and H. Riel, "Impact of Trapassisted Tunneling and Channel Quantization on InAs/Si Hetero Tunnel FETs", 74th IEEE Device Research Conference, pp. 67–68, 2016.

[5] Synopsys Inc., Device Monte Carlo User Guide, V-2015.06.

[6] D. Quang and N. Tung, "A Semiclassical Approach to the Electron Gas in Two-Dimensional Semiconductor Structures, phys. Stat. sol. (b), vol. 207, pp. 111–123, 1998.

[7] T. Ando, "Screening effect and quantum transport in Silicon inversion layer in strong magnetic fields", J. Phys. Soc. Japan, vol. 43, pp. 1616–1626, 1977.

[8] S. Sant and A. Schenk, "Modeling the Effect of Interface Roughness on the Performance of Tunnel FETs", IEEE Electron Device Lett., vol. 38, no. 2, pp. 258–261, 2017.

[9] R. M. Feenstra, D. A. Collins, D. Z.-Y. Ting, M. W. Wang, and T. C. McGill, "Interface roughness and asymmetry in InAs/GaSb superlattices studied by scanning tunneling microscopy, Phys. Rev. Lett. vol. 72, pp. 2749–2752, 1994.

[10] M. Naidenkova, M. S. Goorsky, R. Sandhu, R. Hsing, M. Wojtowicz, T. P. Chin, T. R. Block, and D. C. Streit, "Interfacial roughness and carrier scattering due to misfit dislocations in In$_{0.52}$Al$_{0.48}$As/In$_{0.75}$Ga$_{0.25}$As/InP structures", J. Vac. Sci. Technol. B, vol. 20, no. 3, pp. 1205–1208.

[11] A. Schenk, S. Sant, K. Moselund, and H. Riel, "III-V-Based Hetero Tunnel FETs: A Simulation Study Focussed on Non-ideality Effects", Proc. ULIS-EUROSOI, pp. 9–12, 2016

Material Selection and Device Design Guidelines for Two-Dimensional Materials Based TFETs

Tarun Agarwal[1,2], Bart Soree[1,2], Iuliana Radu[2], Praveen Raghavan[2], Gianluca Fiori[4], Marc Heyns[2,3] and Wim Dehaene[1,2]

[1] Department of Electrical Engineering, Katholieke Universiteit Leuven, Heverlee, 3001 Belgium

[2] Interuniversity Microelectronics Centre (imec), Heverlee, 3001 Belgium

[3] Department of Materials Engineering, Katholieke Universiteit Leuven, Heverlee, 3001 Belgium

[4] Dipartimento di Ingegneria dell'Informazione, Universita' di Pisa, Via Caruso 16, 56122 Pisa, Italy

Email: Tarun.AgarwalKumar@imec.be

Abstract—In this paper, we study the impact of different device architectures and material properties on the performance of two-dimensional tunnel FETs (2D TFETs). We show that single-gate (SG) device architecture in case of monolayer and few layers two-dimensional materials perform better than double-gate (DG) architecture. Due to sharper band bending at the tunneling junction, SG device offers shorter tunneling lengths resulting in larger ON currents. With physical insight into the device structure, we show that the gate-to-source outer fringing fields play a significant role in 2D TFETs performance. In order to reduce the effect of outer fringing fields, we propose a device structure with an interfacial layer (IL) between High-k and 2D material, resulting in a 3-4x increase in ON current. Further, we show that ON currents can be further boosted by increasing the channel thickness. In the end, it is shown that a bi/tri-layer anisotropic 2D material based SG/DG TFET using IL can provide up to 400% higher ON current in comparison to monolayer 2D material based SG FET for equivalent effective mass and bandgap.

I. Introduction

Tunnel FETs have been investigated extensively as they promise sub-60 mV/dec sub-threshold swing (SS) at room-temperature, a key to scale supply voltages below 0.5 V [1]. However, ON currents in TFETs are well below that of MOSFET. To achieve reasonable ON currents in TFETs, we need higher transmission probability using lower effective mass and lower bandgap materials with good electrostatics. III-V material based gate-all-around TFETs offer both lower effective mass/bandgap, and excellent electrostatics. Alternatively, 2D materials are considered for TFETs due to their atomic thickness, which offer better scalability in comparison to Si and III V [2]. Moreover, an extensive list of 2D materials with different effective masses and bandgap openings provide multiple alternatives for material selection [3].

For high-performance 2D material based TFETs, it is imperative to choose the right 2D material with optimum bandgap and effective mass [4]. A lower bandgap and higher effective mass 2D material has been reported to offer higher interband tunneling currents than lower effective mass and higher bandgap 2D materials [5]. Moreover, a circuit level evaluation with delay and energy-delay product (EDP) reported lower bandgap (above 0.5 eV) and higher effective mass 2D materials to show lower EDP for the given performance [6]. The performance of 2D TFETs can be further boosted by designing the device for good electrostatics and enhanced

transmission probabilities. Therefore, a 2D material based double-gate (DG) device is expected to have higher ON currents than single-gate (SG) device, as in the case of bulk materials [7]. However, in case of atomically thin 2D channels, outer fringing fields are reported to play a significant role in electrostatics of 2D material based devices [8]. Therefore, to provide guidelines for device selection of 2D TFETs, a study comparing both single-gate and double-gate device options is required.

We believe that this is the first work that shows the comparison of SG and DG 2D material TFETs for chosen material parameters. In the first part, we provide a detailed explanation on modeling of an anisotropic effective mass 2D material. Using this model, we study the effect of anisotropicity in the effective mass of a small bandgap 2D material on the performance of 2D TFETs. Next, SG and DG device options with 2D materials are compared and thoroughly analyzed. New device solutions are proposed to further boost the performance of 2D TFETs. In the second part, we provide material and device design guidelines by combining different solutions.

II. Modeling and Simulation

The electrical characteristics of 2D TFETs are calculated using two-band tight binding (TB) Hamiltonian with a quantum transport simulation framework based on self-consistent solution of Poisson and Schrödinger equation in non-equilibrium Green's function framework [9].

A. Material Modeling

The two-band Hamiltonian for an anisotropic effective mass two-dimensional material with hexagonal lattice can be written as a 2x2 Hamiltonian matrix:

$$H_{2D} = \begin{bmatrix} E_{cm} & f(k) \\ f^*(k) & E_{vm} \end{bmatrix} \quad (1)$$

where E_{cm}, and E_{vm} denote the bottom of conduction band, and top of the valence band. Further, bandgap (E_G) of the material can be expressed as: $E_G = E_{cm} - E_{vm}$. Here, the $f(k)$ function, due to nearest neighbors, can be written as:

$$f(k) = t_1 e^{(ik_x a/\sqrt{3})} + 2t_2 e^{(-ik_x a/2\sqrt{3})} cos(\frac{k_y a}{2}). \quad (2)$$

Here, t_1, t_2 represents hopping energies in x and y direction respectively, which are calculated using the effective masses

978-1-5090-5979-9/17 $31.00 © 2017 IEEE

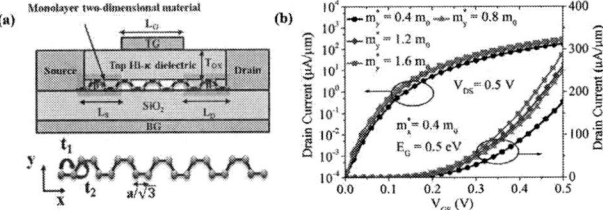

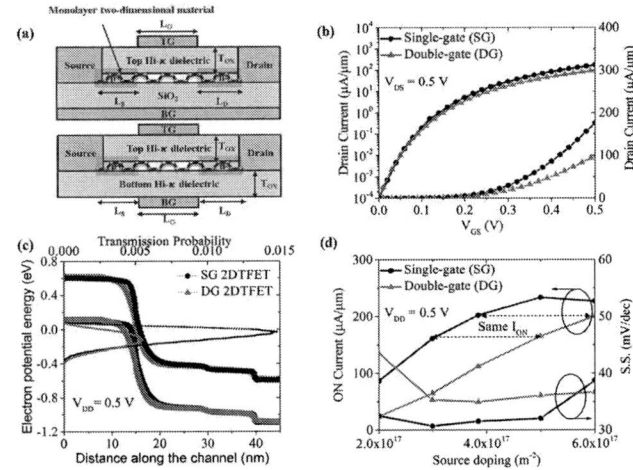

Fig. 1. Effect of anisotropic effective mass in a) 2D material based single-gate TFET, b) Transfer characteristics with different anistropicity (i.e. m_y^*/m_x^*) obtained using two-band quantum simulations, with material and device parameters taken from [6].

in x and y directions and bandgap of 2D material. Here, k_x, and k_y are wave vectors in x & y directions, while a denotes the lattice constant of the two-dimensional hexagonal lattice. Further, using secular equation, we obtain the dispersion relation for the two-band model given as:

$$E^{\pm}(k) = \frac{(E_{cm} + E_{vm}) \pm \sqrt{(E_{cm} - E_{vm})^2 + 4|f(k)|^2}}{2}.$$
(3)

In order to calculate t_1 and t_2 for given effective masses in x and y direction, we use the parabolic effective mass approximation with the two-band model as:

$$\frac{1}{m_x^*} = \frac{1}{\hbar^2} \cdot \frac{\partial^2 E^{\pm}(k)}{\partial k_x^2} \qquad \frac{1}{m_y^*} = \frac{1}{\hbar^2} \cdot \frac{\partial^2 E^{\pm}(k)}{\partial k_y^2}$$
(4)

where m_x^* and m_y^* denotes the reduced effective mass in x and y direction. Using Eq. 2,3,4, and by taking limit of the second derivative at the minimum energy k-point, we can calculate t_1 and t_2 for a given m_x^*, m_y^*, and E_G as:

$$|t_1|^2 = \frac{2\hbar^2 E_G}{3a^2 m_x^*} \qquad |t_2|^2 = \frac{\hbar^2 E_G}{2a^2}\left[\frac{1}{3m_x^*} + \frac{1}{m_y^*}\right]$$
(5)

Using the above framework, we understand the effect of anisotropic effective mass in 2D material based lateral TFET, as shown in Fig. 1. The material and device parameters are taken from our previous study, consisting of optimum effective mass and bandgap of 2D material [6]. Here, source and drain extensions are p^+ doped (3.85×10^{13} cm^{-2}), and n^- doped (3.85×10^{12} cm^{-2}) respectively. While the channel is intrinsic with L_G = 15 nm, source extension (L_S) = 15 nm, and drain extension (L_D) = 15 nm. t_{HfO_2} is chosen to be 3 nm with ϵ_{HfO_2} = 20, and t_{SiO_2} is chosen to be large enough not to affect the electrostatics. As shown in Fig. 1(b), by introducing an anisotropy in the y-direction effective mass (m_y^*) with respect to the effective mass in x-direction (m_x^*), the ON current increases with increasing anisotropy. Increasing m_y^* (i.e. density-of-states) leads to a decrease in the source degeneracy for a given source doping, resulting in wider transmission energy window. It is shown that an anistropicity of 4x in orthogonal direction effective mass can result up to 50% increase in the ON current.

B. Device Modeling

1) Single-gate and Double-gate: The electrical characteristics of single-gate (SG) and double-gate (DG) monolayer (ML) 2D material based TFETs (2DTFET) are shown in

Fig. 2. Comparison of SG and DG 2D TFETs a) Schematic of SG and DG monolayer 2D TFETs, b) Transfer characteristics at V_{DS} = 0.5 V, c) Band-diagrams with transmission probabilities at ON state for SG and DG 2DTFETs, d) ON current (I_{ON}) and sub-threshold slope (SS) with different source extension doping concentrations.

Fig. 2. Contrary to our belief, it is shown that SG ML 2DTFET offers higher ON currents in comparison to DG ML 2DTFET. We see from Fig. 2(b) that although the sub-threshold characteristics are similar for SG and DG ML 2DTFET, the ON-state characteristics of SG ML 2DTFET shows higher ON currents indicating higher tunneling probabilities. Fig. 2(c) shows the tunneling probabilities of SG and DG ML 2DTFETs along with the band diagrams. It is clearly observed that a sharper source-to-channel junction in SG ML 2DTFET results in higher tunneling probability with respect to DG 2DTFET. Further, in order to study the effect of source extension doping, Fig. 2(d) shows behavior of the ON current and SS with different source doping concentrations. It is shown that although the ON currents in DG 2DTFET can be matched with SG 2DTFET by increasing the source doping, the ON currents in single-gate 2DTFET are consistently higher than the ON current in DG 2DTFET for chosen source doping concentrations.

In order to get a physical insight into higher tunneling probabilities observed in SG TFETs, a 2D potential profile for both SG and DG device structures are shown in Fig. 3. In ON state (i.e. at high V_{GS} = 0.5 V), Fig. 3 shows the gate-to-source potential distribution or the outer fringing fields originating from gate edge. In ideal condition when the gate electric field lines are assumed to be perpendicular to the channel, it results in nearly zero outer fringing width, and a steep source/channel junction [8]. While in realistic situation (i.e. Fig. 3), we can clearly see that the outer fringing width at source/channel junction is lower for single-gate device structure, resulting in a higher electric field at source/channel junction, thus higher tunneling probabilities.

2) Proposed device structure: We propose a device structure with a low-k interfacial layer (IL) in between High-k and 2D material to reduce the outer fringing fields, as shown in Fig. 4(a). The gate stack with IL needs to be carefully designed to mitigate gate tunneling through the gate stack. Fig. 4(b) shows that introducing an IL layer improves both

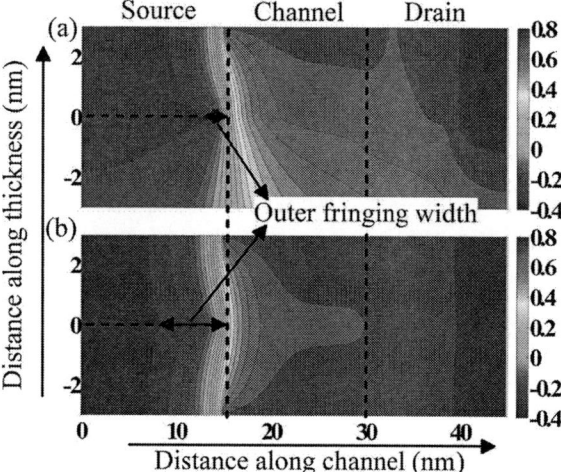

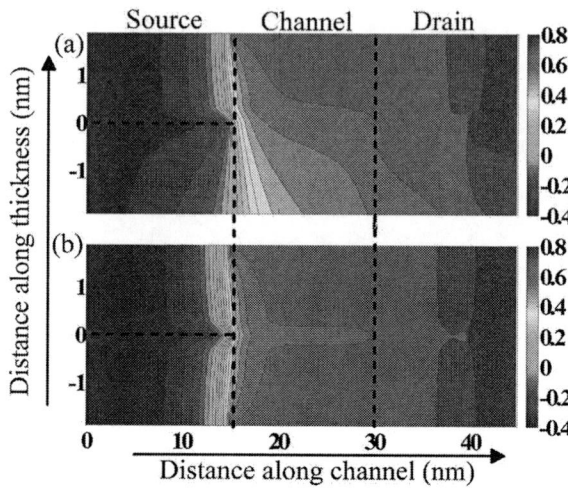

Fig. 3. Effect of outer fringing fields a) 2D Potential distribution in SG (with partial buried oxide), b) 2D Potential distribution in DG, showing the gate-induced potential spread at source-channel junction (i.e. outer fringing fields) in ON state (V_{GS} = 0.5 V).

Fig. 5. Effect of outer fringing fields in devices with IL a) 2D Potential distribution in SG (with partial buried oxide), b) 2D Potential distribution in DG, showing the improved 2D rectilinear potential near source-channel junction (i.e. reduced outer fringing fields) at ON state.

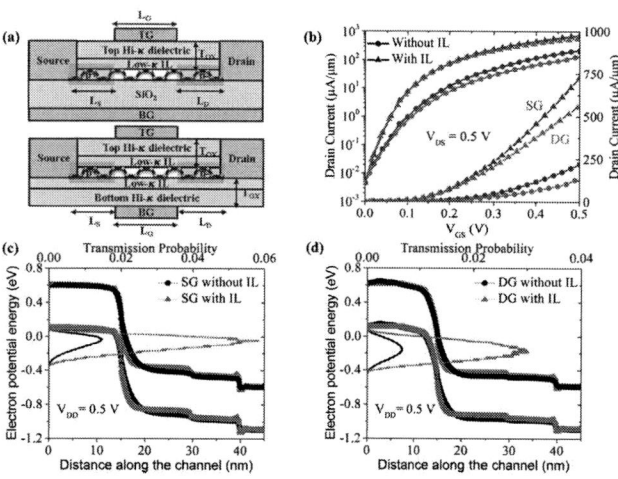

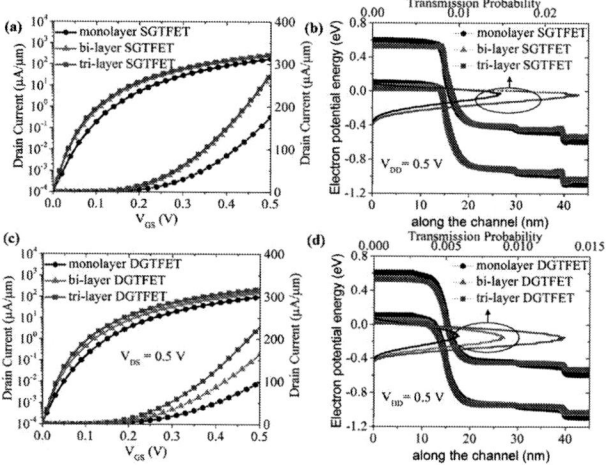

Fig. 4. Proposed device, a) Schematic of SG and DG monolayer 2D TFETs with low-k interfacial layer between High-k and 2D material, b) Transfer characteristics comparing both SG and DG devices with/without IL, c), and d) Band-diagrams with transmission probabilities for SG and DG device structures with/without IL.

Fig. 6. Effect of channel thickness for equivalent effective mass and bandgap, a,c) Transfer characteristics of mono-, bi-, and tri-layer 2D material based SG and DG TFETs, b,d) Band-diagrams with transmission probabilities for multi-layer SG and DG 2D TFETs.

onset voltage (i.e. sub-threshold characteristics) and the ON state (i.e. tunneling probability). For an equivalent effective-oxide-thickness (EOT), introducing IL results in around 3-4x increase in the ON current. Fig. 4(c) and (d) shows an equivalent boost in the transmission probability as the ON current in both SG and DG devices with IL. A significant reduction in the tunneling distances can be seen in the device structures with IL. As shown in Fig. 5, shorter tunneling lengths at source/channel junction are achieved by shaping the potential to be steeper at the junction. Fig. 5 exhibits a near ideal case, where the potential distribution is highly rectilinear (i.e. lesser outer fringing fields) in contrast to Fig. 3 where the potential distribution is elliptical.

3) Effect of the channel thickness: With physical insight on the fringing fields in monolayer based SG and DG de-

vice structures, we further study the effect of multi-layer 2D material on the fringing fields with number of layer, thus the performance. Using the framework of multi-layer 2D materials as given in [9] with t_p=0 eV, performance of mono-, bi-, and tri-layer 2D TFETs are compared in Fig. 6(a).

It is shown that ON current increases with number of layers for given material parameters ($m_x^* = m_y^* = 0.4\ m_0$, and E_G = 0.5 eV). Although in reality increasing the number of 2D layers also change the material parameters such as E_G decreases with increasing number of layers. But, to restrict ourselves in order to understand the effect of increasing channel thickness we use a multi-layer tight binding Hamiltonian with same material parameters. For single-gate case, we can observe a significant increase in the ON current when making a transition from monolayer to bi-layer due to significant reduction in fringing fields, effectively increasing the transmission probability, as shown in Fig. 6(b). While the ON current doesn't increase appreciably when transitioning from bi-layer to tri-layer as the degradation in 2D electrostatistics with increasing number of layers nullifies the boost in ON current due to the reduction in fringing fields. On the other hand, Fig. 6(c), and (d) show that increasing ON current with number of 2D layers for DG 2D TFET which signifies the dominance of the reduction in the fringing fields over the degradation in 2D electrostatistics.

III. MATERIAL SELECTION AND DEVICE DESIGN GUIDELINES

Combining different features to boost I_{ON} as discussed in previous section, we quantify the improvement in the ON current as the percentage with respect to the reference device from [6]. For a single-gate device, anisotropic effective mass ($m_y^* = 4m_x^*$) can result in 35-50 % increase in I_{ON}, while using IL with High-k can boost the ON current by 200-300 %. On the other hand, a DG 2DTFETs can show 50-60 % increase with anisotropic effective mass 2D material and 300-350 % increase with IL. Moreover, increasing the channel thickness by adding an atomic 2D layer can boost I_{ON} by 35-65 %. Combining SG/DG FET with IL, and bi/tri-layer 2D material with anisotropic effective mass can result up to 400 % increase in the ON current.

IV. CONCLUSION

We have studied multiple device options for lateral 2D TFETs. With monolayer 2D material as channel, single-gate TFETs provide higher ON currents in comparison to double-gate 2D TFETs. The performance of 2D TFETs are shown to be limited by outer fringing fields. To mitigate the effect of fringing fields, a device structure with low-k interfacial layer is proposed, resulting in 3-4x performance boost. We show that a bi-/tri-layer anisotropic effective mass 2D material based SG/DG TFET can result in 400% increase in the ON current with respect to single-gate 2D TFET for same material parameters.

REFERENCES

[1] Seabaugh, A. C., and Zhang Q., "Low-Voltage Tunnel Transistors for Beyond CMOS Logic," *Proceedings of the IEEE*, vol.98, no.12, pp.2095-2110, Dec. 2010

[2] Liu, L., Kumar, S.B., Ouyang, Y. and Guo, J., "Performance limits of monolayer transition metal dichalcogenide transistors," *IEEE Transactions on Electron Devices*, vol.58, no.9, pp.3042-3047, 2011

[3] Rasmussen, F. A., and Kristian S. T., "Computational 2D materials database: Electronic structure of transition-metal dichalcogenides and oxides," *The Journal of Physical Chemistry C*, vol.119, no.23, pp.13169-13183, Jun. 2015

[4] Ilatikhameneh, H., Tan, Y., Novakovic, B., Klimeck, G., Rahman, R., and Appenzeller, J., "Tunnel field-effect transistors in 2-D transition metal dichalcogenide materials," *Exploratory Solid-State Computational Devices and Circuits, IEEE Journal on*, vol.1, pp. 12-18, Dec. 2015

[5] Ma, N., and Jena, D., "Interband tunneling in two-dimensional crystal semiconductors," *Applied Physics Letters*, vol.102, no.13, pp.132102, Apr. 2013

[6] Agarwal, T., Radu, I., Raghavan, P., Fiori, G., Thean, A., Heyns, M. and Dehaene, W., "Effect of material parameters on two-dimensional materials based TFETs : An energy-delay perspective," *IEEE European Solid-State Device Research Conference (ESSDERC)*, pp. 47-50, Sept. 2016

[7] Verhulst, A.S., Sore, B., Leonelli, D., Vandenberghe, W.G. and Groeseneken, G., "Modeling the single-gate, double-gate, and gate-all-around tunnel field-effect transistor," *Journal of Applied Physics*, 107(2), pp.024518, 2010

[8] Ilatikhameneh, H., Klimeck, G., Appenzeller, J. and Rahman, R., " Scaling theory of electrically doped 2D transistors," *IEEE Electron Device Letters*, 36(7), pp.726-728, 2015

[9] NanoTCAD ViDES, http://vides.nanotcad.com/vides

Nanometer CMOS Characterization and Compact Modeling at Deep-Cryogenic Temperatures

R. M. Incandela*, L. Song*[†], H.A.R. Homulle*, F. Sebastiano*, E. Charbon*[¶][||] and A. Vladimirescu*[‡][§]

*TU Delft, Delft, The Netherlands, [†]Tsinghua University, Beijing, China, [‡]ISEP, Paris, France,
[§]UC Berkeley, Berkeley, CA, USA, [¶]EPFL, Lausanne, Switzerland, [||]Intel, Hillsboro, OR, USA

Abstract—The characterization of nanometer CMOS transistors of different aspect ratios at deep-cryogenic temperatures (4 K and 100 mK) is presented for two standard CMOS technologies (40 nm and 160 nm). A detailed understanding of the device physics at those temperatures was developed and captured in an augmented MOS11/PSP model. The accuracy of the proposed model is demonstrated by matching simulations and measurements for DC and time-domain at 4 K and, for the first time, at 100 mK.

I. INTRODUCTION

Quantum computers are based on operations performed on quantum bits (qubits), which generally operate well below 1 K [1]. Cryogenic electronics working at temperatures near that of qubits is of utmost importance for performing operation on these quantum devices [2]. The liquid-helium temperature (LHT), i.e. 4.2 K, is the lowest practical temperature at which state-of-the-art refrigerators can offer sufficient cooling power. A few submicron and nanometer CMOS technologies from 350 nm to 14 nm have been characterized above 50 K [3]–[11], these temperatures are however too high for quantum computing applications. Several prior works are targeted to the temperature range of interest down to 4 K [12]–[22], with technologies ranging from 3 μm to 32 nm in feature size. A visual summary, provided in Fig. 1, compares this work with previous publications.

Although the physics of cryogenic CMOS devices is generally understood [23] and enhancing the BSIM4 model was attempted [24], no general compact MOSFET models exist for nanometer CMOS at deep-cryogenic temperatures. In this paper, we present the characterization of nanometer CMOS transistors and propose a SPICE MOSFET model based on MOS11/PSP [25], [26] valid at 4 K and 100 mK in addition to the common operating range (-55 °C to 125 °C).

II. CRYOGENIC MEASUREMENTS

Measurements were performed on CMOS transistors with both thin and thick oxide in two different technologies (160 nm, 40 nm) and with a wide range of feature sizes ranging from minimum-size to long and wide devices, as shown in Table I.

The measured drain current I_D of several thin- and thick-oxide 160 nm NMOS transistors (Fig. 2(a,b)) is generally higher at LHT than at room temperature (RT) due to increased carrier mobility (μ). For the minimum-length thick-oxide NMOS, an obvious kink in I_D is observed for large V_{DS}.

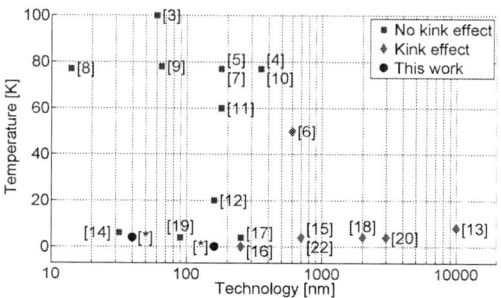

Fig. 1. Summary of CMOS technologies measured at cryogenic temperatures. Devices exhibiting a kink effect (see Section II) are indicated with red diamonds. The devices presented in this paper are indicated with black dot.

TABLE I
SUMMARY OF CHARACTERIZED DEVICES.

Technology	SSMC 0.16 μm		ST 40 nm	
Oxide	Thick	Thin	Thick	Thin
W/L [μm / μm]	2 / 1.61	2.32 / 1.6	1.6 / 1.35	1.2 / 0.4
	2 / 0.322	2.32 / 0.16	1.6 / 0.27	1.2 / 0.04
	0.4 / 1.61	0.232 / 1.6	0.32 / 1.35	0.12 / 0.4
	0.4 / 0.322	0.232 / 0.16	0.32 / 0.27	0.12 / 0.04

For the 40 nm NMOS devices, the I_D also increases at LHT (Fig. 2(c,d)), however, it does not show a kink up to the maximum operating V_{DS}. Measurements at 100 mK of a 40 nm PMOS also show proper transistor operation (Fig. 3(a)).

The I_D-V_{GS} characteristics of the minimum-length 160 nm thin-oxide NMOS show a 3.8× steeper subthreshold slope (SS) and 0.15 V higher threshold voltage V_T at LHT with respect to RT (Fig. 2(e)). The SS of the 40 nm MOSFET is 3.2× steeper and the V_T is 0.1 V higher with respect to RT (Fig. 2(f)) while an increase of 9× was measured at 100 mK (Fig. 3(b)).

III. DISCUSSION ON CRYOGENIC BEHAVIOR

At cryogenic temperatures there is an increase of μ, V_T and SS as expected. The SS, of great interest for leakage minimization, is expected to decrease proportionally with T, since the drain current in subthreshold and the SS can be expressed, respectively, as

$$I_D \approx I_0 \cdot \exp\left[\frac{q(V_{GS} - V_T)}{nkT}\right] \quad (1)$$

$$SS(T) = \left[\frac{\partial \log(I_D)}{\partial V_{GS}}\right]^{-1} = \ln(10)\frac{nkT}{q} \quad (2)$$

978-1-5090-5979-9/17 $31.00 © 2017 IEEE

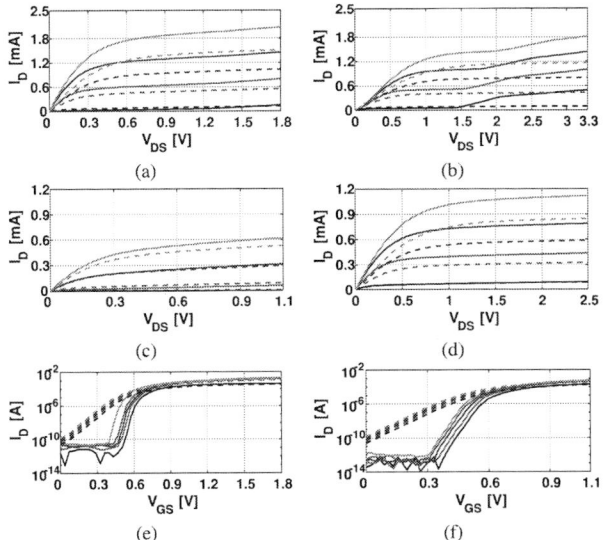

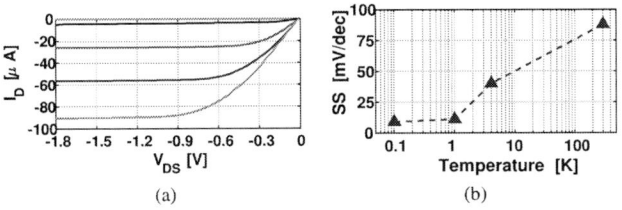

Fig. 2. $I_D(V_{DS})$ of NMOS in 160 nm (a-b) and 40 nm (c-d) thin- (a-c) and thick-oxide (b-d) and $I_D(V_{GS})$ of thin-oxide NMOS in 160 nm (e) and 40 nm (f) at RT (dashed lines) and LHT (solid lines). W/L in μm.
(a) W/L = 2.32/0.16, V_{GS} = 0.68 → 1.8 V; (b) W/L = 2/0.322, V_{GS} = 1.05 → 3.3 V; (c) W/L = 1.2/0.04, V_{GS} = 0.43 → 1.1 V; (d) W/L = 1.6/0.270, V_{GS} = 0.85 → 2.5 V;
(e) V_{DS} = 0.1 → 1.8 V, W/L as (a); (f) V_{DS} = 0.1 → 1.1 V, W/L as (c).

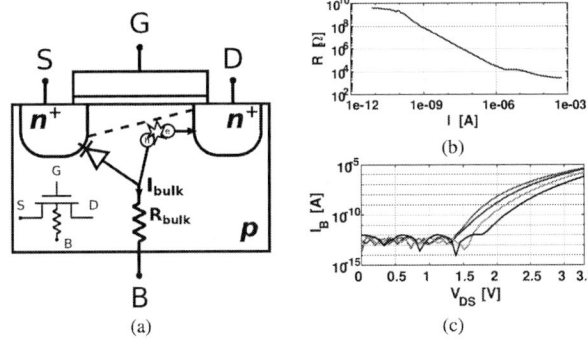

Fig. 3. (a) $I_D(V_{DS})$, V_{GS} = −0.3 → −1.8 V at 100 mK and (b) SS versus temperature of PMOS, W/L = 2.32 μm / 1.6 μm.

Fig. 4. (a) Cross-section and schematic representation of the cause of kink effect at LHT; (b) N-well resistance versus current at LHT; (c) $I_{bulk}(V_{DS})$ of thick-oxide NMOS, W/L = 2.32 μm / 322 nm at LHT.

where k is Boltzmann constant, q the electron charge, T the absolute temperature and n is the subthreshold slope factor. n

can be expressed as

$$n(T) = 1 + \frac{C_{dep}(T)}{C_{ox}} \quad (3)$$

where $C_{dep}(T)$ is the depletion capacitance between channel and bulk and C_{ox} the gate oxide capacitance. However, SS is not proportional to T at cryogenic temperature as shown in Fig. 2(e,f), which can be attributed to a large variation of $C_{dep}(T)$ due to freeze-out at cryogenic temperature and incomplete ionization of dopants.

The kink can be explained as follows. Due to substrate freeze-out at 4 K, the bulk resistance R_{bulk} (Fig. 4(a)) increases drastically [23], as shown by measuring the n-well resistance in the 160 nm process (Fig. 4(b)). Impact ionization due to the high electric field along the channel leads to a significant bulk current (I_{bulk}) (Fig. 4(c)) that raises the internal body potential V_{BI}, thus, in turn, locally lowering V_T and resulting in a jump in I_D. Eventually, V_{BI} becomes positive enough to turn on the Drain-Bulk-Source bipolar transistor, diverting most of the I_D due to ionization and leading to a gradual saturation of I_{bulk} (Fig. 4(c)). The kink current at higher V_{DS} saturates as well, due to the steep turning on of the Bulk-Source (BS) diode pinning V_{BI} at ≈1 V [23]. Unlike the 160 nm-process thick-oxide devices, the thin-oxide MOS in the same process show no kink (Fig. 2(a)), but only the early onset of the substrate-current body effect (SCBE), common in nanometer CMOS and occurring at a lower V_{DS} at LHT.

The different behavior between the thin- and thick-oxide device is attributed to the stronger surface scattering and the higher surface doping in the thin-oxide MOS. The lateral electric field in thin- and thick-oxide MOS is identical, while the vertical field is higher in the thin-oxide MOS devices. The resulting higher surface scattering results in a reduced mobility, which leads to a lower carrier velocity, thus reducing impact ionization and, hence, causing a lower substrate current. Combined with a lower R_{bulk} due to higher doping in thin-oxide MOS, this results in a lower V_{BI}. The reduction in mobility at a higher transversal electric field in a thick-oxide transistor is confirmed by the kink moving to a V_{DS} beyond the measurement range, as V_{GS} increases, see Fig. 2(d). Similarly, no kink is observed in 160 nm thick-oxide PMOS, as their enclosure in a higher-doped n-well and lower ionization current due to lower μ, prevents this effect. In sub-100 nm CMOS, higher surface doping for V_T control and lightly doped drain (LDD) reduce the occurrence of the kink [21].

Device performance at RT and LHT are compared and summarized in Table II. The large increase in SS at LHT is beneficial not only for digital logic performance, but also results in a larger transconductance in weak inversion for a given current, thus resulting in more power-efficient sub-threshold analog circuits. In addition, transistor parameters that strongly affect the performance of analog circuits are not degraded at LHT, and they show even an improvement. The transconductance (g_m) increases more than 27% for 40 nm CMOS and up to 32% for 160 nm CMOS, which will result in circuits with wider bandwidth for the same power

978-1-5090-5979-9/17 $31.00 © 2017 IEEE

TABLE II
COMPARISON OF PERFORMANCE AT RT AND LHT. g_m, g_{ds} AND GATE DELAY ARE MEASURED WITH V_{GS} OF 1.1, 1.8 V (40 NM, 160 NM).

Technology	40 nm		160 nm	
Temperature	4 K	300 K	4 K	300 K
Device W/L [μm / μm]	1.2 / 0.04		2.32 / 0.16	
V_T [V]	0.65	0.55	0.70	0.55
SS [mV/dec]	27.70	88.20	22.80	87.00
g_m ($V_{DS} = 1$ V) [mS]	1.30	1.02	1.61	1.22
g_{ds} ($V_{DS} = 1$ V) [mS]	0.17	0.13	0.23	0.12
Gate delay [ps]		-	30.60	38.30

budget. On the other hand, the concomitant increase of the output conductance (g_{ds}) results in a reduction of the intrinsic transistor DC gain by 2.5% for 40 nm CMOS and 31% for 160 nm.

IV. MODELING

Existing compact models, such as PSP and MOS11, are only certified down to -55 °C; they can however be modified for LHT by extracting the temperature-dependent model parameters from the LHT measurements, and augmenting the intrinsic model with additional electrical components, which account for the device physics at LHT. For the 160 nm CMOS devices, the MOS11 model from the foundry was modified at LHT by changing the values of the parameters listed in Table III and by adding a non-linear resistor in series with the bulk terminal to emulate the increase in R_{bulk} (Fig. 4(a)). The non-linear resistor has an I-V characteristic proportional to that in Fig. 4(b) and was sized to match the measured V_{DS} of the kink. For the 40 nm CMOS devices, only adapting the parameter values in the foundry-provided PSP model (Table III) was sufficient due to the absence of any kink.

The parameter-fitting procedure starts by setting the temperature of the simulator to -200 °C in order to match the SS, which is directly proportional to T in the standard compact models. The parameters with most impact on I_D according to both physics and measurements are extracted next, i.e. V_T (VFB for MOS11, DELVTO for PSP), μ (BETSQ for MOS11, FACTUO for PSP). Additional model tuning is achieved by including mobility degradation effects at LHT due to surface scattering (THESRR for MOS11, THEMUO for PSP), as explained in Section III. Subsequently, velocity saturation was increased (THESATR for MOS11, THESATO for PSP), to match the change of saturation voltage at LHT visible in Fig. 2. The increased impact ionization at higher V_{DS} highlighted in Fig. 4(c) is emulated by parameters A1R, A2R, A3R for the 160 nm technology. The remaining parameters listed in Table III have a minor influence but lead to a better fitting when modified.

The transistor dynamic behavior is set in addition to its DC bias by resistive and capacitive parasitics, among which the gate resistance (R_g) and capacitance (C_g). In particular, R_g and C_g are the poly-gate resistance and the Poly–SiO$_2$–Si capacitance for 160 nm CMOS, respectively, and several LHT physical phenomena, such as partial ionization of dopants, higher mobility and carrier freeze-out, affect their variation with temperature [19]. Generally, R_g and C_g are smaller at

TABLE III
LIST OF MODIFIED PARAMETERS FOR COMPACT MODEL AT LHT.

MOS11 parameters for 160 nm CMOS					
BETSQR	VFBR	THESRR	SDIBLO	ALPR	KOR
THESATR	THERR	A1R	A2R	A3R	
PSP parameters for 40 nm CMOS					
FACTUO	DELVTO	THEMUO	THESATO	RSW1	CFL
ALPL	MUEO	FBET1			

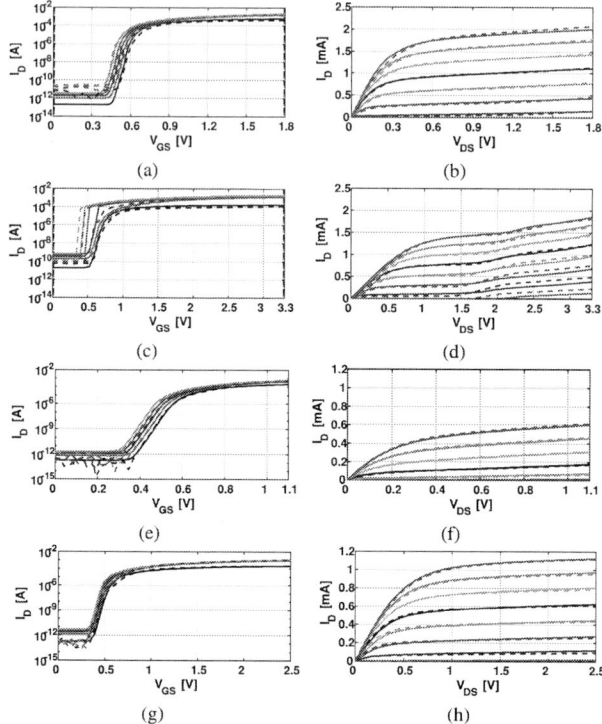

Fig. 5. Models (solid) and measurements (dashed) at LHT. W/L in μm. (a,b) $I_D(V_{GS})$ and $_D(V_{DS})$ of 160 nm thin-oxide NMOS, W/L = 2.32/0.16; (c,d) $I_D(V_{GS})$ and $I_D(V_{DS})$ of 160 nm thick-oxide NMOS, W/L = 2/0.322; (e,f) $I_D(V_{GS})$ and $I_D(V_{DS})$ of 40 nm thin-oxide NMOS, W/L = 1.2/0.040; (g,h) $I_D(V_{GS})$ and $I_D(V_{DS})$ of 40 nm thick-oxide NMOS, W/L = 1.6/0.270.

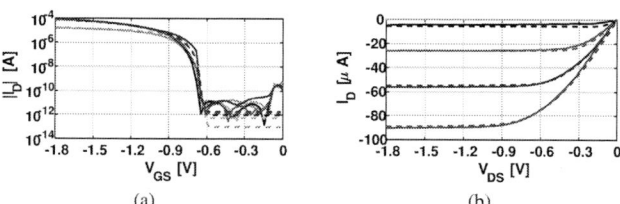

Fig. 6. Models (dashed) and measurement (solid) of 160 nm thin-oxide PMOS at 100 mK. (a) $I_D(V_{GS})$, $V_{DS} = -1.8 \rightarrow -0.09$ V; (b) $I_D(V_{DS})$ $V_{GS} = -0.3 \rightarrow -1.8$ V.

LHT for polysilicon gates leading to a higher intrinsic cut-off frequency [19]. However, variations in parasitic capacitances and R_g have not been accounted for in the proposed model.

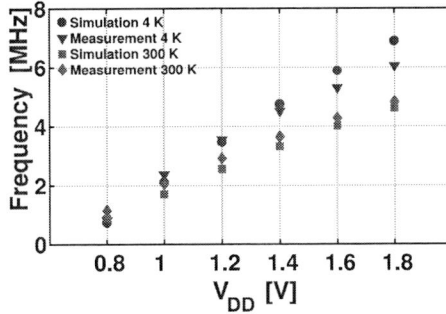

Fig. 7. Ring Oscillator frequency versus supply voltage at 300 K and 4 K. Comparison between measurement and simulation with compact model.

V. SIMULATION VERSUS MEASUREMENT

Very good agreement with DC measurements is achieved for both 160 nm (Fig. 5(a-d)) and 40 nm devices at 4 K (Fig. 5(e-h)) as well as at 100 mK (Fig. 6) over the full voltage range. The measured and simulated output frequency, f_{osc}, of a 160 nm CMOS 2703-stage ring oscillator is plotted in Fig. 7 at 300 K and 4 K showing good agreement over a wide range of supply voltages. Faster saturation of f_{osc} at 4 K at high V_{DD} is observed experimentally with respect to simulations. This is attributed to a number of factors relating to the polysilicon parasitic resistance and capacitances affected by freeze-out, voltage drop on poly and incomplete ionization of dopants not included in the model.

VI. CONCLUSION

A thorough characterization of nanometer CMOS transistors and circuits operating at 4 K was presented and, for the first time, operation at 100 mK of nanometer CMOS devices was proven. An extended MOS11/PSP compact model for nanometer CMOS transistors operating at 4 K and below was conceived and validated at DC and in time domain via both device and circuit characterization. Both subthreshold and strong-inversion regions are modeled with a single set of parameters valid at LHT.

The proposed modeling approach is extensible to any CMOS process node. It ensures high accuracy for nanometer CMOS processes at common bias values while it also predicts the kink effect for older technologies. MOS11 and PSP models, being based on the surface potential, facilitate their extension to cryogenic temperatures. The proposed models have already been used for the design of functional cryogenic CMOS circuits [27], and will enable the future design of even more complex cryogenic CMOS circuits and systems.

ACKNOWLEDGMENT

The authors would like to thank Intel Corp. for funding, and NXP and ST for chip fabrication.

REFERENCES

[1] L. Vandersypen, "Quantum computing - the next challenge in circuit and system design," in *ISSCC*. IEEE, 2017, pp. 24–29.

[2] E. Charbon *et al.*, "Cryo-CMOS for quantum computing," in *IEDM*. IEEE, 2016, pp. 13.5.1–13.5.4.

[3] Z. Chen *et al.*, "Temperature dependences of threshold voltage and drain-induced barrier lowering in 60nm gate length MOS transistors," *Microelectronics Reliability*, vol. 54, no. 6, pp. 1109–1114, 2014.

[4] G. S. Fonseca *et al.*, "Extraction of static parameters to extend the EKV model to cryogenic temperatures," in *SPIE Defense+ Security*. SPIE, 2016, pp. 98 192B–98 192B.

[5] G. D. Geronimo *et al.*, "Front-end ASIC for a liquid argon TPC," in *NSS*. IEEE, 2010, pp. 1658–1666.

[6] W.-Y. Liu, Q. Feng, and R.-J. Ding, "Impact of kink effect on CMOS readout circuits for cryogenic operation," *Laser & Infrared*, vol. 37, pp. 990–992, 2007.

[7] P. Martin *et al.*, "MOSFET modeling for design of ultra-high performance infrared CMOS imagers working at cryogenic temperatures: Case of an analog/digital 0.18 μm CMOS process," *Solid-State Electronics*, vol. 62, no. 1, pp. 115–122, 2011.

[8] M. Shin *et al.*, "Low temperature characterization of 14nm FDSOI CMOS devices," in *WOLTE*, 2014, pp. 29–32.

[9] A. Siligaris *et al.*, "High-frequency and noise performances of 65-nm mosfet at liquid nitrogen temperature," *IEEE Transactions on Electron Devices*, vol. 53, no. 8, pp. 1902–1908, 2006.

[10] H. Zhao and X. Liu, "Modeling of a standard 0.35 μm CMOS technology operating from 77K to 300K," *Cryogenics*, vol. 59, pp. 49–59, 2014.

[11] Z. Zhu *et al.*, "Design applications of compact MOSFET model for extended temperature range (60–400K)," *Electronics letters*, vol. 47, no. 2, pp. 141–142, 2011.

[12] A. Akturk *et al.*, "Compact and distributed modeling of cryogenic bulk MOSFET operation," *IEEE Transactions on Electron Devices*, vol. 57, no. 6, pp. 1334–1342, 2010.

[13] F. Balestra, L. Audaire, and C. Lucas, "Influence of substrate freeze-out on the characteristics of mos transistors at very low temperatures," *Solid-state electronics*, vol. 30, no. 3, pp. 321–327, 1987.

[14] A. H. Coskun and J. C. Bardin, "Cryogenic small-signal and noise performance of 32nm SOI CMOS," in *IMS*. IEEE, 2014, pp. 1–4.

[15] Y. Creten *et al.*, "A cryogenic ADC operating down to 4.2 K," in *ISSCC*. IEEE, 2007, pp. 468–616.

[16] S. R. Ekanayake *et al.*, "Characterization of SOS-CMOS FETs at low temperatures for the design of integrated circuits for quantum bit control and readout," *IEEE Transactions on Electron Devices*, vol. 57, no. 2, pp. 539–547, 2010.

[17] Y. Feng *et al.*, "Characterization and modelling of MOSFET operating at cryogenic temperature for hybrid superconductor-CMOS circuits," *Semiconductor science and technology*, vol. 19, no. 12, p. 1381, 2004.

[18] H. Hanamura *et al.*, "Operation of bulk CMOS devices at very low temperatures," *IEEE Journal of Solid-state Circuits*, vol. 21, no. 3, pp. 484–490, 1986.

[19] S.-H. Hong *et al.*, "Low-temperature performance of nanoscale MOSFET for deep-space RF applications," *IEEE Electron Device Letters*, vol. 29, no. 7, pp. 775–777, 2008.

[20] E. Simoen, B. Dierickx, and C. L. Claeys, "Low-frequency noise behavior of Si NMOSTs stressed at 4.2 K," *IEEE Transactions on Electron Devices*, vol. 40, no. 7, pp. 1296–1299, 1993.

[21] E. Simoen and C. Claeys, "Impact of CMOS processing steps on the drain current kink of NMOSFETs at liquid helium temperature," *IEEE Transactions on Electron Devices*, vol. 48, no. 6, pp. 1207–1215, 2001.

[22] E. Simoen *et al.*, "Impact of irradiations performed at liquid helium temperatures on the operation of 0.7 μm CMOS devices and read-out circuits," in *ESA Special Publication*, vol. 536, 2004, p. 369.

[23] L. Deferm, E. Simoen, and C. Claeys, "The importance of the internal bulk-source potential on the low temperature kink in NMOSTs," *IEEE Transactions on Electron Devices*, vol. 38, no. 6, pp. 1459–1466, 1991.

[24] A. Akturk *et al.*, "Impact ionization and freeze-out model for simulation of low gate bias kink effect in SOI-MOSFETs operating at liquid he temperature," in *2009 Conference on Simulation of Semiconductor Processes and Devices (SISPAD)*. IEEE, 2009, pp. 1–4.

[25] R. Langevelde, A. J. Scholten, and D. B. M. Klaassen, "Physical background of MOS Model 11."

[26] X. Li, W. Wu, G. Gildenblat, G. D. J. Smit, A. J. Scholten, D. B. M. Klaassen, and R. Langevelde, "PSP 102.3."

[27] E. Charbon *et al.*, "Cryo-CMOS circuits and systems for scalable quantum computing," in *ISSCC*. IEEE, 2017, pp. 264–265.

Cryogenic Characterization of 28 nm Bulk CMOS Technology for Quantum Computing

Arnout Beckers[†], Farzan Jazaeri[†], Andrea Ruffino[‡], Claudio Bruschini[†‡], Andrea Baschirotto[§], and Christian Enz[†]

[†]Integrated Circuits Laboratory (ICLAB), Ecole Polytechnique Fédérale de Lausanne (EPFL), Switzerland,
[‡]Advanced Quantum Architecture Lab. (AQUA), Ecole Polytechnique Fédérale de Lausanne (EPFL), Switzerland,
[§]INFN & University of Milano-Bicocca, Milano, Italy,
arnout.beckers@epfl.ch

Abstract—**This paper presents the first experimental investigation and physical discussion of the cryogenic behavior of a commercial 28 nm bulk CMOS technology. Here we extract the fundamental physical parameters of this technology at 300, 77 and 4.2 K based on DC measurement results. The extracted values are then used to demonstrate the impact of cryogenic temperatures on the essential analog design parameters. We find that the simplified charge-based EKV model can accurately predict the cryogenic behavior. This represents a main step towards the design of analog/RF circuits integrated in an advanced bulk CMOS process and operating at cryogenic temperature for quantum computing control systems.**

Keywords—**28 nm bulk CMOS, EKV, slope factor, quantum computing, cryogenic, 4.2 K**

I. INTRODUCTION

Quantum computing promises a rapid enhancement of the available computational power for selected algorithms while transistor scaling is slowing down. Not long ago, it has been proposed that quantum bits ("qubits") can be implemented in the electron spins in silicon [1] and in conformity with industry-standard CMOS technology [2]. This triggered growing interest in the co-integration of the qubits and their control system, requiring the development of dedicated analog/RF CMOS electronics for the initialization, manipulation and read-out of the qubits [3]. However, the theoretical exponential increase of the computational power is only accessible if the qubits' entangled superpositions can be preserved during computation. This is achieved in practice by operating the qubits at deep cryogenic temperature (mK-range), which reduces the thermal noise. In a co-integrated system also the control system operates at cryogenic temperature, which can further reduce the thermal noise by removing the need for direct interconnections from a room temperature (RT) control system to the qubits. In [2] the silicon qubits are implemented on silicon oxide and operate at mK-temperature. Nonetheless, the exact operating conditions of the control system are still unclear to date. For instance, a 3D-integrated system with a temperature gradient from top to bottom [3, Fig. 13], where the control system is at a higher cryogenic temperature (e.g. 77 K, 4.2 K) than the qubits, may be a better solution in terms of analog performance, heat removal or various non-thermal noise sources. In a later stage,

This project has received funding from the European Union's Horizon 2020 Research & Innovation Programme under grant agreement No. 688539 MOS-Quito. The 28 nm bulk technology was provided by ScalTech28, INFN Milano-Bicocca.

TABLE I.   MEASURED DEVICES (28 NM BULK CMOS PROCESS)

Symbol	Type	W/L
●	nMOS	3 μm / 1 μm
▲	pMOS	3 μm / 1 μm
■	nMOS	1 μm / 90 nm
▼	nMOS	3 μm / 28 nm
◆	nMOS	300 nm / 28 nm

co-integration also has the advantage of scalability to large qubit arrays. Compact transistor models extended to cryogenic temperatures will then be a must to increase the chance of first silicon right quantum computing systems. As we will bring forward as the key finding of this work, the simplified charge-based EKV model presents an interesting first step towards the development of such a cryogenic compact model for advanced technology nodes.

II. CRYOGENIC CMOS ELECTRONICS

The control system involves the following analog/RF building blocks, which are to be designed in cryogenic CMOS electronics ("cryo-CMOS"): multiplexers to control multiple qubits at once, low-noise amplifiers and oscillators. For the design of these building blocks, the advanced technology nodes (below 100 nm) are the most important because of their potential for co-integration, low bias operation and very high transit frequency F_t, reaching several hundreds of GHz even at RT. This high F_t can be traded with power consumption by shifting the bias point to weak inversion, where F_t reaches tens of GHz, which is still high enough for qubit manipulation and read-out [2]. On top of that, at cryogenic temperatures we expect to have an additional increase in F_t in weak inversion, allowing for even more current savings.

Previous research shows cryogenic measurement results for the following advanced technology nodes: a 40 nm bulk CMOS process at liquid helium temperature (4.2 K) [3] and a 28 nm FDSOI process at liquid nitrogen temperature (77 K) [4]. In this work, based on cryogenic measurement results of a 28 nm bulk CMOS technology at 77 K and 4.2 K, we quantify the impact of cryogenic temperature on the essential analog design parameters, namely i) the transconductance G_m and transconductance efficiency G_m/I_D used for low power analog design; ii) the intrinsic gain G_m/G_{ds}, important for amplifying the weak signals coming from the qubits, and iii) the transit frequency F_t. This will allow to design low power analog/RF control building blocks that work properly at cryogenic temperature and have a minimal effect on the qubits.

978-1-5090-5979-9/17 $31.00 © 2017 IEEE

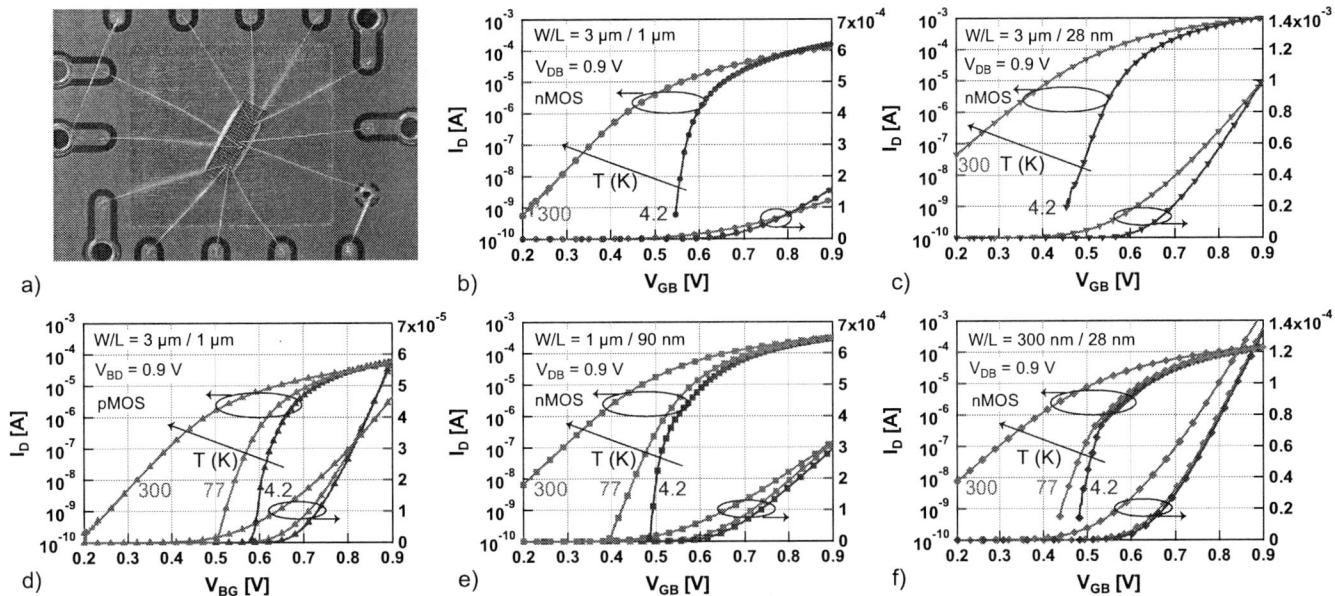

Fig. 1. Cryogenic measurements on a 28 nm bulk CMOS technology, a) Au-wire bonded sample chip glued to a standard PCB, before covering with a globtop, b)-f) Measured transfer characteristics at 300, 77 and 4.2 K in saturation (V_{DB} = 0.9 V). In all measurements the gate voltage is referred to the bulk (V_{GB}) and swept from 0.2 V to 0.9 V with a step size of 1 mV. Marker symbols refer to the device type (as shown in table I). Colors indicate the temperature, red: room temperature (300 K), green: liquid nitrogen temperature (77 K) and blue: liquid helium temperature (4.2 K).

III. CRYOGENIC MEASUREMENTS

A. Measurement Set-up

The measurements were conducted on the devices presented in Table I, fabricated in a 28 nm bulk CMOS process [5], [6]. The sample chips were first wire bonded to standard PCBs using Au-wire bonds (Fig. 1a) and then covered with a globtop. The PCBs were immersed into liquid nitrogen (77 K) and liquid helium (4.2 K) by means of a dipstick. The results were acquired with a Keysight B1500A semiconductor device parameter analyzer. Using this set-up, we measured transfer characteristics in the linear (V_{DB} = 10 mV) and the saturation regions (V_{DB} = 0.9 V), as well as output characteristics for different gate voltages.

B. Measurement Results and Discussion

Figs. 1b-f show the measured transfer characteristics in saturation for the devices in Table I. Clearly, the subthreshold swing SS, defined as $nU_T \ln 10$ [mV/dec] with $U_T \triangleq kT/q$ the thermal voltage, decreases drastically at 77 K and 4.2 K for all devices. The slope factor n expresses the deviation of the SS from the asymptote corresponding to an ideal device with $n = 1$. As illustrated in Fig. 2a, the SS decreases to 11 mV/dec at 4.2 K (-85%) for long channel nMOS ($L = 1\,\mu$m), although this decrease is less than the expected value from the ideal case corresponding to ≈ 0.8 mV/dec, pointing out the much higher slope factor n than unity at 4.2 K. This is attributed to the incomplete dopant ionization at cryogenic temperatures. In short channel nMOS ($L = 28$ nm), the SS reaches 28 mV/dec (-68%). The on-state current decreases at 77 K and 4.2 K for the short devices, while it increases for the long devices (Fig. 2b). This can be explained from the expression of the (electron) current density $\boldsymbol{J}_n = qn\mu_n\boldsymbol{E} + qD_n\nabla n$ with $D_n = \mu_n U_T$. At low bias, J_n is proportional to the temperature

and field-assisted ionization is weak, whereas at high bias the current density increases with decreasing temperature through the mobility and field-assisted ionization is strong in the channel [8]. Additionally, the threshold voltage V_{th} shifts to higher voltages at 77 K and 4.2 K due to incomplete ionization. Indeed, a higher voltage is needed to attract sufficient charges to the surface to create the inversion layer. The shift in threshold voltage ΔV_{th} with respect to RT in Fig. 2c was extracted from the transconductance in saturation and shows a significant increase in the order of 0.1 V. It should be noted that the maximum threshold voltage variation is observed in long channel pMOS devices ($\Delta V_{th} = 0.2$ V). Moreover, relying on the Y-function method [7] the low-field mobility (μ_0) was extracted for different gate lengths. Fig. 2d shows a significant improvement ($\times 3$ for nMOS, $L = 1\,\mu$m) at 4.2 K due to the phonon scattering reduction, although the Coulomb scattering is becoming dominant at low temperature. The transconductance in saturation $G_{m,sat}$ improves at 4.2 K ($\times 1.3$ for nMOS, $W/L = 3\,\mu$m/28 nm). Fig. 2f shows that at 77 K the slope factor n remains close to the value at RT, e.g. 1.6 compared to 1.47 for nMOS $W/L = 300$ nm/28 nm. This slight increase at 77 K was also observed in [9]. Here we report for the first time on the strong increase of n at 4.2 K (Fig. 2f), which was evident from the extracted values of the SS in Fig 2a. The extraction procedure of the slope factor is demonstrated in Fig. 4a at 300 K and 4.2 K. For $n = 33$ (nMOS, $W/L = 3\,\mu$m/28 nm), the modified theoretical trend $nU_T \ln 10$ indicates a value of 27.5 mV/dec, in accordance with the measured value in Fig. 2a.

IV. CRYOGENIC ANALOG DESIGN PARAMETERS

To extract the analog design parameters, we use the simplified and normalized EKV model, described in detail in [10]. This model captures all the changes of the temperature reduction to cryogenic temperatures in only four parameters.

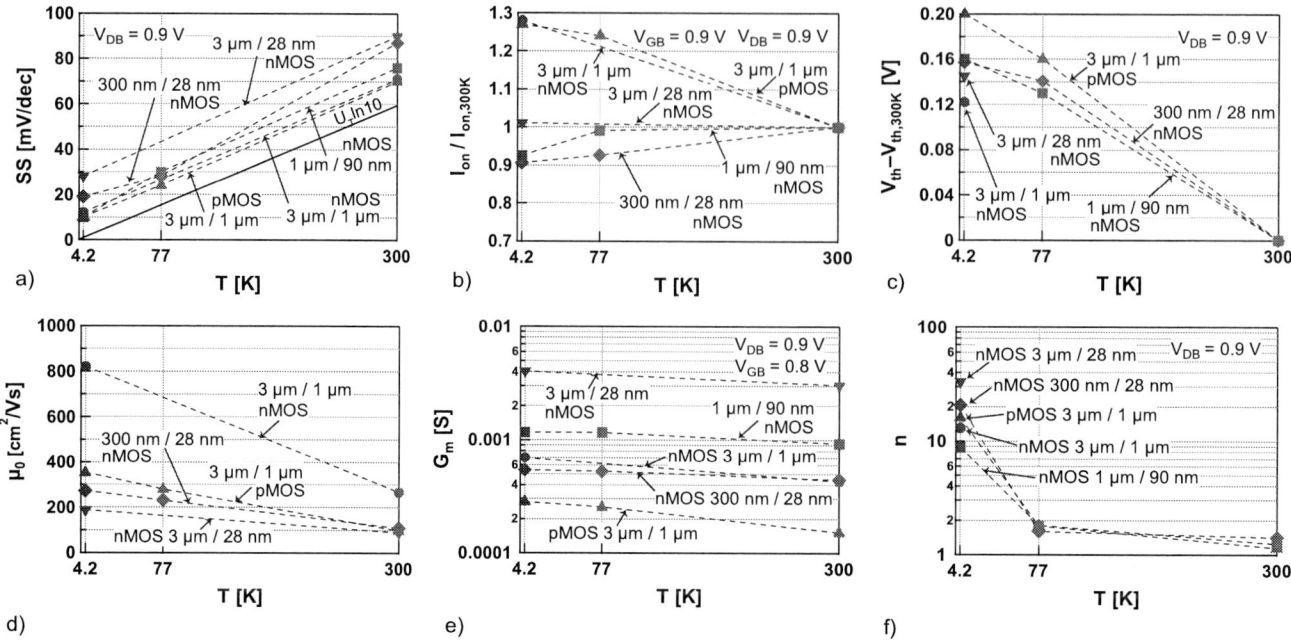

Fig. 2. Extraction of the fundamental physical and technology parameters at 300, 77 and 4.2 K, a) Subthreshold swing versus temperature. The results at 4.2 K show a strong deviation from the theoretical trend ($U_T \ln 10$), which is expressed by an increase in the slope factor n, b) On-state current normalized to RT, c) Shift in threshold voltage at 77 K and 4.2 K with respect to RT, extracted from the transconductance in saturation ($V_{DB} = 0.9$ V) at $V_{GB} = 0.9$ V, d) Low field mobility versus temperature, extracted using the Y-function method [7], e) Transconductance in saturation versus temperature, f) Slope factor versus temperature.

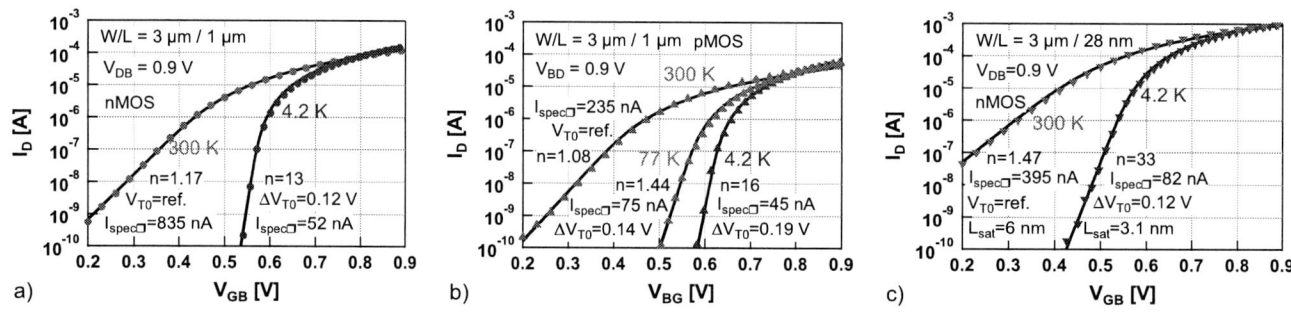

Fig. 3. Transfer characteristics: measured (markers) and modeled (solid lines) with the simplified EKV long/short channel model for a) nMOS $W/L = 3$ μm / 1 μm at RT and 4.2 K, b) pMOS $W/L = 3$ μm / 1 μm at RT, 77 K and 4.2 K and c) nMOS $W/L = 3$ μm / 28 nm at RT and 4.2 K. For each curve, the extracted model parameters are shown.

The long channel model relies on three model parameters, i.e. the slope factor n, the threshold voltage V_{T0} and the specific current ($I_{spec} = I_{spec\square}W/L = 2(W/L)n\mu C_{ox}U_T^2$). The current is normalized to I_{spec} to obtain the inversion coefficient $IC \triangleq I_{D,sat}/I_{spec}$. Initial guesses for the model parameters can be estimated from the extracted values. The short channel model also adds the saturation length L_{sat} as a fourth model parameter, indicating the part of the channel in full velocity saturation. As shown in Fig. 3, both the RT and cryogenic measurements can be accurately predicted by the proposed model for long (Figs. 3a-b) and short (Fig. 3c) devices. It is worth mentioning that the model parameters obtained here confirm the extracted results from Figs. 2c and 2f. Furthermore, I_{spec} decreases by one order of magnitude from RT to 4.2 K due to its quadratic dependency on U_T in the model, which is explained physically by incomplete ionization. In Figs. 4b-f

the impact of cryogenic temperatures on the essential analog design parameters is analyzed. The transconductance efficiency $G_m n U_T / I_D$ in Fig. 4b and the normalized transconductance at the source $g_{ms} = nG_m/G_{spec}$ with $G_{spec} = I_{spec}/U_T$ in Fig. 4c, both show a lower impact of velocity saturation at 4.2 K in strong inversion ($IC > 10$). The saturation length L_{sat} is reduced from 6 nm at RT to 3 nm at 4.2 K ($L = 28$ nm). As can be seen in the output characteristics plotted in Fig. 4d for long channel and in Fig. 4e for short channel, the output conductance G_{ds} remains practically constant with respect to temperature. This results in an increased intrinsic gain G_m/G_{ds} at 77 K ($\times 1.2$) and 4.2 K ($\times 1.3$, for nMOS, $W/L = 300$ nm / 28 nm), which is promising for cryogenic amplifier design (Fig. 4f). Assuming the capacitances do not change significantly going down in temperature [11], the transit frequency F_t follows the increase in the transconductance

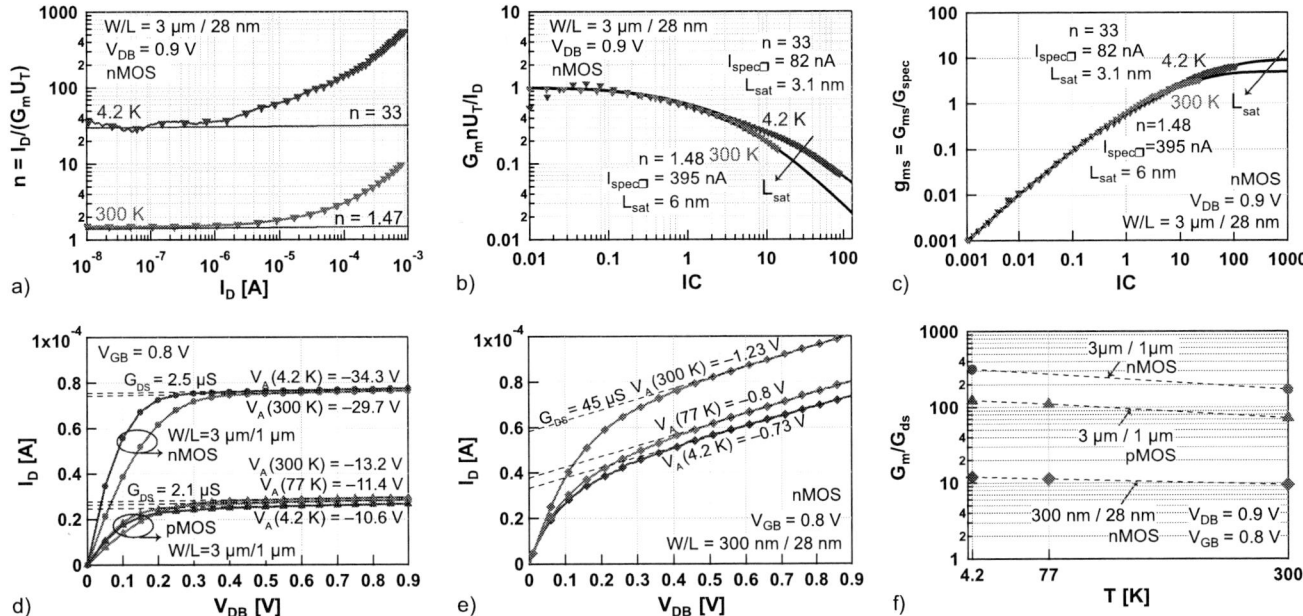

Fig. 4. Impact of cryogenic temperatures on analog design parameters, a) Slope factor versus the drain current at 300 K and 4.2 K for nMOS $W/L = 3\,\mu\text{m}/28\,\text{nm}$, b) Normalized transconductance efficiency versus the inversion coefficient for nMOS $W/L = 3\,\mu\text{m}/28\,\text{nm}$, showing a decreased velocity saturation effect at 4.2 K. Solid lines: model, c) Normalized source transconductance versus the inversion coefficient. Solid lines: model, d) Measured output characteristics for long channel nMOS and pMOS $W/L = 3\,\mu\text{m}/1\,\mu\text{m}$ with extracted values for the output conductance and the Early voltage V_A, e) Measured output characteristics for short channel nMOS $W/L = 300\,\text{nm}/28\,\text{nm}$ with extracted values for the output conductance and V_A, f) Intrinsic gain at 300, 77 and 4.2 K.

shown in Fig. 2e. This increase can be traded-off for a lower power consumption, which is also beneficial in terms of heat dissipation from the control system to the qubits. The current saved at cryogenic temperature to reach the same F_t as at RT, can be evaluated in weak inversion as

$$\frac{I_D}{I_{D0}} = \frac{n}{n_0} \cdot \frac{T_{cryo}[K]}{300},$$

where n_0 and I_{D0} are the RT-values. Ideally, if n would stay the same at RT as at 4.2 K, the current reduction factor would be 71. Unfortunately, due to the strong increase of n at 4.2 K mentioned above, the current reduction is only 5.2 for $W/L = 300\,\text{nm}/28\,\text{nm}$ ($n = 20$), and even 3.2 for $W/L = 3\,\mu\text{m}/28\,\text{nm}$ ($n = 33$) due to lower electrostatic control of wider channels. Since the change of n at 77 K is only minor (typically 1.6), the current reduction at 77 K is a factor of 3.6. In other terms, the n-factor mitigates the expected current savings moving from 77 K to 4.2 K for reaching the same F_t.

V. CONCLUSION

This work presents the influence of cryogenic temperature on a 28 nm bulk CMOS technology for quantum computing control systems. Starting from a detailed analysis of the physical device properties at cryogenic temperatures, encouraging trends in the essential analog design parameters are obtained, although the strong increase in the slope factor at 4.2 K mitigates the expected current savings. The proposed analysis demonstrates, by means of the simplified EKV model, that the cryogenic behavior in advanced CMOS can be accurately predicted. This represents an interesting solution for further implementation in cryogenic compact models for silicon-based quantum computing systems.

ACKNOWLEDGEMENT

The authors would like to thank G. Boero and A. Matheoud for sharing their expertise in cryogenic measurements, G. Corradini for the wire-bonding and P. Van der Wal for providing the liquid nitrogen (all EPFL).

REFERENCES

[1] J. J. Pla, K. Y. Tan et al., "A single-atom electron spin qubit in silicon," Nature, vol. 489, no. 7417, pp. 541–545, 2012.

[2] R. Maurand, X. Jehl et al., "A CMOS silicon spin qubit," Nature Communications, vol. 7, 2016.

[3] E. Charbon, F. Sebastiano et al., "Cryo-CMOS for quantum computing," in Electron Devices Meeting (IEDM), 2016 IEEE, pp. 343–346.

[4] M. Shin, M. Shi et al., "Low-T characterization of 14 nm FDSOI CMOS devices," in Low Temperature Electronics (WOLTE), 2014, pp. 29–32.

[5] C. Zhang, F. Jazaeri et al., "GigaRad Total Ionizing Dose and Post-Irradiation Effects on 28 nm Bulk MOSFETs," in IEEE NSS 2016.

[6] A. Pezzotta, C.-M. Zhang et al., "Impact of GigaRad Ionizing Dose on 28 nm bulk MOSFETs for future HL-LHC," in ESSDERC, 2016.

[7] G. Ghibaudo, "New method for the extraction of MOSFET parameters," Electronics Letters, vol. 24, no. 9, pp. 543–545, 1988.

[8] E. A. Gutierrez-D, J. Deen, and C. Claeys, Low temperature electronics: physics, devices, circuits, and applications. Academic Press, 2000.

[9] D. M. Binkley, "Tradeoffs and optimization in analog CMOS design," in Mixed Design of Integrated Circuits and Systems, 2007. MIXDES'07. IEEE, 2007, pp. 47–60.

[10] C. Enz, M.-A. Chalkiadaki, and A. Mangla, "Low-power analog/RF circuit design based on the inversion coefficient," in European Solid-State Circuits Conference (ESSCIRC), 2015, pp. 202–208.

[11] A. Akturk, N. Goldsman et al., "Effects of cryogenic temperatures on small-signal MOSFET capacitances," in Semiconductor Device Research Symposium. IEEE, 2007, pp. 1–2.

A New Method for Junctionless Transistors Parameters Extraction

Renan Trevisoli, Rodrigo T. Doria, Michelly de Souza
and Marcelo A. Pavanello
Electrical Engineering Department, Centro Universitário FEI
Sao Bernado do Campo, Brazil
renantd@fei.edu.br

Sylvain Barraud
CEA, LETI, Minatec Campus and
University Grenoble Alpes, 38054
Grenoble, France

Abstract—**This work proposes a new method for the extraction of the flatband voltage, effective nanowire width and doping concentration of junctionless nanowire transistors. The accurate extraction of such parameters is essential for the understating of the device behavior and for the prediction of its performance in circuits through analytical models. The method is validated using 3D numerical simulations and has been applied to experimental short-channel devices proving its applicability.**

Keywords—Nanowires, Extraction Method, Effective Width

I. INTRODUCTION

The downscaling of MOSFETs dimensions has led to the loss of the gate control on the channel charges, due to the so-called short-channel effects (SCEs), degrading the devices performance. To overcome this limitation, multiple gate technology has been developed [1] and adopted by the industry for advanced technological nodes [2] due to its better electrostatic control. Junctionless nanowire transistors (JNTs) have been proposed aiming to simplify the fabrication process owing to its constant doping profile from source to drain [3-4], such that ultra-sharp junctions are not necessary. In the literature, fabrication and operation of junctionless nanowires with channel lengths (L) down to 3 nm have been reported, demonstrating the viability of the application of such devices for extremely scaled technological nodes [5].

Analytical models for JNTs to predict their behavior and allow the design of circuits have been the focus of several works [6-11]. It is worth noting that the behavior of JNTs is strongly dependent on the devices physical characteristics [4]. Therefore, it is important to be able to extract their main physical parameters accurately for the calibration of the analytical models. A method for the determination of the flatband voltage (V_{FB}) and doping concentration (N_D for n-type devices and N_A for the p-type ones) of JNTs has been proposed in [12]. However, this method is based on the capacitance curves, which requires very long and/or wide JNTs in order to result in measurable values. In [13], the linear dependence of the drain current (I_D) on the width is used to determine the effective width (W_{eff}) of the transistors. However, this relationship can be affected by the JNTs series resistance, which might influence significantly the device performance.

Therefore, the aim of this work is to propose a new method for effective width, V_{FB} and $N_{D,A}$ extractions in triple-gate

JNTs, using only current-voltage characteristics, such that the method can be applied to short-channel transistors.

II. METHOD FORMULATION

The proposed extraction method explores the threshold voltage (V_{TH}) dependence on the devices dimensions and substrate bias (V_{BS}) as described in [14]. For an n-type transistor, the threshold voltage can be described by

$$V_{TH} = V_{FB} - N_D \frac{qWH}{C_{ox}} - N_D \frac{q}{\varepsilon_{Si}} \left(\frac{WH}{2H+W} \right)^2 = V_{FB} - N_D \alpha - N_D \beta \quad (1)$$

where W and H are the device width and height, respectively, C_{ox} is the gate capacitance per unit of length. The terms $\alpha = qWH/C_{ox}$ and $\beta = (q/\varepsilon_{Si})(WH/(2H+W))^2$ are related to the depleted charges and surface potential, respectively.

According to [14], when a positive (negative) substrate bias is applied to an nMOS (pMOS) transistor, there is an increase in the amount of charges in the channel region proportional to the applied bias. This increase can be accounted by the sum of a substrate-induced charge term (ΔQ_{sub}) to the doping concentration in (1) following [14], resulting in

$$V_{TH,Vbs} = V_{FB} - (N_D + \Delta Q_{sub})(\alpha + \beta) = V_{FB} - \left(N_D + \frac{fC_{Box}(V_{BS} - V_{FBs})}{qH} \right)(\alpha + \beta) \quad (2)$$

where V_{FBs} is the flatband voltage of the silicon/buried oxide interface, $f = (\alpha + 2\beta)/(\alpha + \beta) + (\gamma^2/N_D)\beta/(\alpha + \beta)$ and C_{Box} is the buried oxide capacitance.

In Fig. 1, the simulated I_D of nMOS JNTs is presented as a function of the gate voltage for $L = 100$ nm, demonstrating the V_{BS} influence. The symbols indicate the extracted V_{TH} using the method of [15]. The simulations have been performed using Synopsys tools [16], considering drift-diffusion transport mechanism together with models for the mobility dependence on transversal electric field, bandgap narrowing and doping-dependent generation/recombination lifetimes. The devices physical characteristics are shown in the figure.

The V_{TH} variation with V_{BS} increases when W is incremented [14]. The threshold voltage variation due to V_{BS}

This work was supported by São Paulo Research Foundation (FAPESP) grant #2014/18041-8, CAPES and CNPq.

978-1-5090-5979-9/17 $31.00 © 2017 IEEE

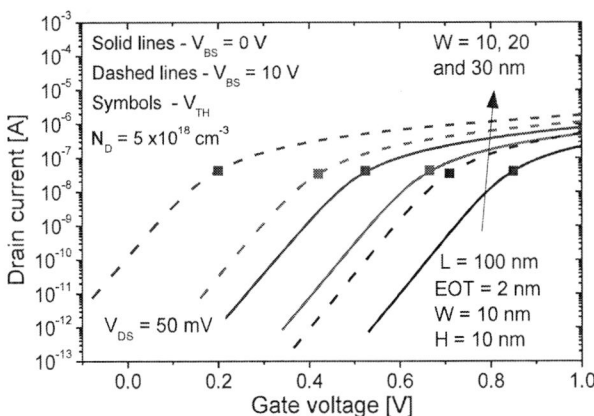

Fig. 1. Drain current as a function of the gate voltage for simulated devices considering different substrate biases and nanowire widths.

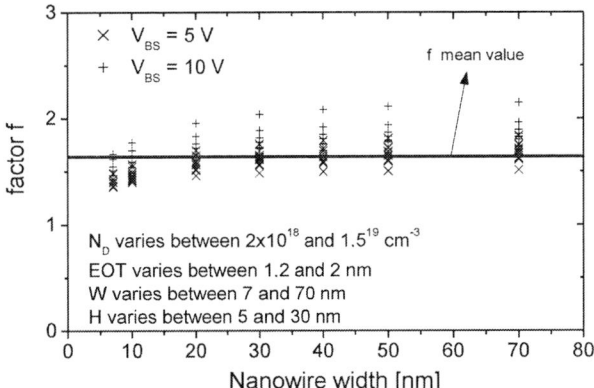

Fig. 2. Factor f versus the nanowire width for several devices of different doping concentrations, heights and effective oxide thicknesses.

influence ($\Delta V_{TH} = V_{TH} - V_{TH,Vbs}$) is given by $\Delta V_{TH} = \Delta Q_{sub}(\alpha + \beta)$. Therefore, V_{TH} can be written as a function of ΔQ_{sub}

$$V_{TH} = V_{FB} - N_D \, \Delta V_{TH} / \Delta Q_{sub} \qquad (3)$$

It is worth noting that f depends on the dimensions of the device. Its first term varies between 1 and 2 whereas its second term depends on the doping concentration. However, when the substrate induced charges are significantly lower than the doping concentration ($\gamma << N_D$), this latter term becomes negligible. In order to provide a deeper analysis on the term f, it has been calculated for several transistors varying the height between 5 and 30 nm, the width between 7 and 70 nm, the effective oxide thickness between 1.2 and 2 nm, the doping concentration between 2×10^{18} and 1.5×10^{19} cm^{-3} and the substrate bias between 5 and 10 V. All the values of f are presented in Fig. 2 as a function of the device width. For N_D extraction, a cascade of JNTs with W is necessary, where each of them presents a slight different value of f. The mean value of f for all analyzed devices is 1.64 with a standard deviation of 0.16. As the mean value has been determined for a wide range of transistors, f has been roughly assumed as 1.64. Usually, narrow JNTs show f lower than its mean value whereas wider ones present higher f. Therefore, the substrate induced charge can be approximated by $\Delta Q_{sub} = 1.64 \, C_{Box} (V_{BS} - V_{FBs})/(qH)$.

Generally, devices fabricated on a same wafer present similar height and buried oxide thickness, which means that the variation of the substrate bias would induce a similar ΔQ_{sub} independently of the devices dimensions. By measuring a device at two different V_{BS}, one can calculate the threshold voltage variation. If the dimensions of the JNT is changed, ΔV_{TH} also changes, since it depends on α and β. When measuring a cascade of JNTs with different widths, each device results in a different ΔV_{TH}, whereas $N_D / \Delta Q_{sub}$ remains the same. It is worth noting that when W is incremented, ΔV_{TH} increases which reduces V_{TH} according to (3). When the width tends to zero, ΔV_{TH} becomes negligible such that $V_{TH} = V_{FB}$. Therefore, by plotting the threshold voltage as a function of ΔV_{TH} considering devices with different widths, V_{FB} is obtained through the linear extrapolation at the y-axis. According to (3),

the slope of the V_{TH} versus ΔV_{TH} curve is given by $-N_D/\Delta Q_{sub}$, such that the doping concentration can be obtained by

$$N_D = 1.64 \, . S_{Vth} . (V_{BS} - V_{FBs}) C_{Box} / (qH), \qquad (4)$$

where S_{Vth} is the slope of the V_{TH} versus ΔV_{TH} curve.

To obtain the channel width reduction (ΔW), i.e. the difference between the mask width and the effective one, V_{TH} can be plotted as a function of the mask width. As expressed by (1) and experimentally demonstrated in [17], one can observe that V_{TH} presents a parabolic dependence on the width. When the width tends to zero, all the square bracket term in (1) also tends to zero such that $V_{TH} = V_{FB}$. Therefore, by fitting the V_{TH} versus W curve as a parabola, ΔW is obtained when $V_{TH} = V_{FB}$.

III. METHOD VALIDATION

In Fig. 3, simulated V_{TH} for n-type transistors is presented as a function of ΔV_{TH} (V_{BS} varied between 0 and 5 V) for different N_D and channel lengths. The V_{TH} has been extracted as described in [15]. It can be seen that the threshold voltage varies linearly with ΔV_{TH} for all simulated N_D. The channel length reduction does not affect the slope of the curves, such that the method can be applied to short-channel JNTs. Following (3), the interception with y-axis represents the flatband voltage as indicated in the figure. One can also observe that the absolute value of the slope increases with N_D.

Through the slopes obtained of Fig. 2, N_D were calculated using (4), resulting in the values shown in Table I. The error has been lower than 3% for any doping concentration. The flatband voltages extracted from Fig. 2 are also presented in Table I, indicating an error lower than 50 mV for any N_D in the analyzed range. It is worth noting that the error in V_{FB} increased with the doping concentration, which might be related to the larger gate voltage interval between the threshold voltage and V_{FB} for the heavier doped devices.

In Fig. 4, the V_{TH} is presented as a function of W. The curves have been fitted by a parabola as mentioned in Section II. When the threshold voltage tends to the flatband voltage, the

978-1-5090-5979-9/17 $31.00 © 2017 IEEE

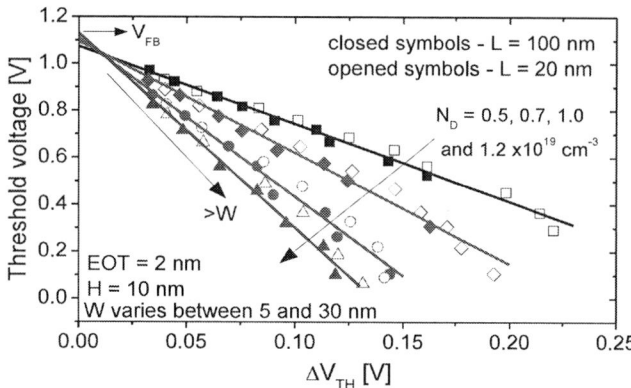

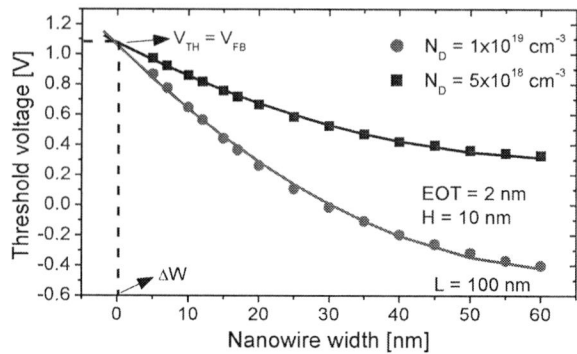

Fig. 3. Threshold voltage *versus* ΔV_{TH} (V_{BS} varied between 0 and 5 V) for devices of different doping concentrations and channel lengths.

Fig. 4. Threshold voltage *versus* nanowire width for devices of different N_D, demonstrating the extraction of the channel width reduction.

effective width of the device should be zero. Therefore, the nanowire width (when $V_{TH} = V_{FB}$ in the fitted curve) represents the channel width reduction during the fabrication process. In the simulated devices, ΔW has been close to zero, as expected.

TABLE I. COMPARISON BETWEEN THE SIMULATED AND EXTRACTED VALUES OF V_{FB} AND N_D FROM FIG. 2.

Doping concentration [cm^{-3}]		Error [%]	Flatband Voltage [V]	
Simulated	Extracted		Simulated	Extracted
5.0×10^{18}	4.92×10^{18}	1.68	1.07	1.07
7.0×10^{18}	7.04×10^{18}	0.57	1.07	1.09
10.0×10^{18}	10.04×10^{18}	0.40	1.08	1.11
12.0×10^{18}	12.31×10^{18}	2.60	1.08	1.13

Both Figs. 3 and 4 consider triple-gate JNTs. To analyze the impact of the gate structure on the method, simulations have been performed considering Pi-gate devices, i.e. when the gate material not only surrounds the silicon, but also deepens into the buried oxide. In Fig. 5, the V_{TH} vs. ΔV_{TH} and ΔV_{TH} vs. W curves comparing different gate structures are shown. The substrate influence is reduced for the Pi-gate device, which affects both V_{FB} and N_D extractions. To overcome the influence of the Pi-gate structure on the parameters extraction, the ΔV_{TH} vs. W curve can be shifted as indicated by the opened symbols in Fig. 5, which results in the similar values for V_{FB} and N_D. The amount of shifting is determined by the maximum of the derivative of the ΔV_{TH} vs. W curve. Therefore, the method can be applied to either triple gate or Pi-gate devices similarly.

IV. EXPERIMENTAL DEVICES

The proposed method has also been applied to Pi-gate devices fabricated in CEA-LETI according to [18] on Silicon-On-Insulator wafers with buried oxide thickness of 145 nm. Both n- and p-type devices have been measured considering different channel lengths (20 and 100 nm) and mask widths (varying between 30 and 240 nm). The gate stack is composed by HfSiON with an effective oxide thickness of 1.3 nm. The nanowire height is 10 nm whereas the targeted doping concentration is in the order of 1-2×10^{19} cm^{-3}.

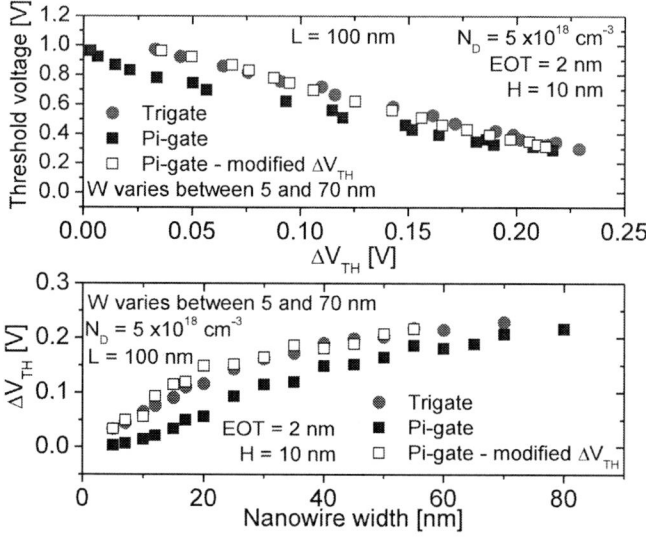

Fig. 5. Threshold voltage as a function of ΔV_{TH} (top) and ΔV_{TH} *versus* W (bottom) for simulated devices considering different gate structures.

In Fig. 6, the devices V_{TH} is presented as a function of ΔV_{TH} for both n- and p-type transistors (V_{BS} varied between 0 and 10V for nMOS and between 0 and −10V for pMOS JNTs). Following the method description in Section III, the curves have been fitted by a line, which results in the V_{FB} at the intersection with the y-axis and a doping dependent slope. The extracted flatband voltage for the n-type JNTs is 0.52V whereas $V_{FB} = -0.49$V for the p-type ones. Since the gate material used in the measured devices is a midgap one, the flatband voltage is expected to be close to the Fermi potential, which is in the order of |0.5| V for heavily doped devices. From the slopes of Fig. 6, the doping concentration was calculated using (4), leading to $N_D = 7.7 \times 10^{18}$ cm^{-3} and $N_A = 1.4 \times 10^{19}$ cm^{-3} for n- and p-type JNTs, respectively.

Fig. 7 shows V_{TH} as a function of the width for both n- and p-type devices. By using the extracted flatband voltage, the channel width reduction has been estimated as indicated in the figure. The extracted values of V_{FB}, $N_{D,A}$ and ΔW have been used together in the drain current analytical model of triple-gate JNTs proposed in [8] aiming to reproduce the I_D vs. V_{GS} JNTs curves. The results are shown in Fig. 8 and indicate that

978-1-5090-5979-9/17 $31.00 © 2017 IEEE

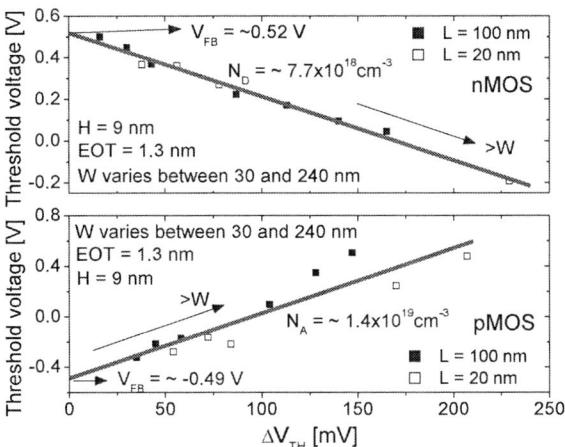

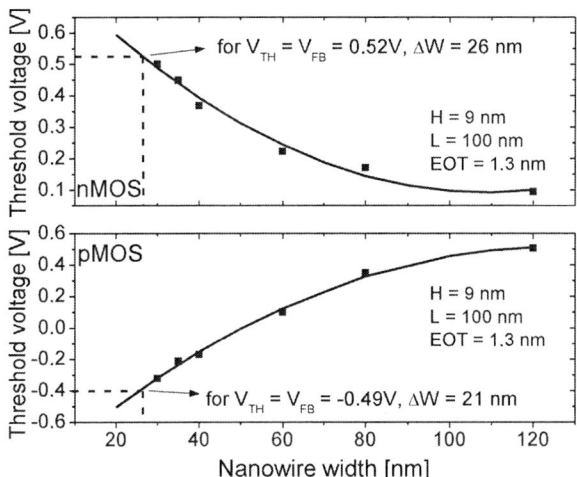

Fig. 6. V_{TH} versus ΔV_{TH} (V_{BS} varied between 0 and 10V for n- and between 0 and −10V for p-type JNTs) for experimental transistors of different L.

the extracted values can be used to calibrate analytical models based on experimental data.

V. CONCLUSIONS

This work proposes a new method for the extraction of the doping concentration, flatband voltage and effective width of junctionless nanowire transistors. Three dimensional numerical simulations were used to validate the method. Its applicability has been demonstrated to both n- and p-type experimental short-channel devices. The extracted values were used together with an analytical model from the literature to predict the JNTs drain current characteristics, demonstrating the applicability of the proposed method in calibrating analytical models.

ACKNOWLEDGMENT

The authors thank M. Vinet and O. Faynot for supplying the devices fabricated by CEA-Leti.

REFERENCES

[1] J.P. Colinge, FinFETs and other multi-gate transistors, 1st ed.. Springer, 2007, p. 340.

[2] http://newsroom.intel.com/docs/DOC-5677.

[3] J.P. Colinge, C.W. Lee, A. Afzalian, N. Dehdashti et al., "SOI gated resistor: CMOS without junctions," In: IEEE Int. SOI Conference, 2009.

[4] J.P. Colinge, A. Kranti, R. Yan, C.W. Lee, I. Ferain et al., "Junctionless Nanowire Transistor (JNT): Properties and design guidelines", Solid-State Electronics, vol. 65-66, pp. 33-37, Nov.-Dec. 2011.

[5] S. Migita, Y. Morita, M. Masahara, and H. Ota, "Electrical Performances of Junctionless-FETs at the Scaling Limit (L_{CH} = 3 nm)", In: IEEE International Electron Devices Meeting (IEDM), 2012.

[6] J.M. Sallese, N. Chevillon, C. Lallement, B. Iñiguez, and F. Prégaldiny, "Charge-based modeling of junctionless double-gate field-effect transistors," IEEE Trans. Elec. Dev., vol. 58, p. 2628-2637, Aug. 2011.

[7] E. Gnani, A. Gnudi, S. Reggiani et al., "Theory of junctionless nanowire FET," IEEE Trans. Electron Devices, vol. 58, pp. 2903-2910, 2011.

[8] R. Trevisoli, R.T. Doria, M. de Souza et al., "Surface-Potential-Based Drain Current Analytical Model for Triple-Gate Junctionless Nanowire Transistors", IEEE Trans. Electron Dev., vol. 59, pp. 3510-3518, 2012.

[9] J.P. Duarte et al., "A compact model of quantum electron density at the subthreshold region for double-gate junctionless transistors," IEEE Trans. Electron Devices, vol. 59, pp. 1008–1012, 2012.

Fig. 7. Threshold voltage versus nanowire width for experimental n- and p-type transistors, used for the extraction of the channel width reduction.

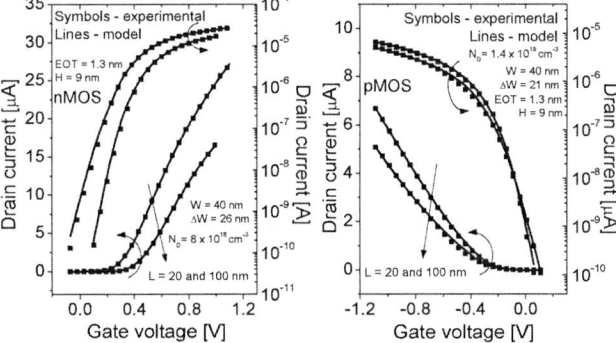

Fig. 8. Comparison between measured data and the model of [8] using the extracted values of flatband voltage, doping concentration and channe width reduction for both n- and p-type transistors.

[10] T. Holtij, M. Graef, F.M. Hain, A. Kloes, B. Iniguez, "Compact Model for Short-Channel Junctionless Accumulation Mode Double Gate MOSFETs," IEEE Trans. Electron Dev., vol. 61, pp. 288 – 299, 2014.

[11] T.-K. Chiang, "A quasi-two-dimensional threshold voltage model for short-channel junctionless double-gate MOSFETs," IEEE Trans. Electron Devices, vol. 59, pp. 2284–2289, 2012.

[12] T. Rudenko, R. Yu, S. Barraud, K. Cherkaoui, and A. Nazarov, "Method for Extracting Doping Concentration and Flat-Band Voltage in Junctionless Multigate MOSFETs Using 2-D Electrostatic Effects", IEEE Electron Device Letters, vol. 34, pp. 957-959, 2013.

[13] D.-Y. Jeon, S.J. Park, M. Mouis, M. Berthomé et al., "Revisited parameter extraction methodology for electrical characterization of junctionless transistors", Solid-State Elec., vol. 90, pp. 86-93, 2013.

[14] R. Trevisoli, R.T. Doria, M. de Souza, and M.A. Pavanello, "Substrate Bias Influence on the Operation of Junctionless Nanowire Transistors", IEEE Trans. Electron Devices, vol. 61, pp. 1575-1582, 2014.

[15] R. Trevisoli, R.T. Doria, M. de Souza, and M. Pavanello, "A physically-based threshold voltage definition, extraction and analytical model for junctionless nanowire transistors", Solid State Elec., vol. 90, pp. 12, 2013.

[16] Sentaurus Device User Guide, Synopsys, USA, 2016.

[17] R.D. Trevisoli, R.T. Doria, M. de Souza, and M.A. Pavanello, "Threshold voltage in junctionless nanowire transistors," Semiconductor Science and Technology, vol. 26, pp. 105 009, 2011.

[18] S. Barraud, M. Berthomé, R. Coquand, M. Cassé, T. Ernst et al., "Scaling of trigate junctionless nanowire MOSFET with gate length down to 13 nm", IEEE Elec. Dev. Lett., vol. 33, pp. 1225-1227, 2012.

978-1-5090-5979-9/17 $31.00 © 2017 IEEE

Avalanche Compact Model Featuring SiGe HBTs Characteristics up to BV$_{CBO}$

Mathieu Jaoul[(1)(2)], Didier Céli[(1)], Cristell Maneux[(2)], Michael Schröter[(3)(4)], Andreas Pawlak[(3)]

[(1)] STMicroelectronics, 850 rue Jean Monnet, 38926 Crolles, France.
[(2)] IMS, Université Bordeaux I, 33405 Talence, France.
[(3)] CEDIC, Dresden University of Technology, D-01062 Dresden, Germany.
(4) University of California, San Diego, La Jolla, CA, USA.

Abstract—**The cut-off frequencies of silicon-germanium hetero-junction bipolar transistors (SiGe HBTs) have entered the THz range at the cost of high current density and relatively low breakdown voltages. Typically, the common-emitter breakdown voltage with open base (BV$_{CEO}$) is used to indicate the allowed breakdown voltage related operation limit. However, an open base (i.e. an infinite source impedance) is rarely encountered in actual circuits, so that BV$_{CEO}$ may be exceeded to a certain extent, maximal up to the open-emitter breakdown voltage BV$_{CBO}$. Therefore, compact HBT models need to be accurate beyond BV$_{CEO}$ up to BV$_{CBO}$. In this paper, the enhancement of the avalanche current implemented in the latest version of HICUM/L2 is presented. The model has been validated for different types of advanced SiGe:C HBTs over a wide range of collector-base voltages and temperatures.**

Keywords— HICUM/L2, strong avalanche, parameter extraction, breakdown voltage, HBT

I. INTRODUCTION

SiGe HBTs have recently been fabricated with transit frequencies f$_T$ beyond 500 GHz and maximum oscillation frequencies f$_{max}$ beyond 700 GHz [1], which confirms the roadmap related predictions in [2]. Such high performance comes at the expense of a relatively low breakdown voltage typically characterized by the open-base common-emitter breakdown voltage BV$_{CEO}$. For maximizing the output power at high frequencies though, circuit operation beyond BV$_{CEO}$ up to the open-emitter collector-base breakdown voltage BV$_{CBO}$ is desired. Utilizing this operation region is possible under specific bias conditions [3]. Therefore, compact models have to be accurate even beyond BV$_{CEO}$ up to BV$_{CBO}$. However, present compact models cover only the weak collector avalanche effect and do not accurately describe the voltage range up to BV$_{CBO}$ [4].

In this paper we present an extended formulation of the avalanche current that enhances the model accuracy up to BV$_{CBO}$ as described in section II. The formulation has been implemented in the latest release of HICUM/L2 which is used as vehicle for the experimental verification. All the measurements shown hereafter feature the STMicroelectronics advanced 55nm node BICMOS technology [5]. Two different device types were investigated: a high speed (HS) NPN HBT (f$_T$=320GHz, BV$_{CEO}$=1.55V) and a high voltage (HV) NPN

HBT (f$_T$=60GHz and BV$_{CEO}$=3.5V) with an effective emitter width of 0.1μm and an effective emitter length of 4.42μm. These two devices differ mainly by their collector doping profile. The extraction of the new model parameters is presented in section III and the validation of the new formulation with measurements is displayed in section IV.

II. MODEL FOMULATION

At high electric field within the base-collector space charge region (BC SCR), electrons can acquire sufficient kinetic energy to create electron-hole pairs by impact ionization [6], [7]. Subsequently, the holes flow into the base inducing a negative base current beyond BV$_{CEO}$. Further increase of the electric field, increases the amount of ionization (as long as the SCR width w$_{BC}$ is not too short [2]) up to the point when every carrier creates a new one within the BC SCR.

Assuming that the ionization rate of electrons is the same as for holes, the multiplication factor is given by the ratio of the electron current leaving the BC SCR over the one entering into the SCR (e.g [8]).

$$M = \frac{I_T + I_{AVL}}{I_T} = \frac{1}{1 - \int_0^{W_{BC}} \alpha dx} \qquad (1)$$

with I$_T$ the transfer current, I$_{AVL}$ the avalanche current and α the ionization rate, empirically determined by Chynoweth [9] as

$$\alpha = a\, e^{-\frac{b}{|E|}} \qquad (2)$$

Where a is the avalanche coefficient and b is the critical electric field. Rearranging (1) and inserting (2) yields

$$1 - \frac{1}{M} = \int_0^{W_{BC}} a\, e^{-\frac{b}{|E|}} dx \qquad (3)$$

Assuming an abrupt junction, sufficiently low carrier densities and neglecting the contribution of the electric field in the base, the electrical field E in the above relation can be obtained from Poisson's equation as

$$E(x) = E_{jc}\left(1 - \frac{x}{w_{BC}}\right) \qquad (4)$$

with E$_{jc}$ as the maximum electric field at the BC junction. Assuming local impact ionization, significant avalanche multiplication occurs only in the vicinity of the maximum electric field. Hence, the inverse value of the electric field can be approximated by

978-1-5090-5979-9/17 $31.00 © 2017 IEEE

$$\frac{1}{|E(x)|} = \frac{1}{E_{jc}}\left(1 + \frac{x}{w_{BC}}\right) \tag{5}$$

Therefore, the ionization integral in (3) becomes

$$1 - \frac{1}{M} = \frac{a}{b}\,E_{jc}\,w_{BC}\,e^{\left(-\frac{b}{E_{jc}}\right)}\left(1 - e^{\left(-\frac{b}{E_{jc}}\right)}\right) \tag{6}$$

Since b is larger than E_{jc}, $e^{\left(-\frac{b}{E_{jc}}\right)} \ll 1$, (6) can be rewritten

$$M - 1 = \frac{\frac{a}{b}\,E_{jc}\,w_{BC}\,e^{-\frac{b}{E_{jc}}}}{1 - \frac{a}{b}\,E_{jc}\,w_{BC}\,e^{-\frac{b}{E_{jc}}}} \tag{7}$$

E_{jc} and w_{BC} can be determined from the BC depletion capacitance as $w_{BC} = \frac{\varepsilon_{Si}A_E}{C_{jci}}$ and $E_{jc} = \frac{2(V_{DCi}-V_{B'C'})C_{jci}}{\varepsilon_{Si}\,A_E}$. Therefore, (7) becomes

$$M - 1 = \frac{f_{AVL}(V_{DCi} - V_{B'C'})e^{-\frac{q_{AVL}}{C_{jci}(V_{DCi}-V_{B'C'})}}}{1 - f_{AVL}(V_{DCi} - V_{B'C'})e^{-\frac{q_{AVL}}{C_{jci}(V_{DCi}-V_{B'C'})}}} \tag{8}$$

with $f_{AVL} = \frac{2\,a}{b}$ and $q_{AVL} = b\,\varepsilon_{Si}\,\frac{A_E}{2}$ as compact model parameters.

The new HICUM/L2 version 2.4.0 [10] is based on (8) with a new parameter k_{AVL} introduced for fine tuning and for being able to turn off the strong avalanche model ($k_{AVL}=0$).

$$M - 1 = \frac{g}{1 - k_{AVL}g} \tag{9}$$

where $g = f_{AVL}(V_{DCi} - V_{B'C'})e^{-\frac{q_{AVL}}{C_{jci}(V_{DCi}-V_{B'C'})}}$.

Physically, k_{AVL} compensates for the simplifications made during the derivation. In order to avoid non-physical negative values of the denominator, a smoothing function is applied

$$M - 1 = \frac{2\,g}{1 - k_{AVL}\,g + \sqrt{(1 - k_{AVL}\,g)^2 + 0.01}} \tag{10}$$

Note that, setting k_{AVL} to zero yields to the expression for weak avalanche,

$$M - 1 = g \tag{11}$$

which is commonly used in many models (MEXTRAM [11], previous HICUM release [8] and VBIC [12]).
In bipolar transistor compact models, the collector impact ionization is represented by a controlled current source I_{AVL} as shown in Fig. 2.a, which is calculated from the multiplication factor defined in (10) and I_{AVL} expressed in (1).

For a better understanding of the limited accuracy of the weak avalanche model at voltages beyond BV_{CEO}, the multiplication factor was studied. When approaching BV_{CBO} at low V_{BE} without self-heating or pinch-in effects [7], the factor $1-1/M$ approaches 1 as shown in Fig. 1.b, or M approaches infinity. This corresponds to a self-sustaining impact ionization process which leads to the destruction of the transistor if the current is not limited. As shown in Fig. 1.a and b, the weak avalanche formulation remains quite accurate beyond BV_{CEO} but start to deviate from the actual data at about 70% of BV_{CBO} whereas the new model shows excellent accuracy up to BV_{CBO} and thus allows to predict the actual breakdown.

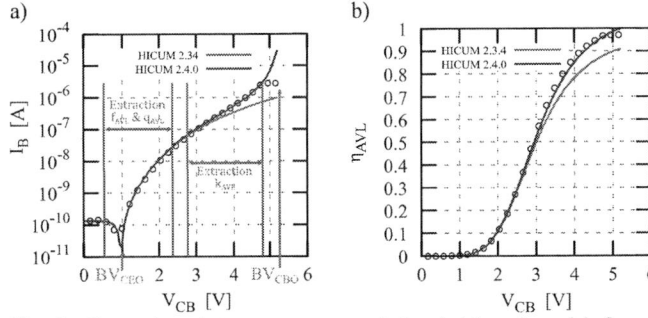

Fig. 1. Comparison between measurement (symbols) and model for HICUM/L2 version 2.3.4 and 2.4.0 at T=25°C.
a) Base current, I_B as a function of collector-base voltage, V_{CB} at a constant base-emitter voltage, V_{BE}=0.6V, showing the f_{AVL}, q_{AVL}, k_{AVL} parameter extraction ranges
b) Avalanche factor η_{AVL}=1-1/M as a function of collector-base voltage, V_{CB}.

III. MEASUREMENT SETUP AND PARAMETER EXTRACTION

To ensure accurate and scalable parameter values representative of process fabrication variations, a dedicated physics based parameter extraction procedure has been developed. For extracting the avalanche effect related model parameters f_{AVL}, q_{AVL} and k_{AVL}, the values of the multiplication factor M are required from a specific measurement setup. As expressed in (1), M can be easily calculated from the avalanche current and the transfer current.

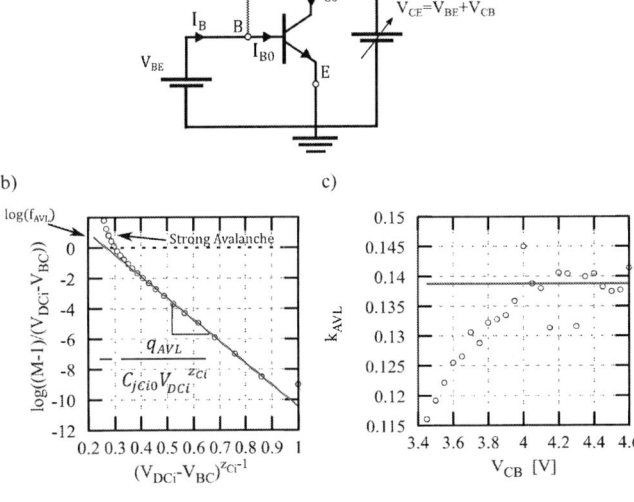

Fig. 2. a) Measurement setup allowing extracting the multiplication factor, M from the avalanche current I_{AVL} represented as a controlled current source.
b) Log-scale representation of the multiplication factor allowing a direct extraction of weak avalanche parameters (f_{AVL} and q_{AVL}). Comparison between measurement (symbols) and model (line) at T=25°C
c) Extraction of the k_{AVL} parameter at high V_{CB} from measured mean value (line).

Fig. 2.a shows the measurement setup and highlights the direct contribution of the avalanche current within the base current. To avoid high injection, self-heating and pinch-in effects, the voltage V_{BE} is fixed at a low value, typically between 0.6V and 0.7 V. Under that condition, I_{C0} and I_{B0} are defined respectively as the collector and the base current at $V_{CB}=0$ V. Therefore, as shown in Fig. 2.a, the avalanche current can be written as

$$I_{AVL} = I_B - I_{B0} \tag{12}$$

Since the DC collector current equals the transfer current I_T, the collector current can be written

$$I_C = M\,I_{C0} = I_{C0} + I_{AVL} \tag{13}$$

From (12) and (13), the multiplication factor M can be expressed as

$$M = \frac{I_C}{I_C + I_B - I_{B0}} \tag{14}$$

Knowing M, the extraction procedure can be performed. To begin with, the parameters V_{DCi}, C_{jCi0} and z_{Ci} used in (9) are already known from the prior depletion capacitance measurement [13]. Then, the first step is to extract the weak avalanche parameters f_{AVL} and q_{AVL} assuming that, under the associated bias conditions, k_{AVL} equals 0. Afterwards, k_{AVL} is extracted under strong avalanche conditions.

A. Weak avalanche regime

As stated in section II, weak avalanche occurs at voltages close to BV_{CEO}. Thus, k_{AVL} is supposed to have no impact on weak avalanche, implying that its contribution can be disabled ($k_{AVL}=0$). Then, by taking (9) in its logarithmic form leads to

$$\log\left(\frac{M-1}{V_{DCi} - V_{BC}}\right) = \log(f_{AVL}) - \frac{q_{AVL}(V_{DCi} - V_{BC})^{z_{ci}-1}}{C_{jCi0}\,V_{DCi}^{z_{ci}}} \tag{15}$$

Here, it is assumed that at low current densities, V_{BC} is close to $V_{B'C'}$ defined as the potential difference between the internal base and collector nodes. Equation (15) provides a linear dependency as a function of $(V_{DCi} - V_{BC})^{z_{ci}-1}$. So, the two parameters f_{AVL} and q_{AVL}, respectively, can be determined directly respectively from the intercept and the slope of the obtained straight line. As shown in Fig. 2.b, the measurements are well aligned (until strong avalanche occurs) which validates both model formulation and measurements.

B. Strong avalanche regime

Once q_{AVL} and f_{AVL} are known, the next step is to extract k_{AVL}, which is associated with the strong avalanche regime. Providing that the voltage range of V_{CB} for this extraction is chosen close before BV_{CBO}, the smoothing equation described in (10) reduces M-1 to

$$M - 1 = \frac{g}{1 - k_{AVL}\,g} \tag{16}$$

Therefore, k_{AVL} can be determined from

$$k_{AVL} = \frac{1}{g} - \frac{1}{M-1} \tag{17}$$

The extraction range of V_{CB} corresponding to the strong avalanche regime is shown in Fig. 1.a. In Fig. 2.c, k_{AVL} reaches 0.14 at high V_{CB}. The parameter value becomes constant between 3.8 and 4.6 V. This corresponds to the suitable and hence selected extraction range, where k_{AVL} is determined as the mean value (line).

C. Avalanche thermal coefficients

The avalanche parameters f_{AVL}, q_{AVL} and k_{AVL} exhibit a

conventional temperature dependence that can be expressed as

$$X_{AVL}(T) = X_{AVL}(T_0)\,e^{AL_{XAV}(T-T_0)} \tag{18}$$

where X_{AVL} is the generic term that represents the avalanche parameters f_{AVL}, q_{AVL}, k_{AVL}, T_0 is the reference temperature (in the present case 25°C), and AL_{XAV} the corresponding thermal coefficient.

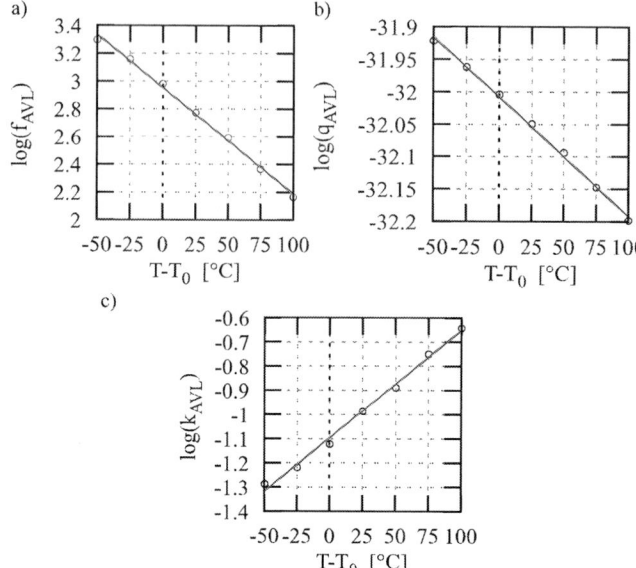

Fig. 3. Comparison between model (line) and measurement (symbols) for a) f_{AVL} temperature parameter extraction b) q_{AVL} temperature parameter extraction c) k_{AVL} temperature parameter extraction

Therefore, the previous extraction procedure is repeated for several temperatures in order to determine the associated thermal coefficients. The first step is to extract at each temperature the respective parameter as shown previously. Using (18) in logarithm form,

$$\log\big(X_{AVL}(T)\big) = \log\big(X_{AVL}(T_0)\big) + AL_{XAV}(T - T_0) \tag{19}$$

Equation (19) presents a linear dependence as a function of the temperature. Therefore, the corresponding thermal coefficients can be directly determined from the slope of this straight line. As shown in Fig. 3 the measurements show a linear behavior, which validates the model equations and the extraction procedure.

IV. EXPERIMENTAL RESULTS AND VALIDATION

This section presents a comparison between the measurements and the compact model formulations (9) and (11), i.e. taking into account or omitting the strong avalanche effect. This comparison is performed over a wide temperature range and for two collector doping profiles representative of HS and HV NPN transistors.

A. Room temperature

Fig. 4 presents the measurement results at 25°C of the base (left) and the collector (right) current as a function of the base-collector voltage in comparison with the simulation results from HICUM/L2 version 2.3.4 and 2.4.0. As expected, the new formulation of the avalanche current is in better agreement with

978-1-5090-5979-9/17 $31.00 © 2017 IEEE

the measurement than the previous one. The new model has been validated for two different device types: NPN HS HBT as in Fig. 4.a and NPN HV HBT as shown in Fig. 4.b. In both cases, the new model predicts the correct trend close to the device BV_{CBO}. Data close to BV_{CBO} were not taken due to measurement instabilities, most likely caused by pinch-in and self-heating.

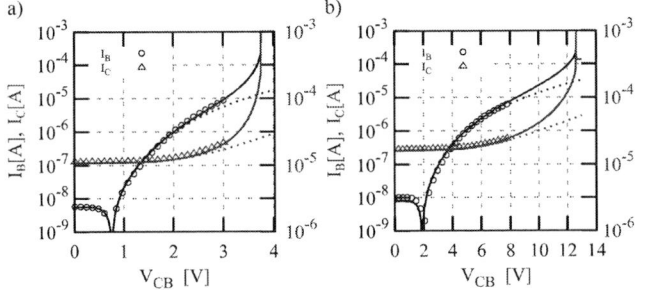

Fig. 4. I_B and I_C at 25°C and at V_{BE}=0.7V – Comparison of the new (line)/old (dashed line) avalanche model of HICUM/L2 with measurement (symbols)
a) For NPN HS HBT (BV_{CEO}=1.5V, BV_{CBO}=5.3V)
b) For NPN HV HBT (BV_{CEO}=3.5V, BV_{CBO}=13.5V)

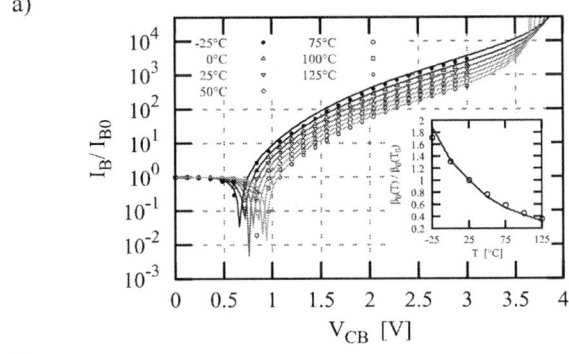

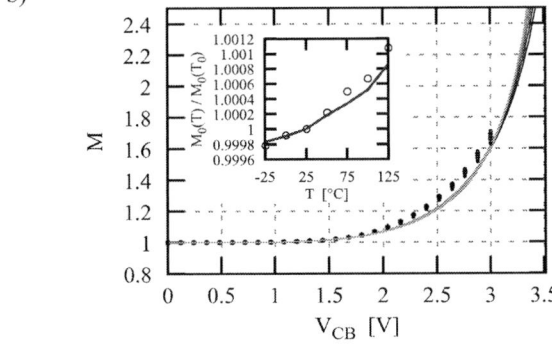

Fig. 5. Comparison between measurement (HS NPN transistors) and HICUM/L2 version 2.4.0 (lines) for a temperature range of [-25,125°C]:
a) I_B/I_{B0} vs. V_{CB} at constant V_{BE} (V_{BE}=0.7V at 25°C and at each temperature V_{BE} is adjusted to keep I_{B0} close to the value at 25°C), the inset shows the normalized current gain I_{C0}/I_{B0} at V_{BC}=0V vs. Temperature.
b) M vs. V_{CB} at constant V_{BE}, the inset shows the normalized multiplication factor M_0 at the breakdown voltage BV_{CEO} vs. Temperature.

B. Temperature range [-25,125°C]

Fig. 5.a shows the normalized current I_B/I_{B0} as a function of V_{CB} according to temperatures in the [-25,125°C] range. This

figure highlights the very good agreement between measurement and simulation results over a wide temperature and voltage range. To understand the increase of the breakdown voltage BV_{CEO} as a function of temperature, the M factor is plotted Fig. 5.b for different temperatures. A zoom at BV_{CEO} shows the small variation of M with the temperature (0.1% of variation between -25°C and 125°C). This, associated with the decrease of the current gain shown in Fig. 5.a, explains the increase of the BV_{CEO} with the temperature.

However, self-heating may cause a different trend of the BV_{CBO} thermal dependency. In order to evaluate this trend, additional pulsed measurements close to BV_{CBO} are being performed to avoid the impact of self-heating on the measured data and to be able to evaluate the model at higher V_{BE} and V_{CB}.

V. CONCLUSION

A new avalanche current model that enhances the accuracy of collector and base current approaching BV_{CBO} has been presented. The new formulation has been implemented in HICUM/L2 version 2.4.0 and has been validated for a wide range of voltages and temperatures for advanced SiGe HBTs featuring different breakdown voltages. A method for extracting the corresponding avalanche model parameters has been developed.

REFERENCES

[1] P. Chevalier et al., "Si/SiGe:C and InP/GaAsSb Heterojunction Bipolar Transistors for THz Applications," *Proc. of the IEEE, 2017.* doi: 10.1109/JPROC.2017.2669087.

[2] M. Schroter et al., "Physical and electrical performance limits of high-speed SiGeC HBTs: Part I: Vertical scaling,," *IEEE Trans. Electron Devices,* vol. 58, no. 11, p. 3687–3696, Nov. 2011..

[3] M. Rickelt, H.-M. Rein and E. Rose, "Influence of Impact-Ionization-Induced Instabilities on the Maximum Usable Output Voltage of Si-Bipolar Transistors," *IEEE Journal of Solid-State Circuits,* vol. 48, no. 4, pp. 774-783, April 2001.

[4] D. Céli and M. Jaoul, "A study of HICUM avalanche model beyond BVCEO," HICUM Workshop - Munich, May 2016. [Online]. Available: https://www.iee.et.tu-dresden.de/iee/eb/forsch/Models/workshop_2016/contr_2016.html.

[5] P. Chevalier et al., "A 55nm Triple Gate Oxide 9 Metal Layers SiGe BiCMOS Technology Featuring 320 GHz fT / 370 GHz fmax HBT and High-Q Millimeter-Wave Passives," Proceedings IEDM, December 2014.

[6] S. L. Miller, "Avalanche Breakdown in Germanium," *Physical Review,* vol. 99, no. 4, pp. 1234-1241, May 1955.

[7] G. Sasso and N. Rinaldi, "Avalanche Multiplication and Pinch-In Models for Simulation Electrical Instability Effect in SiGe HBTs," *Microelectronics Reliability,* pp. 1577-1580, 2010.

[8] M. Schröter and A. Chakravorty, "Compact hierarchical modeling of bipolar transistor with HICUM," World Scientific, Singapore, 2010.

[9] A. G. Chynoweth, "Ionization Rates for Electrons and Holes in Silicon," *Physical Review,* vol. 109, no. 5, pp. 1537-1540, 1957.

[10] M. Schroter and A. Pawlak, "HICUM version 2.4.0 Release Notes," March 2017. [Online]. Available: https://www.iee.et.tu-dresden.de/iee/eb/hic_new/hic_doc.html.

[11] W.J. Kloosterman, J.C.J. Paasschens and R.J. Havens, "A comprehensive bipolar avalanche multiplication compact model for circuit simulation," *IEEE Trans. Electron Devices,* pp. 172-175, 2002.

[12] "VBIC - Vertical Bipolar Intercompany Model," [Online]. Available: http://www.designers-guide.org/VBIC/.

[13] N. Gambetta, B. Cialdella, D. Céli and M. Depey, "A New Extraction Method For Unit Bipolar Junction Transistor Capacitance Parameters," *Proc. ICMTS,* pp. 161-165, 1995.

978-1-5090-5979-9/17 $31.00 © 2017 IEEE

Utilizing *I-V* Non-Linearity and Analog State Variations in ReRAM-Based Security Primitives

G.C. Adam[1,3*&], H. Nili[1*†], J. Kim[2], B. D. Hoskins[1], O. Kavehei[2], D.B. Strukov[1]

[1] UC Santa Barbara, Santa Barbara, CA 93106, USA
[2] RMIT University, Melbourne, Victoria 3000, Australia
[3] National Institute for R&D in Microtechnologies, Bucharest 077190, Romania
[&]gina_adam@engineering.ucsb.edu, [†]hnili@ece.ucsb.edu

Abstract— The underlying variability in the ReRAM device operation, while undesired in many applications, can be advantageous for hardware security primitives. ReRAM devices also come with the advantage of having non-linear multi-state operation. By comparison with previous reported ReRAM PUFs, which utilized spatial variations in the devices' binary ON/OFF states, we proposed to use sneak path currents and device / network nonlinearity as its main source of randomness to implement robust, reconfigurable and dense security primitives. In particular, in this work we present an in-depth discussion of how device non-linearity is affected by the read bias and the thermal stresses applied to the ReRAM crossbar. For the experimental demonstration, we used a three-dimensional stack of two 10x10 Al_2O_3/TiO_{2-x} –based ReRAM crossbars with good uniformity for the memristors in both crossbar layers. The results highlight the utility of device non-linearity to extract more complex and more reliable one-way functions from relatively small ReRAM crossbar arrays.

Keywords—ReRAM, PUF, security, metal-oxide memristor, passive crossbar circuits

I. INTRODUCTION

Hardware-based security primitives have recently attracted significant attention due to their potential to be physically embedded with their cryptographic data thanks to unique and unpredictable structural features. Physically Unclonable Functions (PUFs) are an example of such cryptographic primitives and represent the hardware implementation of a mathematical one-way function that maps an input (challenge) to an ideally unique and unpredictable output (response). PUFs put to use the otherwise disadvantageous device variabilities that naturally arise during the manufacturing process [1-2]. A PUF should ideally be unclonable against a wide range of adversarial attacks [3-4], but also have a stable and reliable operation. Unfortunately, PUF implementations using traditional CMOS technology have been shown to be susceptible to a range of modeling attacks related to their need to reprogram the volatile SRAM memory [5].

ReRAM crossbar arrays are a promising emerging technology for PUF implementation, due to their simple and relatively low-cost manufacturing, scalability, compatibility with CMOS monolithic integration and most importantly, their stochastic process-induced variations in their *I-V* characteristics. ReRAM devices are based on a mixed electronic-ionic transport, utilizing amorphous materials as the switching medium. These amorphous materials are prone to have compositional inhomogeneities, defects, grain boundaries, etc., which together with the inherent imperfections of the electrodes all can lead to variability in the device behavior [6-7].

Previously reported ReRAM-based PUFs utilize the spatial (device-to-device) variations of high and/or low resistance states (HRS/LRS) to construct a unique key [8-12]. In this approach, the ReRAM crossbar is reduced to a static resistive network, needing a large crossbar size for a high reliability. Since the average distributions of HRS and LRS are typically constant across the array, such method is not suitable for implementing multiple different PUF instances.

We proposed to use ReRAM device nonlinearity as the source of intrinsic variations to implement PUFs with increased security metrics [13] (Fig. 1a). Our approach utilizes a crossbar with non-linear devices that can be tuned to a resistive state anywhere in their dynamic range. The variations in the individual device conductances and their respective non-linearities are shown to be ubiquitous toolbox with a high degree of complexity for constructing robust PUF architectures that are difficult to model and predict [14].

This work focuses on the impact of read bias and of the thermal stress to the device non-linearity variation across the ReRAM crossbar. The results show that by using device non-linearity variability as a source of randomness, more complex and more reliable one-way functions can be obtained from relatively small ReRAM crossbar arrays. Since in this approach the ReRAM state can be tuned to any resistance in the range, it is important to choose a median array state that can satisfy both the requirement of high network nonlinearity and the one of minimum power requirements. Higher resistance states typically satisfy both these requirements.

In particular, in order to mitigate the effect of state variations on the PUF response, the devices need to be tuned randomly to a narrow pseudo-normal distribution (Fig. 1b). These device states need to be randomly distributed across the array so that the average column and row resistances are uniformly distributed (Fig. 1c). In this work, we investigated how the read bias affects the apparent non-linearity distribution across the ReRAM crossbar and how stable the state maps are with respect to applied thermal stresses.

---

* These authors contributed equally to this work.

This work was supported by the AFOSR under the MURI grant FA9550-12-1-0038, by NSF CCF-1528250 grant, by the Australian Research council under ARC DP140103448 grant and by the U.S. Department of State under the International Fulbright Science and Technology Award.

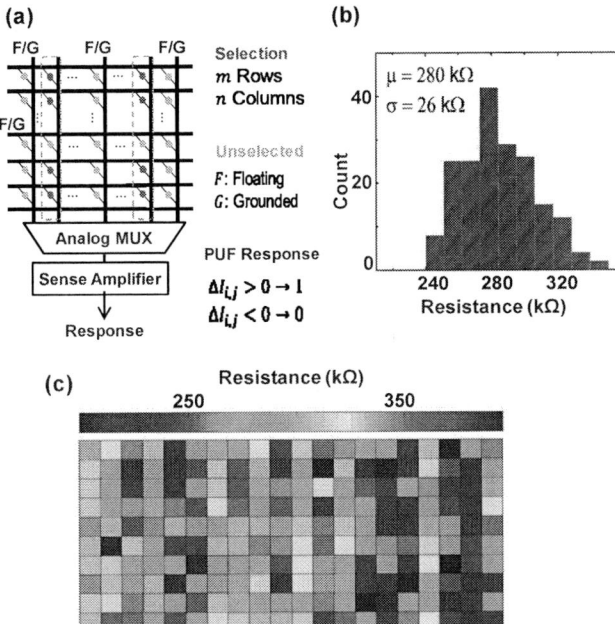

Fig. 1. Proposed ReRAM-based PUF (a) Security primitive selection scheme (b) Tuned resistance pseudo-normal distribution ($R_0 = 280 \pm 60$ KΩ measured 200 mV) (c) Spatial distribution of resistances across the 2x10x10 ReRAM crossbar.

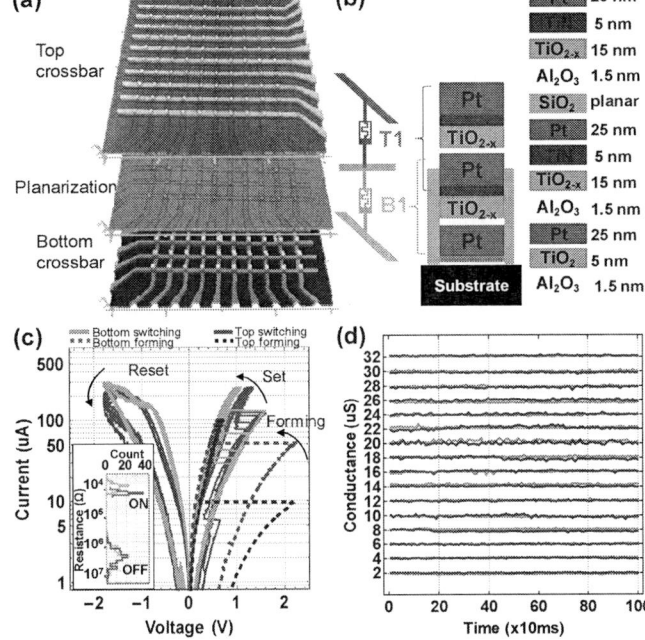

Fig. 2. Fabrication details: (a) AFM images of the two-layer crossbar, and (b) Cartoon of its cross-section showing the material layers and their corresponding thicknesses. (c) Typical I-V forming and switching characteristics for bottom and top devices (inset shows ON and OFF resistance distributions) (d) Bottom and top device analog tuning to 16 states (between 2μS and 32μS).

II. DEVICE FABRICATION AND CHARACTERIZATION

Two fully passive TiO$_{2-x}$ 10×10 memristor crossbars with an active device area of ~350-nm×350-nm were monolithically integrated in a conventional (horizontal) configuration using an in-situ low temperature reactive sputtering deposition and a precise planarization step. The middle electrodes are shared between the bottom and top layers (Fig. 2a and b). These crossbars have a similar fabrication process and electrical parameters as the one described in [15].

The detailed electrical characterization reveals very good uniformity and analog programmability of the fabricated 3D crossbar circuits. All electrical testing was performed at room temperature, unless otherwise specified, with an Agilent B1500A semiconductor device parameter analyzer and an Agilent B1530A waveform generator / fast measurement unit.

The I-V characteristics of the bottom and their corresponding top devices measured in a floating configuration in a small 2×2 portion of the crossbar show good similarity in device behavior between the bottom and top devices. An ON/OFF ratio of ~10 was observed when a conservative reset voltage of -1.8 V and a set current of 300 μA were used (Fig. 2c). The ON/OFF ratio can be further increased to ~100 when a more aggressive reset voltage of -2.4 V is used (Fig. 2c inset).

The devices in both layers show good analog tunability (Fig. 2d) to 16 clearly distinguishable states equally spaced in the 2 μS – 32 μS range. The devices were tuned to 1% precision using the tuning algorithm presented in [16] with 500 μS long voltage pulses of maximum amplitudes ±2.6 V and a step of ±0.01 V.

The variations in effective voltage switching thresholds are sufficiently low for a precise tuning of the devices within array, while still very substantial to be utilized in the considered application – see below. The device I-V is nonlinear, especially at higher resistance states (Fig. 2d).

III. DEVICE NON-LINEARITY IN PUF OPERATION

We investigated the device non-linearity for all the 200 devices in the fabricated stacked crossbar, together with its variation to the read voltage and the applied temperature. The nonlinearity factor was calculated as:

$$NLF = |1 - G_0/G(V_B)| \times 100\% \qquad (1)$$

In Fig. 1a, all the 200 devices were tuned to $R_0 = 280 \pm 60$ KΩ at 200 mV, using the write-verify algorithm. The resistances of these devices show a Gaussian distribution. Their non-linearity factors have a narrower distribution at lower read voltages (300 mV) by comparison with higher read voltages (600 mV).

In Fig. 3a, one can notice how at higher voltages, the nonlinearity changes the conductance map in its totality. Then the spatial distribution of the device nonlinearity over the crossbar is calculated according to Eq. 1. The maps show, as expected, higher nonlinearity at higher voltages as compared to lower biases. The relative nonlinearities of the devices at higher voltages are very different despite the fact the resistance state are so close. These variations in device non-linearity are due to individual device variations which are heavily

978-1-5090-5979-9/17 $31.00 © 2017 IEEE

predicated on the nanometric parameters such as local interfacial states, roughness and doping variations (vacancies, passivation centers, etc.). Therefore even knowing the nominal linear resistances and a working model of average device nonlinearity does not lead total knowledge about the dynamics. This in effect means that nonlinearity can be used both as an additional space of entropy and as a tool for tuning the security by trading off higher operational power.

The larger entropy space at higher voltages will result in more robust adaptive circuits or security primitives. De facto, the operation at higher bias means operating a different crossbar both in terms of absolute conductance value and relative distribution of the map. So the same instance and distribution can be re-purposed for different functions (e.g. unique security primitives) by just tuning the operational bias.

As it can be observed from Fig. 3b, the variability at higher biases means a wider distribution of currents (both absolute and relative measures). Assuming we have a detection limit of 50 nA, more current differentials at 600 mV are above the detection limit than the differential at 200 mV. This means less chance for errors and consequently, this leads to a higher confidence margin for different "differential" operations.

Since the PUF security hardware that we proposed needs tight resistance distributions, it is important for the crossbar

system to maintain a similar response independent of the temperature stress. In terms of the effect of the temperature on the device resistance, Fig. 4 shows that the crossbar has higher temperature swing resilience at higher biases. Firstly, the crossbar was tuned to 280 kΩ±60 kΩ as shown in Fig. 1b, after tuning having a conductance map at 200 mV as shown in Fig. 1c. The tuned crossbar was then stressed thermally to 90°C for 3 hours. After the stress heating, the conductance maps at 200 mV and at 600 mV were measured. As can be seen from Figs. 4a and 4b, there is a smaller change in rest resistance of individual states at 600 mV by comparison with the 200 mV. This smaller change at higher biases means lower susceptibility to thermally stress induced changes in operational characteristics and consequently, a more stable and reliable operation.

IV. IMPLICATIONS FOR PUF DESIGN

A prototype PUF network was implemented in the 2x10x10 ReRAM stacked crossbar structure, by taking advantage of the variations in the devices' nonlinearity and state tuning capabilities. Each device in the stacked crossbar was tuned to the specific pre-calculated values required by the PUF using the write-verify algorithm. The details of the implementation can be found elsewhere [13]. This first experimental demonstration showed good security metrics, such as uniformity, diffuseness and uniqueness.

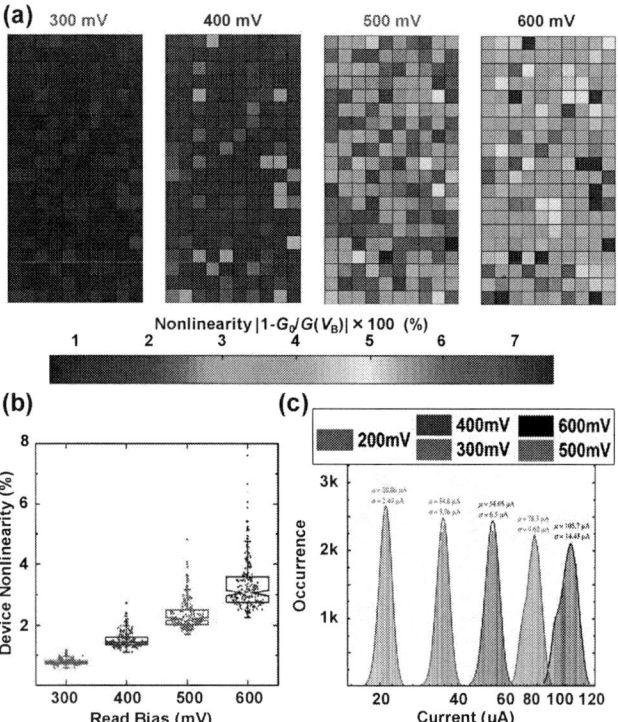

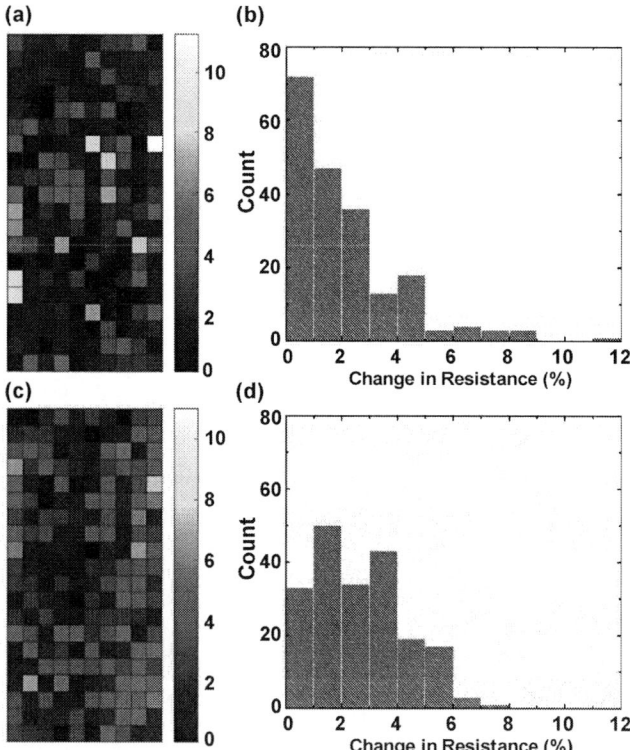

Fig. 3. Nonlinearity and Read bias dependence. (a) Spatial distribution of nonlinearity over the crossbar - calculated at the difference between conductance at V_{bias} and linear conductance at 200 mV normalized to the linear conductance. All devices are tuned to a state between 250-300 kΩ with very tight distribution.(b) Nonlinearity distribution showing a bigger relative space for higher biases; (c) Distribution of READ currents for a selection of ($m = 5$) rows and reading one column multiple times while the rest of the array is floating. The aggregate number of READ operations for each distribution (at each bias) is 128k. At higher biases, a wider current distribution is observed.

Fig. 4. Nonlinearity and resilience to temperature stresses dependence. (a-b) Absolute change in resistance map (a) and distribution (b) at 200 mV after heating the tuned stacked crossbar to 90°C for 3 hours; (c-d) Absolute change in resistance map (c) and distribution (d) at 600 mV after heating the tuned stacked crossbar to 90°C for 3 hours. The changes are measured by comparison with the initial conductance map and distribution at 200mV (Fig.1) where all devices are tuned to a state between 280±60 kΩ.

As expected, increasing by voltage bias from 200 mV to 600 mV improves the uniformity from already decent 49.5 ±6.25% to nearly ideal 50.1 ± 6.26%, while another PUF randomness metric, diffuseness is also close to ideal ~50 ± 6.25% for all voltage cases.

As discussed in the previous section, the better performance PUF at higher voltages can be attributed to the stronger nonlinearity in the device I-Vs at higher voltages. The proposed PUF shows robust functional performance passing the NIST randomness test, while having the additional advantages of reconfigurability, high integration density of ReRAM stacked crossbars and their suitability for monolithical integration onto CMOS chips.

V. DISCUSSION AND SUMMARY

The demonstrated utility of conductance state nonlinearity in enriching the randomness (entropy) space of ReRAM-based PUF architectures and improving the overall network resilience to environmental stress (e.g. temperature variations) highlights the promise of such higher order variations in design and implementation of hardware-intrinsic security primitives. Thus employing such variations along with the first order variations in conductance states across the ReRAM array can enable ubiquitous design for secure and robust security primitives.

The experimental results were obtained on relatively large 350×350 nm devices. However, through aggressive scaling it would be possible to achieve very dense and complex PUF primitives. Previous studies have shown similar performance metal-oxide memristors with dimensions around 8 nm to 10 nm [17, 18], though apparently utilizing slightly different nanometer scale filamentary switching mechanism [19, 20].

In summary, we have experimentally investigated the impact of device non-linearity and analog state variations on a ReRAM implemented PUF system. The three-dimensionally stacked crossbar used for the experiment showed low enough variation as required for the tuning, but high enough for the PUF entropy. Our results show that higher read biases are useful for a higher entropy space and a more stable operation with respect to thermal stresses. We believe that this work is an important step toward ReRAM-based robust implementation of the PUFs. In addition, there are many reservations for improvement, e.g. security metrics can be improved by employing multi-level responses [13].

ACKNOWLEDGMENT

We acknowledge useful discussions with B. Thibeault, M. Prezioso and F. Merrikh-Bayat. This work was funded through the NSF grant CCF-1528205 and through the Maric Sklodowska-Curie grant No. 705957.

All device fabrication was performed in the UCSB cleanroom facility, which is part of the NSF-funded National Nanotechnology Infrastructure Network.

REFERENCES

[1] B. Gassend *et al.*, "Silicon physical random functions," in: *Proc. ACM CCS'02*, 2002, pp. 148-160.

[2] J. Guajardo *et al.*, "FPGA intrinsic PUFs and their use for IP protection," in: *Proc. CHES'07*, 2007, pp. 63-80.

[3] U. Rührmair *et al.*, "Modeling attacks on physical unclonable functions," in: *ACM CCS'10*, 2010, pp. 237-249.

[4] U. Rührmair, J. Sölter, F. Sehnke, X. Xu, A. Mahmoud, V. Stoyanova, *et al.*, "PUF modeling attacks on simulated and silicon data," *IEEE Transactions on Information Forensics and Security*, vol. 8, pp. 1876-1891, 2013.

[5] M. Cortez *et al.*, "Modeling SRAM start-up behavior for Physical Unclonable Functions," in: *Proc. DFTS'12*, 2012, pp. 1-6.

[6] R. Waser, R. Dittmann, G. Staikov, and K. Szot, "Redox-based resistive switching memories- nanoionic mechanisms, prospects, and challenges", *Advanced Materials*, vol. 21, pp. 2632-2663, 2009.

[7] J. J. Yang, D. B. Strukov, and D. R. Stewart, "Memristive devices for computing", *Nature Nanotechnology*, vol. 8, pp. 13-24, 2013.

[8] P.-Y. Chen *et al.*, "Exploiting resistive cross-point array for compact design of physical unclonable function," in: *Proc. HOST'15*, 2015, pp. 26-31.

[9] R. Liu *et al.*, "Experimental characterization of physical unclonable function based on 1 Kb resistive random access memory arrays," *IEEE Electron Device Letters*, vol. 36, pp. 1380-1383, 2015.

[10] Y. Gao *et al.*, "Obfuscated challenge-response: A secure lightweight authentication mechanism for PUF-based pervasive devices," in: *Proc. PerCom'16*, 2016, pp. 1-6.

[11] Y. Gao *et al.*, "Memristive crypto primitive for building highly secure physical unclonable functions", *Scientific reports*, vol. 5, 2015.

[12] L. Gao, P.-Y. Chen, R. Liu, and S. Yu, "Physical unclonable function exploiting sneak paths in resistive cross-point array", *IEEE Transactions on Electron Devices*, vol 63, pp. 3109-3115, 2016.

[13] H. Nili *et al.* "Highly-Secure Physically Unclonable Cryptographic Primitives Using Nonlinear Conductance and Analog State Tuning in Memristive Crossbar Arrays", *arXiv preprint arXiv:1611.07946*, 2016.

[14] J. Kim, *et al.* "A Physical unclonable function with redox-based nanoionic resistive memory", *arXiv:1611.04665*, 2016.

[15] G.C. Adam, *et al.* 3-D memristor crossbars for analog and neuromorphic computing applications. *IEEE Transactions on Electron Devices* vol. 64, pp. 312-318, 2017.

[16] F. Alibart, L. Gao, B.D. Hoskins, and D.B. Strukov, "High precision tuning of state for memristive devices by adaptable variation-tolerant algorithm", *Nanotechnology*, vol. 23, 075201, 2012.

[17] B. Govoreanu *et al.*, "10×10 nm² Hf/HfOx crossbar resistive RAM with excellent performance, reliability and low-energy operation", in: *Proc. IEDM'11*, Washington, DC, Dec. 2011, pp. 31.6.1-31.6.4.

[18] S. Pi, P. Lin, Q. Xia, "Cross point arrays of 8 nm× 8 nm memristive devices fabricated with nanoimprint lithography", *Journal of Vacuum Science & Technology B, Nanotechnology and Microelectronics: Materials, Processing, Measurement, and Phenomena*, vol. 31, no. 6, 06FA02, 2013.

[19] J. P. Strachan *et al.*, "Direct identification of the conducting channels in a functioning memristive device", *Advanced Materials*, vol. 22, pp. 3573-3577, 2010.

[20] D.-H. Kwon *et al.*, "Atomic structure of conducting nanofilaments in TiO₂ resistive switching memory", *Nature Nanotechnology*, vol. 5, pp. 148-153, 2010.

Negative Capacitance Field Effect Transistors; Capacitance Matching and non-Hysteretic Operation

Ali Saeidi[*], Farzan Jazaeri[†], Francesco Bellando[*], Igor Stolichnov[*], Christian C. Enz[†], and Adrian M. Ionescu[*]

[*]NANOLAB, Ecole Polytechnique Fdrale de Lausanne
[†]ICLAB, Ecole Polytechnique Fdrale de Lausanne
Email: {ali.saeidi,adrian.ionescu}@epfl.ch

Abstract—This work experimentally demonstrates negative capacitance MOSFETs in hysteretic and non-hysteretic modes of operation. A PZT capacitor is externally connected to the gate of commercial nMOSFETs fabricated in 28nm CMOS technology to explore the negative capacitance effect. In hysteretic devices, subthreshold slope as steep as 10mV/dec is achieved in the region where the ferroelectric represents an S-shape polarization. In addition, a matching condition is achieved between a PZT capacitor and the gate capacitance of MOSFETs fabricated on SOI substrates. For the first time, we achieve a non-hysteretic switch configuration in our fabricated MOSFETs, suitable for analog and digital applications, for which a reduction in the subthreshold swing is obtained down to 20mV/dec.

I. INTRODUCTION

As CMOS technology continuing its relentless downscaling, power dissipation has become the most important roadblock for future nanoelectronic circuits and systems [1]. It is well known that the power dissipation would be lowered significantly if FETs could be operated at lower voltages [2]. An average subthreshold swing (SS) smaller than the thermal limit of MOSFET swing at room temperature would enable the scaling of the supply voltage, V_{dd}. A sub-thermal subthreshold swing ($< \ln(10)KT/q$, which is $60mV/dec$ at $T = 300K$) can be obtained by decreasing the device body factor, $m = 1+C_s/C_{ins}$, to a value smaller than 1 (where C_s and C_{ins} are the semiconductor and the gate oxide capacitances) [3], [4]. This can be achieved by using the recently proposed negative capacitance (NC) effect of ferroelectric materials to the gate stack of conventional MOSFETs [5], [6]. It has been suggested that a Metal-Ferroelectric-Semiconductor (MFS) can provide a feasible solution to step-up the semiconductor surface potential (ψ_s) above the gate voltage (V_g) which leads to a reduction in the subthreshold swing [7], [8]. The underlying idea consists of exploiting the negative capacitance region of ferroelectric materials, defined as $C_{FE} = dQ/dV_{FE}$, where Q and V_{FE} refer to the charge density and the voltage drop over the ferroelectric, respectively [9], [10]. A ferroelectric capacitor (FE) in series with a dielectric capacitor (DE) of a proper value can be biased in the negative capacitance region, providing a larger capacitance than the constituent DE capacitor [11]. In order to have a non-hysteretic NCFET, the ferroelectric NC and DE capacitor should be well matched to provide a positive total capacitance in the whole range of the operation [12], [13] while the slope of the charge line is smaller than the negative slope of the FE polarization [14], [15].

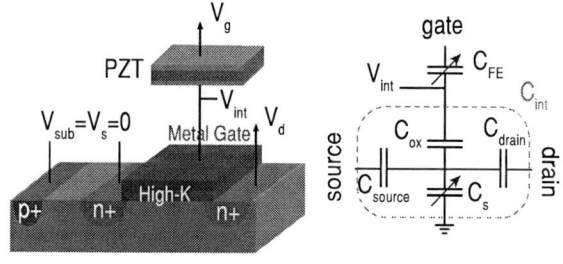

(a)

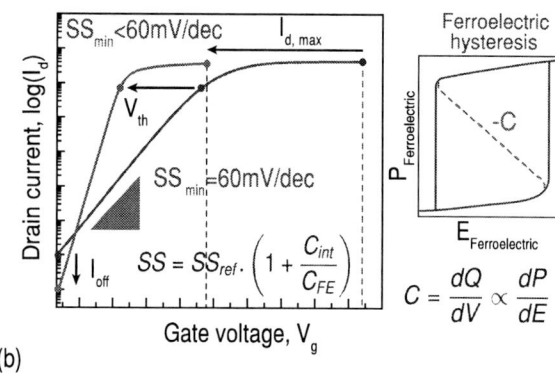

(b)

Fig. 1. The investigated experimental configuration of the NCFET (left) and the capacitance model of the structure (right). C_{int} is the total capacitance of the reference MOSFET looking into the gate (a). Transfer characteristics of a non-hysteretic NCFET versus the base MOSFET highlighting the gain of using the ferroelectric negative capacitance in terms of SS improvement and threshold voltage reduction (b).

In this context, we experimentally investigate the impact of the ferroelectric NC on DC electrical behavior of commercial MOSFETs fabricated in $28nm$ CMOS technology. A matching condition is proposed between the ferroelectric and MOS capacitances in order to obtain the non-hysteretic operation of the device. The subthreshold swing below the theoretical limit at room temperature ($60mV/dec$) over many decades of current on hysteretic NCFETs is achieved in $28nm$ bulk CMOS process. Moreover, the non-hysteretic operation of an NCFET with a matched PZT capacitor and sub-thermal swing is demonstrated.

978-1-5090-5979-9/17 $31.00 © 2017 IEEE

II. HYSTERETIC AND NON-HYSTERETIC OPERATION: CAPACITANCE MATCHING

One may consider the NCFET as a conventional transistor with an added amplifier. Considering a simple capacitance model (Fig. 1-a), the amplification factor of the NC effect can be expressed as

$$\beta = \partial V_{int} / \partial V_g = C_{FE}/(C_{FE} + C_{int}). \quad (1)$$

It should be noted that the following conditions are required to provide a sufficient amplification together with a non-hysteretic behavior in an NCFET: (i) the absolute value of the ferroelectric negative capacitance ($|C_{FE}|$) and the intrinsic gate capacitance (C_{int}) need to be relatively close, and (ii) the total capacitance should remain positive in the whole range of operation; $C_{total}^{-1} = C_{FE}^{-1} + C_{int}^{-1} > 0$. Using the amplification factor, β, the subthreshold swing of an NCFET, SS_{nc}, can be indicated as:

$$SS_{nc} = \left(\frac{\partial Log I_d}{\partial V_g} \right)^{-1} = \frac{\partial V_{int}}{\partial Log I_d} \times \frac{\partial V_g}{\partial V_{int}} = \frac{SS_{ref}}{\beta}. \quad (2)$$

In the subthreshold region, providing an effective NC by the ferroelectric leads to $\beta > 1$ and the SS will be reduced [5]. The hysteretic NCFETs and also a non-hysteretic device which fulfills above conditions will be discussed in the following sections.

III. HYSTERETIC NCFET

The experimental configuration of the NCFET including the capacitance model of the device is schematically depicted in Fig. 1-a, where a PZT capacitor is externally connected to the gate of a conventional MOSFET. Such external electrical connection offers the flexibility of testing tens to hundreds of PZT capacitor values and MOSFETs until the best matching is obtained. The impact of the ferroelectric negative capacitance on DC electrical performance of MOSFETs is schematically illustrated in Fig. 1-b.

High-performance commercial n-type MOSFETs fabricated in $28nm$ CMOS technology have been employed as the reference devices and Pb(Zr$_{46}$,Ti$_{57}$)O$_3$ (PZT) is exploited as the ferroelectric material. The PZT ferroelectric with a thickness of $50nm$ has been grown via the chemical solution deposition root on a Pt-coated silicon wafer [16], [17]. Pt top electrode is deposited at room temperature and patterned by shadow masking. The deposited PZT film has a polycrystalline structure. High-quality epitaxial ferroelectric layers are commonly considered suitable for NC devices due to their ability to form a mono-domain state characterized by a simple coercive field [18]. In this study, a repetitive bipolar voltage is applied to the ferroelectric capacitors, so that the mono-domain behavior is achieved in the polycrystalline ferroelectric material. The characterized electrical parameters of the thin film PZT (i.e. relative permittivity, coercive field, and remanent polarization) are depicted in Fig. 2. Red curve illustrates the relative permittivity (epsilon) with respect to the imposed

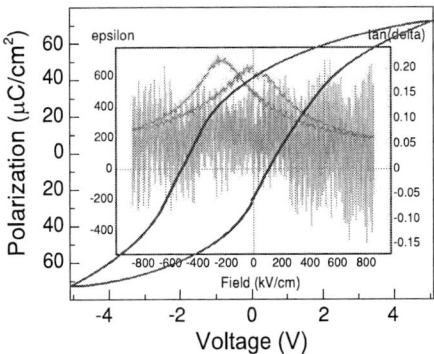

Fig. 2. The film polarization, permittivity, and the phase angle of the capacitance measurement hysteresis loops regarding the applied voltage (electric field) on the ferroelectric layer. The relative permittivity, coercive field, and remanent polarization of $220 - 240$, $260kV/cm$, and $30.2\mu C/cm^2$ have been measured.

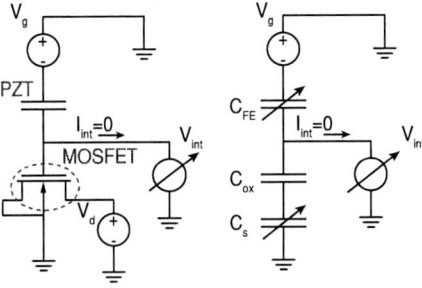

Fig. 3. Measurement setup utilized for probing the internal voltage, V_{int}, using an Agilent 4156C semiconductor analyzer. This probing has a negligible influence of the extracted SS.

electric field which varies from 220 to 240. The coercive field ($\pm 260kV/cm$) and remanent polarization ($\pm 30.2\mu C/cm^2$) of the PZT are extracted from the blue curve, devoted to the polarization-voltage hysteresis.

The experimental setup used for the electrical characterization of NCFETs is represented in Fig. 3. The internal contact is probed (V_{int}) while a voltage is applied to the top gate (a zero current is injected in the internal node). As reported in [6], this probing has a negligible impact on the measurement results of the SS.

Fig. 4 shows experimental characteristics data measured on an nMOSFET with $W = 100nm$, $L = 1\mu m$ (NCFET-a), and a $20\mu m \times 20\mu m$ PZT capacitor connected to the gate with a drain voltage of $100mV$. Fig. 4-a illustrates the hysteretic transfer characteristic of NCFET-a with an improved I_{on}/I_{off} ratio. A sub-thermal swing over 4 decades of current is illustrated in Fig. 4-b. The voltage amplification in the regions corresponding to the device subthreshold slope is depicted in Fig. 4-c. Imposing displacement vector continuity, a unique S-shape behavior in the polarization, P, is obtained and illustrated in Fig. 4-d. A negative slope of the polarization with respect

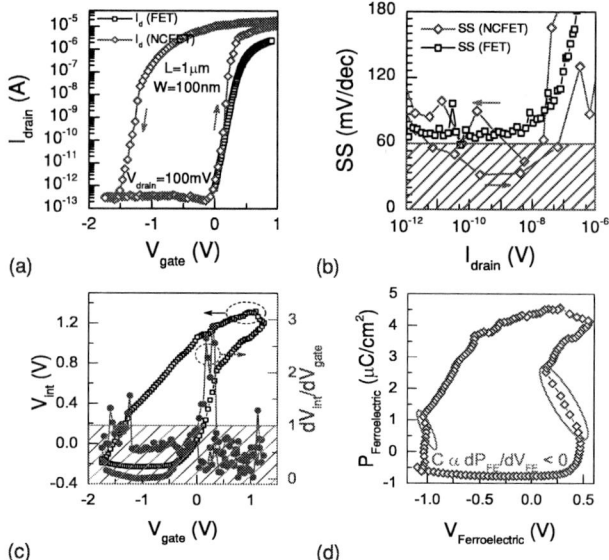

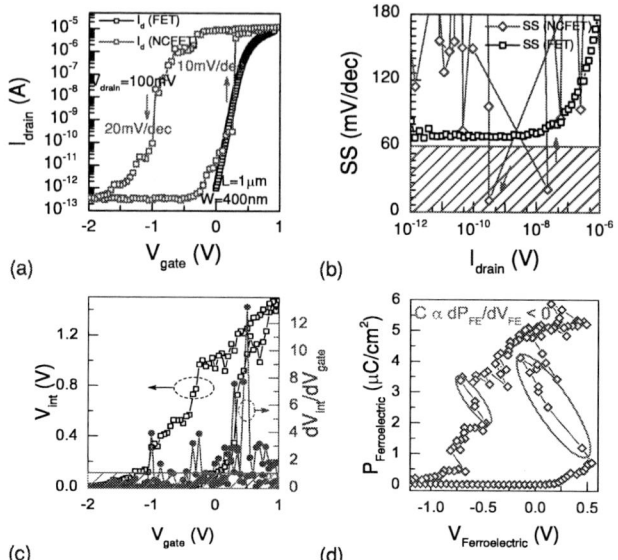

Fig. 4. Transfer characteristic of NCFET-a, where a $20\mu m \times 20\mu m$ PZT capacitor with a thickness of $50nm$ is connected to the gate of a commercial nMOSFET (a). Sub-thermal swing (b) is obtained over many decades of current due to the differential voltage amplification of NC. Amplification factor greater than 1 is achieved (c) in the regions corresponding to the negative slope of the ferroelectric polarization (d).

Fig. 5. The impact of the NC on DC electrical behavior of NCFET-b (b). Subthreshold slope as steep as $10mV/dec$ is demonstrated with a better matching of capacitances (b). The internal voltage represents β greater than 1 up to 12 (c). The polarization of the ferroelectric confirms the existence of NC in both positive and negative going branches (d).

to the applied electric field is observed in a certain range of the polarization, corresponding to the subthreshold region of the device.

The device operation is hysteretic due to the relatively small value of C_{int} where $C_{total}^{-1} = C_{FE}^{-1} + C_{int}^{-1} > 0$ is not fulfilled in the whole range of the gate voltage. Device characterizations of an architecture (NCFET-b) with a better matching of capacitances (higher C_{int} due to the larger gate) is illustrated in Fig. 5. Subthreshold swing of $10mV/dec$ with a $1V$ hysteresis is obtained (Fig. 5-b). The representation of the voltage amplification of this architecture is shown in Fig. 5-c with $\beta > 1$ in both branches (having a peak of above 12). The extracted P-V hysteresis of the ferroelectric capacitor clearly demonstrates the negative slope of the polarization, confirming the NC effect.

IV. NON-HYSTERETIC NCFET

Due to the relatively small intrinsic gate capacitance of commercial MOSFETs, the full capacitance matching between PZT capacitors and MOS capacitors is challenging.

Here, we fabricated MOSFETs on a SOI silicon wafer in relatively large dimensions which fulfill the condition for non-hysteretic NCFET. The devices are built on a p-type SOI substrate with $88nm$ of epitaxial silicon and $145nm$ BOX. A cycle of dry oxidation plus HF-based etching is used to thin down the Si layer to $30nm$, improving the gate electrostatic control. After the source and drain phosphorus implantation and annealing, the FET body is shaped using photolithography and selective plasma etching. HfO$_2$ with $3nm$ thickness has been deposited by ALD on an ultra thin

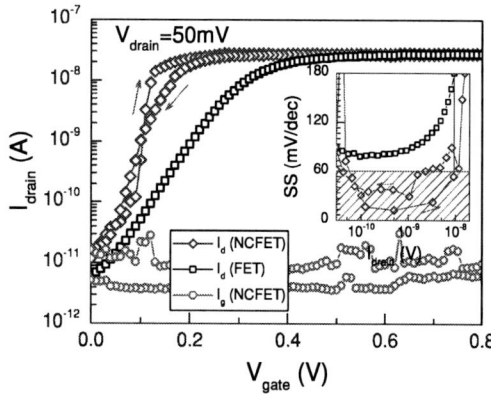

Fig. 6. $I_d - V_g$ characteristic of a non-hysteretic NCFET using a $10\mu m \times 10\mu m$ PZT capacitor comparing the reference MOSFET ($W = 19\mu m$, $L = 2\mu m$) with $V_{drain} = 50mV$. Sub-thermal swing down to $20mV/dec$ is obtained due to the differential amplification effect of the ferroelectric NC.

layer of SiO$_2$ as the gate dielectric. As reported in [19], SiO$_2$ is used in order to provide a proper interface with the Si channel. Photolithography is used to define the source/drain dielectric openings for electrical contact, which are then created by BHF etching. AlSi1% metal contacts have been created by sputtering and lift-off process.

In order to obtain non-hysteretic negative capacitance, A $10\mu m \times 10\mu m$ PZT capacitor is connected to the gate of a MOSFET, figured out above, with a gate length of $2\mu m$ and a gate width of $19\mu m$. Fig. 6 shows $I_d - V_g$ characteristics of the reference MOSFET and the NCFET at $50mV$ drain

978-1-5090-5979-9/17 $31.00 © 2017 IEEE

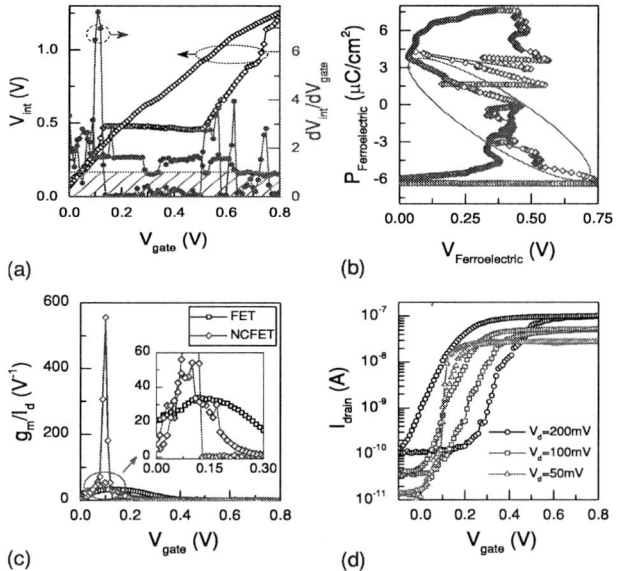

(a) **(b)**

(c) **(d)**

Fig. 7. Electrical characteristics of the proposed non-hysteretic NCFET. The internal voltage measurement depicts an amplification factor greater than 1 in the subthreshold region (a). The polarization of the ferroelectric demonstrates a negative slope in a wide range of the polarization (b). The calculated g_m/I_d of the NCFET represents a significant boosting in the subthreshold region of the device (c). Increasing the drain voltage results in limiting the improvement of the NC effect and increasing of the hysteresis (d).

voltage. The recorded gate leakage confirms that its level is systematically lower than I_{on} and the leakage and charge trapping mechanisms can be neglected in the reported effects. A significant improvement in the SS of the device is obtained when the ferroelectric and gate capacitances are matched so that a non-hysteretic NC operation can be achieved over the whole range of the gate voltage. A sub-thermal swing, down to $20mV/dec$, is observed due to the voltage amplification of the ferroelectric effective NC (the inset figure of Fig. 6). The internal voltage measurement represents the amplification factor up to 7 (Fig. 7-a) in the subthreshold region. Fig. 7-b reports the ferroelectric polarization showing an effective NC in a wide range of the polarization. Fig. 7-c compares g_m/I_d (g_m is the transconductance) of the NCFET with the reference MOSFET, indicating a significant improvement in the subthreshold region. The impact of the drain voltage on the NC effect is exploited in Fig. 7-d. Increasing the drain voltage performs hysteresis in the device operation and reduces the amplification factor.

V. CONCLUSIONS

The hysteretic and non-hysteretic behaviors of negative capacitance-MOSFETs are investigated in this paper. It has been experimentally proved that a significant reduction of the subthreshold swing can be obtained by the impact of the NC on field-effect transistors. A $10mV/dec$ subthreshold slope is achieved with $1V$ of hysteresis in a commercial 28nm CMOS technology. A matching condition between the ferroelectric and MOS capacitances is obtained for the non-

hysteretic operation in MOSFETs fabricated on SOI substrates leading to a subthreshold swing of $20mV/dec$.

ACKNOWLEGMENT

The authors acknowledge Swiss National Science Foundation (Grant NO. 149495) for providing the financial support of this research.

REFERENCES

[1] S. Takagi *et al.*, "Carrier-transport-enhanced channel cmos for improved power consumption and performance," *IEEE Transactions on Electron Devices*, vol. 55, no. 1, pp. 21–39, 2008.

[2] A. M. Ionescu *et al.*, "Ultra low power: Emerging devices and their benefits for integrated circuits," in *Electron Devices Meeting (IEDM), 2011 IEEE International*. IEEE, 2011, pp. 16–1.

[3] S. Salahuddin and S. Datta, "Can the subthreshold swing in a classical fet be lowered below 60 mv/decade?" in *Electron Devices Meeting, 2008. IEDM 2008. IEEE International*. IEEE, 2008, pp. 1–4.

[4] G. A. Salvatore, A. Rusu, and A. M. Ionescu, "Experimental confirmation of temperature dependent negative capacitance in ferroelectric field effect transistor," *Applied Physics Letters*, vol. 100, no. 16, p. 163504, 2012.

[5] S. Salahuddin and S. Datta, "Use of negative capacitance to provide voltage amplification for low power nanoscale devices," *Nano letters*, vol. 8, no. 2, pp. 405–410, 2008.

[6] A. Rusu *et al.*, "Metal-ferroelectric-meta-oxide-semiconductor field effect transistor with sub-60mv/decade subthreshold swing and internal voltage amplification," in *Electron Devices Meeting (IEDM), 2010 IEEE International*. IEEE, 2010, pp. 16–3.

[7] G. A. Salvatore, D. Bouvet, and A. M. Ionescu, "Demonstration of subthrehold swing smaller than 60mv/decade in Fe-FET with $P(VDF - TrFE)/SiO_2$ gate stack," in *Electron Devices Meeting, 2008. IEDM 2008. IEEE International*. IEEE, 2008, pp. 1–4.

[8] J. Jo *et al.*, "Negative capacitance in organic/ferroelectric capacitor to implement steep switching mos devices," *Nano letters*, vol. 15, no. 7, pp. 4553–4556, 2015.

[9] D. J. Appleby *et al.*, "Experimental observation of negative capacitance in ferroelectrics at room temperature," *Nano letters*, vol. 14, no. 7, pp. 3864–3868, 2014.

[10] W. Gao *et al.*, "Room-temperature negative capacitance in a ferroelectric–dielectric superlattice heterostructure," *Nano letters*, vol. 14, no. 10, pp. 5814–5819, 2014.

[11] A. Saeidi *et al.*, "Double-Gate Negative-Capacitance MOSFET With PZT Gate-Stack on Ultra Thin Body SOI: An Experimentally Calibrated Simulation Study of Device Performance," *IEEE Transactions on Electron Devices*, vol. 63, no. 12, pp. 4678–4684, 2016.

[12] C. W. Yeung, A. I. Khan, S. Salahuddin, and C. Hu, "Device design considerations for ultra-thin body non-hysteretic negative capacitance fets," in *Energy Efficient Electronic Systems (E3S), 2013 Third Berkeley Symposium on*. IEEE, 2013, pp. 1–2.

[13] A. I. Khan *et al.*, "Ferroelectric negative capacitance mosfet: Capacitance tuning & antiferroelectric operation," in *Electron Devices Meeting (IEDM), 2011 IEEE International*. IEEE, 2011, pp. 11–3.

[14] A. Rusu, A. Saeidi, and A. M. Ionescu, "Condition for the negative capacitance effect in metal–ferroelectric–insulator–semiconductor devices," *Nanotechnology*, vol. 27, no. 11, p. 115201, 2016.

[15] A. Jain and M. A. Alam, "Stability constraints define the minimum subthreshold swing of a negative capacitance field-effect transistor," *IEEE Transactions on Electron Devices*, vol. 61, no. 7, pp. 2235–2242, 2014.

[16] J. Lee *et al.*, "Built-in voltages and asymmetric polarization switching in $Pb(Zr, Ti)O_3$ thin film capacitors," *Applied physics letters*, vol. 72, no. 25, pp. 3380–3382, 1998.

[17] D.-J. Kim *et al.*, "Evaluation of intrinsic and extrinsic contributions to the piezoelectric properties of $Pb(Zr_{1-X}T_X)O_3$ thin films as a function of composition," *Journal of Applied Physics*, vol. 93, no. 9, pp. 5568–5575, 2003.

[18] P. Zubko *et al.*, "Negative capacitance in multidomain ferroelectric superlattices," *Nature*, 2016.

[19] S. Rigante *et al.*, "Sensing with advanced computing technology: Fin field-effect transistors with high-k gate stack on bulk silicon," *ACS nano*, vol. 9, no. 5, pp. 4872–4881, 2015.

Buried Multi-Gate InAs-Nanowire FETs

Thomas Grap, F. Riederer, C. Gupta and J. Knoch

Institute of Semiconductor Electronics, RWTH Aachen University, 52056 Aachen, Germany

Email: grap@iht.rwth-aachen.de

Telephone: +49-241-8027890

Abstract—We present a study on multi-gate field-effect transistors that allow adjusting the potential landscape in semiconducting nanowires/tubes on the nanoscale. To this end, a damascene-like process is employed that allows fabricating a large number of gate structures that are contacted individually and exhibit lengths and inter-gate distances well below 10nm enabling to realize potential landscapes within a device that exploit quantum effects. The functionality of the multi-gate structures is shown experimentally with InAs nanowire FETs.

I. INTRODUCTION

During the last decades, a performance increase of highly integrated circuits (ICs) has been achieved by increasing the number of transistors onto the same chip area. The scalability of transistors, however, will soon come to an end. Therefore, an attractive alternative to further increase the performance of ICs is to add functionality to the devices as has recently been demonstrated [1], [2], [3]. In order to control the functionality, a potential landscape within a device must be tunable and hence, doping needs to be replaced by additional gate electrodes [1], [4]. Recently, we demonstrated a reconfigurable WSe$_2$ device based on buried triple gate substrates that could be tuned to show n-type and p-type device behaviour as well as functionality as band-to-band tunnel FET [5]. However, for more complex devices multiple, lateral gates are required [1], [6]. These gates, however, need to be scaled down as far as possible in order to guarantee a device with smallest lateral dimensions and to allow for device functionalities that rely on quantum effects such as a resonant tunneling, for instance. Experimental demonstrations of multi-gate transistors so far exhibited gates with lateral dimensions on the order of 50 nm and larger [7], [8], [9], [10] which is not small enough to facilitate such functionalities at room temperature. Here, we present a buried multi-gate platform with 10 nm gates with an 25 nm inter-gate distance that can be scaled down to ~5nm gate length and inter-distance. We demonstrate the functionality with InAs nanowire field-effect transistors and show simulations regarding different device functionalities.

II. MULTI-GATE FETS

Figure 1 (a) shows a schematics of a multi-gate substrate consisting of eight metallic gate electrodes (made e.g. of TiN), each with a gate length of d_g, an inter-gate distance of d_i embedded into SiO$_2$. The gate oxide thickness is d_{ox}. On top of the substrate a nanowire of diameter d_{nw} is placed and contacted with metallic source/drain contacts. In order to study such a multi-gate substrate, particularly with respect to the

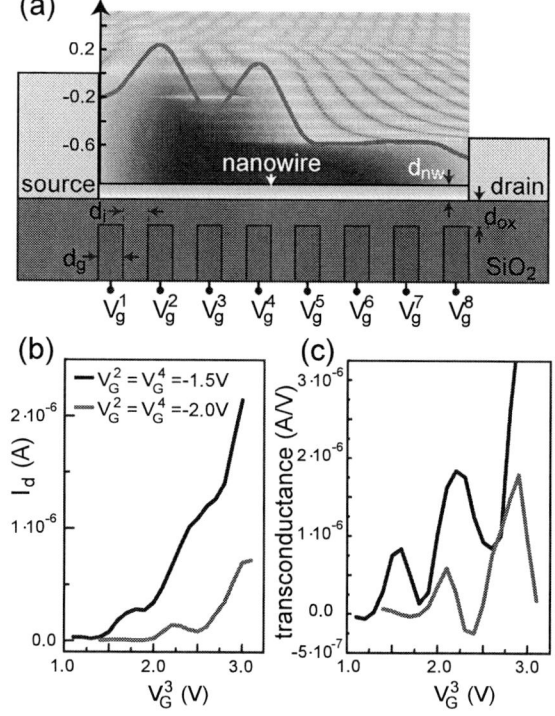

Fig. 1. (a) Schematics of the multi-gate substrate with local density of states. (b) Transfer characteristics as a function of the dot voltage V_G^3 for constant $V_G^{2,4} = -1.5$V (black) and -2V (red); all other gate voltages are constant at $V_G^{1,5} = 2$V, $V_G^{6,7,8} = 1.5$V. (c) Transconductance as a function of the dot voltage; all other parameters are as in (b).

potential landscape, self-consistent Poisson-Schrödinger simulations using the non-equilibrium Green's function formalism (NEGF) were carried out. To keep the computational burden as small as possible, a 1D model for the electronic transport was chosen. To take the electrostatics into consideration a 1D modified Poisson equation is used and the inter-gate regions are taken into account with a conformal mapping technique (see e.g. [11]).

A device with a resonant tunneling diode in front of a conventional transistor is investigated. To this end, we consider a device with the following parameters: $d_{nw} = 5$nm, $d_{ox} = 5$nm, $d_g = 5$nm, $d_i = 5$nm, $m^\star = 0.023m_0$ and a Schottky-barrier of -0.2eV at the nanowire-metal interfaces. A quantum dot is formed by biasing gate no. 2 and 4 at $V_G^{2,4} = -2$V and gate no. 3 is varied while keeping all

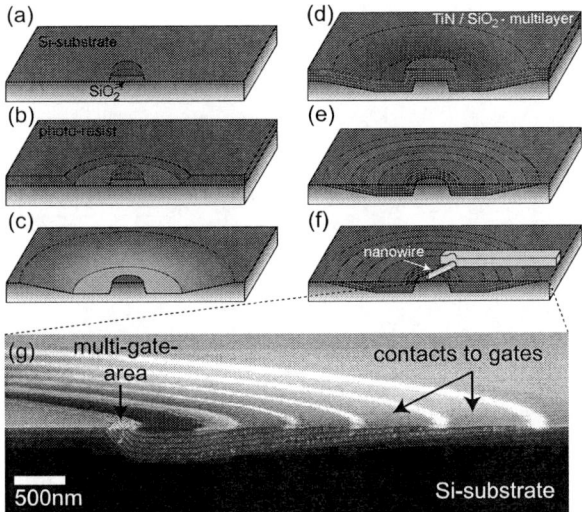

Fig. 2. Fabrication process (a)-(f) and electron micrograph (g) of a multi-gate substrate.

other gates at a constant voltage (see Fig. 1 (b)). Figure 1 (a) shows a local density of states plot of such a device where the quantum dot is clearly visible. Varying V_G^3 leads to a peak structure in the transfer characteristics as shown in (b) with negative differential resistance. Figure 1 (c) displays the transconductance as a function of V_G^3, all other parameters are as in (b).

III. SAMPLE FABRICATION

In order to realize device functionalities discussed in the preceding section we fabricated buried multi-gate substrates. The fabrication starts with the deposition of a 65nm thick SiO_2 hard mask layer using remote plasma-enhanced chemical vapor deposition (rPECVD). Next, the hard mask is patterned with optical lithography and wet chemical etching in buffered oxide etch (BOE) that yields circular shaped SiO_2-mesas (Fig.2 (a)). Afterwards, a photoresist mask (positive process) is patterned, surrounding the SiO_2-mesas (cf. Fig. 2 (b)). The photoresist is baked at 300°C for 5 minutes on a hot plate in order to obtain rounded resist flanks. Thus, in the subsequent SF_6/O_2 plasma etching, the resist mask is eroded while the SiO_2 hard mask is hardly attacked by the etching. This results in a donut-shaped mold with very shallow etch flanks at the outer edge and steep flanks at the inner edge (cf. Fig. 2 (c)). After etching of the Si substrate, alternating layers of 10nm of TiN and 50nm of SiO_2 are deposited with ALD and rPECVD, respectively, in-situ in a cluster-tool (cf. Fig. 2 (d)). The TiN is deposited at 300°C using Tetrakis(dimethylamido)titanium (TDMAT) as precursor with a deposition rate of 0.07nm per cycle [12]. The SiO_2 is deposited in the rPECVD at 350°C using silane and oxygen. While the mold in the Si-substrate is covered conformally by the ALD deposited TiN (i.e. $d_g = 10$nm), the rPECVD SiO_2 exhibits a lower deposition rate on the steep flank compared to the shallow one (ratio ~0.5). Hence, the 50nm SiO_2 reduces to

$d_i = 25$nm SiO_2 inter-gate insulation in the multi-gate area. Recently, we also fabricated substrates with $d_g = 5$nm and $d_i = 7$nm (not shown here). After deposition the samples are polished using a Logitech polisher PM5 (cf. Fig. 2 (e)). Polishing the samples leads to the exposition of the multi-gate-area and the contact area to the gates. Due to the particular shape of the mold, the TiN layers are strongly widened in the contact areas, facilitating to individually contact the gates using electron beam lithography (EBL) and lift-off (cf. Fig. 2 (g)). Finally, a $d_{ox} = 30$nm SiO_2-layer is deposited that serves as top gate dielectric finishing the fabrication of the multi-gate substrate.

VLS-grown InAs nanowires with a diameter of 65nm are randomly deposited onto the buried multi-gate substrates. Subsequently, EBL with PMMA is used to pattern Ti/Al (5nm/50nm) contact leads to the InAs nanowire. Before depositing the metal on the nanowires, the nanowires are dipped in a solution of $HCl:H_2O$ (1:10) in order to remove the oxide film of the InAs nanowire [13]. Afterwards, a second EBL-step is carried out in the gate contact area to etch small stripes uncovering the top gate dielectric (BOE for 12sec). Eventually, individual contacts to the gates are realized with EBL and lift-off (5nm Ti/50nm Al).

Figure 3 (a) shows an electron micrograph of the fabricated buried multi-gate substrate where the multi-gates and the NW are contacted. The NW is located directly above the multi-gates as depicted in Fig. 3 (b).

IV. DISCUSSION

To check the functionality of the multi-gate substrates we first measured the conductivity of the TiN-layers as well as the electrical insulation of the SiO_2 in-between the TiN-layers. To this end, two contacts per multi-gate are realized. The results of these measurements are shown in Fig. 4. The IV-characteristics do not show ohmic behaviour and a rather high resistivity of about 1MΩ. However, if the radius r_{MG} of the multi-gates is increased, the resistivity decreases yielding a higher current through the TiN-layer. The reason for the non-ohmic behavior is still unclear but may stem from an insufficient nitrogen incorporation during the TiN deposition

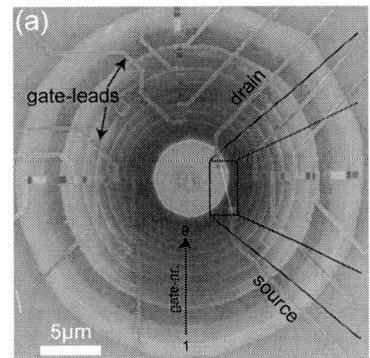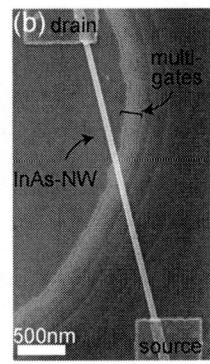

Fig. 3. (a) SEM-image of an InAs-NW multi-gate device and (b) the InAs-NW with the multi-gate area underneath.

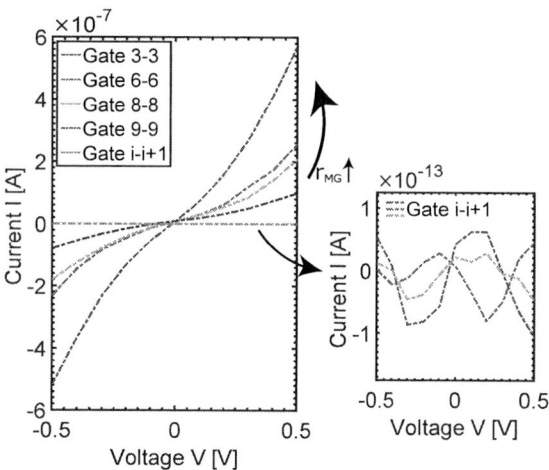

Fig. 4. IV-measurements for as-deposited multi-gates; (a) shows the conductivity for single gates and (b) shows the resistivity between two neighbouring gates.

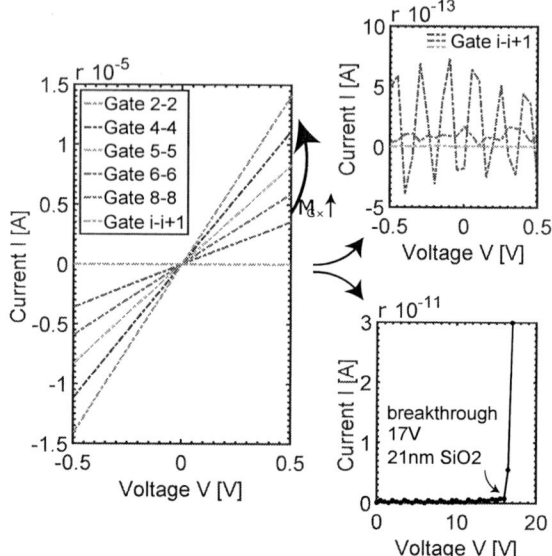

Fig. 5. IV-measurements for tempered multi-gates. (a) shows the conductivity for single gates, (b) shows the resistivity between two neighbouring gates and (c) depicts the electrical stability of the SiO_2.

[12]. Diffusion of oxygen into the TiN-layer during the SiO_2-process might also affect the IV-behaviour as well as oxidation of the TiN-layers.

The insulation (SiO_2) between the TiN-layers is stable up to $\pm 10V$ and with rather small leakage currents of about $10^{-13} A$. The low level of leakage ensures that the voltages applied in the contact area at the various gates in the gate contact area can be assumed to be present in the multi-gate area.

In order to improve the properties of the TiN-layers a RTP-step is carried out before polishing the samples [14]. To this end, the samples are annealed at 500°C for 15min in an Ar-atmosphere. The IV-results of the samples with an additional annealing are shown in Fig. 5 (a). The conductivity is clearly improved by a factor of ≈ 24 and the TiN-layers now show ohmic behaviour [15]; an increase of conductivity of TiN-layers due to annealing has also been reported in [14]. Figure 5 (b) displays the IV-characteristics of the inter-gate SiO_2 layers after annealing. Similar to the non-annealed samples, the leakage through the SiO_2 layers is as small as $5 \cdot 10^{-13} A$ and lower.

After optimizing the multi-gate properties, first devices with VLS grown InAs-NW are realized to demonstrate the functionality of the multi-gate structure. At first, each multi-gate is used separately to sweep through the gate voltage in order to manipulate the conductivity of the NW (see Fig. 6 (measurement set-up 1.1), while all other multi-gates are floating. The transfer characteristics of a single multi-gate (gate-no. 5) is depicted in Fig. 6 (blue line). Clearly, the multi-gate is able to shift the conduction band, hence change the transport properties of the NW. However, a single multi-gate with a gate length of 10nm and $d_{ox} = 30$nm is not able to switch off an InAs-NW with a diameter of 65nm, yielding an I_{on}/I_{off} ratio of ≈ 3.5 for $V_{ds} = 0.15V$. The small gate length, which leads to short channel effects in the NW, as well as the Fermi-Level pinning of InAs yield

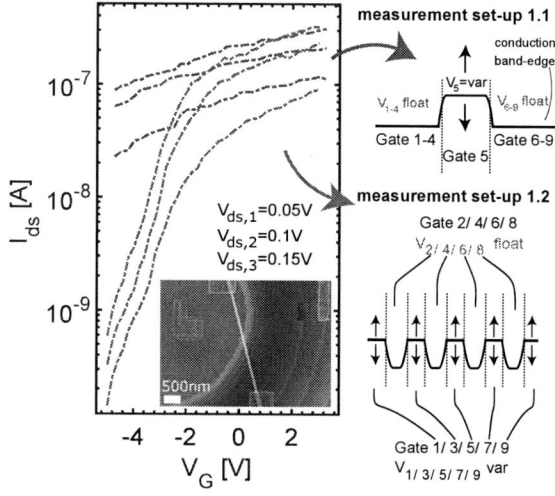

Fig. 6. Transfer characteristics (a) of a single InAs-NW for two different measurement set-ups (b) and (c).

a deteriorated off-state performance. In order to improve the off-state, additional multi-gates are used to manipulate the conduction band (yellow line in Fig. 6 (measurement set-up 1.2) increasing the effective length of the potential barrier in the channel. The measurement presented here corresponds to the use of five multi-gates (gate-no. 1, 3, 5, 7, 9); the remaining gates are again floating. As expected, the off-state behaviour improves by a reduction of I_d ($I_{d,off} = 5.9 \cdot 10^{-10}$A). As expected, the on-state performance is unchanged, compared to measurement set-up 1.1.

In a second measurement set-up (see Fig. 7), the influence and interaction of neighbouring multi-gates on the measure-

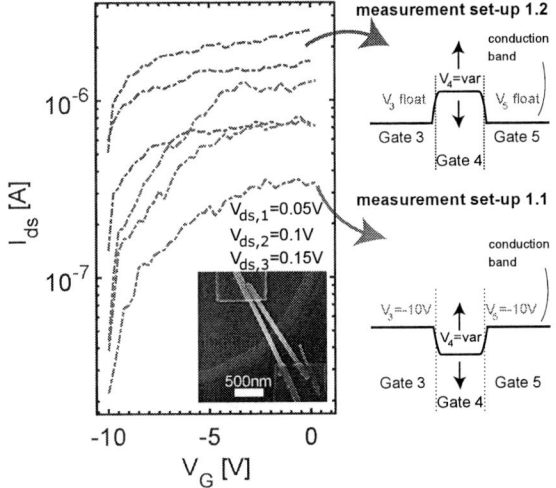

Fig. 7. Transfer characteristics (a) of two InAs-NW for two different measurement set-ups (b) and (c).

ments is studied. To do so, a single multi-gate FET measurement is performed as reference (Fig. 7, measurement set-up 2.1). Note, that this time two NW are contacted, yielding a higher I_d than in the first set-up. Again, an obvious influence of the multi-gate is visible. In the second part of the set-up (measurement set-up 2.2) neighbouring gates are held at a constant gate voltage of $V_{G3,5} = -10V$, while the center gate no. 4 is used to vary the conduction band. All other gates are again floating. In the off-state a lower current is expected in comparison to the single multi-gate device, due to the larger barrier because of an effectively longer gate. If the center-gate had no effect on the neighbouring multi-gates no. 3 and 5 (kept constant at $V_{G3,5} = -10V$) it would be expected, that the on-state behaviour is deteriorated. A similar on-state performance compared to the reference sample, however, indicates an impact of a specific gate on the potential in the adjacent gate areas (also confirmed by simulations). This is important for the engineering of more complex conduction band structures. As the measurements show, an off-current of $I_d = 4 \cdot 10^{-8}A$ compared to $I_d = 5 \cdot 10^{-7}A$ (for $V_{ds} = 0.15V$) is achieved and - as presented in Fig. 7 - the on-state of the second set-up is also decreased by a factor of two ($I_d = 1 \cdot 10^{-6}A$ to $I_d \approx 2 \cdot 10^{-6}A$). Nevertheless, these measurements indicate, that the multi-gates set-ups necessary to define, e.g., quantum dots or resonant tunnel diodes can be realized.

V. CONCLUSION

A fabrication technique for multi-gate substrates that allows the realization of gate lengths and inter-gate distance well below 10nm has been presented. InAs-NW were placed on top of a multi-gate substrate to show the general functionality of the substrates for the realization of NW-FETs. Due to short channel effects (short gates, large d_{nw} and d_{ox}) and Fermi level pinning of InAs, the potential landscape within the InAs NW could only be manipulated in a limited range.

Thinner gate dielectrics and for instance carbon nanotubes will allow to set-up devices which rely on quantum effects such as resonant tunnel transistors or devices with an energy filter for the realization of steep slope transistors.

ACKNOWLEDGMENT

The authors would like to thank Deutsche Forschungsgemeinschaft for financial support under grant no. KN 545/11-1.

REFERENCES

[1] A.M. Burke, D.J. Carrad, J.G. Gluschke, K. Storm, S. Fahlvik Svensson, H. Linke, L. Samuelson, and A.P. Micolich, "InAs Nanowire Transistors with Multiple, Independent Wrap-Gate Segments ", *Nano Lett.*, **15** (2015), pp. 2836-2843.

[2] A. Heinzig, T. Mikolajick, J. Trommer, D. Grimm, and W.M. Weber, "Dually Active Silicon Nanowire Transistors and Circuits with Equal Electron and Hole Transport", *Nano Lett.*, **13** (2013), pp. 4176-4181.

[3] L. Liu, V. Saripalli, E. Hwang, V. Narayanan and S. Datta, "Multi-Gate Modulation Doped In0.7Ga0.3As Quantum Well FET for Ultra Low Power Digital Logic", *ECS Transactions*, **35** (2011), pp. 311-317.

[4] A.V. Thathachary, N. Agrawal, L. Liu, and S. Datta, "Electron Transport in Multigate InxGa1-x As Nanowire FETs: From Diffusive to Ballistic Regimes at Room Temperature", *Nano Lett.*, **14** (2014), pp. 626633.

[5] M.R. Müller, R. Salazar, S. Fathipour, H. Xu, K. Kallis, U. Künzelmann, A. Seabaugh, J. Appenzeller and J. Knoch, "Gate-Controlled WSe2 Transistors Using a Buried Triple-Gate Structure", *Nanoscale Res. Lett.*, **11** (2016), pp. 512.

[6] J. Trommer, A. Heinzig, U. Mühle, M. Löffler, A. Winzer, P.M. Jordan, J. Beister, T. Baldauf, M. Geidel, B. Adolphi, E. Zschech, T. Mikolajick, and W.M. Weber, "Enabling Energy Efficiency and Polarity Control in Germanium Nanowire Transistors by Individually Gated Nanojunctions", *ACS Nano*, **11** (2017), pp. 1704-1711.

[7] A. Fuhrer, C. Fasth, and L. Samuelson, "Single electron pumping in InAs nanowire double quantum dots", *Appl. Phys. Lett.*, **91** (2007), pp. 052109.

[8] M. J. Biercuk, S. Garaj, N. Mason, J. M. Chow, and C. M. Marcus, "Gate-Defined Quantum Dots on Carbon Nanotubes", *Nano Lett.*, **5** (2005), pp. 1267-1271.

[9] S. d'Hollosy, M. Jung, A. Baumgartner, V. A. Guzenko, M. H. Madsen, J. Nygård, and C. Schönenberger, "Gigahertz Quantized Charge Pumping in Bottom-Gate-Defined InAs Nanowire Quantum Dots", *Nano Lett.*, **15** (2015), pp. 4585-4590.

[10] S. Heedt, N. Traverso Ziani, F. Crpin, W. Prost, St. Trellenkamp, J. Schubert, D. Grützmacher, B. Trauzettel and Th. Schäpers, "Signatures of interaction-induced helical gaps in nanowire quantum point contacts", *Nature Phys.* (2017), doi:10.1038/nphys4070.

[11] J. Knoch and J. Appenzeller, "Modeling of High-performance p-type III-V heterojunction tunnel FETs", *IEEE Electron Dev. Lett.*, **31** (2010), pp. 305-307.

[12] J. Musschoot, Q. Xie, D. Deduytsche, S. Van den Berghe, R.L. Van Meirhaeghe, C. Detavernier, "Atomic layer deposition of titanium nitride from TDMAT precursor", *Microelectron. Engin.*, **86** (2009), pp. 7277.

[13] M. Yamaguchi, A. Yamamoto, H. Sugiura and C. Uemura, "Thermal Oxidation Of InAs And Characterization Of The Oxide Film", *Thin Solid Films*, **92** (1982), pp. 361-369.

[14] N.K. Ponon, D.J.R. Appleby, E. Arac, P.J. King, S. Ganti, K.S.K. Kwa, and A. O'Neill, "Effect of deposition conditions and post deposition anneal on reactively sputtered titanium nitride thin films", *Thin Solid Films*, **578** (2015), pp. 3137.

[15] H. Van Bui, A.Y. Kovalgin, J. Schmitz, and R.A.M. Wolters, "Conduction and electric field effect in ultra-thin TiN films", *Appl. Phys. Lett.*, **103** (2013), pp. 051904.

978-1-5090-5979-9/17 $31.00 © 2017 IEEE

Equivalent Circuit Model for the Electron Transport in 2D Resistive Switching Material Systems

E. Miranda and J. Suñé

Departament d'Enginyeria Electrònica
Universitat Autònoma de Barcelona
Cerdanyola del Valles, Spain
enrique.miranda@uab.cat

C. Pan, M. Villena, N. Xiao, and M. Lanza

Institute of Functional Nano & Soft Materials
Soochow University
Suzhou, China
mlanza@suda.edu.cn

Abstract—**A compact model for the low and high resistance state conduction characteristics of electroformed capacitors with hexagonal boron nitride (*h*-BN) as insulator material and with multi-layer graphene and metal electrodes is presented. The model arises from an approximation of the expression for multi-filamentary electron transport with parabolic shaped constrictions. The model takes into account the parallel contribution of partially and fully formed localized current pathways spanning the two-dimensional (2D) film characterized by transmission coefficients $T<1$ and $T=1$, respectively. It is shown how the resulting physical equation for a highly asymmetric constriction can be linked to an equivalent electrical circuit. The proposed approach unveils the connection between filamentary electron transport and diode-like conduction in resistive switching (RS) devices.**

Keywords—Resistive Switching, Graphene, Boron Nitride

I. INTRODUCTION

Electron devices based on two-dimensional (2D) materials as a way to enhance their range of applicability have attracted a great deal of attention in the last years. This strategy has been not only applied for field-effect transistors [1] but also pursued in the field of memory devices such as the resistive RAMs (RRAMs) [2]. Recent publications point out the remarkable physical properties of hexagonal boron nitride (*h*-BN) as insulator material: bandgap between 5.2 and 5.9 eV, relative permittivity between 2 and 4, large tunneling barrier (~3 eV), and dielectric strenght of about 7 MV/cm [3] as well as the capability of reversibly switching its resistance by means of ramped voltage signals [2]. *h*-BN is chemically and thermally stable and, because of its 2D nature, free of dangling bonds and surface charge traps. Moreover, *h*-BN is an isomorph of graphene (conductor material) with a lattice mismatch of ~1.7% [4]. The graphene/*h*-BN material system has not only interesting quantum properties in connection with the capacitance of the system [5] but has also been shown to play a role in the reduction of power consumption in RRAMs [4]. In this work, the conduction characteristics of electroformed *h*-BN sandwiched between multi-layer graphene (MLG) and metal (M) electrodes are investigated. The application of a forming voltage (initial ramp) generates single or multifilamentary current pathways across the *h*-BN, mainly at the grain boundaries, that are able to switch under bias of opposite sign between a high (HRS) and a low (LRS)

resistance state. The Landauer expression for multi-filamentary conduction with parabolic shaped constrictions is used as the starting point [6]. A similar quantum approach has been followed to model the post-breakdown *I-V* characteristics of SiO_2 [7], HfO_2 [8], SiO_x [9] and NiO-based [10] RS devices. We show that the physical model can be represented in terms of an equivalent circuit model solvable using the Lambert W function [11]. The proposed approach sheds light on the connection between filamentary and diode-like conduction models for RS devices.

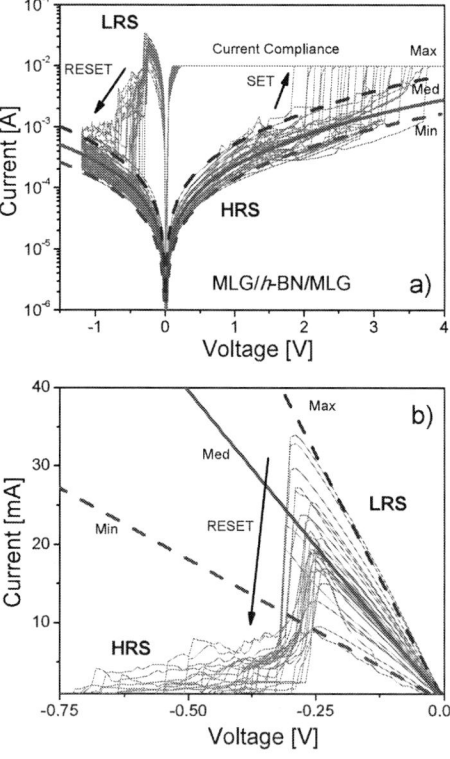

Fig. 1: a). Experimental and model results for the HRS *I-V* curves using (11). The grey lines are experimental data. The colour lines are the fitting results to the minimum ($N=40, R_S=600\Omega$), median ($N=65, R_S=350\Omega$) and maximum ($N=145, R_S=120\Omega$) HRS *I-V* curves. In all cases $n=0, \varphi=3.5eV, \alpha=1eV^{-1}$, and $\beta=1$. b) Detail of the experimental and model results for the LRS characteristics. The dashed lines are fitting results ($N=n\gg1, R_S=7.3,12,27\Omega$).

978-1-5090-5979-9/17 $31.00 © 2017 IEEE

II. Experimental Details

Ti/MLG/*h*-BN/MLG/Au RRAM devices were fabricated with assistance of standard transfer process. The process is as follows: a wafer of 300 nm SiO$_2$/Si was coated with 10 nm Ti (first) and 50 nm Au (second). The Au surface was functionalized via oxygen plasma etching process for 1.5 minutes in O$_2$ atmosphere using a power of 200 W to increase the roughness of the substrate. This favours the adhesion of the 2D material and reduces the amount of wrinkles. A MLG sheet, a thin *h*-BN sheet (previously grown on Ni-doped Cu), and another MLG sheet were sequentially transferred, and Au/Ti electrodes were evaporated on top by shadow mask and e-beam evaporator. For comparison, devices with metal electrodes (Ti and Cu) were also fabricated by chemical vapour deposition (CVD). The devices were characterized using a Cascade M150 probe station and a Keithley 4200-SCS semiconductor parameter analyzer. The devices require an initial electroforming step (not shown). After this, they were voltage ramped with positive and negative biases using a current compliance of 10 mA for the set process. Figure 1.a illustrates a typical RS experiment for MLG/*h*-BN/MLG devices. Figure 1.b shows the details of the *I-V* curves in the LRS regime. Further experimental details about the devices investigated here can be found in [12].

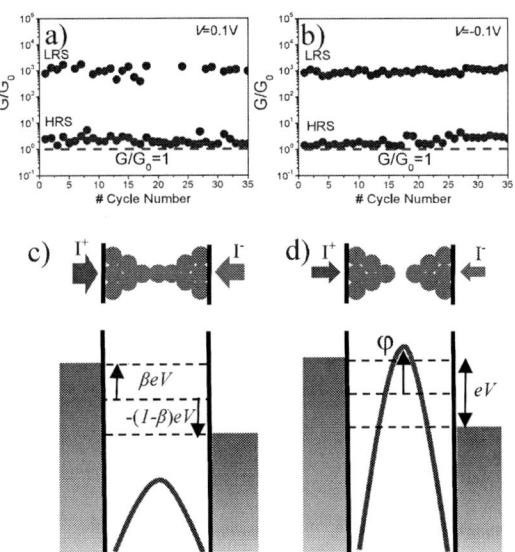

Fig.2: a) Normalised conductance (positive voltage) as a function of the cycle number (data from Fig. 1.a), b) Normalised conductance (negative voltage) as a function of the cycle number (data from Fig. 1.a), c) energy diagram for a wide constriction (R_S=0), d) energy diagram for a narrow constriction (R_S=0). β is the the fraction of the applied bias that drops at the source side of the constriction and φ is the potential barrier height measured from the equilibrium Fermi energy.

In this work, the attention will be focused exclusively on the stable conduction states. Figures 2.a and 2.b shows HRS and LRS conductance ($G=I/V$) values measured at 0.1V and -0.1V, respectively, for the curves shown in Fig. 1. Notice that the HRS conductances for both positive and negative voltages are

close to the quantum conductance unit $G_0=2e^2/h$=77.5μS, where e is the electron charge and h the Planck's constant. This is the signature of electron transport through atomic-sized constrictions and what motivates the use of the Landauer approach [13,14].

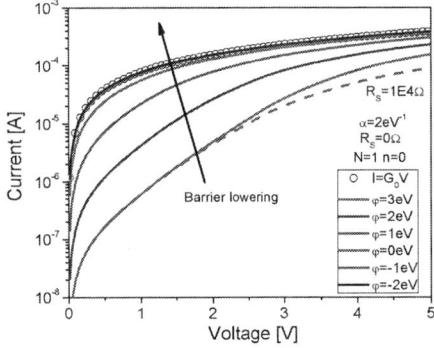

Fig.3: Transition from a partially to a completely formed filament as a consequence of the confinement barrier decrease. The curves were calculated using expression (3). Notice the effect of the series resistance at the largest currents. Negative barrier means below the equilibrium Fermi energy.

III. Model Equations and Fitting Results

For N independent conducting filaments (CF) connecting two electron reservoirs, the current can be expressed as [13]:

$$I = \frac{2e}{h}\int_{-\infty}^{+\infty} \Sigma T \{H[E-\beta e(V-R_SI)] - H[E+(1-\beta)e(V-R_SI)]\}dE \quad (1)$$

where

$$\Sigma T = n + (N-n)T(E) \quad (2)$$

V is the applied voltage, $0\leq\beta\leq1$ the fraction of V-IR_S that drops at the source side of the constrictions, n refers to the number of fully formed CFs (Fig.2.c) and H is the Heaviside function that replaces the Fermi-Dirac distribution function at zero temperature. This is an effective approach which assumes two separate sets of CFs: those with a large cross section (Fig. 2.c: potential barrier height well under the anode quasi-Fermi energy level) and those with a small cross section (Fig. 2.d: potential barrier height above the cathode quasi-Fermi energy level). This distinction corresponds to fully (T=1) and partially formed (T<1) CFs, respectively, where T is the transmission coefficient for a single filament. For an inverted parabolic confinement barrier, (1) yields:

$$I = \frac{N}{R_0+NR_S}V + \frac{(N-n)}{e\alpha(R_0+NR_S)}\ln\left[\frac{exp\{\alpha[\varphi-\beta e(V-IR_S)]\}+1}{exp\{\alpha[\varphi+(1-\beta)e(V-IR_S)]\}+1}\right] \quad (3)$$

where α and φ are parameters related to the width and height of the potential barrier, respectively. $R_0=G_0^{-1}$ and R_S is the series resistance (likely associated in this case with non-broken *h*-BN layers, electrodes and external circuit). (3) is an implicit equation for the current that can be explicitly solved in a few particular cases. Figure 3 shows some numerical results using (3). Notice that the potential barrier depicted in

Fig. 2 is a not a material barrier but a consequence of the confinement effect on the electron wavefunction when passing through the narrowest point along the filament [7]. The barrier represents the bottom of the first conduction subband. (3) was obtained using the transmision coefficient:

$$T(E,V) = \frac{1}{1+exp[-\alpha(E-\varphi)]} \qquad (4)$$

where E is the energy measured from the equilibrium Fermi level. For a large tunneling barrier, i.e. a very narrow constriction (Fig.2.d), (4) can be approximated as:

$$T(E,V) \approx exp[\alpha(E-\varphi)] \qquad (5)$$

Disregarding for the moment the fully formed filaments (n=0), and using (5), (1) can be decomposed as:

$$I = I^+ - I^- \qquad (6)$$

where

$$I^+ = \frac{2eN}{\alpha h}\exp(-\alpha\varphi)\,exp\{\alpha e\beta(V-IR_S)\} \qquad (7)$$

$$I^- = \frac{2eN}{\alpha h}\exp(-\alpha\varphi)\,exp\{-\alpha e(1-\beta)(V-IR_S)\} \qquad (8)$$

which correspond to the right and left-going current components shown in Figs. 2.c and 2.d. (7) and (8) hold for $\beta e(V-IR_S)<\varphi$. The best fitting results for our HRS I-V experimental data is achieved using $\beta \approx 1$, which is an extreme case of the theory and corresponds to a highly asymmetric constriction. A similar behavior was reported in [15,16]. Then, from (7) and (8), and taking into account now the presence of n out of N fully formed CFs, the current reads:

$$I = G_0 n(V-IR_S) + \frac{2e(N-n)}{\alpha h}\exp(-\alpha\varphi)\{exp[\alpha e(V-IR_S)]-1\} \quad (9)$$

which is still an implicit function of I but can be regarded as a diode with parallel and series resistances (see Fig. 4). Using the Thévenin transformation, (9) can be explicitly solved as:

$$I = \frac{VG_P - I_0}{1+R_S G_P} + \frac{1}{\alpha e R_S}W\left\{\frac{\alpha e I_0 R_S}{1+R_S G_P}exp\left[\frac{\alpha e(V+R_S I_0)}{1+R_S G_P}\right]\right\} \quad (10)$$

where W is the Lambert function [11], i.e. the solution of the transcendental equation $We^W = x$ and:

$$G_P = G_0 n \qquad (11)$$

and

$$I_0 = \frac{2e(N-n)}{\alpha h}\exp(-\alpha\varphi) \qquad (12)$$

We can say that (10) establishes the connection between filamentary conduction at a mesoscopic level with the circuital approach (diode-like behavior) [10,16,17]. Model results for the minimum, median, and maximum experimental I-V curves are shown in Figs. 1 and 5. Notice that for the experimental curves shown in Fig. 5, the voltage range is larger than that in Fig. 1 and the current dispersion in the positive HRS region significantly higher. The system MGL/h-BN/MGL exhibits a

high variability not only from cycle-to-cycle but also from device-to-device.

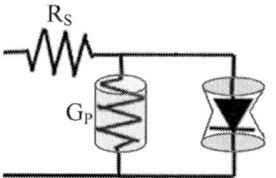

Fig.4: Equivalent circuit model for filamentary conduction under positive bias. For negative bias the diode needs to be inverted.

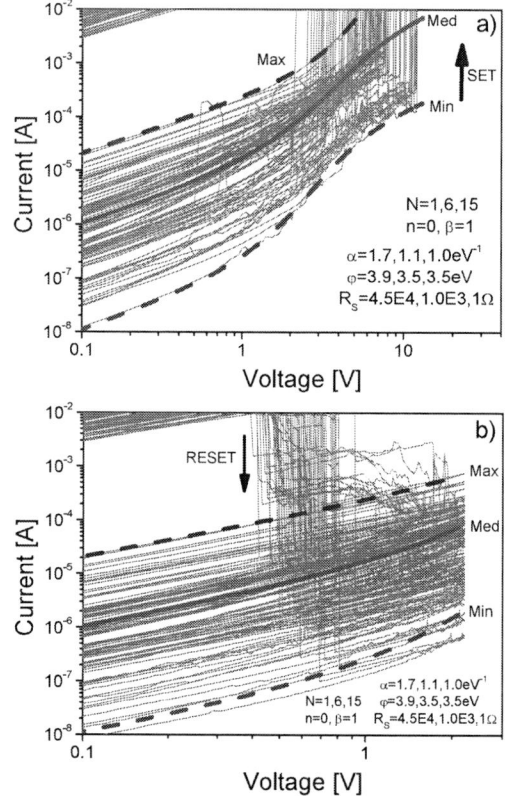

Fig.5: Experimental and model results using (11). a) log-log scale positive voltage, (b) log-log scale negative voltage.

For comparison, Figures 6 and 7 show full RS cycles and partial reset switchings in Ti/h-BN/Cu devices. The transition voltages are much lower than in the previous case. In Fig.7, N=1 and n=0 are considered (no fully formed filament). The issue whether the reset jumps in the I-V curve correspond to a self-healing process of the individual 2D planes of h-BN still needs confirmation [18]. A barrier change of 0.3-0.4eV per jump is detected. Knowing α and φ, the barrier width t at the equilibrium Fermi level (E=0 in Fig.2.c) can be calculated from:

$$t = \frac{\alpha h}{\pi^2}\sqrt{\frac{\varphi}{2m^*}} = 0.24\alpha\sqrt{\varphi}\ [nm] \qquad (13)$$

where $m^*=0.26m$ is the electron effective mass in h-BN [19] and m the free electron mass. The obtained results are compatible with the distances between a few number of h-BN atomic planes. The interlayer distance in h-BN is in the range 0.36nm [19]-0.43nm [20]. In addition, the radius of the filament's bottleneck can be calculated assuming a tube-like constriction with a hard-wall potential [8]:

$$r = \frac{hZ_0}{2\pi\sqrt{2m^*\varphi}} = \frac{0.92}{\sqrt{\varphi}} \ [nm] \qquad (14)$$

where $Z_0=2.404$ is the first zero of the J_0 Bessel function. The obtained results for r are in the nm range (see Fig.7).

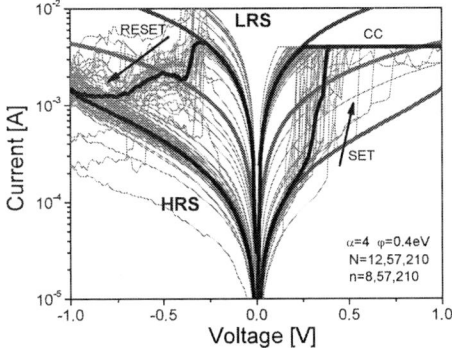

Fig.6: RS cycles in Ti/h-BN/Cu devices. The black heavy solid line is the median curve. Colour lines are simulations.

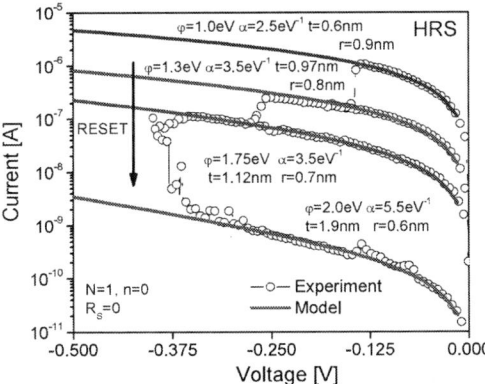

Fig.7: Partial resets in Ti/h-BN/Cu devices. Symbols and solid lines are experimental and model results using (10), respectively.

IV. CONCLUSIONS

A compact model for the low and high resistance states of capacitors with h-BN as insulating layer (graphene and metal electrodes) was presented. The model is based on the Landauer approach for mesoscopic systems. The model is able to account for the linear and non-linear I-V characteristics as well as the their large spread (variations in the number of filaments and shape of the confinement potential barrier).

ACKNOWLEDGMENT

This work was supported in part by the PANACHE EU Project and the DURSI through the Generalitat de Catalunya under Grant 2014SGR384.

REFERENCES

[1] Q. Weng, X. Wang, Y. Bando, D. Golberg, Functionalized hexagonal boron nitride nanomaterials: emerging properties and applications, Chem Soc Rev 45, 3989 (2016)

[2] K. Qian, R. Tay, V. Nguyen, J. Wang, G. Cai, T. Chen, E. Teo, P. Lee, Hexagonal boron nitride thin film for flexible resistive memory applications, Adv Funct Mater 26, 2176 (2016)

[3] G. Lee, Y. Tu, C. Lee, C. Dean, K. Shepard, P. Kim, J. Hone, Electron tunneling through atomically plat and ultrathin hexagonal boron nitride, App Phys Lett 99, 243114 (2011)

[4] Y. Ji, C. Pan, M. Zhang, S. Long, X. Lian, F. Miao, F. Hui, Y. Shi, L. Larcher, E. Wu, M. Lanza, Boron nitride as two dimensional dielectric: reliability and dielectric breakdown, Appl Phys Lett 108, 012905 (2016)

[5] G. Shi, Y. Hanlumyuang, Z. Liu, Y. Gong, W. Gao, B. Li, J. Kono, J. Lou, R. Wajtai, P. Sharma, P. Ajayan, Boron nitride-graphene nanocapacitor and the origins of anomalous size-dependent increase of capacitance, Nano Lett 14, 1739 (2014)

[6] S. Datta, in Electronic Transport in Mesoscopic Systems, Cambridge University Press, 1995

[7] E. Miranda, J. Suñé, Electron transport through broken down ultra-thin SiO$_2$ layers in MOS devices, Mic Rel 44, 1 (2004)

[8] E. Miranda, C. Walczyk, C. Wenger, T. Schroeder, Model for the resistive switching effect in HfO$_2$ MIM structures based on the transmission properties of narrow constrictions, IEEE Elect Dev Lett 31, 609 (2010)

[9] E. Miranda, A. Mehonic, J. Suñé, A.J. Kenyon, Multi-channel conduction in redox-based resistive switch modelled using quantum point contact theory, Appl Phys Lett 10, 222904 (2013)

[10] S. Oliver, J. Fairfield, A. Bellew, S. Lee, J. Champlain, L. Ruppalt, J. Boland, P Vora, Quantum point contacts and resistive switching in Ni/NiO nanowire junctions, Appl Phys Lett 109, 203101 (2016)

[11] R. Corless, G. Gonnet, D. Hare, D. Jeffrey, D. Knuth, On the Lambert W function, Adv Comp Math 5, 329 (1996)

[12] C. Pan, E. Miranda, M. Villena, N. Xiao, X. Jing, X. Xie, T. Wu, F. Hui, Y. Shi, M. Lanza, Model for multi-filamentary conduction in graphene/hexagonal-boron-nitride/graphene based resistive switching devices, 2D Mater 4, 025099 (2017)

[13] J. van Ruitenbeek, M. Morales Masis, E. Miranda, Quantum point contact conduction, in *Resistive switching: From fundamentals of nanoionic redox processes to memristive device applications*, Wiley, 197-224 (2016)

[14] W. Yi, S. Savel'ev, G. Medeiros-Ribeiro, F. Miao, M. Zhang, J. Joshua Yang, A. Bratkovsky, R. Stanley Williams, Quantized conductance coincides with state instability and excess noise in tantalum oxide memristors, Nature Communications 7, 11142 (2016)

[15] X. Lian, X. Cartoixà, E. Miranda, L. Perniola, R. Rurali, S. Long, M. Liu, J. Suñé, Multi-scale quantum point contact model for filamentary conduction in resistive random access memory devices, J App Phys 115, 244507 (2014)

[16] A. Grossi, C. Zambelli, P. Olivo, A. Crespo-Reyes, J. Martin-Martinez, R. Rodríguez, M. Nafría, E. Perez, C. Wenger, Electrical characterization and modeling of 1T-1R RRAM arrays with amorphous and poly-crystalline HfO$_2$, Solid-St Electron 128, pp 187-193 (2017)

[17] J. Blasco, N. Ghenzi, J. Suñé, P. Levy, E. Miranda, Equivalent circuit modeling of the bistable conduction characteristics in electroformed thin dielectric films, Mic Rel 55, pp 1-14 (2015)

[18] Y. Hattori, T. Taniguchi, K. Watanabe, K. Nagashio, Layer-by-layer dielectric breakdown of hexagonal boron nitride, ACS Nano 9, pp 916-921 (2015)

[19] G. Lee, Y. Yu, C. Lee, C. Dean, K. Shepard, P. Kim, J. Hone, Electron tunneling through atomically flat and ultrathin hexagonal boron nitride, Appl Phys Lett 99, 213114 (2011)

[20] L. Britnell, R. Gorbachev, R. Jalil, B. Belle, F. Schedin, M. Katsnelson, L. Eaves, S. Morozov, A. Mayorov, N. Peres, A. Castro Neto, J. Leist, A. Geim, L. Ponomarenko, K. Novoselov, Electron tunneling through ultrathin boron nitride crystalline barriers, Nano Lett 12, pp 1707-1710 (2012)

Analytical Drain Current Model for Non-Ballistic Schottky-Barrier CNTFETs

Igor Bejenari[*§], Michael Schröter [*†‡] and Martin Claus[*‡]

*Chair for Electron Devices and Integrated Circuits, Department of Electrical and Computer Engineering
Technische Universität Dresden, 01062 Dresden, Germany
†Dept. of Electrical and Computer Engineering, University of California at San Diego, La Jolla, CA 92093 USA
‡Center for Advancing Electronics Dresden (Cfaed), Technische Universität Dresden, 01062 Dresden, Germany
§Institute of Electronic Engineering and Nanotechnologies, Academy of Sciences of Moldova, MD 2028 Chisinau, Moldova
Email: igor.bejenari@tu-dresden.de, michael.schroeter@tu-dresden.de, and martin.claus@tu-dresden.de

Abstract—A new analytical static drain current model based on the WKB approximation has been developed for Schottky-Barrier CNTFETs. Electron scattering by acoustic and optical phonons in the channel has been taken into account. By using a simple approximation of both the Fermi-Dirac distribution function and transmission probability, an analytical expression for the drain current in the Landauer–Büttiker formalism has been obtained. This allows to overcome the limitations of existing models and to extend their applicability toward high bias voltages as needed for analog applications. The model results agree well with experimental data.

I. Introduction

High-mobility CMOS channel materials, gate-all-around (nanowire) structures, supply voltages lower than 0.6 V, controlling source/drain series resistance within tolerable limits, and fabrication of advanced non–planar multi–gate and nanowire MOSFETs with gate lengths below 10 nm represent milestones of the International Roadmap for Devices and Systems (IRDS) in the development of CMOS technology for the next decades [1]. Carbon–nanotube FETs (CNTFETs) may satisfy these requirements [2], [3], [4]. Due to a low channel resistance, especially for short-channel devices, the metal-CNT contact resistance can significantly affect the performance of CNTFETs. Many different factors like the type of metal contact, the surface preparation, annealing conditions, the CNT diameter, and the electrostatic potential define the contact resistance [5], [6]. For analytical calculations, it is commonly assumed that the metal-CNT interface can be described as a Schottky barrier (SB) [7], [8].

Defined by the interface quality and physical contact length, the contact–CNT interface resistance can be estimated by simple fitting of the SB characteristics [9], [10]. Along with SB tunneling, electron–phonon scattering becomes important for long-channel devices [11], [12]. Based on the nonequilibrium Greens function (NEGF) method, Wigner transport equation, and Boltzmann equation formalism, computations of the SB-CNTFET transport properties are extremely time consuming [2], [7]. For integrated circuit design, compact models based on analytical expressions are needed for simulations in a SPICE–like environment. Assuming an energy independent transmission probability [7], analytical models agree with experimental data only in a limited bias range. Simplified semi–analytical models [13] solving the current integral involved in the transport calculations numerically are

a compromise between accuracy and computational efficiency, but are still not suitable for circuit simulation and optimization.

To consider both SB tunneling and electron–phonon scattering, a semi-analytical model based on the numerical solution of the current integral taking into account SB contacts and Büttiker's probe approach has been proposed in [14]. This rather sophisticated method has not been validated with experimental data yet. While the semi–empirical CNTFET model [10] and virtual source models [15] have been experimentally verified, they cannot be used for predictions.

In this paper, we propose an analytical drain current model for dissipative transport in SB-CNTFETs, which is free from numerical integrations. The pseudo-bulk approximation is exploited to calculate the channel potential variation under applied bias with respect to channel charge [16], [17]. The model captures a number of features such as transmission through the SB contacts, electron scattering by acoustic and optical phonons, and ambipolar conduction. The proposed model combines the analytical drain current models for both the ballistic transport in SB-CNTFETs [18] and dissipative transport in ohmic CNTFETs [19]. It can significantly improve already existing empirical compact models by providing analytical physics-based transport equations.

II. Transport Model

A. Energy Band Model

We adopted the evanescent mode analysis approach [14] to obtain the band edge profile. In the vicinity of the source and drain contacts, the conduction band bending in the left ($E_{\mathrm{L}}^{\mathrm{c}}$) and right ($E_{\mathrm{R}}^{\mathrm{c}}$) regions are given by exponential functions

$$E_{\mathrm{L}}^{\mathrm{c}}(z) = E_b^s \exp\left(-\frac{z}{\lambda}\right), \qquad (1)$$

$$E_{\mathrm{R}}^{\mathrm{c}}(z) = E_b^d \exp\left(\frac{z - L}{\lambda}\right), \qquad (2)$$

where λ is a characteristic length of the decaying electrostatic potential, L is the total length of the channel, which is assumed to equal the gate length L_g, and $E_b^{s(d)} = \phi_b + q\psi_{cc} - E_{m,0} - qV_{s(d)}$ is the bias dependent potential barrier height with respect to the bottom of the mth conduction subband at the source and drain contacts, respectively. The latter is taken as the energy reference, while

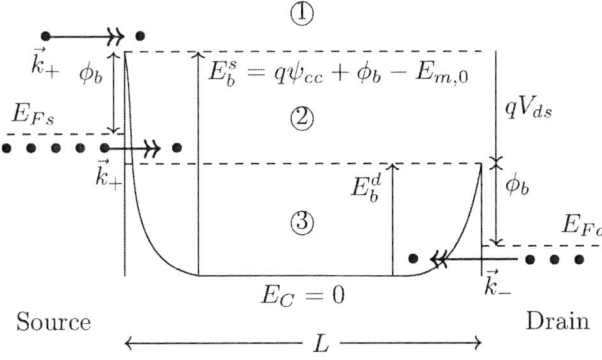

Fig. 1. Energy band diagram describing the thermionic current in region 1, source tunneling in region 2, as well as both source and drain tunneling in region 3. The reference point of energy E_C corresponds to the conduction subband edge in the channel. At equilibrium, the Fermi level E_{Fs} of the source contact is set in the middle of the CNT bandgap E_g.

$E_{m,0} = |3m - n_1 + n_2| V_\pi a_{cc}/d_{CNT}$ corresponds to the conduction band edge at thermal equilibrium [20]. The channel potential, ψ_{cc}, at the channel center, i.e., current control point, is obtained as a solution of the charge conservation equation in the framework of pseudo-bulk approximation [16], [21], [22]. For gate-all-around CNTFETs, the asymptotic value of λ can be estimated by using a simple expression if the CNT diameter d_{CNT} is smaller than the oxide thickness [15].

Based on (1) and (2), the energy band diagram of the SB-CNTFET is shown in Fig. 1. The source Fermi level is defined as $E_{Fs} = q\psi_{cc} - E_{m,0}$. The metal–semiconductor barrier height is determined by the bias independent constant parameter, ϕ_b. For holes, ϕ_b^h equals $E_g - \phi_b$, where $E_g = 2E_{m,0}$ is a band gap. The source and drain Fermi levels E_{Fs} and E_{Fd}, respectively, are related as $E_{Fd} = E_{Fs} - qV_{ds}$, where $V_{ds} = V_d - V_s$ is the drain–source voltage. For the sake of simplicity, we suppose that $V_s = 0$.

Fig. 1 also indicates different injection mechanisms in the SB-CNTFET. Having enough energy to overcome the potential barrier, thermally excited electrons in region 1 are mainly injected from the source over the barrier into the channel without reflection. The electrons with energy belonging to region 2 tunnel from the source through the SB into the channel, where they propagate towards the drain contact. Since these electrons overcome the barrier in the drain region due to their large energy, they are absorbed by the drain contact without reflection. Belonging to region 2, thermally excited electrons injected from the drain freely propagate in the channel and may tunnel through the SB located at the source. Region 3 is limited by the electron subband edge and top of the drain barrier. In region 3, electrons tunnel from both the source and drain through the SBs into the channel. These electrons are absorbed by the source or drain contacts after multiple reflections between the barriers. The electron scattering by optical phonons near the drain contact leads to an accumulation of the back-scattered electrons at the bottom of the conduction subband. This prevents the injection of electrons from the drain into the channel due to the Pauli exclusion principle.

B. Piece-Wise Approximation of Fermi-Dirac Distribution Function

In order to simplify the analytical calculation of the current, we use a piece-wise approximation for the electron distribution function as follows [18].

$$f_{\text{app}}(E) = \begin{cases} 1 - \frac{1}{2}\exp\left(\frac{E - E_F}{c_1 k_B T}\right), & E \le E_F \\ \frac{1}{2}\exp\left(\frac{E_F - E}{c_1 k_B T}\right), & E_F < E < E_F + c_2 k_B T \\ \exp\left(\frac{E_F - E}{k_B T}\right), & E \ge E_F + c_2 k_B T \end{cases}$$
(3)

where $c_1 = 2\ln(2)$ and $c_2 \approx 2.49$.

C. SB Transmission Probability

We use the Wentzel–Kramers–Brillouin (WKB) approximation to obtain the electron transmission probability to tunnel through the SB. In the framework of the parabolic band approach, the approximate transmission probability $T_{\text{SB}}^{s(d)}(E)$ is given by the following expression [18]

$$T_{\text{SB}}^{s(d)}(E) = \exp\left\{-\alpha\sqrt{E_b^{s(d)}}\gamma\left(E/E_b^{s(d)}\right)\right\},$$
(4)

$$\gamma(x) = px - (p+1)\sqrt{x} + 1,$$
(5)

where $p = 0.7113$ and $\alpha = 8\lambda\sqrt{2E_{m,0}}/(\sqrt{3}aV_\pi)$ with the carbon–carbon atom distance $a = 2.49\,\text{Å}$ and carbon $\pi - \pi$ bond energy $V_\pi = 3.033\,\text{eV}$ in the tight binding model [20].

Taking into account electron–phonon scattering, the total transmission probability reads

$$T_{2b}(E) = T_{\text{SB}}^s(E)T_{\text{ch}}(E)T_{\text{SB}}^d(E).$$
(6)

In region 3 at $E_b^s > 0$ and $E_b^d > 0$, both SBs located at the source and drain affect the current. However, the multiple reflections of electrons between the two potential barriers in region 3 of Fig. 1 represent a second order contribution to the net current, which, therefore, is neglected. In region 2, electrons only tunnel through the potential barrier located at the source, hence, $T_{\text{SB}}^d(E) = 1$. The transmission probability $T_{\text{ch}}(E)$ describing the electron-phonon scattering in the channel is given by [23]

$$T_{\text{ch}}^{-1}(E) = T_{\text{ac,low}}^{-1} + \theta(E > \hbar\omega_{\text{op}})(1 - T_{\text{op}})/T_{\text{op}},$$
(7)

where $\theta(x)$ is Heaviside step function. The optical phonon energy is $\hbar\omega_{\text{op}} \approx 0.16$ eV. The electron transmission probability T_{op} is approximated by $T_{\text{op}} = \lambda_{op}/(\lambda_{op} + L)$, where λ_{op} is the mean free path (mfp) of electrons scattered only by optical phonons. Scattered by acoustic phonons, the electrons with a low energy propagate through the channel with the probability $T_{\text{ac,low}} = l_{ac}/(l_{ac} + L)$, where l_{ac} is the effective mfp [23].

D. Total Current

To calculate the total electron current for CNTFETs, we use the Landauer-Büttiker approximation. We neglect band-to-band and direct source-to-drain tunneling. In the case that electrons are only scattered by acoustic phonons, the source and drain components of the current are given by

$$I_1^{s(d)} = \frac{4qe^{\delta C}T_{\text{ac,low}}}{h}\int_0^\infty T_{\text{SB}}^s(E)T_{\text{SB}}^d(E)f_{\text{app}}(E - E_{Fs(d)})dE,$$
(8)

where δ is a correction parameter [18]. If $E_b^d < E < E_b^s$, electrons tunnel only through the SB located at the source, then $C = \alpha\sqrt{E_b^s}$. If $E < E_b^d$, electrons tunnel through both SBs located at the source and drain, then $C = \alpha(\sqrt{E_b^s} + \sqrt{E_b^d})$. The analytical solution of the current integral (8) is given in [18]. In the case that electrons are scattered by both the acoustic and optical phonons, the source and drain components of current are given as

$$
\begin{aligned}
I_2^s = \frac{4qe^{\delta C}}{h} &\left\{ (T_{\mathrm{ch}} - T_{\mathrm{ac,low}}) \right. \\
&\times \int_0^\infty T_{\mathrm{SB}}^s(E) T_{\mathrm{SB}}^d(E) f_{\mathrm{app}}(E - E_{Fs} + \hbar\omega_{\mathrm{op}}) dE \\
&\left. + T_{\mathrm{ac,low}} \int_0^\infty T_{\mathrm{SB}}^s(E) T_{\mathrm{SB}}^d(E) f_{\mathrm{app}}(E - E_{Fs}) dE \right\}, \quad (9)
\end{aligned}
$$

$$
I_2^d = \frac{4qe^{\delta C} T_{\mathrm{ch}}}{h} \int_0^\infty T_{\mathrm{SB}}^s(E) T_{\mathrm{SB}}^d(E) f_{\mathrm{app}}(E - E_{Fd} + q\Delta) dE, \quad (10)
$$

where the parameter Δ is included to take into account the back-scattering of electrons near the drain contact, which prevents the injection of electrons from the drain into the channel [19]. The integrals in (8)-(10) can be solved analytically in terms of complementary error function and Dawson's integral [18]. The total electron current is $I_n(V_{gs}, V_{ds}) = I_n^s(V_{gs}, V_{ds}) - I_n^d(V_{gs}, V_{ds})$.

Using the property of the electron–hole symmetry of the band structure in CNT, we define the total current I_{amb} for ambipolar CNTFETs as

$$
I_{amb}^n(V_{gs}, V_{ds}) = I_n(V_{gs}, V_{ds}) + I_n(V_{ds} - V_{gs}, V_{ds}), \quad (11)
$$

where V_{gs} is a gate voltage. The first (second) term corresponds to a contribution of electrons (holes) to the total current in (11).

The proposed model including (8)–(11) covers all bias regions. The corresponding drain current is smooth and differentiable without application of smoothing functions.

III. RESULTS

Fig. 2 shows the transfer characteristics obtained with the analytical model for a bottom gate n–type SB-CNTFET with a gate length of 100 nm. For the calculations, we use the following parameters: $\Delta = 0.1 V_{ds}$, $\lambda_{ac} = 963$ nm, $\lambda_{op} = 28$ nm. The combination of electron–phonon scattering and SB reflections leads to a significant decrease of the total current. Fig. 3 depicts the calculated dependence of the intrinsic transconductance, $g_m^* = \partial I / \partial \psi_{cc}$, on the tube potential ψ_{cc} at different values of drain–source voltage V_{ds}. The electron scattering by optical phonons causes the sharp variation of g_m^* with ψ_{cc}. Electron reflections at the SB smooth the calculated transconductance.

Fig. 4 compares the output characteristics with experimental data [24] for a bottom gate p–type CNTFET with a gate length of 300 nm. The gate oxide (HfO$_2$) thickness is 10 nm. The relative dielectric constant of HfO$_2$ is about 18.

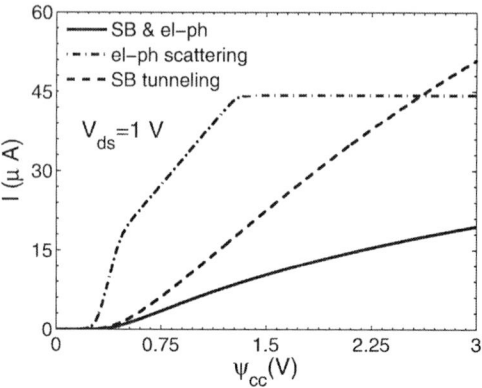

Fig. 2. Total current I calculated analytically as a function of tube potential ψ_{cc} at the drain–source voltage $V_{ds} = 1$ V. The dashed (dot-dashed) line corresponds to the current with only SB tunneling (electron-phonon scattering) included. CNT chirality (19, 0), bandgap $E_g = 0.579$ eV, CNT diameter $d_{CNT} = 1.48$ nm, SB hight $\phi_b = E_g/2$, characteristic length $\lambda = 3$ nm, gate length $L_g = 100$ nm, correction parameter $\delta = 0.0045$ and temperature $T = 300$ K.

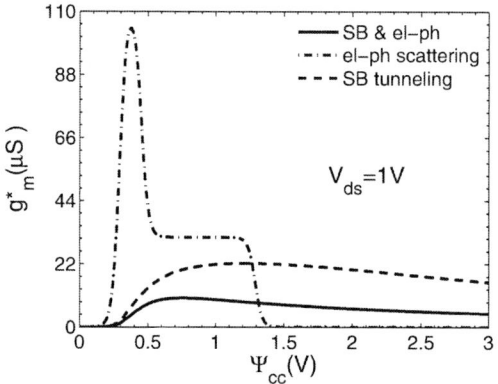

Fig. 3. Intrinsic transconductance g_m^* calculated analytically as a function of tube potential ψ_{cc} at the drain–source voltage $V_{ds} = 1$ V. The dashed (dot-dashed) line corresponds to the current with only SB tunneling (electron-phonon scattering) included. CNT chirality (19, 0), bandgap $E_g = 0.579$ eV, CNT diameter $d_{CNT} = 1.48$ nm, SB height $\phi_b = E_g/2$, characteristic length $\lambda = 3$ nm, gate length $L_g = 100$ nm, correction parameter $\delta = 0.0045$ and temperature $T = 300$ K.

In the experiments, electron scattering by optical phonons is supposed to be negligible [24]. There is only acoustic phonon scattering in this case. To fit the current calculated by using (8) to experimental data, a bias independent SB height $\phi_b = 0.13$ eV and characteristic length $\lambda = 2$ nm are assumed. Also, the effective mfp l_{ac} was chosen to be equal to 62, 152, and 244 nm at different gate voltage $V_{gs} = 1.25$, 1.5, and 1.75 V, which agree with experiment [25], [12], [24]. Due to dependence on the density of states, the electrons with a low energy near the conduction band edge are scattered by acoustic phonons more intensively compared to electrons with a high energy. As a result, the effective mfp l_{ac} of electrons scattered by acoustic phonons should depend on applied bias. The electron–phonon scattering as described in [23], [19] should be modified in order to consider this effect.

978-1-5090-5979-9/17 $31.00 © 2017 IEEE

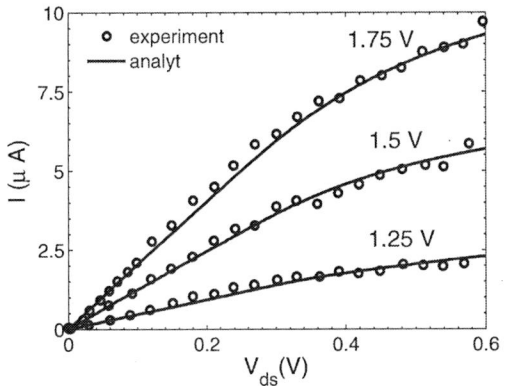

Fig. 4. Source–drain current I calculated with the proposed analytical model as a function of drain–source voltage V_{ds} at a gate voltage V_g equal to 1.25, 1.5, and 1.75 V in comparison with the output characteristics (data are from [24]). CNT chirality (16, 0), bandgap $E_g = 0.687$ eV, CNT diameter $d_{CNT} = 1.25$ nm, SB height $\phi_b = 0.3$ eV, characteristic length $\lambda = 2$ nm, gate length $L_g = 300$ nm, gate oxide capacitance per unit length $C_{ox} = 284$ aF/um, correction parameter $\delta = 0.0045$ and temperature $T = 300$ K. The effective mfp l_{ac} equals 62, 152, and 244 nm at $V_g = 1.25$, 1.5, and 1.75 V, correspondingly. The electron scattering by optical phonons is negligible. The polarity of V_g and V_{ds} are flipped compared to the original data to become n-type CNTFET.

IV. CONCLUSION

An analytical drain current model for SB-CNTFETs has been proposed that takes into account electron–phonon scattering in the channel. To avoid a numerical integration, we have introduced a piece-wise approximation for the Fermi–Dirac distribution function and approximate transmission probability for the barrier, assuming an exponential decay of the potential. The resulting closed-form analytical model formulation significantly reduces the computational effort and simplifies its implementation in Verilog-A. The latter is supported by almost all commercially available circuit simulators and enables fast SPICE-level design and verification of analog, radio frequency (RF), and mixed-signal circuits. Our model can be successfully used for performance predictions of varying SB height, characteristic length, electron mean free path and band gap of the channel material in quasi-1D SB-FETs based on both semiconductor nanowires and CNTs. Hence, the new analytical formulation enables device optimization and variability studies. The comparison with experimental data revealed a deficiency in the treatment of the effective mfp of electrons scattered by acoustic phonons in the existing compact models for CNTFETs.

ACKNOWLEDGMENT

This work was supported in part by a grant from the Cfaed, CAPES project 88881.030371/2013-01, DFG project CL384/2, and DFG project SCHR695/6-2.

REFERENCES

[1] International Roadmap for Devices and Systems. MORE MOORE. White Paper. 2016 Edition. [Online]. Available: http://irds.ieee.org/

[2] J. Guo, S. Datta, and M. Lundstrom, "A numerical study of scaling issues for Schottky-Barrier carbon nanotube transistors," *IEEE Trans. Electron Devices*, vol. 51, no. 2, pp. 172–177, Feb 2004.

[3] C. Qiu, Z. Zhang, M. Xiao, Y. Yang, D. Zhong, and L.-M. Peng, "Scaling carbon nanotube complementary transistors to 5-nm gate lengths," *Science*, vol. 355, no. 6322, pp. 271–276, Jan 2017.

[4] M. Schroter, M. Claus, P. Sakalas, M. Haferlach, and D. Wang, *IEEE J. Electron Devices Soc.*, vol. 1, no. 1, pp. 9–20, Jan 2013.

[5] Z. H. Chen, J. Appenzeller, J.Knoch, Y. M. Lin, and P. Avouris, "The role of metal–nanotube contact in the performance of carbon nanotube field–effect transistors," *Nano Lett.*, vol. 5, no. 7, pp. 1497–1502, Jun 2005.

[6] A. Fediai, D. A. Ryndyk, G. Seifert, S. Mothes, M. Claus, G. Cuniberti, and M. Schröter, "Towards an optimal contact metal for CNTFETs," *Nanoscale*, vol. 8, no. 19, pp. 10 240–10 251, Apr 2016.

[7] C. Maneux, S. Fregonese, T. Zimmer, S. Retailleau, H. N. Nguyen, D. Querlioz, A. Bournel, P. Dollfus, F. Triozon, Y. M. Niquet, and S. Roche, "Multiscale simulation of carbon nanotube transistors," *Solid-State Electron.*, vol. 89, pp. 26–67, Nov 2013.

[8] F. Leonard and A. A. Talin, "Electrical contacts to one- and two-dimensional nanomaterials," *Nature Nanotech.*, vol. 6, no. 12, pp. 773–783, Dec 2011.

[9] A. Pacheco-Sanchez, M. Claus, S. Mothes, and M. Schroter, *Solid-State Electron.*, vol. 125, pp. 161–166, Nov 2016.

[10] M. Schroter, M. Haferlach, A. Pacheco-Sanchez, S. Mothes, P. Sakalas, and M. Claus, "," *IEEE Trans. Electron Devices*, vol. 62, no. 1, pp. 52–60, Jan 2015.

[11] J. Guo and M. Lundstrom, "Role of phonon scattering in carbon nanotube field-effect transistors," *Appl. Phys. Lett.*, vol. 86, no. 19, p. 193103, May 2005.

[12] Z. Zhang, S. Wang, L. Ding, X. Liang, T. Pei, J. Shen, H. Xu, Q. Chen, R. Cui, Y. Li, and L.-M. Peng, "Self-aligned ballistic n-type single-walled carbon nanotube field-effect transistors with adjustable threshold voltage," *Nano Lett.*, vol. 8, no. 11, pp. 3696–3701, Nov 2008.

[13] A. Hazeghi, T. Krishnamohan, and H.-S. P. Wong, "Schottky-barrier carbon nanotube field-effect transistor modeling," *IEEE Trans. Electron Devices*, vol. 54, no. 3, pp. 439–445, Mar 2007.

[14] P. Michetti and G. Iannaccone, "Analytical model of one-dimensional carbon-based Schottky-barrier transistors," *IEEE Trans. Electron Devices*, vol. 57, no. 7, pp. 1616–1625, Jul 2010.

[15] C.-S. Lee, E. Pop, A. D. Franklin, W. Haensch, and H.-S. P. Wong, "A compact virtual-source model for carbon nanotube FETs in the sub-10-nm regime – Part I: Intrinsic elements," *IEEE Trans. Electron Devices*, vol. 62, no. 9, pp. 3061–3069, Sep 2015.

[16] S. Fregonese, H. C. d'Honincthun, J. Goguet, C. Maneux, T. Zimmer, J.-P. Bourgoin, P. Dollfus, and S. Galdin-Retailleau, "Computationally efficient physics–based compact CNTFET model for circuit design," *IEEE Trans. Electron Devices*, vol. 55, no. 6, pp. 1317–1327, Jun 2008.

[17] M. Claus, D. Gross, M. Haferlach, and M. Schroter, "Critical review of CNTFET compact models," in *NSTI-Nanotech (Workshop on Compact Modeling)*, vol. 2, 2012, pp. 770–775.

[18] I. Bejenari and M. Claus. Analytical model of one-dimensional ballistic Schottky-Barrier transistors. [Online]. Available: https://arxiv.org/abs/1703.05092

[19] I. Bejenari and M. Claus, "Electron Back Scattering in CNTFETs," *IEEE Trans. Electron Devices*, vol. 63, no. 3, pp. 1340–1345, Mar 2016.

[20] J. W. Mintmire and C. T. White, *Phys. Rev. Lett.*, vol. 81, no. 12, pp. 2506–2509, Sep 1998.

[21] S. Mothes, M. Claus, and M. Schroter, *IEEE Trans. Nanotechnol.*, vol. 14, no. 2, pp. 372–378, Mar 2015.

[22] A. Rahman, J. Guo, S. Datta, and M. S. Lundstrom, *IEEE Trans. Electron Devices*, vol. 50, no. 9, pp. 1853–1864, Sep 2003.

[23] S. Fregonese, J. Goguet, C. Maneux, and T. Zimmer, "Implementation of electron – phonon scattering in a cntfet compact model," *IEEE Trans. Electron Devices*, vol. 56, no. 6, pp. 1184–1190, Jun 2009.

[24] A. D. Franklin and Z. Chen, "Length scaling of carbon nanotube transistors," *Nature Nanotechnol.*, vol. 5, no. 12, pp. 858–862, Dec 2010.

[25] Z. Yao, C. L. Kane, and C. Dekker, "High-field electrical transport in single-wall carbon nanotubes," *Phys. Rev. Lett.*, vol. 84, no. 13, pp. 2941–2944, Mar 2000.

978-1-5090-5979-9/17 $31.00 © 2017 IEEE

A General Circuit Model for Spintronic Devices Under Electric and Magnetic Fields

Meshal Alawein and Hossein Fariborzi

Computer, Electrical and Mathematical Sciences and Engineering Division
King Abdullah University of Science and Technology
Thuwal 23955-6900, Kingdom of Saudi Arabia
Email: {meshal.alawein,hossein.fariborzi}@kaust.edu.sa

Abstract—**In this work, we present a circuit model of diffusive spintronic devices capable of capturing the effects of both electric and magnetic fields. Starting from a modified version of the well-established drift-diffusion equations, we derive general equivalent circuit models of semiconducting/metallic nonmagnets and metallic ferromagnets. In contrast to other models that are based on steady-state transport equations which might also neglect certain effects such as thermal fluctuations, spin dissipation in the ferromagnets, and spin precession under magnetic fields, our model incorporates most of the important physics and is based on a time-dependent formulation. An application of our model is shown through simulations of a nonlocal spin-valve under the presence of a magnetic field, where we reproduce experimental results of electrical measurements that demonstrate the phenomena of spin precession and dephasing ("Hanle effect").**

I. INTRODUCTION

As the CMOS manufacturing technology approaches sub-10 nm node, the semiconductor chip industry is facing major challenges such as increased power density and degraded energy efficiency that diminish the benefits of further scaling [1]. Fortunately, the past decade has witnessed tremendous research advances on technologies that can possibly replace or augment Si-based CMOS. One possible avenue is the use of the intrinsic electron spin to store and transfer binary information [2]. This emerging technology is known as spintronics and is considered as one of fastest growing "Beyond CMOS" solutions for the advancements of the famous Moore's law. In principle, spintronics can enable faster data transmission and reduced energy per operation and heat dissipation. This technology can also be used in the near future for applications in quantum information processing and cryptography, thus allowing the development of a new paradigm for digital computing [3].

In this paper, we model diffusive spintronic devices with distributed circuit elements based on a circuit-formulation of time-dependent drift-diffusion equations applicable to nonmagnetic and ferromagnetic materials (Fig. 1). The underlying theory considers the presence of both electric and magnetic fields with a few number of assumptions, hence allowing the model to capture most of the known device physics. To be consistent with the notion of the recently established four-component circuit approach (also known as generalized circuit theory) [4], [5], we derive the 4×4 conductance matrices of the nonmagnets and ferromagnets. Finally, the model is benchmarked against experimental results of a nonlocal spin-valve through self-consistent simulations of the stochastic nanomagnet dynamics and the equivalent spin circuits [6].

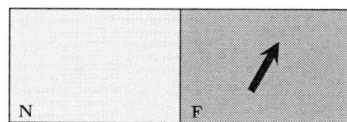

Fig. 1: Schematic of a nonmagnet/ferromagnet (N/F) bilayer.

II. FORMALISM

Below we address the semiclassical model of charge and spin transport based on a modified version of the established drift-diffusion equations [7]. The discussion here is limited to diffusive nonmagnetic materials (the generalization to ferromagnets will be illustrated throughout the development of the circuit model). In addition, in this formalism we assume: (i) High enough temperatures to dispel the question of diffusion, (ii) slow spin relaxation to establish correct equilibrium polarizations, (iii) weak enough external fields such that we can work in the linear response regime (in contrast to Ref. [7]), and (iv) no spin-orbit coupling or spin Coulomb drag [8].

A. Spinor Boltzmann Equation

Let $\Omega \subset \mathbb{R}^3$ be some conducting region. The time-dependent Boltzmann equation which describes the evolution of distribution function can be written in spinor form as [7]

$$\mathcal{L}\hat{f}(t,\mathbf{r},\mathbf{k}) = \mathcal{C}\hat{f}(t,\mathbf{r},\mathbf{k}) \qquad (1)$$

where $\hat{f} \equiv \hat{f}(t,\mathbf{r},\mathbf{k})$ is the distribution function in spin space, $\mathcal{L} = \frac{\partial}{\partial t} + \mathbf{v}\cdot\nabla_{\mathbf{r}} + \frac{1}{\hbar}\mathbf{F}\cdot\nabla_{\mathbf{k}} + \frac{1}{j\hbar}\left[\hat{H},\cdot\right]$ is the Liouvillian operator, $\mathcal{C}$ is the collision operator, $\mathbf{F} = q(\mathbf{E} + \mathbf{v}\times\mathbf{B})$ is the Lorentz force, $q = -e$ is the electron charge, $\mathbf{E}$ is the electric field intensity, $\mathbf{B}$ is the magnetic flux density, $\mathbf{v}$ is the velocity, $\mathbf{k}$ is the wave vector, $\hbar$ is the reduced Planck constant, $\hat{H} = -g\mu_B\boldsymbol{\sigma}\cdot\mathbf{B}/2$ is the spin-dependent Hamiltonian in spin space, μ_B is the Bohr magneton, g is g-factor (typically $g \approx 2$), and $\boldsymbol{\sigma} = (\sigma_x, \sigma_y, \sigma_z)$ is the vector of Pauli spin matrices.

B. Drift-Diffusion Equations

Consider a one-dimensional transport along the z-axis through a uniform cross section A of a nonmagnet. We will assume a homogeneous electric field $\mathbf{E} = E_z \mathbf{a}_z$, where $E_z = -\partial V_C / \partial z$, and set $(\mathbf{v} \times \mathbf{B}) \cdot \nabla_\mathbf{v} \hat{f} = 0$ for simplicity. As mentioned in Ref. [7], the latter (orbital) term should be included if one is interested in ferromagnetic resonance (FMR) applications. By using relaxation time approximation (RTA) for the collision term (for both momentum and spin-flip scattering) and writing the distribution as the sum of equilibrium and nonequilibrium distributions $\hat{f}(t, \mathbf{k}, \mathbf{r}) = \hat{f}_0(t, \mathbf{k}) - \frac{\partial \hat{f}_0}{\partial \varepsilon} \left[f_1(t, \mathbf{k}, \mathbf{r}) \hat{I} + \mathbf{f}_1(t, \mathbf{k}, \mathbf{r}) \cdot \boldsymbol{\sigma} \right]$, the first two moments of the resulting equations yield the following balance and flux equations of charge

$$\frac{\partial n}{\partial t} - \frac{1}{e}\frac{\partial J_C}{\partial z} = 0 \qquad (2)$$

$$J_C = -\tau_m \frac{\partial J_C}{\partial t} + eD\frac{\partial n}{\partial z} + \sigma E_z \qquad (3)$$

and spin

$$\frac{\partial \mathbf{s}}{\partial t} - \frac{1}{e}\frac{\partial \mathbf{J}_S}{\partial z} = -\frac{\delta \mathbf{s}}{\tau_{sf}} - \mathbf{s} \times \boldsymbol{\omega}_L \qquad (4)$$

$$\mathbf{J}_S = -\tau_m \frac{\partial \mathbf{J}_S}{\partial t} + eD\frac{\partial \mathbf{s}}{\partial z} - \tau_m \mathbf{J}_S \times \boldsymbol{\omega}_L \qquad (5)$$

where we have only kept the terms linear in the electric field. The above equations are modified versions of the near-equilibrium (low-field or linear) transport equations with spin precession taken into account [7]. Here τ_m is the momentum relaxation time, $\tau_{sf} = l_{sf}^2/D$ is the spin-flip relaxation time, l_{sf} is the spin-diffusion length, $D = \tau_m \overline{v_z^2}$ is the diffusion coefficient, $\sigma = e\tilde{\mu}n$ is the conductivity, $\tilde{\mu} = e\tau_m/m^*$ is the mobility, m^* is the effective mass of the band, and $\boldsymbol{\omega}_L = \gamma \mathbf{H}$ is the Larmor frequency with $\gamma > 0$ being the Gilbert gyromagnetic ratio (usually taken to be the gyromagnetic ratio associated with the electron spin, i.e. $\mu_0 |\gamma_S|$). In contrast to Eq. (2), the right-hand side of Eq. (4) appears due to the nonconserving spin scattering events (e.g. electron-magnon collisions or spin-orbit interaction on defects and impurities) [9] characterized by τ_{sf}, and spin precession, characterized by $\tau_L = 2\pi/\omega_L$.

III. CIRCUIT MODELING

Below we show how a reformulated version of the above theory along with temporal and spatial discretization schemes results in simple circuit models for the nonmagnetic and ferromagnetic materials. The goal is to describe every region of an arbitrary ferromagnet/nonmagnet multilayer with a 4×4 conductance tensor that allows us to use the four-component spin circuit approach [4], [5] whereby the total current and total voltage are related using the generalized Ohm's law $\mathbf{I} = G\mathbf{V}$.

A. Current-Voltage Relations

1) Nonmagnet:
Equations (2)-(5) can be further manipulated and discretized using simple finite-differences to obtain

$$\Delta I_C = C\frac{\partial V_C}{\partial t} \qquad (6)$$

$$I_C = G\Delta V_C \qquad (7)$$

and

$$\Delta I_{S,k} = \left(C_Q \frac{\partial}{\partial t} + G_Q \right) V_{S,k} + C_Q V_{H,k} \qquad (8)$$

$$I_{S,k} = G\Delta V_{S,k} + \tau_m I_{H,k} \qquad (9)$$

where $k \in \{x, y, z\}$, $\Delta X(t, z) = X(t, z) - X(t, z + \Delta z)$ is the difference in quantity $X \in \{I_C, I_{S,k}, V_C, V_{S,k}\}$ over Δz, C_ℓ is the capacitance per unit length, $C = C_\ell \Delta z$ is the total capacitance, $G = \sigma A/\Delta z$ is the conductance, $C_q = e^2 \partial n/\partial \mu$ is the quantum capacitance per unit volume [10], $C_Q = C_q A\Delta z$ is the quantum capacitance, $G_Q = C_Q/\tau_{sf}$ is the quantum conductance, and $s_k = -e\frac{\partial n}{\partial \mu} V_{S,k}$ is the spin density. Here we have denoted $V_{H,k} = \varepsilon_{kji} V_{S,j} \omega_{L,i}$ and $I_{H,k} = \varepsilon_{kji} \omega_{L,j} I_{S,i}$, where ε_{kji} is the Levi-Civita symbol. Moreover, we have neglected the wavelike behavior since τ_m is typically much less than the time scales of interest. Equations (6)–(9) can be represented with the T-networks shown in Fig. 2.

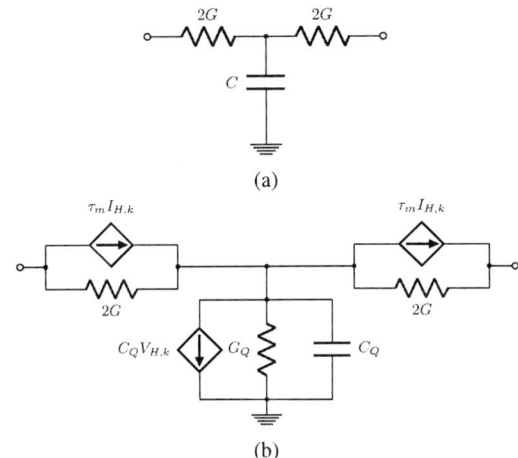

Fig. 2: Distributed T-networks for charge and spin transport along Δz of a nonmagnet. (a) Charge circuit. (b) Spin circuit.

2) Ferromagnet:
For ferromagnets, one has to use the spin-resolved properties in the transport problem [9]. Assuming the magnetization is along the x-axis (taken as the spin quantization axis) and parameterizing the conductivity and diffusion coefficient as $\sigma_{\alpha(\bar{\alpha})} = \frac{\sigma}{2(1\mp\beta)}$ and $D_{\alpha(\bar{\alpha})} = \frac{D}{1\mp\beta'}$, where β and β' are spin-asymmetry coefficients, and $\alpha = \uparrow, \downarrow$ is the spin index, the discretized transport equations in the metallic ferromagnet reads

$$\Delta I_C = C\frac{\partial V_C}{\partial t} \qquad (10)$$

$$I_C = P_\beta G\Delta V_C + \beta' P_{\beta'} G\Delta V_{S,k} \qquad (11)$$

978-1-5090-5979-9/17 $31.00 © 2017 IEEE

and

$$\Delta I_{S,k} = \left(C_Q \frac{\partial}{\partial t} + G_Q \right) V_{S,k} \tag{12}$$

$$I_{S,k} = \beta P_\beta G \Delta V_C + P_{\beta'} G \Delta V_{S,k} \tag{13}$$

where the contribution of $\mathbf{H}$ has been neglected and charge neutrality has been assumed (i.e. $\delta n_{\alpha(\bar{\alpha})} = \pm \delta s_k / 2$). Here we have denoted $P_\beta = 1/1 - \beta^2$ and $P_{\beta'} = 1/1 - \beta'^2$. Equations (10)–(13) yields the circuits shown in Fig. 3.

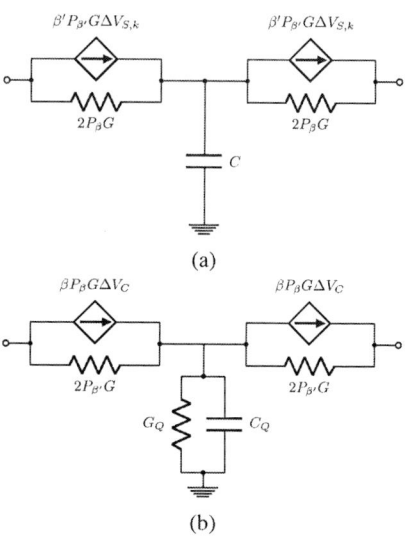

(a)

(b)

Fig. 3: Distributed T-network for charge and spin transport along Δz of a ferromagnet. (a) Charge circuit. (b) Spin circuit.

B. Conductance Matrices

Now that we found the circuit representations of charge and spin currents conduction, we employ finite-differences again, but now for the temporal derivatives, which allows us to solve resistive circuits sequentially at each time step. This framework is similar to the ones implemented in our previous publications [6], [11]. Here we will employ the backward Euler scheme.

1) Nonmagnet: A nonmagnet can be modeled with the T-network shown in Fig. 4, where the circuit is comprised of diagonal series and shunt conductance matrices

$$G_{se}^N = 2G^N \begin{bmatrix} 1 & 0 & 0 & 0 \\ 0 & 1 & 0 & 0 \\ 0 & 0 & 1 & 0 \\ 0 & 0 & 0 & 1 \end{bmatrix} \tag{14}$$

$$G_{sh}^N = G_Q^N \begin{bmatrix} \alpha_G^N & 0 & 0 & 0 \\ 0 & 1 & 0 & 0 \\ 0 & 0 & 1 & 0 \\ 0 & 0 & 0 & 1 \end{bmatrix} \tag{15}$$

along with the currents vectors

$$\mathbf{I}_0^{N,n} = -\frac{1}{\Delta t} \begin{bmatrix} C^N V_C^{N,n} \\ C_Q^N V_{S,x}^{N,n} \\ C_Q^N V_{S,y}^{N,n} \\ C_Q^N V_{S,z}^{N,n} \end{bmatrix} \tag{16}$$

$$\mathbf{I}_1^{N,n} = C_Q^N \begin{bmatrix} 0 \\ V_{H,x}^{N,n} \\ V_{H,y}^{N,n} \\ V_{H,z}^{N,n} \end{bmatrix} \quad \mathbf{I}_2^{N,n} = \tau_m \begin{bmatrix} 0 \\ I_{H,x}^{N,n} \\ I_{H,y}^{N,n} \\ I_{H,z}^{N,n} \end{bmatrix} \tag{17, 18}$$

where $\alpha_G^N = G_C^N / G_Q^N$, $G_C^N = C^N / \Delta t$, $G_Q^N = C_Q^N / \Delta \tau_{sf}^N$, $\Delta \tau_{sf}^N = \left(1/\Delta t + 1/\tau_{sf}^N \right)^{-1}$, and n is the time index.

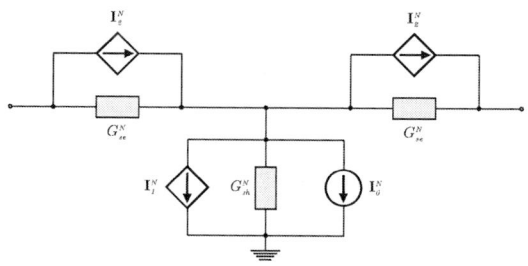

Fig. 4: Equivalent circuit model of the nonmagnet.

2) Ferromagnet: To obtain the conductance matrix of the ferromagnet, we start by considering a simple case in which the magnetization coincides with a single axis, say $\mathbf{m}_0 = \mathbf{a}_x$. After setting β and β' to zero for the y- and z-components of spin, the transport equations in the ferromagnet will result in the T-network shown in Fig. 5, where the circuit is comprised of

$$G_{se}^F (\mathbf{m}_0) = 2G^F \begin{bmatrix} P_\beta & \beta' P_{\beta'} & 0 & 0 \\ \beta P_\beta & P_{\beta'} & 0 & 0 \\ 0 & 0 & 1 & 0 \\ 0 & 0 & 0 & 1 \end{bmatrix} \tag{19}$$

where we note that $G_{sh}^F (\mathbf{m}_0)$ and $\mathbf{I}_0^{F,n}$ have equivalent forms to Eqs. (15) and (16), respectively.

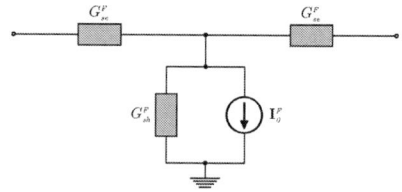

Fig. 5: Equivalent circuit model of the ferromagnet.

3) Ferromagnet/nonmagnet interface: The interface conductances tensor can be found by extending the results of the two-current model [12]. It has been shown that the tensor in fact takes the form [4], [5]

$$G^{F/N}\left(\mathbf{m}_0\right) = G^{F/N} \begin{bmatrix} 1 & P^{F/N} & 0 & 0 \\ P^{F/N} & 1 & 0 & 0 \\ 0 & 0 & \eta_R & \eta_I \\ 0 & 0 & -\eta_I & \eta_R \end{bmatrix} \quad (20)$$

where $G^{F/N} = G_\uparrow + G_\downarrow$ is the interface conductance, $\eta_R = 2\,\mathrm{Re}\{G_{\uparrow\downarrow}\}/G^{F/N}$ is the reduced real-part of the mixing conductance, and $\eta_I = 2\,\mathrm{Im}\{G_{\uparrow\downarrow}\}/G^{F/N}$ is the reduced imaginary-part of the mixing conductance [13].

IV. MODEL BENCHMARKING

In order to demonstrate our model's utility in an application, a nonlocal spin-valve is simulated and afterwards compared to experiment [14]. Using the spin circuit model introduced in Section III and the device parameters in [14], we invoke the self-consistent simulation framework [6], coupling the stochastic magnetization dynamics and spin circuit model. First, we examine our model's ability to capture the spin transport. Figure 6(a) shows the nonlocal resistance versus in-plane magnetic field $B_\parallel$ for both forward and reverse sweeps. Second, to validate the spin precession aspect, we plot Hanle precession curves in Fig. 6(b) where a perpendicular magnetic $B_\perp$ is applied to the device, causing the spins to precess around the magnetic field, resulting in a decrease in the measured signal due to incoherent spin precession and relaxation. Both of the curves generated closely match the experiment.

V. CONCLUSION

In this paper we derived general circuit models for spintronic devices via a circuit-formulation of drift-diffusion equations. In addition to the time-dependent formulation, the model considers the presence of both electric and magnetic fields, and thus can capture important device physics such as spin transport, spin precession, and ultra-fast spin signals. To adapt with the notion of the four-component circuit approach, we derived conductance matrices for the device. To validate the model, we simulated a nonlocal spin-valve, and the results match well with experiments. We believe this model will allow accurate and easy assessment of future spintronic devices and enables circuit level simulation of spin based logic.

ACKNOWLEDGMENT

The authors would like to thank Aurelien Manchon from King Abdullah University of Science and Technology for the helpful comments and discussions.

REFERENCES

[1] R. Gonzalez, B. M. Gordon, and M. A. Horowitz, "Supply and threshold voltage scaling for low power CMOS," *IEEE Journal of Solid-State Circuits*, vol. 32, no. 8, pp. 1210–1216, 1997.

[2] I. Žutić, J. Fabian, and S. D. Sarma, "Spintronics: Fundamentals and applications," *Reviews of modern physics*, vol. 76, no. 2, p. 323, 2004.

[3] A. Hirohata and K. Takanashi, "Future perspectives for spintronic devices," *Journal of Physics D: Applied Physics*, vol. 47, no. 19, p. 193001, 2014.

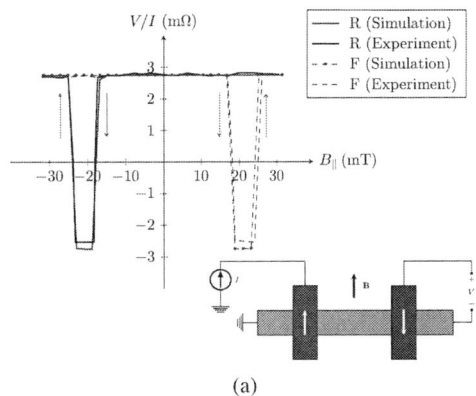

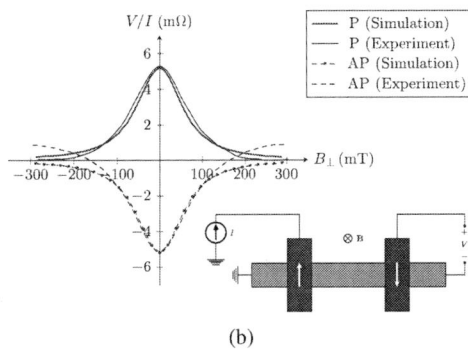

Fig. 6: Simulation of a nonlocal spin-valve using the compact circuit models introduced in this work. Nonlocal resistance signal as a function of (a) in-plane magnetic field at $T = 293$ K, and (b) perpendicular magnetic field ("Hanle effect") at $T = 4.2$ K. The device parameters and experimental data are taken from [14].

[4] B. Behin-Aein, A. Sarkar, S. Srinivasan, and S. Datta, "Switching energy-delay of all spin logic devices," *Applied Physics Letters*, vol. 98, no. 12, p. 123510, 2011.

[5] S. Manipatruni, D. E. Nikonov, and I. A. Young, "Modeling and design of spintronic integrated circuits," *IEEE Transactions on Circuits and Systems I: Regular Papers*, vol. 59, no. 12, pp. 2801–2814, 2012.

[6] M. Alawein and H. Fariborzi, "Improved circuit model for all-spin logic," in *Nanoscale Architectures (NANOARCH), 2016 IEEE/ACM International Symposium on*. IEEE, 2016, pp. 135–140.

[7] Y. Qi and S. Zhang, "Spin diffusion at finite electric and magnetic fields," *Physical Review B*, vol. 67, no. 5, p. 052407, 2003.

[8] I. DAmico and G. Vignale, "Theory of spin coulomb drag in spin-polarized transport," *Physical Review B*, vol. 62, no. 8, p. 4853, 2000.

[9] T. Valet and A. Fert, "Theory of the perpendicular magnetoresistance in magnetic multilayers," *Physical Review B*, vol. 48, no. 10, p. 7099, 1993.

[10] P. Bonhomme, S. Manipatruni, R. M. Iraei, S. Rakheja, S.-C. Chang, D. E. Nikonov, I. A. Young, and A. Naeemi, "Circuit simulation of magnetization dynamics and spin transport," *IEEE Transactions on Electron Devices*, vol. 61, no. 5, pp. 1553–1560, 2014.

[11] M. Alawein and H. Fariborzi, "Dynamic circuit model for spintronic devices," *Procedia Engineering*, vol. 168, pp. 966–970, 2016.

[12] M. Alawein, "Circuit simulation of all-spin logic," Master's thesis, 2016.

[13] A. Brataas, G. E. Bauer, and P. J. Kelly, "Non-collinear magnetoelectronics," *Physics Reports*, vol. 427, no. 4, pp. 157–255, 2006.

[14] F. Jedema, H. Heersche, A. Filip, J. Baselmans, and B. Van Wees, "Electrical detection of spin precession in a metallic mesoscopic spin valve," *Nature*, vol. 416, no. 6882, pp. 713–716, 2002.

Compact Physical Model of a-IGZO TFTs for circuit simulation

Matteo Ghittorelli*, Fabrizio Torricelli*, Carmine Garripoli[†], Jan-Laurens J.P. van der Steen[‡], Gerwin H. Gelinck[‡§], Sahel Abdinia[†], Eugenio Cantatore[†], and Zsolt M. Kovács-Vajna*.

*Department of Information Engineering, University of Brescia, 25123 Brescia, Italy.
Email: m.ghittorelli@unibs.it
[†]Mixed-Signal Microelectronics Group, Eindhoven University of Technology, Eindhoven 5600 MB, The Netherlands.
[‡]Holst Centre/TNO, Eindhoven 5656 AE, The Netherlands.
[§]Department of Applied Physics, Technical University of Eindhoven, Eindhoven 5612, The Netherlands

Abstract—**Amorphous InGaZnO (a-IGZO) is a candidate material for thin-film transistors (TFTs) owing to its large electron mobility. The development of high functionality circuits requires accurate and efficient circuit simulation that, in turn, is based on compact physical a-IGZO TFTs models. Here we propose a compact physical-based and analytical model of the drain current of a-IGZO TFTs. The model accounts for both trapped and free charges by means of an effective density of states that accurately approximate the actual a-IGZO density of states in the energy range relevant for the TFT operation. The model is implemented in a circuit simulator and it is validated with the measurements of both coplanar and staggered a-IGZO TFTs fabricated on flexible substrates.**

I. INTRODUCTION

Amorphous Indium Gallium Zinc Oxide (a-IGZO) is an ideal semiconductor for large area, low temperature electronics, showing electron mobilities as high as 25 $cm^2V^{-1}s^{-1}$ [1]. Thus, a-IGZO enables the fabrication of flexible thin film transistors (TFTs) with good characteristics such as high field effect mobility and steep subthreshold slope, good device lifetime, high flexibility, transparency, and large-area uniform integration even if deposited on flexible and lightweight substrates [2]. Typically, a-IGZO TFTs are used in the backplane switch matrix of organic light emitting diodes (OLEDs) high resolution displays. In recent years several groups have demonstrated sensors and circuits fabricated with a-IGZO TFTs [1], [2].

The high mobility of a-IGZO originates from its peculiar chemical bonding. In a-IGZO the overlapping spherical *s* orbitals of the metal cations form the electron transport paths, enabling band transport and mobilities comparable to those of crystalline metal oxide semiconductors. However, the spatial disorder due to the material amorphousness gives rise to localized states in the sub-gap. [1], [3]. The spherical symmetry of the metal *s* orbitals partially compensates the material disorder. This results in a total number of localized states in a-IGZO of the order of $10^{18}cm^{-3}$, that is orders of magnitude lower than that of covalent oxide semiconductors (i.e. a-Si:H) [1], nano-crystalline ZnO [4] or organic semiconductors [5]. As a consequence, in a-IGZO TFTs, at small gate voltage the trap states are playing the key role in the transport mechanism,

while at large gate voltages both trapped and free charges have to be accounted for.

In the last few years, the a-IGZO TFTs electrical characteristics were extensively investigated [3], [6]–[10]. The electrical characteristics were modelled with empirical mobility functions accounting for trap states only [6], and surface-potential based models were developed, in which free and trap states were considered by means of regional approaches [3], [7]. Furthermore, technology computer-aided design (TCAD) device simulators were used to reproduce the measured current-voltage characteristics and to extract the density of states [8], [9]. Unfortunately, state-of-art analytical models [6] do not take into account the specific transport properties of a-IGZO and therefore are not able to predict the a-IGZO TFTs electrical characteristics in the whole range of transistor operation with a unique and simple formulation. On the other hand, physical models [3], [7], [10] include the device physics but are not suitable for circuit design because their implementation in CAD simulator is typically cumbersome [3], [7], [10].

The possibility to describe the unique charge transport properties of a-IGZO through a compact, analytical and physical-based model that can be easily implemented in a standard CAD software is of paramount of importance to push further the development of a-IGZO TFTs technologies and for the design of robust and high functionality circuits.

In this work we propose a new physical based compact model of a-IGZO TFTs. The a-IGZO density of states (DOS) is approximated by means of an effective DOS that accounts for both localized and delocalized states. The model is implemented in a circuit simulator, it is validated with the measurements of both coplanar and staggered a-IGZO TFTs, and it is currently used for the design of flexible a-IGZO circuits.

II. TRANSISTOR FABRICATION

Coplanar a-IGZO TFTs: Gate, drain and source were deposited by evaporation of gold. A 100 nm thick gate dielectric was formed by atomic layer deposition of Al_2O_3. A 30 nm thick InGaZnO semiconductor was deposited using sputtering from a target with 2:2:1 atomic ratio for In:Ga:Zn. All layers were patterned using standard photolithographic techniques.

978-1-5090-5979-9/17 $31.00 © 2017 IEEE

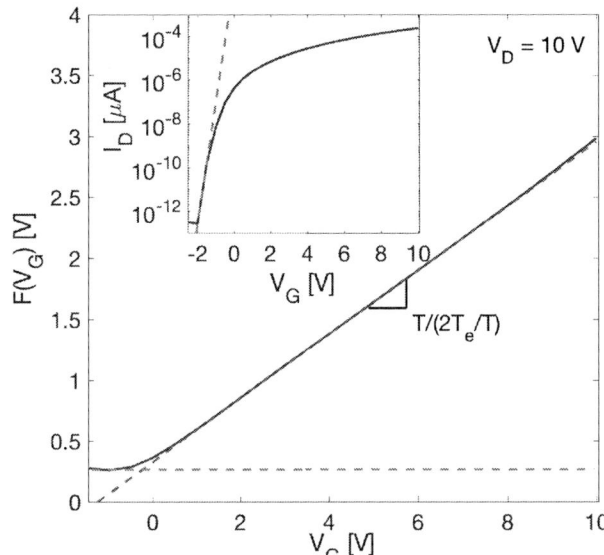

Fig. 1. Main Panel. F functional calculated from the measured transfer characteristic of coplanar a-IGZO TFT with $V_D = 10\ V$ (blue line). The functional is almost flat when $V_G < 0\ V$ (dashed green line), while it can be approximated by a straight line when $V_G > 0\ V$ (dashed red line). Inset. Measured transfer characteristic of the fabricated coplanar a-IGZO TFT, $V_D = 10\ V$. The dashed green line is the least square approximation of $log_{10}(I_D) - V_G$ in the subthreshold regime.

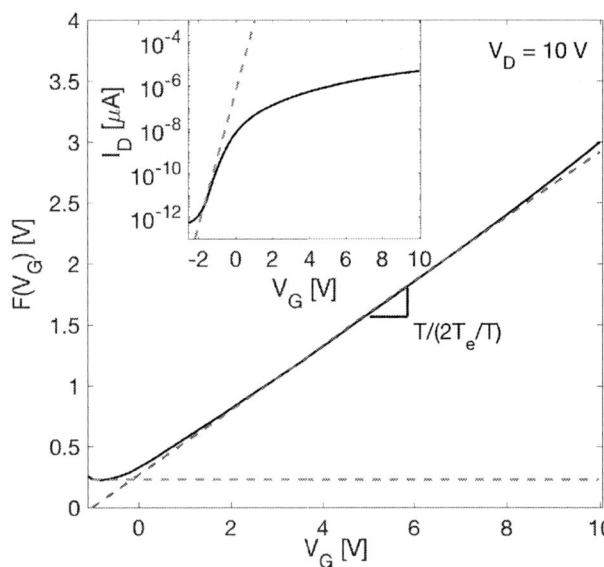

Fig. 2. Main Panel. F functional calculated from the measured transfer characteristic of staggered a-IGZO TFT with $V_D = 10\ V$ (blue line). The functional is almost flat when $V_G < 0\ V$ (dashed green line), while it can be approximated by a straight line when $V_G > 0\ V$ (dashed red line). Inset. Measured transfer characteristic of the fabricated staggered a-IGZO TFT, $V_D = 10\ V$. The dashed green line is the least square approximation of $log_{10}(I_D) - V_G$ in the subthreshold regime.

The measurements were performed in darkness and in vacuum at about 10^{-6} mbar. The measured technological parameters are listed in Tab. I.

Staggered a-IGZO TFTs: A Mo-Cr gate metal was sputtered and patterned using photolithography. A 200 nm thick SiO_2 gate dielectric was formed by a 180 °C plasma-enhanced chemical vapour deposition (PECVD). A 12 nm thick IGZO layer was deposited using sputtering. A 100 nm thick SiO_2 etch stop layer was deposited using PECVD at 200 °C on top of the patterned semiconductor and it was patterned by dry etching process. Mo-Cr is used for the source, drain, and metal lines. The metal lines were patterned using standard photolithography techniques. The measurements were performed in darkness and in vacuum at about 10^{-6} mbar. The measured technological parameters are listed in Tab. II.

III. PRELIMINARY ANALYSIS

The inset of Fig. 1 shows the measured transfer characteristic of a coplanar a-IGZO TFT at $V_D = 10\ V$. In order to gather information on the a-IGZO DOS, the functional $F = \int_0^{V_G} I_D(V_g)dV_g / I_D(V_G)$ is calculated on the measured characteristics [11]. Fig. 1 shows the calculated F as a function of V_G. The functional is almost flat when $V_G < 0\ V$, while it can be approximated by a straight line when $V_G > 0\ V$. A flat F corresponds to an exponential subthreshold current (inset of Fig. 1) that, in turn, indicates a surface density of states (DOS) located at the insulator/semiconductor interface, while a straight line indicates an exponential DOS [11]. The exponential DOS is defined as follows:

$$g_e(E) = \frac{N_e}{k_B T_e} \exp\left(\frac{E - E_c}{k_B T_e}\right) \quad (1)$$

where N_e is the total number of states, T_e is the characteristic temperature of the DOS, k_B is the Boltzmann constant, and E_c is the conduction band energy. By exploiting the functional F it is possibile to estimate the characteristic temperature of the a-IGZO DOS. When the a-IGZO TFT is operated in the saturation regime, i.e. $V_G - V_T < V_D$, the functional can be calculated to be $F = T/(2T_e + T)(V_G - V_T - V_S)$, where T is the temperature, and V_G, V_T, V_D and V_S are the gate, threshold, drain and source voltages, respectively. From the linear least square approximation of F (Fig. 1, red dashed line) results $T_e = 415\ K$ and $V_T = -1.25\ V$. As a confirmation, the analysis is applied to staggered a-IGZO TFTs. The inset of Fig. 2 shows the measured transfer characteristic of a staggered a-IGZO TFT at $V_D = 10\ V$. Fig. 2 shows that the functional F is almost flat when $V_G < 0\ V$, while it can be approximated by a straight line when $V_G > 0\ V$. In the case of staggered a-IGZO TFTs, results $T_e = 409K$ and $V_T = -1.00\ V$.

It is worth noting that in a-IGZO the characteristic temperature of the trap states is typically in the range $500-650K$ [3], [7]–[10] while, interestingly, we obtained $T_e \simeq 410K$ for both coplanar and staggered a-IGZO TFTs. This can be explained as follows: the a-IGZO DOS comprises both localized (trapped) and delocalized (free) states [3], [7]–[10] and, in the above threshold regime (strong accumulation regime), the charge transport depends on the interaction between trapped and free charges. Since the total number of localized and delocalized states are similar, of the order of $10^{18} - 10^{19} cm^{-3}$, the extracted exponential DOS is an effective density of states that describes the interaction between trapped and free charges.

978-1-5090-5979-9/17 $31.00 © 2017 IEEE

IV. COMPACT DRAIN CURRENT MODEL

Taking advantage of the preliminary analysis, a compact drain current model suitable for circuit simulation is derived as follows. The integral form of the drain current, based on the Sao-Pah model and accounting for the multiple trapping and release (MTR) theory, reads [12]:

$$I_D = \frac{W}{L} \int_{V_S}^{V_D} \int_{V_{ch}}^{\varphi_s} \frac{\sigma_0 \exp\left[\frac{q(\varphi - V_{ch})}{k_B T}\right]}{\sqrt{\frac{2q}{\epsilon_s} \int_{V_{ch}}^{\varphi} n(\varphi', V_{ch}) d\varphi'}} d\varphi \, dV_{ch} \quad (2)$$

where W and L are the channel width and length, respectively, V_{ch} is the channel potential (pseudo Fermi energy), φ the electrostatic potential and φ_s is the surface potential at the insulator-semiconductor interface, $\sigma_0 = q\mu_B N_b \exp(\Delta E_{Fi}/k_B T)$, μ_B is is the band mobility, N_b is the total number of delocalized states, $\Delta E_{Fi} = E_{gap}/2$, ϵ_s is the a-IGZO semiconductor permittivity, and n is the carrier concentration that accounts for the exponential DOS (Eq. 1) that reads:

$$n = N_e \frac{\pi T/T_e}{\sin(\pi T/T_e)} \exp\left[\frac{q(\varphi - V_{ch}) + \Delta E_{Fi}}{k_B T_e}\right] \quad (3)$$

since in a-IGZO $T_e > T$. By plugging Eq. 3 in Eq. 2, and solving that with respect to the electrostatic potential φ one obtains:

$$I_D \simeq \frac{W}{L} \omega_0 \int_{V_S}^{V_D} \exp\left[\frac{q\xi_b}{k_B T}\left(\frac{2T_e - T}{2T_e}\right)\right] dV_{ch} \quad (4)$$

where $\omega_0 = \frac{\sigma_0}{k_e} \frac{2k_B T_e T}{2T_e - T}$, $\xi_b = \varphi_s - V_{ch}$ is the band bending, and $k_e = \sqrt{2N_e k_B \pi T/[\epsilon_s \sin(\pi T/T_e)]} \exp[\Delta E_{Fi}/(k_B T_e)]$. By applying Gauss law to the insulator / semiconductor interface and accounting for the interfacial trap density, the continuity of the electric field imposes that:

$$\frac{\varrho_i \xi_b}{\epsilon_s} + k_e e^{\frac{q\xi_b}{2k_B T_e}} = \frac{C_i}{\epsilon_s}(V_G - V_{fb} - \xi_b - V_{ch}) \quad (5)$$

where C_i is the gate capacitance per unit area, V_G and V_{fb} are the gate and the flatband voltages, respectively, and ϱ_i is the surface charge density located at the gate insulator/a-IGZO semiconductor interface. Eq. 5 is a transcendental equation. In order to work out an analytical solution for the band bending, it is useful to evaluate its asymptotic behaviours.

When the transistor works in the weak accumulation regime, i.e. $V_G - V_{fb} < V_{ch}$ the Fermi energy level is located far from E_c and lies within the interfacial traps. As a consequence, the band bending turns out to be:

$$\xi_b = \frac{V_G - V_{fb} - V_{ch}}{1 + \varrho_i/C_i} \quad (6)$$

On the other hand, when the transistor works in the strong accumulation regime, i.e. $V_G - V_{fb} > V_{ch}$ the Fermi energy level is located close to E_c and lies within the exponential DOS. The band bending reads:

$$\xi_b = \frac{2k_B T_e}{q} \log\left[\frac{C_i(V_G - V_T - V_{ch})}{\epsilon_s k_e}\right] \quad (7)$$

where $V_T = V_{fb} + \Delta_{fb}$, and Δ_{fb} takes into account the variation of the threshold voltage due to interfacial traps [9], [13].

TABLE I.   PHYSICAL AND GEOMETRICAL PARAMETERS OF COPLANAR a-IGZO TFTs.

W $[\mu m]$	1000	N_e $[cm^{-3}]$	3.75×10^{18}
L $[\mu m]$	100	T_e $[K]$	415
C_i $[nF/cm^2]$	70	V_T $[V]$	-1.25
μ_b $[cm^2/Vs]$	15	ϱ_i $[cm^{-2}V^{-1}]$	4.3×10^{11}

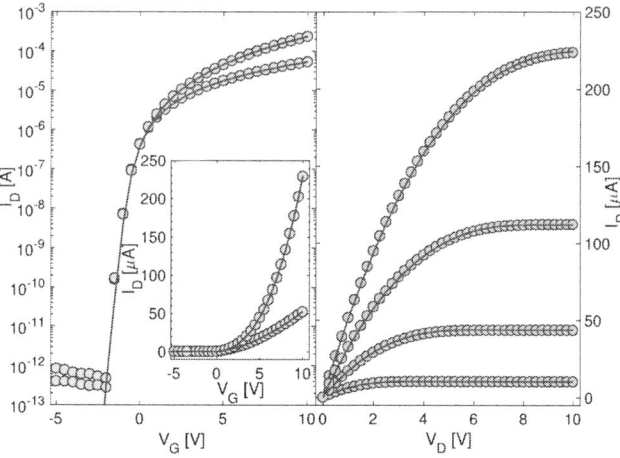

Fig. 3.   Left Panel. Measured (symbols) and modeled (red lines) transfer characteristics of coplanar a-IGZO TFTs. The bottom curves refer to drain voltage $V_D = 1$ V, the top curves to $V_D = 10$ V. Right panel. Measured (symbols) and model (red lines) output characteristics of coplanar a-IGZO TFTs at various gate voltages: from bottom to top $V_G = 2.5, 5, 7.5, 10$ V.

In the weak accumulation regime (below threshold), the band bending has linear dependence on the overdrive voltage. By plugging Eq. 6 in Eq. 4 one obtains:

$$I_{D_{BT}} = \frac{W}{L} I_0 \left\{\left[\exp\left(\frac{V_G - V_{fb} - V_{ch}}{V_i}\right)\right]^{\frac{2T_e - T}{T}}\right\}_{V_D}^{V_S} \quad (8)$$

where $V_i = (2k_B T_e/q)(1 + \varrho_i/C_i)$, $I_0 = \omega_0 V_i T/(2T_e - T)$, and $[f(x)]_a^b = f(b) - f(a)$.

On the other hand, in the strong accumulation regime (above threshold) the band bending has a logarithmic shape. By plugging Eq. 7 in Eq. 4 one obtains:

$$I_{D_{AT}} = \frac{W}{L} \Gamma \left[(V_G - V_T - V_{ch})^{\frac{2T_e}{T}}\right]_{V_D}^{V_S} \quad (9)$$

where $\Gamma = \omega_0 \frac{T}{2T_e}(\epsilon_s k_e/C_i)^{1 - \frac{2T_e}{T}}$. Eq. 8 and Eq. 9 describe the drain current for the below threshold and the above threshold regimes, separately. A continuous and analytical mathematical equation that enables to match these two different shapes can be chosen to be:

$$I_D = \frac{W}{L} \Gamma \left\{\left[V_I \mathcal{F}\left(\frac{V_{GS}}{V_I}\right)\right]^{\frac{2T_e}{T}} - \left[V_I \mathcal{F}\left(\frac{V_{GD}}{V_I}\right)\right]^{\frac{2T_e}{T}}\right\} \quad (10)$$

where $V_{GS} = V_G - V_T - V_S$, $V_{GD} = V_G - V_T - V_D$, $\mathcal{F}(x) = \log[1 + \exp(x)]$, and $V_I = V_i 2T_e/(2T_e - T)$.

V. DISCUSSION

In order to validate the drain current model, the measurements of coplanar bottom gate bottom contact a-IGZO TFTs

TABLE II.   PHYSICAL AND GEOMETRICAL PARAMETERS OF
STAGGERED a-IGZO TFTs.

W $[\mu m]$	200	N_e $[cm^{-3}]$	2.95×10^{18}
L $[\mu m]$	200	T_e $[K]$	409
C_i $[nF/cm^2]$	25	V_T $[V]$	-1.00
μ_b $[cm^2/Vs]$	15	ϱ_i $[cm^{-2}V^{-1}]$	3.9×10^{11}

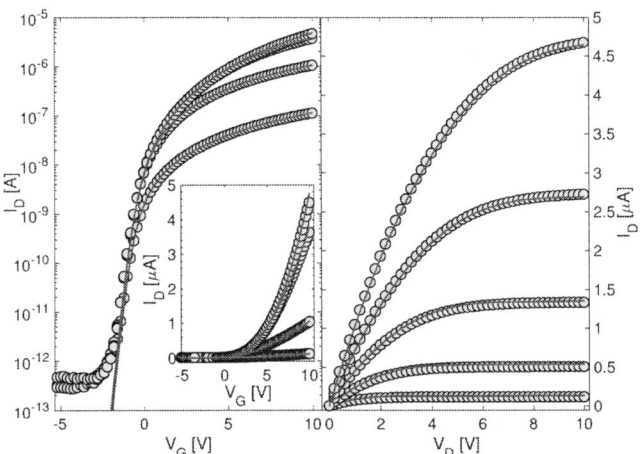

Fig. 4.   Left Panel. Measured (symbols) and modeled (red lines) transfer characteristics of staggered a-IGZO TFTs at various drain voltages: from bottom to top $V_D = 0.1, 1, 5, 10$ V. Right panel. Measured (symbols) and model (red lines) output characteristics of staggered a-IGZO TFTs at various gate voltages: from bottom to top $V_G = 2, 4, 6, 8, 10$ V.

are reproduced. The physical parameters were extracted from a single transfer characteristics, with $V_D = 10$ V, and reported, together with the geometrical a-IGZO TFTs parameters, in Tab. I. The comparison between the analytical drain current model and the measurements of coplanar a-IGZO TFTs are shown in Fig. 3. There is a very good agreement between the experimental data and the model in the whole range of biasing conditions.

As a further confirmation we modeled the measurement of staggered bottom gate top contact a-IGZO TFTs. The extracted physical parameters and the geometrical parameters of staggered a-IGZO TFTs are reported in Tab. II. Fig. 4 shows that, also in the case of staggered a-IGZO TFTs, there is a good agreement between the experimental data and the proposed model. By comparing the extracted parameters in the case of coplanar and staggered a-IGZO TFTs we found that the effective temperature $T_e \simeq 410$ K is the same in both cases, while coplanar a-IGZO TFTs shows a slightly higher N_e, T_e and N_{ss}. It is worth to note that the slightly higher total number and disorder of traps of coplanar a-IGZO TFTs with respect to staggered a-IGZO TFTs can be ascribed to the absence of a capping (ESL) layer in coplanar a-IGZO TFTs according to [1].

VI.   CONCLUSION

In this work a new compact model of the drain current of a-IGZO TFTs is proposed. The physical basis of the model allows to extract from a single transfer characteristic an effective charge carrier concentration: an interfacial density of traps

that represents the deep trapped charges and an exponential carrier concentration that reflects the interaction between the tail trapped and the free charges. The model was implemented in a circuit simulator, and validated with the measurements of both coplanar and staggered a-IGZO TFTs, and it is currently used for the design of flexible a-IGZO circuits.

REFERENCES

[1]   T. Kamiya, K. Nomura, and H. Hosono, "Present status of amorphous In-Ga-Zn-O thin-film transistors," *Sci. Technol. Adv. Mater.*, vol. 11, no. 4, pp. 044305, Aug. 2010.

[2]   C. Garripoli; J.-L. P. J. van der Steen; E. Smits; G. H. Gelinck; A. H. M. Van Roermund; E. Cantatore, "An a-IGZO asynchronous delta-sigma modulator on foil achieving up to 43dB SNR and 40dB SNDR in 300Hz bandwidth," *IEEE International Solid-State Circuits Conference (ISSCC)* , pp. 260-261, Feb. 2017.

[3]   S. Lee, K. Ghaffarzadeh, A. Nathan, J. Robertson, S. Jeon, C. Kim, I-H. Song, and U-I. Chung, "Trap-limited and percolation conduction mechanisms in amorphous oxide semiconductor thin film transistors (TFTs)," *Appl. Phys. Lett.*, vol. 98, no. 20, pp. 203508-1-203508-3, May 2011.

[4]   F. Torricelli, E. C. P. Smits, J. R. Meijboom, A. K. Tripathi, G. H. Gelinck, L. Colalongo, Z.M. Kovacs-Vajna, D. de Leeuw, and E. Cantatore,"Transport physics and device modeling of zinc oxide thin-film transistors - Part II: Contact resistance in short channel devices," *IEEE Trans. Electron Devices*, vol. 58, no. 9, pp. 3025-3033, Sep. 2011.

[5]   F. Torricelli, "Charge Transport in Organic Transistors Accounting for a Wide Distribution of Carrier Energies - Part I: Theory," *IEEE Trans. Electron Devices*, vol. 59, no. 5, pp. 1514-1519, May 2012.

[6]   K. Abe, N. Kaji, H. Kumomi, K. Nomura, T. Kamiya, M. Hirano, and H. Hosono, "Simple analytical model of on operation of amorphous InGaZnO thin-film transistors" *IEEE Trans. Electron Devices*, vol. 58, no. 10, pp. 3463-3471, Oct. 2011.

[7]   W. Deng, J. Huang, X. Ma, and T. Ning, "An Explicit Surface-Potential-Based Model for Amorphous IGZO Thin-Film Transistors Including Both Tail and Deep States," *IEEE Electron Device Lett.*, vol. 35, no. 1, pp. 78-80, Jan. 2014.

[8]   T.-C. Fung, C.-S. Chuang, C. Chen, K. Abe, R. Cottle, M. Townsend, H. Kumomi, and J. Kanicki, "Two-dimensional numerical simulation of radio frequency sputter amorphous InGaZnO thin-film transistors," *J. Appl. Phys.* vol. 106, no. 8, pp. 084511-1-084511-10, Oct. 2009.

[9]   H. Hsieh, T. Kamiya, K. Nomura, H. Hosono, and C. Wu, "Modeling of amorphous InGaZnO$_4$ thin film transistors and their subgap density of states," *Appl. Phys. Lett.*, vol. 92, no. 13, pp. 133503-1-133503-3, Apr. 2008.

[10]   M. Ghittorelli, F. Torricelli, L. Colalongo, and Z. M. Kovács-Vajna, "Accurate analytical physical modeling of amorphous InGaZnO thin-film transistors accounting for trapped and free charges," *IEEE Trans. Electron Devices*, vol. 61, no. 12, pp. 4105-4112, Dec. 2014.

[11]   F. Torricelli, M. Ghittorelli, M. Rapisarda, A. Valletta, L. Mariucci, S. Jacob, R. Coppard, E. Cantatore, Z. M. Kovács-Vajna, and L. Colalongo, "Unified drain-current model of complementary p- and n-type OTFTs," *Org. Electron.*, vol. 22, pp. 5-11, Mar. 2015.

[12]   M. Ghittorelli, F. Torricelli, and Z. M. Kovács-Vajna, "Physical modeling of amorphous InGaZnO thin-film transistors: the role of degenerate conduction," *IEEE Trans. Electron Devices*, vol. 63, no. 6, pp. 2417-2423, Jun. 2016.

[13]   B. Kim, E. Chong, D. H. Kim, Y.W. Jeon, D. H. Kim and S. Y. Lee, "Origin of threshold voltage shift by interfacial trap density in amorphous InGaZnO thin film transistor under temperature induced stress," *Appl. Phys. Lett.*, vol. 99, no. 6, p. 062108-1-062108-3, Aug. 2011.

Complementary Black Phosphorous FETs by workfunction engineering of pre-patterned Au and Ag embedded electrodes

Nicolò Oliva[1], Emanuele A. Casu[1], Wolfgang A. Vitale[1], Igor Stolichnov[1] and Adrian M. Ionescu[1]

[1]EPFL, Nanoelectronic Devices Laboratory (NanoLab), 1015, Lausanne, Switzerland

Abstract—**We propose and experimentally demonstrate top-gated complementary n- and p-type black phosphorous FETs by engineering the workfunction of pre-patterned electrodes embedded in a SiO₂ layer. Pre-patterned electrodes offer the possibility of reducing the exposure time of exfoliated flakes to oxidant agents with respect to top-contacted devices and maximize the accessible area for sensing applications. The devices are realized by exfoliating multilayer black phosphorous flakes on top of pre-patterned embedded source and drain contacts. A capping layer consisting of 15 nm thick Al₂O₃ is used to prevent flakes degradation and serves as top gate dielectric. We deposited both Au and Ag contacts to investigate the impact of the electrode workfunctions on BP FETs polarity. Au contacted devices showed p-type conduction with ON/OFF current ratio 140 and holes mobility up to 40 cm²V⁻¹s⁻¹. Devices with Ag contacts showed prevalent n-type conduction with ON/OFF ratio 1700 and electron mobility 2 cm²V⁻¹s⁻¹. The reported results represent a substantial improvement with respect to reported alternative implementations of black phosphorous FETs with pre-patterned, non-embedded electrodes. Moreover, we demonstrate that Ag is a promising metal for electron injection in black phosphorous FETs.**

Keywords—Black phosphorous, pre-patterned electrodes, 2D materials, field-effect devices, atomic layer deposition.

I. INTRODUCTION

Since the discovery and experimental isolation of graphene, two-dimensional (2D) materials have attracted a huge research interest. Despite the large mobility shown by graphene, its lack of an electronic band gap prevents the realization of field effect devices (FETs) with reasonable I_{ON}/I_{OFF} ratios [1]. Transition metal dichalcogenides (TMDCs) are characterized by large band gaps but low carrier mobilities [2-4]. Black phosphorous (BP) exhibits a finite direct band gap (0.3 eV in bulk up to 1.5 eV in monolayer) together with high hole mobility, which make it a promising alternative to TMDCs for the realization of 2D channel FETs and sensors [5-8]. However, BP layers are not inert in ambient conditions, and the electrical properties of not passivated samples are rapidly degraded to such an extent that the functionality of the field effect devices is compromised [9]. Effective passivation of exfoliated BP flakes against ambient degradation using Al₂O₃ or HfO₂ capping layers has been reported in several works, leading to the demonstration of field effect devices with promising and stable performance [10-11].

BP FETs are usually realized exfoliating black phosphorous flakes on SiO₂ and depositing by lift-off the electrical contacts [5-8]. BP FETs realized with pre-patterned source and drain electrodes have been proposed to reduce the air exposure time of BP during fabrication [9]. Such electrodes allow maximizing the accessible area of black phosphorous flakes for sensing applications, in particular for light detection [12].

However, the reported devices showed an I_{ON}/I_{OFF} ratio lower than 10 and reduced carriers mobilities [9]. Here, we report the implementation of top gated BP FETs with pre-patterned embedded source and drain contacts, realized in Au or Ag. With respect to previously reported pre-patterned electrodes, embedding electrodes in a SiO₂ layer allows reducing the amount of mechanical stress on black phosphorous flakes due to topography steps with height comparable or larger than the flakes thickness. Both n and p-type devices have been demonstrated with I_{ON}/I_{OFF} ratios larger than 2 orders of magnitude.

II. DEVICE FABRICATION

The proposed process flow is summarized in Fig. 1. As a first step, a 280 nm SiO2 layer was grown on a Si wafer using wet oxidation. The areas for embedded source and drain contacts were lithographically defined using a direct writing laser tool, VPG200, and a direct resist, ECI 3027. We then etched 45 nm deep boxes in the SiO2 layer exploiting a combination of dry and wet etching, and the electrodes were formed by evaporation and lift-off of either Ti (5 nm) and Au (40 nm), or Ti (5 nm) and Ag (40 nm). The resulting step between the electrodes top surface and the SiO2 layer is reduced to few nanometers as measured with a mechanical profilometer. Embedded bottom contacts allows to achieve much smaller topography steps with respect to pre-patterned non-embedded contacts.

- Resist patterning
- SiO₂ Dry etching
- Au evaporation and lift-off
- Resist patterning
- SiO₂ Dry etching
- Ag evaporation and lift-off
- BP exfoliation
- ALD Al₂O₃
- PMMA/MMA patterning
- Ni/Au evaporation and lift-off

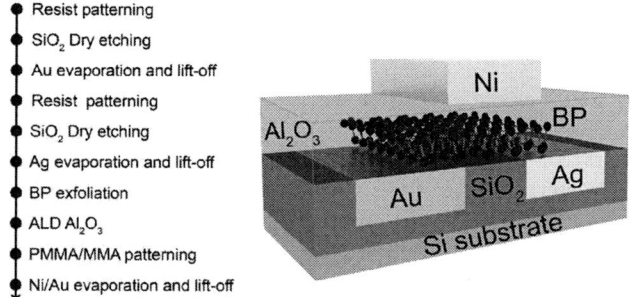

Fig. 1. Schematic view of the realized devices and summary of the proposed process flow. The embedded pre-patterned electrodes have been realized in Au and/or Ag, while Ni has been selected for the top gate. The exfoliated black phosphorus flakes have been passivated with a 15 nm thick Al₂O₃ layer.

978-1-5090-5979-9/17 $31.00 © 2017 IEEE

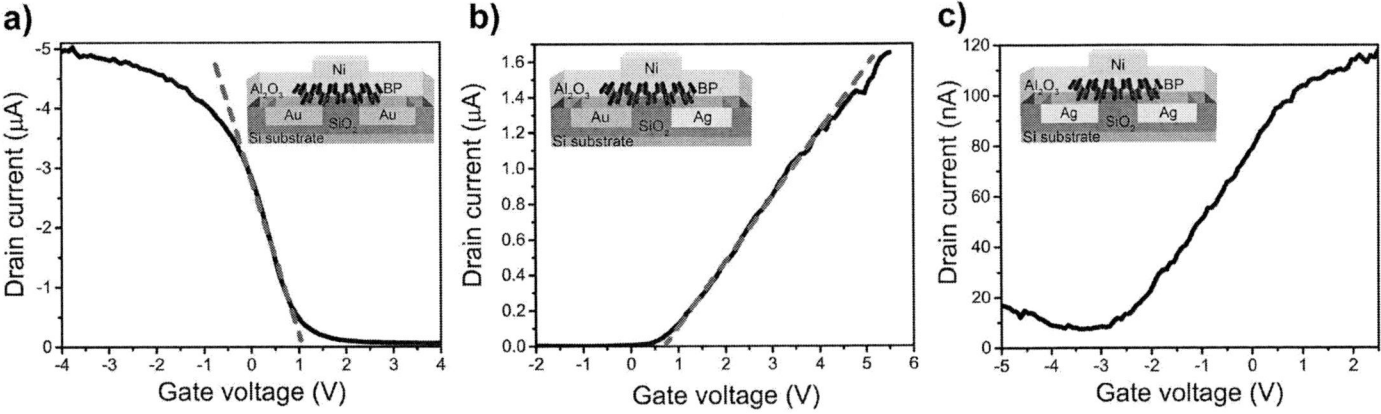

Fig. 2. Comparison of the transfer characteristic $I_D(V_G)$ of BP FETs realized with different embedded bottom contacts. The results here reported have been obtained applying a drain to source voltage of absolute value equal to 750 mV for the Au-Au contact device and 500 mV for the Ag contacted devices. a) Au contacted devices exhibit a clear p-type conduction. b) and c) Devices with Ag contacts or Au-Ag electrodes with the Ag used as transistor source show n-type polarity. The red dotted line in a) and b) represents the linear fit performed for the mobility extraction. Insets: schematic representation of the BP FETs with the different sets of embedded electrodes.

Next, black phosphorous flakes were exfoliated from a bulk crystal sample (synthetized by *Smart Elements*) according to the mechanical scotch taping technique. In order to reduce the exposure to ambient oxygen and water vapor, the samples were passivated shortly after the exfoliation using atomic layer deposition (ALD) of 15 nm of Al_2O_3, which is used also as top gate dielectric. . The ALD was performed in a BENEQ TFS 200 reactor at 200 °C, using TMA and H_2O as precursors. We then identified flakes favorably connected to the embedded electrodes using optical microscopy. Finally, the top gate electrode was obtained by patterning a PMMA/MMA bilayer using electron beam lithography and then evaporating and lifting off Ni (50 nm) and Au (10 nm). BP flakes thickness was determined using atomic force microscopy (AFM). The results reported in the following have been obtained from BP FETs realized with 30 nm thick flakes.

III. RESULTS

All the electrical measurements were performed at ambient conditions and room temperature using an HP 4156C semiconductor parameter analyzer.

A. Impact of eletrodes workfunction on BP FETs polarity

In order to study the impact of the different metals used for the pre-patterned electrodes, Au, Au-Ag and Ag-Ag contacted devices have been characterized. The transfer characteristics in linear scale of three devices with these electrode combinations are shown in Fig. 2a, 2b and 2c. The reported $I_D(V_G)$ curves have been obtained applying a drain to source bias of absolute value equal to 750 or 500 mV. Au contacted devices exhibit p-type conduction, while both Ag and Au-Ag devices, where the Ag contact is used as FET source, show prevalent n-type conduction. The role played by contacts workfunction in the determination of BP FETs polarity can be understood with the band diagram reported in Fig. 3. Multilayer BP flakes have a band structure close to bulk one, characterized by a 0.3 eV band gap and a 4.15 eV workfunction [13]. Polycrystalline Au and

Ag exhibit respectively a 5.1 and a 4.26 eV workfunction. As shown in Fig. 3, Ag is expected to offer a small barrier for electron injection to BP conduction band, explaining the n-type conduction enabled by Ag contacts to BP FETs. Vice versa, Au contacts favor hole injection while providing a high barrier for thermionic injection in BP conduction band. An analogous BP FETs polarity-control based on low workfunction metals (Al and Ti) has been reported for top contacted devices [7,14].

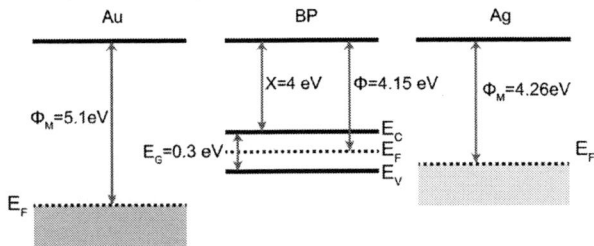

Fig. 3. Band diagram in vacuum of Au, multilayer BP flakes and Ag. Multilayer BP flakes show a band diagram similar to bulk samples. Polycrystalline Ag presents a workfunction close to BP one, so it is promising for electron injection in BP conduction band. Au on the contrary has a larger workfunction, favoring hole conduction in BP rather than electron one.

B. Transfer and output characteristc of n and p type BP FET

In the following we present a detailed characterization of the realized FETs, focused on representative p and n-type devices. We fabricated both the devices on the same wafer with the same processing conditions, and we performed the dielectric deposition simultaneously on the two FETs. The p-type device is contacted with Au electrodes and presents gate length L = 2.6 μm and width W = 1.5 μm. The n-type FET exhibits Au drain and Ag source, with L ≈ W = 1.6 μm. The output characteristics of these two devices, measured at different gate biases, are reported in Fig. 4a and 4b. The $I_D(V_D)$ characteristics of both the Au contacted device (Fig. 4a) and the Ag-Au one (Fig. 4b) show no saturation for the considered gate biases and a good ohmic behavior.

This work was financially supported by the European Research Council (ERC) under the ERC Advanced Grants Milli-Tech (ERC-2015-AdG-695459)

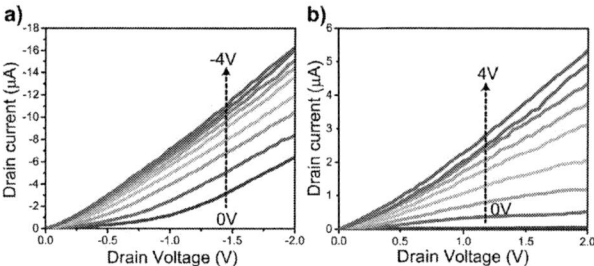

Fig. 4. Output characteristic $I_D(V_D)$ of Au a) and Au-Ag b) contacted BP FETs measured at different gate biases. For both the devices and set of contacts, there is no saturation of the output characteristic and a good ohmic behavior is observed.

We report in Fig. 5a and 5b the transfer characteristics of the same devices in semilogarithmic scale, obtained measuring $I_D(V_G)$ at different drain-to-source biases. The p-type FET (Fig. 5a) shows a I_{ON}/I_{OFF} current ratio larger than two orders of magnitude. The n-type device instead (Fig. 5b) reaches a I_{ON}/I_{OFF} of 1700. These results outperform the ON/OFF current ratios reported for pre-patterned non-embedded contacts [9], and could be further improved in BP FETs realized with thinner flakes and bottom gated, thanks to the larger band gap and higher quality of the dielectric layer.

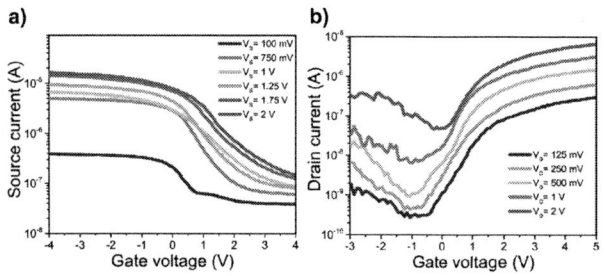

Fig. 5. Transfer characteristic $I_D(V_G)$ at different drain-to-source biases in semilog scale of the characterized p-type a) and n-type b) BP FETs. The p-type device exhibits a I_{ON}/I_{OFF} ratio larger than 10^2, while the n-type FET provides a ON/OFF current ratio larger than 10^3.

Fig. 6 shows the extracted transconductance curves $g_m(V_G)$ for the two devices, measured at different drain-to-source biases. The Au contacted device g_m (Fig. 6a) exhibits a relatively narrow peak in the gate bias range from -1 V to 1 V. This suggests that the contact resistance is limiting the transconductance preventing the saturation of the curve at the peak value [14]. At large negative gate biases, the series resistance provided by the contacts becomes larger than the BP channel resistance, determining a change of the $I_D(V_G)$ curve slope and the consequent degradation of the transconductance. Conversely, the n-type device transconductance, reported in Fig. 6b, saturates to its peak value over a large window of gate voltages suggesting a lower impact of the contact resistance on the FET conduction.

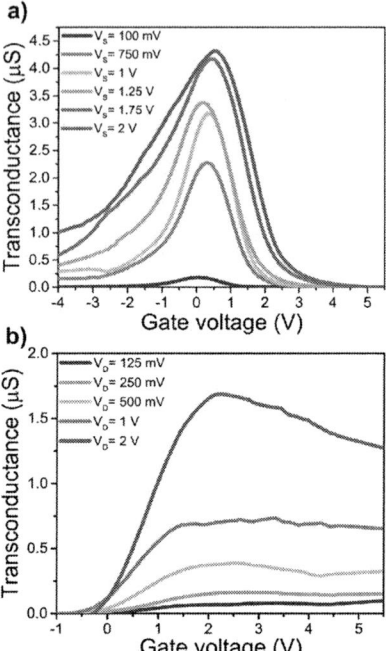

Fig. 6. Gate transconductance $g_m(V_G)$ at different drain-to-source biases of the characterized p-type a) and n-type b) BP FETs. The Au contacted device shows a peak of the transconductance, while the n-type g_m curve saturates to the maximum value over a wide range of the gate bias.

C. Field effect mobility extraction

We then extracted the two-terminal effective field effect mobility, μ_{FE}, using $\mu_{FE} = g_m L / W C_{ox} V_{DS}$, where g_m is the transconductance extracted from the linear part of the $I_D(V_G)$ curve, W is the channel width and C_{ox} is the gate capacitance per unit area for 15 nm of Al_2O_3. The relative permittivity of the dielectric layer has been extracted to be equal to 6.9 by characterizing MIM structures included on the same wafer of the devices. The $I_D(V_G)$ curves of the two devices are shown in Fig. 1a and Fig. 1b together with the fit of the linear part, whose slope is used for the mobility extraction. The obtained hole mobility for the Au contacted transistor is 10 cm^{2}V^{-1}s^{-1} at V_{SD} = 750 mV, while the Au-Ag FET presents an electron mobility of 1.8 cm^{2}V^{-1}s^{-1} at V_{DS} = 500 mV.

In order to obtain a more accurate estimation of the carriers' mobility by decoupling the impact of the contact resistance, we applied the so-called (Ghibaudo) Y function method which provides a robust evaluation of the low-field carrier mobility in 2D channel field effect devices [2]. Assuming that the contact resistance is not dependent on the gate bias, the Y function expression reduces to:

$$Y = \frac{I_D}{\sqrt{g_m}} = \sqrt{\frac{\mu C_{ox} |V_{DS}| W}{L}} (V_G - V_T) \qquad (1)$$

where V_T is the threshold voltage.

Fig. 7 shows the Y function computed for the two reported

devices, and the corresponding fit of the linear region from which it is possible to extract the low field mobility. The obtained mobilities are 40 cm²V⁻¹s⁻¹ for holes in the Au contacted FET (Fig. 7a) at V_{SD} = 750 mV and 2 cm²V⁻¹s⁻¹ for electrons in the Au-Ag device (Fig. 7b) at V_{DS} = 500 mV. The large enhancement of the estimated hole mobility for the p-type device with the Y function method together with the gate dependence of the transconductance shown in Fig. 6a, suggests that the contact resistance is playing a relevant role in limiting the Au contacted device performance. Vice versa, the n-type device shows a limited increase of the estimated mobility when applying the Y function method, suggesting that Ag constitutes a good contact for electron injection in BP.

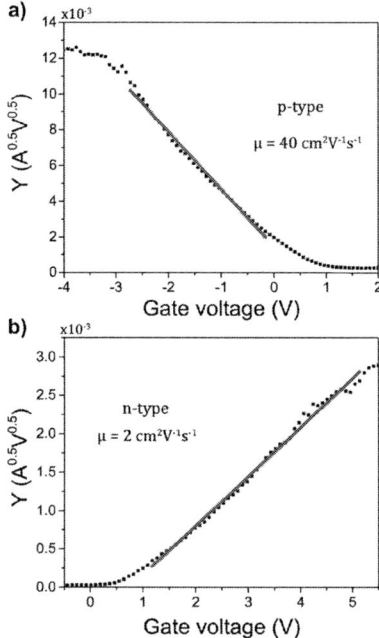

Fig. 7. Extracted Y function vs gate voltage of a) the p-type FET at V_{SD}= 750 mV and b) the n-type device at V_{DS}= 500 mV. The mobility can be extracted from the fitting of the linear part of the curve, shown in red in the figures.

The extracted mobility values and I_{ON}/I_{OFF} ratios are larger than the ones reported for pre-patterned non-embedded contacts [9], but are still lower than other results published for top contacted BP FETs [5,7]. Other sources of mobility reduction not considered here could be the interface roughness scattering, the high density of oxide traps in the SiO_2 and Al_2O_3 dielectric layers and the anisotropic mobility distribution in the 2D plane of black phosphorous [11, 15]. The performance of p-type devices could also be negatively affected by the choice of the top gate dielectric, since Al_2O_3 has been reported to impact considerably the p-dominant conduction mechanism in BP [11].

IV.   CONCLUSION

In summary, we have demonstrated for the first time complementary top gated BP FETs by the workfunction engineering of embedded pre-patterned contacts, exhibiting enhanced performance with respect to alternative implementations of pre-patterned electrodes FETs. Both n and p-type devices have been demonstrated, with electron and hole mobilities respectively 2 cm²V⁻¹s⁻¹ and 40 cm²V⁻¹s⁻¹ extracted applying the Y function method. Moreover, we proved that Ag is a promising choice for electron injection in black phosphorous FETs, whose polarity can be controlled by properly selecting the contacts workfunction. The proposed fabrication approach minimizes the exposure of unprotected flakes to ambient oxidants and contaminants, with reduced topography. Further enhancement of the performance could be obtained by reducing the contact resistance thanks to annealing and by using different high-k top or bottom gate dielectrics [11].

REFERENCES

[1] F. Schwierz, "Graphene transistors," *Nat. Nanotechnol.*, vol. 5, no. 7, pp. 487–496, 2010.

[2] H. Y. Chang, W. Zhu, and D. Akinwande, "On the mobility and contact resistance evaluation for transistors based on MoS2 or two-dimensional semiconducting atomic crystals," *Appl. Phys. Lett.*, vol. 104, no. 11, 2014

[3] J. Kang, W. Liu, and K. Banerjee, "High-performance MoS2 transistors with low-resistance molybdenum contacts," *Appl. Phys. Lett.*, vol. 104, no. 9, pp. 2–7, 2014.

[4] S. Das and J. Appenzeller, "WSe2 field effect transistors with enhanced ambipolar characteristics," *Appl. Phys. Lett.*, vol. 103, no. 10, 2013.

[5] L. Li, Y. Yu, G. J. Ye, Q. Ge, X. Ou, H. Wu, D. Feng, X. H. Chen, and Y. Zhang, "Black phosphorus field-effect transistors.," *Nat. Nanotechnol.*, vol. 9, no. 5, pp. 1–17, 2014.

[6] A. Castellanos-Gomez, et al., "Isolation and characterization of few-layer black phosphorus," *2D Mater.*, vol. 1, no. 2, p. 25001, 2014.

[7] D. J. Perello, S. H. Chae, S. Song, and Y. H. Lee, "High-performance n-type black phosphorus transistors with type control via thickness and contact-metal engineering.," *Nat. Commun.*, vol. 6, p. 7809, 2015.

[8] N. Haratipour and S. Koester, "Ambipolar Black Phosphorus MOSFETs with Record n-Channel Transconductance," *IEEE Electron Device Lett.*, vol. 37, no. 1, pp. 1–1, 2015.

[9] J. O. Island, G. A. Steele, H. S. J. Van Der Zant, and A. Castellanos-gomez, "Environmental instability of few-layer black phosphorus," *2D Mater.*, vol. 2, no. 1, p. 11002.

[10] J. Pei, X. Gai, J. Yang, X. Wang, Z. Yu, D.-Y. Choi, B. Luther-Davies, and Y. Lu, "Producing air-stable monolayers of phosphorene and their defect engineering," *Nat. Commun.*, vol. 7, p. 10450, 2016.

[11] H. Liu, A. T. Neal, M. Si, Y. Du, and P. D. Ye, "The effect of dielectric capping on few-layer phosphorene transistors: Tuning the schottky barrier heights," *IEEE Electron Device Lett.*, vol. 35, no. 7, pp. 795–797, 2014.

[12] M. Buscema, D. J. Groenendijk, S. I. Blanter, G. A. Steele, H. S. J. Van Der Zant, and A. Castellanos-Gomez, "Fast and Broadband Photoresponse of Few-Layer Black Phosphorus Field-Effect Transistors," *Nature materials*, pp. 3–8, 2014.

[13] M. T. Edmonds, A. Tadich, A. Carvalho, A. Ziletti, K. M. O'Donnell, S. P. Koenig, D. F. Coker, B. Özyilmaz, A. H. C. Neto, and M. S. Fuhrer, "Creating a stable oxide at the surface of black phosphorus," *ACS Appl. Mater. Interfaces*, vol. 7, no. 27, pp. 14557–14562, 2015.

[14] S. Das, M. Demarteau, and A. Roelofs, "Ambipolar phosphorene field effect transistor," *ACS Nano*, vol. 8, no. 11, pp. 11730–11738, 2014.

[15] J. Qiao, X. Kong, Z.-X. Hu, F. Yang, and W. Ji, "High-mobility transport anisotropy and linear dichroism in few-layer black phosphorus.," *Nat. Commun.*, vol. 5, p. 4475, 2014

Tunneling Transistors based on $MoS_2/MoTe_2$ Van der Waals Heterostructures

Yashwanth Balaji [†], Quentin Smets, Cesar J. Lockhart de la Rosa[†], Anh Khoa Augustin Lu[†],
Daniele Chiappe, Tarun Agarwal[†], Dennis Lin, Cedric Huyghebaert, Iuliana Radu,
Dan Mocuta, Guido Groeseneken

IMEC, Kapeldreef 75, B-3001 Leuven, Belgium
[†]KU Leuven, Kasteelpark Arenberg 44, B-3001, Leuven, Belgium
Email: Yashwanth.balaji@imec.be

Abstract—**Two-dimensional transition metal dichalcogenides (TMD's) are promising materials for CMOS application[1], [2] due to their ultra- thin channel with excellent electrostatic control. TMD's are especially well suited for Tunneling Field-Effect Transistors (TFETs) due to their low dielectric constant and their promise of atomically sharp and self-passivated interfaces[3]–[5]. Here we experimentally demonstrate for the first-time band-to-band tunneling (BTBT) in Van der Waals (VdW) heterostructures formed by MoS_2 and $MoTe_2$. Density functional theory (DFT) simulations of the band structure show our MoS_2-$MoTe_2$ heterojunctions have a staggered band alignment, which boosts BTBT compared to a homojunction configuration. Low-temperature measurements and electrostatic simulations provide understanding towards the role of schottky contacts and the material thickness on device performance. This work provides the prerequisites and challenges required to overcome at the contact region to achieve a steep subthreshold slope and high ON-currents with 2D-based TFETs.**

Index Terms—**2D materials, TMD, TFET, Band-to-band tunneling, Heterostructures, Schottky contacts.**

I. INTRODUCTION

Tunnel FET's are a promising candidate for low-power logic applications, due to the reduced supply voltage enabled by a steep subthreshold slope less than 60mV/dec[3], [4]. There is increasing research interest in two dimensional TMD heterostructures based TFET's [5], [6], because the 2D layers of these materials are self-passivated and the interaction between layers are only through VdW forces. Therefore, 2D TFETs are expected to have a lower defect concentration and hence lower parasitic leakage current compared to their III-V counterparts[7].

Previous experimental works have been shown on 2D tunneling based devices exhibiting BTBT with different material stacks such as MoS_2/WSe_2[8][9], $SnSe_2/WSe_2$[10], $SnSe_2/BP$[11]. In another work, a steep-slope TFET with polymer electrolyte gating was realized using p-type Ge/MoS_2 3D-2D VdW heterojunction[12]. Though these works have experimentally demonstrated BTBT, not much focus was given to understanding the role of the schottky barriers at the contacts and the recombination and generation currents on the device characteristics. And moreover, to understand the tunneling current operation with respect to the number of layers and gate electrostatics.

In this work, we address the issues mentioned above. The TMD materials used to realize the TFET are Molybdenum disulfide (MoS_2) and Molybdenum ditelluride ($MoTe_2$). MoS_2 is the most widely studied 2D material after graphene and has shown to be an n type semiconductor[1] whereas $MoTe_2$ has shown to be ambipolar with a high p-type current[13][14]. This material stack is chosen due to a staggered band alignment shown by the subsequent DFT simulations.

II. DFT SIMULATIONS

DFT simulations are performed to analyze the band alignment of the $MoS_2/MoTe_2$ stack (Fig.1). We use the quantum espresso software package[15] with the Perdew-Burke-Ernzerhof exchange-correlation functional (PBE)[16] combined with project-augmented waves (PAW) pseudopotentials. Corrections for VdW forces introduced by Grimme[17] are also included. The simulated stacks are firstly 1 layer MoS_2 + 1 layer $MoTe_2$ and secondly 3 layers MoS_2 + 3 layers $MoTe_2$. These two different stacks are chosen to understand the effect of the thickness of the layers on the band alignment at the heterojunction. Fig.1(a) shows the 1+1 layers' band structure, where the band gaps are 1.63eV for MoS_2 and 1.14eV for $MoTe_2$. Together they form a straddled band alignment (type

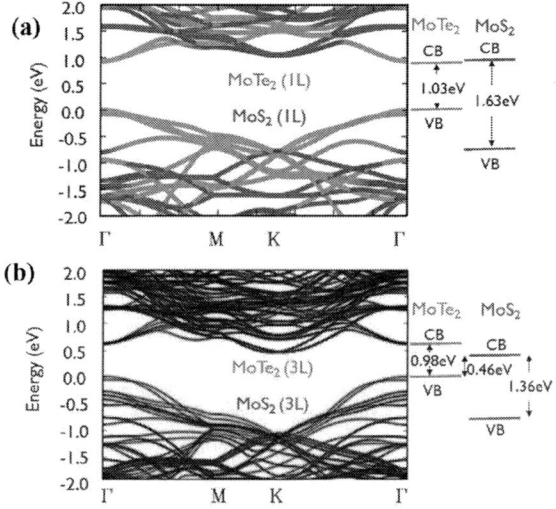

Fig 1. Band structure of overlapping MoS2-MoTe2 stack obtained using DFT simulations along with the band alignment and the bandgaps for (a)1+1 layers and (b) 3+3 layers heterostack.

978-1-5090-5979-9/17 $31.00 © 2017 IEEE

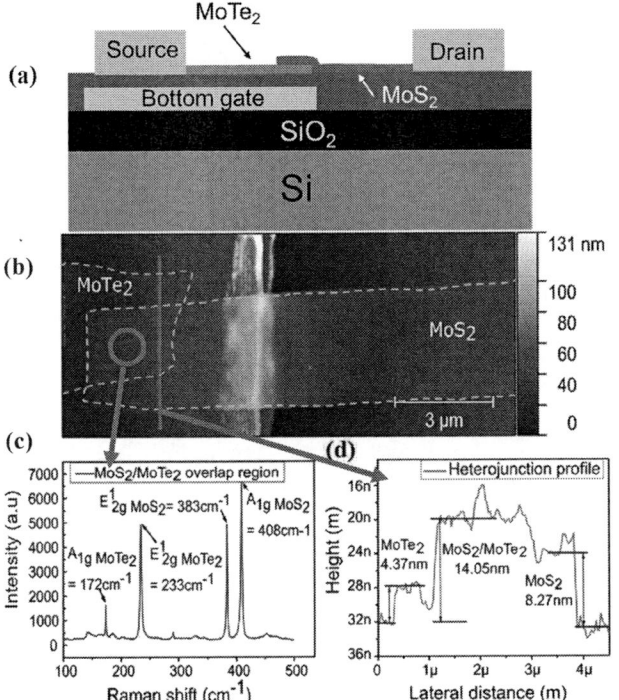

(a) Source — MoTe₂ — Drain
Bottom gate — MoS₂
SiO₂
Si

(b) MoTe₂ — MoS₂ — 3 µm — (131 nm, 100, 80, 60, 40, 0)

(c) MoS₂/MoTe₂ overlap region
E^1_{2g} MoS₂ = 383cm⁻¹
A_{1g} MoS₂ = 408cm⁻¹
A_{1g} MoTe₂ = 172cm⁻¹
E^1_{2g} MoTe₂ = 233cm⁻¹
Raman shift (cm⁻¹)

(d) Heterojunction profile
MoTe₂ 4.37nm — MoS₂/MoTe₂ 14.05nm — MoS₂ 8.27nm
Lateral distance (m)

Fig. 2 (a) schematic cross-section of the device architecture. (b) AFM showing the thickness of various parts of the heterostructure. (c) Raman spectra of the heterojunction depicting the characteristic peaks of MoS₂ and MoTe₂. (d) AFM height profile from the AFM on the MoTe₂, Overlap region and MoS₂.

1). Fig. 1(b) shows the 3+3 layers' band structure, where the MoS₂ bandgap decreases strongly to 1.36eV, and the MoTe₂ bandgap decreases only slightly to 0.98eV. These form a staggered band alignment (type 2) with an effective tunneling bandgap of 0.48eV. The band alignments are in good accordance with literature [18], [19]. When moving from the 1+1 to the 3+3 layer heterostructure, the decrease in effective band gap is indicative of orbital interactions between the two materials, consistent with the branch-overlaps when the bands of the MoS₂ and MoTe₂ overlap at several points in the bandstructure. Our DFT simulations show that thicker layers have more staggered band alignment, which is beneficial for a TFET. However thicker layers also degrade the electrostatics, which will be discussed more in detail later.

III. DEVICE FABRICATION

Since our device configuration uses a back gate and the bottom layer is MoTe₂, we target a thin MoTe₂ layer and thicker MoS₂ layer, to achieve both the staggered alignment and good electrostatic control. A schematic side view of the device configuration is shown in Fig. 2(a). Initially a local bottom gate is fabricated using palladium as metal followed by the deposition of 30nm Al₂O₃ as the gate dielectric. The devices are fabricated using exfoliated flakes where the MoS₂ and MoTe₂ flakes are transferred onto the substrate using a Poly-methyl methacrylate (PMMA) dry transfer technique[20]. During the transfer, the flakes are aligned on the substrate in such a way that the entire MoTe₂ flake and the MoS₂/MoTe₂ overlap region are located above the bottom gate, whereas the MoS₂ flake is ungated as shown in Fig. 2(a). The Silicon back gate is nonfunctional. Gold was deposited as contacts for the source and drain by optical lithography. Fig. 2(d) shows an AFM image of the fabricated device along with the height profile shown along the indicated red line. Fig. 2(c) shows the Raman spectroscopy at the overlap region. The Raman characteristics obtained in the overlap region is a combined signal of the MoS₂ and MoTe₂ peaks, as shown in literature [21], [22].

IV. RESULTS AND DISCUSSION

The devices are electrically characterized in vacuum at temperatures ranging from 77K to 300K. The MoS₂ contact is defined as the drain and the MoTe₂ contact as the source. Fig 3(a) shows the output characteristics at 77K and we shall discuss them at different gate biases.

At Vg= -3V, the MoTe₂ layer is strongly accumulated with holes. Therefore, the valence band minimum of MoTe₂ aligns above the conduction band maximum of MoS₂ creating a window for tunneling, as shown in the sketched band diagram of Fig.5(c). The tunneling current increases for increasing reverse bias (V_D>0). In forward bias (V_D<0), the current is below the noise level (1pA) as the electrostatic barrier for diffusion is very high.

As the V_g is increased to -2V, the electric field at the heterojunction is reduced and the BTBT current decreases. A significant forward bias current is observed at V_D> -0.5V as the electrons acquire enough energy for diffusion. At higher gate biases of Vg > -1V, the tunneling current in the reverse bias is completely suppressed into the noise level (Fig. 3(a)) as there is no tunneling window. At Vg = 0V and above, the reverse bias is dominated by the reverse bias saturation current due to the drift in minority carrier concentration of MoTe₂.

We observe that when both the gate and drain biases are increased by 1V, the BTBT current remains identical. In order

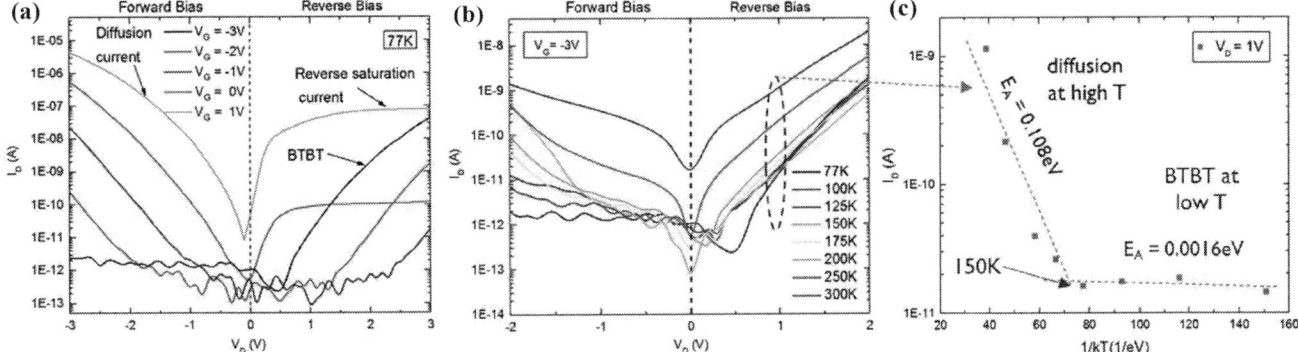

Fig. 3. (a) ID-VD characteristics with varying bottom gate measured at 77K.(b) Temperature dependence of tunneling current in the reverse bias at Vg = -3V.(c) Arrhenius plot extracted from (b) at V = 1V showing the change of slope from diffusion to tunneling at low temperatures.

978-1-5090-5979-9/17 $31.00 © 2017 IEEE

to understand this effect, which is also observed for other 2D TFETs [8], [9], we self-consistently calculate the 1-D electrostatic potential in the vertical direction using the Poisson equation. The electron and hole concentrations are calculated semi-classically. These simulations do not consider BTBT or other transport mechanisms, and do not consider interaction between electronic orbital positions of the different layers but the bandgaps and band alignment are taken from DFT in Fig.1(b). The electrostatics are simulated for 6-layer MoS₂ on 6-layer MoTe₂ on 30nm SiO₂ with 0.3nm VdW spacing considered between each layer. n-type doping is considered in the MoS₂ flake, which is commonly observed[1]. The band diagram in Fig.4(a) show this doping fixes the Fermi level position near the conduction band edge. Hence the back-gate voltage controls mainly the MoTe₂ bands. In Fig.4(b) we observe that decreasing both the gate and the drain bias by 0.25V shifts the bands diagram up in energy, and near-identical BTBT paths are obtained, thus explaining the effect observed experimentally.

The simulations also show that the layers closer to the oxide are modulated more effectively compared to the ones further away. At low electric field, the electrons tunnel from the farthest layers creating long tunneling paths. As the electric field is increased, the tunneling paths are shortened. This transition from long to short tunnel paths is not ideal for a TFET and results in a poor subthreshold swing. To have ideal tunneling characteristics with steep switching, ideally monolayers are required for optimum electrostatic control.

When we compare our results with other work on MoS₂/WSe₂ TFETs[8], [9], the back-gate dependence of the BTBT and diffusion currents in the output characteristics are in good agreement, despite the absence of a top gate on our devices. However, we do not observe a Negative Differential resistance (NDR) in the forward bias. We interpret the lack of NDR in our devices as a lack of degenerate Fermi level in the top flake. The condition of Fermi level degeneracy is required for Esaki diode behavior, but is not required for TFET operation

The temperature dependent I_d-V_d characteristics are plotted in Fig. 3(b) at a bias of V_g = -3V where the tunneling current is dominant. It is observed that the current values do not change significantly at low temperatures, but above 150K the current increases with temperature. To further understand this behavior, Arrhenius plots are extracted as shown in fig.3(b) at V_D = 1V. The plots show two different slopes with the transition taking place at 150K. The low activation energy of E_A=1.6meV below

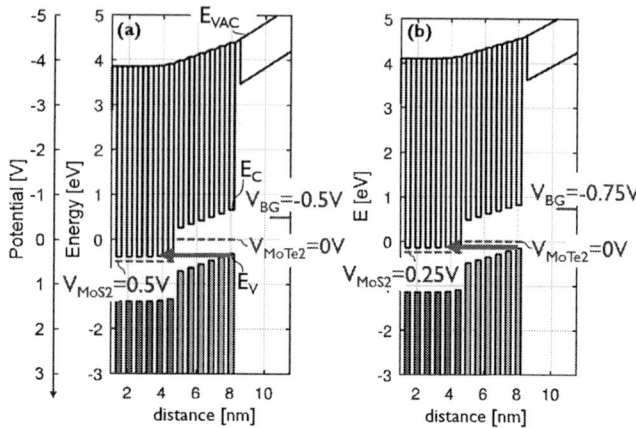

Fig. 4 (a) Simulated band diagrams of 6-layer MoS₂ on 6-layer MoTe₂ on 30nm SiO₂ on a bottom gate. The MoS₂ layers are doped 5×10^{11} cm⁻². Filled electron states in are shown in red for MoS₂ and green for MoTe₂. (b) Simultaneously decreasing VMoS₂ and V_{BG} shifts the band diagram up and near-identical BTBT paths are obtained.

150K is indicative of tunneling-limited current. Since the MoTe₂ flake is gated, the strongly negative V_g thins down the schottky barrier for holes. As a result, at low temperatures the thermionic injection over the barrier is suppressed and the dominant mechanism is schottky tunneling. Above 150K, generation currents become dominant and the carriers acquire enough energy for diffusion above the barrier indicated by the increase in E_A=108meV.

To further understand the different transport mechanisms such as generation and recombination effects and the role of schottky contacts in the forward and reverse biases, we need to measure below 60K to see the effects of the schottky contacts and its influence on the diffusion and BTBT currents. Thus, temperature measurements were done on the same device from 10K to 300K and E_A's were extracted. Fig 5(a) shows the output characteristics where the device is measured in a tunnel diode configuration where the device is left floating. Applying no gate bias prevents in gating of the contact regions, to understand how the ungated schottky barriers play a role in the current transport at different temperatures. At Vg= 0V, the device functions as a diode undergoing diffusion at forward bias (V_D<0V) and BTBT at reverse bias(V_D>3V). Between 0<V_D<3V, we also observe a reverse saturation current similarly seen in Fig. 3(a) and (b) due to the minority carrier which increases in current with increase in temperature.

Arrhenius plots are extracted from the diffusion and

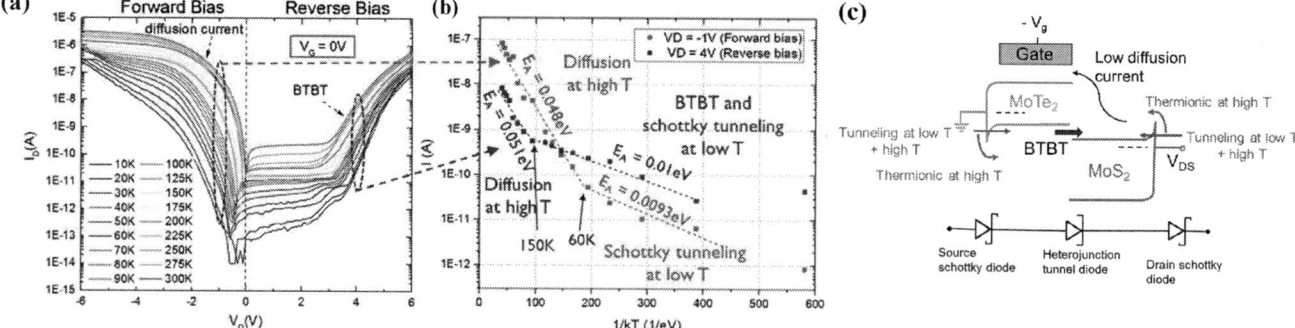

Fig. 5.a) I_D- V_D characteristics at temperatures from 10K – 300K depicting diffusion mechanism at forward bias and BTBT at reverse bias. b) Arrhenius plots for the temperature measurement extracted at the diffusion and tunneling regimes. c) Schematic representation of the different transport mechanisms taking place on the device at different temperatures along showing the different diode behaviors at the schottky contacts and tunnel junction.

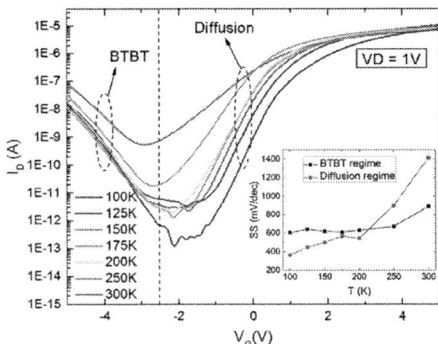

Fig. 6. I_D-V_G characteristics shown at different temperatures with the dependence of the diffusion and tunneling currents with temperature shown in the inset.

tunneling regimes as shown in Fig. 5(b) by the red and blue data points respectively. The Arrhenius plot of the forward bias currents are extracted at V_D= -1V as shown by the red data points. Below 60K, the E_A is very low of 9.3meV which is indicative that the electrons do not have enough energy to thermionically inject across the schottky barrier, therefore they undergo tunneling through the barrier. The high E_A of 48meV above 60K indicates that the electrons start to undergo thermionically assisted tunneling across the schottky barrier along with increased carrier recombination and generation effects with increasing temperature.

Similarly, the Arrhenius plot for the tunneling regime was extracted at reverse bias of V_D = 4V shown by the blue data points in Fig. 5(b). The E_A obtained from 150K to 300K is similar to the E_A in the diffusion regime (60K to 300K) with E_A = 51meV. As observed previously in fig 3(b) and (c), above 150K, the carriers acquire enough energy to undergo diffusion due to the increased thermionic injection from the contacts and generation effects thereby dominating the BTBT current to vary with temperature. Below 150K, it was observed from the previous Arrhenius plot in Fig. 3(c) that the E_A for the BTBT current is very low with a value of 1.6 meV. The reason for the very low E_A obtained is due to the thinning down of the barrier by the applied gate bias. But for Fig. 5(a), there is no gate bias applied, therefore the barriers at the contacts are dominantly present. The E_A obtained in Fig. 5(c) is 10meV, which is 6 times higher and very close to the E_A in the diffusion regime below 60K. This can be understood by the fact that the schottky contacts are more dominant on the current transport and the tunneling in the schottky contacts are in series with the BTBT current.

Therefore, the study shows that the schottky contacts have a current limiting contribution towards the device. We see that, schottky barriers act as individual schottky diodes that are coupled along with our heterojunction tunnel diode. Fig. 5(c) shows a schematic of the different transport mechanisms in the device at different temperatures. The E_A obtained for these barriers are higher than that of the BTBT current which affect the device performance in terms of low subthreshold swing and low ON currents, hence an ohmic contact is required to eliminate the influence at the contacts to solely study the behavior at the heterojunction tunneling region.

Fig. 6. shows the transfer characteristics for the device at different temperatures. The device behaves as a p-TFET for Vg<-2V or a n-MOSFET for Vg>-2V. In the p-TFET mode, the BTBT current does not significantly vary with T in the range 100-200K and the subthreshold swing is 600mV/dec and

constant with temperature (inset of Fig. 6), indicative of BTBT. However, for T>200K, the current increases with temperature due to recombination and schottky contacts undergo thermionic injection. In the n-MOSFET mode the SS increases linearly with temperature, indicative of diffusion current. The subthreshold swing of these modes is very high, which is due to the large estimated oxide thickness (EOT) present thereby preventing fast switching.

V. CONCLUSION

In summary, we have demonstrated band-to-band tunneling in a TMD based MoS_2/$MoTe_2$ heterostructures. DFT simulations result in a staggered band alignment and 1-D electrostatic potential simulations depict the need for working with thinner layers for optimum device performance. The low temperature measurements clearly distinguish the different mechanisms in the forward bias and reverse bias, and that the diffusion and tunneling currents are largely contact limited. The transfer characteristics show the device as a p-TFET for negative V_g and n-MOSFET for positive V_g. In conclusion, a key challenge in achieving a steep subthreshold swing in 2D hetero TFETs apart from a low EOT and thinner layers are having ohmic contacts with a low contact resistances.

REFERENCES

[1] B. Radisavljevic et al., "Single-layer MoS2 transistors.," *Nat. Nanotechnol.*, vol. 6, no. 3, pp. 147–150, 2011. [2] G. Fiori *et al.*, "Electronics based on two-dimensional materials," *Nat. Nanotechnol.*, vol. 9, no. 10, pp. 768–779, 2014. [3] A. C. Seabaugh et al., "Low-Voltage Tunnel Transistors for Beyond CMOS Logic," *Proc. IEEE*, vol. 98, no. 12, pp. 2095–2110, Dec. 2010. [4] A. M. Ionescu et al., "Tunnel field-effect transistors as energy-efficient electronic switches.," *Nature*, vol. 479, no. 7373, pp. 329–37, Nov. 2011. [5] D. Jena, "Tunneling Transistors Based on Graphene and 2-D Crystals," *Proc. IEEE*, vol. 101, no. 7, pp. 1585–1602, Jun. 2013. [6] a K. Geim et al., "Van der Waals heterostructures.," *Nature*, vol. 499, no. 7459, pp. 419–25, Jul. 2013. [7]H. Lu et al., "Tunnel Field-Effect Transistors: State-of-the-Art," *IEEE J. Electron Devices Soc.*, vol. 2, no. 4, pp. 44–49, Jul. 2014.[8]T. Roy *et al.*, "Dual-Gated MoS2 /WSe2 van der Waals Tunnel Diodes and Transistors," *ACS Nano*, vol. 9, no. 2, pp. 2071–2079, 2015.[9] A. Nourbakhsh et al., "Transport properties of a MoS2/WSe2 heterojunction transistor and its potential for application," *Nano Lett.*, vol. 16, no. 2, pp. 1359–1366, 2016.[10] T. Roy et al., "2D-2D tunneling field-effect transistors using WSe2/SnSe2 heterostructures," *Appl. Phys. Lett.*, vol. 108, no. 8, 2016.[11] R. Yan *et al.*, "Esaki Diodes in van der Waals Heterojunctions with Broken-Gap Energy Band Alignment.," *Nano Lett.*, vol. 15, no. 9, pp. 5791–8, Sep. 2015.[12] D. Sarkar *et al.*, "A subthermionic tunnel field-effect transistor with an atomically thin channel," *Nature*, vol. 526, no. 7571, pp. 91–95, 2015.[13] S. Fathipour *et al.*, "Exfoliated multilayer MoTe 2 field-effect transistors," vol. 192101, pp. 19–21, 2014.[14] F. R. Mote *et al.*,"Field-Effect Transistors Based on," no. 6, pp. 5911–5920, 2014.[15] P. Giannozzi *et al.*, "QUANTUM ESPRESSO: a modular and open-source software project for quantum simulations of materials.," *J. Phys. Condens. Matter*, vol. 21, no. 39, p. 395502, Sep. 2009.[16] J. Perdew et al., "Generalized Gradient approximation Made Simple.," *Phys. Rev. Lett.*, vol. 77, no. 18, pp. 3865–3868, Oct. 1996.[17] S. Grimme et al, "Semiempirical GGA-Type Density Functional Constructed with a Long-Range Dispersion Correction," vol. 16, 2006.[18]P. Miró et al., "An atlas of two-dimensional materials.," *Chem. Soc. Rev.*, vol. 43, no. 18, pp. 6537–54, Aug. 2014.[19] C. Gong et al., "Band alignment of two-dimensional transition metal dichalcogenides: Application in tunnel field effect transistors," *Appl. Phys. Lett.*, vol. 103, no. 5, p. 053513, 2013.[20]T. Roy *et al.*, "Field-effect transistors built from all two-dimensional material components," *ACS Nano*, vol. 8, no. 6, pp. 6259–6264, 2014.[21]T. U. S.Sugai, "High- pressure Raman spectroscopy in the layered materials 2H-MoS2,2H-MoSe2 and 2H-MoTe2," vol. 26, no. 12, pp. 6554–6558, 1982.[22] H. Li *et al.*, "From Bulk to Monolayer MoS2: Evolution of Raman Scattering," *Adv. Funct. Mater.*, vol. 22, no. 7, pp. 1385–1390, Apr. 2012.

Temperature Dependence of Contact Resistance for Gold-Graphene Contacts

A. Gahoi, [a] S. Kataria, [a, b] M. C. Lemme [a, b*]

[a] University of Siegen, School of Science and Technology, Hölderlinstr.3, 57076 Siegen, Germany
[b] RWTH Aachen University, Faculty of Electrical Engineering and Information Technology, Otto-Blumenthal-Str. 25, 52074 Aachen, Germany
Corresponding author: max.lemme@uni-siegen.de

Abstract — We report temperature dependent transport properties of back-gated graphene TLM structures in a wide temperature range from 35 K to 450 K. We use gold as the contact material and find that the contact resistance exhibits a strong temperature dependence, dropping considerably to a value of 315±127 Ωμm at 35 K as compared to 957±210 Ωμm at 450 K measured at the Dirac point. This significant drop in R_C is attributed to an increase in carrier mean free path in graphene which enhances the transmission efficiency through the metal - graphene junction. At 35 K, the carrier mean free path in graphene is calculated to be ~ 41.6 nm compared to ~ 12.67 nm at 450 K.

Keywords— contact resistance; temperature; graphene; TLM

I. INTRODUCTION

Graphene has attracted worldwide attention for post silicon nanoelectronics due to its unprecedented electronic properties including high charge carrier mobility (μ) [1], [2] [3]. Now, large-area graphene synthesis using chemical vapor deposition (CVD) technique using hydrocarbon molecules in the presence of metal catalyst has reached a pre-commercialization stage [4]. In addition, CVD graphene can be easily transferred to arbitrary substrates [5] [6]. Numerous electronic applications have been demonstrated such as radio frequency (RF) analog transistors [2] [7] [8], photodetectors [9], [10] [11], nanoelectromechanical systems [12], [13], terahertz modulators [14] using CVD graphene. However, contacts are unavoidable for charge injection and collection from graphene for any desired application and hence contact resistance (R_C) between graphene and contact material plays a crucial role. Although the metal-graphene (M-G) contact is ohmic in nature, the low density of states in graphene generally leads to high R_C values [15] and is one of the major challenges towards industrial adoption of graphene [15]–[18]. The R_C requirement for silicon state-of-the-art metal oxide field effect transistors (MOSFETs) is 80 Ωμm per contact [19]. Reported values of R_C for M-G contacts is scattered in the literature and varies greatly from a few tens (69 Ωμm) [20] to mega ohms (10^6 Ωμm)[16]. This large variation is attributed due to the measurement technique, fabrication procedure, quality of graphene layer, and measurement conditions.

High R_C at the M-G interface stems due to the unique physical interaction that takes place between contact metal and atomically thin graphene. High R_C limits the on-current of graphene field-effect transistors (FETs) which deters the performance of analog and RF devices [21], [22]. High transconductance (g_m) is of utmost importance in RF devices as it may lead to an improved intrinsic voltage gain [21]. Unfortunately, extrinsic g_m is usually suppressed by series resistance which are not gateable such as lead, access and R_C. While lead and access resistance can be technologically reduced [23], [24], R_C is mostly determined by the electrical contact properties at M-G interface. High R_C is indeed a big bottleneck in high-frequency GFETs devices and reports suggest it could be comparable to gateable channel resistance thus suppressing extrinsic transconductance and voltage gain [21]. Real world applications require operation of the graphene devices in a wide range of temperatures. For example, spintronics applications at low temperature [25] and detection of fuel leaks in space craft at high temperatures [26]. Therefore, it is utmost important to understand the electrical properties of M-G interfaces, i.e. R_C, at various temperatures in order to develop graphene based electronics. However, such temperature dependent studies are scattered in the literature.

In this work, transmission line method (TLM) structures were fabricated on monolayer CVD graphene with gold (Au) as the contact material. Au was selected as the contact material as it is shown to exhibit low R_C [27]. Electrical characterization was carried out at Lakeshore Cryotronics using a Keithley SCS4200 parameter analyzer. Transport measurements were performed in a wide temperature range from low (35 K) to extremely high temperature (450 K) and subsequently R_C was extracted at different back-gate bias.

II. EXPERIMENTAL

Thermally grown silicon wafers (p dope) with resistivity (1-20 Ωμm) were used as starting substrates. The thickness of SiO_2 achieved was around 85 nm. The oxidized silicon wafers was diced into sample size of 1.3 cm x 1.3 cm using advanced dicing saw. Large area graphene was grown on a copper (Cu) foil in a NanoCVD (Moorfield, UK) rapid thermal processing tool [4]. After graphene growth polymethyl methacrylate (PMMA) was spin coated on graphene grown on copper foil and was baked at 180°C. Graphene was then transferred on the SiO_2/Si substrate using electrochemical delamination technique [5]. Optical micrograph of a graphene transferred on a SiO_2/Si substrate is shown in Figure.1. Detailed description about the graphene transfer is mentioned in [28].

978-1-5090-5979-9/17 $31.00 © 2017 IEEE

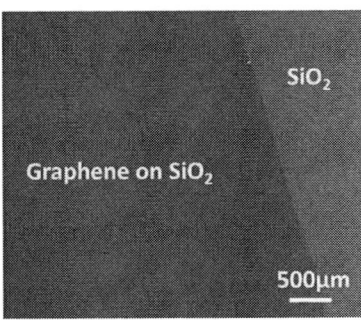

Figure 1. CVD monolayer graphene transfer on SiO₂ substrate by electrochemical delamination method.

Later on, optical lithography and oxygen (O₂) plasma etching were used to define graphene channels in the presence of O₂ (80 sccm), pressure (80 mtorr) and power (50 W). Au was thermally evaporated, and a lift-off process was carried out in order to define source-drain contact pads. Figure 2 shows the optical image of the fabricated TLM structures with contact spacing (L_{CH}) varying from 5 μm to 30 μm for a defined channel width (W_{CH}) of 10 μm. TLM was used as a measurement technique in order to extract R_C due to the fact that channel length was much larger than the carrier mean free path (λ) in the graphene channel [17] [29].

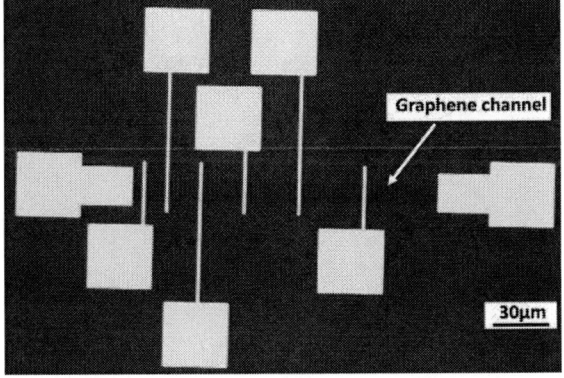

Figure 2. Optical image of the TLM structures with gold contacts.

III. RESULTS & DISCUSSION

Electrical characterization was carried out at Lakeshore Cryotronics using a Keithley SCS4200 parameter analyzer. Figure 3a-3e shows the transfer characteristics (I_{SD}-V_{BG}) of TLM structures measured at 35 K, 150 K, 300 K, 400 K and 450 K, respectively. All these measurement were carried out under a high vacuum of around 10^{-6} mbar. Fig. 3f shows the transfer curves for a channel length of 5 μm in the studied temperature range. It can be seen that with increase in temperature, the transfer curves exhibit significant changes like decrease in I_{SD} for a particular channel length and an obvious broadening which points towards the increased charge carrier scattering at high temperatures [30]. A shift in Dirac

voltage (V_{Dirac}), point of minimum conductivity, towards negative voltage is also observed and it may indicate that extrinsic factors are still affecting the transport properties of our devices.

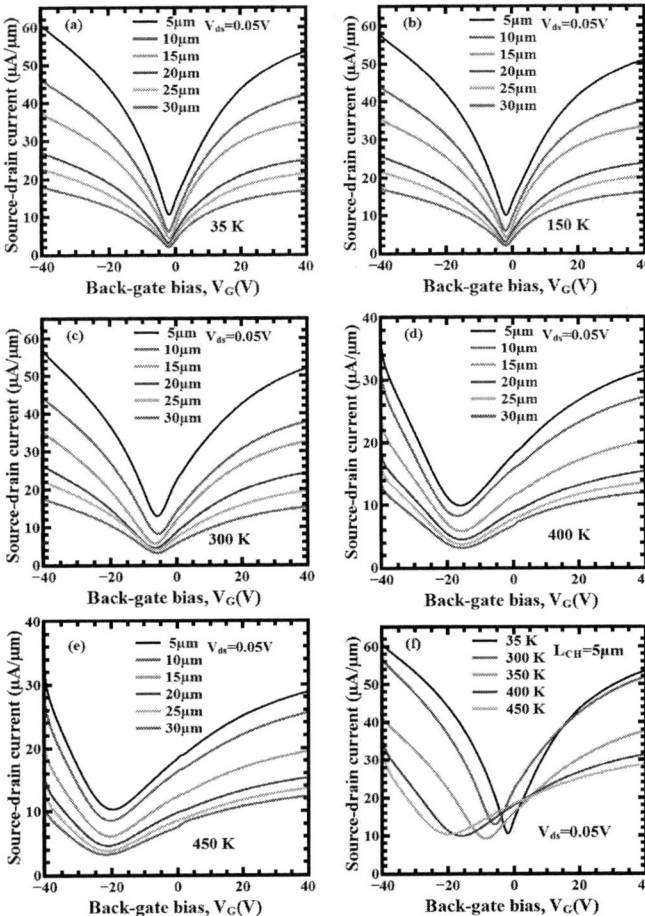

Figure 3. Transfer characteristics of a graphene FET on a TLM structures with contact separations from 5 μm to 30 μm with a contact width of 10 μm measured at (a) 35 K, (b) 150 K, (c) 300 K, (d) 400 K, (e) 450 K and (f) a fair comparison of transfer characteristics for L_{CH} = 5 μm measured at 35 K, 300 K, 350 K, 400 K and 450 K..

Figure 4 shows the source-drain resistance as a function of back-gate bias measured at 35 K. Total source-drain resistance is the combination of the R_C, length dependent channel resistance and resistance of metal and leads which is generally neglected. From the figure it is quite evident that R_C in the n branch is always larger than that in the p branch [15]. This is due to the fact that when a positive back gate bias (V_G) is applied to the device it form a p-n-p junction near the drain contact which leads to additional resistance [15]. Figure 5 shows R_C as a function of gate bias for the TLM devices measured at 35 K. R_C exhibits clear gate dependence as it peaks around V_{Dirac} and decreases as the channel is

978-1-5090-5979-9/17 $31.00 © 2017 IEEE   111

electrostatically field doped by the back gate indicating a strong doping dependence of R_C.

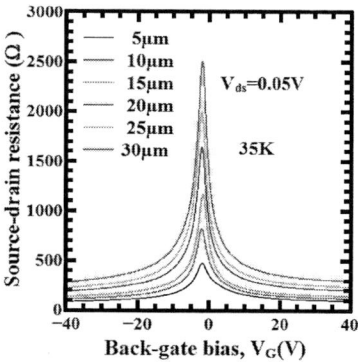

Figure 4. Source-drain resistance in a back-gated GFET as a function of back-gate bias (V_{bg}) at 35 K.

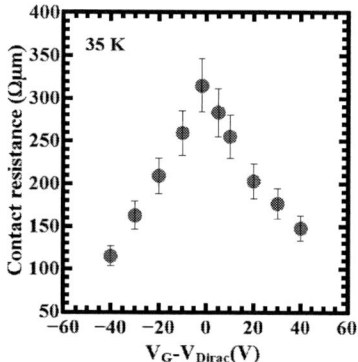

Figure 5. Measured contact resistance (R_C) as a function of a back gate bias at 35 K.

Figure 6 shows the measured R_C at V_{Dirac} (primary Y-axis) and electron mobility (μ_e) in graphene (secondary Y-axis) for the TLM structures measured at different temperatures. R_C is found to increase with rise in temperature, being minimum at 35 K with a value of 315±127 Ωμm as compared to 957±210 Ωμm at 450 K. Error bars indicate the upper and lower limits while fitting the data. A mobility of 1232 cm²/V.s is extracted at 35 K as compared to 652 cm²/V.s at 450 K at a carrier density of 6.54×10^{12} cm⁻². Low R_C could be attributed to an increase in carrier mean free path in graphene (λ) at low temperatures [17]. Current transmission efficiency (T_{MG}) in diffusive limit is given as $\sqrt{\lambda}/(\lambda_m + \lambda)$ where λ_m is the metal graphene coupling length and with the increase of λ, T_{MG} increases [17]. In order to evaluate the improvement in graphene devices at low temperature, λ is calculated by $\lambda = (h/e).\mu.(\pi n_s)^{1/2}$ where h is the reduced Planck's constant and e is the elementary charge [31]. This resulted in a λ of 41.6 nm at 35 K and 12.67 nm at 450 K as shown in Figure 7. At low temperature, calculated λ in graphene increases under the metal contact which results in an increase in the current transmission efficiency through the metal-graphene junction and, hence a lower R_C is achieved.

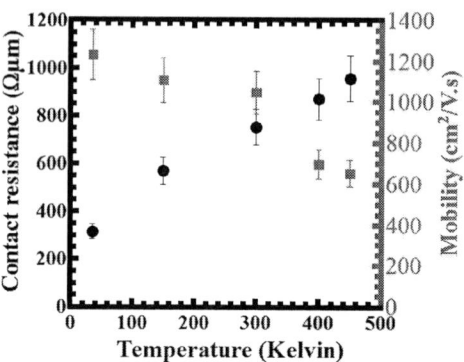

Figure 6. Measured gold-graphene contact resistance (R_C) and electron mobility (cm²/V.s) as a function of temperature at V_{Dirac}.

Enhancement in μ_e is due to the increase in λ in graphene, which in turn depends on various carrier scattering mechanisms. Higher the λ in graphene, less is the scattering due to collision and higher the μ_e. Generally, scattering of electrons with LA (longitudinal-acoustic) phonons is suppressed at low temperature in graphene which leads to high mobility [32]. From the above experimental study one important thing is pretty evident that λ in graphene plays a dominating role in improving electrical characteristics of graphene based devices. Low R_C leads to high transconductance (g_m) in graphene device. Transconductance (g_m) is of paramount importance in RF applications. Graphene amplifier gain characteristics is a function of g_m of the graphene FET [24], [33].

Figure 8 shows g_m with respect to back gate bias for 35 K, 300 K and 450 K. It can be seen that at 35 K high g_m is

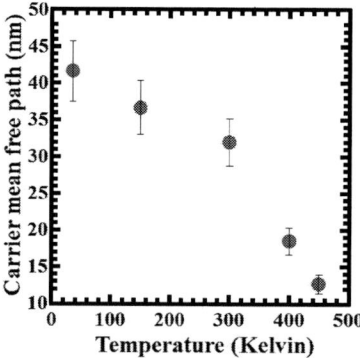

Figure 7. Calculated carrier mean free path in graphene channel with respect to temperature.

achieved as compared to 450 K. Table 1 shows the summary of the results for the temperature dependent measurement on Au-graphene contacts. The result shows low R_C of 315±127 Ωμm and high mobility of 1232 cm²/V.s is achieved at 35 K due to increase in λ of 41.6 nm in graphene as compared to measurement taken at 450 K. The λ decreases to 12.67 nm at 450 K leading to highly diffusive transport due to

978-1-5090-5979-9/17 $31.00 © 2017 IEEE

increased charge carrier scattering, ultimately leading to high R_C and low mobility.

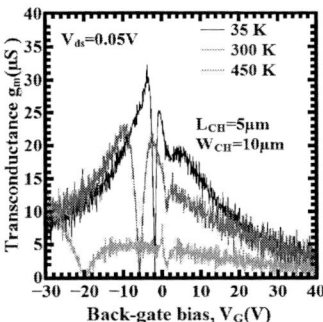

Figure 8 Transconductance in monolayer graphene devices as a function of back gate bias for 35 K,300 K and at 450 K.

Table 1 Summary of results			
Temp (K)	Mobility (cm^2/V.s)	Carrier mean free path (nm)	Contact resistance ($\Omega.\mu m$)
35	1232	41.6	(315±127)
150	1107	36.67	(569±154)
300	1048	32	(753±276)
400	697	18.49	(870±243)
450	652	12.67	(957±210)

IV. CONCLUSION

In this work systematic measurements were carried out on Au-graphene contact in wide range of temperatures from 35 K to 450 K in order to study its effect on R_c. From the transfer characteristics R_c, μ_e, λ, g_m were extracted. Carrier mean free path in graphene (λ) calculated at 35 K was around ~ 41.6 nm as compared to around ~ 12.67 nm at 450 K. Increased in the λ lead to a low R_c value of 315±127 $\Omega\mu m$ at 35 K as compared to 957±210 $\Omega\mu m$ at 450 K as it enhances the transmission efficiency through the M-G interface. High mobility of 1232 cm^2/V.s is achieved at 35 K as compared to 652 cm^2/Vs at 450 K. Overall performance of graphene devices is enhanced at low temperature as compared to high temperature. The results presented above provides a clear outlook of R_C measured in a wide range of temperatures which is of great importance for the development of graphene based hostile environmental electronics.

References

[1] K. S. Novoselov et al., "Electric field effect in atomically thin carbon films," Science, vol. 306, no. 5696, pp. 666–669, 2004.
[2] M. C. Lemme, L.-J. Li, T. Palacios, and F. Schwierz, "Two-dimensional materials for electronic applications," MRS Bull., vol. 39, no. 08, pp. 711–718, 2014.

[3] M. C. Lemme, "Current status of graphene transistors," Solid State Phenom., vol. 156, pp. 499–509, 2010.
[4] S. Kataria et al., "Chemical vapor deposited graphene: From synthesis to applications," Phys. Status Solidi A, vol. 211, no. 11, pp. 2439–2449, 2014.
[5] Y. Wang et al., "Electrochemical delamination of CVD-grown graphene film: toward the recyclable use of copper catalyst," ACS Nano, vol. 5, no. 12, pp. 9927–9933, 2011.
[6] X. Li et al., "Transfer of large-area graphene films for high-performance transparent conductive electrodes," Nano Lett., vol. 9, no. 12, pp. 4359–4363, 2009.
[7] J. S. Moon et al., "Epitaxial-graphene RF field-effect transistors on Si-face 6H-SiC substrates," Electron Device Lett. IEEE, vol. 30, no. 6, pp. 650–652, 2009.
[8] Y.-M. Lin et al., "100-GHz transistors from wafer-scale epitaxial graphene," Science, vol. 327, no. 5966, pp. 662–662, 2010.
[9] M. C. Lemme et al., "Gate-activated photoresponse in a graphene p–n junction," Nano Lett., vol. 11, no. 10, pp. 4134–4137, 2011.
[10] F. Xia, T. Mueller, Y. Lin, A. Valdes-Garcia, and P. Avouris, "Ultrafast graphene photodetector," Nat. Nanotechnol., vol. 4, no. 12, pp. 839–843, 2009.
[11] S. Riazimehr et al., "Spectral sensitivity of graphene/silicon heterojunction photodetectors," Solid-State Electron., vol. 115, pp. 207–212, 2016.
[12] A. D. Smith et al., "Electromechanical piezoresistive sensing in suspended graphene membranes," Nano Lett., vol. 13, no. 7, pp. 3237–3242, 2013.
[13] A. D. Smith et al., "Resistive graphene humidity sensors with rapid and direct electrical readout," Nanoscale, vol. 7, no. 45, pp. 19099–19109, 2015.
[14] B. Sensale-Rodriguez et al., "Broadband graphene terahertz modulators enabled by intraband transitions," Nat. Commun., vol. 3, p. 780, 2012.
[15] K. Nagashio and A. Toriumi, "Density-of-states limited contact resistance in graphene field-effect transistors," Jpn. J. Appl. Phys., vol. 50, no. 7, p. 0108, 2011.
[16] K. Nagashio, T. Nishimura, K. Kita, and A. Toriumi, "Metal/graphene contact as a performance killer of ultra-high mobility graphene analysis of intrinsic mobility and contact resistance," in Electron Devices Meeting (IEDM), 2009 IEEE International, 2009, pp. 1–4.
[17] F. Xia, V. Perebeinos, Y. Lin, Y. Wu, and P. Avouris, "The origins and limits of metal-graphene junction resistance," Nat. Nanotechnol., vol. 6, no. 3, pp. 179–184, 2011.
[18] S. Russo, M. F. Craciun, M. Yamamoto, A. F. Morpurgo, and S. Tarucha, "Contact resistance in graphene-based devices," Phys. E Low-Dimens. Syst. Nanostructures, vol. 42, no. 4, pp. 677–679, 2010.
[19] L. Wilson, "International Technology Roadmap for Semiconductors (ITRS)," Semicond. Ind. Assoc., 2013.
[20] H. Zhong et al., "Realization of low contact resistance close to theoretical limit in graphene transistors," Nano Res., vol. 8, no. 5, pp. 1669–1679, 2015.
[21] T. Palacios, A. Hsu, and H. Wang, "Applications of graphene devices in RF communications," IEEE Commun. Mag., vol. 48, no. 6, 2010.
[22] A. Hsu, H. Wang, K. K. Kim, J. Kong, and T. Palacios, "Impact of graphene interface quality on contact resistance and RF device performance," Electron Device Lett. IEEE, vol. 32, no. 8, pp. 1008–1010, 2011.
[23] L. G. Rizzi et al., "Cascading wafer-scale integrated graphene complementary inverters under ambient conditions," Nano Lett., vol. 12, no. 8, pp. 3948–3953, 2012.
[24] E. Guerriero, L. Polloni, L. G. Rizzi, M. Bianchi, G. Mondello, and R. Sordan, "Graphene audio voltage amplifier," Small, vol. 8, no. 3, pp. 357–361, 2012.
[25] A. C. Neto, F. Guinea, N. M. R. Peres, K. S. Novoselov, and A. K. Geim, "The electronic properties of graphene," Rev. Mod. Phys., vol. 81, no. 1, p. 109, 2009.
[26] V. K. Nagareddy et al., "High temperature measurements of metal contacts on epitaxial graphene," Appl. Phys. Lett., vol. 99, no. 7, p. 073506, 2011.
[27] A. Gahoi, V. Passi, S. Kataria, S. Wagner, A. Bablich, and M. C. Lemme, "Systematic comparison of metal contacts on CVD graphene," in Solid State Device Research Conference (ESSDERC), 2015 45th European, 2015, pp. 184–187.
[28] A. Gahoi, S. Wagner, A. Bablich, S. Kataria, V. Passi, and M. C. Lemme, "Contact resistance study of various metal electrodes with CVD graphene," Solid-State Electron., 2016.
[29] S. Venica, F. Driussi, P. Palestri, L. Selmi, A. Gahoi, V. Passi, M. C. Lemme, " Detailed characterization and critical discussion of series resistance in graphene – metal contacts, " in 2017 International Conference on Microelectronic Test Structures (ICMTS), 2017
[30] H. Wang, Y. Wu, C. Cong, J. Shang, and T. Yu, "Hysteresis of electronic transport in graphene transistors," ACS Nano, vol. 4, no. 12, pp. 7221–7228, 2010.
[31] S. Weingart, C. Bock, U. Kunze, F. Speck, T. Seyller, and L. Ley, "Low-temperature ballistic transport in nanoscale epitaxial graphene cross junctions," Appl. Phys. Lett., vol. 95, no. 26, p. 262101, 2009.
[32] E. H. Hwang and S. D. Sarma, "Acoustic phonon scattering limited carrier mobility in two-dimensional extrinsic graphene," Phys. Rev. B, vol. 77, no. 11, p. 115449, 2008.
[33] G. Fiori, D. Neumaier, B. N. Szafranek, and G. Iannaccone, "Bilayer graphene transistors for analog electronics," IEEE Trans. Electron Devices, vol. 61, no. 3, pp. 729–733, 2014.

Radical Oxidation Process for Hybrid SAM/HfO$_x$ Gate Dielectrics in MoS$_2$ FETs

Takamasa Kawanago, Ryo Ikoma, Tomoaki Oba, and Hiroyuki Takagi

Quantum Nanoelectronics Research Center, Institute of Innovative Research, Tokyo Institute of Technology,
2-12-1, Ookayama, Meguro-ku, Tokyo, 152-8552, Japan,
Tel.: +81-3-5734-2542, E-mail: kawanago.t.ab@m.titech.ac.jp

Abstract—This study describes the fabrication of hybrid SAM/HfO$_x$ gate dielectrics by the radical oxidation in molybdenum disulfide (MoS$_2$) field-effect transistors (FETs). The fabrication process involves the radical oxidation to form HfO$_x$ at the surface of metallic HfN, SAM formation by immersion, and deterministic transfer of MoS$_2$ flakes. A subthreshold slope (SS) of 75 mV/dec and small hysteresis were demonstrated with the hybrid SAM/HfO$_x$ gate dielectrics accompanied by a low gate leakage current. TEM observation revealed a uniform formation of HfO$_x$ at the surface of HfN. This study opens up intriguing possibilities of radical oxidation process for research in the applications and developments for functional electronic devices.

Keywords— Radical oxidation; Self-assembled monolayer; MoS$_2$; FETs; Hafnium oxides;

I. INTRODUCTION

A radical oxidation process can provide ultrathin and highly reliable gate oxides at the surface of silicon (Si) and germanium (Ge) substrates [1] − [3]. The prominent features of the radical oxidation are low process temperature, low process damage and high oxidative reaction [1] − [3]. This radical oxidation using the highly reactive oxygen radicals motivates the formation of new classes of gate oxides such as high-k gate dielectrics for improving electrostatics in electronic devices [4].

This paper describes a method to fabricate hybrid self-assembled monolayer (SAM)/hafnium oxides (HfO$_x$) gate dielectrics in molybdenum disulfide (MoS$_2$) FETs. The FETs are constructed by radical oxidation, SAM formation and transfer printing of MoS$_2$ flakes. Semiconducting MoS$_2$ has been considerable attention for functional FET applications because of free from dangling bonds that originates from layered crystal structure [5]. Previous studies demonstrated that the SAM/aluminum oxides (AlO$_x$) gate dielectrics can be applied to low voltage operation in MoS$_2$ FETs since the molecule monolayer enables the low density of traps accompanied by low gate leakage current [6], [7]. In this study, the HfO$_x$ layer are fabricated by the radical oxidation at the surface of metallic hafnium nitride (HfN). Previous study reveal that the phosphonic acid-based SAM can be organized at the surface of HfO$_2$ [8]. The gate capacitance is increased with high dielectric constant of HfO$_x$ layer, resulting in the improvement of electrostatics. It is speculated that the highly reactive oxygen radicals makes it possible to form the HfO$_x$ at the surface of HfN. Furthermore, the metallic HfN underneath the HfO$_x$ is applicable to the bottom gate electrodes in MoS$_2$ FETs. The effectiveness of our strategy is demonstrated in this study through the fabrication and characterization of devices.

This study was supported by JST CREST (Grant No. JPMJCR16F4), a JSPS Grant-in-Aid for Research Activity Start-up (Grant No. 15H06204), and Yazaki Memorial Foundation for Science and Technology.

II. DEVICE FABRICATION

Firstly, HfN (60nm) was deposited by RF sputtering in Ar ambient with HfN target. HfN was patterned by wet etching with HF : H$_2$O$_2$: H$_2$O = 1 : 2 : 40 using photoresist mask. 30 nm of gold (Au) was deposited on HfN as contact pads. The radical oxidation (microwave power: 500 W, frequency: 2.45 GHz, process pressure: 0.1 Torr, oxygen flow: 5 sccm) was performed to form HfO$_x$ at the surface of HfN with various substrate temperature and oxidation time. Next, 10 nm of Al and 40 nm of Au were deposited on the substrate as source and drain contacts, respectively. Al serves as an adhesion layer between Au and SiO$_2$. Subsequently, the substrate was exposed to oxygen plasma (plasma power: 300 W, process pressure: 300 mTorr, oxygen flow: 100 sccm, duration: 10 min) to form hydroxyl groups on the surface of HfO$_x$. Then, the substrate was immersed into 2-propanol containing 5 mM n-octadecylphosphonic acid (ODPA) for 3 hours at room temperature [7]. After that, the substrate was rinsed with 2-propanol, dried with nitrogen, and annealed at 100 °C in N$_2$ for 30 min to stabilize ODPA. Metal-insulator-metal (MIM) capacitors were also prepared to evaluate the leakage current and capacitance. Mechanically exfoliated MoS$_2$ flakes were transferred to the substrate by the poly(dimethylsiloxane) (PDMS) elastomer and a micromanipulator [7]. Finally, devices were subjected to annealing in forming gas (H$_2$: N$_2$ = 3% : 97%) at 150 °C for 30 min. The fabrication process and device structure are summarized in Fig. 1.

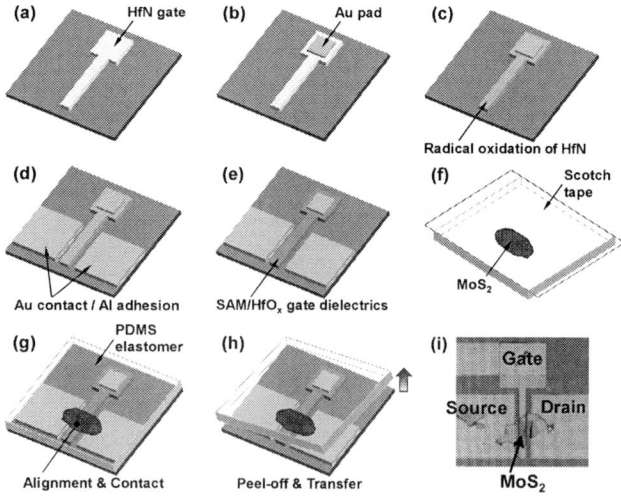

Fig. 1. Schematic of the fabrication process and device structure

III. RESULTS AND DISCUSSIONS

The most important experimental results are shown in Fig. 2. Fig. 2 shows the $I_d - V_g$ characteristics of MoS_2 FETs with hybrid SAM/HfO$_x$ gate dielectrics. A low gate current can be demonstrated using hybrid SAM/HfO$_x$ gate dielectrics. The HfO$_x$ was prepared by radical oxidation at 200 °C for 30 min. The hybrid SAM/HfO$_x$ gate dielectrics is found to be applicable to the fabrication of the low-voltage MoS_2 FETs. The threshold voltage (V_{th}) was extracted by linear extrapolation with the drain current measured as a function of gate voltage at a low drain voltage (V_{ds}) of 0.05 V [9]. This method is based on the linear extrapolation of the $I_d - V_g$ curve at the point of maximum transconductance to the gate voltage axis intercept (i.e., $I_d = 0$). The transconductance (G_m), which is defined as $G_m = dI_d/dV_g$, was evaluated from the $I_d - V_g$ curve at $V_{ds} = 0.05$ V. The tangent to the $I_d - V_g$ curve at the maximum G_m was extrapolated to $I_d = 0$. V_{th} was determined by $V_{th} = V_{gsi} - V_{ds}/2$, where V_{gsi} is the intercept gate voltage. V_{th} of -0.55V and SS of 75 mV/dec were obtained from the $I_d - V_g$ characteristics, as shown in Fig. 2. The small hysteresis was clearly observed, as shown in Fig. 2. The SS of 75 mV/dec and the small hysteresis in the $I_d - V_g$ characteristics indicate a low defect density at the MoS_2/SAM interface. The representative $I_d - V_d$ characteristics are shown in Fig. 3. These $I_d - V_d$ characteristics indicate the fabrication and operation of MoS_2 FETs with hybrid SAM/HfO$_x$ gate dielectrics. Fig. 2 and Fig. 3 demonstrate the effectiveness of presented approach.

Next, cross-sectional transmission electron microscopy (TEM) and Energy Dispersive X-ray Spectroscopy (EDX) were employed to obtain detailed information on the gate structure. Fig. 4 shows a TEM image of the gate structure in the MoS_2 FETs. The MoS_2 channel layer, gate dielectrics, and gate electrode can be distinguished in the TEM image. This TEM image highlights the uniform formation of the HfO$_x$ by radical oxidation process. The previous study employed the Al gate and plasma oxidation for constructing the SAM-based gate dielectrics [7]. Large thickness fluctuation of plasma-oxidized AlO$_x$ was observed from cross-sectional TEM image in previous study [7]. It was found that the uniform HfO$_x$ at the surface of HfN can be formed with radical oxidation method. The multilayer structure of MoS_2 is clearly observed from the TEM image. In this study, no attempt was made to control the thickness of the MoS_2 flakes. The physical thickness of the transferred MoS_2 was 56 nm. Fig. 5 shows EDX profile in gate structure. The surface of HfN was oxidized using radical oxidation process, as shown in Fig. 5. It was concluded that the HfO$_x$ was fabricated at the surface of metallic HfN by radical oxidation.

The capacitance and leakage current of hybrid SAM/HfO$_x$ gate dielectrics were evaluated through MIM capacitors in which the gate area is precisely defined. Fig. 6 shows the capacitance at 1V between the single layer of HfO$_x$ and the dual layer of the ODPA/HfO$_x$ gate dielectrics as a function of frequency ranging from 100 Hz to 1MHz. The capacitance of the ODPA/HfO$_x$ gate dielectrics is lower than that of the single layer of HfO$_x$ because of the presence of ODPA on HfO$_x$. The capacitance equivalent thickness (CET) at 100 kHz for the single layer of HfO$_x$ and the ODPA/HfO$_x$ are 2.5 nm and 5.5 nm, respectively. The physical thickness and dielectric constant

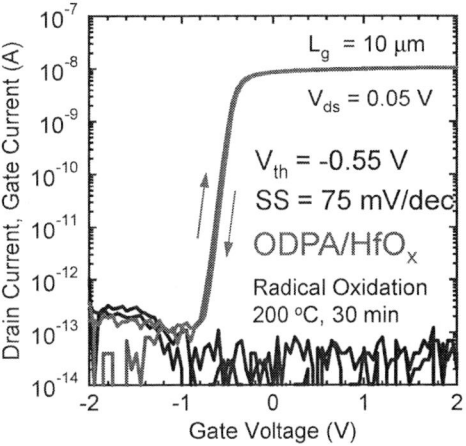

Fig. 2. I_d-V_g and I_g-V_g characteristics of MoS_2 FETs with hybrid SAM/HfO$_x$ gate dielectrics.

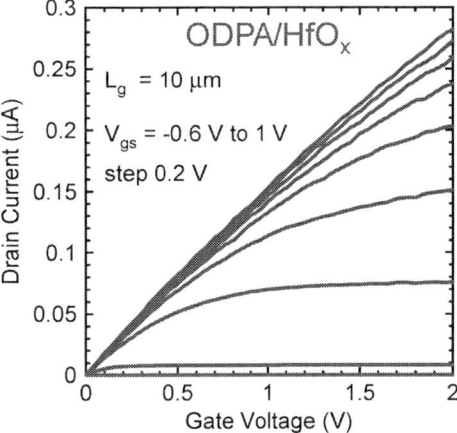

Fig. 3. Representative I_d-V_d characteristics of MoS_2 FETs with hybrid SAM/HfO$_x$ gate dielectrics.

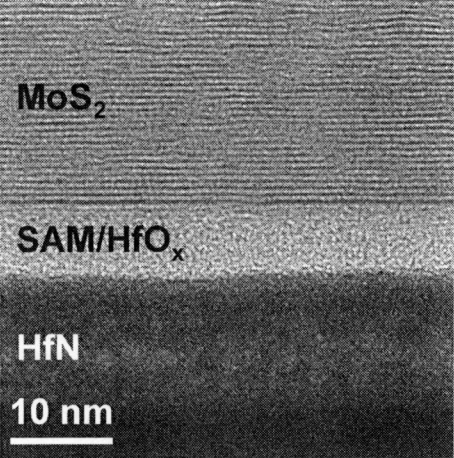

Fig. 4. Cross-sectional TEM image of gate strucure in MoS_2 FET.

of ODPA are 2.1 nm and 2.5, respectively [6]. Therefore, the dielectric constant of HfO_x using radical oxidation corresponds to 12−14 from TEM image and capacitance measurement.

Fig. 7 shows the gate leakage current between the single layer of HfO_x and the dual layer of the $ODPA/HfO_x$ gate dielectrics. The breakdown voltage of the hybrid $ODPA/HfO_x$ gate dielectrics is higher than that of the single layer of HfO_x, while the leakage current through the single layer of HfO_x at low voltage is comparable to that of the hybrid $ODPA/HfO_x$ gate dielectrics. The presence of ODPA on the HfO_x contributes to the improvement of the breakdown voltage. This is consistent with previous study with $ODPA/AlO_x$ gate dielectrics [7].

The substrate temperature and process duration of radical oxidation affect the physical thickness and the gate leakage current [1] − [3]. Fig. 8 shows the capacitance at 1V for single layer of HfO_x gate dielectrics with various condition of radical oxidation as a function of frequency ranging from 100 Hz to 1MHz. The capacitance was increased with decreasing the substrate temperature and process duration of radical oxidation. Increase in the capacitance is thought to be associated with the decrease in physical thickness of HfO_x layer. Fig. 9 shows the gate leakage current for single layer of HfO_x gate dielectrics with various condition of radical oxidation. Clearly, the gate leakage current is increased with decreasing the substrate temperature and process duration of radical oxidation. Furthermore, the breakdown voltage is decreased. These results indicate that the physical thickness of HfO_x layer is dependent on the process temperature or process duration during radical oxidation. These results are consistent with radical oxidation of Si and Ge substrate [1] − [3].

Finally, the impact of SAM on interfacial properties were experimentally investigated. As previously shown in Fig. 7, the current density through the single layer of HfO_x gate dielectrics was 10^{-7} A/cm^2 order at an applied voltage of 2V. This leakage current is small enough to construct MoS_2 FET with the single layer of HfO_x gate dielectrics. The sweep rate of the gate voltage during the I–V measurements was automatically varied with respect to the current value. For the currents of 10 pA, 100 pA, and 1 nA, the times required for measurement were set to 1 s per point, 0.2 s per point, and 0.1 s per point, respectively. The measurement time for currents of 10 nA and greater was set to 0.02 s per point. Fig. 10 shows the hysteresis in $I_d − V_g$ characteristics of MoS_2 FETs with single layer of HfO_x gate dielectrics. The gate current was also shown in Fig. 10. The hysteresis in I_d-V_g characteristics was clearly observed, while the gate leakage current is 10^{-12} A order. The hysteresis in clockwise direction is caused by the electron trapping at defects. This result is the experimental evidence that the presence of ODPA significantly contribute to the superior interfacial properties in MoS_2 FETs. Typically, the hysteresis is increased with increasing applied gate voltage, since high electric field promotes the electron injection from channel layer into gate dielectrics. Fig. 11 shows the hysteresis in $I_d − V_g$ characteristics as a function of sweep range of gate voltage. The hysteresis is found to be increased with increasing the sweep range of gate voltage. This result indicates that a huge quantity of traps exist in HfO_x layer neighboring MoS_2/HfO_x interface.

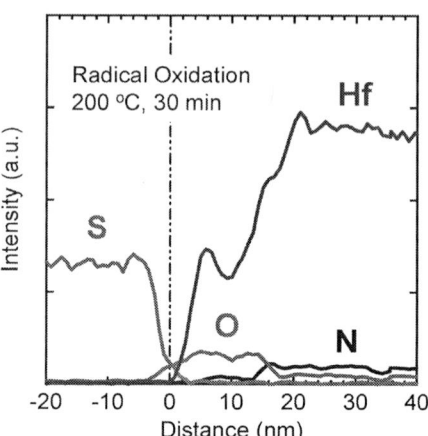

Fig. 5. EDX profile in gate structure of MoS_2 FET.

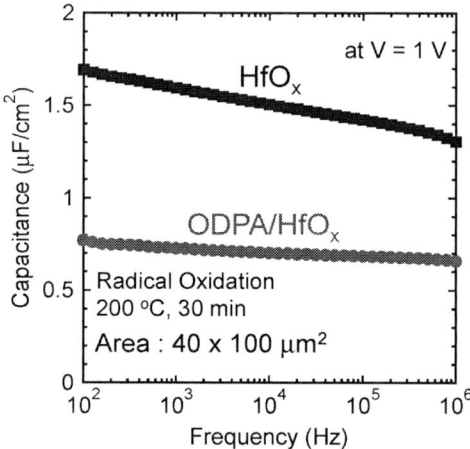

Fig. 6. Capacitance at 1V between the single layer of HfO_x and the dual layer of the $ODPA/HfO_x$ gate dielectrics as a function of frequency ranging from 100 Hz to 1MHz

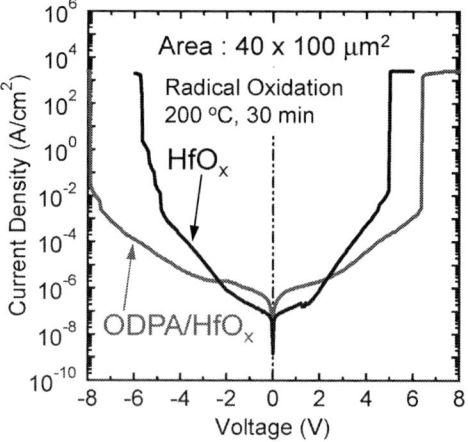

Fig. 7. Gate leakage current between the single layer of HfO_x and the dual layer of the $ODPA/HfO_x$ gate dielectrics.

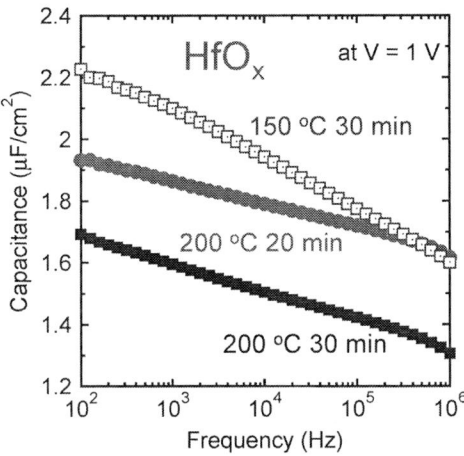

Fig. 8. Capacitance at 1V for single layer of HfOx gate dielectrics with various condition of radical oxidation.

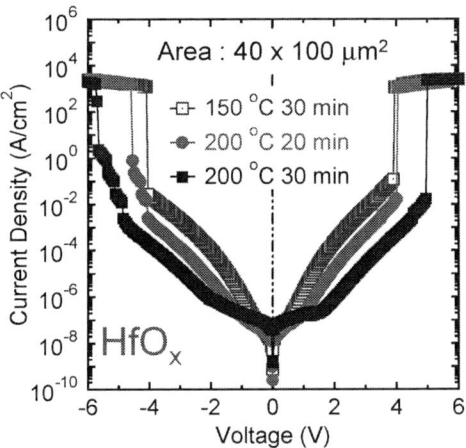

Fig. 9. Gate leakage current for single layer of HfOx gate dielectrics with various condition of radical oxidation.

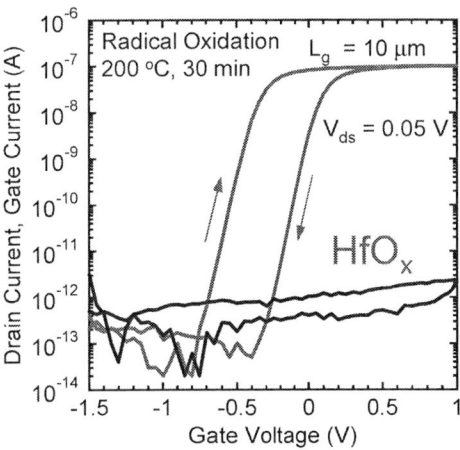

Fig. 10. Hysteresis in $I_d - V_g$ characteristics of MoS2 FETs with single layer of HfOx gate dielectrics.

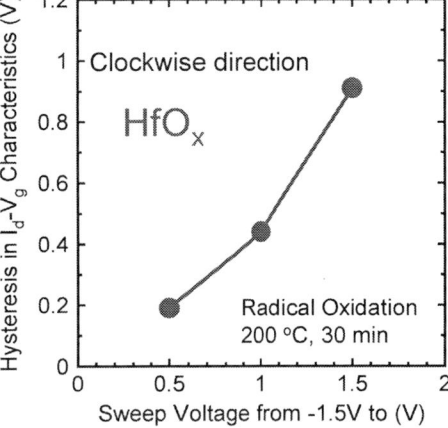

Fig. 11. Relationship between hysteresis and sweep range of gate voltage.

IV. CONCULUSIONS

In this study, the radical oxidation was applied to construct the hybrid SAM/HfOx gate dielectrics in MoS2 FETs. A low gate leakage current was demonstrated using the hybrid SAM/HfOx gate dielectrics. Moreover, it was found that SS is decrease to 75 mV/dec with small hysteresis in electrical characteristics. It was revealed that the radical oxidation can prepare the uniform HfOx at the surface of metallic HfN. The SAM plays an important role in the superior interfacial properties in MoS2 FETs. The utilization of the radical oxidation is not restricted to Si and Ge substrates and is also applicable to the oxidation of metal materials to form high-k gate dielectrics. This study opens up intriguing possibilities of radical oxidation process for research in the applications and developments for functional electronic devices.

ACKNOWLEDGMENT

The authors would like to thank Professor K. Kakushima, Professor H. Wakabayashi, Professor K. Tsutsui and Professor S. Oda of Tokyo Institute of Technology for their continuous support in the experiments.

REFERENCES

[1] M. Nagamine et al., in IEDM Tech. Dig., 593 (1998).
[2] K. Sekine et al., IEEE Trans. Electron Devices 480, 1550 (2001).
[3] M. Kobayashi et al., J. Appl. Phys. 106, 104117 (2009).
[4] K. Henson et al., in IEDM Tech. Dig., 645, (2008).
[5] B. Radisavljevic et al., Nat. Nanotechnol. 6, 147 (2011).
[6] H. Klauk et al., Nature 445, 745 (2007).
[7] T. Kawanago et al., Appl. Phys. Lett. 108, 041605 (2016).
[8] H. Ma et al., Phys. Chem. Chem. Phys. 14, 14110 (2012).
[9] D. K. Schroder, Semiconductor Material and Device Characterization (Wiley, New York, 2005).

CoolSiC™ and major trends in SiC power device development

Roland Rupp

IPC DD

Infineon Technologies AG

Neubiberg, Germany

roland.rupp@infineon.com

Abstract—Silicon Carbide (SiC) diodes are already commercially available since 15 years and have gained significant market share in power supply and solar converter applications. In the last few years, the SiC device family was enriched by switches. They become increasingly more important for differentiation of power converters in size, weight and/or efficiency. The dedicated material properties of SiC enable the design of minority carrier free unipolar devices instead of the charge modulated IGBT devices. As such, they deliver highest efficiency, higher switching frequencies, reduced heat dissipation and space savings – benefits that, in turn, also lead to overall lower cost. Infineon's CoolSiC™ SiC device trademark stands not only for those outstanding properties but also come with the promise of SiC device reliability in the same range as for the traditional Si power devices. One key enabler for this is a unique asymmetric SiC trench MOSFET concept, which allows low $R_{DS,on}$ x A at Gate Oxide field stress in a range what is also typical for Si IGBTs.

Future trends in SiC material and device development will further shrink the cost gap between Si and SiC based products, paving the way for an increasing market penetration not only in small high end market segments but also in automotive, UPS and drives applications.

Keywords—silicon carbide; power devices; energy efficiency; MOSFET; diode

I. INTRODUCTION

Just looking at the material properties of SiC this material besides GaN appears as an ideal semiconductor for manufacturing power devices:

- The high breakdown field strength allows the manufacturing of majority carrier devices up to 3.3kV blocking capability with still very attractive on resistance

- A thermal conductivity close to copper enables high power densities and excellent heat spreading.

- The bandgap of 3.2eV leads to very low leakage current of pn-junctions and negligible intrinsic carrier density even at device temperatures of several hundred degrees centigrade.

However, the world market of SiC power devices today is only in the range of some 100 Million \$. This is mainly related to the technological difficulties of SiC wafer manufacturing and the long lasting problems to achieve attractive and reliable SiC Gate Oxide structures.

For both problems significant progress was achieved in the recent years. SiC wafers with 150mm diameter are used in production now, their cost per wafer are down to the same level as for 100mm wafers 8 years ago with a reasonably broad supplier base. And for the SiC MOSFET modern device concepts allow an excellent reliability in combination with very attractive performance as explained in section B.

II. SiC DIODE HISTORY

A. Basic Device Concepts

When SiC diodes have been commercially released in 2001 (600V and 1200V class) they have been mostly constructed as pure Schottky diodes. Besides the fact, that they allowed for the 1st time in these voltage classes virtually loss less commutation, those devices had a severe draw back with respect to surge current capability [1]. For their most important application (Power Factor correction in high end Power supplies), this required a significant over-dimensioning of the diodes only to cope with occasional AC drop outs and the subsequent peak current event to recharge the output capacitor.

This issue was fixed only few years later by the market introduction of so called MPS (merged-PIN-Schottky) rectifiers [1]. These devices are pure majority carrier devices in normal operation but allow minority carrier based resistivity modulation in surge current operation (> 5000A/cm² for 10ms pulse time for 650V devices). This allowed a significant shrink of the devices and respective cost reduction.

Meanwhile nearly all commercial SiC diodes from more than 10 vendors are based on the MPS or JBS (Junction-Barrier-Schottky) principle. They are considered as the best concept for voltage classes up-to 3.3kV. Only above pure pn diodes will be the better choice taking into account the higher knee voltage (2.7V for pn vs ~1V for MPS) on one side and the lower differential resistance of the pn diodes on the other side.

978-1-5090-5979-9/17 \$31.00 © 2017 IEEE

B. Further differentiation of SiC MPS diodes

Since the introduction of MPS diodes those devices developed further in 3 major steps:

- *Improved power handling capability of discrete devices via reduction of $R_{th,JC}$.* This was enabled by replacing classical soft soldering (~60µm) by AuSn based diffusion solder (1.2µm) [2]. Together with the excellent thermal conductivity of the SiC chip itself this was another big step for chip shrinkage and reduction of device cost.

- *Introduction of thin wafer technology for SiC.* SiC substrates typically have a thickness of 350µm, this is demanded to allow stable wafer yield during the frontend process chain. However, in case of 650V diodes the substrate resistivity is dominating the differential resistance of the diodes. Therefore, thinning down the substrate after having completed all front side technology steps was a big lever for further performance improvement. This was enabled mainly by the capability to generate the necessary ohmic backside contact via laser annealing instead of a RTP furnace process [3].

- *Schottky barrier reduction.* Finally by moving from Ti based Schottky contacts to a Mo based metal system a reduction of the knee voltage from ~ 1V to 0.7V was achieved keeping the same differential resistance for the diode [4]. This approach requested a complete redesign of the MPS cell structure to keep the Schottky leakage current under control also for T_j=175°C and is a big step for the improvement of the low load efficiency.

The sum of these measures made SiC diodes an affordable choice for all compact and fast switching applications. The cost ratio towards fast Si diodes is meanwhile clearly below 3.

III. COOLSIC POWER SWITCH

A. Challenges compared to silicon-based MOSFETs

Si as well as SiC has a thermal oxide which is at a first glance the common way to create an almost ideal MOS interface. But there are some well-known challenges to make a SiC MOSFET. Carbon atoms at the interface tend to form clusters or dangling bonds and cause significant lower field effect channel mobility due to a much higher density of interface states compared to Si. Hence, a much more sophisticated gate oxide process is needed to mitigate the

negative effect of these interface states. The field effect channel mobility is still in the range of 5–50cm²/Vs only which is a poor fraction of the bulk mobility of ~200cm²/Vs (at a bulk doping level being equal to the channel doping). Further defects are located near the interface but in the gate oxide (NIT) and trap electrons. Due to their energy levels which are positioned somewhere within the larger bandgap of SiC they can interact in a larger span of time constants by trapping or emitting electrons. The balance between trapping and emission rates of these NIT states causes higher threshold voltage shifts depending on gate voltage profile and temperature [5]. Furthermore, as SiC devices allow roughly 10 times higher electric fields than their Si counterparts, the electric field in the gate oxide has to be limited in order to maintain a required reliability of the device.

B. Cell concept of the CoolSiC™ MOSFET

The CoolSiC™ MOSFET is, different to the commonly used planar cell, a trench SiC MOSFET based on a novel asymmetric concept [6].

In the trench device, only one side of the trench sidewall is used as MOS channel which is well aligned to the preferred <11$\overline{2}$0> crystal plane by a special process. Making use of this favorite crystal plane is seen as the key in achieving a minimum of interface states. It was shown experimentally that the channel mobility for this crystal plane is about two times better than for other crystal planes [7] and that high channel mobility can be realized. In combination with nitridation techniques to ensure a good interface state passivation, the channel mobility is further improved and shows reduced Coulomb scattering. At the same time, all these measures minimize the amount of threshold voltage shift with temperature and improve the device reliabilities. Thanks to the improved channel properties the device can be driven at a sufficient low gate oxide field in the on-state. The oxide thickness is designed for the commonly used on-state gate source voltage of V_{GS} = + 15V.

Deep p-wells are used in order to limit the electric field in the gate oxide at the bottom and at the corners of the trench. These p-type regions serve as well as emitters of the body diode which can be used for freewheeling operation. This MOSFET structure is very compact resulting in a low on-resistance which is about half the value of typical DMOS cells. This cell construction inherently owns a favorable small ratio of the Miller charge Q_{GD} related to the gate source charge Q_{GS}. Q_{GS} is comparably large since a large part of the trench contributes to it, i.e. the n+-type areas and all p-type areas which are all well connected to source. This allows for a well-controlled switching with very low dynamic losses [8]. In particular this feature is essential to suppress undesirable additional losses caused by a parasitic turn-on in topologies using half bridges.

The cell design is further supportive to obtain an adequate short circuit capability. The JFET region formed by the adjacent p-emitter regions is not only good to limit the oxide field in the trench corner but also lowers the saturation current of the device by adjusting the distance between the p-type regions. A smaller distance supports both a lower saturation

TABLE I.   KEY PARAMETERS OF THE COOLSIC™ MOSFET

Parameter	Value	Unit	Condition
$R_{DS(on), typ}$	45 (75)	mΩ	Single die, T_j = 25°C (175°C) I_D = 20A, V_{GS} = 15V
V_{DSS}	> 1200	V	-55°C < T_j < 175°C
V_{GS}	-5 / +15	V	recommended range
V_{GS}	-10 / +20	V	maximum rating
V_{GSth}	4.5 (3.8)	V	T_j = 25°C (175°C) I_D = 10mA, V_{GS} = V_{DS}
V_{SD}	3.3 (3.1)	V	T_j = 25°C (175°C) I_D = 20A, V_{GS} = 0V

current and lower electric field in the gate oxide of the trench corner but causes an additional contribution in the overall on-state resistance due to the JFET.

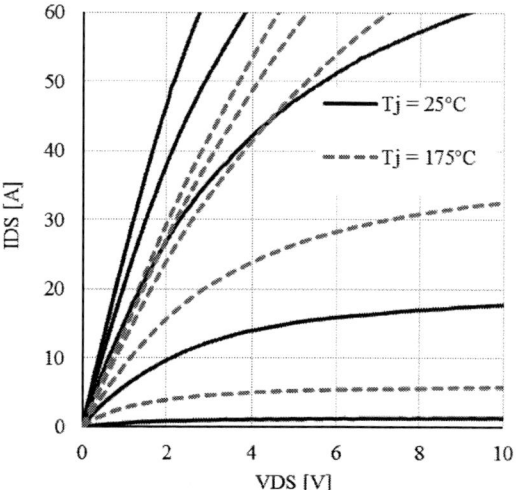

Fig. 1: Typical 1st quadrant output characteristics for gate voltages of V_{GS} = 17, 15, 13, 10, 7V at 25°C (solid) and at 175°C (dashed)

The 1st voltage class of the CoolSiC MOSFET introduced to the market is 1200V addressing e.g. photovoltaic applications and uninterruptable power supplies (UPS). TABLE I. lists key parameters of the device. It is tailored to be long term stable within a gate voltage range between -5V and +15V.

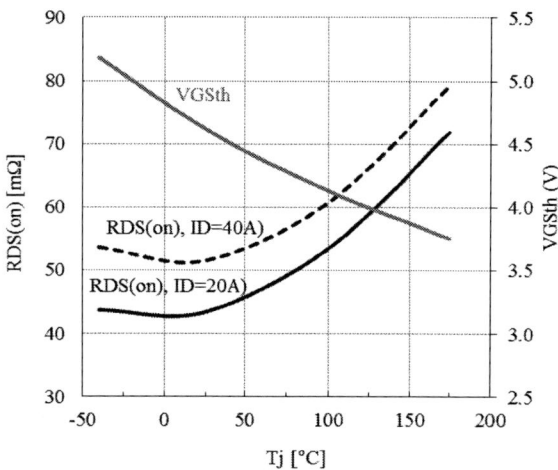

Fig. 2: Typical temperature dependence of $R_{DS(on)}$ at I_D = 20A (solid black), 40A (dotted black) and V_{GSth} (solid red, V_{GS}=V_{DS}, I_D = 10mA)

C. Static Performance

The 1200 V CoolSiC™ MOSFET is optimized for an operation with standard gate driver voltage levels of -5V to 0V for off-state and +15V for on-state. The output characteristics at two temperatures of 25°C and 175°C for selected gate voltages are shown in Fig. 1. The on-state depends on the applied gate voltage, a feature which is common for SiC MOSFETs. However, due to the reduced contribution of the channel to the total resistance the dependence is less severe than in planar channel DMOS device

The body diode shows a low forward voltage drop below 4V (V_{GS} = -5V, channel off, I_D = 20A). In reverse conduction the on-resistance amounts 33mΩ at V_{GS} = 15V. This I_D-V_{DS} curve is more linear than in the 1st quadrant due to the JFET effect formed by the p-wells.

Fig. 2 shows the temperature dependence of the threshold voltage (red) and of the on-resistance (black) which increases by 70% from 25°C to 175°C as expected for a MOSFET with low defect densities in the channel area. This is a reasonable positive temperature coefficient important for paralleled devices. The on-resistance shows a minimum around room temperature and is only slightly increasing towards lower temperatures. The temperature behavior mainly results from the sum of the resistances of the MOS channel, the JFET zone and the drift layer. The MOS channel has a negative temperature characteristic mainly due to the threshold voltage which decreases with temperature whereas the n-type doped drift zone has a positive temperature coefficient. The increase of the resistance with load current is again explainable by the JFET section of the trench MOSFET [6].

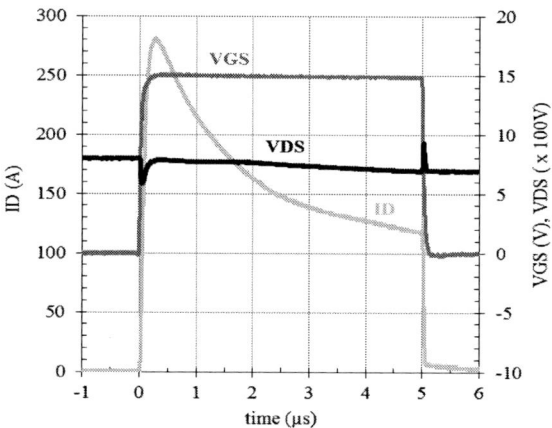

Fig. 3: Typical short circuit oscillography: the device can width-stand 5µs in short circuit. Conditions: T_c = 175°C, R_G = 10Ω, V_{bus} = 800V, V_{GS} = 0V / +15V, package TO-247-4 pin.

The p-wells which form the JFET region are also beneficial in order to limit the saturation current and make the device short circuit rugged. Fig. 3 shows a typical short circuit test. The device is switched on / off between 0V and +15V under 800V bus voltage and a case temperature of 175°C (i.e. short circuit type 1). After 5µs the device is able to turn-off safely.

D. Dynamic behavior

The capacitances and their dependency on drain source voltage are key with respect to switching behavior. Due to the chosen asymmetric p-well structure the Miller capacity C_{rss} is quite small. This makes parasitic return-on losses easy to suppress. The turn-on energies E_{on} dominate the switching

978-1-5090-5979-9/17 $31.00 © 2017 IEEE   120

losses and can be minimized independent of the temperature in combination with a SBD diode (Fig. 4). Compared to E_{on} of the MOSFET in half bridge configuration 30-50 % can be saved. In half-bridge configuration the body diode is active and shows an increasing impact with larger load current as well as higher temperature. This is a bipolar effect which generates a reverse recovery charge. However, the absolute values at the rated current of 20A are still very small compared to 1200V Si IGBTs so that the body diode can easily be used for commutation

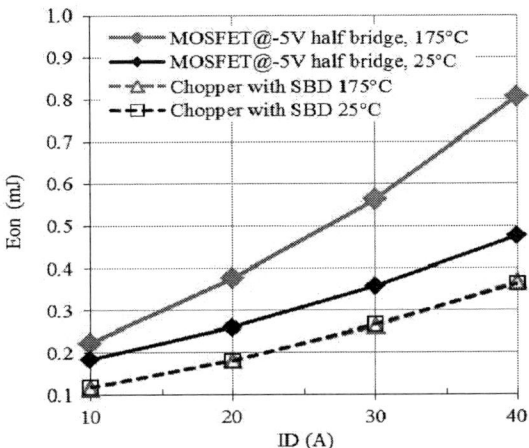

Fig. 4: Turn-on energy as function of drain current, measured in TO-247-4 pin ($V_{DS} = 800V$, $R_g = 2.2\Omega$, $V_{GS} = -5V / +15V$)

E. Reliability

To investigate the gate oxide reliability over the targeted device lifetime of 20 years, long-time gate stress tests were performed with a large number of devices in order to determine the extrinsic gate oxide failure rates. The investigation was done for 2 groups consisting of 1000 discrete devices. The tests were performed at 150°C under constant gate bias stress for 300 days. During this test, the gate source voltage was kept constant for 100 days and then increased by +5V after each 100 day period. The time stamp of each failure was monitored. In case of group G1, the test started at a gate source voltage of +25V with zero fails after 100 days. The test of group G1 ended at +35V, which is +20V above the recommended use voltage of +15V, with in total 2.9% fails after 300 days. The 2nd group started at 30V, continued at 35V and ended at 0V, with 6.5% fails in total (for more details see [6]).

As demonstrated in [9], these failure statistics fit well to the linear E-Model. By extrapolating this result to a life time of 20 years of device operation, the model predicts a failure rate of 0.2ppm. This experiment gives evidence to an IGBT-like reliability of the gate oxide with a failure rate under use conditions which is well below 100ppm.

IV. FUTURE TRENDS IN SiC POWER DEVICE TECHNOLOGY

Besides the foreseeable expansion of the voltage classes of commercial SiC power devices to 3.3kV and above the following major developments are expected:

- Base material
  The diameter increase towards 200mm is clearly foreseeable as 1st sample wafers are already displayed at various opportunities. In addition the defect density continuously decreases whereas the understanding of the impact of material defect on device performance grows.

- Technology
  Together with the increased wafer diameter access to more modern manufacturing tools is given, leading to continuously decreasing defect density, smaller feature size and – as consequence – better performance/cost.

- Application requirements
  Better understanding of the SiC device specific trade-offs (e.g. V_{th} vs $R_{DS,on}$ or short circuit withstand time t_{sc} vs $R_{DS,on}$) trigger innovative system and driver concepts, causing a more efficient utilization of the unique properties of SiC power devices.

These measures together with a continuous gain in confidence in SiC device reliability will be the main driving forces for a significant SiC power device market growth in the near future.

V. ACKNOWLEDGEMENT

The author wants to thank the complete SiC technology development team at Infineon which have enabled this overview paper by their work.

REFERENCES

[1] R. Rupp, M. Treu, S. Voss, F. Dahlquist, T. Reimann, "2nd Generation SiC Schottky diodes: A new benchmark in SiC device ruggedness", Proc. ISPSD 2006.

[2] R. Rupp, F. Björk, G. Deboy, M. Holz, M. Treu, J. Hilsenbeck, R. Otremba, H. Zeichen, "A new generation of SiC Schottky diodes with improved thermal management and reduced capacitive losses" Material Science Forum Vols. 645-648 (2010) pp 885-888.

[3] R. Rupp, R. Kern, R. Gerlach, „Laser backside contact annealing of SiC Power devices: A Prerequisite for SiC thin wafer technology", Proc. ISPSD 2013.

[4] M. Draghici, R. Rupp, R. Elpelt, R. Gerlach, R. Schörner, "A new SiC diode with significantly reduced threshold voltage", Proc. ISPSD 2017 to be published

[5] G. Pobegen, J. Weisse, M. Hauck, H.B. Weber and M. Krieger, "On the origin of threshold voltage instability under operating conditions of 4H-SiC n-channel MOSFETs", Materials Science Forum, Vol. 858, 2016, pp.473-476

[6] D. Peters, R. Siemieniec, T. Aichinger, T. Basler, Romain Esteve†, W. Bergner, D. Kueck, "Performance and Ruggedness of 1200V SiC Trench MOSFETs" Proc. ISPSD 2017 to be published

[7] H. Yano, H. Nakao, T. Hatayama, Y. Uraoka and T. Fuyuki, "Increased channel mobility in 4H-SiC UMOSFETs using on-axis substrates" Materials Science Forum, Vols. 556-557, 2007, pp.807-811

[8] D. Heer, D. Domes and D. Peters, "Switching performance of a 1200 V SiC-Trench-MOSFET in a Low-Power Module", Proc. PCIM Nuremberg 2016

[9] D. Peters, T. Aichinger, T. Basler, W. Bergner, D. Kueck, R. Esteve, "1200V SiC Trench-MOSFET Optimized for High Reliability and High Performance", ECSCRM'16, Sep. 2016, to be published in Materials Science Forum in 2017

978-1-5090-5979-9/17 $31.00 © 2017 IEEE

Gated Base Structure for Improved Current Gain in SiC Bipolar Technology

B. Gunnar Malm, Hossein Elahipanah, Arash Salemi, Mikael Östling
School of Information and Communication Technology
KTH Royal Institute of Technology
Kista, SWEDEN
gunta@kth.se

Abstract—**Silicon Carbide (SiC) bipolar integrated circuits are a promising technology for extreme environment applications. SiC bipolar technology shows stable operation over a wide range of temperature. However, the current gain of the devices is suffering from high surface recombination, due to poor oxide passivation. In this paper we propose a gated base structure that offers improved current gain control. A polysilicon gate is formed on the passivation oxide on top of the base-link region. We investigate the current gain as a function of gate bias and temperature. A negative gate bias improves the gain at low collector current by more than 30% by suppressing the surface recombination. Measurements are presented at temperatures ranging from 300 K to 550 K and the gain is consistently improved. The proposed structure is also useful as a process monitor for the passivation oxide quality.**

Keywords—*current gain; silicon carbide (SiC); bipolar; surface passivation; extreme enviroment; process monitor; test structure*

I. INTRODUCTION

Wide bandgap (WBG) semiconductor technology based on silicon carbide (SiC) or Gallium Nitride (GaN) has several advantages over silicon devices, including higher temperature operation, high voltage and power handling capability and also radiation hardness [1, 2]. However a common issue for WBG technology is the difficulty to find a suitable surface passivation that does not lead to instabilities in threshold voltages or drive current for field effect devices or non-ideal current gain for bipolar devices. Also, long-term reliability during normal or elevated temperature operation needs to be proven [3]. In previous work, the surface passivation on SiC has been improved significantly by using a post oxide deposition step in N_2O [4, 5] and peak current gains in the range 100 to 200 can be achieved. There has also been modelling work based on highly doped structures in order to suppress surface carrier recombination [6, 7] resulting in higher gain. Other work has suggested high resistivity regions, with low carrier concentration, as a remedy [5]. That concept has been experimentally verified, but the process integration, involving additional epitaxial layers, is rather complex. In this work we investigate a process module, based on a polysilicon metal-oxide-semiconductor (MOS) gate that offers direct control of surface potential, and the carrier concentration at the p-type base surface. The module is easily integrated in the bipolar

process flow, since it only requires a single deposition/ pattering/etch step. No electrical properties of the baseline devices are altered. This paper is organized as follows. The experimental section outlines the process flow, with focus on the gate integration. In the results section measurement data is presented. The influence of the MOS gate bias on the current gain is demonstrated for different configurations and then analyzed as a function of base doping concentration and temperature. Finally, an n-channel MOS test structure is demonstrated by capacitance-voltage (C-V) measurements at elevated temperature.

II. EXPERIMENTAL PROCESS FLOW

Our in-house SiC bipolar integrated technology is fabricated using an all-epitaxial process flow. For the samples under study we included a thickness and doping split of the p-type base epitaxial layer according to Table 1. The base design was targeted for a narrow base, high speed RF transistor with a supply voltage of 15 V. A reference design with lower doping concentration but identical base dose (doping concentration × thickness) was included for comparison. The base doping uniformity, across the 100 mm wafers, was monitored by Fourier Transform Infrared Spectroscopy (FTIR) mapping and verified by Secondary Ion Mass Spectroscopy (SIMS).

TABLE I. EPITAXIAL BASE SPLIT

Samples	Base layer epitaxial parameters		
	Thickness (nm)	*Acceptor Doping (cm^{-3})*	*Base Dose (cm^{-2})*
High speed design	200	1.5×10^{18}	3×10^{13}
Reference design	250	1.2×10^{18}	3×10^{13}

The process flow is based on sequential dry-etching of the emitter, base and collector mesas. A 5% overetching is considered in each step and the etch uniformity of ~7% is achieved by employing multi-step controlled etching. A dry sacrificial oxidation is done at 1100 °C for 1 hour. After the final etching step a surface passivation step is needed. In our approach a low-temperature deposited oxide (PECVD) is annealed in N_2O at 1250 °C for 1 hour. The thickness of the oxide layer is about 40 nm and it also serves as the gate oxide in the gated base structures. The oxide thickness was based on previous optimization studies where it was found

---

This work was supported by the KAW foundation project 'Working on Venus' and the Swedish Research Council National Infrastucture Myfab.

978-1-5090-5979-9/17 $31.00 © 2017 IEEE

that sufficient thickness was a key parameter to achieve high gain [4]. Since we did not want to compromise the current gain thinner oxides were not implemented. An in-situ n-type doped polysilicon layer with a thickness of 100 nm was selected as the gate electrode, thanks to the straight forward process integration. Patterning of the gate electrode was done by standard stepper lithography and dry etching, using endpoint. At this point, a Ni layer was deposited and patterned by a lift-off process, and annealed at 950 °C for 1 min in N_2 ambient to form the n-Ohmic contact. For the p-Ohmic contact a Ni/Ti/Al stack was used, annealing conditions were 820 °C for 90 sec in Ar ambient. The rest of process flow includes a thicker dielectric oxide layer, followed by via openings and a two-layer Al/TiW/Ti metallization.

A schematic cross-sectional view and the top view microscopic image of a finished device are shown in Fig. 1. Note that the gate can be independently biased, by a separate probing pad, in our test transistor layout. One could also consider layouts where the gate shares a common potential, with either the emitter or the base. The gate edge was placed close to the emitter mesa with about 0.5 μm separation. The default length of the MOS gated region on top of the base-link region was 2 μm. In a more aggressive layout the gate edge was aligned to the emitter mesa and the gate length was 3 μm. These devices had significantly lower yield, due to gate leakage and will not be further discussed.

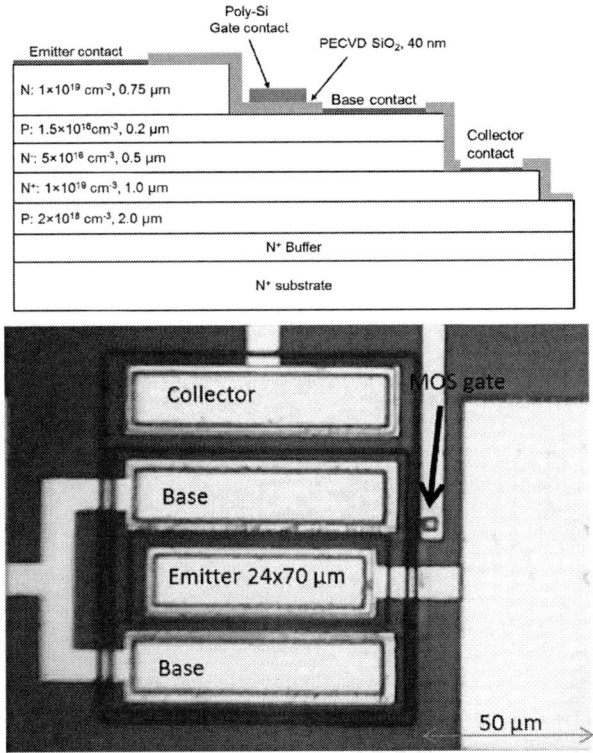

Fig. 1. (a) Schematic cross-sectional view showing Poly-Si gate on the base-link region and (b) top view microscope image of double base bipolar transistor with base surface polysilicon (MOS) gate surrounding the emitter mesa (24 x 70 μm²), lighter colored shape.

III. RESULTS AND DISCUSSION

A. Tied and floating gate electrode biasing

The bipolar transistor, with a gated base structure, becomes a four-terminal device. For a circuit application, it would be preferable to tie the gate potential to either the emitter (*Gate->Emitter*) or to the base terminal (*Gate->Base*), in order to avoid additional circuit complexity. We first analyzed these two options, compared to a *floating gate* reference case. For the floating gate case the exact surface potential on top of base region is varying with lateral position from the emitter contact edge to the base contact edge. Devices with a floating gate behaved identically to standard layout, ungated bipolar transistors on the same die, showing that the addition of a poly-Si gate, on top of the passivation oxide did not cause any performance change. The current-voltage (I-V) Gummel curves at room temperature (300 K) in Fig. 2. show that biasing the gate mainly influences the base current at low forward base-emitter bias. In particular, the *Gate->Emitter* case exhibits a lower base current, with a higher ideality factor (steeper slope). The collector current was almost overlapping for the investigated cases.

The current gain (β) was calculated from the measured base and collector currents as shown in Fig. 3. The *Gate->Emitter* case shows the highest gain, while the *Gate->Base* configuration actually decreases the gain. The peak gain coincides for the *floating gate* and *Gate->Emitter* cases, while the gain at low collector current is very similar in the *floating gate* and *Gate->Base* cases.

To summarize, it is beneficial to apply a lower potential on the base-link surface underneath the gate, since surface inversion can be avoided or suppressed. Hence, the recombination that is proportional to the product of electron and hole concentrations can be minimized. For our particular choice of base doping concentration and passivation oxide thickness gate voltages with larger absolute value must be applied to find the optimum gate bias condition. This is further discussed in the following section.

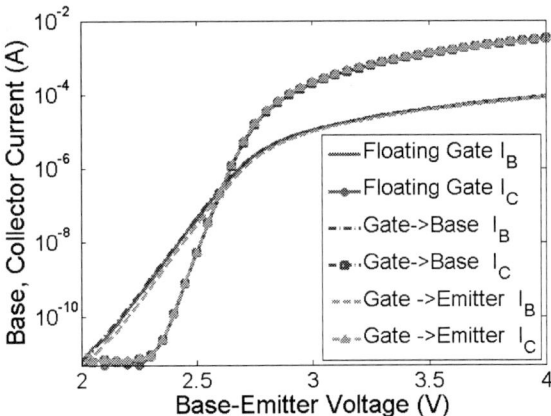

Fig. 2. Base and collector currents for different gate bias conditions, $V_{BC} = 0$ V (200 nm base sample), showing the effect on the non-ideal base current magnitude and slope.

978-1-5090-5979-9/17 $31.00 © 2017 IEEE

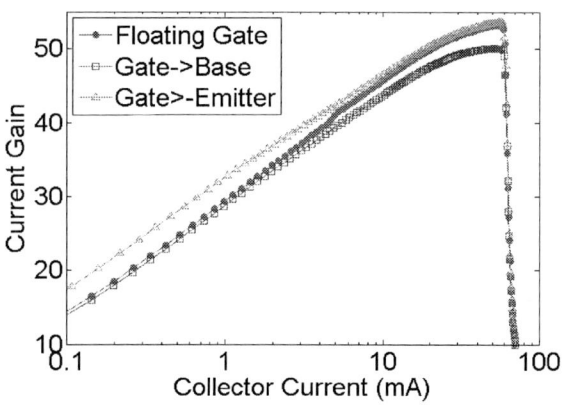

Fig. 3. Current gain for different gate bias conditions, $V_{BC} = 0$ V (200 nm base sample).

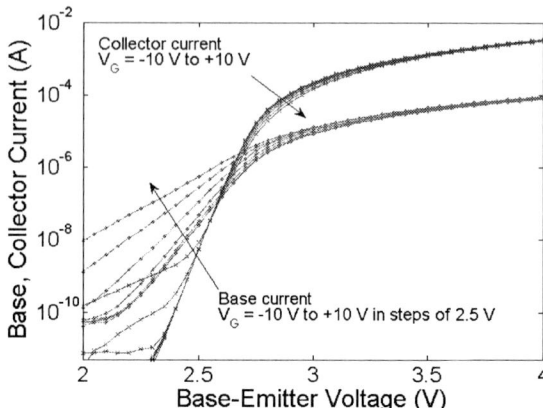

Fig. 4. IV Gummel characteristics for independent gate bias $V_G = -10$ to $V_G = +10$ V, $V_{BC} = 0$ V (200 nm base sample).

B. Independent gate electrode biasing from accumulation to inversion

The purpose of the gate is to modulate the surface potential in the base-link region. There are two important limiting cases here. Applying a positive potential on the gate for a MOS structure on a p-type substrate, e.g. the base region in a bipolar npn-transistor, leads to inversion. While a negative gate potential eventually leads to surface depletion and accumulation. In the former case a high surface electron concentration severely limits the achievable gain, due to increased electron-hole recombination. In the latter case, the accumulation of holes, in the p-type base, slightly reduces the base sheet resistance, and hence the base-link voltage drop. To investigate this, I-V Gummel curve measurements were performed for different independent gate bias conditions, shown in Fig. 4. While in the cases discussed above the gate-bias was not independently controlled, since the gate electrode was tied to either the base or emitter potential. Here, the gate was stepped from a large negative voltage $V_G = -10$ (accumulated interface) to $V_G = +10$ V (inverted interface). By re-measuring the device, with the gate floating, after this gate voltage test sequence we observed a minor change in the base current, as compared to a fresh device. We attribute this base current change to possible slow hole trap filling, during the high negative gate voltage conditions. The response, in the non-ideal part of base current at low forward bias, was pronounced. This is illustrated in Fig. 5., where the gain at two selected base-emitter voltage (V_{BE}) values 2.4 and 4.0 V is shown. The reported values are normalized to the gain at zero V_G. For V_{BE} of 4.0 V the normalized gain remains close to unity while for lower V_{BE} of 2.4 V a gain improvement in the order of 30% can be achieved, since any change in the non-ideal base current directly affects the gain. For comparison this graph also includes data from the sample with thicker base/lower doping. The variation with gate bias becomes stronger (close to 100 %) for lower doping, since the substrate (base layer) is more easily inverted (low threshold voltage).

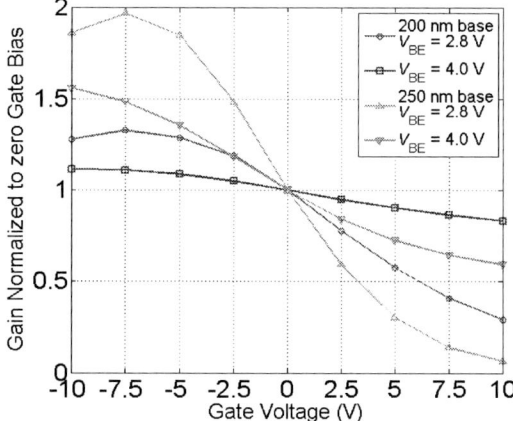

Fig. 5. Normalized gain vs. gate bias for the 200 nm base and 250 nm base samples for low forward bias 2.8 and 4.0 V.

C. Gating effect vs. temperature and dopant activation

The peak current gain of the SiC bipolar is typically decreasing from room temperature up to around 300 °C [8]. At higher temperatures an increase in the gain has been observed which is related to the lifetime dependence on temperature. That has enabled circuit demonstrators to be operational at 500 °C [9, 10]. In our work the focus is on the current gain in the low forward bias region. In Fig. 6. we demonstrate that gating from -10 V to +10 V was effective up to 550 K (277 °C), close to the limit of the wafer prober hot chuck. For the highest temperature data, it is clear that the gating effect mainly acts on the low gain at low collector current, while the peak gain remains relatively unaltered by gate bias. It should be mentioned that the base acceptors dopant are not fully ionized at room temperature and hence the effective doping of the p-type substrate increases with temperature.

978-1-5090-5979-9/17 $31.00 © 2017 IEEE

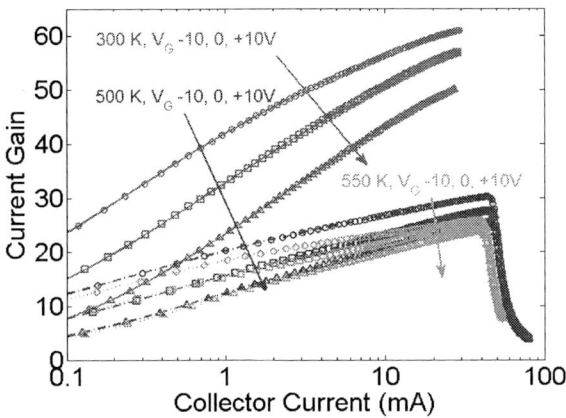

Fig. 6. Current gain for negative, zero and postive gate bias at temperatures 300 (room temperature), 500 and 550 K (limit of hot chuck) for the 200 nm base sample.

D. N-channel MOS test structure

Our layout also included simple n-channel MOS test structures, where the highly doped n-type emitter region was patterned into source and drain contacts, and etched to reach a recessed channel region i.e. the p-type base epi-layer. These structures were analyzed by C-V measurements. The capacitance measured in the gate-to-channel configuration is shown in Fig. 7 for a device with gate length of 5 μm and width of 50 μm. For increasing temperature the threshold voltage is reduced, while the hysteresis in the CV-curve, due to trapped charge is relatively stable, around 200 mV. The measured inversion capacitance, with overlap capacitances subtracted, was in agreement with values calculated from the nitrided passivation oxide thickness.

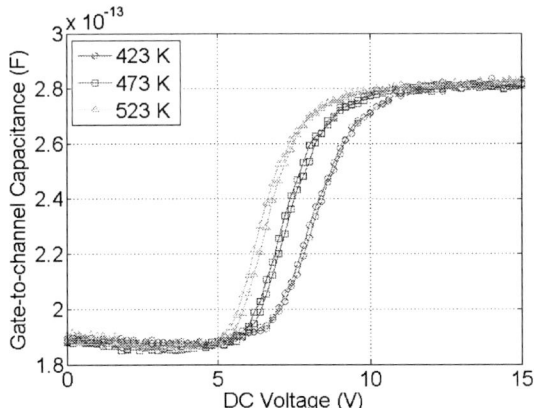

Fig. 7. Gate-to-channel capacitance, including source-drain overlap capacitances, f = 300 kHz vs. temperature for test structure n-channel MOS device, area 5 × 50 μm².

IV. CONCLUSIONS

A novel four-terminal gated base bipolar transistor has been designed to achieve higher current gain, by suppressed carrier recombination at the oxide passivated base-link surface. A polysilicon MOS gate was implemented, using only one additional process step. It was demonstrated that a negative gate bias improves the gain at low collector current by more than 30%, by suppressing the surface recombination. Measurements were presented at temperatures ranging from 300 K to 550 K and the gain was consistently improved. The proposed MOS structure is also useful, as a process monitor for the passivation oxide quality.

ACKNOWLEDGMENT

Doc. Per-Erik Hellström and the EKT group are thanked for sharing their valuable expertise.

REFERENCES

[1] A. S. Kashyap, C. P. Chen, R. Ghandi, A. Patil, E. Andarawis, L. Yin, D. Shaddock, P. Sandvik, K. Fang, Z. Z. Shen, and W. Johnson, "Silicon Carbide Integrated Circuits for Extreme Environments," *2013 1st IEEE Workshop on Wide Bandgap Power Devices and Applications (Wipda)*, pp. 60-63, 2013.

[2] C. M. Zetterling, "Integrated circuits in silicon carbide for high-temperature applications," *MRS Bulletin*, vol. 40, pp. 431-438, May 2015.

[3] L. C. C. Yu, G. T. Dunne, K. S. Matocha, K. P. Cheung, J. S. Suehle, and K. A. Sheng, "Reliability Issues of SiC MOSFETs: A Technology for High-Temperature Environments," *IEEE Transactions on Device and Materials Reliability*, vol. 10, pp. 418-426, Dec 2010.

[4] L. Lanni, B. G. Malm, M. Ostling, and C. M. Zetterling, "Influence of passivation oxide thickness and device layout on the current gain of SiC BJTs," *IEEE Electron Device Letters*, vol. 36, pp. 11-13, 2015.

[5] K. Nonaka, A. Horiuchi, Y. Negoro, K. Iwanaga, S. Yokoyama, H. Hashimoto, M. Sato, Y. Maeyama, M. Shimizu, and H. Iwakuro, "Suppressed surface-recombination structure and surface passivation for improving current gain of 4H-SiC BJTs," *Physica Status Solidi A-Applications and Materials Science*, vol. 206, pp. 2457-2467, Oct 2009.

[6] B. Buono, R. Ghandi, M. Domeij, B. G. Malm, C. M. Zetterling, and M. Ostling, "Influence of emitter width and emitter-base distance on the current gain in 4H-SiC power BJTs," *IEEE Transactions on Electron Devices*, vol. 57, pp. 2664-2670, 2010.

[7] M. Nawaz, "On the assessment of few design proposals for 4H-SiC BJTs," *Microelectronics Journal*, vol. 41, pp. 801-808, Dec 2010.

[8] L. Lanni, B. G. Malm, M. Ostling, and C. M. Zetterling, "500 °C bipolar integrated OR/NOR Gate in 4H-SiC," *IEEE Electron Device Letters*, vol. 34, pp. 1091-1093, 2013.

[9] R. Hedayati, L. Lanni, B. G. Malm, A. Rusu, and C. M. Zetterling, "A 500 °c 8-b Digital-to-Analog Converter in Silicon Carbide Bipolar Technology," *IEEE Transactions on Electron Devices*, vol. 63, pp. 3445-3450, 2016.

[10] R. Hedayati, L. Lanni, S. Rodriguez, B. G. Malm, A. Rusu, and C. M. Zetterling, "A monolithic, 500 °c operational amplifier in 4H-SiC bipolar technology," *IEEE Electron Device Letters*, vol. 35, pp. 693-695, 2014.

On the Understanding of Cathode Related Trapping Effects in GaN-on-Si Schottky Diodes

W. Vandendaele, T. Lorin, R. Gwoziecki,
Y. Baines, J. Biscarrat, M.A. Jaud, C. Gillot,
M. Charles, M. Plissonnier and G. Reimbold
CEA LETI
MINATEC Campus, 17 rue des Martyrs
Grenoble, FRANCE F-38054
Email: william.vandendaele@cea.fr

Abstract—Cathode related current collapse effect in GaN on Si SBDs (Schottky Barrier Diode) is investigated in this paper. Capacitance and current relaxation measurements on diodes and gated-VDP (Van Der Pauw) are associated with temperature dependent dynamic R_{ON} transients analysis showing that the main part of the current collapse at the cathode comes from a combination of electron trapping in the passivation layer and in a carbon related hole trap in the GaN buffer layers ($E_A = E_T - E_V \simeq 0.9eV$). These two parasitic effects can lead to long recovery time ($> 1ks$) after reverse bias stress.

I. INTRODUCTION

Gallium Nitride based power devices have emerged during the last decade as candidates to replace silicon based devices in medium power applications (100-1200V / 10-100A). Recently, promising results have been demonstrated on both HEMTs (High Electron Mobility Transistor) [1]–[3] and SBDs (Schottky Barrier Diode) [4]. However, a "current collapse" effect related to electron trapping can induce an ON-state resistance (R_{ON}) increase and thus a high power dissipation during switching. Consequently, recoverable processes remain a matter of concern at the same level as long-term reliability. Academic and R&D efforts are focused on reducing and even removing this effect [1], [3] and also on understanding its origins [5]–[7]. Optimization of the epitaxial buffer [5], [6], [8] and the introduction of field plates in the device structure [7], [9], [10] can significantly improve the dynamic behavior of GaN-on-Si power devices. In particular, the effect of drain field plates in transistor configuration has been explored in [7] and related to electron trapping under the field plate in interface states between the passivation layer and the front barrier.

In this paper, different cathode field plate (FPC) configurations are studied using dynamic R_{ON} and capacitance measurements and in depth analysis will be presented on associated relaxation transients to give a novel insight into cathode field plate related current collapse. Section II describes the device configurations used in this work both for capacitance and current relaxation transients, and section III focuses on the recovery of capacitance and current characteristics on gated-VDP (Van Der Pauw) structures consistent with a FPC diode structure. Finally, temperature dependent measurements leading to the extraction of the energies of traps are presented in section IV allowing interpretation of cathode related current collapse effects.

II. DEVICE AND SETUP DESCRIPTION

A. Devices description

200mm GaN-on-Si wafers were grown using metalorganic chemical vapor deposition (MOCVD). A structure consisting of "Transition Layers" is grown directly on a 1mm thick lightly doped ($\sim 10 \ \Omega.cm$) p-type Si substrate. A Carbon doped GaN (GaN:C) layer is then grown on top to ensure electrical insulation between the substrate and the top active layer while improving GaN breakdown voltage [11], and the GaN growth is finished with a thin unintentionally doped GaN layer (GaN:UID) which forms the channel. Next, a 1nm AlN spacer is grown, followed by a 24nm AlGaN barrier (23% in Al content) to form the 2DEG (2-D Electron Gas) at the GaN interface, before a final 10nm in-situ SiN passivation layer is grown.

The diode structure is presented in Fig.1.(a). Titanium nitride (TiN) is used as a Schottky contact while an Ohmic contact is formed using a TiAl 875°C annealed stack. Anode to Cathode length is 15μm to ensure a high lateral breakdown voltage. Three anode field plates (FPAs) are embedded to smooth the electrical field at the edge of the anode during reverse bias electrical stress. On the cathode side, a field plate (FPC) is defined on top of the10nm in-situ SiN/140nm LPCVD SiN stack. The FPC length varies between $L_{FPC} = 0.5$ and 3.5μm. The nominal width of the power SBD devices is W = 52mm. The anode and backside contacts are connected in this study.

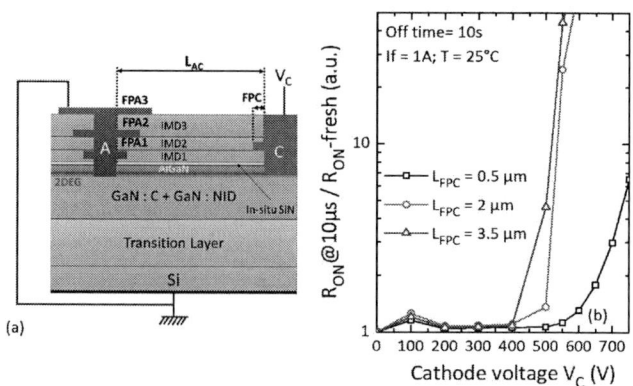

Fig. 1: GaN on Si SBDs structure (a) and evolution of dynamic ON resistance (R_{ON}) under reverse stress up to 750 V for increasing FPC length (L_{FPC})

(b)

B. Characterization Setup

To characterize R_{ON} recovery transients, a microsecond resolved test setup is used [12] giving accurate acquisition in the range of $3\mu s$ to several ks after switching. SBDs were reverse biased up to $V_{REV} = -750V$ for 10s and then a forward current of $I_F = 1A$ is used to characterize ON-state recovery. Fig.1.(b) depicts the evolution of normalized R_{ON} extracted at $10\mu s$ after switching under increasing reverse voltage for diodes with different FPC configuration.

Capacitance/current based measurements are performed woth a Keysight B1505 combined with a high voltage bias-tee. A Cascade TESLA probe station is used to probe the devices while maintaining a fixed chuck temperature between 25 and 150°C.

III. CATHODE FIELD-PLATE CAPACITANCE MEASUREMENTS

As shown in Fig1.(b), a severe R_{ON} increase is observed after reverse stress above -400V for the configuration with a long FPC, revealing strong electron trapping in the cathode field plate area. Diode reverse capacitance in Fig.2 shows the direct relationship between FPC length and the depletion regime that occurs at high reverse voltage ($|Vrev| > 400V$). Depletion occurs under the FPC, which is emphasized by the vertical cathode-substrate capacitance where depletion arises at the same voltage as for the SBD. Nevertheless, this depletion effect on the cathode side is not straightforward since electrons are attracted by the cathode and repulsed by the anode.

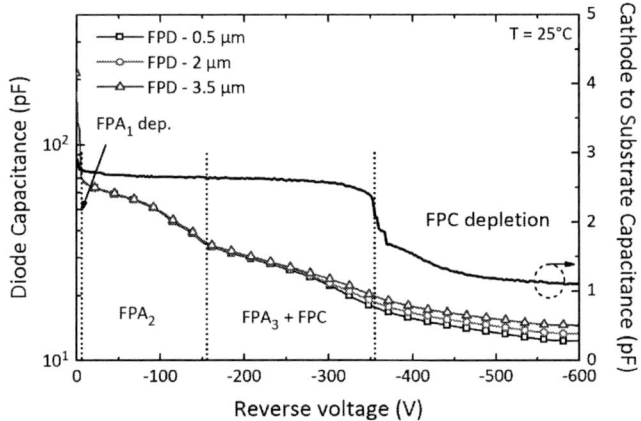

Fig. 2: GaN on Si SBD capacitance versus reverse voltage V_{REV} as a function of FPC length. Temperature is set at 25°C.

In-depth analysis has been carried out through recovery measurements (capacitance and current) on a specific VDP-gated structure (Fig.3) that reproduces the configuration of a diode FPC.

This structure enables us to perform both current and capacitance measurements, and two types of electrical stress have been applied on this gated-VDP. In the first case, a pure gate-to-channel stress is applied where the gate is biased at a given V_G, so that electrons are injected from the 2DEG directly into the FPC equivalent dielectric stack. In the second stress case, a negative bias is applied to the substrate while the gated-VDP structure is grounded, representing the vertical stress on the SBD cathode when polarized in reverse mode.

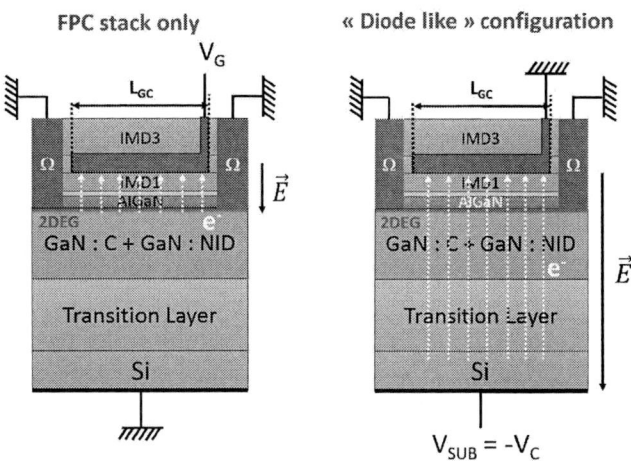

Fig. 3: Gated-VDP structure in gate stress configuration (left) and on substrate stress configuration (right). This latter configuration can also be regarded as very similar to a diode FPC configuration.

A. Gated-VDP capacitance relaxation

Capacitance-voltage C(V) characteristics as a function of relaxation duration after both gate and substrate stresses, $\{60V, 10s\}$ and $\{-500V, 60s\}$ respectively, are shown in Fig.4.

The C(V) characteristics shift towards positive voltages for both stress conditions, which is a sign of electron trapping inside the dielectric stack under the FPC. The independence of conductance peaks related to GaN/AlGaN and AlGaN/in-situ SiN interfaces (not shown) implies that the electron trapping arises in the dielectric volume (SiN LPCVD here). There are significant differences between the two stress conditions, in terms of both amplitude and shape of the curves. An additional capacitance is observed after a substrate stress around the pinch-off voltage of the gated-VDP which tends to disappear after a sufficient recovery duration.

To further analyse these effects, the pinch-off voltage shift ($\Delta V_{po} = V_{po}(t) - V_{po,fresh}$) is plotted as a function of recovery duration for the two stress types and at two temperatures,

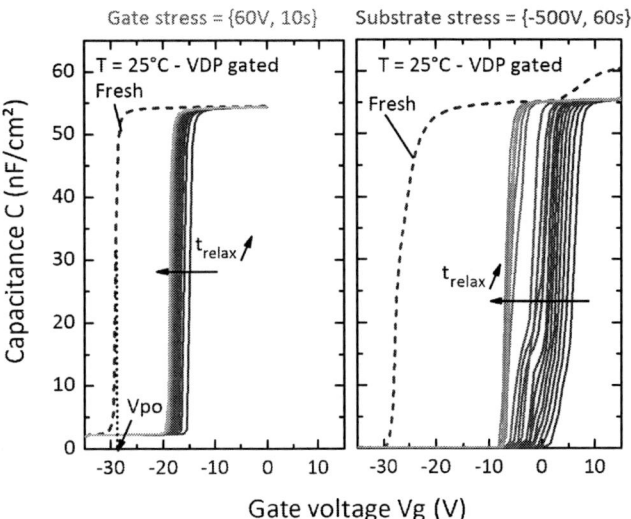

Fig. 4: C(V) characteristics evolution after a $\{60V, 10s\}$ gate stress (left) and a $\{-500V, 60s\}$ substrate stress (right) for increasing recovery durations (4s to 2000s).

25°C and 150°C (Fig.5). Gate stress recovery transients show a power law recovery $\Delta V_{po} = \alpha \cdot t^\beta$, suggesting a self-limiting process, which is well known in the field of BTI (Bias Temperature Instabilities) recovery transients. Substrate stress recovery transients are more complex and exhibit two regimes for both temperatures, with the latter transient following the same trend as a gate stress recovery combined with an additional short time mechanism.

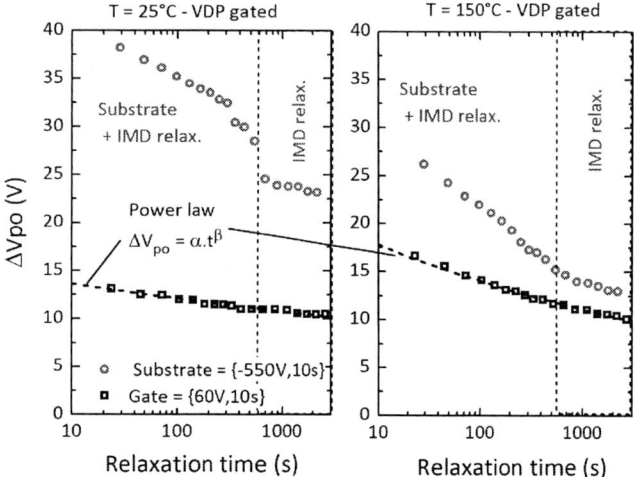

Fig. 5: Recovery transients of ΔV_{po} for gated-VDP structures at 25°C (left) and 150°C (right) after gate and substrate stresses.

This analysis suggests that additional electron trapping towards the substrate under the FPC could act as a second source of R_{ON} increase. We can also consider the so-called FPC depletion in Fig.2 as a V_{po} shift under the FPC region induced by electron injection into the passivation dielectric.

B. Gated-VDP current relaxation

The lateral current relaxation in the gated-VDP structrues after a gate or a substrate stress on the gated-VDP structure was also studied (Fig.6). Sheet resistance was measured after the stress by setting the gate voltage to $V_G = 0V$ with a 100mV lateral voltage to get a realistic image of the 2DEG depletion.

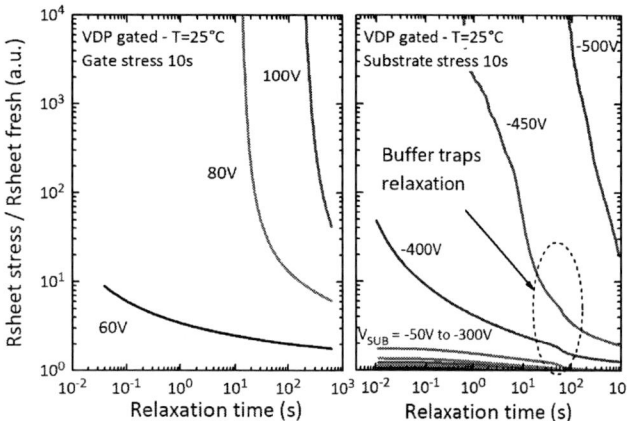

Fig. 6: Sheet resistance R_{sh} recovery transients at 25°C for 10s of gate stress (left) (60,80 and 100V) and (right) after a 10s substrate stress (from -50V to -500V). R_{sh} are normalized to $R_{sh,fresh}$ values.

The amplitude of the gate voltage stress induces a major increase of the R_{sh} underneath the gated-VDP by inducing a

consequent V_{po} shift. It should be noted that thermal detrapping time constant for the dielectric is extremely long (> 1ks). The substrate stress barely has an effect until the bias reaches -400V. The recovery transient shows two superimposed mechanisms after an electrical field is applied between the substrate and the gate of the VDP. A power law similar to the V_{po} transient is superimposed with a buffer relaxation transient. In this latter case, electrons trapped in the buffer layers or GaN:C have a relaxation time constant of approximately 70s.

IV. DYNAMIC R_{ON} AND TRAPS ENERGY EXTRACTION

A temperature dependent dynamic R_{ON} study was performed to confirm the combined effect of trapping in the dielectric under the FPC and in the substrate. The amplitude of the current collapse effect was measured for two designs, with Fig.7 showing R_{ON} recovery between $1\mu s$ and 600s after cathode stress up to 750V. While the amplitude of the current collapse effect is almost null at 600V on a SBD with $L_{FPC} = 0.5\mu m$, the amplitude is multiplied by up to 100 when $L_{FPC} = 2\mu m$. In addition, the relaxation times for the traps appear to be different for these two SBDs structures.

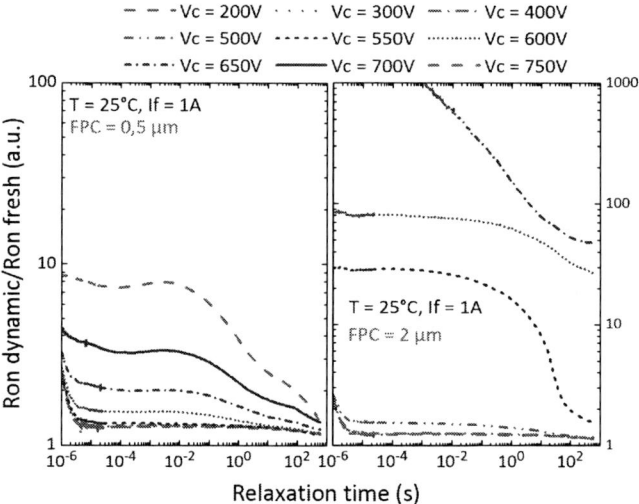

Fig. 7: Dynamic R_{ON} recovery for SBDs diodes with a FPC length of $0.5\mu m$ (left) and $2\mu m$ (right) for different cathode bias up to 750V.

Dynamic R_{ON} transients were performed for temperatures between 25°C and 120°C on SBD with two FPC lengths, 0.5 (not shown here) and $2\mu m$ (Fig.8). The time constants for the traps were extracted using a multiexponential fit and the derivative of the R_{ON} vs log(t) after a smart polynomial fit.

An Arrhenius plot (Fig.9) reveals four traps for the short FPC configuration. The trap extracted at $E_A = 0.56eV$ has been reported to be related to Si-dopants, while the trap extracted at $E_A = 0.75eV$ tends to imply the presence of Nitrogen in an interstitial position in the lattice [13]. Although these two traps are apparent for the short FPC ($0.5\mu m$), they play only a minor role in the amplitude of the trapping. On the contrary, a large amplitude of the current collapse in the short FPC diode is related to a trap located at $E_A = 0.92eV$. This trap is related to carbon impurities introduced in the GaN buffer to ensure electrical insulation by creating a deep acceptor (also called a hole-like trap) [8], [13], [14]. This trap signature is also extracted for the long FPC ($2\mu m$) configuration and is

978-1-5090-5979-9/17 $31.00 © 2017 IEEE

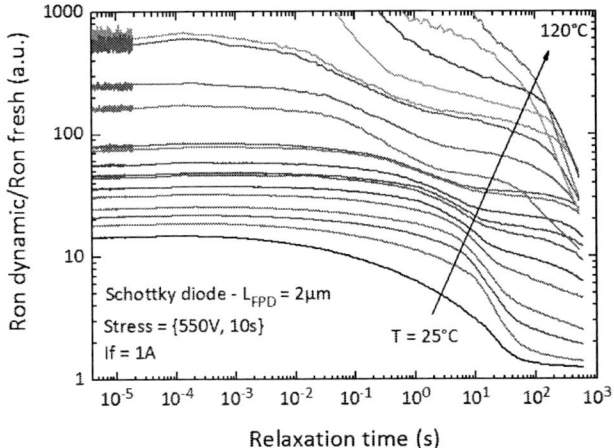

Fig. 8: Temperature dependent R_{ON} transients recorded after a {550V, 10s} reverse stress on a 2μm long FPC SBD.

responsible for the majority of the current collapse effect. This proves that increased trapping in the GaN:C buffers arises in the case of long FPC in diodes caused by electron trapping due to the vertical electrical field. The deviation from the Arrhenius law at low temperatures observed in this case has been recently related to barrier lowering due to the Poole-Frenkel effect [14] when a charge is stored into the buffer layer during a vertical electrical stress which creates a natural electric field in the C-doped region.

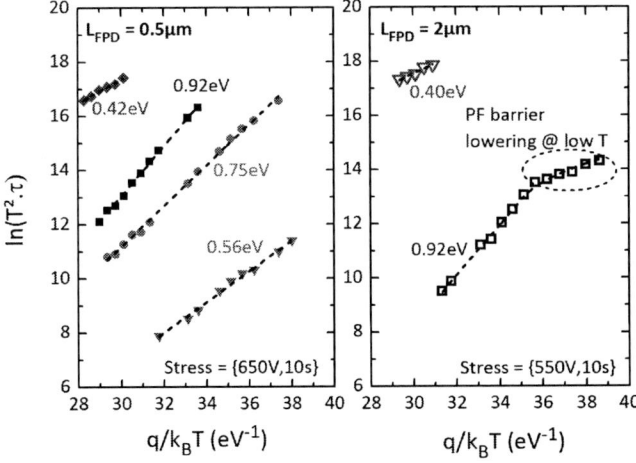

Fig. 9: Arrhenius plot $\ln(T^2 \cdot \tau)$ vs $q/k_B T$ extracted for 0.5μm FPC long SBD (left) and for a 2μm long FPC (right).

The trap extracted at $E_A \simeq 0.4$eV has not been reported in the literature and does not have a suitable cross section to be considered as a semiconductor trap ($\sigma > 10^{-10}$ cm^2). This is in fact an artifact of the R_{ON} transients derivatives analysis and rely on the relaxation of the electrons trapped in the dielectric passivation layer under the FPC. Increased trapping in the GaN:C layer and in the passivation dielectric layer is seen when the FPC gets longer, so a reduced field-plate length combined with a thicker dielectric layer would be an attractive way to move the cathode FP away from the 2DEG and thus to reduce or even avoid electron trapping near the cathode edge region.

V. CONCLUSION

In this paper, we have presented an exhaustive analysis of cathode related current collapse effect in GaN on Si SBDs. Capacitance measurements on power diodes and on gated-VDP confirmed that electron trapping in the passivation dielectric below the FPC can arise under reverse bias stress, while current relaxation measurement on gated-VDP showed that additional trapping occurs in the buffer underneath the FPC. Temperature dependent analysis on diode recovery transients confirmed that a severe trapping arises in the GaN:C layer (on carbon deep acceptor level $E_A = 0.9$eV) as well as in the dielectric under the FPC.

ACKNOWLEDGMENT

This work was performed in the frame of TOURS 2015, project supported by the French "Programme de l'économie numérique des Investissements d'Avenir".

REFERENCES

[1] P. Moens, A. Banerjee, A. Constant, P. Coppens, M. Caesar, Z. Li *et al.*, "(invited) intrinsic reliability assessment of 650v rated algan/gan based power devices: An industry perspective," *ECS Transactions*, vol. 72, no. 4, pp. 65–76, 2016.

[2] P. Moens, C. Liu, A. Banerjee, P. Vanmeerbeek, P. Coppens, H. Ziad *et al.*, "An industrial process for 650v rated gan-on-si power devices using in-situ sin as a gate dielectric," in *Power Semiconductor Devices & IC's (ISPSD), 2014 IEEE 26th International Symposium on*. IEEE, 2014, pp. 374–377.

[3] S. Kaneko, M. Kuroda, M. Yanagihara, A. Ikoshi, H. Okita, T. Morita *et al.*, "Current-collapse-free operations up to 850 v by gan-git utilizing hole injection from drain," in *Power Semiconductor Devices & IC's (ISPSD), 2015 IEEE 27th International Symposium on*. IEEE, 2015, pp. 41–44.

[4] M. Zhu, B. Song, M. Qi, Z. Hu, K. Nomoto, X. Yan *et al.*, "1.9-kv algan/gan lateral schottky barrier diodes on silicon," *IEEE Electron Device Letters*, vol. 36, no. 4, pp. 375–377, 2015.

[5] P. Moens, A. Banerjee, M. Uren, M. Meneghini, S. Karboyan, I. Chatterjee *et al.*, "Impact of buffer leakage on intrinsic reliability of 650v algan/gan hemts," in *Electron Devices Meeting (IEDM), 2015 IEEE International*. IEEE, 2015, pp. 35–2.

[6] G. Meneghesso, M. Meneghini, I. Rossetto, D. Bisi, S. Stoffels, M. Van Hove *et al.*, "Reliability and parasitic issues in gan-based power hemts: a review," *Semiconductor Science and Technology*, vol. 31, no. 9, p. 093004, 2016.

[7] M. Wespel, V. Polyakov, M. Dammann, R. Reiner, P. Waltereit, R. Quay *et al.*, "Trapping effects at the drain edge in 600 v gan-on-si hemts," *IEEE Transactions on Electron Devices*, vol. 63, no. 2, pp. 598–605, 2016.

[8] A. Chini, G. Meneghesso, M. Meneghini, F. Fantini, G. Verzellesi, A. Patti *et al.*, "Experimental and numerical analysis of hole emission process from carbon-related traps in gan buffer layers," *IEEE Transactions on Electron Devices*, vol. 63, no. 9, pp. 3473–3478, 2016.

[9] A. Brannick, N. A. Zakhleniuk, B. K. Ridley, J. R. Shealy, W. J. Schaff, and L. F. Eastman, "Influence of field plate on the transient operation of the algan/gan hemt," *IEEE Electron Device Letters*, vol. 30, no. 5, pp. 436–438, 2009.

[10] M. T. Hasan, T. Asano, H. Tokuda, and M. Kuzuhara, "Current collapse suppression by gate field-plate in algan/gan hemts," *IEEE Electron Device Letters*, vol. 34, no. 11, pp. 1379–1381, 2013.

[11] S. Kato, Y. Satoh, H. Sasaki, I. Masayuki, and S. Yoshida, "C-doped gan buffer layers with high breakdown voltages for high-power operation algan/gan hfets on 4-in si substrates by movpe," *Journal of Crystal Growth*, vol. 298, pp. 831–834, 2007.

[12] T. Lorin, W. Vandendaele, C. Gillot, M. Charles, J. Biscarrat, M. Plissionnier *et al.*, "A microsecond time resolved current collapse characterization test setup dedicated to gan based schottky diode characterization," in *IEEE Internationnl Conference on Microelectronic Test Structures*. IEEE, 2017.

[13] D. Bisi, M. Meneghini, C. de Santi, A. Chini, M. Dammann, P. Brückner *et al.*, "Deep-level characterization in gan hemts-part i: Advantages and limitations of drain current transient measurements," *IEEE Transactions on Electron Devices*, vol. 60, no. 10, pp. 3166–3175, 2013.

[14] H. Yacoub, C. Mauder, S. Leone, M. Eickelkamp, D. Fahle, M. Heuken *et al.*, "Effect of different carbon doping techniques on the dynamic properties of gan-on-si buffers," *IEEE Transactions on Electron Devices*, vol. 64, no. 3, pp. 991–997, 2017.

Temperature Dependent Substrate Trapping in AlGaN/GaN Power Devices and the Impact on Dynamic R_{on}

Arno Stockman[*§], Michael Uren[†], Alaleh Tajalli[‡], Matteo Meneghini[‡], Benoit Bakeroot[§] and Peter Moens[*]

[*]ON Semiconductor, Oudenaarde, Belgium
[†]University of Bristol, Bristol, United Kingdom
[‡]University of Padova, Padova, Italy
[§]CMST imec/Ghent University, Ghent, Belgium
Arno.Stockman@onsemi.com; Arno.Stockman@ugent.be

Abstract—**Temperature dependent substrate ramp measurements on AlGaN/GaN power high-electron-mobility transistors (HEMTs) are used to extract information on charge redistribution in the buffer structure as a function of substrate bias. The measurements are compared to a theoretical model, representing the ideal capacitive buffer stack. It is found that at room temperature some negative charge is stored in the buffer. However, at higher temperatures, positive charge storage occurs as a result of band-to-band tunneling through the GaN channel layer, suppressing bulk trap induced dynamic on-resistance.**

I. INTRODUCTION

AlGaN/GaN based HEMTs show great promise in power applications due to their low on-resistance and high critical electric field [1]. The absence of inversion symmetry in the wurtzite crystal along with the piezoelectric properties of the material causes a high polarization charge at the AlGaN/GaN interface, generating a highly conductive two-dimensional electron gas (2DEG) [2]. When grown on a foreign substrate (such as Si), internal stress leads to a high density of threading dislocations, which, accompanied by (un)intentionally implanted dopants, generate trap states in which dynamic charge storage can occur. In properly passivated structures, buffer charge storage is the main source of dynamic on-resistance R_{on} [3], [4].

In this paper, temperature dependent substrate ramp measurements are used to obtain information about charge buildup inside the buffer structure and results are compared to a theoretical model for the capacitive behavior of the buffer. The capacitive model holds well at room temperature. However, at higher temperature positive charge storage occurs at the bottom of the carbon doped GaN layer, suppressing bulk trap induced dynamic R_{on}.

II. BUFFER STRUCTURE AND DEVICES

A schematic representation of the buffer structure is shown in Fig. 1. A thin AlGaN barrier grown on top of unintentionally doped (UID) GaN provides a positive polarization charge along its heterojunction interface, creating the low-resistive 2DEG. The carbon doped GaN layer is used as a blocking layer for electron current towards the substrate. A superlattice structure acts as a strain relief layer (SRL) for the strain induced by the lattice constant mismatch between AlGaN/GaN layers and the Si substrate.

The devices used in this study are depletion mode 650 V rated AlGaN/GaN power devices. GaN buffers are grown on a Si substrate. The devices have in-situ SiN passivation and several metal layers for field plate design (see [5], [6] for more information). The transistors used are single finger test structures of width $W = 200\,\mu$m.

Temperature dependent substrate bias measurements are performed on a Keysight B1505A Power Device Analyzer. The drain to source voltage V_{ds} is set at 1 V and the gate is grounded. The substrate is ramped from $V_{\mathrm{sub}} = 0$ V down to -650 V and back, with a sweep rate of 2 V/s. The drain current is measured and normalized to its initial value, resulting in a normalized 2DEG conductivity versus substrate bias plot. Pulsed IV measurements are done on a home-built system, details can be found in [7].

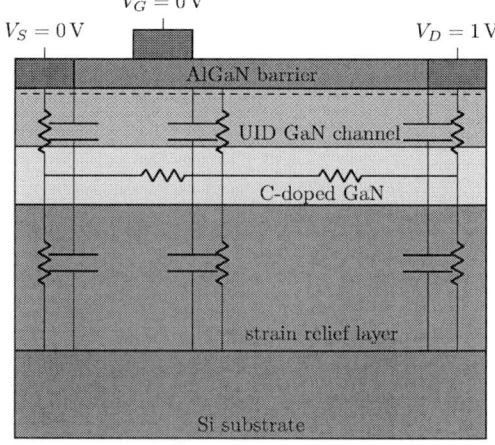

Fig. 1: Schematic representation of an AlGaN/GaN-on-Si buffer stack along with the electrical equivalent circuit under substrate ramp conditions.

978-1-5090-5979-9/17 $31.00 © 2017 IEEE

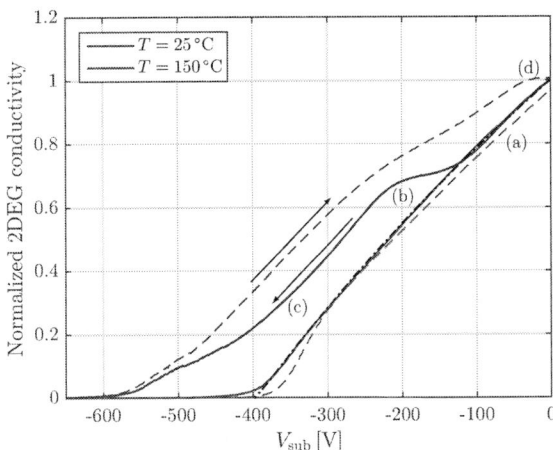

Fig. 2: Normalized 2DEG conductivity at room temperature and $T = 150\,^\circ$C, forward (full lines) and reverse sweep (dashed lines). The black dash-dotted line is the theoretical capacitive behavior as calculated in (1) at room temperature.

III. EXPERIMENTAL DATA

A. Substrate Ramp

Under off-state conditions, the buffer stack is subject to large electric fields (lateral and vertical), causing dynamic charge storage in multiple regions within the buffer. Any negative charge build-up close to the 2DEG partially depletes the transistor channel and will increase the on-resistance. The different layers in the buffer stack cannot be probed separately and not much is known about the exact current transport mechanisms.

Substrate ramp is a technique which is able to give qualitative and quantitative information about the magnitude, the sign and the position relative to the 2DEG of stored charge. By ramping the substrate to a high (negative) potential, mimicking the off-state operation under the drain contact in a transistor, and monitoring the channel conductivity, changes in the electric field close to the 2DEG are observed. Any charge redistribution in the buffer upon reverse bias will change the electric field and if the charge is in close proximity to the 2DEG it will be sensed as a change in the 2DEG conductivity. As such, the buffer charge trapping or storage will be visible in the substrate ramp characteristic.

Fig. 2 shows the normalized conductivity measured at room temperature and $T = 150\,^\circ$C for the buffer stack studied in this paper. The experimental curves are complemented by the theoretically calculated normalized conductivity, depicting the situation when no leakage occurs inside the buffer and thus the buffer stack behaves as a capacitor. The calculated normalized conductivity slope can be expressed as:

$$\frac{d(I_d/I_{d,0})}{dV_{\mathrm{sub}}} = C_{eq}\,\frac{\mu}{I_{d,0}}\,\frac{(V_{ds} - 2I_d R_c/W)^2}{V_{ds}} \qquad (1)$$

with C_{eq} the equivalent capacitance of the complete buffer

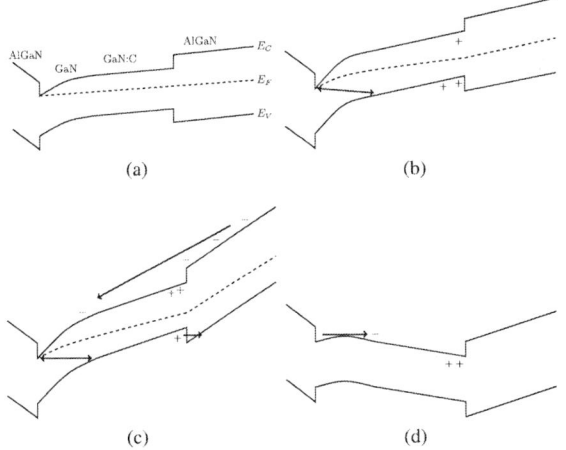

Fig. 3: Schematic band diagrams showing inferred charge storage locations and current flows under different substrate bias conditions. The four graphs represent the different regions labeled in Fig. 2. (a) low V_{sub}, no charge storage; (b) intermediate V_{sub}, leakage of the GaN UID layer; (c) high V_{sub}, buffer behaves resistive; (d) return sweep.

stack, I_d and $I_{d,0}$ the drain current at any given V_{sub} and at $V_{\mathrm{sub}} = 0$ V, respectively, V_{ds} the drain/source voltage, μ the 2DEG mobility (1940 cm^2/Vs at room temperature), R_c the contact resistance (0.6 Ωmm) and W the device width (200 μm). The accuracy of the model will be discussed later.

From Fig. 2, it follows that at room temperature the substrate ramp characteristic (at that ramp rate) behaves almost like an ideal capacitor. At any substrate voltage upon the return sweep, the normalized conductivity is (slightly) below the ideal capacitance curve, which can only be explained by depletion of the 2DEG due to negative stored buffer charge. A possible explanation for this stored negative charge can be found in the loss of 1D field distribution when the channel is fully pinched off, inducing non-uniform trapping in the drain/source gap. At $T = 150\,^\circ$C, the situation is different. Up to $V_{\mathrm{sub}} \approx |-150|$ V, the buffer still behaves capacitively. However, between $V_{\mathrm{sub}} \approx -150$ V and -250 V, the normalized conductivity saturates, i.e. does not change with V_{sub}. This implies that the electric field at the 2DEG does not change with increasing V_{sub}, which is explained in Fig. 3, where the schematic band diagrams for the four corresponding regions in the substrate ramp characteristic of Fig. 2 are schematically depicted. At low V_{sub}, no charge storage occurs and the buffer behaves capacitively (Fig. 3a). At $V_{\mathrm{sub}} \approx -150$ V, the UID GaN channel layer starts to conduct current through a band-to-band tunneling process, initiated along dislocations. The net effect is positive charge storage at the GaN:C/SRL interface. This positive charge builds up the field with increasing V_{sub}, screening the 2DEG from the substrate potential (Fig. 3b). From $|-250|$ V onward, the complete buffer structure starts to leak and the buffer becomes more resistive (Fig. 3c). Upon

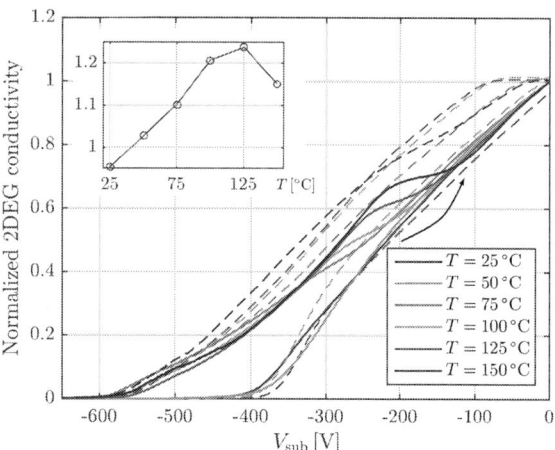

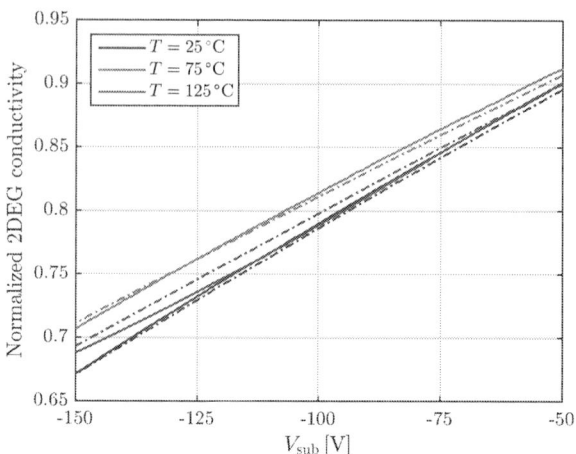

Fig. 4: Normalized 2DEG conductivity from room temperature up to $T = 150\,°C$ in steps of $25\,°C$, forward (full lines) and reverse sweep (dashed lines). Inset: relative conductivity (reverse w.r.t. forward) versus temperature at $V_{\mathrm{sub}} = -100\,V$.

Fig. 5: Normalized 2DEG conductivity at $T = 25\,°C$, $75\,°C$ and $125\,°C$ as measured (full lines) and calculated (dash-dotted lines).

return sweep, the reverse characteristic is situated above the forward characteristic, indicative of positive charge storage (for any given V_{sub}, the normalized conductivity is higher in the reverse than in the forward sweep, indicating a higher 2DEG density). The amount of positive charge can be estimated either from the plateau in the forward characteristic (region (b)) or from the magnitude of the positive hysteresis.

Fig. 4 shows the substrate ramp data for the complete temperature range. Between $T = 50\,°C$ and $75\,°C$, the shape of the curve changes drastically, likely caused by a leakage path between the 2DEG and the bottom of the GaN:C layer, which could indicate a sharp temperature dependent resistivity change of the GaN layers and their inherent dislocations. From $75\,°C$ onward, the substrate ramp at these intermediate temperatures show a gradual shift of the onset of the conductivity plateau towards lower electric fields at higher temperatures. In other words, to obtain the same level of hole conduction, the electric field over the GaN channel and GaN:C layer is lower at higher temperatures, consistent with a trap-assisted band-to-band mechanism [8].

The amount of positive charge is estimated from the magnitude of the positive hysteresis at $V_{\mathrm{sub}} = -100\,V$ and is shown in the inset of Fig. 4. There is a clear trend that the positive charge increases with temperature. Also, the onset of positive buffer charge storage occurs at a lower field with increasing temperature (notice the plateau appearing at lower voltage for increasing temperature, Fig. 4).

Fig. 5 compares the experimental forward substrate bias ramp and the theoretical model (1) at low substrate voltages and at different temperatures. Measurements performed up to $T = 75\,°C$ are in good agreement with the capacitive model. At higher temperatures, the measured curve starts to deviate from the ideal capacitive situation. It is speculated that this is due to the temperature dependence of the contact resistance,

which decreases with temperature and gives rise to a steeper slope, as is indeed observed from Fig. 5.

The vertical leakage currents during substrate ramp are plotted in Fig. 6. The measured currents are the sum of the diffusion current in the structure and the displacement current $I_{\mathrm{disp}} = CdV/dt$. As the displacement current switches sign upon reverse sweep, the leakage current crosses zero at low voltages, where the leakage current is of the same (or smaller) magnitude as the negative displacement current. With increasing substrate voltage, the leakage current (originating via dislocation lines through the nucleation layer) can be fitted to a Poole-Frenkel conduction model with a barrier

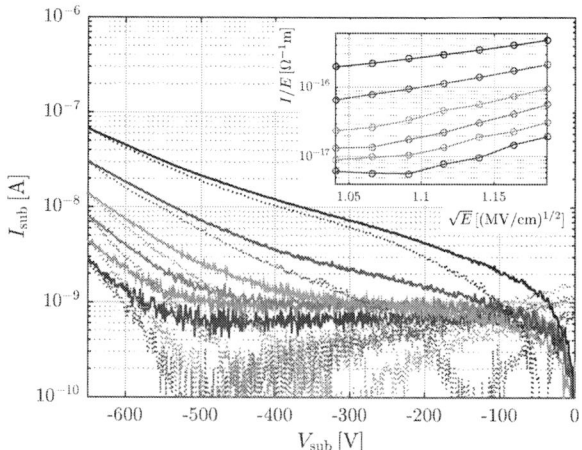

Fig. 6: Vertical leakage current from room temperature up to $T = 150\,°C$ in steps of $25\,°C$, forward (full lines) and reverse sweep (dotted lines). Inset: area times conductivity I/E versus $\sqrt{E}$ for different temperatures.

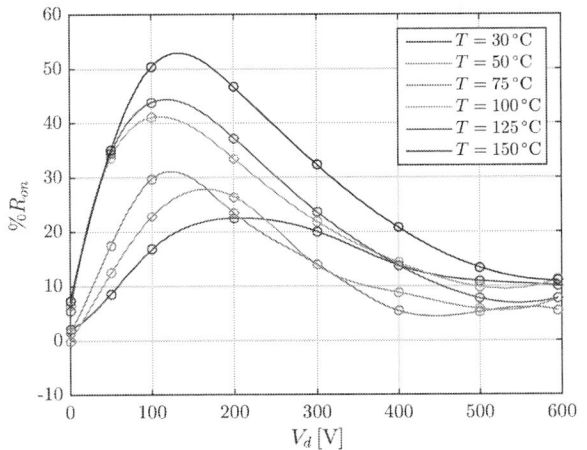

Fig. 7: Pulsed IV measurements from room temperature up to $T = 150\,°C$ in steps of $25\,°C$ and $t_{on} = 20\,\mu s$ and $t_{on} = 2\,ms$.

$\phi = 1.56\,eV$, as can be seen in the inset of Fig. 6, showing the (area times) conductivity I/E versus the square root of the electric field for different temperatures [9]. Note that the model fits better at high rather than low temperature, as at low temperatures and electric fields the substrate leakage is dominated by the displacement current.

B. Pulsed IV

Dynamic R_{on} measurements are performed on the same devices as a function on drain quiescent voltage and temperature, see Fig. 7. The on and off times are $20\,\mu s$ and $2\,ms$, respectively. The gate quiescent voltage is $-20\,V$. At low temperature, the maximum dynamic $R_{on} \approx 25\%$ at 200–300 V. At higher drain voltages, the dynamic R_{on} decreases due to the build-up of positive buffer charge that counteracts the dynamic R_{on} increase, as follows from the substrate ramp data (see Fig. 4). At higher temperatures, the amount of buffer charge increases and the onset of positive charge storage occurs at lower drain quiescent voltage. Hence, at higher temperature the dynamic R_{on} suppression is stronger and occurs at lower fields, as can be clearly seen from Fig. 7. As a result, the dynamic R_{on} at typical operating conditions for these devices ($V_{ds} \approx 400$–$500\,V$) is very low ($< 15\%$) and almost independent of temperature.

IV. CONCLUSION

We have shown temperature dependent substrate ramp and pulsed IV measurements and how they give information about charge storage inside the buffer of an AlGaN/GaN HEMT. The voltage at which positive charge storage is initiated at the bottom of the carbon doped GaN layer shifts to lower substrate voltages at higher temperatures, consistent with a trap-assisted band-to-band mechanism. This voltage can be determined by introducing a temperature dependent capacitive model and observing the point at which the measured characteristic deviates from the ideal capacitive structure. The impact of the amount, the field dependence and the temperature dependence of the positive charge storage on dynamic R_{on} is such that the devices have low dynamic R_{on} at typical operating conditions and over the full temperature range.

ACKNOWLEDGMENT

A. Stockman wants to acknowledge the Baekeland scholarship from the Flanders Innovation and Entrepreneurship (VLAIO). Part of this work is also funded by the Horizon2020 project Innovative Reliable Nitride based Power Devices and Applications (InRel-NPower).

REFERENCES

[1] N. Ikeda, Y. Niiyama, H. Kambayashi, Y. Sato, T. Nomura, S. Kato, and S. Yoshida, "GaN power transistors on Si substrates for switching applications," *Proc IEEE*, vol. 98, no. 7, pp. 1151–1161, July 2010.
[2] O. Ambacher, J. Smart, J. R. Shealy, N. G. Weimann, K. Chu, M. Murphy, W. J. Schaff, L. F. Eastman, R. Dimitrov, L. Wittmer, M. Stutzmann, W. Rieger, and J. Hilsenbeck, "Two-dimensional electron gases induced by spontaneous and piezoelectric polarization charges in N- and Ga-face AlGaN/GaN heterostructures," *Journal of Applied Physics*, vol. 85, pp. 3222–3233, March 1999.
[3] G. Meneghesso, M. Meneghini, E. Zanoni, P. Vanmeerbeek, and P. Moens, "Trapping induced parasitic effects in GaN-HEMT for power switching applications," *IC Design & Technology (ICICDT)*, 2015.
[4] J. Würfl, O. Hirl, E. Bahat-Treidel, R. Zhytnytska, P. Kotara, F. Brunner, O. Krueger, and M. Weyers, "Techniques towards GaN power transistors with improved high voltage dynamic switching properties," *IEDM Technical Digests*, 2013.
[5] P. Moens, C. Liu, A. Banerjee, P. Vanmeerbeek, P. Coppens, H. Ziad, A. Constant, Z. Li, H. De Vleeschouwer, J. Roig-Guitart, P. Gassot, F. Bauwens, E. De Backer, B. Padmanabhan, A. Salih, J. Parsey, and M. Tack, "An industrial process for 650V rated GaN-on-Si power devices using in-situ SiN as a gate dielectric," *ISPSD*, pp. 374–377, 2014.
[6] P. Moens, P. Vanmeerbeek, A. Banerjee, M. Caesar, J. Guo, C. Liu, P. Coppens, A. Salih, M. Meneghini, M. Kuball, M. J. Uren, H. Meneghesso, E. Zanoni, and M. Tack, "On the impact of carbon-doping and channel thickness on the dynamic R_{on} of 650V GaN power devices," *ISPSD*, pp. 37–40, 2015.
[7] D. Bisi, A. Stocco, M. Meneghini, F. Rampazzo, A. Cester, G. Meneghesso, and E. Zanoni, "High-voltage double-pulsed measurement system for GaN-based power HEMTs," *52nd IEEE International Reliability Physics Symposium, IRPS*, pp. CD.11.1–CD.11.4, June 2014.
[8] V. Moroz, H. Y. Wong, M. Choi, N. Braga, R. V. Mickevicius, Y. Zhang, and T. Palacios, "The impact of defects on GaN device behavior: Modeling dislocations, traps and pits," *ECS Journal of Solid State Science and Technology*, vol. 5, pp. 3142–3148, 2016.
[9] Y. Zhang, H.-Y. Wong, M. Sun, S. Joglekar, L. Yu, N. A. Braga, R. V. Mickevicius, and T. Palacios, "Design space and origin of off-state leakage in GaN vertical power diodes," *IEDM*, pp. 899–902, 2015.

978-1-5090-5979-9/17 $31.00 © 2017 IEEE

Material and device innovation impact on reliability for scaled CMOS technologies

T. Nigam^, A. Kerber*, T. Shen*, R. Ranjan* and L. Cao*

Reliability Engineering, GLOBALFOUNDRIES

^ Santa Clara, CA, USA, *Malta, NY, USA

Tanya.nigam@globalfoundries.com

Abstract—Comprehending material, technology and design interaction with key reliability metrics is becoming critical to ensure successful product introduction with continued scaling beyond 10nm. This paper summarizes the key challenges for reliability and its implication for the design/technology co-optimization. Building in reliability requires a detailed understanding of material properties, device architecture and circuit usage model/sensitivity to various failure mechanisms. For logic application, mean Fmax degradation due to BTI/HCI addresses circuit impact while for SRAM memories it is critical to comprehend BTI induced variations. For BEOL the choice of material is driven by EM/TDDB and design complexity/routing strategy with the goal to reduce RC component.

Keywords—reliability; variability; RO degradation; BTI/HCI; TDDB

I. INTRODUCTION

A very diverse product landscape is emerging to meet the future market needs. This divergence in needs is driving different dimensions of technical innovation namely new paradigms of silicon scaling, differentiated silicon and system level integration, Fig.1 [1]. Scaled nodes have to address the high performance sector such as server, graphic chips and high end mobile applications while differentiated silicon solutions are needed for low power applications like IoT, virtual reality and artificial intelligence. Autonomous automotive applications are spread across both spectrums with more stringent reliability requirement.

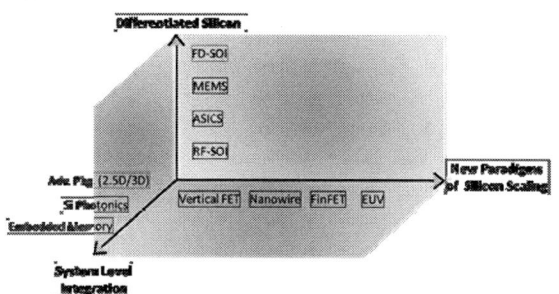

Figure 1 Dimensions of innovation targeted to address emerging market segments.

Scaling minimum feature size and enhancing performance with equivalent level of reliability will continue in the foreseeable future. Moore's law based scaling has contributed

to numerous material, device architecture and design innovation. Each of these innovations led to new reliability challenges/interactions which needed to be addressed to ensure product functionality for the intended lifetime. Introduction of High-K Metal Gate (HK MG) introduced Positive Bias Temperature Instability (PBTI), Stress Induced Leakage current (SILC) and a new mode of Time Dependent Dielectric Breakdown (TDDB) exhibiting a bimodal distribution [2-4]. Some elements such as SiGe/Ge in the channel have helped Negative Bias Temperature Instability (NBTI) by moving the band alignment [6-8]. With the introduction of novel barrier layers for interconnect metallization continuous trade off between (Back End of Line) BEOL TDDB/EM (Electromigration) and (Resistance Capacitance) RC performance has to be made [9]. Introduction of ultra-low K material poses challenges for Chip Package Interaction (CPI), recently discussed in [10]. The move to FINFET or (Fully Depleted Substrate on Insulator) FDSOI introduces the challenges of Self Heating (SH) that occurs during operation but more importantly during assessment of reliability at elevated voltages/temperatures [11-12]. The need to understand the magnitude of SH and modeling its impact is critical. With reduction in feature sizes variability becomes a challenge [13-15]. Time zero variability due to lithography or dopant distribution may get compounded by degradation during operation leading to a significant impact at end-of-life (EOL) operation.

All of these innovations have been coupled with enhancement in measurement/modeling and physical understanding of the various degradation mechanisms. Examples include introduction of ultra-fast measurement capability necessitated by recovery in BTI, use of Ring Oscillators to understand BTI/HCI impact at circuit level as they become coupled during DC degradation studies [16-18], test structure optimization to reduce SH during reliability assessment and array level design to study local vs global variability component. In this paper, how reliability challenges have evolved with scaling so far and where focused effort will be required in future is presented.

The paper is divided into 4 sections. First part covers the aspect of material innovation and the reliability challenges they pose. Guidance is provided for key areas which need focused effort for scaled nodes. Second, we will talk about the reliability challenges for architectural solutions to address the

978-1-5090-5979-9/17 $31.00 © 2017 IEEE

growing market needs for scaled and differentiated silicon. Third aspect of this paper focusses on variability and how to ensure that both intrinsic and extrinsic reliability is comprehended during technology development to shorten the time to market. Final section will focus on device to product reliability correlation using Ring Oscillator (RO) for performance degradation trade off vs process choices, impact of substrate type plus TDDB assessment under AC stress.

II. MATERIAL INNOVATION AND IMPACT ON RELIABILITY

Material innovation has been the bed rock of CMOS scaling. In FEOL, HK MG was introduced to enable gate dielectric thickness scaling and address leakage challenges [19-20]. In BEOL, metallization change from Aluminum (Al) to Copper (Cu) lead to significant improvement in RC and EM limitation while requiring optimization of barrier/seed layer [21]. For sub-7 nm nodes, new channel materials and BEOL material are being evaluated to improve performance and RC delay.

A. HK MG

Introduction of HfO_x/SiO_x introduced PBTI due to electron trapping in HfO_x as shown in Fig.2. Impact of PBTI in technology was reduced with thinner HK and Inter Layer (IL) optimization. Note that similar level of NBTI is observed in HfO_x/SiO_x and Poly/SiON as the IL properties dominate NBTI.

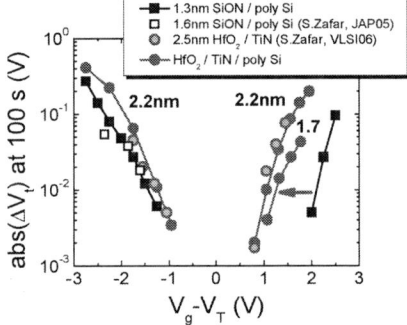

Figure 2 Positive Bias Instability is measured for HK MG unlike Poly/SiON while similar level of Negative Bias Temperature Instability continues to exits.

Additional challenge with HK MG was a lower Weibull slope measured on small area devices. The transition to higher Weibull slope for large area devices as shown in Fig. 3, due to dual layer breakdown statistics allowed the increase of Vmax in HK MG CMOS technologies [4-5]. These findings are independent of Gate first versus Gate last HK MG process. They are related to difference in defect generation rate in dual layer dielectric stack and can be applied to any multi-layer dielectric stack independent of device architecture.

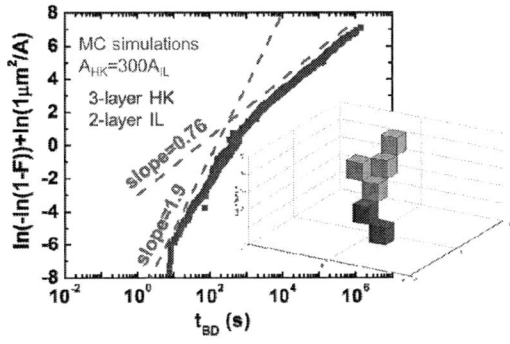

Figure 3 Bi-modal TDDB distribution in HK MG is observed. Using larger area devices it is demonstrated that a higher Weibull slope of 1.5+ can be obtained for NFET devices consistent with percolation theory.

For scaled nodes, enhanced EM lifetime is required to meet the product needs. In Fig. 4a and 4b, Co liner/cap optimization leads to significant improvement in EM with some reduction in voltage acceleration for TDDB while improving the RC component.

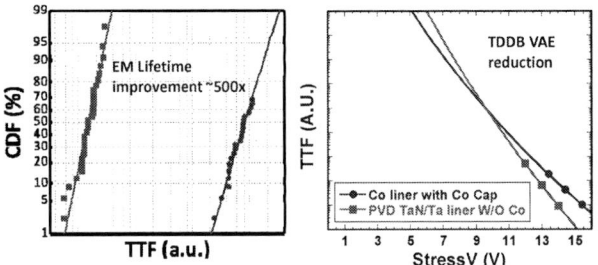

Figure 4a) EM lifetime enhancement using Co liner with Co cap. 4b) Reduction in VAE for TDDB leads to a lower Vmax for EOL. Careful optimization is needed to find the best trade-off between EM and TDDB.

In the Gate first approach, band alignment for V_T required introduction of SiGe in the channel improving NBTI. [6] Ge based FETs also show improved NBTI due to favorable band alignment and separation of carriers from the channel [7-8]. Moving to III-V materials for NFET (Replacement Metal gate) RMG optimization continues to be a challenge along with controlling channel/oxide interface quality [22-23].

III. ARCHITECHTURAL INNOVATION AND IMPACT ON RELIABILITY

Introduction of FINFET in 22nm/14nm node allowed voltage and gate dielectric thickness scaling with better electrostatics, lower T0 variability and reduced field over the gate dielectric [24-25].

978-1-5090-5979-9/17 $31.00 © 2017 IEEE   135

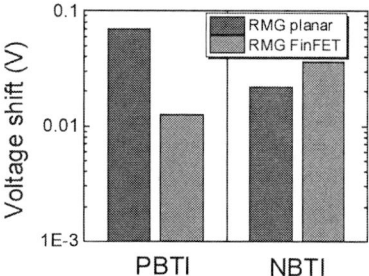

Figure 5 Lower PBTI is measured for FINFET devices as compared to planar while NBTI is slightly enhanced.

Fig.5 shows that PBTI significantly benefits from reduced field for the same HK thickness; while NBTI remained a challenge due to similar IL quality and (110) interface [12].

Similar to BTI, TDDB benefited from lower field and exhibited random defect generation and breakdown. All methodologies/learning from planar technologies continue to drive the Vmax entitlement for TDDB in FINFET geometry and FDSOI. Other mechanisms like HCI are sensitive to device architectural solution and substrate material. FINFET geometry leads to lower heat dissipation than planar technology requiring methodology/test structure optimization to minimize self-heating during HCI assessment. Additional work is needed to validate their extension in Nanowires or Vertical FETs which will be operating in the sub-1.0V range.

A. Bulk vs SOI

From the reliability perspective, choice of substrate has little impact on BTI and TDDB but plays a strong role during HCI assessment.

- Self-Heating: Under typical HCI-like stress conditions when the device is turned on, 5x more heat is generated in SOI based FINFETSs as compared to bulk FINFETs, see Fig 6 [26]. Similar observation is made on FDSOI with thin silicon channel as compared to bulk devices.

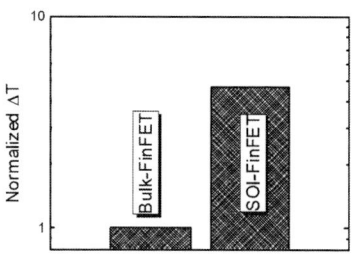

Figure 6 Change in temperature due to self-heating in bulk vs SOI FINFETs.

As shown in Fig. 7, the impact of SH on HCI is a strong function of layout with multiple FINs per active (RX) showing higher levels of HCI degradation in SOI substrate vs bulk. SH to a small extent is present in bulk FINFET as well and is expected to increase as we scale to 7 nm and below. Leveraging the learning from SOI, test structure optimization with fewer FINs per RX and more RX is clearly beneficial. Calibration methodology which already exists in literature for SOI based

technologies can be employed to quantify the level of SH during HCI in scaled bulk FINs [27-28]. As will be shown later in Fig. 12, SH is not present to the same extent under AC operation even in SOI substrates.

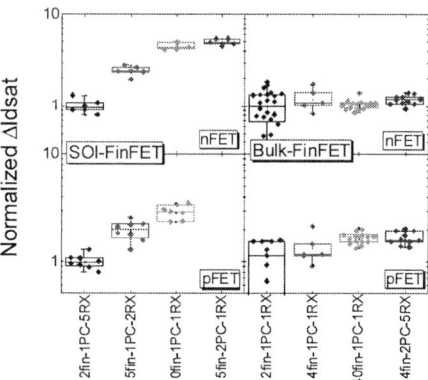

Figure 7 Impact of Number of FINs per RX for SOI vs Bulk FINFET technology subjected to HCI stress.

B. Nanowires, Vertical FETs etc.

Gate length (Lg) scaling has virtually stopped in FINFET geometry as the FIN width scaling is challenging. To continue Lg scaling and ensure good electrostatics moving to Gate-All-Around (GAA) nanowires offers many advantages. Vertical implementation of Nanowires is more likely [29]. Additionally, Vertical FETs using III-V channel material is also being evaluated [30]. Reliability challenges in all these architectures will continue to be determined by the channel/oxide interface quality and choice of materials. Any work function material optimization for V_T will also impact reliability.

IV. VARIABILITY IN SCALED NODES

T0 variability and process control is critical in scaled nodes. The key sources of T0 variation in scaled devices are Random Discrete Dopant, Line Edge Roughness, Gate Edge Roughness, Metal Gate Grain size etc. EOL variability impact is visible either as additional V_T variation due to defect generation/trapping [14-15] or hard TDDB fails due to local thinning.

A. T0 and EOL Variability for Performance/Guard band

Moving to FINFETs or FDSOI based technologies leads to a reduction in T0 V_T variation due to undoped or lowly doped channels as shown for pFET devices in Fig 8 (left panel). This implies that BTI/HCI induced contribution for EOL variability increases. Therefore, impact of NBTI induced variability in scaled SRAM's needs to be comprehended explicitly. As shown in the right panel of Fig. 8, NBTI induced variability is similar for planar and FINFET devices and has similar magnitude as the T0 FINFET variability [31].

In logic applications, mean shift is more critical as EOL variability due to HCI and BTI is averaged out. As observed in Fig. 9, $\Delta f/f$ distribution for ROs stressed to median shift of 5%

show a similar 3-sigma scaling factor (Kwc) value as large area ΔV_T BTI, which is a reflection of process variability.

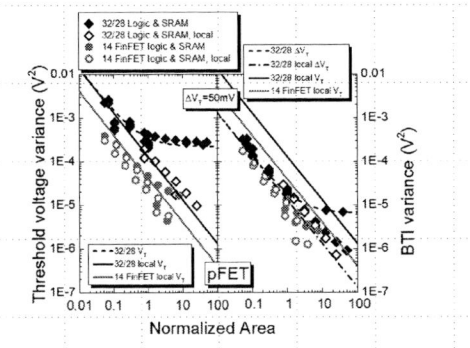

Figure 8 T0 vs NBTI induced (ΔV_T=50mV) variability for planar and FINFET devices.

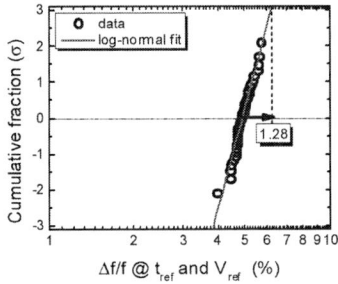

Figure 9 $\Delta f/f$ distribution for a 101 stage RO.

B. T0 vs EOL Variability from Process variation leading to hard fails

BEOL and MOL process are prone to variability and in many cases they are easily visible in T0 Pareto leading to a lower yield. Once a healthy yield distribution is established the variability can unveil itself as lower Vmax due to a reduction in Weibull slope and t63 as shown in Fig. 10 for an un-optimized flow. Similar observations have been made for FEOL TDDB, where no leakage signature was observed at T0 but a lower Weibull slope was measured on thinner buffer layer thickness due to variability introduced during eSiGe process [32]. These variations can be sensitive to layout density necessitating HTOL like stress in logic along with SRAM in scaled nodes.

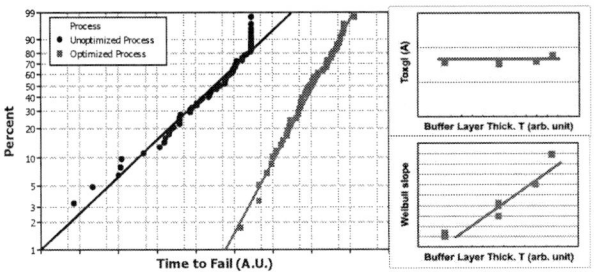

Figure 10 Observation of lower Weibull slope due to process induced variability leading to a lower Vmax in BEOL/FEOL TDDB.

V. CIRCUIT LEVEL ASSESSMENT OF RELIABILITY

With continued scaling reliability margins are getting reduced as voltage scaling is not following the expected reduction. As AC characterization methods have evolved, mimicking circuit level operation using single devices or using ring oscillators (RO)/SRAM arrays to predict product fail modes is becoming critical. In this section ultra-fast RO characterization methodology is leveraged to understand possible impacts on process choices for BTI, substrate impact on HCI and TDDB Vmax entitlement.

A. RO Degradation

Ultra-fast RO characterization methodology as outlined in [18] has been used to understand and correlate impact of device level degradation to a simple logic circuit

- BTI and process choices: In Fig. 11 we observe that process choices to improve T0 performance can lead to different EOL implication for different usage conditions [25]. At typical operating conditions the decision to move to T3 would be beneficial even after taking EOL performance guard band into account. But for overdrive conditions typically used in the industry T3 would not be beneficial compared to T1 after factoring in BTI impact. Such feedback early in the technology development can help quickly drive decisions and reduce technology development time.

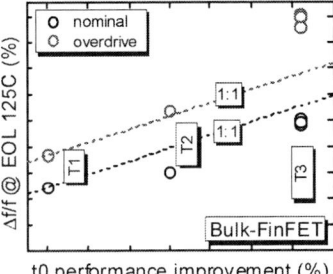

Figure 11 Impact of T0 performance increase on EOL Frequency degradation under nominal and overdrive conditions

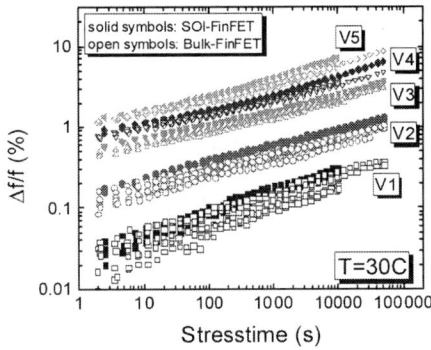

Figure 12 RO frequency degradation on bulk vs SOI substrate for FINFET based technology. Low voltage stress is dominated by BTI while HCI contribution is observed at higher stress bias.

978-1-5090-5979-9/17 $31.00 © 2017 IEEE

- **HCI and impact of SH in Bulk vs SOI FINs**: As discussed in Section III, DC HCI is very sensitive to substrate type and layout geometry due to significant SH during DC stress. Fig 12 shows the weak dependence of RO degradation on bulk vs SOI substrate. For both bulk and SOI, at elevated voltages we can see the HCI signal leading to a change in degradation slope. This assessment clearly demonstrates the need to comprehend SH and test structure optimization for DC HCI modeling to provide the correct DC to AC ratio as the SH contribution between bulk and SOI differs by 5x (Fig. 8). For both types of substrate BTI degradation dominates at typical operating conditions for EOL performance guard band.

- **TDDB and Vmax entitlement under AC stress**: TDDB is considered a hard fail and typically determines the Vmax for a technology. DC characterization for TDDB has been the corner stone for all technology qualifications. For customers looking to over drive the base technology, understanding TDDB benefit if any under AC stress becomes crucial for any design-technology co-optimization. Device level AC characterization has shown such benefit. In Fig. 13 static vs dynamic stress in an RO is compared using a voltage ramp stress (VRS) methodology [33]. Note that in a typical ramped stress both device and RO are limited by NFET TDDB due to the higher VAE for NFET TDDB. The obtained slope for RO VBD is consistent with NFET breakdown and ~100mV enhancement in VBD is observed under AC conditions.

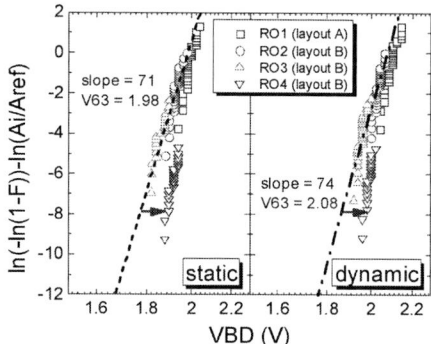

Figure 13 VRS breakdown study on RO with different number of stages demonstrates good area scaling and a VBD slope consistent with SG NFET TDDB.

VI. CONCLUSIONS

Future products require continued silicon scaling in conjunction with differentiated silicon and system level integration. New material and architectural solutions are being explored for sub 7nm. Key challenges for reliability remains maintaining a good quality IL/channel interface, reduced charge trapping and defect generation under operating bias. New challenges like self-heating in FINFET/FDSOI geometry, modeling variability and circuit level assessment require

innovative test structure design to enable accurate modeling of reliability limitations. Process variability for tight metal pitches will dominate failure rate in BEOL TDDB while performance enhancement elements may lead to local weakness and reduced lifetime for FEOL TDDB. For sub-1.0V application, we need to reassess which defect generation mechanisms are likely to dominate EOL.

ACKNOWLEDGMENT

The authors would like to thank QRA team/management at GLOBALFOUNDRIES for their support and guidance.

REFERENCES

[1] J. G. Pellerin, , "Dimensions of Innovation to Enable the Next Era of Intelligent Systems," Electron Device Technology and Manufacturing Conference, , Feburary 2017.

[2] Eduard Cartier and Andreas Kerber, "Stress-induced leakage current and defect generation in nFETs with HfO2/TiN gate stacks during positive-bias temperature stress", IEEE International Reliability Physics Symposium, 2009, pp.486-4923.

[3] A. Kerber; E. Cartier; L. Pantisano; R. Degraeve; T. Kauerauf; Y. Kim; A. Hou; G. Groeseneken; H. E. Maes; U. Schwalke, "Origin of the threshold voltage instability in SiO2/HfO2 dual layer gate dielectrics," Volume: 24, Issue: 2, IEEE Electron Device Letters, 2003, pp. 87-89.

[4] T. Nigam; A. Kerber; P. Peumans, "Accurate model for time-dependent dielectric breakdown of high-k metal gate stacks," IEEE International Reliability Physics Symposium, 2009, pp. 523-530.

[5] A. Kerber; E. Cartier; B. P. Linder; S. A. Krishnan; T. Nigam, "TDDB failure distribution of metal gate/high-k CMOS devices on SOI substrates," IEEE International Reliability Physics Symposium, 2009, pp. 505-509.

[6] S. Krishnan; U. Kwon; N. Moumen; M. W. Stoker; E. C. T. Harley; et al.,, "A manufacturable dual channel (Si and SiGe) high-k metal gate CMOS technology with multiple oxides for high performance and low power applications," International Electron Devices Meeting, 2011, 28.1.1-28.1.4.

[7] P. Srinivasan; J. Fronheiser; S. Siddiqui; A. Kerber; L. F. Edge; R. G. Southwick; E. Cartier, "NBTI in Si0.5Ge0.5 RMG gate stacks — Effect of high-k nitridation," IEEE International Reliability Physics Symposium, 2009, pp. 2F.5.1-2F.5.6.

[8] Jacopo Franco; Ben Kaczer; Jérôme Mitard; María Toledano-Luque; Philippe J. Roussel; Liesbeth Witters; Tibor Grasser; Guido Groeseneken, "NBTI Reliability of SiGe and Ge Channel pMOSFETs With SiO2/HfO2 Dielectric Stack," Volume: 13, Issue: 4, IEEE Transactions on Device and Materials Reliability, 2013, pp. 497-506.

[9] T. Nogami, M. Chae3, C. Penny, T. Shaw1, H. Shobha, J. Li, S. Cohen1, et al., "Performance of Ultrathin Alternative Diffusion Barrier Metals for Next-Generation BEOL Technologies, and their Effects on Reliability," International Interconnect Technology Cconference, 2014.

[10] X. H. Liu; T. M. Shaw; M. W. Lane; E. G. Liniger; B. W. Herbst; D. L. Questad, "Chip-Package-Interaction Modeling of Ultra Low-k/Copper Back End of Line," International Interconnect Technology Conference, June 2007.

[11] C Prasad, L Jiang, D Singh, M Agostinelli, C Auth, P Bai, T Eiles, J Hicks,et al., "Self-heat reliability considerations on Intel's 22nm Tri-Gate technology," IEEE International Reliability Physics Symposium, 2013, pp. 5D.1.1-5D.1.5.

[12] Changze Liu ; Hyun-Chul Sagong ; Hyejin Kim ; Seungjin Choo ; Hyunwoo Lee ; Yoohwan Kim, et al., "Systematical study of 14nm FinFET reliability: From device level stress to product HTOL," IEEE International Reliability Physics Symposium, 2015, pp. 2F3.1-2F3.5.

[13] E. Stewart, I.I.I. "Rauch, Review and reexamination of reliability effects related to NBTI-induced variations, IEEE Trans. Device Mater. Reliab. Vol. 7, No. 4, pp. 524–530, 2007.

978-1-5090-5979-9/17 $31.00 © 2017 IEEE

[14] A. Kerber, "Methodology for Determination of Process Induced BTI Variability in MG/HK CMOS Technologies Using a Novel Matrix Test Structure", IEEE Electron Device Letters, Vol. 35, No. 3, pp. 294-296, 2014.

[15] B. Kaczer, T. Grasser, Ph. J. Roussel, J. Franco, R. Degraeve, L.-A. Ragnarsson, E. Simoen, G. Groeseneken, H. Reisinger, "Origin of NBTI Variability in Deeply Scaled pFETs", Proc. Int. Rel. Phys. Symp., pp. 26-32, 2010.

[16] H. Reisinger, O. Blank, W. Heinrigs, A. Mühlhoff, W. Gustin, and C. Schlünder, "Analysis of NBTI degradation- and recovery-behavior based on ultra fast VT-measurements", in Proc. Int. Rel. Phys. Symp., pp. 448–453, 2006.

[17] A. Kerber, "Assessment of sense measurement duration on BTI projection in MG/HK CMOS technologies using a novel stacked transistor test structure", submitted to Microelectronics Reliability.

[18] A. Kerber, X. Wan, Y. Liu, T. Nigam, "Fast wafer-level stress-and-sense methodology for characterization of Ring-Oscillator degradation in advanced CMOS technologies", IEEE Trans. Electron Devices, vol. 62, no. 5, pp. 1427 - 1432, 2015.

[19] K. Mistry, C. Allen, C. Auth, B. Beattie, D. Bergstrom, M. Bost, M. Brazier, M. Buehler, et al., "A 45nm Logic Technology with High-k+Metal Gate Transistors, Strained Silicon, 9 Cu Interconnect Layers, 193nm Dry Patterning, and 100% Pb-free Packaging", Digest International Electron Device Meeting, pp. 247-250, 2007.

[20] Chudzik, M., et. al. "High-Performance High-K/Metal Gates for 45nm CMOS and Beyond with Gate-First Processing", Technical Digest of the IEEE Symposium on VLSI Technology, 2007, Kyoto, Japan, Session 11A-1.

[21] S. Venkatesan; A.V. Gelatos; S. Hisra; B. Smith; R. Islam; J. Cope; B. Wilson; D. Tuttle; et al., "A high performance 1.8 V, 0.20 /spl mu/m CMOS technology with copper metallization,"Digest International Electron Device Meeting 1997, Page(s):769- 772.

[22] M. Heyns; A. Alian; G. Brammertz; M. Caymax; G. Eneman; J. Franco; et al., "Challenges for introducing Ge and III/V devices into CMOS technologies," IEEE International Reliability Physics Symposium, 2015, pp. 5D1.1-5D1.10.

[23] A. Vais; A. Alian; L. Nyns; J. Franco; S. Sioncke; V. Putcha; et al., "Record mobility (μeff ~3100 cm2/V-s) and reliability performance (Vov~0.5V for 10yr operation) of In0.53Ga0.47As MOS devices using improved surface preparation and a novel interfacial layer," IEEE Symposium on VLSI Technology, 2016.

[24] C. Auth, C. Allen, A. Blattner, D. Bergstrom, M. Brazier, M. Bost, M. Buehler, V. Chikarmane, T. Ghani, et al., "A 22nm High Performance and Low-Power CMOS Technology Featuring Fully-Depleted Tri-Gate Transistors, Self-Aligned Contacts and High Density MIM Capacitors",VLSI Technology Symposium, pp. 131-132, 2012.

[25] S. Ramey, A. Ashutosh, C. Auth, J. Clifford, M. Hattendorf, J. Hicks, R. James, A. Rahman, V. Sharma, A. St Amour, C. Wiegand, "Intrinsic Transistor Reliability Improvements from 22nm Tri-Gate Technology", Proc. Int. Rel. Phys. Symp., pp. 4C.5.1 - 4C.5.5, 2013.

[26] A. Kerber, P. Srinivasan, S. Cimino, P. Paliwoda, S. Chandrashekhar, Z. Chbili, S. Uppal, R. Ranjan, M.-I. Mahmud, D. Singh, P.P. Manik, J. Johnson, F. Guarin, T. Nigam, B. Parameshwaran, "Device reliability metric for End-Of-Life performance optimization based on circuit level assessment", Proc. Int. Rel. Phys. Symp., pp. 2D.3.1 – 2D.3.8, 2017.

[27] L. T. Su, J. E. Chung, D. A. Antoniadis, K. E. Goodson, and M. I. Flik, "Measurement and modeling of self-heating in SOI MOSFET's," IEEE Trans. Electron Devices, vol. 41, no. 1, pp. 69–75, 1994.

[28] S. Mittl and F. Guarin, "Self-Heating and Its Implications on Hot Carrier Reliability Evaluations", Proc. Int. Rel. Phys. Symp., pp. 4A.4.1 - 4A.4.6, 2015.

[29] H. Mertens,; R. Ritzenthaler, A. Chasin, T. Schram, E. Kunnen, A. Hikavyy, L.-Å. RagnarssonGAA et al., "Vertically Stacked Gate-All-Around Si Nanowire CMOS Transistors with Dual Work Function Metal Gates,", Digest International Electron Device Meeting, pp. 247-250, 2015.

[30] A. V-Y Thean ; D. Yakimets ; T. Huynh Bao ; P. Schuddinck ; S. Sakhare ; M. Garcia Bardon, et al., "Vertical device architecture for 5nm and beyond: Device & circuit implications," IEEE Symposium on VLSI Technology, 2015.

[31] A. Kerber, "Reliability of Metal Gate / High-k devices and its impact on CMOS technology scaling" (invited), MRS Spring Symposium, in press, 2017.

[32] R. Ranjan, S. Uppal, H. Yu, B. Parameshwaran, T. Nigam, A. Kerber, and C. LaRow, and M.I. Natarajan, "Impact of e-SiGe S/D processes on PFET TDDB Reliability," Electron Device Technology and Manufacturing Conference, , Feburary 2017.

[33] A. Kerber and T. Nigam, "Application of CVS and VRS method for correlation of logic CMOS wear out to discrete device degradation based on Ring Oscillator circuits", VLSI Technology Symposium, pp. 44-45, 2016.

Carrier lifetime evaluation in FD-SOI layers

K. H. Lee, M. Bawedin, H-J. Park, M. Parihar, S. Cristoloveanu

IMEP-LAHC, Grenoble Institute of Technology, Minatec
Grenoble, France
leek@minatec.grenoble-inp.fr

Abstract— Virtual P-N diodes are emulated in undoped SOI films by biasing the front and back gates such as to induce electrostatically doped regions. The I-V curves are diode-like and can be engineered via gate voltage. We exploit the characteristics of the virtual diode for lifetime characterization, which was considered as a very challenging task in ultrathin SOI films. Two original methods are proposed and compared based on systematic experiments and simulations. It is shown that the carrier lifetime is very short in 7 nm thick SOI layers.

Keywords— Silicon-on-insulator (SOI); p-n diode; virtual doping; electrostatic doping; lifetime

I. INTRODUCTION

The carrier lifetime is a very informative parameter for monitoring the crystal quality and purity of semiconductors. It also governs the operation of a majority of devices: leakage current in MOSFETs, gain of bipolar transistors, switching capability of power devices, performance of photo-detectors and solar cells, etc. In fully depleted SOI (FD-SOI) MOSFETs, the carrier lifetimes also control the activation and amplitude of floating-body mechanisms such as kink effect, parasitic bipolar transistor, transient currents and memory mechanisms [1-3]. It is of paramount importance in the operation of SOI band-modulation devices [4-11] as it defines the sharp switch and hysteresis features.

The difficulty to measure the carrier lifetime in ultrathin devices explains the lack of published data. Pulsing an MOS capacitor into deep depletion (Zerbst method [12]) does not work simply because the depletion depth is restricted by the film thickness. Techniques developed for thicker SOI MOSFETs consist in monitoring the drain current of minority carriers as a function of time during the gradual build-up of a majority carrier layer [13]. These methods fail in sub-10 nm thick films where the super-coupling principle forbids the co-existence of electrons and holes in the transistor body [14]. In addition, the lifetime is very short in FD-SOI, being dominated by the proximity of interfaces

$$\frac{1}{\tau_{eff}} = \frac{1}{\tau_v} + \frac{S_1 + S_2}{t_{si}}$$

(1)

where τ_{eff} is the effective lifetime, τ_v is the volume lifetime and $S_{1,2}$ are surface generation-recombination velocities.

In this work, we introduce a novel method for lifetime measurement in FD-SOI devices. The test device is an astonishing virtual diode shown in Fig. 1. It consists in a PIN diode with undoped body controlled by front and back MOS gates. If the gates are unbiased, the device operates as a regular PIN diode (Fig. 1a). Negative bias V_{GF} on front gate and positive bias V_{GB} on the back gate emulate a P-N junction, out of nothing, in the middle of the body. We have recently reported diode-like I-V characteristics measured on this virtual

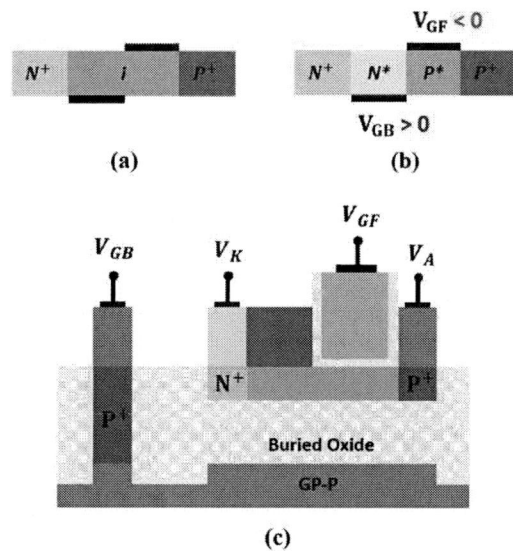

Fig.1 (a) PIN diode without gate biasing. (b) Front-gate and back-gate voltages induce free electrons N* and holes P* that form a virtual P-N diode. (c) Configuration of the fabricated FD-SOI device.

diode [15]. Our diode is an extension of the plasma charge diode proposed earlier [16-18] in the sense that the concentrations of 'electrostatic' doping N* and P* are adjustable on demand via gate biasing. Our goal is to take advantage of the flexibility of the virtual diode in order to extract the carrier lifetime.

II. DEVICE FABRICATION AND TYPICAL CHARACTERISTICS

The devices have been processed with the 28 nm FD-SOI technology at STMicroelectronics [19]. They feature high-

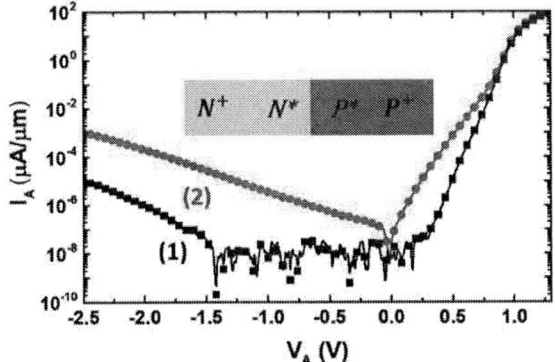

Fig. 2 Typical current-voltage characteristic of the virtual diode. (1) V_{GF} = − 1.2 V, V_{GB} = + 5 V, (2) V_{GF} = − 2 V, V_{GB} = + 10 V. The reverse current is in absolute value. L_g = L_{in} = 500 nm, width = 1 μm, T = 300 K.

K/metal gate stack with 3.7 nm equivalent oxide thickness, 7 nm thick SOI film and 25 nm thick BOX. The thickness of source, drain and gate-underlap regions has been completed by epitaxy to reduce series resistances. The BOX separates the undoped film (residual doping N_A = 10^{15} cm^{-3}) from a highly doped P-type ground plane (N_A = 10^{18} cm^{-3}) acting as a back-gate. We consider relatively long diodes with equal length of gated and ungated regions (L_g = L_{in} = 500 nm).

The reverse and forward currents of virtual diodes were measured for different combinations of gate and ground-plane biases at room temperature. The experimental I-V curves (Fig. 2) look familiar to a conventional diode [20, 21] with ion-implanted dopants. However, a detailed analysis reveals significant differences exposed in [22].

III. LIFETIME EXTRACTION

The reverse current cannot be captured by the classical theory of the P-N diode [20]. Unlike the physical diode, where the doping is fixed, there is a dynamic change in P* electrostatic doping with anode voltage. For a given negative

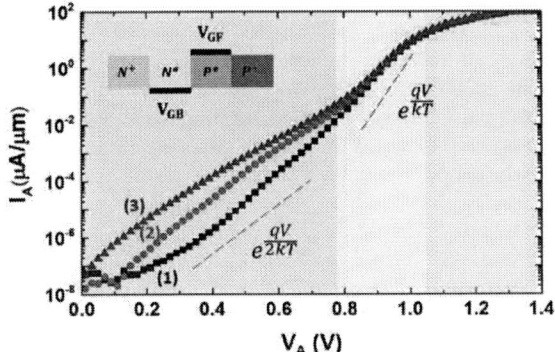

Fig. 3 Forward current-voltage characteristics of the virtual diode measured for variable bias on front and back gates. (1) V_{GF} = − 2 V, V_{GB} = + 5 V, (2) V_{GF} = − 1.2 V, V_{GB} = + 10 V, (3) V_{GF} = − 2 V, V_{GB} = + 15 V.

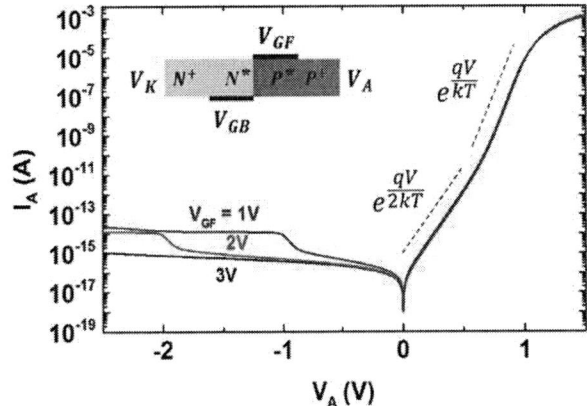

Fig. 4 Simulated current-voltage characteristics of a double-gated virtual diode (inset). The increase in reverse current is due to the full depletion of the P* electrostatic doping. V_{GF} = − V_{GB}.

gate voltage, increasing the anode reverse bias decreases the P* concentration. Ultimately, for $|V_A| \sim |V_{GF}|$, the holes cannot be sustained under the gate and the region becomes 'intrinsic'.

We exploit the forward I-V curves which are more familiar. Fig. 3 shows the typical regions of operation of a P-N diode: recombination current, diffusion current, high-level injection and series resistances [21]. This behavior is reproduced by numerical simulations with TCAD Sentaurus tools, as illustrated in Fig. 4.

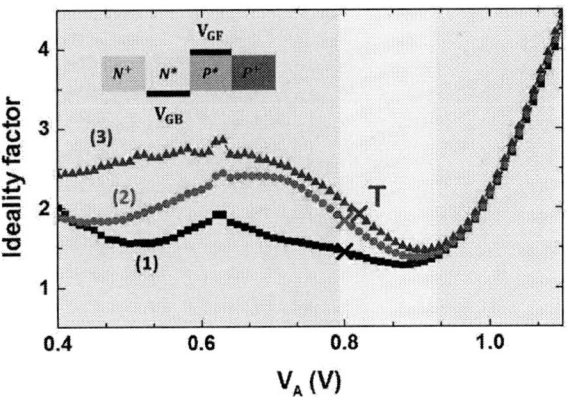

Fig. 5 Ideality factor extracted from Fig. 3 versus forward voltage. Point T indicates the transition from recombination current to diffusion current. (1) V_{GF} = − 2 V, V_{GB} = + 5 V, (2) V_{GF} = − 1.2 V, V_{GB} = + 10 V, (3) V_{GF} = − 2 V, V_{GB} = + 15 V.

At low bias, the recombination current prevails and follows the conventional law:

$$I_R = I_{R0} \cdot \exp(\frac{qV}{2kT})$$

(2)

where $I_{R0} = (q \cdot W_D \cdot n_i)/\tau$ and W_D is the depletion width, n_i the intrinsic carrier concentration and τ the recombination lifetime.

The diffusion current dominates at higher voltage:

$$I_D = I_{D0} \cdot \exp(\frac{qV}{kT}) \tag{3}$$

where $I_{D0} = (q \cdot n_i^2 \cdot L_p)/(\tau \cdot N_D \cdot \tanh(W/L_p))$ and $L_{n,p}$ are the diffusion lengths of electrons or holes.

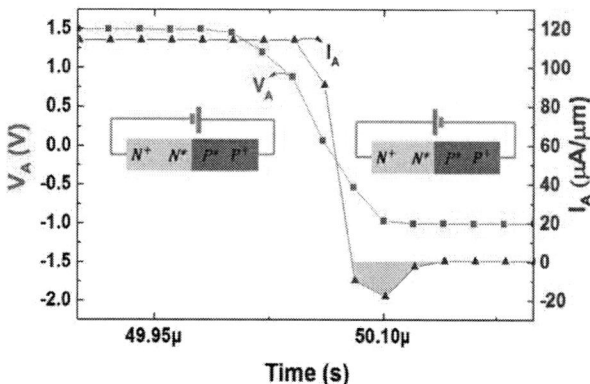

Fig. 6 Switching of a diode from forward bias to reverse bias. $V_{GF} = -2$ V, $V_{GB} = +10$ V.

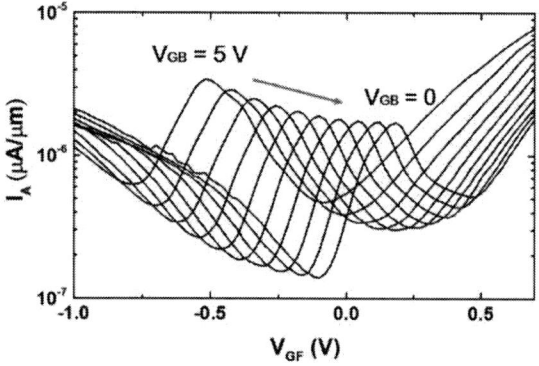

Fig. 7 Experimental forward current versus gate voltage in a FD p–i–n gated diode. $V_A = +0.4$ V, $V_C = 0$.

We focus on the transition between recombination current and diffusion current. Figure 5 shows the variation of the ideality factor n, calculated from $I = I_0 \ (e^{qV/nkT} - 1)$, between $\sim$ 2 (recombination) and $\sim$ 1 (diffusion). The transition voltage V^*, for which the two currents are equal, is defined as the inflection point. Equating Eqs. 2 and 3 yields:

$$\frac{L_p \cdot n_i}{N_D \cdot \tanh(\frac{L_g}{L_p})} \cdot \exp(\frac{V_A}{2kT}) = W_D \tag{4}$$

Eq. 4 is used to determine the lifetime. This method has first been validated by simulating physical diodes with implanted doping. The lifetime extracted from Eq. 4 matches the value introduced in the simulator.

The same methodology is used for the virtual diode. The electrostatic doping needed in Eq. 4 is assumed to be the average concentration of gate-induced free carriers, derived from the MOS theory:

$$N^* = \frac{C_{BOX}(V_{GB} - V_{TB,n})}{qt_{Si}} \tag{5}$$

and

$$P^* = \frac{C_{OX}(V_{GF} - V_{TF,p} - V^*)}{qt_{Si}} \tag{6}$$

For the gate voltage used in Fig. 5, the average doping is in the range of 10^{19} cm^{-3} leading to $\sim$10 nm width of the space charge region. The lifetime extracted from the measurements shown in Fig. 5 is 1–10 ns.

A second method is to monitor the recovery time of the virtual diode. The diode is pulsed from forward mode to reverse mode and the current is recorded as a function of time. Figure 6 shows a signal variation similar to the curves reported in conventional doped diodes. The lifetime is here given by:

$$\frac{Q_B(t)}{Q_B(0)} = (1 + \frac{I_R}{I_F}) \cdot \exp(\frac{-t}{\tau_n}) - \frac{I_R}{I_F} \tag{7}$$

where $Q_B(t)$ is the excess minority charge stored in the base region, $Q_B(0)$ is the excess minority charge just after the diode is switched from forward bias to reverse bias.

According to this experiment, the lifetime is $\sim$10 ns.

IV. DISCUSSION

For validation purposes, we have performed a third experiment by adapting the method proposed by Ernst *et al* [17] to ultrathin films. The test device is here a PIN diode with the gate extending all the way from cathode to anode. The front and back gates are biased such as the in-depth potential profile is quasi-flat and the carrier concentrations are constant in the whole body volume. The forward current is measured at low anode bias (where the recombination mechanism dominates) as a function of gate voltage. Clear peak values of current are visible in Fig. 7. They correspond to maximum carrier recombination which occurs when the quasi-Fermi levels are equidistant from the mid-gap and the concentrations of electrons and holes become equal: $n = p = n_i \cdot \exp(qV/2kT)$.

The excess peak current ΔI_F is obtained by simply multiplying the maximum recombination rate $R_{max} = n/\tau$ by the volume of the body [22, 23]:

$$\Delta I_F = \frac{qn_i}{\tau_n} \cdot \exp\left(\frac{qV_A}{2kT}\right) \cdot \left(L \cdot W \cdot t_{si}\right) \tag{8}$$

The lifetime calculated from Fig. 7 with Eq. 8 is ~10 ns.

All three methods indicate very short lifetime in ultrathin SOI films. Earlier measurements in thicker FD-SOI devices, where the interface contribution is less severe, resulted in lifetime values in the 0.1-1 µs range [22].

The virtual diode appears as a new, highly original tool for ultrathin film characterization. The lifetime has been extracted with first-order models, borrowed from the diode theory. They can be customized by considering the special 2D behavior of the virtual diodes. For example, the in-depth variation of electrostatic doping and extension of the lateral space charge region is not captured in Eqs. 4-6.

V. CONCLUSION

This work proposes a solution to the problem of lifetime evaluation in ultrathin FD-SOI films. We have introduced two parameter extraction methods that take advantage of the flexibility and properties of the gate-controlled virtual diode. The carrier lifetime appears to be very short in sub-10 nm thick SOI devices. Based on our results, a realistic rather than guessed value for the carrier lifetime can now be introduced in TCAD tools for the simulation of switching, memory and power SOI devices.

ACKNOWLEDGMENT

The virtual diode has been discovered during the accomplishment of the European Union project REMINDER. Thanks are due to our partners, especially to Dr. P. Fonteneau, Dr. Ph. Galy and Mr. H. El Dirani.

REFERENCES

[1] Liu, F., Ionica, I., Bawedin, M., and Cristoloveanu, S, "Extraction of the parasitic bipolar gain using the back-gate in ultrathin FDSOI MOSFETs," IEEE Electron Device Letters, pp. 96-98, 36(2), 2015.

[2] Cristoloveanu, S. and Li. S, Electrical Characterization of Silicon-On-Insulator Materials and Devices. Springer, 1995.

[3] Colinge, J.-P., Silicon-on-insulator Technology: Materials to VLSI, Springer Science & Business Media, 2013.

[4] Parihar, M. S., Lee, K. H., Bawedin, M., Lacord, J., Martinie, S., Barbé, J.-C., Xu, Y., Taur, Y., and Cristoloveanu, S. "Impact of Carrier Lifetime on Z²-FET Operation," Joint International EUROSOI Workshop and International Conference on Ultimate Integration on Silicon (ULIS), 2017.

[5] Cristoloveanu, S., Wan, J., and Zaslavsky, A. "A review of sharp-switching devices for ultra-low power applications," IEEE J. Electron Devices Society, pp. 215-226, 4(5), 2016.

[6] Wan, J., Cristoloveanu, S., Le Royer, C., and Zaslavsky, A. "A feedback silicon-on-insulator steep switching device with gate-controlled carrier injection," Solid-State Electronics, pp. 109-111, 76, 2012.

[7] El Dirani, H., Fonteneau, P., Solaro, Y., Legrand, C. A., Marin-Cudraz, D., Ferrari, P., and Cristoloveanu, S. "Sharp-switching band-modulation back-gated devices in advanced FDSOI technology," Solid-State Electronics, pp. 180-186, 128, 2017.

[8] Salman, A. A., Beebe, S. G., Emam, M., Pelella, M. M., and Ioannou, D. E. "Field Effect Diode (FED): A novel device for ESD protection in deep sub-micron SOI technologies," Electron Devices Meeting, pp. 1-4, 2006.

[9] Wan, J., Le Royer, C., Zaslavsky, A., and Cristoloveanu, S. "A compact capacitor-less high-speed DRAM using field effect-controlled charge regeneration," IEEE Electron Device Letters, pp. 179-181, 33(2), 2012.

[10] Wan, J., Le Royer, C., Zaslavsky, A., and Cristoloveanu, S. "Progress in Z²-FET 1T-DRAM: retention time, writing modes, selective array operation, and dual bit storage," Solid-State Electronics, pp. 147-154, 84, 2013.

[11] Solaro, Y., et al. "Innovative ESD protections for UTBB FD-SOI technology," IEEE International Electron Devices Meeting, 2013.

[12] M. Zerbst, "Relaxation effects at semiconductor-insulator interfaces," Z. Angrew. Phys., p.30, 22, 1966.

[13] D. E. Ioannou, Cristoloveanu, S., Mukherjee, M., and Mazhari, B., "Characterization of carrier generation in enhancement-mode SOI MOSFETs," IEEE Electron Device Letters, 11(9), p. 409, 1990.

[14] Eminente, S., Cristoloveanu, S., Clerc, R., and Ghibaudo G, "Ultra-thin fully depleted SOI MOSFETs : special charge properties and coupling effects," Solid-State Electronics, pp. 239-244, 51(2), 2007.

[15] Lee, K. H., Bawedin, Park, H-J., Singh Parihar, M., and Cristoloveanu, S. "A virtual SOI diode with electrostatic doping," Joint International EUROSOI Workshop and International Conference on Ultimate Integration on Silicon (ULIS), 2017.

[16] Hueting, R. J., et al. "The charge plasma pn diode," IEEE Electron Device Letters, pp. 1367-1369, 29(12), 2008.

[17] Rajasekharan, B., Hueting, R. J., Salm, C., van Hemert, T., Wolters, R. A., and Schmitz, J. "Fabrication and characterization of the charge-plasma diode," IEEE Electron Device Letters, pp. 528-530, 31(6), 2010.

[18] Singh, P., and Pandey, S. "Work function engineered charge plasma diodes for enhanced performance." Journal of Physics D: Applied Physics, 48(49), 2015.

[19] El Dirani, H., Solaro, Y., Fonteneau, P., Ferrari, P., and Cristoloveanu, S. "Sharp-switching Z²-FET device in 14 nm FDSOI technology," Solid State Device Research Conference (ESSDERC), pp. 250-253, 2015.

[20] Taur, Y., and Tak H. Ning. Fundamentals of modern VLSI devices. Cambridge University Press, 2013.

[21] Grove, A. S. Physics and Technology of Semiconductor Devices. Wiley, 1967.

[22] Ernst, T., Vandoorren, A., Cristoloveanu, S., Colinge, J.-P., and Flandre, D. "Carrier lifetime Extraction in Fully Depleted Dual-Gate SOI devices," IEEE Electron Device Letters, 20(5), 1999.

[23] Ernst, T., Cristoloveanu, S., Vandooren, A., Rudenko, T., Colinger, J-P. "Recombination current modeling and carrier lifetime extraction in dual-gate fully-depleted SOI devices," IEEE Trans. Electron Devices, pp.1503-1509, 46(7), 1999.

Precise EOT regrowth extraction enabling performance analysis of Low Temperature Extension First devices

J. Micout[1,2], Q. Rafhay[1], X. Garros[2], M. Cassé[2], J. Coignus[2], L. Pasini[1,2,3], C.-M. V. Lu[2], N. Rambal[2], C. Fenouillet-Beranger[2], L. Brunet[2], G. Romano[3], R. Gassilloud[2], P. Batude[2], M. Vinet[2] and G. Ghibaudo[1]

[1]IMEP-LAHC, MINATEC/INPG, Univ. Grenoble Alpes, F-38016, Grenoble, France
[2]CEA, Leti, MINATEC Campus, Univ. Grenoble Alpes, F-38054, Grenoble, France
[3]STMicroelectronics, 850 rue Jean Monnet, F-38926, Crolles Cedex, France
jessy.micout@cea.fr

Abstract— **3D sequential integration requires top FETs processing with a low thermal budget (500°C). The analysis of the origin of the performance difference between Low Temperature (LT) MOSFET and high temperature standard process must take into account a potential EOT modification for short gate lengths. In this work, the difficulty of precise EOT extraction for scaled devices is observed by CV measurements and an alternative methodology using IV measurements is proposed. This methodology has been applied to an extension first integration, and the extraction accuracy is high enough to conclude to an EOT regrowth for the low temperature nFETs only. Thus, the origin of performance degradation between LT and HT, previously attributed to larger access resistance, highlights also a detrimental role of gate stack instability. The origin of this variation is attributed to oxygen ingress, through the thin extension first liner which should be suppressed by minor process optimizations.**

I. INTRODUCTION

3D sequential integration is an alternative approach to conventional scaling. It consists in fabricating stacked layers of devices, sequentially one after the other, and enables to obtain the highest density of contact between the stacked layers (e.g. $10^8/mm^2$ for a 14nm node technology) [1]. To develop this integration, the thermal budget of the top transistor has to be reduced down to 500°C in order to preserve the underlying devices. The main challenge in thermal budget reduction concerns the thermal dopant activation usually made at temperatures around 1050°C. Low Temperature (LT) activation using Solid Phase Epitaxy Recrystallization (SPER) at 600°C has been shown to lead to device performance close from state of the art devices [2], as seen in **Figure 1**.

In this previous study, the performance degradation has been attributed to access resistance degradation, assuming a constant EOT with scaled gate lengths. Indeed EOT has been only extracted on large device in order to have a sufficient signal.

However an EOT variation for the shortest length devices could be observed and this regrowth might be different for the high temperature and low temperature process. As a reminder, a 1 Å difference between HT POR and LT splits is expected to leads to 10% performance degradation. This potential EOT regrowth might extend the original conclusion obtained on the performance degradation of LT process, presented in our previous work [2].

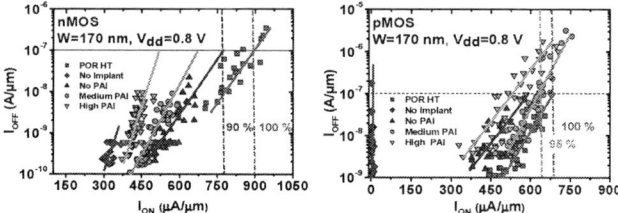

Figure 1: I_{on} I_{off} trade-off for low temperature extension first devices compared to high temperature process of reference [2]

In this in-depth investigation, the eventual EOT regrowth has been evaluated using both CV and IV measurements. The first section describes the device fabrication process; the second and third section presents the CV and IV results and the last section details the interpretation using models of gate current tunneling.

II. DEVICE FABRICATION

The FDSOI devices are fabricated on 6nm Si and SiGe$_{27\%}$ thick channels with 20nm Buried Oxide, followed by HfO$_2$/ TiN/Poly-Si gate stack. Modifications appear afterwards, at the beginning of the junction formation. More specifically, **Figure 2** shows the difference between the HT POR and the extension first (X^{1st}) flow. The Low Temperature devices have been processed using the X^{1st} flow (**Figure 2b**)). This process aims at localizing dopants at the gate edge as this low temperature process allows no dopant diffusion. Indeed, instead of depositing and etching a 6 nm Offset Spacer as for the POR HT case, implantations are carried out directly at the channel entrance through a 3 nm thin nitride liner. After the implantation, a nitride deposition is performed in order to reach the final POR spacer thickness. This thin and implanted (thus damaged) liner can be oxidized and thus could act as an oxygen reservoir very close to the gate edge.

978-1-5090-5979-9/17 $31.00 © 2017 IEEE

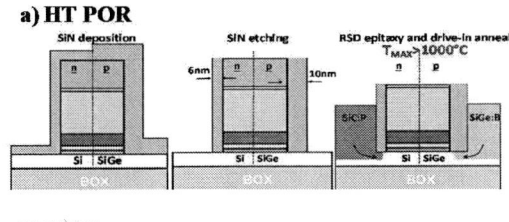

a) HT POR

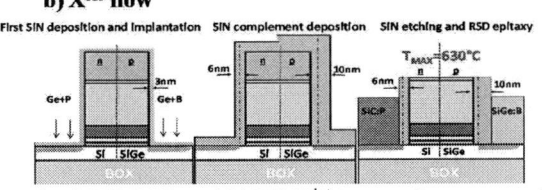

b) X^{1st} flow

Figure 2: a) HT POR and b) LT X^{1st} integration process flow.

It is therefore possible that this process flow could lead to an EOT regrowth for short devices (L$_g$<100nm), by oxygen ingress. To understand if the thin liner and/or the implantation can lead the EOT to increase, three technological splits will be compared, i.e.:

- The HT POR: standard FDSOI integration
- The LT Implant: X^{1st} integration with implanted liner
- The LT No Implant: X^{1st} integration without implanted liner

III.   EOT EXTRACTION BY C-V MEASUREMENTS

The EOT absolute value, extracted by CV measurements at 90 KHz, is obtained for all the gate length by fitting the curve between the measured capacitance and the simulated one, using [3].The differentiation of CV values for two close gate lengths enable to remove parasitic capacitance and to remove gate length uncertainties compared to the mask dimension. Moreover, an interdigitated transistor structure enables to measure the capacitance for the shortest gate lengths. **Figure 3** shows the C-V and the absolute EOT which is deduced for both n&p MOS. The differentiation has been carried out for the following couples of gate lengths: $\Delta L_1 = 150$ nm – 60 nm, and $\Delta L_2 = 26$ nm – 24 nm.

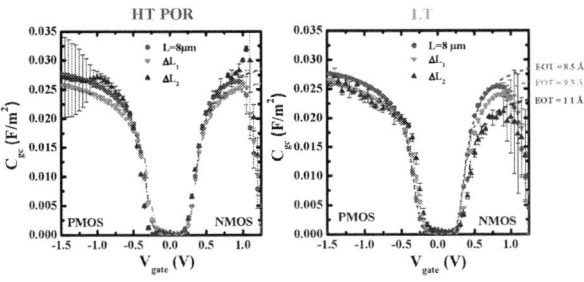

Figure 3: C-V graph and the deduced absolute EOT values

While for long channel the dispersion of the data is suitable for a correct extraction of EOT, the dispersion is too large for short channels. For example, in **Figure 3**, for the nMOS POR HT split, $\Delta EOT = 1$ Å ± 1 Å, and for the nMOS LT No implant

split, $\Delta EOT = 2$ Å ± 1 Å. A conclusion about an EOT variation is therefore impossible using CV measurement. In addition, for pMOS HT POR, the fit for the shortest gate length is impossible due to the large dispersion.

The problem of the large dispersion could be explained by 1/ the linear dependency of the oxide capacitance with the EOT and by 2/ a too large gate leakage.

This variation is thus not satisfying to fully conclude about an EOT regrowth due to the LT process flow, and another method is proposed in this study.

IV.   EOT EXTRACTION BY I-V MEASUREMENTS

In the simple direct Fowler-Nordheim model, the gate current is linked to the EOT following equation (1):

$$I_g = W.L.A.e^{\left(\frac{-EOT}{B}\right)} \quad (1)$$

where A and B are constants depending on material parameters, but independent of the gate length L and the width W. Because of the exponential dependency, the measured gate current is expected to lead to a larger signal in case of EOT regrowth. It is however assumed here that any variation of the material parameters (like gate dielectric constant or effective masses) is interpreted as an EOT variation. The impact will however be the same at the performance level.

To ensure unbiased extraction, outlier data are at first removed by using Grubbs-Smirnov's statistical test [4].Then, to avoid the impact of short channel effects, the gate current is measured at low drain voltage (V$_d$ = 25 mV). In addition, to avoid trap-assisted current leakage, the gate current is taken at high V$_g$. Finally, as no gate resistance can distort the measurements as I$_g^{max}$<100 µA, the highest value of available gate voltage is taken, which is about 1.7 V. **Figure 4** shows the average gate current I$_g^{lin}$ versus the gate voltage V$_g$ measurements, for both HT POR and LT splits and for three gate lengths.

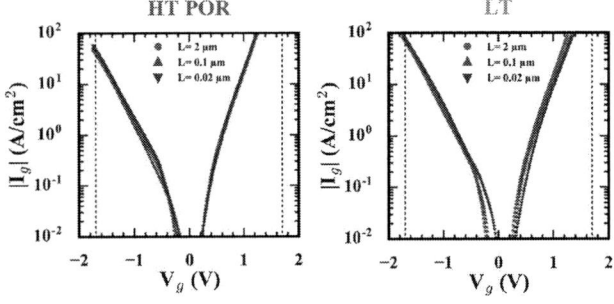

Figure 4: I-V graph

The comparison of the averaged I$_g^{lin}$ as a function of the gate length allows to identify if any degradation occurred. This degradation could be explained by 1/ an EOT variation, 2/ a difference between the mask and the effective channel length or 3/ a difference of the potential between the overlap and the channel, which will impact the gate current in the same manner as 2/. These three hypotheses will be studied in the following paragraphs.

978-1-5090-5979-9/17 $31.00 © 2017 IEEE   145

At first, it has to be noted that with a lithography precision estimated around ±2 nm in the worst case scenario, the hypothesis 2/ would lead at maximum to a 20% degradation of I_g^{lin} for the shortest gate length.

Then, assuming that there is no EOT regrowth and using the very simple direct tunneling model, 2/ and 3/ would influence the gate current according to equation (2):

$$I_g(L) = W.(L + \Delta L).A.e^{\left(-\frac{EOT}{B}\right)} + I_{g,overlap} \quad (2)$$

Using a differential method with respect to the gate length, would lead to equation (3), which becomes independent of the gate length:

$$I_g^{new}\left(\frac{L_2 + L_1}{2}\right) = \frac{\partial I_g}{\partial L} = \frac{I_g(L_2) - I_g(L_1)}{L_2 - L_1} = W.A.e^{\left(-\frac{EOT}{B}\right)} \quad (3)$$

Therefore, if a monotonic gate current variation is caused by an extra gate length ΔL, applying the differential method would give an I_g^{new} parameters independent of L.

On the contrary, if the gate current variation is cause by an EOT variation with L, the differential method would give I_g^{new} values still varying with L, according to:

$$I_g^{new}\left(\frac{L_2 + L_1}{2}\right) = W.A.e^{\left(-\frac{EOT(L)}{B}\right)}\left(1 - \frac{L + \Delta L}{B}\frac{\partial EOT(L)}{\partial L}\right) \quad (4)$$

In that situation, EOT(L) could be extracted more simply from the logarithm of I_g per unit area, if a value of B is known.

a) Classical method

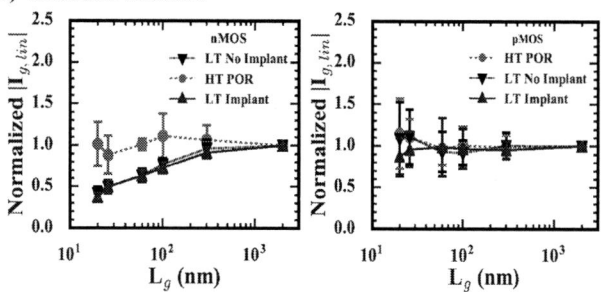

b) Differential method

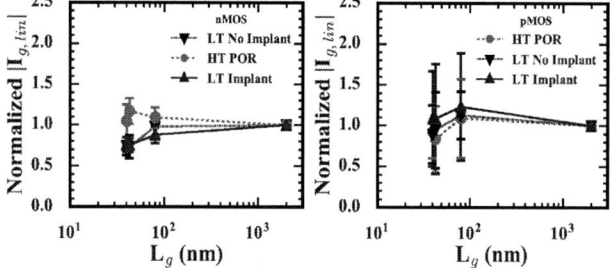

Figure 5: Relative $I_{g,lin}$ vs gate length, for W = 1 μm, |Vd| = 0.025 V and |Vg| = 1.7 V for both methods

Figure 5a) plots the I_g^{lin} per unit area, normalized with respect to the I_g^{lin} per unit area of the longest gate length, as a function of the gate length, for nMOS and pMOS devices of the HT POR and the LT processes. **Figure 5b)** plots the result of the differential method described by equation (3) and (4) and is therefore based on gate current values not divided by the length. The results of the differential methods are then

normalized with respect to the value obtained for the longest gate length.

The following results are shown in **Figure 5a)**:

- For nMOS, the normalized I_g^{lin} of the HT POR is relatively constant, which clearly indicates a stable EOT or a very limited ΔL. A significant degradation however occurs starting from 300 nm for the LT splits, and larger than a factor 2 for the shorter gate length of 20 nm.
- For pMOS, the variation of the normalized I_g^{lin} is far less significant, and the variability in the value of I_g^{lin} is much larger than in the nMOS case, as indicated by the errors bars (standard deviation of I_g^{lin} around its average).

The use of the differential method shown in **Figure 5b)** is then used to clarify the origin of the previous results:

- As expected, the differential of I_g^{lin} for nMOS devices of HT POR is relatively constant, which clearly indicates a stable EOT and very limited ΔL. For the LT splits however, the differential is not constant, which indicates a degradation of the EOT, and not an impact of the ΔL.
- The weakly varying values for pMOS confirm a weaker regrowth of the EOT. Note that the presence of a source of variability strongly impacts the gate current with this method.

These results therefore suggest that it is possible to extract a value of ΔEOT from the logarithm of the gate current per unit area, if a value of B is known:

$$\ln(I_g(L_{Long})) - \ln(I_g(L)) = \frac{\Delta EOT(L)}{B} \quad (5)$$

V. COMPARISON BY MODELISATION

The value of B in equation (5) is highly dependent on the model of gate tunneling current considered. The simple direct tunneling model is convenient to separate the different contribution of the variation of I_g^{lin}, as highlighted in the previous section, but it suffers from too strong assumptions to be applied to the complex SiO_2/HfO_2 gate stack of the devices considered here. To obtain a better estimation of the B parameter, a more complete modeling of the gate current has hence been carried using the scattering matrix formalism, assuming a 3D electron gas in the semiconductor [5][6]. The SiO_2 thickness has been varied from 2 Å to 4 Å and the HfO_2 one from 19 Å to 21 Å. The value of the SiO_2 effective mass has been set to 0.5 m_0, while the one of HfO_2 has been fixed at 0.165 m_0. The tunneling currents obtained with this approach are shown in **Figure 6**.

A value of 1.2 Å has been extracted from these calculations for the B parameter. It will be used in the next section to deduce the resulting ΔEOT observed in LT nMOS devices.

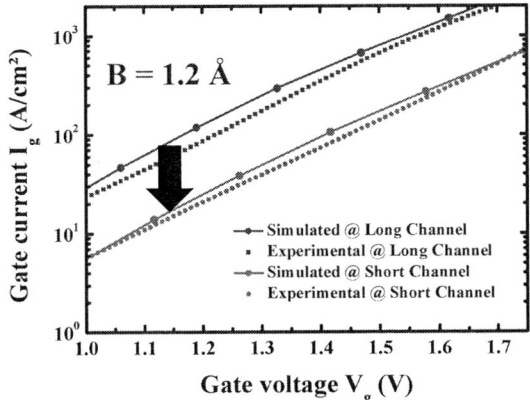

Figure 6: Comparison between simulated and experimental gate current for short and long gate lengths.

VI. ANALYSIS

The ΔEOT value, plotted in **Figure 7**, is hence deduced from equation (5). The value of B is obtained using the scattering matrix formalism.

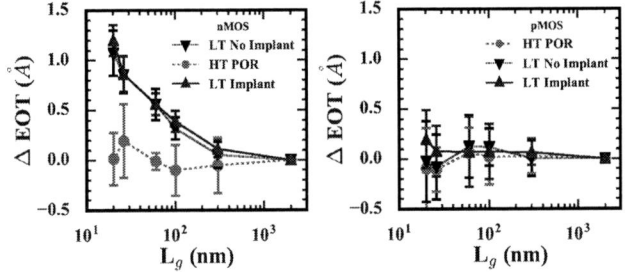

Figure 7: ΔEOT variation as function of the gate length, for W=1µm, |Vd|=0.025 V and |Vg|= 1.7 V

In the case of pMOS, no EOT regrowth is found for HT POR and LT with a high precision of 0.4 Å.

For nMOS, HT POR shows also no EOT regrowth, while for LT, a clear degradation of 1.2 Å ± 0.2 Å between short and long channel is observed. This degradation corresponds to a variation of 10% of the EOT and could thus lead to 10% I_{on} degradation, as I_{on} is inversely proportional to EOT.

As anticipated, this EOT regrowth could be attributed to an oxygen ingress coming from an oxidation of the thin liner. This ingress is expected to occur during the spacer complement deposition, which is made at 630°C during 2 hours. The thin liner oxidation is hence not linked to the implantation as the EOT regrowth is strictly the same of the LT splits, with or without implantation. The difference in EOT regrowth between n and p MOS LT splits can be explained by the fact that the latter is built on SiGe$_{27\%}$ channel and the oxidation kinetics in SiGe is lower than for Si [7].

To resolve this problem, a small reducing treatment could be applied before spacer complement deposition. A queue time

reduction between the first liner deposition and the complement one can also be envisaged.

VII. CONCLUSION

In conclusion, gate current measurement appears as an interesting source of information to see the EOT value evolution for short gate lengths, enabling a higher precision than a capacitance measurement. A precise evaluation of the B parameters (which could be considered as a characteristic tunneling length) is however needed. It has been estimated in this work using the scattering matrix formalism.

These electrical characterizations have shown that EOT regrowth could be observed between LT and HT splits for nMOS only, while from CV measurements the uncertainties is too high to fully conclude about an EOT regrowth. Thanks to this work, the origin of performance degradation between LT and HT, previously only attributed to larger access resistance [2], has been extended to include the detrimental role of gate stack instability.

To avoid EOT regrowth for low temperature process, it is suggested to use a small reducing treatment, which should allow to obtain greater performance.

Acknowledgment

This work is partly funded by the French Public Authorities through NANO 2017 program and EQUIPEX FDSOI11, ST-LETI Alliance program and by LabEx Minos ANR-10-LABX-55-01.

References

[1] P. Batude et al., "3DVLSI with CoolCube process: An alternative path to scaling," in 2015 Symposium on VLSI Technology (VLSI Technology), 2015, pp. T48–T49.

[2] L. Pasini et al., "High performance CMOS FDSOI devices activated at low temperature," in 2016 IEEE Symposium on VLSI Technology, 2016, pp. 1–2.

[3] K. Romanjek, F. Andrieu, T. Ernst, and G. Ghibaudo, "Characterization of the effective mobility by split C(V) technique in sub 0.1 µm Si and SiGe PMOSFETs," Solid-State Electron., vol. 49, no. 5, pp. 721–726, May 2005.

[4] C. Adam, Essential mathematics and statistics for forensic science, Wiley. 2010.

[5] J. Coignus, C. Leroux, R. Clerc, G. Ghibaudo, G. Reimbold, and F. Boulanger, "Experimental investigation of transport mechanisms through HfO2 gate stacks in nMOS transistors," in 2009 Proceedings of the European Solid State Device Research Conference, 2009, pp. 169–172.

[6] D. K. Ferry, Quantum mechanics: an introduction for device physicists and electrical engineers, CRC Press. 2001.

[7] J. Grabowski and R. B. Beck, "Oxidation kinetics of silicon strained by silicon germanium," J. Telecommun. Inf. Technol., vol. nr 3, pp. 30–32, 2007.

Back-gate bias effect on UTBB-FDSOI non-linearity performance

B. Kazemi Esfeh[1], V. Kilchytska[1], B. Parvais[2*], N. Planes[3], M. Haond[3], D. Flandre[1] and J.-P. Raskin[1]

[1]ICTEAM, Université catholique de Louvain, 1348 Louvain-la-Neuve, Belgium
Email: Babak.kazemiesfeh@uclouvain.be, Tel: (+32) 10478386
[2]imec, Kapeldreef 75, 3001 Leuven, Belgium
*Vrije Universiteit Brussel, Dept. of Electronics and Informatics (ETRO), Pleinlaan 2, B-1050 Brussels, Belgium
[3]ST, ST-Microelectronics, 850 rue J. Monnet, 38926 Crolles, France

Abstract—**This work investigates experimentally the non-linearities of FDSOI MOSFETs from DC to RF frequencies. The effect of the back-gate bias on non-linearity of the device is studied by means of 2nd and 3rd harmonic distortions (HD2 and HD3) extracted from dc I-V curves as well as from large-signal RF measurements using 1-dB and IP3 points. It is shown that the non-linearity is reduced by applying a positive back-gate bias. The reasons for this reduction are increasing of "effective body factor" and lesser mobility degradation with increase of the positive back-gate bias.**

Keywords—Fully depleted (FD) SOI; MOSFETs; non-linearity; harmonic distortion; measurements

I. INTRODUCTION

Ultra-thin body (UTB) and ultra-thin buried oxide (UTBB) technology used for fully depleted silicon-on-insulator (FD-SOI) transistors is widely recognized as a promising candidate to continue downscaling trends beyond 28 nm-node low power CMOS applications requested by ITRS [1]. It is also an important node toward RF-digital integration for such applications as, e.g., 5G and Internet of Things (IoT). This technology has already demonstrated not only improved DC but also RF performances. This technology has already demonstrated its outstanding performance with respect to other counterparts in terms of variability [2], electrostatic control and short channel effects [3]-[5], parasitic capacitances and resistances [6]-[8] as well as RF [9] and analogue [10] figures of merit (FoM). One of the most interesting features of this technology is the possibility of back-gate control schemes or double-gate (DG) configuration thanks to ultra-thin buried oxide (BOX). By combining this feature with a highly-doped layer underneath the BOX, so called ground-plane (GP), an architecture with dual gate is achievable by which both static and dynamic behaviors of the device are further controlled and improved [11]. In this work, a 3-port configuration of 30nm High-k/Metal Gate (HKMG) UTBB FD SOI MOSFET is employed to investigate the effect of DC back-gate bias on the RF and non-linearity figures of merit. While the non-linearity of MOSFET is an intrinsic feature that is crucial for RF communication circuits, only few reports on the non-linearity behavior of scaled FDSOI exist. For example in [10], the non-linearity in FDSOI devices is compared to their bulk counterpart, but the work is limited to current-voltage measurements and their derivatives. The present work extends that study to include large-signal harmonic distortions extracted from RF measurements.

II. DEVICE AND MEASUREMENT DESCRIPTION DETAILS

The studied UTBB FD-SOI nMOSFET is fabricated in the 28 nm FDSOI platform of ST-Microelectronics [12]. BOX, Si body and equivalent gate oxide thickness are 25 nm, 7 nm and 1.3 nm respectively. The channel is rotated by 45° from the <100> plane. Studied device features 40 parallel fingers, 30 nm of length and 1 µm of width each.

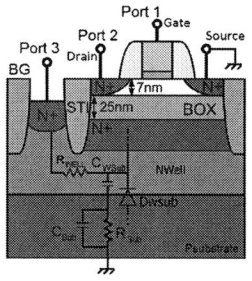

Fig. 1. Simplified cross section of studied NMOS transistor with back-gate scheme

For RF characterization the multi-finger device is designed and embedded in coplanar waveguide access pads. The common-source configuration design of the FD SOI nMOSFET provides the possibility of 3-port measurement in which the front-gate, drain and back-gate are referred to RF port1, 2 and 3 respectively. As shown in Fig. 1 illustrating the simplified cross section, the device layout features a particular access to the back-gate, so called flip-well architecture [13], including a heavily doped n-type back plane located below the buried oxide (BOX) and an n-well which provides a natural substrate insulation between the devices. De-embedding structures including open- and short-pads are implemented on the Si chip. Therefrom, in small-signal measurements the effects of interconnections on extrinsic capacitances and series resistances are withdrawn from S-parameters. In the large-signal measurements, the signal at the fundamental frequency of 900 MHz generated by an Agilent E8267D is amplified up to 4 dBm and injected into the DUT. The output signal is recorded and measured by an Agilent E4440A spectrum analyzer and an Agilent U2000B USB Power Sensor (with a frequency bandwidth of 10 MHz - 18 GHz). The recorded output signal contains the fundamental, 2nd and 3rd harmonics. The HD measurement is performed for various back-gate biases to observe its effect on non-linearity of the transistor (Fig. 2) by investigating of different key parameters e.g. 1-dB

978-1-5090-5979-9/17 $31.00 © 2017 IEEE

compression and third-order interception points which are discussed in next parts.

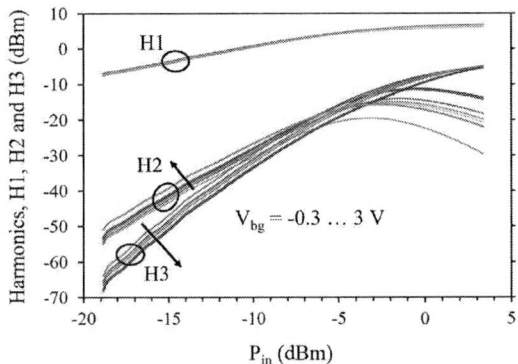

Fig. 2. Experimental Fundamental output (H1), 2nd (H2) and 3rd (H3) harmonics curves of studied FDSOI device at $V_d = 1$ V, constant V_g-V_{th} for various V_{bg} versus P_{in}.

III. NON-LINEAR RESULTS

A. Static analysis

Fig. 3 shows g_{m1}-I_d curves of FDSOI device in saturation, at various V_{bg} where $g_{m1}=dI_d/dV_g$. As can be seen in Fig. 3, positive V_{bg} results in the reduction of g_{m1} maximum and global curve flattening. Similar trends are observed for higher order derivatives.

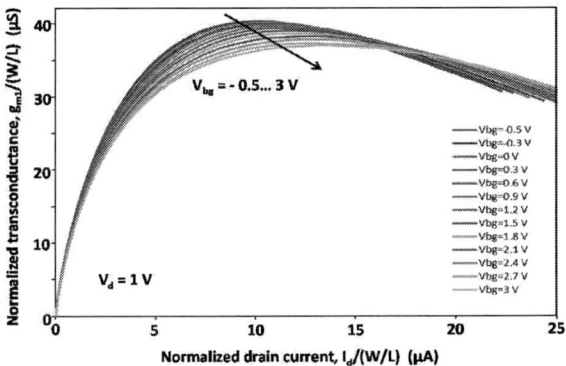

Fig. 3. Experimental g_{m1}-I_d curves of FDSOI device at various V_{bg}. L = 30 nm. W = 40 μm. $V_d = 1$ V.

Next to that, assuming a memoryless circuit excited by a sinusoidal signal with AC amplitude A, we calculated 2nd and 3rd order harmonic distortions as $HD_2=A/2|K_2/K_1|$; $HD_3=A^2/4|K_3/K_1|$ with $K_n=1/n!.d^nI_d/dV_g^n$. It can be seen in Fig. 4 that HD_2 minimum is shifted by V_{bg} to higher I_d (and to lower V_g following V_{th} (V_{bg}) trend, not shown due to lack of space). Thus, depending on the regime (bias, current), HD_2 can be considerably reduced. This is interesting from a design point of view. If one needs a certain current level, the better linearity can be achieved for that current by changing the V_{bg}.

For example, Fig. 4 shows that 10-30 dB improvement can be obtained depending on the V_{bg} at given normalized current $I_d/(W/L) > 12$ μA. In order to remove the effect of HD2 shift

along the I_d axis (due to V_{th} (V_{bg}) dependence) and visualize a pure effect of V_{bg} on HD2, Fig. 5 aligns HD2 minimums and plots HD2 as a function of V_g - $V_{g\_min\,HD2}$).

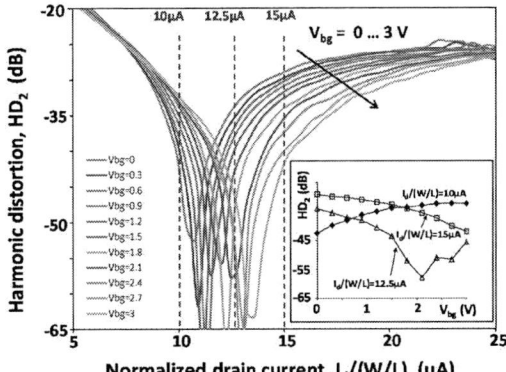

Fig. 4. Experimental HD_2-I_d curves of FDSOI device at various V_{bg}. $V_d = 1$ V. L = 30 nm. W = 40 μm. Insert gives HD_2 vs. V_{bg}.

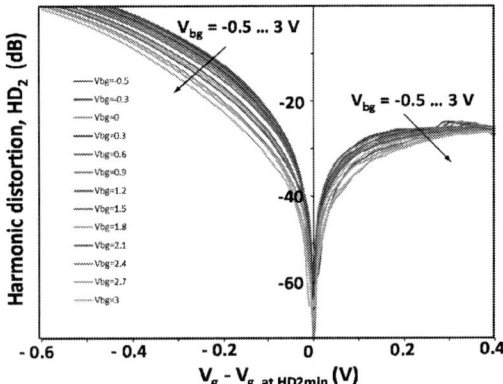

Fig. 5. Experimental HD_2 as a function of V_g - $V_{g\_min\,HD2}$ curves at various V_{bg}. $V_d = 1$ V. L=30 nm. W=40 μm.

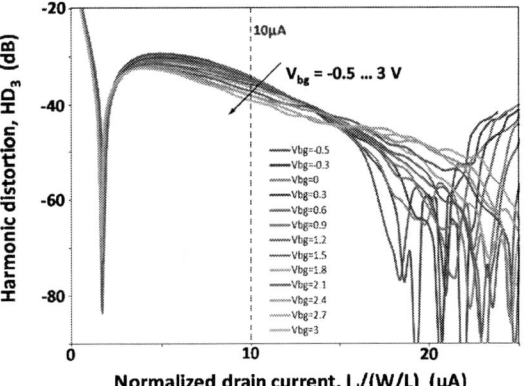

Fig. 6. Experimental HD_3-I_d curves of FDSOI device at various V_{bg}. $V_d = 1$ V. L = 30 nm. W = 40 μm.

Improvement of HD2 by V_{bg} is clearly seen. Similarly for HD3, one can see from Fig. 6 that there exists a current (and

978-1-5090-5979-9/17 $31.00 © 2017 IEEE   149

bias) range where HD$_3$ can also be improved by the positive V$_{bg}$ application (one can win about 3-5 dB).

Improvement of non-linearity with positive V$_{bg}$ application can be related to higher "effective" body factor [14] and lesser vertical electric field mobility degradation.

B. RF Figures of Merit

Fig. 7 illustrates extracted cut-off frequency f$_T$ values as a function of gate bias for various back-gate bias V$_{bg}$ for the high and low drain biases (V$_d$ = 1 V and V$_d$ = 0.6 V, respectively). Maximum f$_T$ values as high as ~360 GHz are reached for the case of V$_d$ = 1 V. These very high f$_T$ values are achievable thanks to the well-optimized (reduced) parasitic elements, as was studied in details in our previous work [15]. It is worth mentioning that even at low V$_d$ of 0.6 V, the device features rather high maximum f$_T$ ~ 280 GHz which is beneficial for low voltage and low power applications. Next to that, it is worth pointing that application of positive V$_{bg}$ results in a slight reduction of maximum achievable f$_T$ and in a shift of V$_g$ at which the maximum is achieved to the lower values. Moreover, global f$_T$ vs. V$_g$ curve flattens. Or in another words, one can say that the gate bias range where f$_T$ and f$_{max}$ stay close to their peak values extends. These features are directly related to the above discussion on the g$_{m\_max}$ and g$_m$ versus V$_g$ behavior as a function of V$_{bg}$ (Fig. 2). Indeed, f$_T$ is directly proportional to g$_m$.

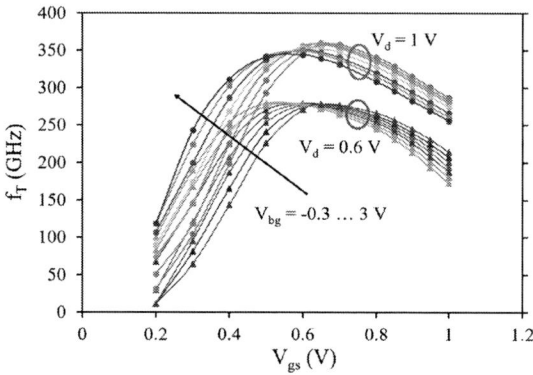

Fig. 7.Experimental cut-off frequency f$_T$ curves versus front-gate bias V$_g$ of studied FDSOI transistor for drain voltages of V$_d$ = 0.6 V and 1 V at various V$_{bg}$.

C. 1-dB compression and IP3 points

The transistor in common source configuration with an input large-signal at the gate introduces non-linearity in the output signal at the drain. Hence, in this mode, the output signal includes harmonics of the fundamental signal. The amount of non-linearity of a transistor in large-signal operation could be evaluated by two parameters, 1-dB compression point and third-order intercept point (TOI) as shown in Fig. 8. According to well-known theory, the 1-dB gain compression point (P$_{-1dB}$) is defined as the power gain at which it lies 1 dB below the linear power gain in small-signal operation due to the non-linearities of the transistor. The dynamic range (DR) shown in Fig. 8 is the range where the transistor in amplifying mode has a linear power gain which is limited by the noise floor level [16]. In this range,

the extrapolation of the fundamental and third-order nonlinearity curves (slope 1 and 3, respectively) will cross at a point called third-order intercept point (IP3). The IP3 usually refers to the input level and it is sometimes denoted IIP3 while the output level corresponding to the IP3 is presented as OIP3 [17]. When assessing the linearity of devices and systems, higher IP3 and 1-dB compression points are desirable. Below we show that the application of positive V$_{bg}$ moves those points to higher values.

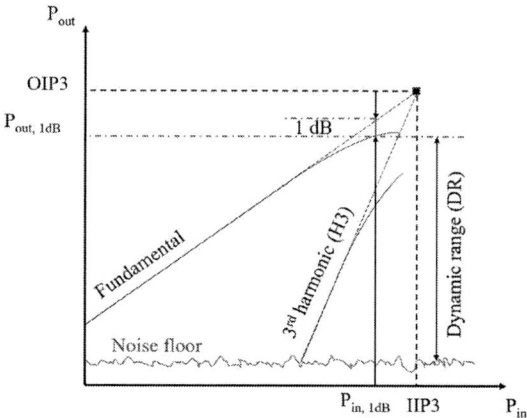

Fig. 8.Output power at 1-dB compression point and third order intercept point (IP3) versus input power on a logarithmic scale. The dynamic range (DR) is also shown.

Fig. 9 shows the input and output referred 1-dB compression point versus V$_{bg}$ from -0.3 to 3 V at 900 MHz. The inset gives the output power, P$_{out}$ and gain, G, versus input power, P$_{in}$ at 1-dB compression point for different V$_{bg}$. The transistor is biased at V$_d$ = 1 V and constant overdrive voltage (V$_g$-V$_{th}$) at which maximum g$_m$ is achieved. As can be observed in Fig. 9, by tuning V$_{bg}$ to higher positive voltages, 1-dB compression point is shifted to higher (P$_{in}$, P$_{out}$) values showing more linearity of the device. By changing V$_{bg}$ from -0.3 to 3 V, 1-dB compression point is improved by 1.2 dB extending linear DR of the device and thus making possible to get 1.2 dB more linear output power from the device. However, as illustrated in Fig. 9, at higher V$_{bg}$, the gain slightly decreases. This behavior is quite consistent with above discussion on the non-linearities extracted from I-V measurements i.e. obtaining lower peak-g$_m$ and flatter g$_m$ curves leading to lower gain and better linearity of the transistor. As shown in Fig. 3, the condition of maximum g$_m$ is achieved for normalized drain current I$_d$ / (W/L) of around 10 µA. As can be seen in Fig. 4 and 6, at this current (bias) condition, a degradation of HD2 and an improvement of HD$_3$ by the positive V$_{bg}$ is observed. The same trend is observable in RF measurement results shown in Fig. 2. The IP3 point is in correlation with HD3 which is expected to be improved.

The improvement of third-order intercept point (IP3) with increase of positive V$_{bg}$ is shown in Fig. 10. It can be seen that by tuning of V$_{bg}$ from 0 V to 3 V, OIP3 and IIP3 increase by about 1 and 1.25 dB, respectively.

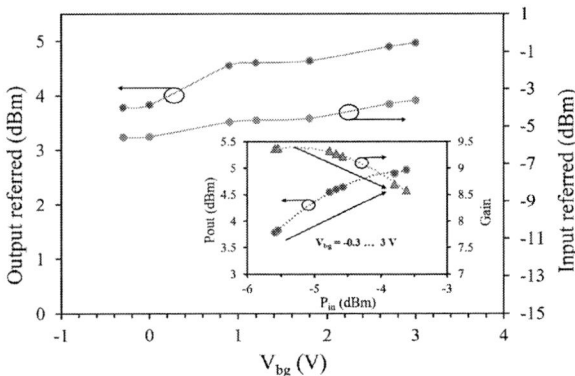

Fig. 9. Input and output referred 1-dB compression point at $V_d = 1$ V, constant over drive voltage $V_g - V_{th}$ at maximum g_m and 900 MHz versus V_{bg} from -0.3 V up to 3 V. Inset gives the gain and output power vs. input power at 1-dB compression point for various V_{bg}.

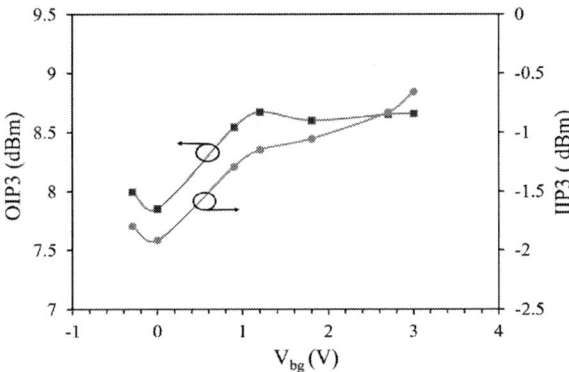

Fig. 10. Input and output referred IP3 point at $V_d = 1$ V and constant over drive voltage $V_g - V_{th}$ at maximum g_m for various V_{bg} from -0.3 V up to 3 V.

IV. CONCLUSION

Non-linear behavior of 3-port FDSOI MOSFETs was studied through dc and RF experimental results. Effect of back-gate bias on non-linearity and RF FoMs was analyzed. Application of positive V_{bg} in FDSOI was shown to open a way for non-linearity reduction (providing e.g. 10-30dB win in HD2, 3-5dB in HD3, depending on bias/current conditions as well as output-referred 1-dB compression and IP3 points increase by 1.2 dB and 1 dB). Non-linearity behavior as a function of V_{bg} revealed by DC and RF measurements agree well. The reduction of non-linearity by the positive V_{bg} application can be understood in terms of "effective" body factor and mobility behaviors as a function of V_{bg}.

ACKNOWLEDGMENT

The authors wish to acknowledge ST Microelectronics and IMEC for fabrication and providing the device. The work was partially funded by Eniac "Places2Be" and Ecsel "Waytogofast" projects.

REFERENCES

[1] International Technology Roadmap for Semiconductor (ITRS), 2011 Edition. <http://www.itrs.com>.

[2] Nobuyuki Sugii *et al.*, 'Comprehensive study on vth variability in silicon on Thin BOX (SOTB) CMOS with small random-dopant fluctuation: Finding a way to further reduce variation,' *2008 IEEE IEDM*, San Francisco, CA, pp. 1-4, 2008.

[3] T. Skotnicki, C. Fenouillet-Beranger, C. Gallon, S. Monfray, F. Payet, A. Pouydebasque, et al., 'Innovative materials, devices, and CMOS technologies for low-power mobile multimedia,' *IEEE Trans Electron Devices*, 55 (1), pp. 96–130, 2008.

[4] Faynot O, Andrieu F, Weber O, Fenouillet-Béranger C, Perreau P, Mazurier J, *et al.*, 'Planar fully depleted SOI technology: a powerful architecture for the 20-nm node and beyond,' In: *IEEE electron devices meeting (IEDM)*, pp. 3–2, 2010.

[5] Gallon C, Fenoujilet-Beranger C, Vandooren A, Boeuf F, Monfray S, Payet F, *et al.*, 'Ultra-thin fully depleted SOI devices with thin BOX, ground plane and strained liner booster,' In: *Proc IEEE inter SOI conference*, pp. 17–8, 2007.

[6] C. L. Chen et al., 'Fully Depleted SOI RF Switch with Dynamic Biasing,' *2007 IEEE Radio Frequency Integrated Circuits (RFIC) Symposium*, Honolulu, HI, pp. 175-178, 2007.

[7] R. Valentin, *et al.*, 'RF small signal analysis of Schottky-Barrier p-MOSFET,' *IEEE Trans Electron Devices*, 55, pp. 1192–1202, 2008.

[8] A. Vandooren *et al.*, 'Mixed-signal performance of sub 100-nm fully-depleted SOI devices with metal gate, high K (HfO2) dielectric and elevated source/drain extensions,' In: *IEEE int. electron devices meeting*, Washington, DC, USA, pp. 11.5.1–11.5.3, 2003.

[9] B. Kazemi Esfeh, V. Kilchytska, V. Barral, N. Planes, M. Haond, D. Flandre, J. -P. Raskin, 'Assessment of 28 nm UTBB FD-SOI Technology Platform for RF Applications: Figures of Merit and Effect of Parasitic Elements,' *Solid-State Electronics*, vol. 117, pp. 130-137, 2016.

[10] V. Kilchytska, B. K. Esfeh, C. Gimeno, B. Parvais, N. Planes, M. Haond, J.-P. Raskin, D. Flandre, 'Comparative study of non-linearities in 28 nm node FDSOI and Bulk MOSFETs,' *proceeding ULIS 2017*.

[11] T. Skotnicki, M. Takayanagi, W. Kleemeier, H. Bu, 'Impact of back bias on ultra-thin body and BOX (UTBB) devices,' In: *Proc IEEE symp on VLSI technology (VLSIT)*, pp. 160–1. 2011.

[12] N. Planes *et al.*, '28 nm FDSOI technology platform for high-speed low-voltage digital applications,' in: *Symposium on VLSI Technology*, pp. 133-134. 2012.

[13] J.-P. Noel, O. Thomas, M.-A. Jaud, C. Fenouillet-Beranger, P. Rivallin, P. Scheiblin, T. Poiroux, F. Boeuf, F. Andrieu, O. Weber, O. Faynot and A. Amara, 'UT2B-FDSOI Device Architecture Dedicated to Low Power Design Techniques,' *ESSDERC 2010*, pp. 210-213.

[14] V. Kilchytska, D. Bol, J. De Vos, F. Andrieu, D. Flandre, 'Quasi-double gate regime to boost UTBB SOI MOSFET performance in analog and sleep transistor applications', *Solid State Electronics*, Vol. 84, 2013, pp. 28–37.

[15] B. Kazemi Esfeh, V. Kilchytska, B. Parvais, N. Planes, M. Haond, D. Flandre and J.-P. Raskin, 'Back-gate bias effect on FDSOI MOSFET RF Figures of Merits and Parasitic Elements,' proceeding ULIS 2017.

[16] G. Gonzalez, Microwave Transistor Amplifiers Analysis and Design. Prentice-Hall, 1984.

[17] D. M. Pozar, Microwave Engineering. John Wiley & Sons, 2005.

Evolution of oxygen vacancies under electrical characterization for HfO$_x$-based ReRAMs

B. Attarimashalkoubeh*, J. Sandrini, E. Shahrabi and Y. Leblebici

Microelectronic System Laboratory (LSM)

Swiss Federal Institute of Technology(EPFL) Lausanne, Switzerland

Corresponding author*: Behnoush.attarimashalkoubeh@epfl.ch

Abstract—Recently, studies on ReRAMs and their reliability have received increased attention. The reliability issue is due to the nature of oxygen vacancies behaviour under biasing conditions which necessitate further studies to achieve an in-depth understanding. In this work, we fabricated several HfOx ReRAM devices with different structure, material, and thickness, followed by a study of their electrical characteristics under DC biasing. We show an improvement in the switching parameters through engineering of the device structure. Moreover, we demonstrate a certain required thickness for the oxide layer for the ease of oxygen vacancies relocations, thinner oxide layer led to the common ReRAMs performance failure in the low resistance state.

I. INTRODUCTION

Memory technologies have significantly changed from Static Random Access Memory (SRAM) to relatively recent nanoscale Non-Volatile (NVM) devices. The need for developing alternative candidates to replace conventional charge-based memories for future computing system has brought noticeable attention to Resistive Random Access Memory (ReRAM) with its comparatively easy fabrication methods and structure [1]. The simple ReRAM structure consists of a metal-oxide layer sandwiched between two electrodes and exhibits appropriate tradeoff between the most important memory specifications such as: density, scalability, high endurance, low power, random access and read/ write throughput [2], [3].

According to ITRS, Oxide based RAMs (OxRAM) are considered as one of the major categories of ReRAM which operate based on the formation and annihilation of conductive paths composed of oxygen vacancies through their oxide layer [4]. In OxRAMs the resistance states could drift from high resistance state (HRS) to low resistance state (LRS) during a process called set and from LRS to HRS during reset process. For OxRAMs, different metal oxides (i.e. HfO$_2$, Al$_2$O$_3$, TaO$_x$, TiO$_2$,....) and electrodes (Hf, Ta, etc.) have been studies as conventional materials [4].

Besides all highlighted advantages of OxRAM, the reliability is thought to be the main hurdle on its way to replace conventional technologies [5]. The reliability issue is due to the nature of switching mechanism that is based on the random formation of conductive filaments -with different shapes and sizes- through insulator layer which necessitates further studies to improve it. Solving reliability issue is possible through a better understanding of ReRAMs switching mechanism. For OxRAMs, it involves deep knowledge about oxygen vacancies as the responsible elements for the resistive switching.

In this study, we fabricated different HfOx-based memory devices and studied their characteristics under DC measurement condition. The HfOx as the oxide material was chosen due to its reported proper trade-off endurance and switching window [6]. Several devices with different structure have been fabricated and their electrical characterisation have been studied.

II. MATERIALS AND METHODS

A. Fabrication process flow

The "in-via" design is implemented to create the memory element(s) stack vertically between the two metal electrodes, Fig. 1 schematically demonstrates the information of patterning steps for the fabrication process including the electrodes size and shape, while the details of fabrication are been thoroughly explained in following. After deposition of a thin layer of Ti on top of 4" conventional test wafers with 500nm silicon oxide, 100nm Pt metal was grown using DC sputtering (1000W). The wafers are spin coated with positive resist of AZ3007, soft baked to reduce the solvent content and increase its adhesion to the substrate, exposed by using an optical lithography tool (20 mW/cm^2) and finally developed. The bottom electrode (BE) etched with STS-ICP dry etch system (Ar/Cl$_2$ chemistry) using laser end-point detection for optimizing the process time. Next, 100nm LTO was grown by LPCVD at 425° C as a passivation layer separating the electrodes and in order to achieve the in-via shape. The second level photolithography process was performed as been described earlier, followed by BHF bath etch to remove LTO inside the vias for the further deposition of memory element materials. Varied thickness of memory elements were deposited for different device, the HfO$_x$ layer for all samples were deposited by ALD (200° C using TEMAH precursors), TiO$_2$, Ti and Pt layers were deposited by DC sputtering using TiO$_2$, Ti and Pt targets at the room temperature. Thick layer of noble Pt metal has been utilised as the capping layer to prevent further oxidation of switching materials after deposition and to protect the electrodes in the contact with the probe tips during the measurement. Next, the top electrode (TE) after resist coating, exposure and development was defined and the resist stripped away through conventional procedure. Fig. 2a shows schematic demonstration of fabrication process flow and optical picture exhibiting access pads for further electrical

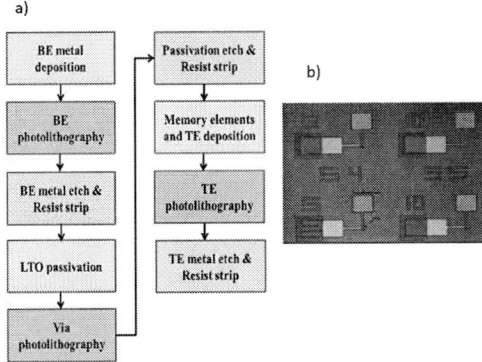

a)

b)

Fig. 1. a) Schamatic summary of the process flow steps, b) Optical picture of the ReRAMs cells before measurement, demonstrating TEs, BEs and vias.

TABLE I
A SUMMARY OF THE FABRICATED DEVICE WITH THEIR STACK STRUCTURES

Device No.	Device Stack
1	Pt(BE)/ **HfO$_2$(5nm)**/ Pt
2	Pt(BE)/ **HfO$_2$(5nm)** / TiO$_2$(3nm)/ Pt(TE)
3	Pt(BE)/ **HfO$_2$(5nm)**/ TiO$_2$(5nm)/ Pt(TE)
4	Pt(BE)/ TiO$_2$(3nm)/ **HfO$_2$(5nm)**/ Pt(TE)
5	Pt(BE)/ **HfO$_2$(5nm)**/ TiO$_2$(3nm)/ Ti(5nm)/ Pt(TE)
6	Pt(BE)/ **HfO$_2$(5nm)**/ TiO$_2$(5nm)/ Ti(5nm)/ Pt(TE)

measurement.Several devices were fabricated with different structure as are listed in Table I.

B. Electrical setup and measurement

The electrical characteristics of the devices were studied by using Agilent B1500 parameter analyzer. The bias was applied on the top electrodes (TE) while the bottom electrodes (BE) kept were grounded during the measurement. To protect the devices from hard breakdown, a compliance current of 150 μA was forced to the samples during measurement. All devices have been measured under the same electrical conditions; when the first bias swept on the TE from 0 to 7V, the current in all the devices appeared to be few μA and gradually increased as the voltage increased, at a specific voltage (V_{TH}), the resistance level of the devices was suddenly reduced, the current reached the compliance currents value and the device reached to its low resistance state (LRS-on state). In addition, applying a negative voltage on the top metal electrode gradually led the devices to the high resistance state (HRS-off state).

III. RESULTS AND DISCUSSION

A. Device engineering

For the device No.1 with HfO$_2$=5nm, when first the fresh device cell is biased by positive sweeps from 0 to 7, the V_{TH} appears to be 4.1V and the device switched back to high

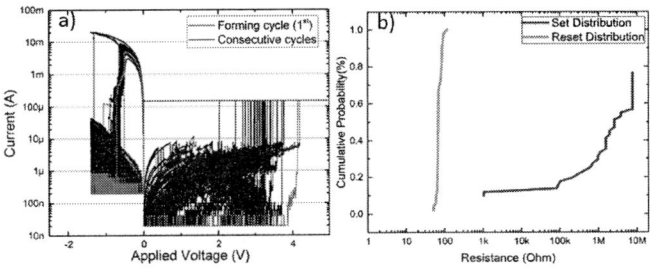

Fig. 2. Device No.1 with Pt(BE)/ HfO$_2$(5nm)/ Pt(TE) structure: a) I-V characteristic for first 100 voltage sweeps, b)LRS and HRS distribution read at 250mV.

resistance state (HRS) simply by negative bias (Fig. 2a). The first cycle (plotted in red and called forming cycle) as expected shows to have higher value (~4.1V) comparing to the next cycles (plotted in black color).

For the device No.1, the wide variation in the switching parameters was observed during the next 100 write operations (Fig.2b), which is related to the random nature of formation of filament inside the insulator layer. To overcome the variation in the switching parameters, the devices No.2 and 3 with the extra layer of TiO$_x$ have been fabricated to suppress the random formation of the filament, as a result they both performed with the lower variation in the switching parameters but gradually their switching window degraded. Next, device No.4 with different order of oxide layer (TiO$_2$/HfO$_2$ rather than HfO$_2$/TiO$_2$)fabricated and measured, the device shows to have high forming voltage (~8V),a large overshoot (~15mA) and within next few cycles the device failed in the LRS. Device No.4 performance is attributed to the different oxide quality obtained from different deposition tools(ALD and sputtering), higher forming voltage is needed to form the first step of the filament in the high quality HfO$_2$ and the device failed quickly regarding to the defective and leaky oxide as the switching material.

In the previous reported studies [7],[8],[9], addition of an extra metal layer has been proven to have significant effect in lowering variation in the key switching materials, hence devices No.5 and 6 with thin layer of Ti (as the most compatible metal with layer underneath has been recommended[10]) were fabricated and their electrical characteristics have been studied as their I-V characteristics for the first 100 sweeps are shown in Fig. 3 and 4. Comparing two devices with 3nm and 5nm TiO$_2$ layer, the device with thicker TiO$_2$ (device No.6) as expected shows higher forming voltage, followed with a big overshoot in the reset as its consequences (Fig.4). Moreover the device No.5 has smaller switching window (on/off ratio) which is a result of oxygen vacancies accumulation in the thin layer of TiO$_2$. Fig. 5 schematically explains the difference between the switching mechanisms of device No.1, No.5 and 6. Device No.1 as explained before has a wide variation in switching mechanism due to uncontrolled formation of filaments with different shape and size (Fig.5a), while in

978-1-5090-5979-9/17 $31.00 © 2017 IEEE   153

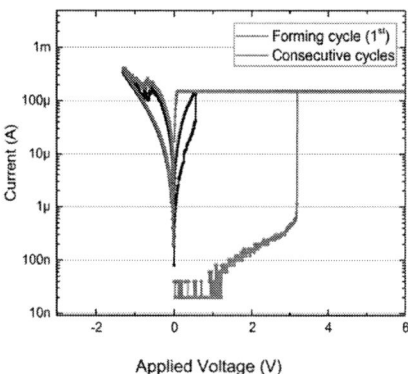

Fig. 3. I-V characteristic of the device No.5 with Pt(BE)/HfO₂(5nm)/TiO₂(3nm)/Ti(5nm)/Pt(TE) structure, the forming cycle is plotted in red and black represents the average switching in the next consecutive cycles.

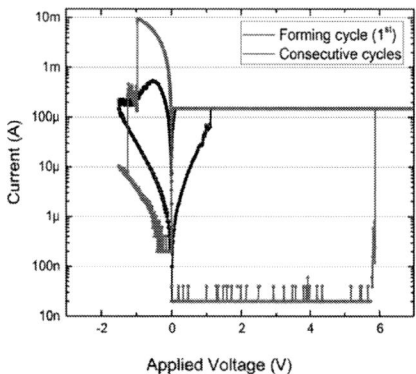

Fig. 4. I-V characteristic of the device No.6 with Pt(BE)/HfO₂(5nm)/TiO₂(5nm)/Ti(5nm)/Pt(TE) structure, the forming cycle is plotted in red and black represents the average switching in the next consecutive cycles

the devices No.5 and 6, this issue has effectively improved by utilizing bi-oxide structure with an added extra layer of reactive metal Ti. During fabrication of the devices No.5 and 6, reactive metal of Ti scavenged oxygen ions from oxide layer below and left permanent oxygen vacancies inside the oxide underneath, during the forming process more oxygen vacancies are created and bridged between TE and BE and caused the drift between HRS and LRS. Negatively biasing of these samples led to back-diffusion of oxygen ions from Ti/TiO₂ to HfO₂/Pt interface and rupture the CF.

B. Electrical setup engineering

In order to achieve stable resistive switching, it is crucial to obtain a sound understanding of the behaviour of oxygen vacancies as the main element responsible for the switching mechanism. Therefore, we implemented different measurement experiments to study oxygen vacancies evolution during electrical measurement. Among fabricated devices, No.5 and 6 have been chosen for further study due to their stable resistive switching. Hence, immediately after the first 100 sweeps with

TE biased and BE grounded (fig. 3 and 4) were finished, the direction of the bias was reversed, as the BE biased for the next 100 consecutive sweeps while TE was grounded (fig. 6a for device No.5 and fig. 7a for device No.6). Next, to investigate the reversibility of the occurred oxygen vacancies relocation, immediately after the reversed 100 sweeps were finished, the direction of the bias was reversed and the next 100 sweeps applied as the same setup with initial original measurement setup (TE biased, BE grounded) to each devices (fig.6b for device No.5 and fig. 7b for device No.6). When the biased was reversed for the devices No.5 and 6, both successfully switched for the next 100 sweeps without any significant change comparing to their performance in the previous original biasing, whereas they exhibited highlighted response to the change in the measurement setup as it turned back to the original one. When the measurement setup for the device No.5 changed back to its original form, the device did not reset and failed at LRS state, while device No.6 successfully switched for the next 100 sweeps (TE biased, BE grounded).

Fig. 8 exhibits proposed evolution of conductive filament (for reversed and back to original measurement) based on the relocation of the oxygen vacancies within the oxide layers. After the bias is reversed for the both devices (BE biased, TE grounded), Hf reduces Ti and creates more oxygen vacancies inside TiO_{2-x} near the TiO_{2-x}/HfO_x interface, the back-diffusion of oxygen ions and their recombination with oxygen vacancies led to changes in the conductive filament shape, therefore application of negative voltage on BE causes the filaments rupture at the TiO_{2-x}/HfO_x interface.

When the setup is reversed back to the original one, the device No.5 failed due to high accumulation of oxygen vacancies in its oxide layer as it is been observed earlier by small switching window (low on/off ratio) of device No.5 in its I-V curve (fig. 3). Implementation of the negative voltage on TE leads to back diffusion of oxygen ions and their

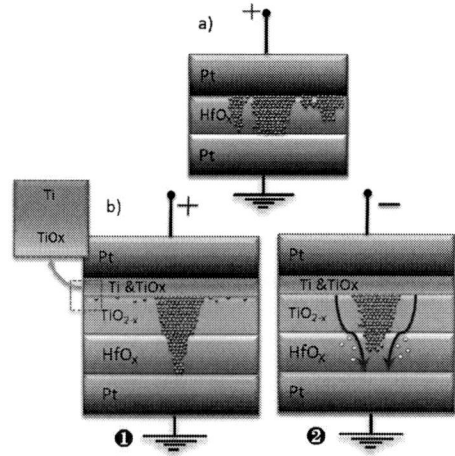

Fig. 5. a) Device No.1 with Pt(BE)/HfO₂(5nm)/Pt(TE), b) Switching mechanism for the devices No.5 and 6 with Pt(BE)/HfO₂(5nm)/TiO₂(3 and 5nm)/Pt(TE) during: 1- set and 2- reset processes.

978-1-5090-5979-9/17 $31.00 © 2017 IEEE

recombination with oxygen vacancies which in the device No.5 due to accumulation of oxygen vacancies, their number is much higher than the oxygen ions, therefore not enough oxygen ions reach to the Hf_x/Pt interface to cause the rupture. Moreover during all bias direction variations, we assume that the TiO_x layer is not thick enough for the relocation of oxygen vacancies, therefore the filaments changed from cone shape to pillar and form robust filaments which are impossible to be dissolved even by high reset voltage ($\sim$-9V), consequently the device No.5 when reversed back to original measurement setup failed and could not been reset.

IV. CONCLUSION

In this work we demonstrated fabrication steps and electrical characterizations of different HfO_2 ReRAMs. In the first part of the work, the impact of changes in device stack structure on the electrical properties of the samples been discussed. In the second part of the work, the devices have been investigated by reversed measurement setup which has been followed immediately by the original measurement setup. The observed I-V characteristics have been explained by proposition of a switching mechanism. We realized the movement of oxygen vacancies is reversible for oxide thickness value that are larger than a certain limit value, which allows proper migration of vacancies. The proposed switching mechanism explains this phenomena through oxygen vacancies movement.

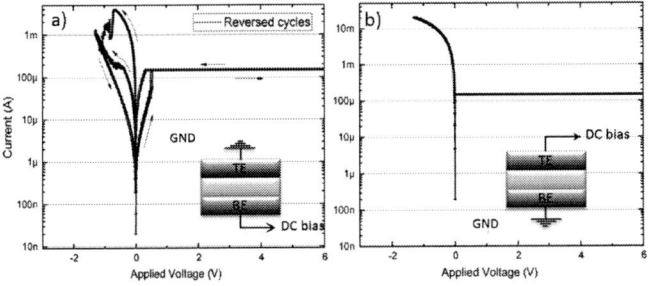

Fig. 6. a) Device No. Pt(BE)/HfO$_2$(5nm)/TiO$_2$(3nm)/Ti(5nm)/Pt(TE) for: a) Reversed measurement setup (BE biased, TE grounded), b) back to original (TE biased, BE grounded)

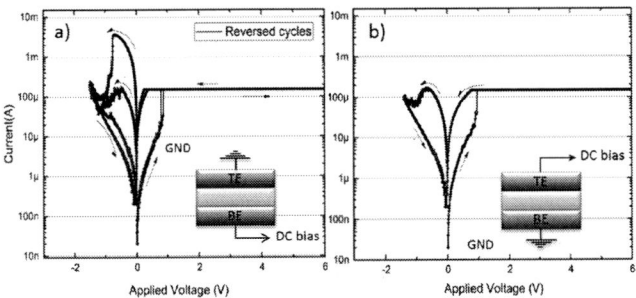

Fig. 7. a) Device No.6 Pt(BE)/HfO$_2$(5nm)/TiO$_2$(5nm)/Ti(5nm)/Pt(TE) for: a) Reversed measurement setup(BE biased, TE grounded), b) back to original (Te biased, BE grounded)

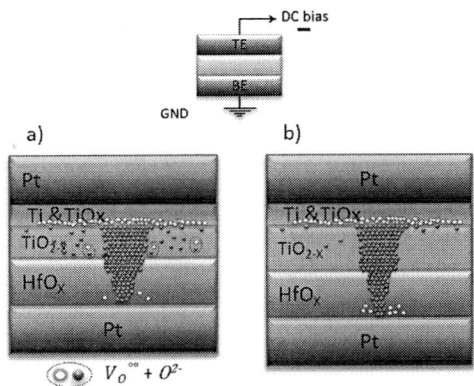

Fig. 8. Schematic demonstration of reset process after switching back to original measurement setup for (TE biased, BE grounded) for: a) Device No.5, b) Device No.6

ACKNOWLEDGMENT

The authors would like to acknowledge the staff of the CMI Clean Room of EPFL, for providing technical advice.

REFERENCES

[1] I. Baek, D. Kim, M. Lee, H.-J. Kim, E. Yim, M. Lee, J. Lee, S. Ahn, S. Seo, J. Lee et al., "Multi-layer cross-point binary oxide resistive memory (oxrram) for post-nand storage application," in Electron Devices Meeting, 2005. IEDM Technical Digest. IEEE International. IEEE, 2005, pp. 750–753.

[2] B. J. Choi, A. C. Torrezan, K. J. Norris, F. Miao, J. P. Strachan, M.-X. Zhang, D. A. Ohlberg, N. P. Kobayashi, J. J. Yang, and R. S. Williams, "Electrical performance and scalability of pt dispersed sio2 nanometallic resistance switch," Nano letters, vol. 13, no. 7, pp. 3213–3217, 2013.

[3] M.-J. Lee, C. B. Lee, D. Lee, S. R. Lee, M. Chang, J. H. Hur, Y.-B. Kim, C.-J. Kim, D. H. Seo, S. Seo et al., "A fast, high-endurance and scalable non-volatile memory device made from asymmetric ta2o5-x/tao2- x bilayer structures," Nature materials, vol. 10, no. 8, pp. 625–630, 2011.

[4] R. Waser and M. Aono, "Nanoionics-based resistive switching memories," Nature materials, vol. 6, no. 11, pp. 833–840, 2007.

[5] G. Meijer, "Who wins the nonvolatile memory race?" Science, vol. 319, no. 5870, pp. 1625–1626, 2008.

[6] E. A. Miranda, C. Walczyk, C. Wenger, and T. Schroeder, "Model for the resistive switching effect in HfO2 mim structures based on the transmission properties of narrow constrictions," IEEE Electron Device Letters, vol. 31, no. 6, pp. 609–611, 2010.

[7] R. Meyer, L. Schloss, J. Brewer, R. Lambertson, W. Kinney, J. Sanchez, and D. Rinerson, "Oxide dual-layer memory element for scalable non-volatile cross-point memory technology," in Non-Volatile Memory Technology Symposium, 2008. NVMTS 2008. 9th Annual. IEEE, 2008, pp. 1–5.

[8] D. Lee, J. Woo, S. Park, E. Cha, S. Lee, and H. Hwang, "Dependence of reactive metal layer on resistive switching in a bi-layer structure ta/hfox filament type resistive random access memory," Applied Physics Letters, vol. 104, no. 8, p. 083507, 2014.

[9] B. Attarimashalkoubeh, J. Sandrini, E. Shahrabi, M. Barlas, and Y. Leblebici, "Effect of hf metal layer on the switching characteristic of hfox-based resistive random access memory," in Ph. D. Research in Microelectronics and Electronics (PRIME), 2016 12th Conference on. Ieee, 2016, pp. 1–4.

[10] C.-Y. Lin, C.-Y. Wu, C.-Y. Wu, T.-C. Lee, F.-L. Yang, C. Hu, and T.-Y. Tseng, "Effect of top electrode material on resistive switching properties of zro2 film memory devices," IEEE Electron Device Letters, vol. 28, no. 5, pp. 366–368, 2007.

Emerging Memory technologies for high density applications

Giorgio Servalli

Micron Technology, Vimercate - Italy
gservall@micron.com

Abstract— **Comparison of most mature and promising emerging memory technologies respect to mainstream NAND and DRAM and challenges for the introduction in the market for high density applications.**

Keyword— **Emerging memory metrics, high density application, storage class memory**

I. INTRODUCTION

In the last 10 years, the memory scenario for high density applications consolidated around two key technologies, NAND and DRAM. The semiconductor industry demonstrated to be able to follow a fast pace of scaling, technology innovation, improvement of performance and overall reduction of cost/bit.

Technology scaling has resulted in ever increasing complexity and consequent increasing cost of innovation. The huge capital investment needed to run this technology race caused also a consolidation process in the industry of high density memory manufacturers. The market of NAND and DRAM is today in the hands of only 4 big companies.

In parallel to the scaling activity on conventional technologies, in the last years multiple different emerging memory (EM) technologies have been studied and developed, each based on different memory cell physics.

The development activity on EM has been focused on finding a solution to the two main problems of consolidated technologies: their scalability limits and the increasing performance gaps between DRAM and NAND.

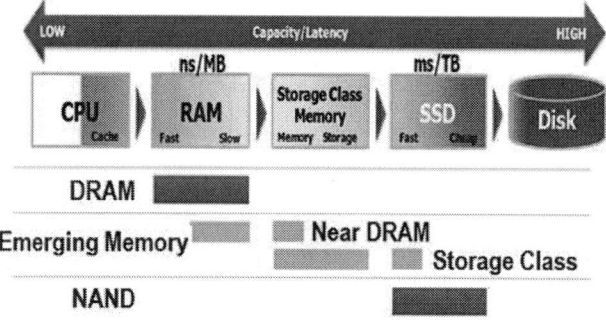

FIGURE 1. MEMORY HIERARCHY FOR EM, FROM DRAM TO NAND

As shown in Figure 1, an EM technology for high density application can be generally classified as "near-DRAM" (high performances, high cost/bit), which aims to compete with DRAM adding non-volatility features and promising scalability beyond DRAM 1T-1C conventional architecture, or as "Storage Class Memory" [1] (intermediate-low performance, low cost/bit), which aims to cover the gap between DRAM and NAND or to challenge NAND in terms of compromise between performance and cost.

The EM technologies reported here, PCM, RRAM and STTRAM, are non-volatile memories (NVM) generally considered as the most mature and promising. But to achieve success in the highly competitive field of high density application, an EM cannot provide only good performance, it must offer an overwhelming advantage over conventional memories to motivate the entire ecosystem to adopt it. This advantage must necessarily include also cost/bit, manufacturability and scalability.

The overall challenge for any EM is to competitively penetrate a consolidated memory market duopoly, where NAND has already migrated to the multilevel (MLC) 3D technology, well beyond 1x nm equivalent planar node, and the DRAM is continuing to scale into the 1x nodes increasing its performance.

II. CELL METRICS FOR EMERGING MEMORIES

For a meaningful comparison among EM technologies and to state of the art conventional technologies, it is necessary to take into account several metrics: switching energy, switching efficiency, switching speed, number of particles, retention and endurance capability. In the following paragraphs each metric is analyzed considering cells at 20nm tech node.

A. Switching Energy and Efficiency

Cell switching energy is widely used as a metric for memories, since it has a large impact at product level.

An EM requiring a large current (e.g. STTRAM and PCM) requires increased voltages and power consumption due to large IR loss in the die. High currents reduce the programming bandwidth since it's not possible to switch a large number of cells simultaneously. An EM requiring a high voltage (like NAND) needs large and inefficient pumps with increased power consumption, as well as large high voltage transistors that result in a larger die size.

978-1-5090-5979-9/17 $31.00 © 2017 IEEE

Besides switching energy, also energy efficiency, defined as the ratio between energy stored in the cell for retention and reading purposes respect to energy used to write data, is a key metric for high density applications.

Table I shows a comparison of switching energy and energy efficiency for different memories. Note that considered EM have significantly larger switching energy and worse energy efficiency compared to NAND and DRAM. PCM needs the highest energy, but at least a reasonable fraction of this is stored in the cell itself. On the opposite, STTRAM has a switching energy closer to conventional memories, but its efficiency is really poor, which gives a very small read signal (read current of high-current state is just 2X-3X larger than low current one [2]).

TABLE I. SWITCHING ENERGY AND ENERGY EFFICIENCY

	Cell Switching Energy (J)	Energy Efficiency	Sources of Energy Loss
DRAM	$E=0.5CV^2=\sim$1E-15	>0.8	Dielectric Relaxation Dielectric Leakage [3]
NAND	$E=0.5CV^2=\sim$1.0E-14	>0.8	Dielectric leakage Parasitic electron trapping in wrong location [4]
PCM	$E=I*V*t=\sim$1.0E-10	~1E-2	Parasitic Heat Loss [5]
RRAM	$E=I*V*t=\sim$1.0E-11	~1E-4	Thermal energy loss Parasitic current
STTRAM	$E=I*V*t=\sim$1.0E-13	~1E-6	Spin related thermal agitation Tunneling Efficiency Stochastic Switching Joule heating of barrier [2]

B. Switching Speed

Together with switching energy, the switching speed of a memory cell, limited by the operating physics of the device, is a key component of the overall bandwidth of a memory chip.

TABLE II. SWITCHING SPEED LIMITATIONS

	Fundamental Switching speed limitation	Typical Switching time	Typical Tradeoffs	Cell programming parallelism and max write bandwidth
DRAM	Capacitor charging time	<10nS	Transistor drive refresh, etc.	64K 6.4GB/s (DDR4)
NAND	RC time constant to reach ~25V (tunneling is fast)	>10µS	Cost and Array size	256K 250MB/s
PCM	Slow temperature ramp down to control crystallization	100-400nS	Separation of states	128b 40MB/s [6]
RRAM	Ion migration time through ~20A oxide	10-100nS	Retention Separation of states	512b [7] 200MB/s
STTRAM	Stochastic switching due to spin precession	10-50ns	TDDB Write BER	512-1024b 2.66GB/s [8]

Table II shows switching speed and limitations, together with the number of cells programmed simultaneously and the highest reported bandwidth for the EM.

Low energy technologies like NAND and DRAM can switch large numbers of cells simultaneously yielding high bandwidth. High current cells typical of EMs forces to switch far fewer bits simultaneously, which may result in a lower bandwidth in spite of high intrinsic performance. Trade off on speed can be adopted to achieve other benefits: e.g. NAND MLC trades off power, energy, time, and switching speed to gain the highest density and the lowest possible cost/bit.

C. Number of particles

The number of particles used to set a state in the memory cell is another key element to compare different technologies, since it has impact in the read signal, in the variability among cells, in the retention capability and in the possibility of storing more than one bit per cell.

The planar NAND scaling limit was essentially defined by decreasing number of electrons that could be stored in a floating gate with a smaller and smaller capacitance [9]. 3DNAND overcomes this limitation using cells that are much bigger than planar devices.

DRAM is currently under pressure because the cell capacitance decreases with scaling due to the lack of space to form a large capacitor and the complexity in creating taller and taller structures.

An EM will have a competitive advantage in terms of signal, noise, scalability, etc., if it uses a large number of particles to set the state of a cell: Table III shows a comparison for different memory technologies.

TABLE III. NUMBER OF PARTICLES IN A 20NM CELL

	Fundamental Particle	Number
DRAM	electron	~5E4 (15fF * 0.6V/q)
NAND	electron	~1E4 (50aF*25V/q)
PCM	Atomic Bond Bond Angle Bond Coordination	~2E4 (DFT on 20nm cube)
RRAM	Cluster of oxygen vacancies or metal ions	10-1000 (DFT [10])
STTRAM	Correlated spins Bohr Magneton	~1E60 (2nm free layer x 20nm diameter)

Among the considered EM, only RRAM can be critical in terms of particles: according to DFT simulation a single cell may work utilizing less than 2 hundred metal oxide vacancies.

D. Retention

DRAM is a volatile memory which requires constant refresh during operation to maintain the data. The refresh operation results in increased power, complexity, and loss of bandwidth and latency [8,9].

NAND typically meets retention requirements of months or years as a function of temperature, cycling, etc.

Most EM under development are NVM. Non-volatility requires the expenditure of more energy to change states than a volatile memory, to overcome a barrier high enough for retention. Generally, a high activation energy for an EM is a good indicator of good retention, but an accurate retention model is needed. As shown in Table I, the energy required to change states for EM is several orders of magnitudes higher than DRAM. Table IV shows a comparison of barrier energies for EM.

TABLE IV. BARRIER ENERGIES

	State Change	Barrier (eV)	Barrier between states	Retention Limitation
DRAM	Stored charge	0.55	Si Bandgap	Junction leakage Transistor leakage GIDL, etc.
NAND	Stored charge	0.3 SILC 1.1 Detrapping 0.2 Interface Trap	Insulator	Insulator leakage [4]
PCM	Amor. Crystal phase	2.4	Gibbs Free Energy	Crystallization of amorphous state
RRAM	Filament	1.4-1.8	Diffusion and electro-chemical barrier for ionization and reduction	Thermal diffusion of ions to or from filament
STTRAM	Spin	1.5	Magnetic Anisotropy of the free layer	Random thermal fluctuation

E. Endurance

For any memory, endurance is a key parameter.

The endurance of DRAM is considered as infinite ($>10^{15}$ cycles) at cell level, because the barrier energy that must be overcome to change states is essentially zero.

On the contrary, any NVM must supply energy to the cell to overcome a barrier to enable retention as shown in Table IV: expending energy causes damage to the cell, limiting its endurance.

Comparing the endurance of EMs is difficult because most publications show single cell data or limited distributions. The performance of a single cell can be misleading and does not represent the true capability of the technology Strong statistics are needed to enable comparison which are generally not available, and the real endurance capability of an application depends on error correction, usage conditions, memory management, etc.

Table V shows the endurance failure modes and the endurance limits for EMs single cell.

TABLE V. ENDURANCE FAILURE MODELS

	Endurance limit (single cell)	Endurance Failure Modes
PCM	1E9	Atom segregation (local stoichiometry variation) Physical separation of electrode due to volume change [11]
RRAM	1E6 – 1E12	TDDB Over-accumulation of atoms in filaments
STTRAM	1E15	TDDB [2]

III. CHALLENGES FOR EM INTRODUCTION IN THE MARKET

With the introduction of 3DNAND technology, the challenges for any EM to compete with NAND in terms of density and cost/bit increased to such a level that it's hard to imagine a realistic solution.

As shown in Figure 2, the density of state of the art 3DNAND technology now overcomes the last planar NAND technology by ~10X [12], with a considerable advantage in terms of cost/bit since this scaling path, even if requiring more and more deposition steps, does not require investments for lithography.

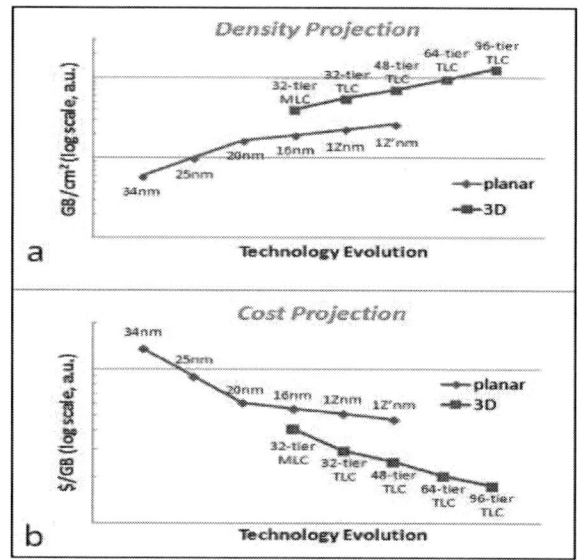

FIGURE 2. 3DNAND DENSITY AND COST VS PLANAR NAND

Moreover, channel area and cell capacitance of 3DNAND technologies are large compared to a planar cell, which gives a relevant reduction of cell Vth variability and noise, opening the path for advanced multilevel storage: 3 bits/cell products are standard today, 4 bits/cell ones are being introduced.

Going to numbers, a planar EM 1bit/cell technology that wants to compete with 3DNAND today [13] needs to have a cell area of less than 230 nm^2, which means a technology node F<8nm in the case of optimal 4F^2 cell layout and ideal 100% array efficiency. Apart from the technical feasibility and the cost to integrate such features, it's a number that is close to what is supposed to be the physical limits of considered EM

978-1-5090-5979-9/17 $31.00 © 2017 IEEE

cells, and it's needed just to start the competition with 3DNAND today.

The alternative is to follow directly the path of NAND scaling, that is going multilevel and integrating vertically. Both solutions appear very complex for EM: multilevel storage concept is hard to be applied to RRAM, PCM and STTRAM because of read noise, of resistance drift and of small read signal; vertical integration is quite challenging because of the series resistance added to cells with high switching current and because of the need of adding a selector to the memory cell or finding a self-selecting solution.

A crosspoint stacked solution is more realistic in the short term than a vertical 3DNAND like integration, but it's not cost competitive because of the large number of critical litho levels that need to be replicated to create a long stack of levels.

Respect to the race with 3DNAND, competition with DRAM conventional 1T-1C architecture looks more realistic for EM memories: since the DRAM scaling pace looks decreasing and the cost/bit is high, the competition is in this case focused on performances and scalability.

However today no EM can achieve DRAM performance (programming bandwidth and endurance). The best performing EM, STTRAM, needs to prove its manufacturability at sub 20nm node [14] and its scalability beyond, since exotic ferromagnetic materials are used in MTJ stack and thermal budget compatibility with CMOS doping is challenging. PCM has proven its manufacturability, but its endurance looks limited to $<10^9$ cycles. Oxide RRAM has excellent process compatibility with mainstream CMOS technology and potentially can give up to 10^{12} cycles, but the variability of switching parameters and the small read signal makes it difficult to reach DRAM performances

While there is little or no chance for EM to compete with DRAM and 3DNAND today, there is the opportunity to close the large performance gap between mainstream technologies, occupying the Storage Class Memory space introduced in fig 1.

Since all EM technologies have characteristics that are in between DRAM and NAND, it may look easy to enter in the market in this way, but practically even this path is challenging: EM manufactures need at first to develop a technology with the right balance between performances and cost, making it attractive respect to the present duopoly in the high density application market, then need to involve the hardware and software ecosystems to support their new memory with dedicated solutions and eventually need to convince customers that the added value of the new technology is worth the risk and cost of adopting it. All while conventional NAND and DRAM technologies are not sitting still and consolidated eco-systems continuously work on them.

However, the recent announcement of Intel-Micron 3D XPoint™ technology [15], with a 128Gb product at 20nm lithography node (2layers of $4F^2$ cells for an effective cell size of $2F^2$, as shown in figure 3), shows the possibility for a new NVM technology to enter into the market with a unique combination of performance and density, enabling high speed and high capacity data storage close to the processor.

FIGURE 3. 3D XPOINT™ CROSSPOINT ARRAY SCHEME

IV. SUMMARY

A huge research effort is expended on EM, looking to find alternatives to 3DNAND and DRAM mainstream technologies.

Even the most promising EM (PCM, RRAM and STTRAM) cannot compete today with the extreme high density and low cost of 3DNAND MLC devices or with the performance of DRAM.

The opportunity to enter in the market in the Storage Class Memory segment is also challenging: right balance of cost and performance and involvement of the eco-system and customers are key factors for success.

3D XPoint™ is the 1st EM showing the possibility to emerge in the high density application scenario.

ACKNOWLEDGMENT

The author gratefully acknowledges the contributions of the Micron R&D organization and the authors of the references.

REFERENCES

[1] R. F. Freitas, Storage-class memory: the next storage system technology, IBM J. of Res. and Dev., Vol. 52, Issue 4.5, 2008, pp. 438-447

[2] D. Tang, et al., Magnetic Memory – Fundamentals and Technology. Cambridge Univ. Press, 2010, pp. 122-165

[3] E. Cartier, et al., Characterization and optimization of charge trapping in high-K dielectrics, IEEE IRPS, 2013, **DOI:** 10.1109/IRPS.2013.6532019

[4] K. Lee, et al., Activation energies (Ea) of failure mechanisms in advanced NAND flash cells for different generations and cycling, IEEE TRED, 2013, Vol. 60, No. 3, pp. 1099-1107

[5] S.M. Sadeghipour, Phase change random access memory, thermal analysis, ITHERM, 2006, **DOI:** 10.1109/ITHERM.2006.1645408

[6] Y. Choi et al. A 20nm 1.8V 8Gb PRAM with 40MB/s program bandwidth, PROC IEEE ISSCC, San Francisco, CA, 2012

[7] R. Fackenthal, et al., A 16Gb reram with 200MB/s write and 1GB/s read in 27nm technology. IEEE ISSCC 2014, pp. 338-339

[8] https://www.everspin.com/ddr3-dram-compatible-mram-spin-torque-technology-0#overlay-context

[9] K. Prall, Scaling non-volatile memory below 30nm, IEEE NVSMW, 2007, pp. 5-10

[10] S. Sills, et al., A copper ReRAM cell for storage class memory applications, IEEE VLSI 2014, DOI: 10.1109/VLSIT.2014.6894368

[11] S. Lee, et al., A Study on the failure mechanism of a phase change memory in write/erase cycling, IEEE EDL, 2009, Vol 30, No. 5, pp. 448-450

[12] P. Cappelletti, Non volatile memory evolution and revolution, IEEE IEDM 2015

[13] T.Tanaka, M.Helm, T.Vali et al, A 768 3b/cell 3D floating gate NAND Flash Memory, IEEE ISSCC 2016

[14] J.Kim et al., Verification on the extreme scalability of STT-MRAM without loss of thermal stability below 15nm MTJ cell, IEEE Symposium on VLSI Technology, 2014

[15] www.micron.com/about/our-innovation/3d-xpoint-technology

Anti-ferroelectric ZrO_2:
An Enabler for Low Power Non-volatile 1T-1C and 1T Random Access Memories

M. Pešić*[†], M. Hoffmann*, C. Richter*, S. Slesazeck*, T. Kämpfe[‡], L. M. Eng[‡], T. Mikolajick*[§], U. Schroeder*

*NaMLab GmbH, Noethnitzer Str. 64 Dresden 01187, Germany
[‡]Institute of Applied Physics, Technische Universität Dresden, 01062 Dresden
[§]Chair of Nanoelectronic Materials, Technische Universität Dresden, 01062 Dresden
[†]E-Mail:Milan.Pesic@namlab.com

Abstract — **Recently it was demonstrated that an asymmetric DRAM capacitor stack can introduce non-volatility and at the same time outperform ferroelectric HfO_2 based FeRAM in terms of cycle endurance. With the present work, we provide an in-depth study of the underlying mechanisms and perform a comprehensive retention study that characterizes ferroelectric memories. Piezoelectric force microscopy is applied to prove the ultimate scalability of the proposed concept beyond capacitor based application. Finally, switching density plots reveal a much lower device to device variability and high switching uniformity with respect to ferroelectric HfO_2 based films.**

Keywords — anti-ferroelectric; non-volatile DRAM; ZrO_2; reliability.

I. INTRODUCTION

The ever-growing market of battery powered electronics together with a transition to data oriented computing continuously pushes the requirements for low power memory solutions. A significant portion of the consumed power can be attributed to DRAM, which, due to its volatility, requires continuous refresh in the non-operating state. Even though many different memory concepts were suggested during the last two to three decades, neither one concept became the solution for the missing link that combines the speed of DRAM with the non-volatility of Flash and hence address the massive memory storage demands of tomorrow. One of the candidates are the long time envisioned ferroelectric (FE) memories. While offering stable remanent polarization at zero electric field together with the fact that the information can be stored as polarization direction which can be altered by an electrical field makes ferroelectric materials a natural choice to realize a non-volatile memory. Even though the concept of FE memory was proposed already in 1950s [1], Ramtron was the first company that initiated the mass production of ferroelectric memories in 1993 [2]. More than a decade later the most aggressively scaled technology node was commercialized by Texas Instruments [3]. These memories are based on lead zirconate titanate (PZT), which is one of the most prominent ferroelectrics, but unfortunately not the best material with respect to scaling and CMOS compatibility. Discovery of the ferroelectric properties [4] within HfO_2 oxide bridged the gap between the state-of-the-art technology nodes and ferroelectric memories [5]. Beside

non-volatility new memory concepts have to ensure sufficient endurance and operation stability.

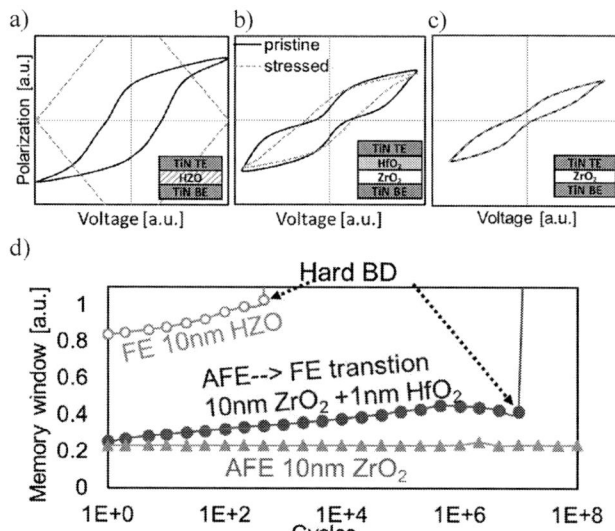

Fig. 1. Comparison of pristine (solid) and cycled (dashed) P-V characteristics measured on the MIM structures comprising a) FE HZO b) AFE ZrO_2 and c) 10nm HfO_2 and 1nm ZrO_2 as a dielectric. d) Field cycling Endurance of the ZrO_2-HfO_2 mixtures. Decrease of the stability with increase of HfO_2 content within the films is observable. A MW above 1 denotes leakage.

Interestingly, depending on the ratio between ZrO_2 and HfO_2 within $Hf_xZr_{1-x}O_2$ a phase transition from ferroelectric (x=0.5; see Fig.1a) to anti-ferroelectric (x=0; see Fig.1b) behavior can be induced. In contrast to FE which are by nature non-volatile materials, anti-ferroelectrics (AFE) do not show any remanent polarization at zero-field. Here, difference between the positive and negative remanent polarization defines the memory window (MW). Nonetheless, these AFE materials reveal a very stable endurance with respect to the FE counterparts which exhibit changes in the MW [6] followed by hard breakdown much earlier than required. Reliability aspects of FE were addressed in [7] and the interfacial layer was identified as the week link. Hence, the interface engineering would be required for the reliability improvements. Due to their endurance strengths, AFE materials would be a preferred choice for integration, its non-volatility is a significant obstacle. Interestingly, this AFE behavior is observable in state-of-the-art DRAM capacitor stacks. In this study, we show how the non-volatility can be introduced into the AFE based MIM capacitors and offer unmatched endurance to FE based NVMs, respectively (Fig.1).

The paper is organized as follows: After the fabrication details, the concept of the anti-ferroelectric non-volatile memory is presented. This is followed by the device performance analysis and its comparison to a FE based random access memories. Finally using microscopic measurements, the applicability of the AFE non-volatile memory was explored for ultimately scaled single transistor memory cells.

II. EXPERIMENTAL

TiN/Hf$_x$Zr$_{1-x}$O$_2$/TiN metal-ferroelectric-metal (MFM) stacks were fabricated on Si substrates. First, a 10 nm thick TiN bottom electrode (BE) and 10 nm Hf$_x$Zr$_{1-x}$O$_2$ dielectrics mixtures were deposited by reactive sputtering at room temperature and atomic layer deposition at 280 °C, respectively. Further sample preparation included reactive sputtering of a 12 nm TiN top electrodes (TE) at room temperature. After completion of the stack, Hf$_x$Zr$_{1-x}$O$_2$ mixtures were crystalized by a 20 s RTP anneal at 800 °C and 450°C in nitrogen atmosphere. Finally, circular hard masks consisting of 10 nm Ti (adhesion layer) and 50 nm Pt were deposited in an electron beam evaporator using a shadow mask defining the area of device of 9500 μm². Discrete capacitor structures were patterned by wet etching the TiN in-between the Pt dots. Furthermore, RuO$_x$ top-electrodes with different workfunction were directly patterned by adoption of the shadow mask. Polarization-voltage (PV) characteristic of the capacitors was recorded using a aixACCT TF Analyzer 3000 by applying triangular pulses at 10 kHz frequency and measuring transient current response, which is subsequently converted into a polarization charge. Retention recording and the device variability investigations of the capacitor structures were performed using a semiautomatic probe station from Cascade Microtech and Keithley 4200 SCS equipped with an internal pulse-measure unit. Ultimate scaling of the device concept was analyzed by piezoelectric force microscopy (PFM) using an Asylum Research Cypher S.

III. CONCEPT AND DESIGN STRATEGIES

Utilizing a capacitor stack with asymmetric electrode work functions (WF) Pešić et al. [8, 9] reported how the generated internal bias field can be utilized to modify the energy landscape of a AFE capacitor and to make a DRAM non-volatile (Fig.2). To materialize the high endurance strength of AFE materials (Figs.1b, d) and ensure that the generated internal bias field is sufficient for inducing the field shift, a branch of the pinched hysteresis of the AFE has to be positioned in the vicinity of zero-field. The internal bias field E$_{built-in}$ generated by using electrodes with different workfunction (WF) values is described by the following equation:

$$E_{built-in} = \frac{1}{t \cdot q} \cdot (WF_{TE} - WF_{BE}), \quad (1)$$

where t represents the thickness of the dielectric, q the elementary charge, WF$_{TE}$ the workfunction value of the top electrode and WF$_{BE}$ the workfunction value of the bottom electrode. The introduced internal bias field centers one branch

of the pinched hysteresis loop and enables non-volatile operation between the polarized (logical binary "0") and un-polarized (logical binary "1") state (see Fig.2).

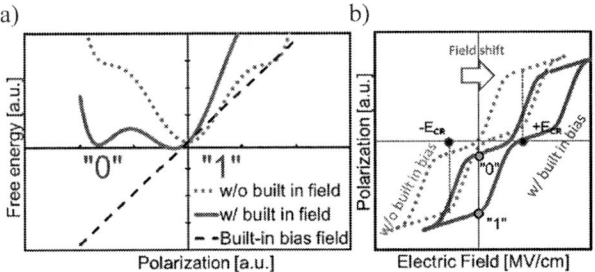

Fig. 2. a) Free energy vs. polarization and corresponding b) polarization hysteresis of AFE materials at zero external fields with (solid) and without (dotted) built-in bias field. The shift in the polarization characteristics along the field axis results in a P$_r$ at zero bias and therefore enables non-volatile data storage.

Besides the asymmetric workfunction values of the electrode as reported in [8] (Fig.3a) two additional methods are suggested. The required internal bias field can be generated by introduction of an interfacial dipole [10] (Fig.3b.) or fixed charges between dielectrics [11]. A surface density difference of oxygen atoms at the interface between two oxide materials can be used as an intrinsic origin for dipole formation. In a similar way, the change of composition from one material to another can induce positive or negative fixed charge [11]. Both effects can be used for the introduction of an internal bias field. It should be noted that the direction of the internal bias field, generated by the asymmetric WFs can be altered combining the top electrode with a higher or lower workfunction material with respect to the bottom electrode. While the dipole or interface charge effects can be controlled by the sequence of the deposited layers.

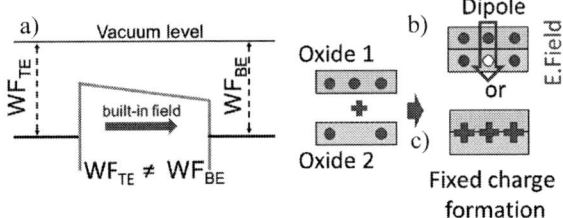

Fig. 3. Built-in bias field introduction by a) asymmetric stack comprising electrodes with different workfuction values and by cobining the heterostructure of two oxide materials with different surfac density of oxygen which result in either b) dipole fromation or c) fixed charge introduction.

According to the WF values from literature [12], it can be expected that a RuO$_x$ or Pt top electrode (TE) induces a shift of about 0.7 to 1.1 MV/cm (for a 10 nm dielectric) if combined with a typical TiN based bottom electrode (BE). This shift is sufficient for the modification of the free energy landscape (Fig.2) and centering of one of the AFE hysteresis branches resulting in a two-state non-volatile memory (Fig.4a) based on the state-of-the art DRAM capacitor.

After a theoretical considerations and simulation (Fig.2), an AFE behavior is stabilized within the TiN/ZrO$_2$/TiN film (see

Fig.1c). Removal of the TiN electrode was followed by a deposition of the RuO$_x$ TE through the shadow mask. Consequently, a polarization measurement resulted in a centered hysteresis curve characteristic for the ferroelectric material (Fig.4a). Extension of the sweep towards more positive values prove that the material is still anti-ferroelectric (Fig. 4a dashed trace) but the hysteresis is shifted along the field axis.

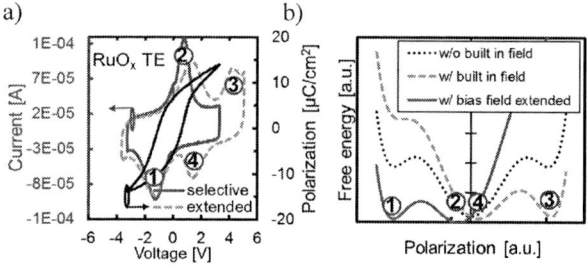

Fig. 4. a) Comparison of AFE TiN/ZrO$_2$/RuO$_x$ NV-DRAM stacks with extended range and selective read-out. b) Free energy and ion position depending on the applied voltage.

IV. PERFORMANCE OF NV-DRAM

After a fabrication consideration and simulation, the device performance is investigated. In the first part, a single transistor single capacitor (1T-1C) memory architecture is investigated, while in the second part ultimate scaling and possibility of the material integration in single transistor (1T) architectures is investigated.

A. 1T-1C Architecture

In contrast to the ferroelectric based memories, a wake-up free stable endurance characteristic was observed (see Figure 5). Small changes in the memory window are caused by an imprint effect. In addition to the imprint based retention tests reported in [8], a detailed retention study was carried out. In case of FeRAM memories, same state (SS), opposite state (OS) and new same state (NSS) retention can be defined [13]. The pulse sequences used for positive and negative state retention tests are illustrated in Figure 6a while the calculation of the respective parameters (e.g. OS) is shown in Figure 5c. To assess all states by retention tests, four discrete capacitors were used and exposed to different pulse sequences shown in Figure 5b. Similar to the PUND tests, the resulting current is dependent on the voltage history and the previous state in which the capacitor resided before the measurement. The first switching pulse comprises dielectric and ferroelectric switching current components, whereas during the second pulse only the dielectric response (including leakage currents and relaxed polarization) is observed. Subtracting the resulting current of the second non-switching pulse from the first one, a normalized switching contribution (or normalized polarization "1" was obtained). The last pulse of each sequence was used to set a defined polarization state, after which the sample was stored at different temperatures for different time intervals. After each bake, the same sequence was applied. Applying previously described methodology, retention was recorded proving that the NV-DRAM is capable of preserving the logical state at 100 °C with negligible loss for at least 75h (Fig.5d). Besides the 1T-1C concept, AFE based films can be utilized in 1T memory cells [8]. In the next section the

integration potential of AFE films in ultimately scaled devices will be explored.

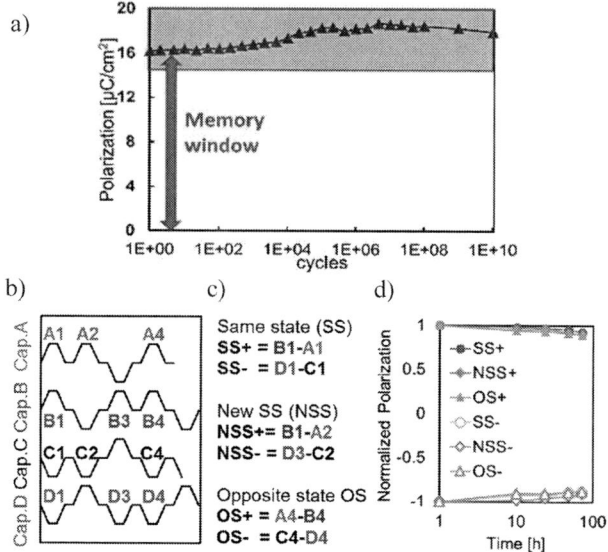

Fig. 5. Reliability assesment of AFE capcitor stack .a) Memory window as a function of field cycles for a 10 nm thick ZrO$_2$ based DRAM stack with TiN top and bottom electrode. b)Waveform transient used for obtaining the retention characteristics measured on four capacitor devices; c) Calculation of the retention values specific for FeRAM from the respective integrated current response; d) Retention of different states (SS,OS,NSS) of the TiN/ZrO$_2$/RuO$_x$ device recorded at 100 °C.

B. 1T Architecture

To ensure the scaling of the devices for a 1T cell architecture and to confirm the material properties on a microscopic level, PFM measurements were performed on a AFE-stack in comparison to FE stacks. The measurement with a Pt-tip on the bare AFE oxide (Fig.6b) resulted in centering of the hysteresis loop (internal bias field between Pt-tip and TiN BE) in the same manner as in the macroscopic experiments. In contrast, the same experiment on the FE Sr doped HfO$_2$ resulted in a biased loop (Fig.6a).

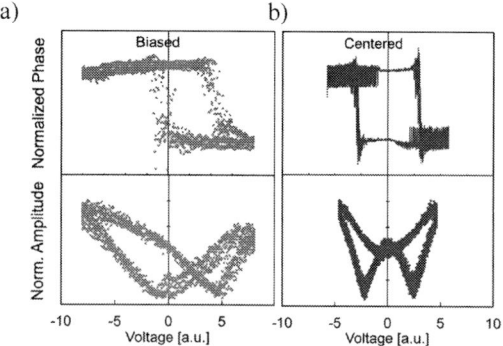

Fig. 6. PFM measurements of the ferroelectric and anti-ferroelectric material on bare oxide. WF difference between the Pt-tip and TiN BE induces the internal bias which consequently a) bias phase and amplitude of the FE and b) centers the phase and amplitude of the AFE, respectively.

978-1-5090-5979-9/17 $31.00 © 2017 IEEE

The polycrystalline nature of both FE and AFE binary oxides and subsequent device patterning may result in a scenario where one discrete device may possess a different number of domains characterized by different distributions of parameters like switching voltage, biasing of the grains with respect to the neighboring. Variability related to the switching domain distribution and the local domain bias is an important parameter of ultra-scaled polarization based devices [14].

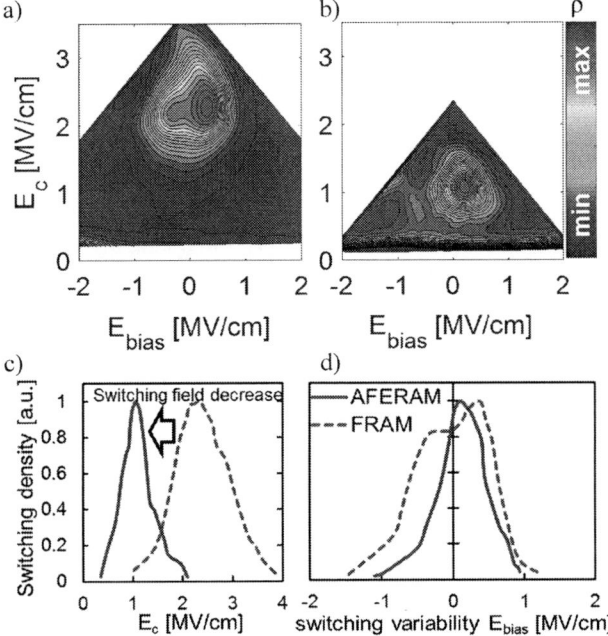

Fig. 7. Preisach switching density $\rho(E_c, E_{bias})$ denoting a switching distribution plots correlated with film variability obtained from first order reversal curves of a) FE and b) AFE-device. c) Comparison of domain switching field distribution and d) Comparisopn of local bias field variability of FE- and AFE- devices.

By applying the first order reversal curve (FORC) [6] method the variability of the device was investigated. The FORC measurement procedure initiates at fields high enough to reach the saturation polarization. Subsequently, the electric field is swept toward the negative direction and back to saturation polarization, stepwise lowering the value of the reversal field until the opposite saturation polarization is reached. This measurement procedure and the derivation of the switching density $\rho(E_r, E)$ utilized for probing the reversal fields enables variability comparison of domain distributions of ferroelectric and anti-ferroelectric material stacks. Out of the switching density $\rho(E_r, E, E_{bias}, E_c)$ the two most important parameters of the film can be evaluated. The local internal bias field E_{bias} and the coercive field E_c. In contrast to the FE case, an AFE-RAM exhibits a much narrower switching distribution indicating a lower device-to-device variability in ultra-scaled devices (Fig.7). Besides the improvement of the variability of local bias field (E_{bias}) a significant decrease of the coercive field (E_c) distribution and magnitude is observed (Fig.7c-d).

V. CONCLUSION

After the recent report of a NV-DRAM concept where asymmetric electrode materials induce an internal bias field, thus enabling non-volatility in state-of-the art DRAM capacitors, we extend our study and focus on the reliability and detailed investigation of the suggested concept. This study sheds light on the important factors for improving the reliability performance by this approach compared to classical ferroelectric based FeRAM. Hence, in order to bypass the need for usage of asymmetric workfunctions, internal bias generation was revisited and two additional paths for internal bias introduction are suggested. Microscopic PFM experiments prove the feasibility of scaling the concept beyond capacitor application towards scaled 1T architectures. The concept enables the reduction of the operating voltage and reduces variability compared to ferroelectric HfO$_2$ based devices.

ACKNOWLEDGMENT

This work is supported by the Free State of Saxony, Germany.

REFERENCES

[1] D. Bondurant et al., "Ferroelectronic RAM Memory Family for Critical Data Storage, Ferroelectrics" 112 (1990), 273-282.

[2] Ramtron. F-ram technology brief. Technical report, Ramtron, 2007. URL: "http://www.digikey.com/-eb%20Export/Supplier%20Content/ramtron-1140/pdf/ramtron-tech-ferroelectric.pdf?redirected=1.," p. 2007, 2007.

[3] H.P. McAdams et al., "A 64-Mb Embedded FRAM Utilizing a 130-nm 5LM Cu/FSG Logic Process", IEEE J. Solid-St. Circ. 39 (2004), 667-677.

[4] T. S. Böscke et al., "Ferroelectricity in hafnium oxide thin films," Appl. Phys. Lett. 99, 102903 (2011); doi: 10.1063/1.3634052.

[5] J Müller et al. "Ferroelectricity in HfO2 enables nonvolatile data storage in 28 nm HKMG," 2012 Symposium on VLSI Technology (VLSIT), Honolulu, HI, 2012, pp. 25-26. doi: 10.1109/VLSIT.2012.6242443.

[6] M. Pešić et al., "(2016), "Physical Mechanisms behind the Field-Cycling Behavior of HfO2-Based Ferroelectric Capacitors". Adv. Funct. Mater., 26: 4601–4612. doi:10.1002/adfm.201600590.

[7] M. Pesic et al., "Root cause of degradation in novel HfO2-based ferroelectric memories," 2016 IEEE International Reliability Physics Symposium (IRPS), Pasadena, CA, 2016, pp. MY-3-1-MY-3-5.doi: 10.1109/IRPS.2016.7574619.

[8] M. Pesic, S. Knebel, M. Hoffmann, C. Richter, T. Mikolajick and U. Schroeder, "How to make DRAM non-volatile? Anti-ferroelectrics: A new paradigm for universal memories," 2016 IEEE International Electron Devices Meeting (IEDM), San Francisco, CA, 2016, pp. 11.6.1-11.6.4. doi: 10.1109/IEDM.2016.7838398.

[9] M. Pešić et al., (2016), "Nonvolatile Random Access Memory and Energy Storage Based on Antiferroelectric Like Hysteresis in ZrO2". Adv. Funct. Mater., 26: 7486–7494. doi:10.1002/adfm.201603182.

[10] K. Kita et al., "Origin of electric dipoles formed at high-k/SiO2 interface" App.Phys.Lett. 94,13, 2008.

[11] DK Simon et al,"On the Control of the Fixed Charge Densities in Al2O3-Based Silicon Surface Passivation Schemes"ACS A.Mat&Int 7, 51, 2015.

[12] HK. Kim et al., ACS App.Mat&Int 2013 5 (4), 1327-1332 DOI: 10.1021/am302604e.

[13] J. Rodriguez et al., "Reliability of Ferroelectric Random Access Memory embedded within 130 nm CMOS," IEEE Int. Reliab. Phys. Symp. Proc., pp. 750–758, 2010.

[14] H. Mulaosmanovic et al., "Evidence of single domain switching in hafnium oxide based FeFETs: Enabler for multi-level FeFET memory cells" IEDM 2015 26.8. 1-26.8.3.

From Planar To Vertical Capacitors : A Step Towards Ferroelectric V-FeFET Integration

Karine Florent[*†·], Simone Lavizzari[†], Luca Di Piazza[†], Mihaela Popovici[†], Goedele Potoms[†], Tom Raymaekers[†], Guido Groeseneken[*†] and Jan Van Houdt[*†]

[*] Department of Electrical Engineering, KU Leuven, Leuven, Belgium
[†] imec, Leuven, Belgium
[·] E-mail : karine.florent@imec.be

Abstract— **Ferroelectric hafnium oxide (HfO₂) attracted a lot of interests since its discovery in 2007. Its scalability and CMOS compatibility are two advantages over conventional ferroelectric materials, favoring new device integration. Doped ferroelectric HfO₂ Metal/Insulator/Metal capacitors have been widely studied for DRAM and FeFET applications. Silicon electrodes have not been discussed in much detail so far, though it provides an input for vertical ferroelectric FET (V-FeFET), based on Flash NAND architecture. In this work, planar and vertical capacitors are presented using silicon as electrodes and aluminum doped HfO₂ as ferroelectric material. The process of the vertical integration is described. Polarization-Voltage measurements show a steep ferroelectric hysteresis loop for the planar as well as vertical capacitors. This demonstrates the conservation of the ferroelectric properties of the dielectric after deposition on an etched vertical wall and lays the first stone toward V-FeFET integration.**

Keywords—ferroelectricity; hafnium oxide; vertical capacitor

I. INTRODUCTION

Ferroelectric (FE) material has recently engendered great attention from the memory community with the discovery of FE-HfO₂. Such material could enable the fabrication of low power devices in standalone semiconductor manufacturing facilities due to its scalability and CMOS compatibility. Already, a one-transistor ferroelectric field effect transistor embedded NVM has been implemented for low power application [1]. This integration demonstrates the potential utilization of such material system for memory applications.

The existence of ferroelectric properties in HfO₂ was discovered in 2007 by Böscke *et al.* using silicon as dopant in a planar Metal/Insulator/Metal (MIM) capacitor [2]. The presence of a small amount of dopant, a capping layer and a thermal anneal facilitate the transformation of the monoclinic phase into a non-centrosymmetric orthorhombic phase which is the origin of this property. FE-HfO₂ has been mainly studied with planar MIM capacitor [3]. 3D FE-HfO₂ capacitor has been reported with Al as dopant and TiN as electrodes [4]. The use of polycrystalline silicon (poly-Si) as

electrodes has already been proposed [5]. Such study provides an important input for 3D applications, in which poly-Si is used as channel [6].

In this work, Al doped HfO₂ Silicon/Insulator/Silicon (Al:HfO₂ SIS) capacitors have been fabricated. After observation of the FE-property on a planar structure using material and electrical analyses, the stack was integrated in a vertical cylindrical device. The process flow is first described followed by electrical measurements.

II. DEVICE FABRICATION AND MATERIAL ANALYSES

A. Planar Capacitor

The devices were fabricated on highly doped p-type silicon substrates ($\sim 10^{20}$ at/cm³). After a HF clean for native oxide removal, FE-Al:HfO₂ was deposited by atomic layer deposition process at 300 °C using hafnium chloride HfCl₄ and trimethylaluminum Al(CH₃)₃ as metal precursors, H₂O as oxidant and with a pulse ratio of Hf:Al = 34:1. The thickness of the dielectric, ~8 nm, was measured by X-Ray Reflectivity. A 50 nm thick amorphous silicon was deposited on top of the dielectric at a temperature of 450 °C by chemical vapor deposition (CVD), followed by an arsenic implant. A rapid thermal anneal step at 850 °C for 1 minute in N₂ ambient at atmospheric pressure was performed subsequently for dopant activation and crystallization. The poly-Si served as contact pad for electrical measurements. After lithography, dry etch was used to define the capacitors structures.

GIXRD was performed using a JVX7300M (Jordan Valley) to study the crystallinity of FE-HfO₂ after the dielectric deposition, top electrode deposition and after the final thermal treatment. The GIXRD results are shown in Fig. 1. After FE-HfO₂ deposition, the dielectric is amorphous as expected. Crystallization happens during the top electrode deposition confirmed by the presence of several intensity peaks. The dielectric seems to have mainly an orthorhombic and/or cubic structure. It is not possible to clearly distinguish between cubic and orthorhombic HfO₂,

978-1-5090-5979-9/17 $31.00 © 2017 IEEE

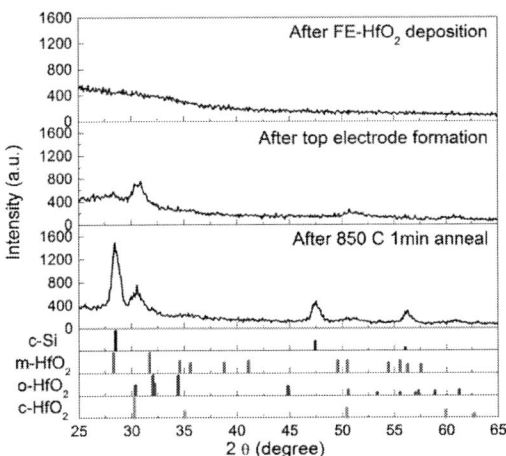

Fig. 1. GIXRD data showing the evolution of the dielectric and top electrode crystallization.

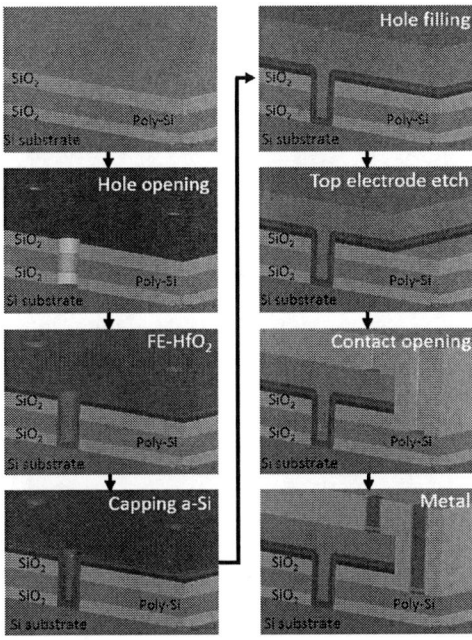

Fig. 2. Process flow of vertical SIS capacitor, showing the important fabrication steps.

as they have similar diffraction lines (common peaks at 2θ of 30.4 ° and 50.5 °). A low intensity peak can be observed at $2\theta = 28.4$ °, corresponding to monoclinic HfO_2. After the final thermal treatment, a large increase in the intensity peak at $2\theta = 28.4$ ° is observed. Intensity peaks at $2\theta = 47.4$ ° and 56.1 ° are also detected. These diffraction lines correspond to the top Si electrode, which crystallized in a cubic structure. Cubic Si has also an intensity peak at $2\theta = 28.4$ °, which is overlapping with monoclinic HfO_2. This explains the high intensity peak at this 2θ value. Therefore, the presence and/or increase of m-HfO2 phase cannot be confirmed. The intensity peak at $2\theta = 30.4$ ° was not significantly altered by the anneal. However, a modification of the HfO_2 crystal structure could still have occurred as orthorhombic and cubic phases cannot be differentiated.

B. Vertical Capacitor

Fig. 2 illustrates the process sequence for the vertical device. In this process, 30 nm thick SiO_2 is first deposited, followed by a 50 nm thick p-doped poly-Si gate and finally 40 nm thick SiO_2 deposition. Vertical cylindrical holes with diameters ranging from 60 nm to 100 nm are etched after dopant activation. An 8 nm thick FE-HfO_2 is then deposited by atomic layer deposition after a HF clean, followed by a thin n-doped Si protective layer deposition. The hole is filled with highly doped a-Si, which serves as electrode. The stack is then annealed at 850 °C for 1 min in N_2 ambient for dopant activation as well as FE-layer crystallization. Finally, the layers are etched to reach the gate electrode, followed by the deposition of metal contacts.

GIXRD was attempted on this device, but the signal intensity was too low to be able to discern the dielectric peaks. Indeed, in this device, HfO_2 is etched to be able to connect the poly-Si gate electrode. Therefore, the FE-material is only present on the small area, while in the planar case, the dielectric is everywhere on the wafer.

High Resolution Transmission Electron Microscope (HRTEM) was performed on the vertical structure and a cross-section of the device with a hole diameter of 90 nm is shown in Fig. 3(a). A small recess in silicon substrate of ~ 15 nm can be seen, resulting from holes overetch. The oxide recess on the side is due to the HF clean, performed before HfO_2 deposition, in order to remove native oxide on the poly-Si. A very good conformity of the ALD layer is visible. Annular Bright-Field and Dark-Field Scanning Transmission Electron Microscope (ABF-STEM and DF-STEM respectively) were used to observe the crystallization of the a-Si and FE-HfO_2 (Fig. 3(b) and Fig. 3(c)). A clear border is noticeable between FE-HfO_2 and poly-Si, without the presence of an amorphous layer at the interfaces. Electrical measurements were achieved with a Keithley 4200-PIV parameter analyzer for both type of capacitors.

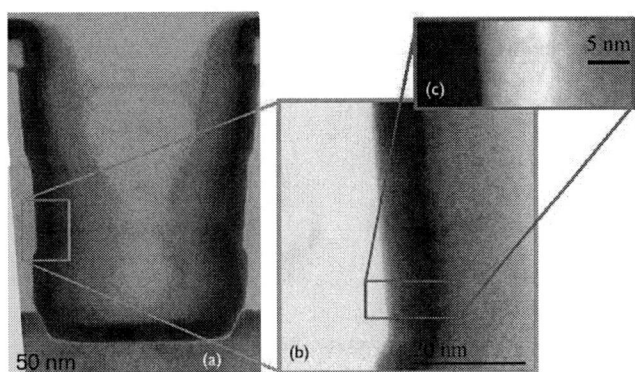

Fig. 3. (a) HRTEM cross-section of the vertical capacitor. (b) ABF-STEM of the sidewall showing crystallization of FE-HfO2 and silicon. (c) DF-STEM exhibiting no clear presence of interfacial layers.

978-1-5090-5979-9/17 $31.00 © 2017 IEEE

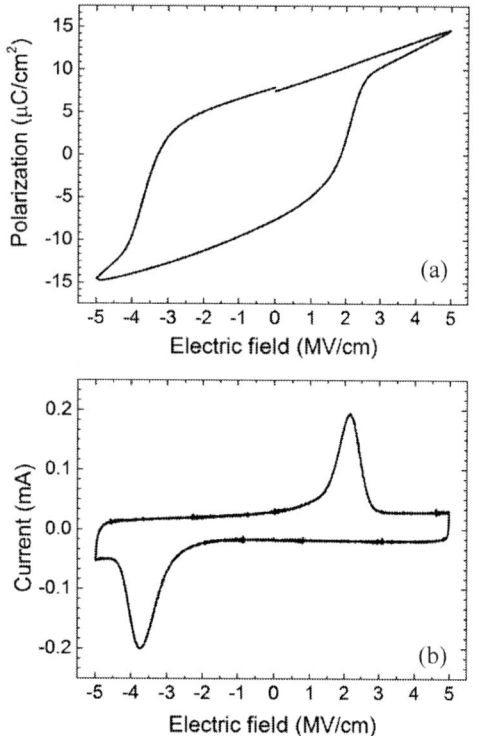

Fig. 4. (a) P-E characteristics of the planar capacitor with an applied electric field of 5 MV/cm after 1000 cycles. (b) Corresponding I-E measurements.

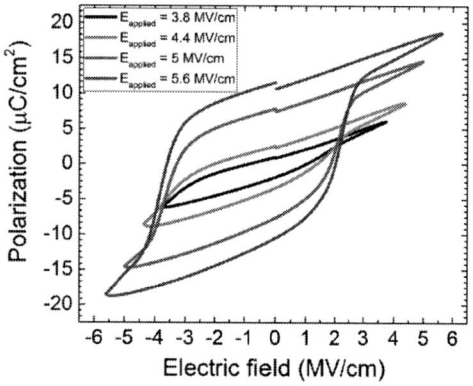

Fig. 5. Example of sub-loops after 1000 cycles, leading to a saturated hysteresis.

III. RESULTS AND DISCUSSION

Polarization-Voltage (P-V), or Polarization-Electric field (P-E) hysteresis was extracted from Current-Voltage (I-V) measurements by applying a triangular pulse with a constant ramp of 10 μs/V. Bipolar square pulses were used to cycle the capacitors.

The planar device exhibits steep ferroelectric switching with a 'square' P-E loop (Fig. 4(a) and Fig. 4(b)). The observed shift towards the left is probably due to difference in electrodes work functions (p-type vs n-type Si) and/or

possible charge accumulation at the thin interfacial layers. In literature, coercive fields (Ec) of around ±1 MV/cm are reported for Al:HfO$_2$ [4], [7]. The larger 2Ec observed here can be the result of polysilicon depletion and presence of interfacial oxide, resulting in a voltage drop.

Fig. 5 presents the evaluation of the P-V hysteresis from sub-loops to saturated loop. A maximum remnant polarization (Pr) of ± 11 μC/cm^2 at an applied electric field of 5.6 MV/cm is extracted. As the applied voltage is close to the breakdown voltage, the endurance is limited in this device below 10000 cycles for the saturated loop. A better approach would be to use electrodes with similar work functions.

P-E measurements were carried out on vertical devices with various diameters, ranging from 60 nm to 100 nm. To get a measurable current, an array of 100k holes in parallel was measured. Fig. 6 presents the results using holes with a diameter of 80 nm. When comparing the currents between Fig. 4(b) and Fig. 6(b), it can be observed that the current is ~8x lower for the vertical capacitor as compared to the planar one: this is perfectly consistent with a ~8x smaller equivalent area in the vertical structure. A 'square' P-E hysteresis is observed with steep switching, confirming that ferroelectricity can be achieved on cylindrical vertical structure with silicon electrodes. A maximum electric field of 5 MV/cm was applied before breakdown of the dielectric.

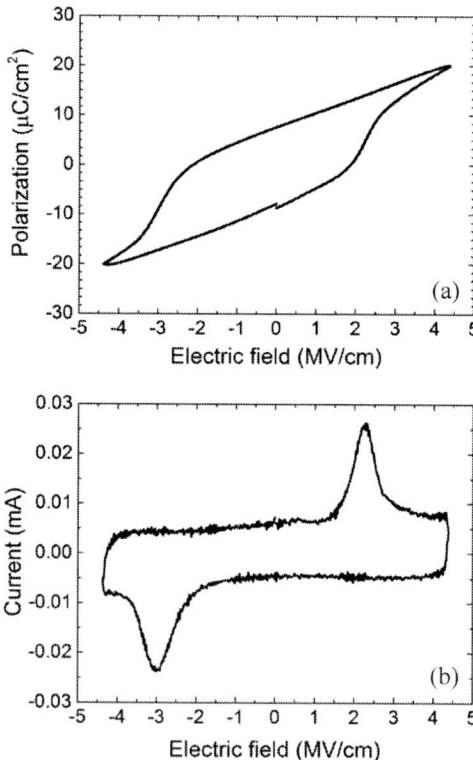

Fig. 6. (a) P-E results of vertical capacitor using an array of 100k holes with a diameter of 80 nm and at an applied electric field of 4.4 MV/cm after 1000 cycles. (b) Corresponding I-E measurements.

978-1-5090-5979-9/17 $31.00 © 2017 IEEE

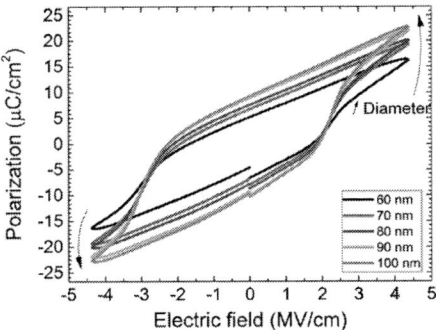

Fig. 7. P-E hysteresis of vertical capacitors with various diameters.

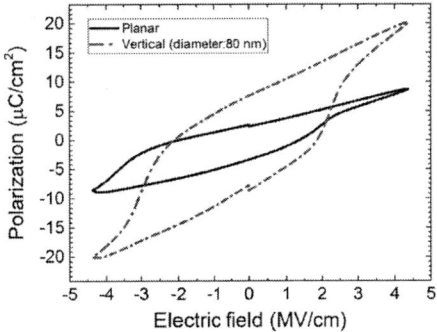

Fig. 8. Comparison between planar and vertical (80 nm diameter hole) capacitors at same applied electric field with vertical capacitor showing larger Pr.

Fig. 7 displays P-E measurements of vertical capacitors with various diameters. A trend is noticeable as a function of the size. It appears that larger diameter results in larger Pr. This is attributed to the stress coming from amorphous silicon during crystallization which depends on the diameter. This seems to enhance the degree of polarization, while maintaining the same coercive field. Since the hole etch was optimized for an 80 nm diameter, the effective area is also a factor not to be neglected, as it might impact the normalization and thus the polarization extraction.

A comparison between planar and vertical capacitors is shown in Fig. 8. A larger Pr is observed for the vertical device. The shapes of the P-E loops, *i.e.* the slope of the flat parts, are different, leading to a dielectric constant of 32 for the vertical device and 18 for the planar capacitor. This indicates that the hole leads to a FE-material with more orthorhombic structure than the planar capacitor, since the dielectric constant of the planar structure corresponds to a monoclinic HfO_2 phase.

IV. CONCLUSION

Al:HfO_2 vertical capacitors with various holes diameter and using silicon as electrodes were fabricated and compared to a planar device. Ferroelectric properties were observed on both cases. 3D structure exhibits steep switching, higher remnant polarization and larger dielectric constant than planar device which is attributed to the crystallization of the FE-HfO_2 in orthorhombic structure with larger amount. Such study opens the door for the fabrication of V-FeFET with 3D NAND architecture.

REFERENCES

[1] M. Trentzsch, S. Flachowsky, R. Richter, J. Paul, B. Reimer, D. Utess, S. Jansen, H. Mulaosmanovic, S. Müller, S. Slesazeck, J. Ocker, M. Noack, J. Müller, P. Polakowski, J. Schreiter, S. Beyer, T. Mikolajick, and B. Rice, "A 28nm HKMG super low power embedded NVM technology based on ferroelectric FETs," in 2016 IEEE International Electron Devices Meeting (IEDM), 2016, p. 11.5.1-11.5.4.

[2] T. S. Böscke, J. Müller, D. Bräuhaus, U. Schröder, and U. Böttger, "Ferroelectricity in hafnium oxide thin films," Appl. Phys. Lett., vol. 99, no. 10, pp. 2–4, 2011.

[3] U. Schröder, S. Müller, J. Müller, E. Yurchuk, D. Martin, C. Adelmann, T. Schlösser, R. van Bentum, and T. Mikolajick, "Hafnium Oxide Based CMOS Compatible Ferroelectric Materials," ECS J. Solid State Sci. Technol., vol. 2, no. 4, pp. N69–N72, 2013.

[4] P. Polakowski, S. Riedel, W. Weinreich, M. Rudolf, J. Sundqvist, K. Seidel, and J. Muller, "Ferroelectric deep trench capacitors based on Al:HfO2 for 3D nonvolatile memory applications," 2014 IEEE 6th Int. Mem. Work. IMW 2014, pp. 1–4, 2014.

[5] K. Florent, M. Popovici, S. Lavizzari, L. Di Piazza, G. Groeseneken and J. Van Houdt, "Impact of Top and Bottom Conductive Layers on Electrical and Material Properties of Ferroelectric Aluminum Doped HfO2", discussed at the 2016 IEEE SISC, San Diego, CA, USA

[6] G. Van Den Bosch, G. S. Kar, P. Blomme, A. Arreghini, A. Cacciato, L. Breuil, A. De Keersgieter, V. Paraschiv, C. Vrancken, B. Douhard, O. Richard, S. Van Aerde, I. Debusschere, and J. Van Houdt, "Highly scaled vertical cylindrical SONOS cell with bilayer polysilicon channel for 3-D nand flash memory," IEEE Electron Device Lett., vol. 32, no. 11, pp. 1501–1503, 2011.

[7] S. Müller, J. Müller, A. Singh, S. Riedel, J. Sundqvist, U. Schröder, and T. Mikolajick, "Incipient ferroelectricity in Al-doped HfO2 thin films," Adv. Funct. Mater., vol. 22, no. 11, pp. 2412–2417, 2012.

Doped GeSe Materials for Selector Applications

Naga Sruti Avasarala[#1,2], B. Govoreanu[1], K. Opsomer[1], W. Devulder[1], S. Clima[1],
C. Detavernier[3], Marleen van der Veen[1], Jan Van Houdt[1,2], Marc Henys[1,2], L. Goux[1] and G.S. Kar[1]

[1]Imec, Kapeldreef 75, B-3001 Leuven, Belgium, [2]also with KU Leuven, Arenbergpark 10, B-3001 Leuven, Belgium, [3]Ghent University, Krijgslaan 281/S1, 9000 Ghent, Belgium [#]Contact: avasar49@imec.be.

Abstract — **We report on the thermal and electrical performance of nitrogen (N) and carbon (C) doped GeSe thin films for selector applications. Doping of GeSe successfully improved its thermal stability to 450°C. N doping led to a decrease in the off-state leakage and an increase in threshold voltage (V_{th}), while C doping led to an increase in leakage and reduced V_{th}. Hence, we show an effective method to tune the electrical parameters of GeSe selectors by using N and C as dopants.**

Keywords— *Selector; Ovonic Threshold Switching (OTS); GeSe; Doping; Threshold voltage; Nitrogen; Carbon.*

I. INTRODUCTION

The emergence of 3D stackable memory in the cross-bar architecture has been receiving attention as a viable candidate for storage class memory. Selectors become essential as the density of arrays increases, to avoid sneak path issues, which hinder memory array operation [1], [2]. Ovonic threshold switching (OTS), which is a field induced volatile switching mechanism, has been demonstrated by Ovshinsky [3] in chalcogenide materials. These materials have gained interest for selector applications due to their high on current and high non-linearity which are required for the implementation of large memory arrays.

Ge_xSe_{1-x} glasses are promising OTS materials. The modulation of OTS switching by varying the Ge concentration in Ge_xSe_{1-x} has been demonstrated [4], [5]. Varying the thickness of the layer also reduces the V_{th} at the cost of increased I_{off} [6]. The effect of dopants, Sb [7], N [8] and Bi [9] on the switching characteristics of Ge_xSe_{1-x} have been studied before for different values of x ranging from 0.4-0.7.

These dopants were used to modulate the threshold voltage of GeSe. In addition to tunable operation parameters, to implement selectors on memory arrays in an integrated process, improvement of the thermal stability of these glasses is essential.

In this work, we study the effect of addition of N and C dopants on the OTS functionality of GeSe. These dopants are interesting due to their high co-ordination number (C=4, N=3) which would increase the average co-ordination number of the network and enhance the network connectivity [5]. This is expected to strengthen the amorphous network and postpone crystallization.

II. EXPERIMENTAL

The selector devices studied in this work are cross-bar structures with the overlap area between a bottom electrode (BE) and a top electrode (TE) defining the active device area. The stack consists of BE Ru/ 20nm GeSe/ 20nm TiN TE as illustrated in Fig. 1 (a). The devices were processed on a 300 mm substrate, with the patterning of the BE, followed by oxide deposition and chemical mechanical planarization (CMP) to planarize the top wafer surface.

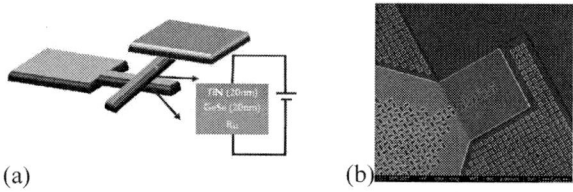

(a)   (b)

Fig. 1(a) Schematic of the selector device (b) Top-down SEM of the cross-bar selector devices which overlap area of 3umx3um. The pattern visible on the substrate is due to dummies used to control the CMP process.

Subsequently, the TE was defined by an e-beam write process as shown in Fig. 2. Poly methyl methacrylate (PMMA) resist was spun on Ru BE and written by ebeam lithography. The resist was developed after exposure for 1 minute in a solution of Mehtyl Iso Butyl Ketone and Iso Propanol Alcohol. GeSe films, 30nm thick, were deposited by co-sputtering from two different targets, at room temperature, to attain the desired Ge content in the GeSe film. The content of N in the deposited GeSe film has been controlled by adjusting the N_2 partial pressure in the gas flow. C was introduced by co-sputtering with a graphite sputter target in a custom-made multitarget sputtering system. The amount of C in the GeSe film has been controlled by adjusting the sputtering power during the deposition. TiN was then deposited in the same system by sputtering from a TiN target avoiding vacuum breaks. Finally, lift-off was performed in acetone solution to achieve the TE definition.

III. RESULTS AND DISCUSSION

A. N doping

Three N dopant concentrations, N1, N2 and N3 are studied in this work with N1<N2<N3. To determine the crystallization temperature (T_c) of the amorphous material, an

978-1-5090-5979-9/17 $31.00 © 2017 IEEE

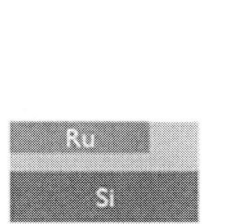 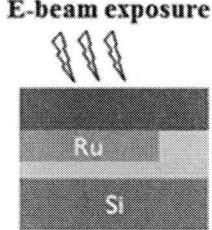 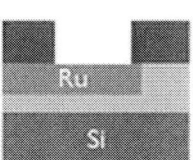 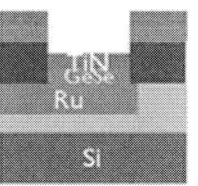 

E-beam exposure **Development of resist** **Deposition of stack** **Lift-off of the resist**

Fig. 2. Schematics of the process flow for the fabrication of the selector devices, starting from a 300mm-processed substrate (leftmost figure).

in situ XRD technique was used, which allowed taking XRD spectra while heating the samples with a specific temperature ramp rate. While the T_c of undoped GeSe is 350°C, the addition of N inhibits crystallization and retains the amorphous state beyond 450°C. Hence N is very effective in delaying crystallization and improving the thermal stability of GeSe.

B. Electrical Characterisation, N:GeSe

The sub-threshold leakage of the pristine samples is plotted in Fig. 3(a) while the averaged and statistical characteristics of the current at 2V across 8 fresh devices for each sample is shown in Fig. 3(b). The initial leakage is symmetric for both polarities despite the asymmetry in the stack, suggesting that the electrode material plays a marginal role, for this stack configuration. The addition of N decreases the I_{off} of the device.

The switching characteristics of the N doped OTS devices are plotted in Fig 3. To initiate switching in the devices, a first fire voltage (V_{ff}) is required for the first cycle which is higher than the stabilized switching voltage observed for the subsequent cycles [10]. The first fire in undoped GeSe, N1 and N2 samples was achieved by applying a triangular pulse of amplitude 7V with rise and fall times of 100ns.to minimize the structural change in the material which occurred when DC sweep was applied. The sample N3 could not be initiated with triangular pulses and needed a DC sweep to initiate it.

Following the initialization step, triangular voltage pulses with a rise and fall time of 500ns were applied across the device while the current across the device is calculated by the voltage drop across a 50Ω load resistor. The devices exhibit good OTS behavior with a transition to lower resistance above the transition/treshold voltage (V_{th}) and regaining the high resistance state when the voltage drops below the hold voltage (V_h). The mean switching curves for each sample are compared in the inset of Fig. 3(d). After the initial cycle, the devices switched at lower voltages for subsequent cycles. Continuing the trend observed for the initial leakage, addition of N increases V_{th} as seen in Fig. 3(c).

C. C doping

Two different doping concentrations of C, C1 and C2 are studied in this work with C1<C2. Addition of C increases T_c with an increase of 150°C for C2. Hence C is also effective in delaying crystallization and consequently improving the thermal stability of GeSe.

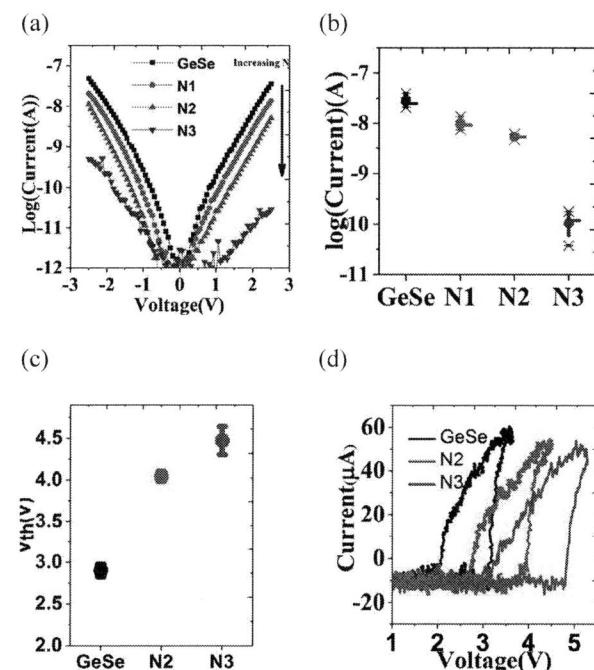

Fig. 3. (a) I-V characteristics of N doepd selector devices selector devices (b) Current@ 2V for different N contents. (c) V_{th} of N doped samples (d) Switching curves at 500ns triangular pulse rise/fall times.

D. Electrical Characterisation, C:GeSe

The initial leakage of the C doped samples is plotted in Fig. 4(a). The addition of C increases the I_{off} of the device in contrast to N. The increase in leakage with increasing C content is insignificant. The V_{th} also reduces with C doping as shown in Fig, 4(b).

The sub-threshold conduction in the samples, was measured at 25°C, 45°C, 65°C and 85°C. The plots for sample

N2(used for representation) are shown in Fig. 5. J-E vs sqrt(E) (Fig. 5(a)) in the voltage range 1.5-2.5 V shows a linear fit indicating that Poole-Frenkel conduction is dominant at this field. The Arrhenius plot for activation energy is shown in Fig 5(b). The trap depth is given by the slope of the plot log(J/E) vs 1/T (inset of Fig 5(b)) while the intercept of the plot gives the trap density, N_T. The extracted parameters are given in Table 1.

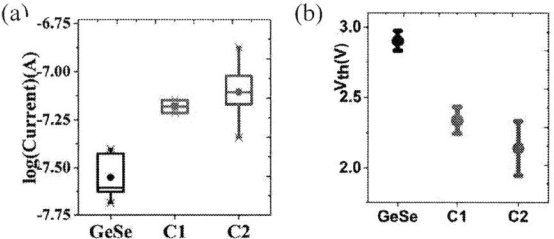

Fig. 4. (a) Current@ 2V for different C contents (b) V_{th} of C doped samples

$$\log\left(\frac{J}{E}\right) = \log(q\mu N_T) + \frac{q}{kT}\left(\sqrt{\frac{qE}{\pi\varepsilon_r\varepsilon_0}} - \varphi_T\right)$$

(a)   (b)

Fig. 5. T dependence of sub-threshold conduction in pristine N2 devices (a) ln(J/E) vs sqrt(E) shows a linear fit confirming Poole-Frenkel conduction (b) Activation Energy for conduction for different field across stacks (Inset: Barrier height lowering vs sqrt(E)). Poole Frenkel model predicts that the barrier height is reduced as a function of the square root of the field with the intercept giving the trap depth.

Table 1: Extracted values from Poole-Frenkel model

Sample	φ_T(eV)	$q\mu N_T$
GeSe	0.54	0.2
N1	0.55	0.019
N2	0.62	0.018
N3	0.7	2e-4
C2	0.53	0.5

IV.   DISCUSSION

To understand the effect of dopant on the thermal and electrical properties of GeSe, it is necessary to understand the bonding in the network and how it is influenced by the dopant. Amorphous GeSe has been shown to consist of corner sharing Ge(Se$_{(1/2)}$)$_4$ tetrahedra units which comply with the 8-N rule for both Ge and Se [11].

The homopolar and heteropolar bond energies have been calculated using the model outlined by Lankhorst [12] and are listed in Table 2.

Table 2: Bond Enthalpies from Lankhorst Model

Bond	Ge-Se	Ge-Ge	Se-Se	
Enthalpy(KJ/mol)	234.5	186	227	
Bond	Ge-N	Se-N	Ge-C	Se-C
Enthalpy(KJ/mol)	354.1	295.7	269.5	262

Both C and N form stronger bonds with Ge than Se, i.e Ge-C and Ge-N being the strongest in the system followed by Ge-Se, Se-Se and Ge-Ge. Hence Ge-N and Ge-C bonds are expected to be formed in the doped GeSe films. For the GeSe composition used in this work, most of the Se co-ordination is expected to be satisfied by Ge, there would be minimal formation of Se-Se chalcogen chain of atoms.

The formation of Valence Alternation Pairs(VAPs) by the lone pair electrons in chains of chalcogen atoms has been identified as responsible for of the traps [15] associated with OTS switching. However, in the equiatomic composition of GeSe, there are expected to be very few homopolar Se-Se bonds as discussed. Hence, another type of defect, probably, Ge dangling bonds, could be responsible for the OTS switching

With the addition of N, the amorphous network which previously consisted of weakly 2-fold coordinated chalcogen atoms, has now an increased 3-fold co-ordination of N, thus increasing cross linking in the network [13]. N is incorporated into the matrix and forms bonds with Ge [14]. The high Ge-N bond energy partially arises from the huge electronegativity difference between Ge and N, which also lends a partial ionic character to the bond. The presence of N also trades weaker homopolar Ge-Ge bonds for the stronger Ge-N bonds. The kinetic barriers to break these bonds would strengthen the network, hinder diffusion and combined with the increased entropy will postpone crystallization

N passivates the Ge dangling bonds and pushes the edge states deeper into the conduction band increasing the mobility gap of the material. This also leads to a reduction in the number of traps contributing to conduction. At the highest N content, Ge$_3$N$_4$ structural units could also be formed [13]. The values extracted in Table 1 also indicate that with the addition of N there is an increase in the trap depth, φ_T and a drop in the density of trap states, N_T. Both contribute to a decrease in leakage and increase in V_{th} [15]. The necessity of a DC first fire for higher N contents, N3, is also a consequence of the increased rigidity of the network.

The addition of C also increases T_c like N. However, the magnitude of increase in T_c is lower for C because the Ge-C bond is weaker than the Ge-N bond. It is to be noted that C, which forms strong homopolar bonds, has the tendency of forming sp^2 hybridized graphitic C chains which could also delay crystallization due to the high energy required to break the C-C bonds. Evidence of C-C [16] bonds was confirmed in C doped GeTe previously.

The values extracted in Table 1 indicate that with the addition of C there is an increase in the density of trap states, N_T. This accounts for the increased leakage and decrease in V_{th} observed with addition of C. The nature and source of these trap states is not clear yet and is under investigation. It is probable there exist triple coordinated C-Ge$_3$ with C 2p states forming gap states or tail states. Another possibility is that C-C chains form states within the band gap and lead to increase in N_T.

V. CONCLUSIONS

We have demonstrated the effect of N and C on the switching behavior of GeSe. N decreases the off-state leakage, which is crucial for application in cross-bar devices while increasing the V_{th} of the selector. In contrast, C increases the off-state leakage while reducing V_{th}. Hence switching parameters of GeSe selector can be tailored by adding C and N. Both dopants increase the T_c of the selector improving the thermal stability of the device.

REFERENCES

[1] G. W. Burr, R. S. Shenoy, K. Virwani, P. Narayanan, A. Padilla and B. K. Hwang, "Access devices for 3D crosspoint memory," *J. Vac. Sci. Technol. B*, vol. 32, no. 4, pp. 40802-1–23, 2014.

[2] B. Govoreanu, L. Zhang, and M. Jurczak, "Selectors for high density crosspoint memory arrays: Design considerations, device implementations and some challenges ahead," *Int. Conf. IC Des. Technol.*, 2015.

[3] S. R. Ovshinsky, "Resistive Electrical Switching Phenomena in Disordered Structures," *Phy. Rev. Lett.*, vol. 21, no. 20, 1968.

[4] S. D. Kim, H. W. Ahn, S. Y. Shin, D. S. Jeong, S. H. Son, H. Lee, B. K. Cheong, D. W. Shin, and S. Lee, "Effect of Ge Concentration in Ge$_x$Se$_{1-x}$ Chalcogenide Glass on the Electronic Structures and the Characteristics of Ovonic Threshold Switching (OTS) Devices," *ECS Solid State Lett.*, vol. 2, no. 10, pp. Q75–Q77, 2013.

[5] B. Govoreanu et al, "Thermally stable integrated Se-based OTS selectors with > 20 MA / cm^2 current drive,>3.10^3 half-bias nonlinearity, tunable threshold voltage and excellent endurance," *VLSI*

2017, in press.

[6] H. W. Ahn, D. S. Jeong, B. K. Cheong, S. D. Kim, S. Y. Shin, H. Lim, D. Kim, and S. Lee, "A Study on the Scalability of a Selector Device Using Threshold Switching in Pt/GeSe/Pt," *ECS Solid State Lett.*, vol. 2, no. 9, pp. N31–N33, 2013.

[7] S. Y. Shin, J. M. Choi, J. Seo, H. W. Ahn, Y. G. Choi, B. Cheong, and S. Lee, "The effect of doping Sb on the electronic structure and the device characteristics of Ovonic Threshold Switches based on Ge-Se," *Sci. Rep.*, vol. 4, p. 7099, 2014.

[8] H. W. Ahn, D. Seok Jeong, B. K. Cheong, H. Lee, H. Lee, S. D. Kim, S. Y. Shin, D. Kim, and S. Lee, "Effect of density of localized states on the ovonic threshold switching characteristics of the amorphous GeSe films," *Appl. Phys. Lett.*, vol. 103, no. 4, 2013.

[9] J. Seo, H. W. Ahn, S. Y. Shin, B. K. Cheong, and S. Lee, "Anomalous reduction of the switching voltage of Bi-doped Ge$_{0.5}$Se$_{0.5}$ ovonic threshold switching devices," *Appl. Phys. Lett.*, vol. 104, no. 15, pp. 0–4, 2014.

[10] S. B. Bhanu Prashanth and S. Asokan, "Composition dependent electrical switching in Ge$_x$Se$_{35-x}$Te$_{65}$ (18 <x <25) glasses-the influence of network rigidity and thermal properties," *Solid State Commun.*, vol. 147, no. 11–12, pp. 452–456, 2008.

[11] Y. G. Choi, S. Y. Shin, R. Golovchak, S. Lee, B. K. Cheong, and H. Jain, "EXAFS spectroscopic refinement of amorphous structures of evaporation-deposited Ge-Se films," *J. Alloys Compd.*, vol. 622, pp. 189–193, 2015.

[12] M. H. R. Lankhorst, "Modelling glass transition temperatures of chalcogenide glasses. Applied to phase-change optical recording materials," *J. Non. Cryst. Solids*, vol. 297, no. 2–3, pp. 210–219, 2002.

[13] Gourong Chen and Jijian Cheng, "Role of Nitrogen in the Crystallization of Silicon Nitride-Doped Chalcogenide Glasses," *J. Am. Chem. Soc.*, vol. 36, no. 190478, pp. 2934–2936, 1999.

[14] J. N. Hart, N. L. Allan, and F. Claeyssens, "Ternary silicon germanium nitrides: A class of tunable band gap materials," *Phys. Rev. B - Condens. Matter Mater. Phys.*, vol. 84, no. 24, pp. 1–7, 2011.

[15] D. Ielmini and Y. Zhang, "Analytical model for subthreshold conduction and threshold switching in chalcogenide-based memory devices," *J. Appl. Phys.*, vol. 102, no. 5, pp. 0–13, 2007.

[16] G. E. Ghezzi, J. Y. Raty, S. Maitrejean, A. Roule, E. Elkaim, and F. Hippert, "Effect of carbon doping on the structure of amorphous GeTe phase change material," *Appl. Phys. Lett.*, vol. 99, no. 15, pp. 10–13, 2011.

Multilevel SOT-MRAM Cell with a Novel Sensing Scheme for High-Density Memory Applications

Behzad Zeinali, Mahsa Esmaeili, Jens K. Madsen, Farshad Moradi
Integrated Circuits and Electronics Lab (ICELAB), Department of Engineering, Aarhus University
Finlandsgade 22, 8200 Aarhus N, Denmark; e-mail: moradi@eng.au.dk

Abstract— **This paper presents a multilevel spin-orbit torque magnetic random access memory (SOT-MRAM). The conventional SOT-MRAMs enables a reliable and energy efficient write operation. However, these cells require two access transistors per cell, hence the efficiency of the SOT-MRAMs can be questioned in high-density memory application. To deal with this obstacle, we propose a multilevel cell which stores two bits per memory cell. In addition, we propose a novel sensing scheme to read out the stored data in the multilevel SOT-MRAM cell. Our simulation results show that the proposed cell can achieve 3X more energy efficient write operation in comparison with the conventional STT-MRAMs. In addition, the proposed cell store two bits without any area penalty in comparison to the conventional one bit SOT-MRAM cells.**

Keywords— SOT-MRAM; p-MTJ; Spin-Hall Effect; Sensing scheme; Multilevel cell.

I. INTRODUCTION

Technology scaling has faced a big challenge due to the ever-increasing process variations and short channel effects (SCEs) leading to a higher leakage power. In addition, today's leading memories such as Dynamic Random-Access Memory (DRAM) and Static Random-Access Memory (SRAM) are volatile and require a constant power supply to hold their data. This further increases leakage and therefore static power consumption [1]. To achieve a memory array with lower leakage and more reliability, there is a need for highly reliable non-volatile memories. Magnetic Random Access Memory (MRAM) is a promising candidate due to its unique features including non-volatility, long endurance, CMOS-compatibility, high-speed access and zero standby power consumption [1]. The MRAM is based on Magnetic Tunnel Junction (MTJ) technology, which has two permanent states with different resistance values which can be utilized to store binaries. The MTJ is composed of two magnetic layers, i.e. pinned layer (PL) and the free layer (FL), and one oxide barrier layer. MTJ resistance is determined by the relative magnetization direction of two ferromagnetic layers. When the magnetization directions of two magnetic layers are parallel (P-state) or anti-parallel (AP-state), MTJ resistance is in low or high states, respectively. Thus, the MTJ can be used as a binary memory cell, low resistance (R_P): logic "0" – high resistance (R_{AP}): logic "1". Based on the easy axis direction, the MTJs can be categorized in either in-plane anisotropy MTJ (i-MTJ) or perpendicular-anisotropy MTJ (p-MTJ). The latter MTJ has attracted a lot of attention in high-density memory applications, mainly due to the higher thermal stability of p-MTJs [2].

The widely used mechanism in scaled technologies recruits a spin-polarized bidirectional current (I_{STT}) by applying appropriate voltages on bit-line (BL) and source-line (SL) and asserting word-line (WL) signal. This charge current applies a spin transfer torque (STT) to enable MTJ

switching. The STT-MRAM can provide a great scalability due to the fact that the critical switching current (I_C) decreases in proportion to the size of the MTJ [1]. However, it also scales thermal stability (Δ) of the MTJ leading to a higher probability of undesirable switching during Read operation. Therefore, it is challenging to reduce I_C while keeping a large Δ leading to a high energy overhead. In addition, passing a high write current through the MTJ imposes severe stress on the oxide layer leading to occasionally the MTJ barrier breakdown and reliability issues. Therefore, the efficiency of the STT-MRAMs can be questioned for sub-22 nm technology nodes [3]. The reliability issue due to high write current can be addressed with emerging spin-orbit torque (SOT) based switching mechanism as shown in Fig. 1 (a) [4]. This technology works based on Spin-Hall effect (SHE) [5] and/or Rashba effect [3] where a charge current (I_{SOT}) flowing through a non-magnetic heavy metal (HM) exerts a torque to switch the FL magnetization. This switching mechanism eliminates the severe stress imposed on the oxide layer due to passing a high write current through the MTJ since the oxide layer is not stressed anymore during write operation. In addition, in comparison to the STT switching mechanism, the SOT enables more energy efficient and faster write operation thanks to higher effective spin injection efficiency [4]. It is shown that an effective spin injection efficiency >100% can be achieved by appropriate sizing of HM [4]. The conventional bit-cell realized by SOT-based MTJ is shown in Fig. 1 (b) where the current paths of read and write are activated by asserting WL_1 and WL_2 signals, respectively. This configuration decouples the read and write paths leading to independent read and write optimization.

Despite all the advantages of SOT-MRAM, the main disadvantage of the cell shown in Fig. 1 (b) is the area, so that two transistors are required for each bit-cell. Therefore, in high-density memory applications, SOT-MRAM cannot be an appropriate choice [4]. To address this obstacle, a well-known technique is to implement a multilevel cell. In this paper, we propose a multilevel design for SOT-MRAM to improve integration density and reduce the cost per bit. In addition, a novel sensing scheme is proposed to read out the stored data in the cell. The remainder of the paper is organized as follows. Section II reviews the previous Multilevel SOT-MRAM designs. The new Multilevel SOT-MRAM design and novel sensing scheme are proposed in

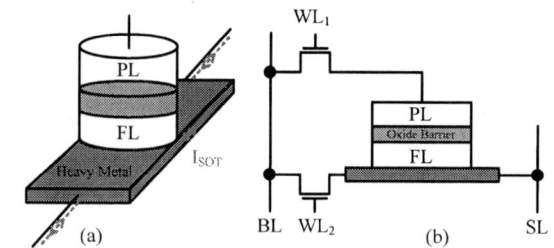

Fig. 1. (a) SOT-based MTJ, (b) Conventional SOT-MRAM.

978-1-5090-5979-9/17 $31.00 © 2017 IEEE

section II as well. The modeling and simulation of Multilevel SOT-MRAM are described in section III. Section IV concludes the paper.

II. MULTILEVEL SOT-MRAM

The integration density of MRAM cell may be increased by storing multiple bits in each memory bit-cell. This can be achieved by including more MTJs in each bit-cell. In the case of SOT-MRAM, multilevel design concepts are studied by Kim et al. [4] where two series MTJ multi-level cell (S-MLC) and Parallel MTJ multi-level cell (P-MLC) shown in Fig. 2, are investigated and demonstrated. In S-MLC, two series MTJs make the storage element of the cell while only the FL of MTJ_2 is in contact with the HM. According to this configuration [Fig. 2 (a)], MTJ_2 can be switched by SOT mechanism while MTJ_1 switching is accomplished by the STT current. During the write operation, the MTJ_1 is written first and then the MTJ_2 is written that leads to Write–disturb failure elimination in this cell. In this case, even if an unwanted switching occurs in MTJ_2 during programming of MTJ_1, the correct data can be written to the MTJ_2 in the following cycle. However, it is obvious that during programming of MTJ_1, large write current flows through the MTJ_2 as well, which degrades the reliability of the MTJ_2. On the other hand, in P-MLC, the FLs of both MTJs are in contact with the HM and the MTJs are connected in parallel to each other as shown in Fig. 2 (b). In this configuration, the write operation of the MTJs is accomplished by the SOT switching mechanism leading to no reliability issue. However, since both MTJs are in the same SOT current path, they should have different critical currents. It may be achieved by increasing cross-sectional area for HM underneath the MTJ_2 leading to higher I_C. In other words, by reducing the effective spin injection efficiency of the SOT mechanism for the MTJ_2, the requirement for different I_C's can be met. In the case of P-MLC, the MTJ with a larger I_C (MTJ_2) is programmed first and then the MTJ_1 is written by passing a lower SOT current than the critical current of MTJ_2. Therefore, in scaled technology node, engineering of the MTJ to have different I_C's and providing a write current lower than I_C of MTJ_2 may be challenging.

The read operation of the S-MLC and P-MLC is enabled by providing four distinguishable resistance states. To this end, the MTJs are realized by different cross-sectional areas, the MTJ_1 is designed with the minimum feature size of technology while the MTJ_2 needs to be larger. Therefore, the MTJ_2 requires a larger write current leading to higher write power dissipation. In addition, the MTJ_1 and MTJ_2 have different thermal stability due to different sizes. Therefore, although the S-MLC and P-MLC can improve the integration density of SOT-MRAMs, the write and read operations are challenging in these cells. To address these obstacles, we propose a new Multilevel SOT-MRAM design and a novel sensing scheme to distinguish resistance states without thermal stability degradation and extra power dissipation.

A. Proposed Multilevel Cell

Figure 3(a) shows the proposed Multilevel SOT-MRAM cell where the storage element contains a series stack of two MTJ with a shared PL. In this bit-cell, FLs of MTJ_1 and MTJ_2 are in contact with different nonmagnetic HMs enabling separate SOT-based switching for each MTJ. During the write operation of the MTJ_1, write current flows through the top HM by asserting WL_1 signal while WL_2 signal is off. It is worth to mention that when MTJ_1 is written, the sneak current passing through the stack MTJ may be eliminated by floating BL_2. On the other hand, switching of the MTJ_2 is enabled by asserting WL_2 signal while WL_1 signal is off and BL_1 is float.

In order to read out stored data of the stack MTJs, we may achieve four distinguishable resistance states. To this end, the conventional multi-level bit-cells have utilized different cross-sectional area (A_{MTJ}) and/or different oxide thickness (t_{MgO}) for MTJ_1 and MTJ_2 [4]. As mentioned earlier these physical designs lead to thermal stability degradation and extra power dissipation. To eliminate this challenge, in this paper, we design the same MTJ_1 and MTJ_2. Therefore, we have three different resistance states for the stack MTJ noted by $R_{00}=2R_L$, $R_{11}=2R_H$ and $R_{01}=R_{10}=R_L+R_H$ where the first letter is resistance state of MTJ_1 and the second letter is for MTJ_2. The challenge in the read operation is to disguise between R_{01} and R_{10} because these states have the same resistance. To deal with this challenge, we propose a two-step sensing scheme to read out the resistance state of the stack MTJs.

When I_{SOT} is larger than $I_{C,SOT}$, e.g. during the write operation, the magnetization direction of the FL is switched. However, when the SOT current is smaller than the critical current, the induced torque exerts little influence on the magnetization direction and put it in an intermediate state between P-state and AP-state. Since the MTJ resistance is determined by θ (the magnetization angle between FL and PL), small I_{SOT} can generate intermediate resistance states between R_L and R_H. When the MTJ is in P-state, applying the SOT current increases the MTJ resistance. On the other hand, when the MTJ is in AP-state, the MTJ resistance reduces by applying the small SOT current. This feature can be recruited during read operation to distinguish between R_{01} and R_{10} as shown in Fig. 4. During read operation, the WL_2 and REN signals are asserted while the WL_1 is turned off. In the first step, a read current (I_{R1}) is passed through the stack MTJ to generate a BL voltage (V_{BL11}). This voltage is stored in a capacitor (C_1) by enabling S_1 signal. The R_{00} and R_{11} states can be read out at the end of this step by comparing the V_{BL11} with the reference voltages. Otherwise, when the resistance state is R_{01} or R_{10}, the second step of the sensing scheme is conducted. In the second step, S_2 signal is asserted along with WL_2 and REN signals are asserted leading to passing I_{R1} and I_{R2} through the stack MTJ and the down HM, respectively. I_{R2} changes the MTJ_2 resistance to an intermediate state, leading to a different BL voltage (V_{BL12}). This voltage is stored in another capacitor (C_2) by enabling S_2 signal. The sensing process is accomplished by comparing V_{BL11} and V_{BL12}. When the MTJ_2 is in P-state, V_{BL11} is smaller than V_{BL12} while in AP-state of the MTJ_2, V_{BL11} is larger than V_{BL12}.

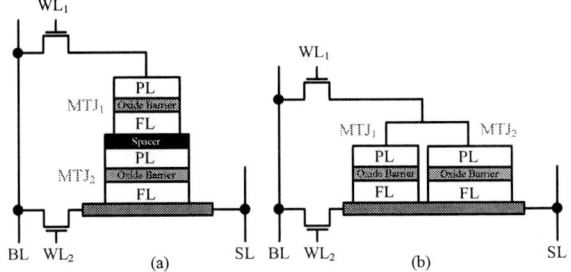

Fig. 2. Conventional multilevel cells, (a) S-MLC, and (b) P-MLC.

978-1-5090-5979-9/17 $31.00 © 2017 IEEE

III. SIMULATION RESULTS AND DISCUSSION

A. MTJ Modeling

In order to analyze the proposed Multi-bit SOT-RAM cell, in this section, we provide a model which describes the behavior of the p-MTJ in the presence of both I_{SOT} and I_{STT} [6]. Although our proposed cell is valid regardless of the MTJ technology, in this paper, we recruit the p-MTJ. The main drawback in the case of the p-MTJ switched by SOT mechanism is stochastic switching determined by the thermal fluctuation. The deterministic switching process can be achieved by using an external magnetic field (H_{ext}) [3]. Therefore, we provide a model which describes the behavior of the MTJ on the presence of H_{ext}, I_{SOT} and I_{STT}. The dynamic behavior of the MTJ free layer is described by the Landau-Lifshitz-Gilbert (LLG) equation, where the torque induced by the injected spin current is expressed in a similar manner to STT. As mentioned earlier, without loss of generality, SHE is considered as the physical origin of the SOT. The variation of the MTJ resistance is described by a voltage-dependent Tunneling magneto-resistance ratio (TMR) equation. Figure 5 shows the block diagram of the modeling containing the equations used to describe the MTJ behavior. The first block is a LLG solver which estimates the magnetization direction of the free layer. In theis block, γ and α are the gyromagnetic ratio and the Gilbert damping constant, respectively. The STT and SOT are modeled by the last two items in the right side of the equation, respectively, where M_s, p, η, $\vec{m}_r$ and $\vec{\sigma}_{SHE}$ represent the saturation magnetization, effective spin polarization, spin Hall angle, PL magnetization and polarization direction of pure spin current, respectively. In addition, the physical parameters of t_{FL}, A_{MTJ} and A_{HM} are FL thickness and cross-sectional areas of the MTJ and HM, respectively. The second block in Fig. 5, captures the MTJ resistance (R_{MTJ}) dependence on the voltage bias of the MTJ (V_{MTJ}) and the magnetization angel between FL and PL (θ). In this equation, R_{p0} and TMR_0 are the R_p and TMR under zero bias voltage while V_h is the voltage at which the TMR_0 is divided by 2.

B. Area

The physical layout of the proposed multilevel SOT-MRAM cell in 28nm UTBB SOI technology is drawn in Fig. 6 to estimate the layout area of the cell in comparison to the conventional SOT-MRAM cell. According to this figure, for the access transistor sizes larger than 370 nm, the area of the proposed multilevel cell is equal to that of the conventional cell. In other words, although the proposed cell needs three vertical metals (BL$_1$, BL$_2$ and SL), when the width of access transistor is larger than 370 nm, the metal spacing between vertical metals is not dominant.

C. Simulation Results

To prove the efficiency of the proposed cell, simulation results are explored in this section. The proposed MRAM cell is designed in 28 nm UTBB SOI technology. The parameters of MRAM cell are shown in Table I where these sizings provide 130 μA and 95 μA during Write-0 and Write-1 operations, respectively. In addition, a permanent magnetic field of 1000 Am^{-1} is generated by biasing a metal layer on top of the MTJs. During the write operation, at the beginning of the write operation, the SOT current along with the fixed external magnetic are applied. The spin torque induced by I_{SOT} rapidly rotates the magnetization direction of the FL from the perpendicular direction (z-axis) to the in-plane orientation (x-axis). However, due to existing the torque induced by H_{ext}, the magnetization passes in-plane direction and is stabilized at a specific orientation between in-plane and z-axis. This process for write-1 and write-0 is sketched in Fig. 7. In this figure m_x, m_y and m_z define the FL magnetization vector ($\vec{m}$) elements along x-, y- and z-axis, respectively.

According to the LLG and MTJ resistance equations in Fig. 5, variation of physical parameters of the MTJ (t_{FL}, A_{MTJ} and A_{HM}) and the transistors leads to considerable behavior changing of the MTJ. To address this concern, the performance of the proposed SOT-MRAM cell in the presence of process variation is studied by the use of Monte-Carlo simulations. In Monte-Carlo, the value of 3σ is equal to 10% of the nominal physical values. The following results are achieved by doing the Monte-Carlo simulation with 1000 samples for each case.

To accomplish the write operation, I_{SOT} should be removed at an appropriate time while H_{ext} continues to achieve the complete switching. Therefore, it is important to

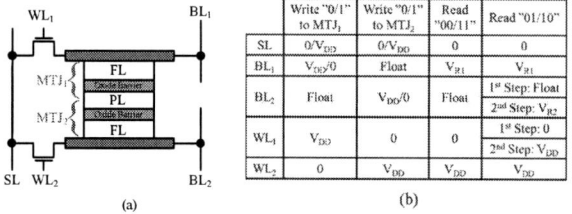

Fig. 3. (a) Proposed Multilevel schematic. (b) Bias conditions during Read and Write for the proposed bit-cell.

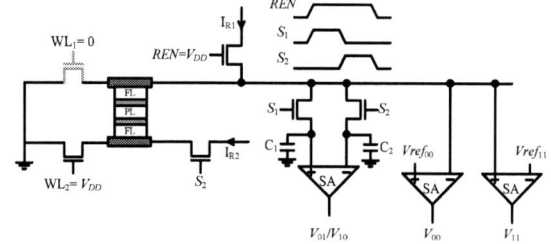

Fig. 4. Conceptual design of the proposed sensing scheme for the multilevel SOT-MRAM.

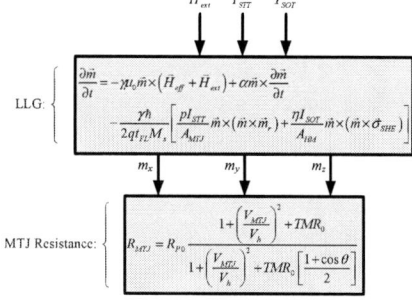

Fig. 5. Block diagram of the MTJ modeling.

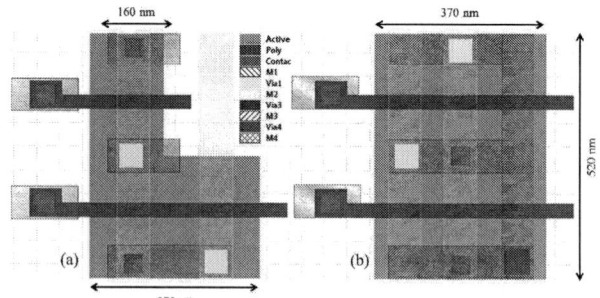

Fig. 6. Layout of (a) the SOT-MRAM cell and (b) the proposed multilevel SOT-MRAM cell.

find the optimum duration of WL signals. Figure 8 (a) and (b) show the distribution of the required time at which the magnetization direction of FL is rotated to the in-plain orientation (T_{SOT}) during Write-0 and Write-1, respectively. According to this figure, the 6σ yield can be supported when the T_{SOT} is 1.2 ns. Figure 8 (c) and (d) show the distribution of overall required time for Write-0 and Write-1 where the switching operation is accomplished in 12.9 ns to achieve the 6σ yield. The distribution of Write-0 and Write-1 energies for the proposed SOT-MRAM in comparison to a conventional STT-MRAM are shown in Fig. 9. According to this figure, the proposed cell enables 3X more energy efficient switching process in comparison to the conventional STT-MRAM.

During Read operation, the proposed sensing scheme is applied to the multilevel SOT-MRAM cell. As shown in Fig. 10(a) which shows the distribution of V_{BL11}, when the stored data in the stacked MTJ is "00" or "11", different BL voltages are generated. During sensing "01" and "10", the BL voltages in the first step are the same for both stored data. However, by applying the SOT current, two distinguishable resistance states can be generated as shown in Fig. 10(b).

TABLE I. SOT-MRAM CELL PARAMETERS

Parameter	Value
Areas of the MTJ (A_{MTJ})	25 nm×25π nm
Free Layer Thickness (t_{FL})	1.4 nm
Oxide Thickness (t_{MgO})	0.85 nm
TMR_0	120 %
Spin Hall angle (η)	0.3
Areas of the HM (A_{HM})	50 nm×60 nm
Supply Voltage (V_{DD})	0.75V
Access transistor (W/L)	1 μm/30 nm
H_{ext}	1000 Am^{-1}

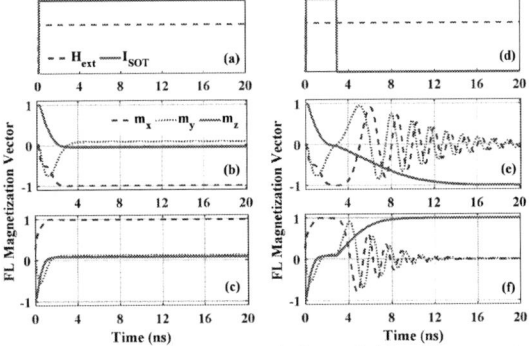

Fig. 7. (a) When I_{SOT} is not removed during switching, FL magnetization is stabilized at a specific orientation between in-plane and z-axis during (b) Write-1 and (c) Write-0. (d) When I_{SOT} is removed during switching, FL magnetization aligned to the easy axis during (e) Write-1 and (f) Write-0.

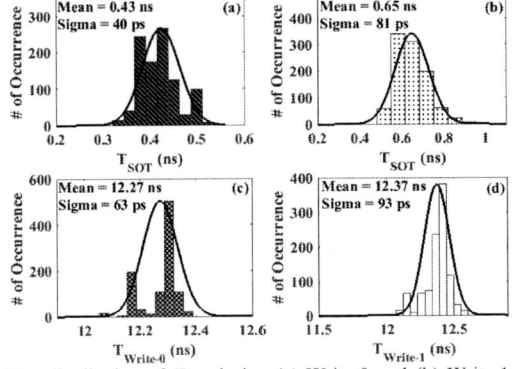

Fig. 8. The distribution of T_{SOT} during (a) Write-0 and (b) Write-1. The distribution of switching time during (c) Write-0 and (d) Write-1

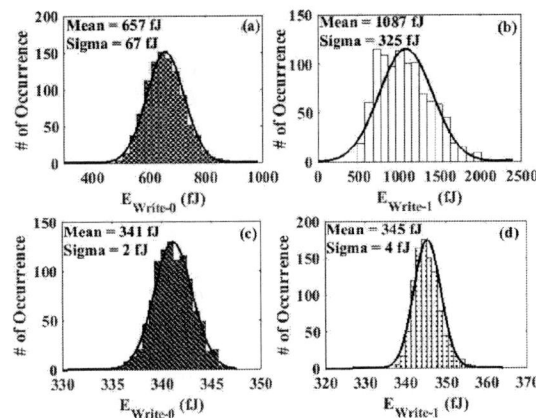

Fig. 9. The distribution of a STT-MRAM switching energy during (a) Write-0 and (b) Write-1. The distribution of the proposed multilevel SOT-MRAM switching energy during (c) Write-0 and (d) Write-1.

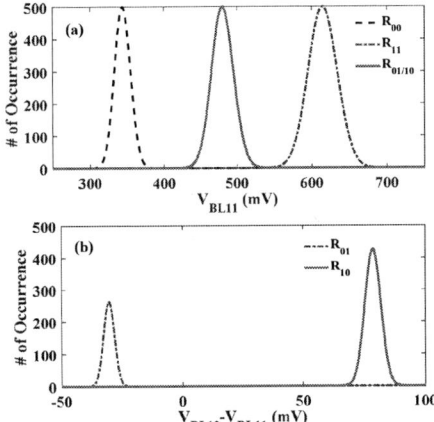

Fig. 10. The distribution of (a) V_{BL12} and (b) V_{BL12}-V_{BL11} during read operation.

IV. CONCLUSION

In this paper, we proposed a multilevel SOT-MRAM for high-density memory application where it stored two bits per memory cell. In addition, we proposed and evaluated a novel sensing scheme to read out the stored data in the multilevel SOT-MRAM cell. To evaluate the efficiency of the proposed, cell and sensing scheme, we provided simulation results showing 3X more energy efficient write operation in comparison with the conventional STT-MRAMs without any area penalty in comparison to the conventional one bit SOT-MRAM cells.

REFERENCES

[1] Y. Huai, "Spin-transfer torque MRAM (STT-MRAM): Challenges and prospects," *AAPPS Bull.*, vol. 18, no. 6, pp. 33–40, Dec. 2008.

[2] S. Ikeda et al., "A perpendicular-anisotropy CoFeB–MgO magnetic tunnel junction," *Nature Materials*, vol. 9, pp. 721-724, Sept. 2010.

[3] G. Prenat et al., "Ultra-Fast and High-Reliability SOT-MRAM: From Cache Replacement to Normally-Off Computing," *IEEE Trans. Multi-Scale Computing Syst.*, vol. 2, no. 1, pp. 49-60, Jan. 2016.

[4] Y. Kim, X. Fong, K.-W. Kwon, M.-C. Chen, and K. Roy, "Multilevel Spin-Orbit Torque MRAMs", *IEEE Tran. on Elec. Devices*, vol. 62, no. 2, pp. 561-568, Feb. 2015.

[5] Y. Seo, K.-W. Kwon, and K. Roy, "Area-Efficient SOT-MRAM with a Schottky Diode," *Accepted for inclusion in Device Lett.*.

[6] Z.Wang,W.Zhao, E. Deng, J. Klein, and C. Chappert, "Perpendicular-anisotropy magnetic tunnel junction switched by spin-Hall-assisted spin-transfer torque," *J. Phys.: Appl. Phys.*, vol. 48, 2015.

On the Ballistic Ratio in 14nm–Node FinFETs

F. M. Bufler[*†], K. Miyaguchi[*], T. Chiarella[*], N. Horiguchi[*], and A. Mocuta[*]

[*]IMEC, Kapeldreef 75, 3001 Leuven, Belgium

[†]Institut für Integrierte Systeme, ETH Zürich, Gloriastrasse 35, CH-8092 Zürich, Switzerland

Email: Fabian.Bufler@imec.be

Abstract—Ballisticity in 14nm–node FinFETs is investigated by Monte Carlo device simulation. Analytic doping profiles are reverse–engineered to measured transfer characteristics of FinFETs from literature and from this work and good agreement between Monte Carlo simulations and measurements is achieved without any device–parameter calibration. The ballistic ratio, defined as the ratio of the on–current with and without scattering in the channel region, varies between 60 % and 76 % at a metallurgical gate length of about 20 nm, the higher ratio being obtained for devices with gate overlap. This shows that scattering in the channel is still important in current FinFET technology with doping and geometry details having an influence on ballisticity beyond only the gate length. As a particular result, it is found that valley occupancies change in the channel also in the absence of scattering which is attributed to free propagation leading to valley changes.

I. INTRODUCTION

As the channel length currently becomes still smaller in new CMOS technology nodes and the applied drain voltage is not reduced accordingly, carriers suffer fewer scattering events while traveling through the channel in the presence of an increased driving field. This has motivated simulation studies to estimate the performance of future devices by purely ballistic transport, e.g. using the nonequilibrium Green's function approach [1] or simply the injection velocity [2]. Also compact models for use in circuit simulation have been introduced which rely on ballistic transport and account for scattering with models for backscattering calibrated to numerical reference simulations [3] or simply multiply a ballistic current with a ballistic ratio from literature [4]. This raises the question how close to the ballistic limit state–of–the–art devices actually operate.

Ballistic ratios were determined by Monte Carlo (MC) device simulations [5]–[9] or subband Boltzmann solvers [9]. However, it is not clear how accurately these simulations describe electrical characteristics of real state–of–the–art devices in view of either simplified device structures and/or approximations involved in these microscopic device simulation approaches. In most cases also only one, in part unspecified, doping profile was used, although it was already found that different doping can lead to significantly different ballistic ratios [5].

It is the purpose of this work to investigate to what extent MC device simulation can predict performance of current FinFET devices, to assess the influence of doping and to extract corresponding realistic ballistic ratios. To this end electrical characteristics of FinFETs from literature [10] and new FinFET measurements are used to reverse–engineer analytical doping profiles. Based on these profiles and geometry details obtained from TEM images self–consistent single–particle MC

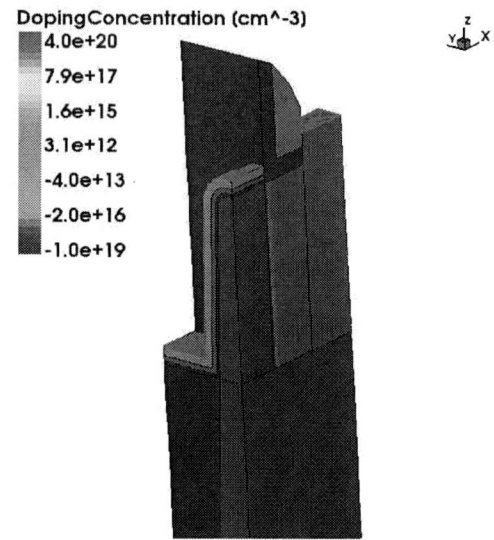

Fig. 1. Geometry and doping of the *n*–type 21nm–FinFET corresponding to Ref. [10].

device simulations [11] are performed without any calibration or change of MC–model parameters. The good agreement achieved between measurement and MC simulation finally allows to extract realistic ballistic ratios for current FinFET technology. The MC transfer characteristics, the ballistic ratio and the reflection coefficient, defined as the ratio between the currents of backward– and forward–flying carriers at the injection point [6], can then be used to predict FinFET performance and for calibration of compact models for circuit simulation.

II. SIMULATION APPROACH

For MC electron transport, an analytic two–band model with anisotropic nonparabolicity is used for the three Δ–valley pairs [12]. Nevertheless a full–band MC algorithm and the full Brillouin zone is used for the simulation. The analytic formulas only serve to compute band energies and group velocities at the wave–vector points of the discretized Brillouin zone. This involves — in contrast to a pseudopotential table — some discontinuities between the regions in the Brillouin zone associated with the different Δ–valleys, but does not require the computation and disk storage of a pseudopotential table. Heavy–hole, light–hole and split–off valence bands are described by an analytic 6–band $\mathbf{k} \cdot \mathbf{p}$ model. Scattering comprises acoustic and optical phonon scattering, ionized impurity scattering according to the Brooks–Herring (BH) model and a combination of specular and diffusive surface scattering.

978-1-5090-5979-9/17 $31.00 © 2017 IEEE

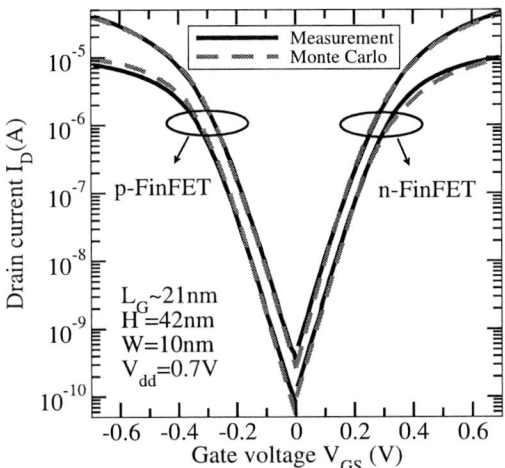

Fig. 2. Transfer characteristics at high (V_{DS}=0.7 V) and low (V_{DS}=0.05 V) drain bias of the *n*–type and *p*–type 21nm–FinFETs according to measurements [10] and Monte Carlo device simulation. W corresponds to the fin width at half of the fin height H.

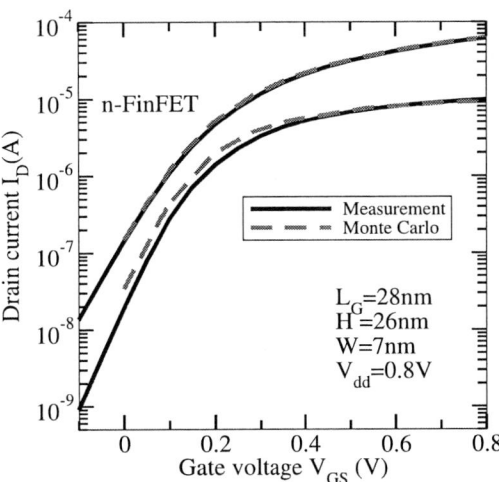

Fig. 3. Transfer characteristics at high (V_{DS}=0.8 V) and low (V_{DS}=0.05 V) drain bias of the *n*–type 28nm–FinFET according to measurements of this work and Monte Carlo device simulation.

A ratio of 85 % for specular surface scattering, governed by energy and parallel–momentum conservation, was found to accurately reproduce measured FinFET electron and hole mobilities for both (110) and (100) sidewall orientations [11]. Quantization is taken into account via modified oxide permittivity and workfunctions extracted from density–gradient simulation in the fin cross–section [13] since it allows to use specular/diffusive surface scattering shown to be in very good agreement with measured FinFET mobilities. Fermi statistics is only used in the computation of the inverse screening length for impurity scattering and the corresponding Fermi energy is determined numerically using the MC density–of–states. The rate for impurity scattering is multiplied by a doping–dependent prefactor which is calibrated to compensate the overestimation of the bulk mobility by the BH model at high phosphorus and boron concentrations with respect to measurements [14].

A quarter of the FinFET geometry of Ref. [10] is shown in Fig. 1. The corresponding simulation setup can be obtained from the Applications Library in Sentaurus Workbench [15]. The corresponding electrical characteristics of the *n*–type and *p*–type 21nm–FinFETs are shown in Fig. 2, where the measured drain currents per footprint width of Ref. [10] have been converted into amperes per fin by using the specified fin pitch of 42 nm. The measurements of this work for the *n*–type 28nm–FinFET can be seen in Fig. 3. The FinFETs have ⟨110⟩ sidewall and ⟨110⟩ channel orientations. No stress is considered for the *n*–type FinFETs, while mechanical stress simulation of the *p*–type FinFET with $Si_{0.5}Ge_{0.5}$ source/drain pockets leads to a compressive channel stress of 660 MPa in channel direction and a tensile channel stress of 500 MPa in fin height direction.

For the reverse–engineering of the doping profiles only the transfer characteristics at high drain bias are used. Essentially, extension doping, channel doping and gate overlap are cali-

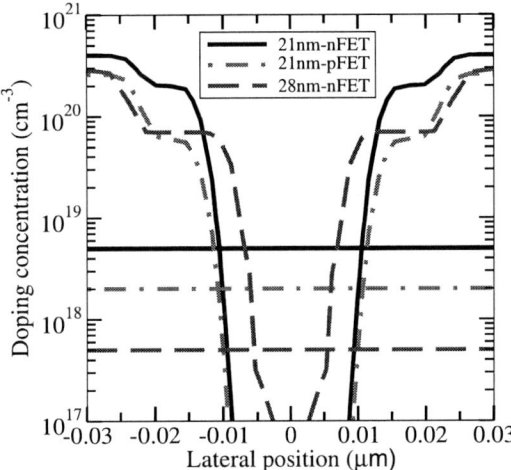

Fig. 4. Source/drain and extension doping as well as the constant channel doping as a function of lateral position used for the *n*–type and *p*–type 21nm–FinFETs [10] as well as for the *n*–type 28nm–FinFET from this work. Note that the 21nm–FinFETs and the 28nm–FinFET feature gate underlap and gate overlap, respectively.

brated to reproduce with drift–diffusion (DD) simulation the measured subthreshold–slopes (SS), since SS is the quantity which is similar between DD and MC. Then the source/drain doping is chosen such that the measured on–current is captured by MC simulation. No calibration at low drain bias is done; drain–induced barrier lowering (DIBL) and linear current (Idlin) are well captured without calibration with some deviation for the 28nm–FinFET. The doping profiles resulting from this procedure can be seen in Fig. 4. Most notably, the 21nm–FinFETs feature gate–underlap and the 28nm–FinFET gate–overlap. Of course, this approach is affected by uncertainties and the result is not necessarily unique. But reproducing

978-1-5090-5979-9/17 $31.00 © 2017 IEEE

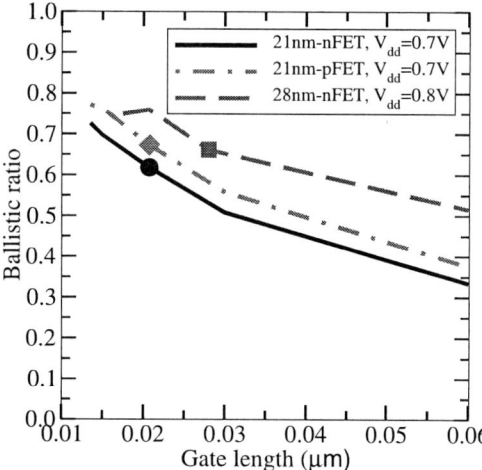

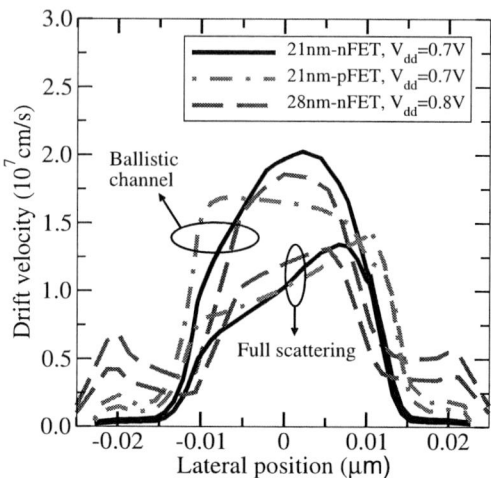

Fig. 5. Ballistic ratio, defined as the ratio of the on–current with scattering and the on–current with scattering switched–off in the channel region, as a function of the metallurgical gate length for the n–type and p–type 21nm–FinFETs [10] as well as for the n–type 28nm–FinFET from this work according to Monte Carlo device simulation. The gate lengths of 21nm and 28nm, marked by symbols, belong to the real devices the measurements of which are reported in Figs. 2 and 3, respectively. Upon scaling, only the gate length is changed.

Fig. 6. Drift velocity profiles, averaged with the carrier density over the fin cross–section, along the channel for the n–type and p–type 21nm–FinFETs [10] as well as for the n–type 28nm–FinFET from this work. The Monte Carlo results are shown for the case with scattering and the case where scattering is switched off in the channel region.

with reasonable doping profiles the electrical characteristics of different FinFETs by MC device simulation without any calibration of the MC model gives confidence that realistic ballistic ratios can be extracted from MC simulation.

III. Ballistic Channel Transport

In this section we investigate the influence of scattering in the channel on FinFET performance. Scaling the gate length reduces scattering in the channel, but not in the highly–doped source/drain and extension regions the relative importance of which increases due to the reduced channel resistance. Scattering in the access region also ensures the thermalization of carriers as already pointed out in Ref. [6] and affects the potential profile along the channel. Therefore the ballistic ratio should be defined as the ratio between the on–current with scattering in the whole device and the on–current where scattering is switched off only in the channel region (as was also done in Refs. [5], [6], [8]). The corresponding ballistic ratios for the three FinFETs are shown in Fig. 5 as a function of the metallurgical gate length. Note that only the gate length is changed for this scaling which is in contrast to Ref. [8] where upon scaling also other geometry and bias parameters were changed according to technology–node specifications. At the metallurgical gate lengths of 21 nm and 28 nm, respectively, of the actual FinFETs ballistic ratios of 62 % for the n–type 21nm–FinFET, 67 % for the p–type 21nm–FinFET under mechanical stress and 66 % for the n–type 28nm–FinFET are found (switching off scattering in the whole device would change the ballistic ratio e.g. from 67 % to 56 % in case of the p–type 21nm–FinFET). But changing the gate length of the 28nm–FinFET to 21 nm increases the ballistic ratio to 76 % compared to 62 % of the 21nm–FinFET, which shows that gate overlap enhances the ballistic ratio (also the higher drain

voltage of 0.8 V compared to 0.7 V contributes to the higher ballistic ratio, of course).

Figure 6 shows the drift velocity profiles along the channel for the three FinFETs with their actual metallurgical gate lengths. It can be seen that the steep velocity increase occurs for the 28nm–FinFET at the pn junction where also the driving field starts to increase strongly (not shown). This observation together with the similar ballistic ratios of 62 % and 66 % for the two n–type FinFETs suggests that the ballistic ratio is a function of the effective gate rather than of the metallurgical gate length. The comparison of the velocity profiles with and without scattering in the channel also demonstrates — in addition to the corresponding ballistic ratios of around 65 % — that transport in current FinFET technology is still significantly away from the ballistic limit.

Finally, we investigate valley repopulation effects when scattering is switched off in the channel. Figure 7 shows the corresponding occupations of the three Δ–valley pairs along the channel. It turns out that despite scattering being switched off in the channel the occupation of the Δ_z–valley differs from those of the two equivalent Δ_x– and Δ_y–valleys. The effect is similar if the pseudopotential band structure or the standard analytical ellipsoidal band model is used. Specular surface scattering principally allows by default valley changes, but excluding this possibility changes the effect only slightly. Therefore these valley changes are caused by the free flight according to Newton's equations of motion where electrons move from regions in the Brillouin zone associated with the Δ_z–valley to a region associated with another valley and vice versa. This effect is also present if scattering is present in the channel, but it is less pronounced in this case (not shown). This means that scattering in the channel tends to keep electrons in the same valley–pair. Consequently valleys should not be treated isolatedly in microscopic device simulation since this

978-1-5090-5979-9/17 $31.00 © 2017 IEEE

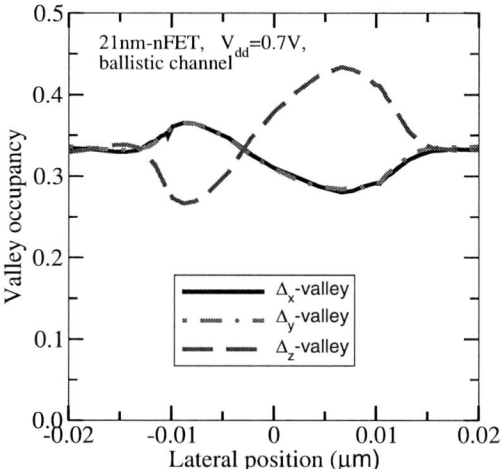

Fig. 7. Occupancies of the three Λ–valley pairs along the channel in the on–state of the n–type 21nm–FinFET.

would prevent electrons from changing the valley via free propagation.

IV. CONCLUSION

MC device simulation of complete transfer characteristics has been found to be in good agreement with corresponding FinFET measurements without any calibration of MC–model parameters making it the method of choice for the prediction of the performance of FinFETs and as reference for compact model calibration. Ballistic ratios in 14nm–node FinFETs depend in particular on the doping profile and are in the order of 65 %, i.e., FinFETs operate still far from the ballistic limit.

ACKNOWLEDGMENT

We would like to thank G. Eneman and A. De Keersgieter for useful discussions.

REFERENCES

[1] S. H. Park, Y. Liu, N. Kharche, M. S. Jelodar, G. Klimeck, M. S. Lundstrom, and M. Luisier, "Performance comparisons of III–V and strained–Si in planar FETs and nonplanar FinFETs at ultrashort gate length (12 nm)," *IEEE Trans. Electron Devices*, vol. 59, no. 8, pp. 2107–2114, 2012.

[2] D. Connelly, P. Zheng, and T.-J. K. Liu, "Channel stress and ballistic performance advantages of gate-all-around FETs and inserted–oxide FinFETs," *IEEE Trans. Nanotechnology*, vol. 16, no. 2, pp. 209–216, 2017.

[3] A. Rahman and M. S. Lundstrom, "A compact scattering model for the nanoscale double–gate MOSFET," *IEEE Trans. Electron Devices*, vol. 49, no. 3, pp. 481–489, 2002.

[4] U. K. Das, M. G. Bardon, D. Jang, G. Eneman, P. Schuddinck, D. Yakimets, P. Raghavan, and G. Groeseneken, "Limitations on lateral nanowire scaling beyond 7–nm node," *IEEE Electron Device Lett.*, vol. 38, no. 1, pp. 9–11, 2017.

[5] C. Jungemann, N. Subba, J.-S. Goo, C. Riccobene, Q. Xiang, and B. Meinerzhagen, "Investigation of strained Si/SiGe devices by MC simulation," *Solid–State Electron.*, vol. 48, pp. 1417–1422, 2004.

[6] P. Palestri, D. Esseni, S. Eminente, C. Fiegna, E. Sangiorgi, and L. Selmi, "Understanding quasi–ballistic transport in nano-MOSFETs: part I–scattering in the channel and in the drain," *IEEE Trans. Electron Devices*, vol. 52, no. 12, pp. 2727–2735, 2005.

[7] W. Guo, M. Choi, A. Rouhi, V. Moroz, G. Eneman, J. Mitard, L. Witters, G. V. der Plas, N. Collaert, G. Beyer, P. Absil, A. Thean, and E. Beyne, "Impact of 3D integration on 7nm high mobility channel devices operating in the ballistic regime," in *IEDM Tech. Dig.*, 2014, pp. 168–171.

[8] A. T. Elthakeb, H. A. Elhamid, and Y. Ismail, "Scaling of TG–FinFETs: 3–D Monte Carlo simulations in the ballistic and quasi-ballistic regimes," *IEEE Trans. Electron Devices*, vol. 62, no. 6, pp. 1796–1802, 2015.

[9] M. Choi, V. Moroz, L. Smith, and J. Huang, "Extending drift–diffusion paradigm into the era of FinFETs and nanowires," in *Proc. SISPAD*, Washington, DC (USA), Sep. 2015, pp. 242–245.

[10] S. Natarajan, M. Agostinelli, S. Akbar, M. Bost, A. Bowonder, V. Chikarmane, S. Chouksey, A. Dasgupta, K. Fischer, Q. Fu, T. Ghani, M. Giles, S. Govindaraju, R. Grover, W. Han, D. Hanken, E. Haralson, M. Haran, M. Heckscher, R. Heussner, P. Jain, R. James, R. Jhaveri, I. Jin, H. Kam, E. Karl, C. Kenyon, M. Liu, Y. Luo, R. Mehandru, S. Morarka, L. Neiberg, P. Packan, A. Paliwal, C. Parker, P. Patel, R. Patel, C. Pelto, L. Pipes, P. Plekhanov, M. Prince, S. Rajamani, J. Sandford, B. Sell, S. Sivakumar, P. Smith, B. Song, K. Tone, T. Troeger, J. Wiedemer, M. Yang, and K. Zhang, "A 14nm logic technology featuring 2nd-generation FinFET, air-gapped interconnects, self-aligned double patterning and a 0.0588 μm^2 SRAM cell size," in *IEDM Tech. Dig.*, 2014, pp. 71–73.

[11] F. M. Bufler, F. O. Heinz, and L. Smith, "Efficient 3D Monte Carlo simulation of orientation and stress effects in FinFETs," in *Proc. SISPAD*, Glasgow (UK), Sep. 2013, pp. 172–175.

[12] F. M. Bufler, F. O. Heinz, A. Tsibizov, and M. Oulmane, "Simulation of ⟨110⟩ nMOSFETs with a tensile strained cap layer," *ECS Trans.*, vol. 16, no. 10, pp. 91–100, 2008.

[13] F. M. Bufler and L. Smith, "3D Monte Carlo simulation of FinFET and FDSOI devices with accurate quantum correction," *J. Comput. Electron.*, vol. 12, pp. 651–657, 2013.

[14] G. Masetti, M. Severi, and S. Solmi, "Modeling of carrier mobility against carrier concentration in arsenic–, phosphorus–, and boron–doped silicon," *IEEE Trans. Electron Devices*, vol. 30, no. 7, pp. 764–769, 1983.

[15] Synopsys Inc., "Applications Library in Sentaurus Workbench (Applications_Library/AdvancedTransport/MonteCarlo)," Release M-2016.12, Mountain View (USA), 2016.

978-1-5090-5979-9/17 $31.00 © 2017 IEEE

Three-dimensional Multi-subband Simulation of Scaled FinFETs

L. Donetti, C. Sampedro, F.G. Ruiz, A. Godoy, F. Gámiz

Departamento de Electrónica and CITIC, Universidad de Granada, Granada, Spain. e-mail: donetti@ugr.es

Abstract—We simulated the static behavior of scaled FinFETs employing a self-consistent Multi-Subband Ensemble Monte Carlo simulator for non-planar devices. To be able to take into account the three-dimensional device structure, the 2D Schrödinger equation is solved in several cross sections; the coupled solution of the 3D Poisson equation and the 1D Boltzmann transport equation through the ensemble Monte Carlo method guarantees self-consistency. This simulator allowed us to study the effects of different fin aspect ratios while properly taking into account quantum confinement in the plane perpendicular to the transport direction. We studied the transfer characteristic as a function of different geometric parameters, the effect of quantum confinement on the charge distribution and the scaling behavior, down to a channel length of 10 nm.

I. INTRODUCTION

Non-planar, multi-gate structures provide a better electrostatic control of the channel of MOS transistors with respect to planar devices, providing, as a consequence, improved scaling properties [1]. Despite the fabrication challenges due to their three-dimensional device structure, FinFETs have already entered mass production [2], [3] and are expected to be able to meet the requirements for the next technology nodes.

From the simulation point of view, FinFETs represent a demanding task because they need a full 3D description and require the inclusion of quantum effects, especially for very narrow fins. These aspects are not fully taken into account in the different tools employed up to now to study these devices. Indeed, several simulators (including commercial ones) allow the analysis of 3D devices; however, quantum effects are only taken into account using quantum corrections algorithms which need proper calibration for each considered structure and crystal orientation (see for example [4], [5]). Another approach is represented by 2D Multi-Subband Ensemble Monte Carlo (MS-EMC) simulators which properly take into account quantum confinement in one dimension. They have been employed to study FinFETs, but only in the high aspect ratio limit, when the fin height is large and a planar approximation is assumed [6], [7]. Alternatively, the self consistent solution of 2D Schrödinger and Poisson equations in a cross section of the device has been employed to study the in-plane charge distribution and the low field mobility [8], [9], but this approach does not provide the device current nor information about the Short-Channel Effects (SCEs).

Here, we employ a self-consistent MS-EMC simulator for 3D devices where two-dimensional confinement is fully accounted for thanks to the solution of the 2D Schrödinger equation in several device cross sections. Self-consistency is achieved by the iterative solution of the 3D Poisson equation, the 2D Schrödinger equation and the 1D Boltzmann transport equation through the Monte Carlo (MC) method.

This work is organized as follows. Section II is devoted to the description of the simulator, then Section III defines the structure of the simulated FinFETs and reports the obtained results. Finally, conclusions are drawn in Section IV.

II. SIMULATOR DESCRIPTION

The simulator employed in this work is based on the space-mode approach [10], where the 2D Schrödinger equation is solved in several cross sections of the device perpendicular to the source-drain axis, and carrier transport is simulated in the perpendicular direction. The solution of the Schrödinger equation provides the electron eigen-energies, $E_{i,\nu}(z)$, and wave functions, $\xi_{i,\nu}(x,y,z)$, for different values of the position z along the transport direction, where i and ν are the valley and subband indices, respectively. After the solution of the Schrödinger equation, energy levels and subband structure are corrected to take into account the non-parabolic behavior of Si conduction band as described in [11]. Transport in each subband is simulated employing the MC method in 1D, where the driving force is proportional to the derivative of $E_{i,\nu}(z)$ with respect to z. The carrier density, $n(x,y,z)$ is obtained from the MC simulation by sampling the population of each subband at the different z positions, $n_{i,\nu}(z)$, multiplied by the distribution function given by the corresponding wave functions $|\xi_{i,\nu}(x,y,z)|^2$. This is employed to compute the total charge density $\rho(x,y,z)$, which is fed into the 3D Poisson equation. In this way the Schrödinger equation, MC transport and the Poisson equation are solved repeatedly in a loop until self-consistency is achieved. The MS-EMC method has been employed for 3D devices only in [12], [13], while it has been widely and successfully used for planar devices [14], [15], [16].

The structure of the considered device is described employing a 3D Finite Element (FE) mesh obtained by extrusion of a 2D triangular mesh. FEs are chosen to allow an accurate description of complex geometries (e.g. cylindrical nanowires, FinFETs with rounded corners and leaning sidewalls) and to formulate the Poisson and Schrödinger equations near material boundaries in a natural way. In the MC simulation, scattering events are implemented in such a way that Pauli exclusion principle is taken into account [17]. The scattering mechanisms currently implemented include acoustic and optical phonons [9]. Due to the stochastic nature of the MC method, it is

978-1-5090-5979-9/17 $31.00 © 2017 IEEE

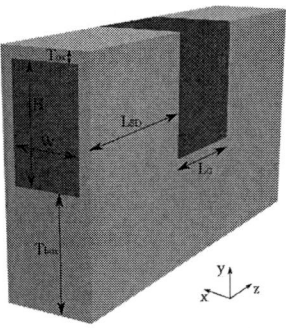

Fig. 1. FinFET structure used for the simulations of this work.

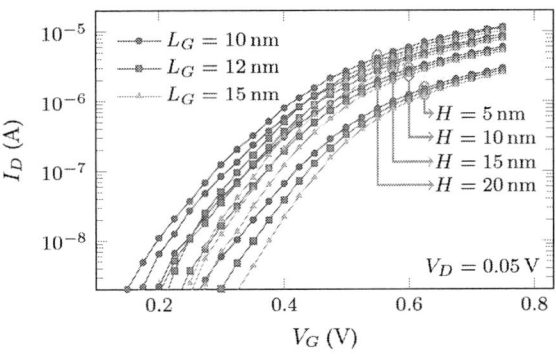

Fig. 2. Drain current for devices with $W = 6\,\text{nm}$ and different values of fin height H and channel length L_G, computed at $V_D = 50\,\text{mV}$.

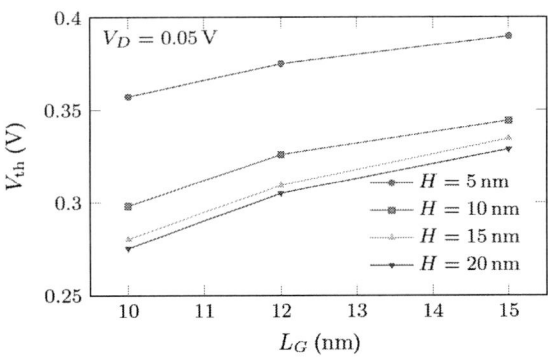

Fig. 3. Threshold voltage V_{th} for devices with $W = 6\,\text{nm}$ and different values of H as a function of channel length, computed at $V_D = 50\,\text{mV}$.

difficult to study the sub-threshold behavior of transistors since the low current levels are easily masked by the statistical noise. To mitigate this effect, the simulator includes a variance reduction technique based on non-uniform super-particle weight: the weight is determined based on energy [18].

The simulation of large 3D devices with a MS-EMC simulator is quite demanding from a computational point of view. To be able to simulate relatively large structures, it is essential to make the computations as efficient as possible. The MC method, in principle, allows for a large degree of parallelization, since the flights of the different super-particles are independent; however, particular care is needed for synchronization to inject and delete particles at contacts and to gather particle statistics. For the same purpose, the solution of the Poisson and Schrödinger equations is obtained employing specialized and optimized sparse matrix parallel routines.

III. RESULTS

The structure of the simulated FinFETs is shown in Figure 1, along with the definition of the relevant length parameters: the orientation is the standard one, where the vertical direction is $\langle 100 \rangle$ and the channel direction is $\langle 110 \rangle$. We employ two values for the fin width, $W = 5\,\text{nm}$ and $W = 6\,\text{nm}$, and fin heights H ranging from $5\,\text{nm}$ to $20\,\text{nm}$: in most of the cases the aspect ratio H/W is not large enough to justify an infinite height approximation. To analyze scaling effects we consider three different channel length values: $L_G = 10\,\text{nm}$, $12\,\text{nm}$ and $15\,\text{nm}$. In all devices we consider source and drain regions with length $L_{SD} = 14\,\text{nm}$. The thickness of the gate oxide (SiO_2) is $T_{\text{ox}} = 1\,\text{nm}$ for the top and lateral oxides, and the one of the buried oxide is $T_{\text{box}} = 10\,\text{nm}$. The channel is considered undoped ($N_A = 10^{15}\,\text{cm}^{-3}$) and midgap metal is assumed for the gate (with work function $4.56\,\text{eV}$). The source and drain doping density is $N_{SD} = 10^{20}\,\text{cm}^{-3}$, with an underlap of $L_{sp} = 2\,\text{nm}$ and Gaussian distribution with $\sigma = 0.8\,\text{nm}$. The boundary condition applied to the lower oxide/substrate interface is equivalent to a strongly n-doped ground plane with $V_B = 0\,\text{V}$. The mesh used for the smallest device considered employs 68573 nodes, 385020 tetrahedra and includes 1828 triangles in the corresponding 2D mesh for the Schrödinger equation. For the largest device simulated in

this work, these numbers grow up to 336243, 1942920 and 10446, respectively.

Figure 2 shows the transfer characteristics (I_D–V_G curves) of the simulated devices with $W = 6\,\text{nm}$: a gate bias up to $V_G = 0.7\,\text{V}$ is considered with a drain bias of $V_D = 50\,\text{mV}$. We can notice that, as expected, the current is higher for taller FinFETs (larger H) due to the bigger channel and for shorter L_G. In Figure 2 it is also evident that the threshold voltage varies between different devices. Even if the statistical noise inherent to the MC procedure noise is reduced, it is nonetheless too large to derive the threshold voltage V_{th} as the maximum of the second derivative of I_d with respect to V_G. Therefore we compute V_{th} with the constant current method; we employ a normalized current per unit length (the part of the perimeter covered by the gate $P = 2H + W$) of $1\,\mu\text{A}/\mu\text{m}$ for the longest device and scale it inversely with channel length: $I_{D,\text{th}} = 1\,\mu\text{A}/\mu\text{m} \times (2H + W) \times (15\,\text{nm}/L_G)$.

The results are shown in Figure 3 for $W = 6\,\text{nm}$. In the device with $H = 5\,\text{nm}$, the fin is almost square so that quantum confinement is noticeable also in the vertical direction: as a consequence, V_{th} is substantially larger than in the taller devices. The threshold voltage always gets smaller for increasing H, but the variations also decrease once the effects of vertical confinement become negligible. An important V_{th}

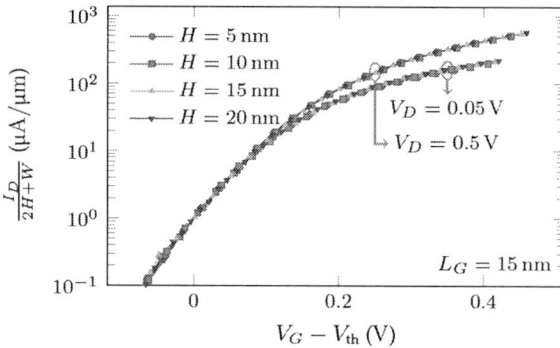

Fig. 4. Drain current normalized by $I_D/(2H+W)$ vs. gate overdrive $V_G - V_{th}$ for $W = 6$ nm and $L_G = 14$ nm. Two sets of overlapping curves can be observed for the two values of $V_D = 0.05$ V and $V_D = 0.5$ V.

roll-off is observed for all devices, although slightly smaller in the case of $H = 5$ nm due to the better electrostatic control of the channel. If we plot the normalized drain current $I_D/(2H+W)$ versus the gate overdrive $V_G - V_{th}$, we can observe that the curves corresponding to devices with different diameters collapse. This can be seen, for example, in Figure 4, for the case $W = 6$ nm and $L_G = 15$ nm: the curves corresponding to different values of the fin height H coincide.

Figure 5 shows the electron distribution obtained in a cross section of the FinFETs with $W = 6$ nm and different fin heights in the middle of the channel at $V_G = V_{th} + 0.3$ V. In such narrow fins the carrier density does not show two different charge regions near the lateral walls: electrons are spread across the whole device, except for the region very near to the interfaces due to the quantum effects. In the smallest device ($H = 5$ nm), the peak of the charge distribution essentially takes up the whole channel. For the taller devices a peak near the top of the channel is observed: it is caused by the joint effect of the top and the lateral gates.

In the standard FinFET orientation, the six valleys of Si conduction bands split into two sets: two valleys, Δ_2, with $m_x = 0.19\,m_0$, $m_y = 0.916\,m_0$, $m_z = 0.19\,m_0$ (see Figure 1 for the definition of x, y and z directions) and four valleys, Δ_4, with $m_x = 0.315\,m_0$, $m_y = 0.19\,m_0$, $m_z = 0.553\,m_0$ [19]. The energy profiles along the channel for few low-energy subbands are shown in Figure 6 for $W = 6$ nm, $L_G = 15$ nm and different fin heights at $V_D = 0.05$ V, $V_G = V_{th}$. The lowest energy subband belongs to the Δ_2 valleys for all the considered devices because its m_y is by far the largest value among all confinement masses. However, the second subband (if we order them according to energy) can belong to either the Δ_2 or Δ_4 valleys depending on the position z along the channel and on the fin height. In most cases it is the second Δ_2 subband, but for $H = 5$ nm and 10 nm in the channel region below the gate it is the first Δ_4 subband. The reason is that stronger quantum confinement in the y direction, caused by a smaller H or by the gate field, affects the Δ_2 subbands more than Δ_4 subbands because of the larger m_y. Subbands corresponding to the Δ_4 valleys present smaller variations

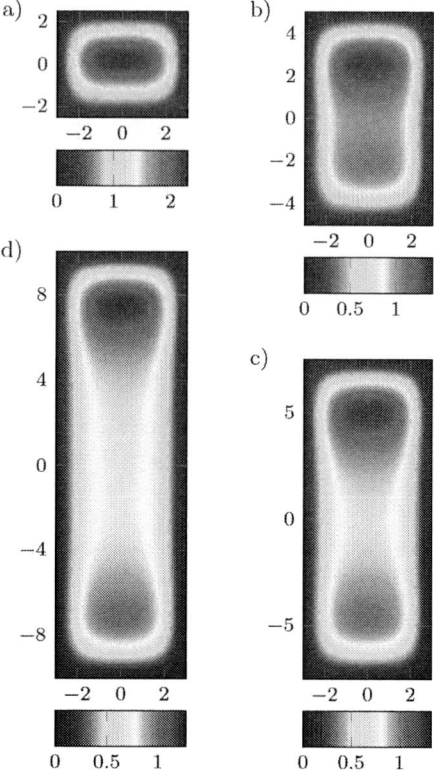

Fig. 5. Electron density in a cross section in the middle of the channel for $V_G = V_{th} + 0.3$ V and devices with $W = 6$ nm and a) $H = 5$ nm, b) $H = 10$ nm, c) $H = 15$ nm, d) $H = 20$ nm. Dimensions in the cross sections are in nm and electron density are in 10^{19} cm^{-3}.

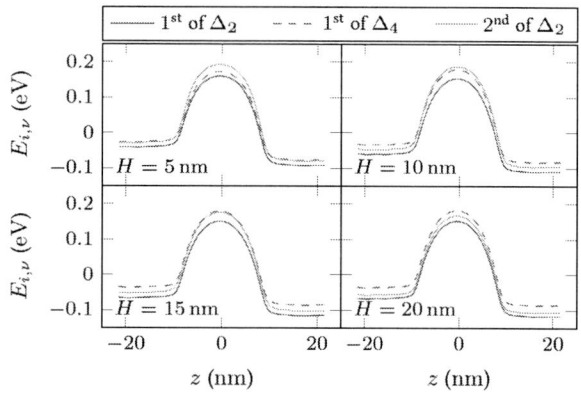

Fig. 6. Energy profile of the first two subbands of Δ_2 valleys and the first subband of Δ_4 valleys, for $W = 6$ nm, $L_G = 15$ nm, $V_D = 0.05$ V and $V_G = V_{th}$.

with H, and can be more populated than subbands of the Δ_2 valleys with equal or lower energy, as can be seen in Figure 7. Two reasons combine to produce this behavior: the larger degeneracy factor (4 vs. 2) and the larger m_z [20].

Finally, we turn to the analysis of SCEs and in particular of the Drain Induced Barrier Lowering (DIBL). Figure 8 shows

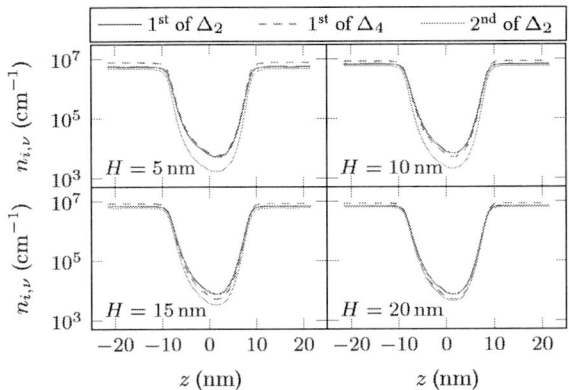

Fig. 7. Population of the first two subbands of Δ_2 valleys and the first subband of Δ_4 valleys along the channel, for $W = 6\,\text{nm}$, $L_G = 15\,\text{nm}$, $V_D = 0.05\,\text{V}$ and $V_G = V_{\text{th}}$.

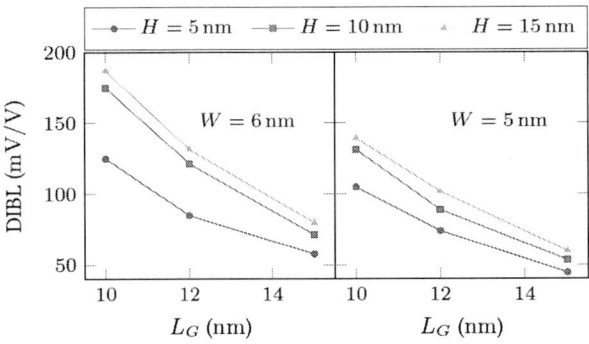

Fig. 8. DIBL as a function of L_G for different values of W and H.

the DIBL (computed between $V_D = 0.05\,\text{V}$ and $V_D = 0.5\,\text{V}$) as a function of the channel length L_G, for different fin geometries. As expected, in all cases the DIBL increases for decreasing L_G. For $W = 6\,\text{nm}$, we can observe a large difference between devices with $H = 5\,\text{nm}$ and $H = 10\,\text{nm}$, while the further move to $H = 15\,\text{nm}$ only produces a small variation. On the contrary, for $W = 5\,\text{nm}$ the changes due to different fin heights are more uniform, thanks to the improved electrostatic control of the lateral gates.

IV. CONCLUSION

In this paper, we described the MS-EMC simulator recently developed by our group for arbitrary 3D device geometries and its application to the simulation of scaled FinFETs. The solution of 2D Schrödinger equation in cross sections of the device perpendicular to the source-drain direction is self-consistently coupled to the solution of the 3D Poisson equation and the 1D Boltzmann transport equation along the channel through an ensemble MC scheme. We showed the importance of the two-dimensional quantum confinement effects and the relevance of the fin height when the fin aspect ratio is small enough. Drain current curves for channel lengths down to $10\,\text{nm}$ have been computed, and the effect of the FinFET cross-section on the SCEs has been evaluated.

ACKNOWLEDGMENT

The authors would like to thank the financial support of Spanish Government (TEC2014-59730-R) and EU H2020 program (REMINDER, grant agreement No 687931 and WAY-TOGO FAST, ECSEL-2014-2-662175).

REFERENCES

[1] J.-P. Colinge, *FinFETs and Other Multi-Gate Transistors*, 1st ed. Springer, Nov. 2007.

[2] C. Auth, C. Allen *et al.*, "A 22nm high performance and low-power CMOS technology featuring fully-depleted tri-gate transistors, self-aligned contacts and high density MIM capacitors," in *2012 Symposium on VLSI Technology (VLSIT)*, Jun. 2012, pp. 131–132.

[3] S. Natarajan, M. Agostinelli *et al.*, "A 14nm logic technology featuring 2nd-generation FinFET, air-gapped interconnects, self-aligned double patterning and a 0.0588 μm2 SRAM cell size," in *2014 IEEE International Electron Devices Meeting*, Dec. 2014, pp. 3.7.1–3.7.3.

[4] F. M. Bufler and L. Smith, "3d Monte Carlo simulation of FinFET and FDSOI devices with accurate quantum correction," *Journal of Computational Electronics*, vol. 12, no. 4, pp. 651–657, Dec. 2013.

[5] M. A. Elmessary, D. Nagy *et al.*, "Anisotropic Quantum Corrections for 3-D Finite-Element Monte Carlo Simulations of Nanoscale Multigate Transistors," *IEEE Transactions on Electron Devices*, vol. 63, no. 3, pp. 933–939, Mar. 2016.

[6] D. Lizzit, P. Palestri *et al.*, "Analysis of the Performance of n-Type FinFETs With Strained SiGe Channel," *IEEE Transactions on Electron Devices*, vol. 60, no. 6, pp. 1884–1891, 2013.

[7] C. Medina-Bailon, C. Sampedro *et al.*, "Confinement orientation effects in S/D tunneling," *Solid-State Electronics*, vol. 128, pp. 48–53, Feb. 2017.

[8] F. Garcia Ruiz, A. Godoy *et al.*, "A Comprehensive Study of the Corner Effects in Pi-Gate MOSFETs Including Quantum Effects," *Electron Devices, IEEE Transactions on*, vol. 54, no. 12, pp. 3369–3377, 2007.

[9] A. Godoy, F. Ruiz *et al.*, "Calculation of the phonon-limited mobility in silicon Gate All-Around MOSFETs," *Solid-State Electronics*, vol. 51, no. 9, pp. 1211–1215, Sep. 2007.

[10] R. Venugopal, Z. Ren *et al.*, "Simulating quantum transport in nanoscale transistors: Real versus mode-space approaches," *Journal of Applied Physics*, vol. 92, no. 7, pp. 3730–3739, Sep. 2002.

[11] S. Jin, M. V. Fischetti, and T.-w. Tang, "Modeling of electron mobility in gated silicon nanowires at room temperature: Surface roughness scattering, dielectric screening, and band nonparabolicity," *Journal of Applied Physics*, vol. 102, no. 8, p. 083715, Oct. 2007.

[12] C. Sampedro, L. Donetti *et al.*, "3d multi-subband ensemble Monte Carlo simulator of FinFETs and nanowire transistors," in *2014 International Conference on Simulation of Semiconductor Processes and Devices (SISPAD)*, Sep. 2014, pp. 21–24.

[13] L. Donetti, C. Sampedro *et al.*, "Multi-Subband Ensemble Monte Carlo simulation of Si nanowire MOSFETs," in *2015 International Conference on Simulation of Semiconductor Processes and Devices (SISPAD)*, Sep. 2015, pp. 353–356.

[14] J. Saint-Martin, A. Bournel *et al.*, "Multi sub-band Monte Carlo simulation of an ultra-thin double gate MOSFET with 2d electron gas," *Semiconductor Science and Technology*, vol. 21, no. 4, p. L29, 2006.

[15] E. Sangiorgi, P. Palestri *et al.*, "The Monte Carlo approach to transport modeling in deca-nanometer MOSFETs," *Solid-State Electronics*, vol. 52, no. 9, pp. 1414–1423, Sep. 2008.

[16] C. Sampedro, F. Gámiz *et al.*, "Multi-Subband Monte Carlo study of device orientation effects in ultra-short channel DGSOI," *Solid-State Electronics*, vol. 54, no. 2, pp. 131–136, Feb. 2010.

[17] P. Lugli and D. K. Ferry, "Degeneracy in the ensemble Monte Carlo method for high-field transport in semiconductors," *IEEE Transactions on Electron Devices*, vol. 32, no. 11, pp. 2431–2437, Nov. 1985.

[18] A. Pacelli and U. Ravaioli, "Analysis of variance-reduction schemes for ensemble Monte Carlo simulation of semiconductor devices," *Solid-State Electronics*, vol. 41, no. 4, pp. 599–605, Apr. 1997.

[19] M. Bescond, N. Cavassilas, and M. Lannoo, "Effective-mass approach for n-type semiconductor nanowire MOSFETs arbitrarily oriented," *Nanotechnology*, vol. 18, no. 25, p. 255201, Jun. 2007.

[20] F. Gámiz, J. Roldán *et al.*, "Electron mobility in extremely thin single-gate silicon-on-insulator inversion layers," *Journal of Applied Physics*, vol. 86, no. 11, pp. 6269–6275, Nov. 1999.

Study of Strained Effects in Nanoscale GAA Nanowire FETs Using 3D Monte Carlo Simulations

Muhammad A. Elmessary*†, Daniel Nagy‡, Manuel Aldegunde§, Antonio J. García-Loureiro‡ and Karol Kalna*

*NanoDeCo Group, College of Engineering, Swansea University, Swansea SA1 8EN, Wales, United Kingdom
†Engineering Math & Physics Department, Faculty of Engineering, Mansoura University, Mansoura 35516, Egypt
‡CITIUS, Universidade de Santiago de Compostela, 15782 Santiago de Compostela, Galicia, Spain
§Dept. Aerospace and Mechanical Engineering, University of Notre Dame, Notre Dame, IN 46556, USA
Email: M.A.A.Elmessary.716902@swansea.ac.uk, Phone: +44 (0) 1792 602816

Abstract—3D Finite Element ensemble Monte Carlo simulations with integrated 2D Schrödinger Equation quantum corrections are employed to forecast the performance of scaled Si gate-all-around (GAA) nanowire (NW) FETs with unstrained/strained channel. The results from the 3D MC toolbox were compared against experimental I-V characteristics of a 22 nm gate length GAA NW FET with excellent agreement. The NW FET is then scaled to a 10 nm gate length, studying the interplay of the pre-existing quantum confinement in $\langle 100 \rangle$ and $\langle 110 \rangle$ channel orientations with uniaxial strain engineering, with a strength of 0.5%, 0.7% and 1.0%. We found that increasing the uniaxial strain in the channel is largely limited by the quantum confinement which weakens the strain induced drive current increase to about 7% in the $\langle 100 \rangle$ channel and to less than 5% in the $\langle 110 \rangle$ channel.

I. Introduction

Gate-all-around (GAA) nanowire (NW) FETs have superior electrostatics, suppressing the short channel effects and making them the most promising candidates for the sub-10 nm digital technology [1]. However, the strong quantum confinement might limit the application of strain engineering in the Si channel aiming to increase the drain current by driving more electrons into the valleys with a lighter effective mass in the transport direction [2]. The question arises as to how much the quantum confinement induced valley splitting counteracts the effectiveness of strain [3] in these nanoscaled devices?

In this work, we investigate the effects of uniaxial tensile strain in nanoscale Si GAA NW FETs with gate lengths of 22 nm and 10 nm with different channel orientations. The 10 nm gate length GAA NW FET has an elliptical cross-section ($R = 5.7/7.17$ nm) with an effective oxide thickness (EOT) of 0.8 nm (Fig. 1) and the further dimensions are summarized in Table I. We simulate $\langle 100 \rangle$ and $\langle 110 \rangle$ channel orientations and two types of tensile strain: uniaxial $\langle 100 \rangle$

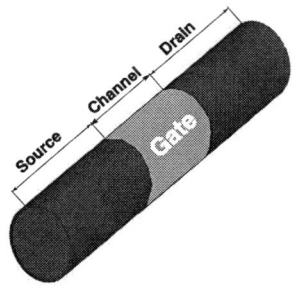

Fig. 1.   3D schematic of the investigated n-channel Si GAA nanowire FET.

and uniaxial $\langle 110 \rangle$ with strengths of 0.5%, 0.7% and 1.0%. To show the effect of quantum confinement on strain, the 10 nm gate length NW is compared to a bigger device with an elliptical cross-section ($R = 11.3/14.22$ nm) and a gate length of 22 nm with an EOT of 1.5 nm [1].

II. 3D FE MC Simulation Toolbox

This study uses an in-house 3D Finite Element (FE) Monte Carlo (MC) toolbox with calibration-free Schrödinger Equation based Quantum Corrections (SEQC) [4]–[6] allowing to account for the exact nanoscale device geometry and thus accurate quantum confinement. The 3D FE mesh contains predefined 2D planes perpendicular to transport direction to solve 2D Schrödinger Eq. separately for the three/six Δ valleys [4]:

$$-\frac{\hbar^2}{2}\nabla_\perp \cdot \left[(\mathbf{m}^*)^{-1} \cdot \nabla_\perp \psi(y,z)\right] + U(y,z)\psi(y,z) = E\psi(y,z),$$

where $(\mathbf{m}^*)^{-1}$ is the inverse effective mass tensor [4] $U(y,z) = -[qV(y,z) + \chi(y,z)]$ is the potential energy, $\chi(y,z)$ is the electron affinity, $\psi(y,z)$ is the wavefunction penetrating into oxide, and E is the energy [5]. Note that I-V characteristics obtained from the 3D MC toolbox were verified against experimental data for the 22 nm GAA NW FET with the $\langle 110 \rangle$ channel orientation with excellent agreement [1], [7] as shown in Fig. 2.

The strain is modeled by shifting the Si conduction valleys by ΔE_C according to the strain type and strength (Table II) [8]. For the uniaxial $\langle 110 \rangle$ strain, the transverse effective masses in $\Delta 3$ valley split due to a band-structure warping, resulting in a lighter m_t in the transport direction [8].

TABLE I.   Dimensions and Design Parameters of Ellipsoidal 22 and 10 nm Gate Length GAA FETs.

Gate Length [nm]	22	10
Perimeter [nm]	40.21	20.29
Area [nm^2]	126.14	32.1
Major Diameter [nm]	14.22	7.17
Minor Diameter [nm]	11.3	5.7
EOT [nm]	1.5	0.8
$N_\mathrm{D,max}$ [cm^{-3}]	5×10^{19}	5×10^{19}
Gaussian σ_x [nm]	7.1	3.23
High V_D [V]	1.0	0.7

978-1-5090-5979-9/17 $31.00 © 2017 IEEE

TABLE II. ENERGY-EDGE SHIFTS FOR SI Δ-VALLEYS WITH DIFFERENT TYPES AND STRENGTHS OF TENSILE STRAIN [8].

Strain	Uniaxial $\langle 100 \rangle$			Uniaxial $\langle 110 \rangle$		
Strength	0.5%	0.7%	1.0%	0.5%	0.7%	1.0%
$\Delta 1$ [eV]	+0.03	+0.042	+0.06	-0.01	-0.014	-0.02
$\Delta 2$ [eV]	-0.045	-0.063	-0.09	-0.01	-0.014	-0.02
$\Delta 3$ [eV]	-0.045	-0.063	-0.09	-0.065	-0.091	-0.13

TABLE III. V_T AND SUB-THRESHOLD SLOPE (SS) AT $V_D = 0.05$ V (LOW) AND 1.0/0.7 V (HIGH) FROM THE DD, DIBL FROM THE DD AND FROM THE MC, AND DRIVE CURRENTS (I_{MC}) AT $V_G = 1.0$ V COMPARING THE 22/10 NM GATE LENGTH GAA FETS.

Method	Gate length [nm]	22	10
MC	V_T [V]	0.3	0.35
DD	SS_{LOW} [mV/dec]	74	67
DD	SS_{HIGH} [mV/dec]	76	68
MC	$DIBL_{\langle 100 \rangle}$ [mV/V]	81	66
MC	$DIBL_{\langle 110 \rangle}$ [mV/V]	64	39
MC	$I_{\langle 100 \rangle}$ [$\mu A/\mu m$]	1222	1320
MC	$I_{\langle 110 \rangle}$ [$\mu A/\mu m$]	1000	1100

III. PERFORMANCE OF SCALED GAA NW FETS

Fig. 3 shows I_D-V_G characteristics at $V_D = 0.05$ V and $V_D = 0.7$ V (the current is normalized by the nanowire perimeter) for the scaled 10 nm gate length NW FET with the $\langle 100 \rangle$ and $\langle 110 \rangle$ channel orientations obtained from the 3D FE SEQC MC. Table III compares device operating characteristics predicting that the scaling to the 10 nm gate will ensure superior electrostatic integrity of a nearly ideal sub-threshold slope of 68 mV/dec and a DIBL of 39 mV/V with satisfactory ON-current (the ON-current increase is $\sim$ 8/10% for the $\langle 100 \rangle$/$\langle 110 \rangle$ orientation with respect to the 22 nm gate NW).

Figs. 4(a) and 4(b) illustrate the average electron density cross-sections in the middle of the channel for the 10 nm and 22 nm gate length devices, respectively, at $V_G = 0.8$ V and $V_D = 0.7/1.0$ V. In the smaller device, volume inversion occurs due to strong quantum confinement providing a good

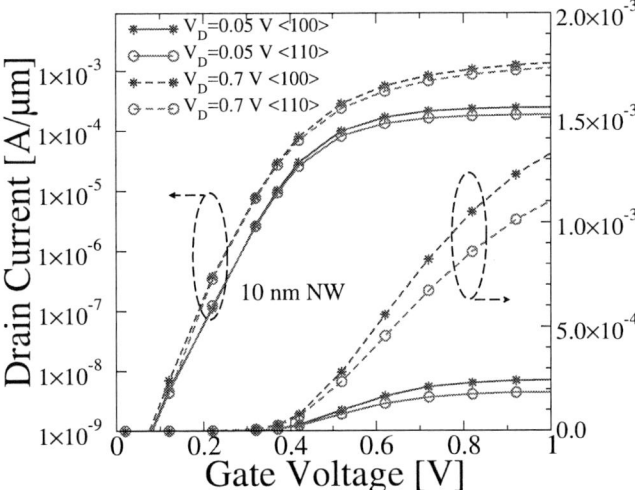

Fig. 3. I_D-V_G characteristics at $V_D = 0.05$ V and 0.7 V for the scaled 10 nm gate length GAA NW FET predicted by the 3D FE SEQC MC along the $\langle 100 \rangle$ and $\langle 110 \rangle$ channel orientations.

gate control over the channel. In the bigger device, the electron density becomes more side-walled with a subsequent weaker quantum confinement.

IV. PERFORMANCE OF STRAINED GAA NW FETS

Fig. 5 shows I_D-V_G characteristics at $V_D = 0.7$ V for the 10 nm gate length NW in the $\langle 100 \rangle$ channel orientation under three uniaxial $\langle 100 \rangle$ strengths delivering increase in the drive-current by 6%, 6.9% and 7.3%, respectively. The sub-threshold slope exhibits a small deterioration under the strain, except for 1.0% case. Fig. 6 plots the 3 Δ valleys contributions under uniaxial strain of the current in the $\langle 100 \rangle$ device. Under strain, $\Delta 1$ valley is shifted up (with the largest effective transport mass, m_l) reducing its contribution to the current. $\Delta 2$ and $\Delta 3$ valleys (Table II) are shifted down (with the smallest effective transport mass, m_t) increasing their current contributions.

Since a larger confinement (perpendicular-to-transport) mass results in a lower bound state, the $\Delta 3$ with a larger confinement effective mass (m_l) contributes more to the current than $\Delta 2$. The reduction in the strain effectiveness caused by the quantum confinement is seen in the less confined 22 nm device

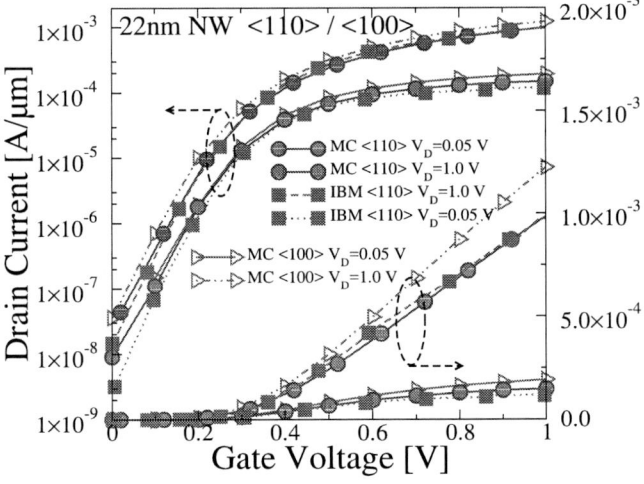

Fig. 2. I_D-V_G characteristics at $V_D = 0.05$ V and 1.0 V for the 22 nm GAA NW FET in the $\langle 110 \rangle$ (circles) and $\langle 100 \rangle$ (triangles) channel orientations from the 3D FE MC with anisotropic SEQC [1], [7] compared against experimental data (IBM) in the $\langle 110 \rangle$ (squares) orientation [1].

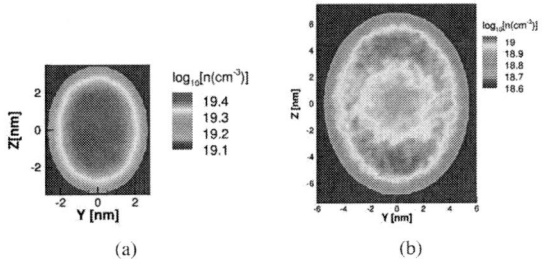

Fig. 4. Electron density across the middle of the $\langle 110 \rangle$ channel for (a) the 10 nm gate length at $V_D = 0.7$ V and (b) the 22 nm gate length at $V_D = 1.0$ V, both NWs at $V_G = 0.8$ V. Note a different colour scale in (a) an (b) to show the contrast between high and low areas of density.

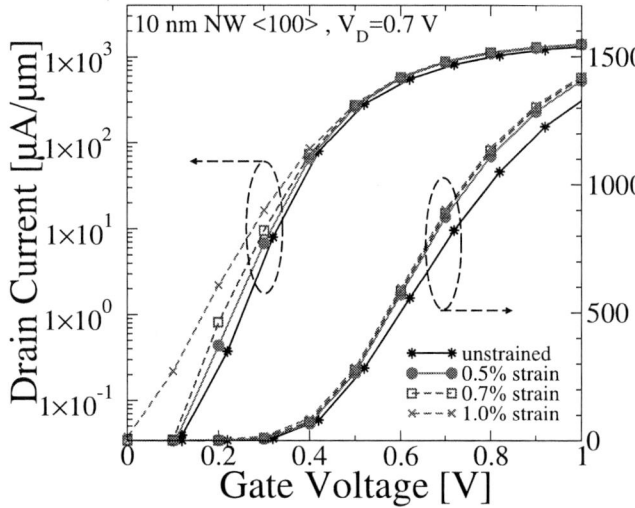

Fig. 5. I_D-V_G characteristics at $V_D = 0.7$ V for the 10 nm gate length $\langle 100 \rangle$ channel GAA NW FET under increasing uniaxial $\langle 100 \rangle$ strain strengths.

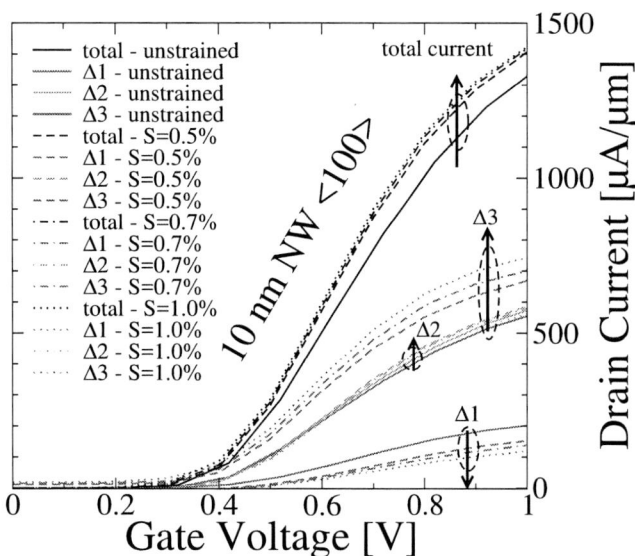

Fig. 6. Valley contributions to the current at $V_D = 0.7$ V for the 10 nm gate length $\langle 100 \rangle$ channel GAA-NW under indicated uniaxial $\langle 100 \rangle$ strains.

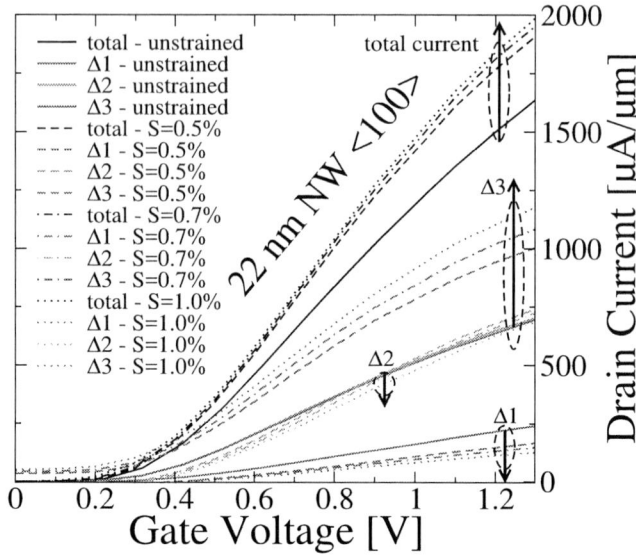

Fig. 7. Valley contributions to the current at $V_D = 1.0$ V for the 22 nm gate length $\langle 100 \rangle$ channel GAA-NW under indicated uniaxial $\langle 100 \rangle$ strains.

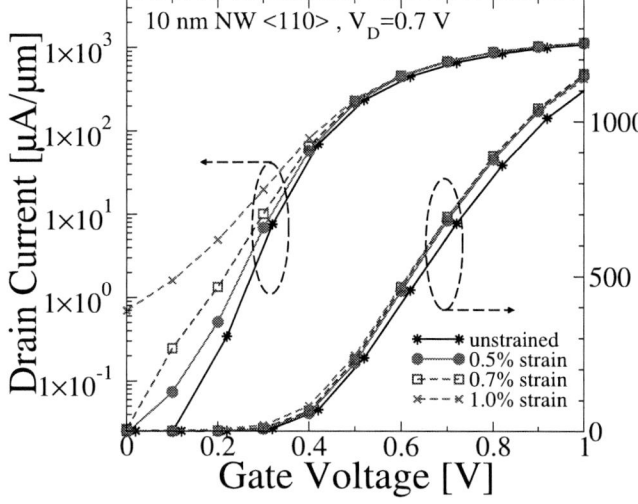

Fig. 8. I_D-V_G characteristics at $V_D = 0.7$ V for the 10 nm gate length $\langle 110 \rangle$ channel GAA NW FET under indicated uniaxial $\langle 110 \rangle$ strains.

(Fig. 7). The 0.5/0.7/1.0% uniaxial $\langle 100 \rangle$ strain increases the drive current (at $V_G = 1.3$ V) by 16.6/19.3/21.4% compared to only 6/6.9/7.3% for the 10 nm gate length device.

Fig. 8 shows I_D-V_G characteristics for the 10 nm gate length NW FET with a $\langle 110 \rangle$ channel orientation at $V_D = 0.7$ V under uniaxial $\langle 110 \rangle$ strain. Without applying any strain, the $\langle 100 \rangle$ channel device delivers more current (20.4%) than the $\langle 110 \rangle$ channel device due to enhanced mobility (lighter effective transport mass). When applying uniaxial $\langle 110 \rangle$ strain, the drive-current at $V_G = 1.0$ V increases by 4.2%, 4.8%, and 3.3% at increasing strain strengths, respectively. The drive-current drop at 1.0% strain is due to the swap of valley contributions to the current with increasing strain. To see the effect of confinement, Fig. 10 shows for comparison the less-confined 22 nm gate length NW in which the 0.5/0.7/1.0% uniaxial $\langle 110 \rangle$ strain increases the drive current (at $V_G = 1.0$ V) by 11.6/12.7/7.6% compared to only 4.2/4.8/3.3% in the 10 nm gate device, see Fig. 9.

Figs. 11 and 12 compare the average electron velocity at V_G-V_T = 0.7/1.0 V and V_D = 0.7/1.0 V for the unstrained/strained 10/22 nm gate length GAA NW FET in the $\langle 100 \rangle$ and the $\langle 110 \rangle$ channel orientations. The average velocity is larger in the $\langle 100 \rangle$ channel orientation than in the $\langle 110 \rangle$ one. The peaks of average electron velocity along the channel of the 22 nm gate NW FET increase more than those of the 10 nm gate length NW FET due to a smaller quantum confinement.

978-1-5090-5979-9/17 $31.00 © 2017 IEEE

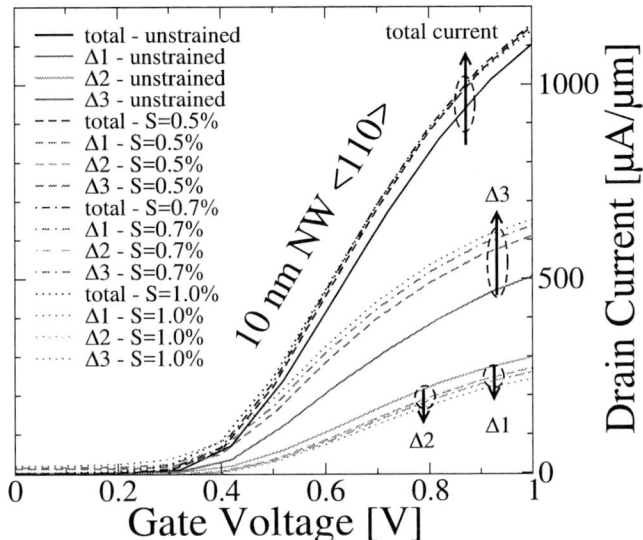

Fig. 9. Valley contributions to the current at $V_D = 0.7$ V for the 10 nm gate length $\langle 110 \rangle$ channel GAA NW FET under indicated uniaxial $\langle 110 \rangle$ strains.

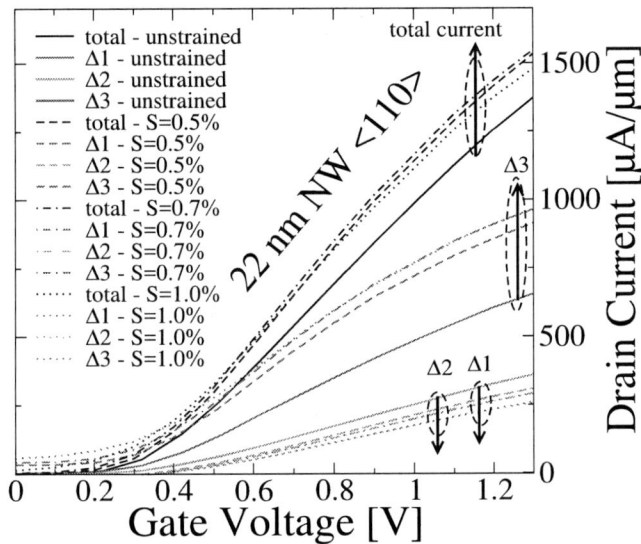

Fig. 10. Valley contributions to the current at $V_D = 1.0$ V for the 22 nm gate length $\langle 110 \rangle$ channel GAA-NW under different uniaxial $\langle 110 \rangle$ strains.

V. CONCLUSION

The drive current in the 10 nm GAA $\langle 100 \rangle$ channel NW FET can be more affected by the strain engineering when compared to the $\langle 110 \rangle$ channel one. With increasing the strain strength, the quantum confinement pre-existing valley splitting starts to weaken the strain effect especially in the $\langle 110 \rangle$ channel. Therefore, the 0.5/0.7/1.0% uniaxial $\langle 100 \rangle$ strain can increase the drive current only by 6/6.9/7.3% in the strongly confined 10 nm gate length FET compared to 16.6/19.3/21.4% in the less confined 22 nm gate length FET; and the uniaxial $\langle 110 \rangle$ strain by only 4.2/4.8/3.3% in the 10 nm gate length FET as compared to 11.6/12.7/7.6% increase in the 22 nm gate length transistor.

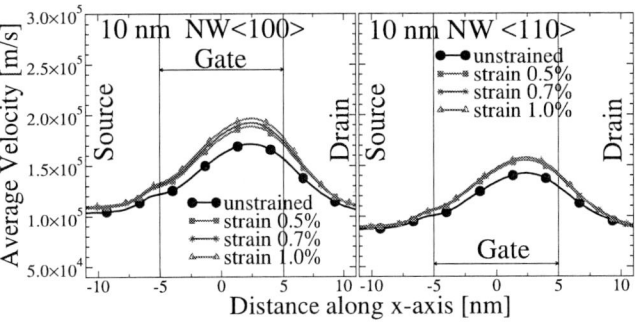

Fig. 11. Average electron velocity at $V_G = 1.0$ V and $V_D = 0.7$ V for the unstrained/strained 10 nm gate length GAA NW FET in the $\langle 100 \rangle / \langle 110 \rangle$ channel orientation. The zero is set in the middle of the gate.

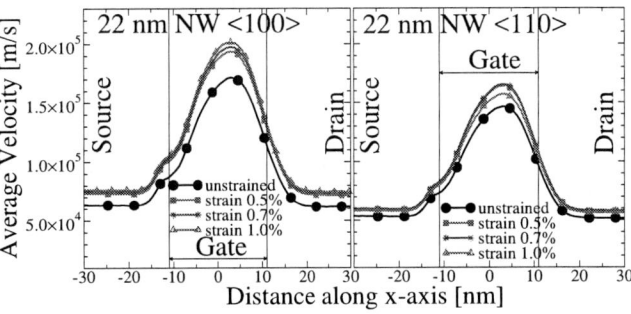

Fig. 12. Average electron velocity at $V_G = 1.3$ V and $V_D = 1.0$ V for the unstrained/strained 22 nm gate length GAA NW FET in the $\langle 100 \rangle / \langle 110 \rangle$ channel orientation. The zero is set in the middle of the gate.

REFERENCES

[1] S. Bangsaruntip et al., "Density scaling with gate-all-around silicon nanowire MOSFETs for the 10 nm node and beyond," *IEDM Tech. Dig.*, pp. 526–529, 2013.

[2] K. M. Liu, L. F. Register, and S. K. Banerjee, "Quantum Transport Simulation of Strain and Orientation Effects in Sub-20 nm Silicon-on-Insulator FinFETs," *IEEE TED*, vol. 58, no. 1, pp. 410, 2011.

[3] M. V. Fischetti, F. Gamiz, and W. Hansch, "On the enhanced electron mobility in strained- silicon inversion layers," *J. Appl. Phys.*, vol. 92, no. 12, pp. 73207324, 2002.

[4] M. A. Elmessary et al., "Anisotropic Quantum Corrections for 3-D Finite-Element Monte Carlo Simulations of Nanoscale Multigate Transistors," *IEEE Trans. Electron Devices*, vol. 63, no. 3, pp. 933-939, 2016.

[5] J. Lindberg et al., "Quantum Corrections Based on the 2D Schrödinger Equation for 3D Finite Element Monte Carlo Simulations of Nanoscaled FinFETs," *IEEE Trans. Electron Devices*, vol. 61, no. 1, pp. 423-429, Jan. 2014.

[6] M. Aldegunde, A. J. García-Loureiro, and K. Kalna, "3D Finite Element Monte Carlo Simulations of Multi-Gate Nanoscale Transistors," *IEEE Trans. Electron Devices*, vol. 60, no. 5, pp. 1561-1567, 2013.

[7] M. A. Elmessary et al., "Scaling/LER study of Si GAA nanowire FET using 3D Finite Element Monte Carlo simulations," *Solid-St. Electron.*, vol. 128, pp. 17-24, 2017.

[8] K. Uchida et al., "Physical mechanisms of electron mobility enhancement in uniaxial stressed MOSFETs and impact of uniaxial stress engineering in ballistic regime," *IEDM Tech. Dig.*, pp. 129-132, 2005.

[9] A. Rahman et al., "Generalized effective-mass approach for n-type metal-oxide-semiconductor field-effect transistors on arbitrarily oriented wafers," *J. Appl. Phys.*, vol. 97, no. 5, pp. 053702, 2005.

978-1-5090-5979-9/17 $31.00 © 2017 IEEE

Investigation of Electrically Gate-All-Around Hexagonal Nanowire FET (HexFET) Architecture for 5 nm Node Logic and SRAM Applications

Jeffrey A. Smith[1]*, Kai Ni[1], Ram Krishna Ghosh[1],
Jeff Xu[2], Mustafa Badaroglu[2], PR. Chidi Chidambaram[2] and Suman Datta[1]

[1]University of Notre Dame, Notre Dame, IN, USA; [2]Qualcomm Technologies Incorporated, San Diego, CA, USA;
*Email: jsmith80@nd.edu

Abstract—This work investigates, in detail, the electrically gate-all-around (eGAA) Hexagonal NW FET (HexFET) which combines the high current drive of FinFETs with the excellent electrostatic robustness of conventional Gate-All-Around Nanowire (GAA NW) FETs. We evaluate HexFET as a potential successor to FinFET for 5nm node logic and SRAM applications using first principles atomistic-based modeling, calibrated 3D numerical device simulations, and circuit-level benchmarking. From this, we conclude that the eGAA HexFET architecture offers superior performance to both FinFET and GAA NW FET for 5nm node applications.

Keywords—5 nm Node; FinFET; Gate-All-Around Nanowire; Electrically Gate-All-Around Hexagonal Nanowire;

I. INTRODUCTION

As CMOS technology approaches the 5nm node, various optimization pathways to extend FinFET technology are currently being explored. FinFET is currently the primary transistor architecture of choice for 14nm [1], 10nm and 7nm technology nodes [2-3] due to its high current drive and effective control of short channel effects (SCE). However, at the 5nm technology node (L_G=11nm), deterioration of SCE observed in FinFET has led to the evaluation of circular gate-all-around nanowire (GAA NW) FET, due to its superior electrostatic control [4-5]. To further increase drive currents in GAA NWs, stacking of conventional NWs [6] and diamond-shaped single NW FETs exhibiting higher hole mobility [7-8] have also been proposed.

In this work, we evaluate the concept of electrically GAA (eGAA) Hexagonal NW FET (HexFET) architecture that combines the high current drive of FinFET and the robust electrostatics of GAA NW FET. The device and circuit level performance of HexFET is benchmarked against FinFET and GAA NW FET for 5nm node logic and SRAM applications. This manuscript is organized as follows: Section II describes the inherent carrier transport benefits of a hexagonal-shaped channel, and sections III-A and III-B discuss the device and circuit level performance of all candidates before concluding remarks are given in section IV.

II. IMPROVING CARRIER TRANSPORT WITH CONFINED SILICON CHANNELS

Starting at the 90nm node, various strain-induced mobility enhancement techniques have been applied to improve the Silicon CMOS drive current [9-11]. To further complement strain engineering, confined silicon channels exhibiting inherently higher carrier mobilities may also be incorporated [12]. To verify this claim, first principles tight-binding band structure calculations are performed for the three silicon channels (FinFET, circular GAA NW and hexagonal NW) as shown in Figs. 1(a-b). Due to increased confinement, circular GAA NW and hexagonal NW display an indirect to direct band gap transition, causing carrier repopulation to the Γ-valley. This carrier relocation reduces the electron transport mass by a factor of two in circular and hexagonal NW structures over a rectangular FinFET (Fig. 1(c)). In the case of hexagonal NW, the reduced transport effective mass is further complemented by a reduction in scattering due to confined phonon modes, providing a remarkable 3.2x increase in the electron-phonon limited mobility over a rectangular FinFET. With the inherent transport improvement in confined channels, the device-level implications of each channel geometry remains questionable and is further investigated through 3D TCAD numerical simulations.

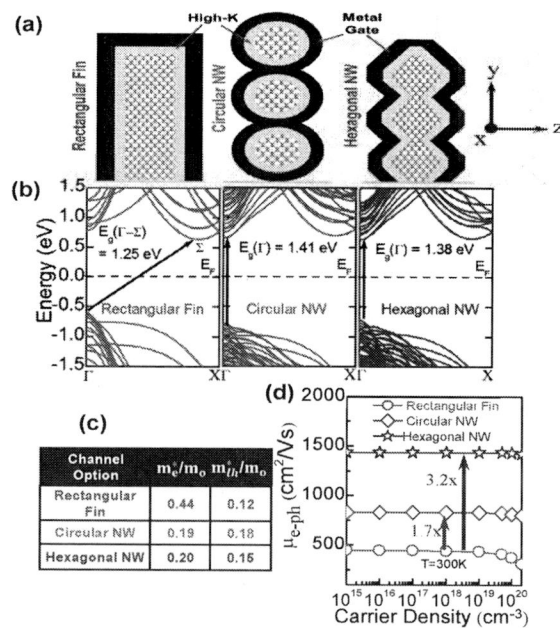

Fig. 1: (a) Atomistic configuration and (b) Tight-Binding (sp³d⁵s*) calculated band structures for Silicon FinFET, circular GAA NW and Hexagonal NW FET. (c) Transport effective mass and (d) electron-phonon limited mobility as a function of carrier density.

This work was supported by Qualcomm Technologies, Inc., San Diego, CA, USA and Lam Research Corporation, Fremont, CA, USA.

III. 5NM NODE TRANSISTOR ARCHITECTURE

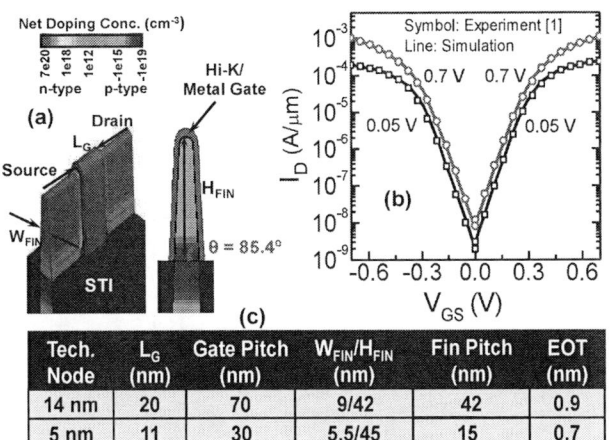

Fig. 2: (a) 14nm node 3D FinFET schematic showing channel and source/drain doping profiles. Sub-fin doping is included to suppress sub-surface leakage paths. (b) The 3D numerical model calibrated to 14nm experimental transfer characteristics. (c) Summary of key dimensions for logic transistors at 14nm and 5nm technology nodes (Courtesy, Lam Research Corporation).

Tech. Node	L_G (nm)	Gate Pitch (nm)	W_{FIN}/H_{FIN} (nm)	Fin Pitch (nm)	EOT (nm)
14 nm	20	70	9/42	42	0.9
5 nm	11	30	5.5/45	15	0.7

Fig. 2(a) shows the constructed 3D FinFET model, with channel and source/drain doping profiles. The device model is then calibrated to experimental FinFET data in [1] using a modified drift-diffusion approach, incorporating models for non-equilibrium transport via field-dependent mobility and velocity overshoot. The density gradient method is utilized to treat quantum confinement in the narrow fin, as carrier confinement will increase with the reduced fin width (Fig. 2(c)) at the 5nm node. Device structures of the 5nm node candidates from Section II are shown in Figs. 3(a)-(c). The HexFET channel is shaped by selective etching of the <111> crystal plane to create the <111>/<110> facets in a fin [8].

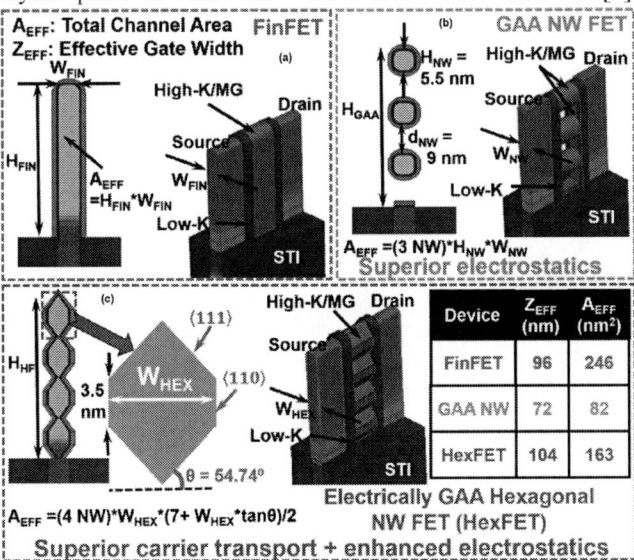

Device	Z_{EFF} (nm)	A_{EFF} (nm²)
FinFET	96	246
GAA NW	72	82
HexFET	104	163

Fig. 3: Comparison of 5nm technology candidates: (a) FinFET (b) GAA NW FET and (c) HexFET. Table summarizes effective gate width (Z_{EFF}) and channel area (A_{EFF}). HexFET offers increased Z_{EFF} over FinFET and GAA NW FET, and increased A_{EFF} over GAA NW FET.

This is in contrast to conventional GAA NW formation, which requires SiGe/Si/SiGe epitaxy and NW release etch steps in a conventional FinFET process flow [4-6].

A. Device Level Implications

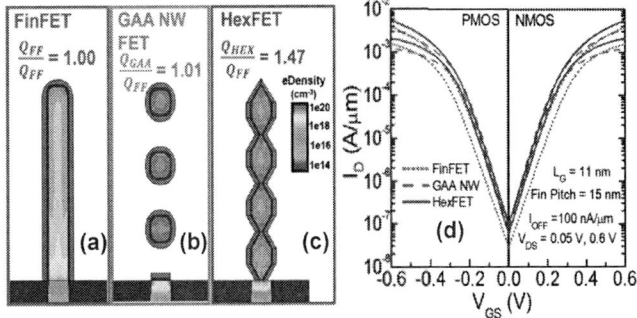

Fig. 4: Distribution of electron density in (a) FinFET (b) GAA NW and (c) HexFET at $V_{GS}=V_{DS}=0.6V$ and $I_{OFF}=100nA/um$. HexFET and GAA NW experience enhanced volume conduction to offer higher mobile charge than FinFET. (d) Simulated transfer characteristics for 5nm node devices. Current is normalized to the layout density per 1µm; Fin Pitch = 15 nm.

Through faceting of the channel of a FinFET, the HexFET increases the effective gate width (Z_{EFF}) over both FinFET and GAA NW FET at matched fin height. As inter-NW spacing is eliminated in the HexFET, a stacked hexagonal NW channel also increases the total channel cross-sectional area (A_{EFF}) over stacked GAA NW FET. This resulting Z_{EFF}/A_{EFF} trade-off manifests itself in terms of the carrier density in the channel. Figs. 4(a)-(c) show the electron density distribution at each device's respective virtual source for $V_{DD} = 0.6V$ and $I_{OFF} = 100$ nA/µm. HexFET exhibits volume conduction along the entire height, circumventing the loss of conduction area in GAA NW FET, to offer 47% higher mobile charge than FinFET. The resulting 5nm node transfer characteristics are shown in Fig. 4(d). At 0.6V, HexFET shows 42% higher $I_{D,SAT}$ than FinFET, while GAA NW FET is lower than FinFET due to lower A_{EFF}. Additionally, drain-induced barrier lowering (DIBL) is suppressed in HexFET due to the screening of the drain field lines that results from the increased channel charge. Table I summarizes the key device metrics: linear threshold voltage, $V_{T,LIN}$, subthreshold slope (SS), DIBL and effective drive current (I_{EFF}) of each device. HexFET exhibits a remarkable SS of 68 mV/dec, 28% lower DIBL and 49% higher I_{EFF} than FinFET. This result highlights that an eGAA channel simultaneously combines excellent SCE immunity and high current drive capability for the 5nm technology node.

Table I: Summary of Electrical Parameters for 5nm Node Candidates at $V_{DD} = 0.6V$ and $I_{OFF} = 100$ nA/µm

Device	$V_{T,LIN}$ (mV)	SS (mV/dec)	DIBL (mV/V)	I_{EFF} (mA/µm)
FinFET	211	78	82	1.98
GAA NW	151	65	18	1.94
HexFET	180	68	29	2.92

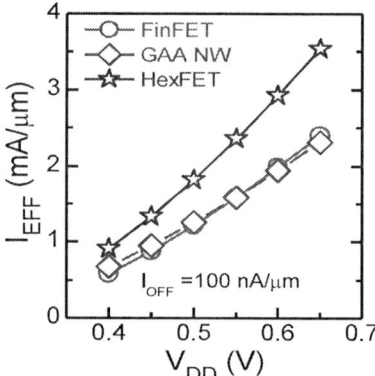

Fig. 5: Effective drive current, I_{EFF}, vs V_{DD} for FinFET, GAA NW FET and HexFET. HexFET offers I_{EFF} gain over FinFET for 0.4-0.65V supply voltages.

In order for a proposed transistor architecture to contend with FinFET, it must offer higher drive current at not just one, but over a range of supply voltages (V_{DD}). Fig. 5 shows the I_{EFF} vs V_{DD} for FinFET, GAA NW FET and HexFET. Due to excellent SCE control and increased mobile carriers in the channel, HexFET retains I_{EFF} gain over FinFET for a range of supply voltages from 0.4-0.65V.

However, the observed I_{EFF} gain in HexFET comes at the cost of 1.74x higher total effective gate capacitance (C_{gg}) than FinFET, as shown in Fig. 6. In addition to increased charge accumulation in the HexFET channel, fringing field-lines terminating at the source/drain extension regions result in additional capacitance contributions that increase C_{gg}. At low V_{GS}, the total gate capacitance of GAA NW FET is higher than both FinFET and HexFET, which results from the high-k dielectric and metal gate deposition on the spacer region after NW release etch. Therefore, the transient response of prototypical logic circuits requires further investigation, as the observed C_{gg}/I_{EFF} tradeoff in HexFET may negate all observed DC benefits.

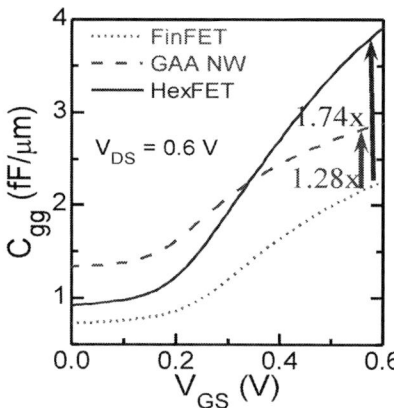

Fig. 6: Total gate capacitance (C_{gg}) vs V_{GS} for each device. HexFET exhibits 1.74x higher C_{gg} than FinFET due to increased Z_{EFF}.

B. Circuit-Level Benchmarking

To investigate the circuit level implications of each architecture, a look-up table based Verilog-A compact model is constructed using the I-V and C-V characteristics obtained from 3D numerical device simulations as in [13], capturing

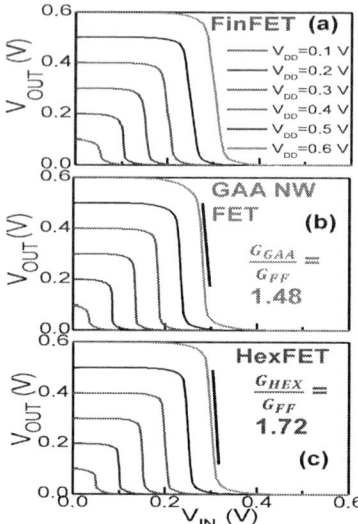

Fig. 7: Inverter voltage transfer characteristics (VTCs) for (a) FinFET, (b) GAA NW FET and (c) HexFET.

both the DC and transient behavior of each device. DC inverter operation is investigated first, with the inverter voltage transfer characteristics (VTCs) of each architecture shown in Fig. 7. Each PMOS/NMOS transistor is comprised of a single fin. GAA NW FET and HexFET show 1.48x and 1.72x higher inverter voltage gain, respectively, over FinFET.

Nine-stage ring-oscillator simulations are performed to investigate the transient response, as shown in Fig. 8. Ring-oscillator frequencies of 29.9, 24.4 and 30.7 GHz are obtained for FinFET, GAA NW and HexFET, respectively, with the energy-delay curves extracted for benchmarking (in Fig. 8(c)). HexFET exhibits a 17% reduction in energy at iso-delay (29% delay reduction at iso-energy) over FinFET and 30% reduction in energy at iso-delay (47% delay reduction at iso-energy) over GAA NW FET. Therefore, it is concluded that the increased I_{EFF} benefits the circuit-level performance of HexFET, and is not degraded below that of FinFET, despite the higher total gate capacitance in the former.

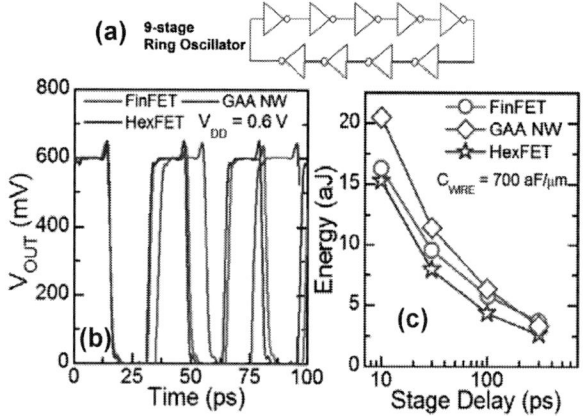

Fig. 8: (a) Schematic of a nine-stage ring oscillator. (b) Ring oscillator transient characteristics for FinFET, GAA NW FET and HexFET. (c) Energy versus delay curves for all devices.

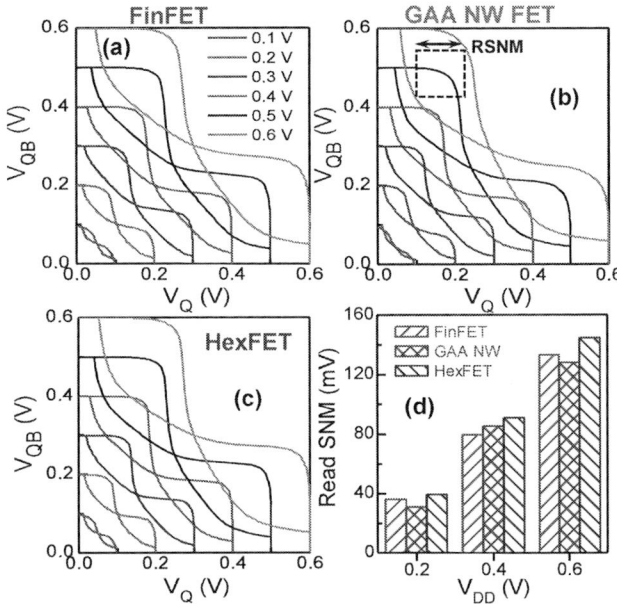

Fig. 9: Butterfly curves of 6T-SRAM cells constructed from (a) FinFET (b) GAA NW FET and (c) HexFET. (d) Read Static Noise Margin (RSNM) vs supply voltage, V_{DD} For all devices.

To investigate the role of transistor architecture on the read stability in memory cells, 6T-SRAM cells are constructed based on the approach in [14]. Butterfly curves for FinFET, GAA NW FET and HexFET are shown in Figs. 9(a)-9(c), respectively. HexFET shows high read static noise margin (RSNM) for supply voltage operation at 0.6V and below. Fig. 9(d) shows the RSNM vs V_{DD} for all devices. At 0.6V, HexFET offers an 8.5% and 13% increased RSNM than FinFET and GAA NW FET, respectively. The higher RSNM of HexFET is retained at lower supply voltages, down to 0.2V.

IV. CONCLSIONS

In this work, we investigate in detail, an electrically gate-all-around hexagonal NW (HexFET) architecture that combines the superior electrostatic control of GAA NW FET and high current drive capability of FinFET for 5nm technology node and beyond. Through first-principles atomistic modeling, 3D numerical device simulations and circuit-level benchmarking, we conclude that HexFET is a promising candidate to replace FinFET at the 5nm technology node. Under iso-I_{OFF} considerations, the increased Z_{EFF} in HexFET produces 47% higher charge than FinFET at virtual-source, giving rise to 49% higher I_{EFF}. Despite higher C_{gg} from increased Z_{EFF}, the HexFET architecture delivers a 17%

reduction in energy at iso-delay (or 29% delay reduction at iso-energy) when compared to FinFET.

REFERENCES

[1] S. Natarajan, et al., "A 14nm logic technology featuring 2nd-Generation FinFET transistors, air-gapped interconnects, self-aligned double patterning and a 0.0588 μm^2 SRAM cell size", International Electron Devices Meeting Technical Digest, pp. 71-74, 2014.

[2] R. Xie et al., "A 7nm FinFET technology featuring EUV patterning and dual strained high mobility channels", International Electron Devices Meeting Technical Digest, pp. 47-50, 2016.

[3] S.-Y. Wu, et al., "A 7nm CMOS platform technology featuring 4th eneration FinFET Transistors with a 0.027 um² high density 6-T SRAM cell for mobile SoC applications", International Electron Devices Meeting Technical Digest, pp. 43-46, 2016.

[4] A.Veloso et al., Gate-All-Around NWFETs vs. triple-gate FinFETs: junctionless vs. extensionless and conventional junction devices with controlled EWF modulation for multi-VT CMOS", Symp. VLSI Technology, pp. 138-139, 2015.

[5] I. Lauer et al., "Si nanowire CMOS Fabricated with minimal deviation from RMG FinFET technology showing record performance", Symp. VLSI Technology, pp. 140-141, 2015.

[6] H. Mertens et al., "Gate-All-Around MOSFETs based on vertically stacked horizontal Si Nanowires in a replacement metal gate process on bulk-Si substrates", Symp. VLSI Technology, pp. 158-1459, 2016.

[7] Y. Lee et al., "Diamond-shaped Ge and $Ge_{0.9}Si_{0.1}$ Gate-All-Around nanowire FETs with four {111} facets by Dry etch technology", International Electron Devices Meeting Technical Digest, pp. 382-385, 2015.

[8] F. Hou et al., "Suspended diamond-shaped nanowire with four {111} facets for high-performance Ge Gate-All-Around FETs", IEEE Trans. Elec. Dev., vol. 63, no. 10, p. 3837-3843, 2016.

[9] S. Thompson et al., "A 90 nm logic technology featuring 50nm strained silicon channel transistors, 7 layers of Cu interconnects, low k ILD, and 1 um² SRAM Cell", International Electron Devices Meeting Technical Digest, pp. 61-64, 2002.

[10] P. Bai et al., "A 65nm logic technology featuring 35nm gate lengths, enhanced channel strain, 8 Cu interconnect layers, low-k ILD and 0.57 um² SRAM cell", International Electron Devices Meeting Technical Digest, pp. 6657-660, 2004.

[11] K. Mistry et al., "A 45nm logic technology with high-k+metal gate transistors, strained silicon, 9 Cu interconnect layers, 193nm dry patterning, and 100% Pb-free packaging", International Electron Devices Meeting Technical Digest, pp. 247-250, 2007.

[12] S. Jin, M.V. Fischetti, and T.W. Tang, "Modeling of electron mobility in gated silicon nanowires at room temperature: surface roughness scattering, dielectric screening, and band nonparabolicity", Journal of Applied Physics, 102, 083715 (2007).

[13] H. Liu et al., "Technology assessment of Si and III-V FinFETs and III-V Tunnel FETs from soft error rate perspective", International Electron Devices Meeting Technical Digest, pp. 577-580, 2012.

[14] N. Agrawal et al. "Impact of variation in nanoscale silicon and non-silicon FinFETs and Tunnel FETs on device and SRAM performance", IEEE Trans. Elec. Dev., vol. 62, no. 6, 2015.

978-1-5090-5979-9/17 $31.00 © 2017 IEEE

Modeling of Dynamic Trap Density Increase for Aging Simulation of Any MOSFET Circuits

M. Miura-Mattausch, H. Miyamoto, H. Kikuchihara, D. Navarro, T. K. Maiti, N. Rohbani*, C. Ma**, H. J. Mattausch

Hiroshima University, Kagamiyama 1-3-1, Higashi-Hiroshima 739-8530, Japan
*Sharif University of Technology, Tehran, Iran; **ASTRI, Science Park, Shatin, Hong Kong

A. Schiffmann, A. Steinmair, E. Seebacher
ams AG, Graz A8141, Austria

Abstract—**A compact aging model for circuit simulation has been developed by considering all possible trapped carriers within MOSFETs. The hot carrier effect and the N(P)BTI effect are modeled by integrating the substrate current as well as the oxide field change due to the trapped carriers. Additionally, the carriers trapped within the highly resistive drift region are included for high-voltage (HV)-MOSFET modeling. The aging model considers the dynamic trap-density increase as a function of circuit-operation time with dynamically varying stress conditions for each individual MOSFET. A self-consistent solution is obtained by iteratively solving the Poisson equation including the trap density. The model is verified to be applicable for any type of MOSFETs covering advanced technologies as well as HV-MOSFETs.**

Keywords—*aging simulation; carrier traps; dynamic change; self-consistent solution*

I. INTRODUCTION

Device degradation is remaining a serious problem due to aggressive circuit design for achieving high performance as well as advanced technologies utilizing ultimately scaled device dimensions. To maintain sufficient safe operating area for circuit design even after long-term usage, accurate prediction of circuit aging is a key. However, circuit-aging prediction is still a severe task due to limitations of measurement and simulation capabilities. It is known that the trap-density increase is mainly responsible for the device aging in addition to inherent trap states [1-3]. Thus compact models for circuit-aging prediction must consider the carrier-trapping events during dynamic stress/relaxation repetitions.

Conventionally, aging is modeled as the threshold-voltage (V_{th}) shift, which fits the measured V_{th} shift as a function of time and stress conditions [4]. In the present investigation, the aging effect is modeled based on the trap-density increase due to its two physical origins. One is the trap increase by hot carriers (the hot-carrier (HC) effect) [5], and the other is the trap increase by the high electric field in the oxide (the N(P)BTI effect) [6]. Carrier trapping, even in the highly resistive drift region, is considered additionally

for power MOSFETs [7, 8]. Therefore, the developed model is applicable for any MOSFET application from advanced down-scaled to high-voltage MOSFETs. The surface-potential-based compact MOSFET model HiSIM [9], solving the Poisson equation explicitly with an iterative approach, is applied for this purpose. To model the long-term circuit aging, integrated physical quantities such as the substrate current are utilized to model the trap-density increase. These integrated physical quantities are determined as a function of stress duration and enable description of dynamic stress-condition changes as well as dynamic device reactions. Trap-time constants for trapping and detrapping events determine the frequency dependence of the aging characteristics observed mostly for pMOSFETs.

II. MODELING MOSFET AGING

A. Basic Equations for Aging Modeling

Fig. 1 shows two different stress conditions to be considered for circuit simulation. One induces the HC effect and the other the high oxide-field effect investigated as the N(P)BTI effect [10]. Modeling these effects is done separately based on their individual origins as depicted in Fig. 2. The developed model equations are summarized in Table 1. For the N(P)BTI effect it is known that trapping occurs during and detrapping after gate-voltage stress, which is modeled with different time constants for trap and detrap events. The trap density N_{trap} is included in the Poisson equation below to self-consistently capture its overall effect.

$$\nabla^2 \phi = -\frac{q}{\varepsilon_s}\left(p - n + N_D - N_A + N_{trap\_D} - N_{trap\_A}\right)$$

Here $N_{trap\text{-}D}$ and $N_{trap\text{-}A}$ are donor-like and the acceptor-like trap densities, respectively. Since the carrier densities p and n as well as the trap density N_{trap} are functions of the electrostatic potential ϕ_s at the surface for bulk MOSFETs, accurate calculation of ϕ_s is a prerequisite. The N(P)BTI is measured as the threshold voltage (V_{th}) shift, referring to the homogeneous trap density distribution within the bandgap (see Fig. 3). Thus modeling is done as the V_{th} shift. The

978-1-5090-5979-9/17 $31.00 © 2017 IEEE

Poisson equation is solved iteratively without simulation time penalty in HiSIM together with the Gauss law, where ΔV_{th} is included as the gate voltage V_{gs} shift. Since the trap density as well as the V_{gs} shift change the electrostatic potential distribution, all device characteristics such as carrier density and mobility are automatically influenced by N_{trap} in the physically correct way. Thus no additional fitting parameters for aging are required.

B. Long-Term Aging Modeling

Two independent trap-density distributions (shallow and deep) are considered as shown in Fig. 2, where the deep-trap density is responsible for the long-term HC-effect aging. The dynamic trap-density increase is written as a function of I_{subt} ($=I_{\text{sub}}$*stress duration), because the substrate current I_{sub} is a universal measure for the hot-carriers strength. The long-term N(P)BTI effect is modeled by the trap-density increase as a function of time as well. However, it is translated to the observed threshold voltage shift (ΔV_{th}), which changes the oxide field E_{ox}. However, dynamically changing I_{sub} as well as E_{ox} are not known before calculating N_{trap} as well as ΔV_{th}. Therefore, initial values are taken and solved iteratively to achieve a consistent solution. To simplify the simulation procedure, we have developed a method to implement our aging model into the compact model HiSIM in two ways, for the DC simulation and the transient simulation as shown in Fig. 4. The DC simulation is expected to be used for the parameter extraction, where the iteration is performed for each bias condition to achieve a self-consistent solution. For circuit simulation, the time step for the simulation is very small (the order of nano seconds), which can be considered in the iteration procedure by using the previous time step solution as the initial value for the next time step. The circuit simulation is performed for a certain time length and the self-consistent solution is obtained within the simulation time DEGTIME0. In practical applications, the convergence to the consistent aged physical values can be achieved very fast (see Fig. 8).

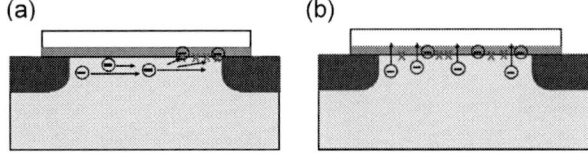

Fig. 1. Schematics of carrier trap events, (a) due to hot carriers, (b) due to high oxide field.

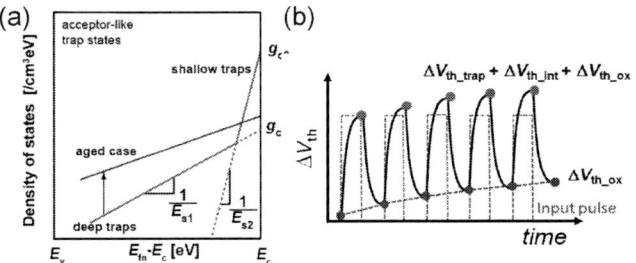

Fig. 2. (a) Trap state density as a function of carrier energy, (b) threshold voltage shift ΔV_{th} due to trap/detrap events measured in pMOSFETs, where long-term aging is shown by $\Delta V_{\text{th\_ox}}$.

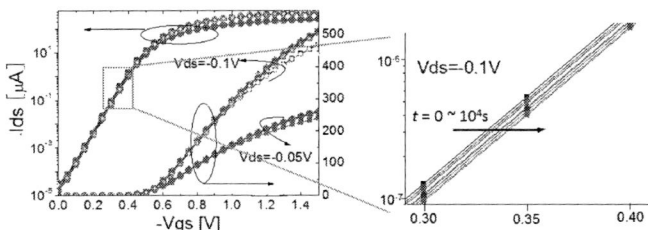

Fig. 3. Measured pMOSFET I_{ds} vs. V_{gs} characteristics under different stress durations and bias conditions.

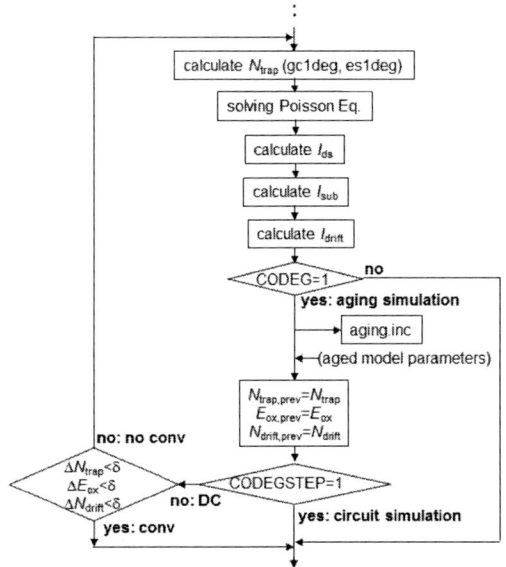

Fig. 4. Flowchart of aging-model implementation into HiSIM for both DC and transient simulations. Three key physical quantities (N_{trap}, E_{ox}, N_{drift}) are obtained iteratively.

Table 1. Model equations for aging.
[HC-effect]

$$N_{\text{trap-A}} = N_0 \exp\left(\frac{E_{\text{f}} - E_{\text{C}}}{E_{\text{s}}}\right)$$

$$E_{\text{f}} - E_{\text{c}} = -\left(qV_{\text{ds}} - E_{\text{v}} - kT\ln\frac{N_{\text{v}}}{N_{\text{A}}}\right) + (q\phi_{\text{s}} - E_{\text{c}})$$

$$N_{01} = g_{c1\text{deg}}(t) \cdot E_{s1\text{deg}}(t) \cdot \frac{\dfrac{kT}{E_{s1\text{deg}}(t)}}{\sin(\dfrac{kT}{E_{s1\text{deg}}(t)})}$$

$$g_{c1\text{deg}}(t) = \textbf{TRAPGC1} + \frac{\textbf{GC1MAX}}{gc\_time1}\exp\left[-\frac{1}{2}\left(\frac{gc\_time - gc\_time2}{gc\_time1}\right)^2\right]$$

$$E_{1\text{deg}}(t) = \textbf{TRAPE1} + \frac{\textbf{E1MAX}}{e1\_time1}\exp\left[-\frac{1}{2}\left(\frac{gc\_time - e1\_time2}{e1\_time1}\right)^2\right]$$

$$gc\_time = \ln\left(I_{\text{subt}}/W\right)$$

gc_time1 & $e1\_time1$: stress time when aging starts
gc_time2 & $e1\_time2$: stress time when aging saturates

[N(P)BTI]

$$\Delta V_{\text{th}} = N_{\text{trap}} \cdot (1.0 - \exp(-\frac{t}{taus\_trap}))$$

$$N_{\text{trap}} = \textbf{ATRAP} \cdot \exp(\textbf{BTRAP} \cdot E_{\text{ox}})$$

$$taus\_trap = \textbf{ATAUTRAP} \cdot \exp(\textbf{BTAUTRAP} \cdot E_{\text{ox}})$$

$$E_{\text{ox}} = \frac{V_{\text{gs}} - (V_{\text{fb}} + \Delta V_{\text{th}}) - \phi_{\text{s}}}{T_{\text{OX}}}$$

III. SIMULATION RESULTS

DC- stress measurements are used to extract aged model parameters as a function of stress duration under different stress conditions. Fig. 5 compares measured and calculated I_{ds}-V_{gs} degradation for low and high V_{ds} after a constant DC-bias stress with different stress durations. From the reduction of the aging effect with increased V_{ds}, it is concluded that the carrier traps are located at the drain side. Fig. 6 shows the fitting capability of the N(P)BTI effect for different stress conditions. Fig. 7 shows aging simulation results of a ring-oscillator depicted in the upper left graph of Fig. 7a. Each MOSFET within the circuit is stressed differently according to the individual dynamic bias changes as can be seen in the 2 lower graphs of Figs. 7. Fig. 8 depicts the convergence waveform to self-consistent N_{trap} and ΔV_{th} solutions as a function of circuit simulation time DEGTIME0 for the ring-oscillator (see Fig. 7, upper left). The time DEGTIME0 is the input condition for circuit-simulation duration, determined in the netlist to predict circuit performance. Though the convergence takes longer time for high-frequency circuits, as in the studied case, it requires usually only a few deca-nano seconds in conventional cases.

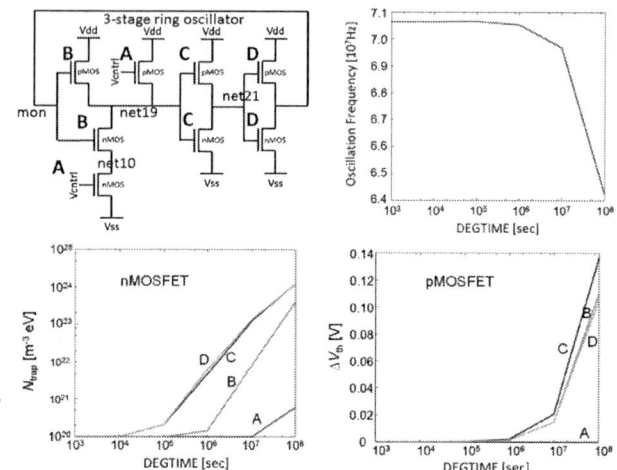

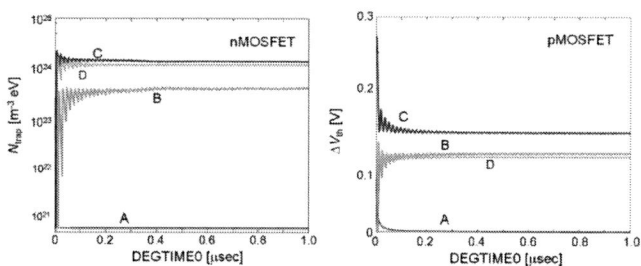

Fig. 7. Studied ring-oscillator (upper left), and simulated aged oscillation frequency after DEGTIME stress duration (upper right). Aging features of each transistor within the ring-oscillator circuit are compared in lower figures.

Fig. 8. Convergence features of N_{trap} in (a) four nMOSFETs and ΔV_{th} in (b) four pMOSFETs of the ring-oscillator shown in Fig. 7 at DEGTIME=10^8sec during circuit simulation for the duration of DEGTIME0.

High-voltage (HV) MOSFETs have an additional highly-resistive drift region, where high applied V_{ds} is mostly sustained (see Fig. 9). Modeling of the HV-MOSFET is done by considering the resistor current in the drift region (I_{ddp}) as indicated in Fig. 10, where the node potential at DP (see Fig. 9) is calculated iteratively to maintain the current-continuity law between I_{ds} and I_{ddp} [11]. There are several regions where carrier trapping occurs within such high-power devices. One is near at the drain contact (position A in Fig. 9) as usual, and another one is at the STI corner (position B) [7, 8]. For aged power MOSFETs, the threshold voltage shift due to the carrier traps in the A region is relatively small as observed in measurements. However, measured drain current shows a non-monotonous feature of I_{ds} as a function of stress time. An I_{ds} reduction occurs first as expected, and I_{ds} starts to recover after certain time, around 10^4s (see Fig. 11). The charge-pumping measurement reveals that the first current reduction is due to interface-state generation in the A region. However, the interface-state generation saturates after approximately 10^3 to 10^4s. The subsequent I_{ds} modification is explained by the fact that the carrier trapping occurs in the B region as well [7, 8]. The non-monotonous feature is dependent on the technologies applied for the fabrication as

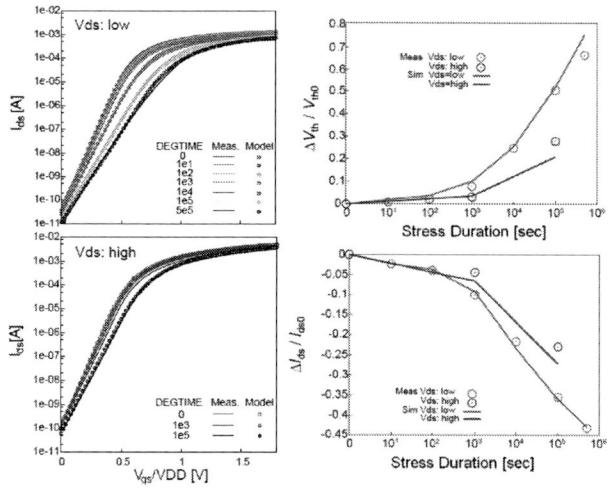

Fig. 5. Comparison of *I-V* characteristics between measurements and calculation results with the developed aging model for two different V_{ds} values. The DC-stress condition is V_{gs}=0.8*VDD and V_{ds}=2*VDD for different durations. I_{ds0} and V_{th0} denote drain current and threshold voltage before stress, respectively.

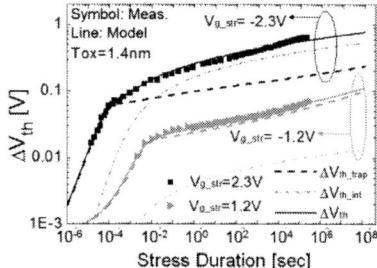

Fig. 6. Comparison of calculation results of threshold voltage shift ΔV_{th} with measurements as a function of DC stress duration. Results for two stress conditions (V_{gs}=-1.2V and -2.3V) are depicted.

well as the stress conditions. Measurements demonstrate that the $I_{ds,lin}$ recovery is much enhanced with increased V_{gs}. However, $I_{ds,sat}$ then recovers much slower. Modeling the carrier traps within the drift region is done with the electric field change, which modulates the carrier density within the drift region. The developed aging equations for high-voltage MOSFETs are summarized in Table 2. Simulated current characteristics as a function of stress duration DEGTIME are depicted in Fig. 12 for the studied high-voltage nMOSFET, which verifies that the HC-effect in the channel is mostly responsible for the aging degradation. The measured non-monotonous features of I_{ds} as a function of stress duration are well reproduced.

ACKNOWLEDGMENT

The model development was originally supported by STARC. The authors express sincere thanks for this support.

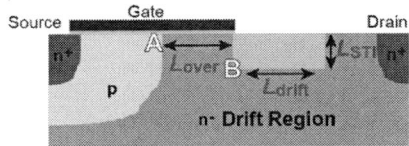

Fig. 9. Schematic of a typical high-voltage MOSFET with highly resistive drift region.

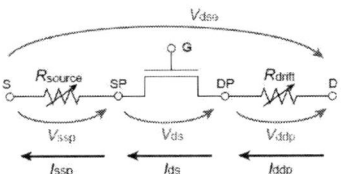

Fig. 10. Modeling approach of HiSIM_HV, where the internal node DP is iteratively solved preserving the continuity law between I_{ds} and I_{ddp}.

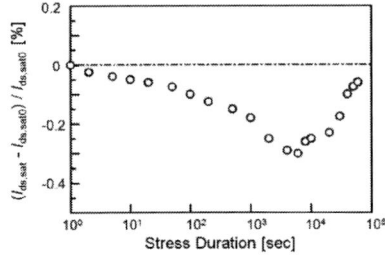

Fig. 11. Measured aging of saturation drain current $I_{ds,sat}$ of a LDMOS transistor as a function of stress duration [7].

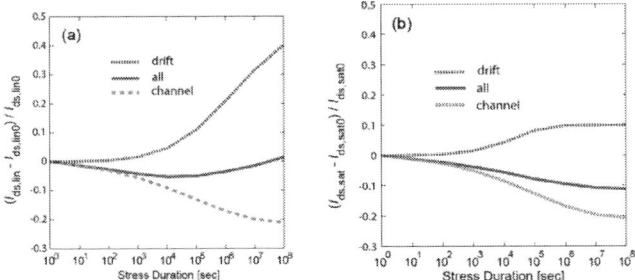

Fig. 12. Calculated aging of (a) linear drain current $I_{ds,lin}$ and (b) the saturation current $I_{ds,sat}$ as a function of stress duration. Two separate aging contributions are depicted separately together with the total aging.

Table 2. Model equations for high-power MOSFETs. All variables for Iddp are described in the HiSM_HV manual available as open source.

$$I_{ddp} = W \cdot X_{ov} \cdot N_{drift} \cdot \mu_{drift} \cdot \frac{V_{ddp}}{L_{drift}}$$

$$N_{drift} = \mathbf{NOVER} \cdot \left[1 + \mathbf{RDRCAR} \cdot (1 - \frac{1}{1 + \mu_{drift,max}}) \right] + \mathbf{NOVER} \cdot D_{vddp}$$

$$\mu_{drift,max} = \frac{\mu_{drift0}}{v_{max}} \cdot \frac{V_{ddp}}{L_{drift}}$$

$$D_{vddp} = D_{vddp,deg} \cdot \exp(-\frac{V_{dseff} - \phi_{SL} + \phi_{S0}}{\mathbf{TRAPDLX}})$$

$$D_{vddp,dep} = \mathbf{TRAPDVDDP} + \frac{\mathbf{TRAPD1MAX}}{time1} \cdot \exp\left[-\frac{1}{2}(\frac{time - time2}{time1})^2\right]$$

$time = \ln(I_{ds} \cdot \mathbf{DEGTIME} / W)$
$time1$: stress time when aging starts
$time2$: stress time when aging saturates

REFERENCES

[1] P. J. Caplan, E. H. Poindexter, B. E. Deal, R. R. Razouk, "ESR centers, interface states, and oxide fixed charge inthermally oxidized silicon wafers," J. Appl. Phus., vol. 50, no. 9, pp. 5847-5854, 1979.

[2] P. M. Lenahan, J. J. Mele, J. P. Campbell, A. Y. Kang, R. K. Lowry, D. Woodbury, S. T. Liu, and R. Weimer, "Direct experimental evidence linkinh silicon dangling bond defects to oxide leakge current," Proc. IEEE IRPS, pp. 150-155, 2001.

[3] T. Grasser, H. Reisinger, P. J. Wagner, and B. Kaczer, "Time-dependent defect spectroscopy for characterization of border traps in metal-oxide-semiconductor transistors," Phys. Review B, vol. 82, no. 24, pp. 245318-1-10, Dec. 2010.

[4] T. L. Tewksbury and H.-S. Lee, "Characterization, modeling, and minimization of transient threshold voltage shifts in MOSFET's," IEEE J. Solid-State Circuits, vol. 29, no. 3, pp. 239-252, 1994.

[5] H. Tanoue, A. Tanaka, Y. Oodate, T. Nakahagi, D. Sugiyama, C. Ma, H. J. Mattausch, and M. Miura-Mattausch, "Compact modeling of dynamic MOSFET degradation due to hot-electrons," IEEE Trans. Device and Material Reliability, vol. 17, no. 1, march, pp. 52-58, 2017.

[6] C. Ma, H. J. Mattausch, K. Matsuzawa, S. Yamaguchi, T. Hoshida, M. Imade, R. Koh, T. Arakawa, and M. Miura-Mattausch, "Universal NBTI compact model for circuit aging simulation under any stress conditions," et al., IEEE Trans. Device and Materials Reliability, vol. 14, no.3, Sept. pp. 818-825, 2014.

[7] P. Moens, G. Van den Bosch, and G. Groeseneken, "Hot-carrier degradation phenomena in lateral and vertical DMOS transistos," IEEE Trans. Electron Devices, vol. 51, no. 4, pp. 623-628, 2004.

[8] J. M. Park, M. Knaipp, Y. Shi, and N. Feilchenfeld, "Hot-carrier behavior and Ron-BV trade-off optimization for p-channel LDMOS transistors in a 180nm HV-CMOS technology," Proc. Int. Symp. Power Semiconductor Devices & IC's, 2012.

[9] M. Miura-Mattausch, N. Sadachika, D. Navarro, G. Suzuki, Y. Takeda, M. Miyake, T. Warabino, Y. Mizukane, R. Inagaki, T. Ezaki, H. J. Mattausch, T. Ohguro, T. Iizuka, M. Taguchi, S. Kumashiro, S. Miyamoto, "HiSIM2: Advanced MOSFET Model Valid for RF Circuit Simulation," IEEE Trans. Electron Devices, Vol. 53, No. 9, pp. 1994-2007, 2006.

[10] S. Mahapatra, V. D. Maheta, A. E. Islam, and M. A. Alam, "Isolation of NBTI tress generated interface trap and hole-trapping components in PNO p-MOSFETs," IEEE Trans. Electron Devices, vol. 56, no. 2, pp. 236-242, Feb. 2009.

[11] A. Tanaka, Y. Oritsuki, H. Kikuchihara, M. Miyake, H. J. Mattausch, M. Miura-Mattausch, Y. Liu, and K. Green, "Quasi-2-dimensional compact resistor model for the drift region in high-voltage LDMOS devices, " IEEE Trans Electron Devices, vol. 58, no. 7, pp. 2072-2080, 2011.

Comprehensive Compact Electro-Thermal GaN HEMT Model

M. Alshahed, M. Dakran, L. Heuken, M. Alomari and J. N. Burghartz

Institut für Mikroelektronik Stuttgart
Stuttgart, Germany
www.ims-chips.de
Alshahed@ims-chips.de

Abstract—In this work we demonstrate a semi-empirical GaN HEMT model implemented in Verilog-A format. The model captures accurately the DC operation of test devices fabricated and measured at IMS CHIPS including the thermal effects. In addition, the off-state leakage current is physically modeled as a space-charge limited current prior to the onset of the physical breakdown. The dynamic current recovery of the transistor after stress bias is physically included by implementing the model in the form of a finite state machine, capturing the memory effect of the device. The drain current is modified to be a summation of multiple exponential terms, each corresponding to a given trapping center. This model can be used to predict the overall performance of the GaN HEMTs based on the epitaxial material composition or to infer the material composition and quality based on the measured device characteristics.

Keywords—Compact modeling; GaN HEMT; Pulsed current; Off-state leakage current; Current recovery; Charge trapping

I. INTRODUCTION

Gallium Nitride (GaN) based electron devices have attracted a lot of attention for more than three decades as viable candidates for power electronics and high frequency applications [1]. Having a wide bandgap, high electron saturation velocity and high stability at elevated temperatures, GaN-based semiconductor devices, especially High Electron Mobility Transistors (HEMTs), proved to be alternatives to the traditional Silicon (Si)-based switches. Their superior intrinsic material properties allowed the GaN HEMT to possess higher breakdown voltage (> 1 kV), higher current density and much higher operational temperature as compared to the currently available commercial Si-based devices [2].

In order to utilize GaN based devices in any application, one has to be able to predict the performance of GaN HEMTs using device compact models, thus, enabling circuit and system level simulations and design. Unlike the well-established and standardized compact modeling process of Si-based devices, the development of compact models of GaN HEMTs is challenging due to the lack of a readily available material model. The unintentional incorporation of a wide span of species (dopants) during GaN epitaxy introduces several DC and dynamic effects which vary from supplier to supplier. In addition, commercially available GaN wafers are always grown on foreign substrates and, thus, feature a relatively high defect density (> 10^9 cm^{-2} for GaN/Si) which varies from supplier to supplier. Such defects are electrically active and influence both the DC and dynamic properties of the device, namely the

leakage current and the dynamic on-resistance (current dispersion) [3].

Several attempts have been made in order to develop accurate GaN HEMT compact models as summarized in [4]. Among these is Angelov's model which has been modified to be used to model GaN HEMTs. This model is classified as an empirical equivalent circuit model. Given an accurate parameter extraction methodology, the Angelov model can accurately represents the GaN HEMT's electrical characteristics. However, accurate conclusions about the epitaxial material quality can be barely inferred. The Curtice3 model is also available for GaN HEMTs; it is also an empirical model. This model is neither an electro-thermal model nor geometry scalable. However, many models have been developed since then to account for its shortcomings like the CMC (Curtice/Modelithics/Cree), CFET and EEHEMT. Other models have been developed to account for self-heating effects in GaN HEMTs and the dynamics of the drain current. In [5] an empirical model is proposed, which is able to capture self-heating effects as well as charge trapping/de-trapping kinetics in GaN HEMTs. Hou, Bilbro and Trew [6] proposed a multi-region physical model where the conducting channel in the HEMT is divided into several regions. These multi-region models, however, did not account for the physical leakage current or the dynamics of charge trapping/de-trapping in the HEMT.

In this work, a semi-empirical compact model with emphasis on the physics of the off-state leakage current and the kinetics of charge trapping and de-trapping is developed. This model can serve as a platform for GaN HEMT developers (suppliers, foundries and designers) to directly infer the effect of the epitaxial layers quality and composition on the performance of the GaN HEMTs. The model is implemented in Verilog-A format and, thus, suitable for circuit simulations.

II. PROPOSED GAN HEMT COMPACT MODEL

A. Proposed model background

A HEMT structure is characterized by a Two-Dimensional Electron Gas (2DEG) channel at the hetero-interface between a barrier layer and a channel, which is a result of the charge balance between the polarization discontinuity at the interface and the counter charge at the surface (Fig. 1). The presence of unintentional n-type dopants such as Oxygen and Si hinders the high voltage operation of the devices. As such, a compensation-doped layer is typically grown in order to

---

In addition to the internal funding, this work has been partially sponsored by IMS Mikro-Nano Produkte GmbH.

978-1-5090-5979-9/17 $31.00 © 2017 IEEE

achieve the high resistivity needed for high voltage and low leakage current applications. This compensation is achieved by doping the layer with deep acceptors such as Iron (Fe) or Carbon (C). These deep acceptors together with the dislocations in the material act as trapping centers. During the device operation, particularly during the off-state stress bias, charge injection into surface or buffer traps causes a reduction in the channel density and, hence, a higher on-resistance. This coupling between the channel and the different trapping centers can be visualized as an RC circuit with a specific time constant as depicted in Fig. 1. These equivalent circuits can be unfolded into multiple circuits, each representing a certain species in the material. Specific for each of the species are the capture and emission time constants. These time constants are a function of the trap density of states, trap energy level and the capture cross-section. In this work, the model puts emphasis on capturing the dynamics imposed by this wide span of trapping centers in the material. The charge transport dynamics to and from the trapping centers are assumed to be negligible in comparison to the dynamics of trapping and de-trapping.

B. Proposed model structure

The proposed model consists of four main modules: (1) the DC core module, (2) the off-state high-voltage module, (3) the charge trapping and de-trapping kinetics module and (4) the thermal network module. The non-linear AC parameters module is not considered in this work.

Starting with the DC module, the threshold voltage can be physically expressed as:

$$V_{th}(x) = \frac{1}{q}[\phi_B(x) - \Delta E_c(x)] - \frac{\sigma(x)}{C_{AlGaN}(x)} \quad (1)$$

where ϕ_B is the energy barrier height between the gate metal and the semiconductor, ΔE_c is the conduction band discontinuity between the AlGaN barrier and the GaN buffer, σ is the polarization discontinuity between the AlGaN and GaN, C_{AlGaN} is the effective barrier capacitance and finally x is the Aluminum (Al) composition in the barrier layer [7]. In case of a HEMT with an additional gate dielectric layer and considering the presence of dielectric charges, the threshold voltage is adjusted as:

$$V_{th,MIS} = V_{th} \cdot \left(1 + \frac{C_{AlGaN}}{C_{dielectric}}\right) - \frac{Q_{it} + Q_{dielectric}}{C_{dielectric}} \quad (2)$$

where $C_{dielectric}$ is the capacitance of the gate dielectric and Q_{it} & $Q_{dielectric}$ are the interface and dielectric charges respectively. The charge function, and hence the current equation, are adopted from the MIT Virtual Source Model (VSM) [8]. The drain current can be represented as:

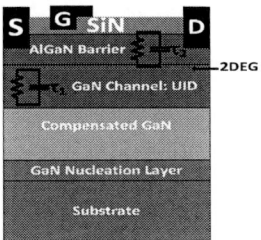

Fig. 1: Cross-section of a GaN HEMT epitaxial stack. The channel coupling to surface and buffer traps characterized by time constants τi.

$$I_{DS} = W \cdot Q \cdot v \quad (3)$$

where W is the total device's width, Q is the channel density function defined by (4) and v is the electron velocity function defined by (5).

$$Q = C_{inv} \cdot n \cdot \varphi_t \cdot \ln(1 + exp\frac{V_{gs} - V_{th}}{n \cdot \varphi_t}) \quad (4)$$

$$v = F_s \cdot v_{SAT} = \frac{V_{ds}/V_{ds,SAT}}{(1 + (V_{ds}/V_{ds,SAT})^\beta)^{1/\beta}} \cdot v_{SAT} \quad (5)$$

where n is the sub-threshold slope factor, φ_t is the thermal voltage, $V_{ds,SAT} = \frac{L_g \cdot v_{SAT}}{\mu_n}$ (L_g is the channel length and μ_n is the 2DEG electron mobility) is the drain voltage determining the onset of current saturation, v_{SAT} is the saturation velocity and β is a fitting parameter to determine the transition profile from linear to saturation regimes.

The second aspect, namely the off-state leakage current modeling, can be categorized into two partitions: 1) the forward and reverse physical breakdown of the gate-drain as well as the gate-source section modelled as Schottky diode and 2) the leakage current profile prior to the onset of physical breakdown.

In the forward direction (positive gate bias), the forward breakdown can be described by the Schottky diode equation as:

$$I_{gs,Forward} = I_0 \cdot \exp\left(\frac{-\varphi_B}{\eta \cdot \varphi_t}\right) \cdot \left[\exp\left(\frac{V'_{gs}}{\eta \cdot \varphi_t}\right) - 1\right] \quad (6)$$

where: $V'_{gs} = V_{gs} - V_{dielectric} - V_{TO}$ where $V_{dielectric}$ is the forward voltage drop on the gate dielectric, V_{TO} is the turn-on voltage of the respective diode, I_0 is the pre-exponent diode current derived from the thermionic emission, φ_B is the energy barrier height collected experimentally and η is the diode ideality factor. The reverse break down can be calculated as:

$$I_{gs,Reverse} = I_0 \cdot (1 + K_{bd,s} \cdot exp\frac{V'_{sg} - V_{bd,s}}{\varphi_{bd}}) \quad (7)$$

where $V_{bd,s}$ is the reverse breakdown voltage of the gate-source diode extracted from the device geometry design and $K_{bd,s}$ & φ_{bd} are technology related fitting parameters accounting for the surface electric field distribution effects [9]. It is worth mentioning that the gate-drain diode equations have the same form as in (6) and (7) used for the gate-source diode.

Before the onset of the physical breakdown of the HEMT, the vertical leakage current is said to follow a Space-Charge Limited Current (SCLC) profile in the presence of trapping centers and background doping concentration. As will be shown in Section III, in the case of a SCLC the current starts with an ohmic profile at low applied voltages until a certain voltage where most of the background compensation acceptor dopants and/or trapping centers are filled with electrons from the background doping and/or from the injected electrons from the electrodes. At this voltage the ohmic trend develops to a quadratic dependence of the current on the applied voltage which is a characteristic of the SCLC [10]. This transition voltage can be calculated as:

$$V_{transition} \approx \frac{q.(N_t - N_D)}{2 \cdot \varepsilon} \quad (8)$$

where N_t is the effective density of free acceptor traps, N_D is the effective density of n-type donor dopants responsible for the background n-type carriers and ε is the dielectric constant of the material. However, in this work this transition voltage is calculated dynamically during the run time of the model as a function of the state of the device as will be shown in Section III. The SCLC equation is then approximated by [10]:

$$J_{SCLC} \approx \frac{9}{8} \cdot \mu \cdot \varepsilon \cdot \theta \cdot \frac{V^2}{L^3} \cdot \exp\left[\frac{\left(\frac{q^3 \cdot V}{\pi \cdot \varepsilon \cdot L}\right)^{0.5}}{K \cdot T}\right] \quad (9)$$

where θ is the ratio between free electrons and trapped electrons, L is the length that has to be travelled with the charge carriers (in this case the thickness of the epitaxial layer) and T is the absolute temperature. The exponential term accounts for Poole-Frenkel (PF) emission and its importance is highlighted in Section III.

The last module is the dynamic charge trapping and de-trapping in the buffer layer which is also related to the SCLC discussed above. For each of the species that act as trapping centers, the trap signature has to be identified. This trap signature consists of the capture cross-section, the trap energy level and its degeneracy and the density of states of that trapping level. If known, one can calculate the capture and emission time constants, and accordingly, the dynamics of the HEMT. The stress-time dependent density of trapped charges, the specific capture time constant and the emission time constant are given as ([11]):

$$n_t(t) = \sigma_n \cdot v_{th} \cdot n \cdot \tau_{capture} \cdot N_t \cdot \ln\left[1 + \frac{t}{\tau_{capture}}\right] \quad (10)$$

$$\tau_{capture} = \frac{K \cdot T \cdot n_{T0}}{q \cdot \varphi_0 \cdot N_t \cdot \sigma_n \cdot v_{th} \cdot n} \quad (11)$$

$$\tau_{emission}^{-1} = A \cdot \sigma_n \cdot v_{th} \cdot N_c \cdot \exp\left[\frac{-q \cdot (E_t - \beta \cdot \sqrt{\xi})}{K \cdot T}\right] \quad (12)$$

where σ_n is the capture cross-section for electrons, v_{th} is the thermal velocity of electrons in the semiconductor, n is the background free carrier concentration, N_t is the total trap density with an energy level E_t, ξ is the applied electric field, $\beta = \sqrt{\frac{q^3}{\pi \cdot \varepsilon}}$, φ_0 is the initial energy barrier at the trapping site and A is a fitting parameter which is between 0.9 and 1 in this work. The equilibrium density of trapped charges is calculated as:

$$n_{T0} = \frac{N_t}{1 + \frac{1}{g}\exp\left[\frac{E_t + \beta \cdot \sqrt{\xi} - E_{FQ}}{K \cdot T}\right]} \quad (13)$$

with E_{FQ} as the quasi Fermi level for electrons in the presence of an applied external field and g as the degeneracy of the trapping level.

The time-dependent on-state drain current of the transistor is then modified as:

$$I_{ds}(t) = I_{ds,0} \cdot \left(1 - \sum_{i=1}^{J} \alpha_i \cdot e^{\frac{-t}{\tau_{emission,i}}}\right) \quad (15)$$

where $I_{ds,0}$ is the original drain current without charge trapping, J is the total number of trapping species in the material and α is the ratio between the density of trapped electrons to the total density of the conducting channel.

Finally, a one-pole thermal node is added to the model where the threshold voltage, electron mobility and saturation velocity are a function of the calculated device's temperature [12]. In addition, parasitic source and drain linear resistances are added to account for the metal-semiconductor contact resistances and the drain and source access regions.

C. Model implementation

The proposed model is implemented in Verilog-A format in the form of a Finite State Machine (FSM). The FSM concept is employed so that the memory of the device operation can be tracked. Fig. 2 depicts a simplified flow chart of the algorithm (not all states are included for simplicity).

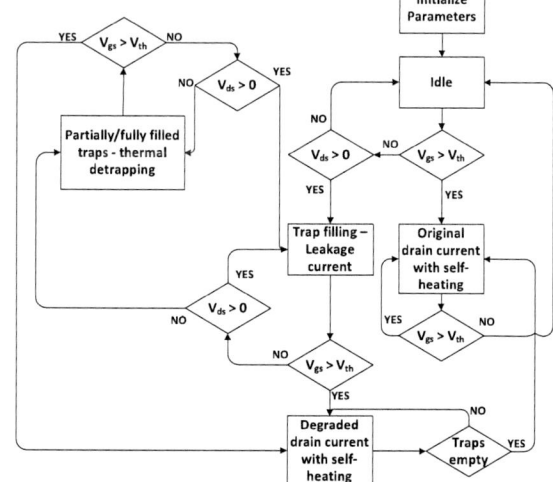

Fig. 2: Selected scenario of the FSM implementation of the model

If the device is biased in the on-state directly after initialization, the maximum device's current is generated (no charge trapping). If the device is initialized in the off-state at a finite drain bias, the charge trapping module is activated. Depending on the level of the off-state stress bias and the amount of time spent in this condition, the degraded on-state current dynamics are determined and used when the device state changes to on-state.

III. MEASUREMENT RESULTS AND MODEL VERIFICATION

In order to verify the model, three test cases are considered for the fabricated devices in this study: 1) The DC output characteristics including the device's self-heating, 2) the off-state leakage current as a function of the applied bias and 3) the dynamic current recovery after off-state stress bias. This dynamic measurement is carried out with a resolution of millisecond using two Keithley 24xx series synchronized together by using hardware trigger links. The composition of the material used is given by the material supplier to be: Oxygen = 1.5E16 cm^{-3}, Silicon = 3E16 cm^{-3} and Carbon 1E17 cm^{-3}. The corresponding trap signatures are acquired from [11].

Starting with the DC characteristics, Fig. 3a depicts the simulated and measured output characteristics of an AlGaN/GaN HEMT with 4 μm gate length and 100 μm gate width. The model captures the measured data with high accuracy in strong inversion. However, in the moderate and

978-1-5090-5979-9/17 $31.00 © 2017 IEEE

weak inversion, the model deviates from the measured current. This could be corrected using a more accurate charge function.

The dynamic current recovery after stress biasing the HEMT in off-state is depicted in Fig. 3b where the voltage is pulsed with a period of 100 ms (with 50 % duty cycle) in which 100 measurements are conducted in order to capture the current recovery. The stress bias is applied at a drain voltage of 100 V at gate voltage of -5 V. In the on-state the drain voltage is set to 1 V and the gate is set to 0 V. The model is able to capture accurately the global trend of the drain current (blue line in Fig. 3b). However, there is a discrepancy between the measured and modeled data regarding the starting point of the current directly after the stress bias. This is due to the fact that the trapping centers included in the model are only those typically reported by the wafer supplier, disregarding those originating from process variations or threading dislocations. This fitting can be improved once all the species are known. Alternatively, the model can be used to infer the composition and quality of the material based on the measured data.

The final aspect to be verified, namely the off-state three terminal leakage current, is then measured and the results are compared to the model in Fig. 4a. Depending on the state of occupancy of the traps, which is dynamically calculated in the model, the leakage current is selected to either follow a SCLC or a conventional linear behavior of a lightly doped semiconductor. In Fig. 4b, the device is stressed in off-state for 1 hr and then the leakage current is measured again. In this case, the trapping sites are filled from the beginning of the stress (charge de-trapping time constant > measurement time) and thus, a SCLC behavior is seen across the entire voltage range. In addition, the SCLC model including the PF emission is compared with that without the PF emission. From this comparison it is obvious that the model without the PF effect

underestimates the current at higher drain biases since electron emission from trapping sites is not included.

IV. CONCLUSIONS

In this work a comprehensive GaN HEMT semi-empirical Verilog-A model with a focus on the off-state leakage current and the kinetics of charge trapping and de-trapping is presented and experimentally verified. The model is able to capture the DC characteristics, the off-state leakage characteristics as well as the dynamic current recovery of the fabricated GaN HEMTs. As such, the model directly relates to the structure and properties of the GaN epitaxial layers and the carrier substrate. The model can, therefore, be used to fit the measured data in order to deduce information about the species incorporated during the growth based on the dynamics extracted by the model.

V. REFERENCES

[1] S. Keller *et al.*, "Gallium nitride based high power heterojunction field effect transistors: process development and present status at UCSB," in IEEE EDL, vol. 48, no. 3, pp. 552-559, Mar 2001.

[2] C. Armbruster, A. Hensel, A. H. Wienhausen and D. Kranzer, "Application of GaN power transistors in a 2.5 MHz LLC DC/DC converter for compact and efficient power conversion," EPE'16 ECCE Europe, Karlsruhe, 2016, pp. 1-7.

[3] M. Alshahed *et al.*, "600 V, low-leakage AlGaN/GaN MIS-HEMT on bulk GaN substrates," 46th European Solid-State Device Research Conference (ESSDERC), Lausanne, 2016, pp. 202-205.

[4] L. Dunleavy, C. Baylis, W. Curtice and R. Connick, "Modeling GaN: powerful but challenging," in IEEE Microwave Magazine, vol. 11, no. 6, pp. 82-96, Oct. 2010.

[5] A. Birafane, P. Aflaki, A. Kouki, and F. Ghannouchi, "Enhanced DC model for GaN HEMT transistors with built-in thermal and trapping effects, " *in* Solid-State Electronics, *vol. 76, pp. 77–83, 2012.*

[6] G. L. Bilbro and R. J. Trew, "A five-parameter model of the AlGaN/GaN HFET," in IEEE Transactions on Electron Devices, vol. 62, no. 4, pp. 1157-1162, April 2015.

[7] O. Ambacher, et al., „Pyroelectric properties of Al(In)GaN/GaN hetero- and quantum well structures", in Journal of Physics: Condensed Matter, Vol. 14, No. 13, 2002.

[8] U. Radhakrishna, T. Imada, T. Palacios, and D. Antoniadis, "MIT Virtual Source GaNFET-High Voltage (MVSG-HV) model: A physics based compact model for HV-GaN HEMTs," in Physica status solidi (c), vol. 11, no. 3-4, pp. 848–852, 2014.

[9] O. Hilt, F. Brunner, E. Cho, A. Knauer, E. Bahat-Treidel and J. Würfl, "Normally-off high-voltage p-GaN gate GaN HFET with carbon-doped buffer," 2011 IEEE 23rd International Symposium on Power Semiconductor Devices and ICs, San Diego, CA, 2011, pp. 239-242.

[10] S. Takeshita, "Modeling of space-charge-limited current injection incorporating an advanced model of the Poole-Frenkel effect", M. S. thesis, electrical and computer engineering dept., Clemens University, Clemens, S. C., USA, 2008.

[11] D. Bisi et al., "Deep-level characterization in GaN HEMTs-Part I: advantages and limitations of drain current transient measurements," in IEEE Transactions on Electron Devices, vol. 60, no. 10, pp. 3166-3175, Oct. 2013.

[12] M. Florovič, R. Stoklas, and P. Kordoš, "Temperature dependence of the threshold voltage in GaN-based HFETs and MOSHFETs", in WOCSDICE EXMATEC, 2016.

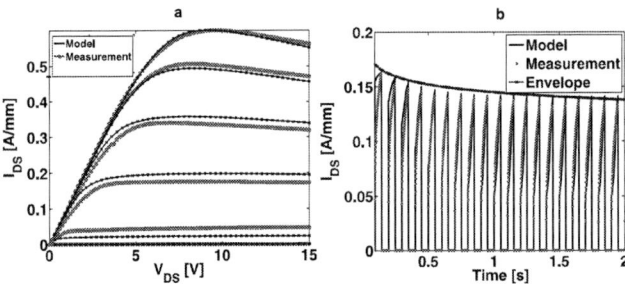

Fig. 3: (a) Measured vs. simulated DC output characteristics of the HEMT ($V_{gs} = 0$ V : -5 V), (b) measured vs. simulated pulsed I_{ds} after stress-bias

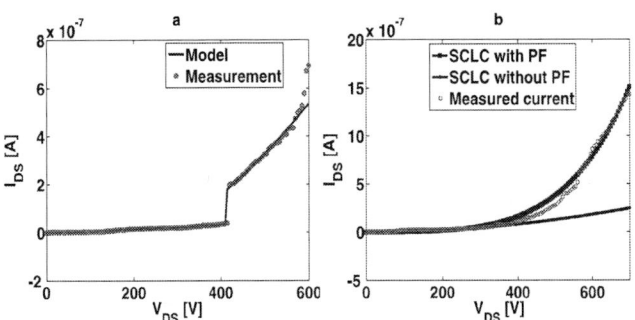

Fig. 4: Off-state 3-terminal leakage current of (a) fresh device and (b) 1 hr. stressed device with and without PF emission

Trap-Assisted Carrier Transport through the Multi-stack Gate Dielectrics of HKMG nMOS Transistors: A Compact Model

Apoorva Ojha and Nihar R. Mohapatra*

Department of Electrical Engineering, IIT Gandhinagar, Palaj, 382355, Gandhinagar, India
*Email: nihar@iitgn.ac.in

Abstract—**In this paper, a compact model for the gate current in HKMG nMOS transistors is presented. The carrier transport through the multi-stack gate dielectrics of HKMG MOS transistors is shown to be dominated by the Trap Assisted Tunneling and Poole-Frenkel conduction mechanisms. Both these mechanisms occur simultaneously and each is dominant in a particular gate voltage range. The interdependence and simultaneity of both the mechanisms is modeled to get a compact gate current formulation. The model is valid for all gate voltages and for different temperatures. The model also includes the formulation of inelastic TAT in a compact format. The accuracy of the model is validated with the measurement data.**

Index Terms—**gate tunneling, HKMG, Inelastic, Poole Frenkel, TAT**

I. INTRODUCTION

Carrier transport through the gate dielectrics is an important leakage mechanism in modern MOS transistors which needs to be accounted for and modeled properly [1-5]. The adoption of High-K Metal Gate (HKMG) CMOS technology with multilayer dielectric stack (HfO_2 on top of a SiO_2 interfacial layer) has given the advantage of increased physical oxide thickness. But, the transition metal oxide films like HfO_2 have large amounts of positively charged oxygen vacancies (V_0^+, V_0^{2+}) and these oxygen vacancies create defect states (trap energy levels) located ~0.6eV (V_0^+) and 1.5eV (V_0^{2+}) below the conduction band edge of HfO_2 [6]. Since the energy levels of these traps are aligned to the conduction band edge of silicon during normal operating conditions, they lead to increased trap assisted conduction of carriers through the gate dielectric stack.

Traditionally, the commercial compact models for gate current [7-9] are described by direct tunneling mechanism. However, for a HKMG MOS transistor, the direct tunneling component is very small. Therefore the existing models do not fit properly with the measured data from HKMG transistors (see Fig. 1). Note, the trap assisted conduction in a dielectric has two components: (a) Trap Assisted Tunneling (TAT) and (b) Poole-Frenkel (PF) conduction. The relative contribution of TAT or PF conduction to the gate current depends on factors like the density of traps, position of traps relative to the conduction band of the dielectric, the tunneling barrier and the electric field. There are separate formulations of TAT and PF conduction mechanism in literature [10-15]. But, there are none which captures the simultaneity of the two. Since, both of these mechanisms together determine the current transport through the gate dielectrics (gate current) of a MOS transistor

and the gate current is an important component of the circuit standby power, there needs to be a compact model that takes into account both the mechanisms and predicts the gate current accurately.

In this work, we have developed the same. Our model is a physics based compact model suitable for describing carrier transport through multi stack gate dielectrics of the HKMG MOS transistors. It captures the simultaneity and interdependence of both PF and TAT mechanisms. The developed model is validated through comparison with the measured data (from HKMG nMOS transistors fabricated using a 28nm CMOS technology) and found to be very accurate.

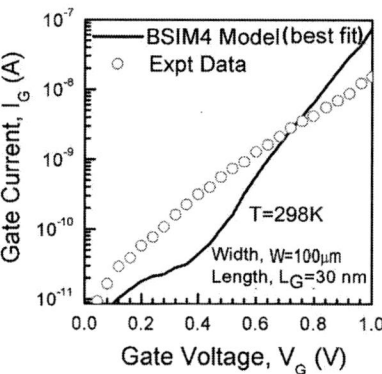

Fig. 1. Comparison of the BSIM4 gate current model (best fit) with the measurement data. The experimental data is obtained form the nMOS transistors fabricated using 28nm HKMG CMOS technology. The BSIM4 model could not predict the gate current in HKMG MOS transistors accurately.

II. METHODOLOGY

As mentioned in the previous section, the gate current in HKMG MOS transistors includes both TAT (elastic and inelastic) and PF conduction components. There are two defect states available in HfO_2 through which the TAT and PF conduction can happen [5]. In our analysis, these two defect levels are termed as E_{T1} and E_{T2}. The E_{T1} is 0.6eV and E_{T2} is 1.5eV below the conduction band edge of HfO_2. Fig. 2 (a-c) shows the band diagram of $TiN/HfO_2/SiO_2/Si$ system at different gate voltages (V_G). The band diagram is extracted from the "Multi-Dielectrics Energy Band Diagram Program"

978-1-5090-5979-9/17 $31.00 © 2017 IEEE

developed by nanoscale materials and device group of Boise State University [16]. As shown, for positive V_G, the carrier transport through the gate stack could be divided into three distinct parts,

(a) Low V_G – The Si conduction band is aligned to E_{T2}. So, the electrons tunnel to the SiO_2/HfO_2 interface and then tunnel through ϕ_2 to the gate.

(b) Medium V_G – The Si conduction band is not aligned to E_{T1} or E_{T2}. The electrons tunnel to the SiO_2/HfO_2 interface then to E_{T2} through an inelastic process (phonon assisted tunneling). Finally they tunnel through E_{T2} to the gate.

(c) High V_G – The Si conduction band is aligned to E_{T1}. The electrons tunnel to the SiO_2/HfO_2 interface and then tunnel through E_{T1} to the gate. Some of the electrons move to the conduction band from E_{T1} (higher transverse field (F_T)) and subsequently transported to gate through PF conduction.

In this work, we have obtained a compact expression for inelastic transition probability and then formulated an expression for trap-assisted conduction capturing simultaneity of TAT and PF conduction.

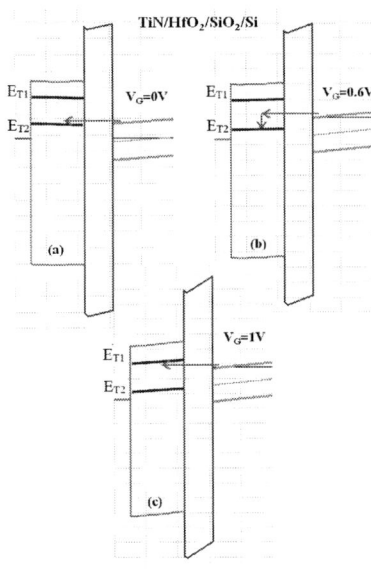

Fig. 2. The energy band diagram of the $TiN/HfO_2/SiO_2/Si$ system at (a) $V_G = 0V$, (b) $V_G = 0.6V$ and (c) $V_G = 1V$. The band diagram graphically depicts the carrier transport mechanism for different V_Gs.

III. MODEL

A. Modeling of Inelastic Trap Assisted Tunneling (TAT)

As discussed in section III, the TAT process at medium V_G consists of inelastic transitions that involve multi-phonon transition. The carrier loses energy in form of phonons, drops down to a trap level E_T (see Fig. 3) and then tunnels to the gate through the trap.

Fig. 3 schematically depicts the whole process. The transport of electron from A to B involves two steps, (a) Tunneling

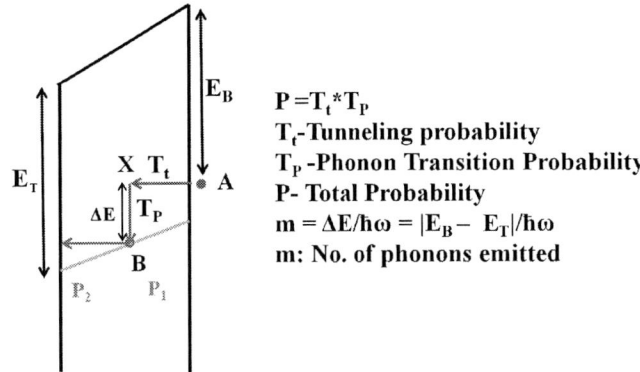

Fig. 3. Schematic diagram showing the inelastic phonon assisted TAT mechanism in HKMG MOS transistors. The inelastic TAT is dominant at medium V_Gs.

from A to X with a probability T_t and (b) Inelastic transition from X to B with a probability T_P. The number of phonons emitted during the transition process (X → B) depends on ΔE and energy of a single phonon.

The multiphonon transition probability has already been discussed in literature [17-18] and can be written as,

$$T_P = c_0 \left(\frac{f_B + 1}{f_B} \right)^{m/2} exp\left(-2\left(2f_B + 1\right)\right)$$
$$I_m \left(2S \left(\sqrt{f_B\left(f_B + 1\right)} \right) \right) \quad (1)$$

In (1), c_0 is capture cross section which is proportional to F_T^2 (F_T = Transverse electric field in the dielectric), S is the Huang Rhys factor (17 for HfO_2 [18]), f_B is the Bose function that gives phonon population statistics $\left(f_B = 1/\left(exp\left(\frac{\hbar\omega}{kT}\right) - 1\right)\right)$, m is the no. of phonons emitted and I_m is the modified Bessel function of the first kind of order m. Note that it is difficult to implement I_m in circuit simulators and hence a compact expression is needed for (1). Also note that, for a particular dielectric, m and T are the only variables (others being material parameters are constant). The variation of (1) with m and T is analyzed. Fig. 4(a) shows the variation of (1) with m and T. It is observed that the variation of $ln\left(T_P\right)$ with T is quadratic for all values of m. It is also observed that the coefficients of this quadratic formalism can be scaled progressively as m increases. Based on these observations, a simplified and empirical formula of T_P is developed as a function of T and m as described below.

$$T_P = A_{ITAT} F_T^2 exp\left(k^m \left(-aT^2 + bT - c\right)\right) \quad (2)$$

Here k, a, b and c are positive and are constants for a particular dielectric. A_{ITAT} is a fitting parameter dependent on the capture cross section of the trap. Fig. 4(b) shows the comparison between the model and actual formula (1) as a function of T for different m. As shown the model could accurately predict T_P. The k, a, b and c used to fit (2) are summarized in Table I. Note that as shown in Fig. 3, we have considered P_1 as the carrier transition probability from silicon channel to trap and P_2 as the carrier transition probability from trap to gate.

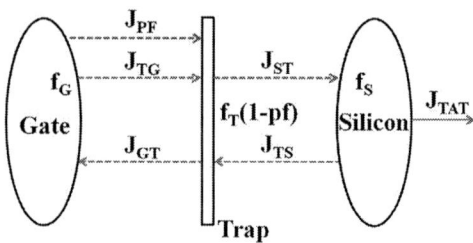

Fig. 6. Schematic representation of gate current components in the dielectric system with a trap level. In this work, gate current is analyzed taking forward and reverse currents into account.

Fig. 6 schematically represents the current components between silicon, trap and gate. J_{ST} (electron tunneling from silicon to trap) and J_{TG} (electron tunneling from trap to gate) are forward currents. J_{GT} (electron tunneling from gate to trap) and J_{TS} (electron tunneling from trap to silicon) are reverse currents. J_{PF} is the current component due to PF conduction and J_{TAT} is the current component due to TAT. These current components can be mentioned as follows,

$$J_{ST} = \frac{qN_T}{\tau}\left(1 - f_T\left(1 - pf\right)\right)f_S P_1 \tag{3}$$

$$J_{GT} = \frac{qN_T}{\tau}\left(1 - f_T\left(1 - pf\right)\right)f_G P_2 \tag{4}$$

$$J_{TS} = \frac{qN_T}{\tau}\left(af_S\right)f_T(1 - pf)P_1 \tag{5}$$

$$J_{TG} = \frac{qN_T}{\tau}\left(af_G\right)f_T(1 - pf)P_2 \tag{6}$$

Here, f_S and f_G are the occupation probabilities in silicon and gate respectively and $af_S = 1 - f_S$, $af_G = 1 - f_G$. They can be written as,

$$f_S = \frac{1}{\left(1 + exp\left(\frac{\phi_B + (E_G/2) - \psi_S}{v_T}\right)\right)} \tag{7}$$

$$f_G = \frac{1}{\left(1 + exp\left(\frac{V_g + \phi_B + (E_G/2) - \psi_S}{v_T}\right)\right)} \tag{8}$$

$$pf = exp\left(\frac{-\left(E_T - \beta\sqrt{F_T}\right)}{v_T}\right) \tag{9}$$

pf is the fraction of carriers released from the trap to the conduction band of HfO_2 and is defined as (9) [10-12]. $f_T(1 - pf)$ is the occupation probability of trap, τ is the time constant for transition to the trap and N_T is the trap density. Combining all the current components, we could write,

$$J_{TS} + J_{TG} = J_{ST} + J_{GT} \tag{10}$$

Using (3)-(6) in (10),

$$\begin{aligned}\left(1 - f_T\left(1 - pf\right)\right)\left(P_1 f_s + P_2 f_g\right) = \\ f_T(1 - pf)\left(P_1 af_S + P_2 af_G\right) \quad (11)\end{aligned}$$

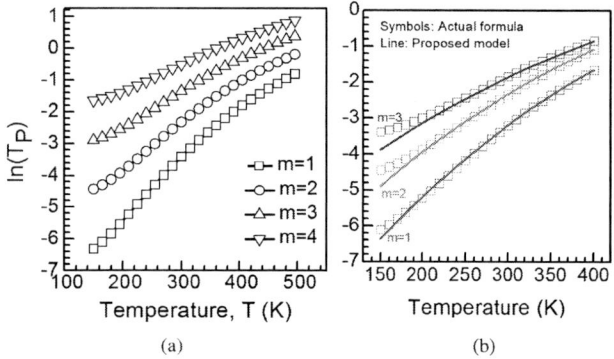

(a)   (b)

Fig. 4. (a) Variation of T_P with temperature T for different m. $ln\left(T_P\right)$ varies quadratically with T for all values of m. (b) Comparison of the developed compact expression (2) with the actual formula. The proposed compact expression is accurate and computationally workable.

TABLE I
THE FITTING PARAMETERS USED IN (2).

Parameter	Value
k	0.892
a	2.587e-5 K^{-2}
b	3.54e-2K^{-1}
c	19.611

B. Consolidated formulation of TAT and PF conduction

Fig. 5 depicts the simultaneity of TAT and PF conduction in SiO_2/HfO_2 dielectric stack with a single trap. Note that in a multiple trap system, the trap energy level close to the conduction band is responsible for both TAT and PF conduction. The other trap levels are only responsible for TAT.

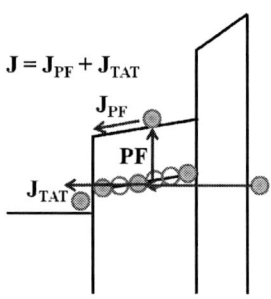

Fig. 5. Consolidated representation of PF and TAT conduction mechanisms.

$$f_T = \frac{(P_1 f_S + P_2 f_G)}{(1 - pf)(P_1 + P_2)} \quad (12)$$

The total TAT current density can be written as,

$$\begin{aligned} J_{TAT} &= J_{ST} - J_{TS} = \\ &\frac{q N_T P_1}{\tau} \left[(1 - f_T) f_s - (1 - pf)(1 - f_s) f_T \right] \\ &= A_{TAT} P_1 \left[(1 - f_T) f_s - (1 - pf)(1 - f_s) f_T \right] \quad (13) \end{aligned}$$

Here, A_{TAT} is a fitting parameter dependent upon trap density N_T and τ. The current component due to PF conduction depends on the electric field [13-14]. So, J_{pf} can be written as,

$$J_{pf} = A_{PF} * F_T * f_T * pf * \left(\frac{T_0}{T} \right)^{\alpha} \quad (14)$$

Here, $f_T * pf$ is the probability of electron transition from trap to the conduction band of HfO_2. A term $(T_0/T)^{\alpha}$ is added to empirically model mobility degradation that exists in the dielectrics at higher temperature. Note that T_0 is 298K and α is a fitting parameter between 1-3. A_{PF} is also a fitting parameter which is dependent on N_T. Finally, the total gate current density is written as,

$$J_G = J_{TAT} + J_{PF} \quad (15)$$

Fig. 7 shows the comparison between the model and measured $I_G (J_G \times W \times L)$ for different temperatures. The measurement data is represented as symbols. As shown, the model successfully predicts I_G for different V_Gs. The A_{TAT}, A_{pf}, a used to fit the measured data are summarized in Table II. Note, the PF component of gate current is dominant at higher V_G and the TAT mechanism is more relevant otherwise.

TABLE II
TYPICAL VALUES OF PARAMETERS USED FOR THE MODEL

Parameter	Value
A_{TAT}	$0.67\ Acm^{-2}$
A_{PF}	$2.52e\text{-}07\ \Omega^{-1} cm^{-1}$
α	2.5

IV. CONCLUSIONS

To summarize, it was observed that the gate current in HKMG MOS transistors is mainly dominated by TAT and PF conduction mechanisms. A multiple trap level formulation is proposed to explain trap assisted conduction at all gate voltages. The phonon assisted inelastic TAT is modeled in a compact mathematical form which is responsible for temperature dependence of carrier transport. Finally, a compact and composite model is proposed which predicts the gate current in HKMG system accurately. The accuracy of the model is

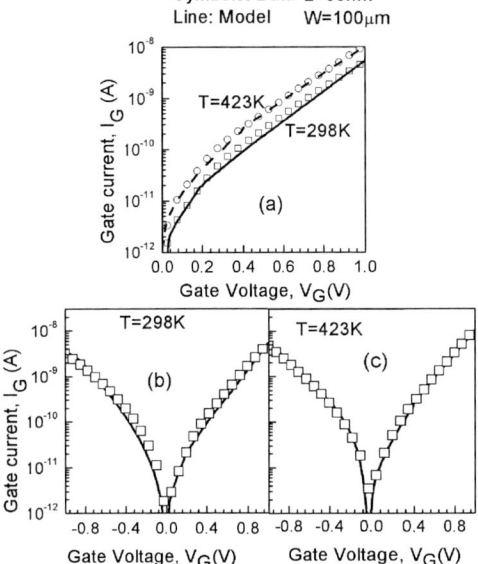

Fig. 7. (a) Comparison of model (solid line) with the measurement data (symbols) for two different temperatures. The model predicts the gate current accurately. (b) Comparison for all V_Gs at 298K and (c) Comparison for all V_Gs at 423K.

verified by comparing it with the experimental data obtained from the transistors fabricated using 28nm HKMG CMOS technology at different temperatures.

REFERENCES

[1] K F. Schuegraf et. al., Symposium on VLSl Technology, pp. 18-19, 1992.
[2] W.C. Lee and C. Hu, IEEE Trans. Electron Devices, vol. 48, no.7, pp 1366-1373, Jul 2001.
[3] Y.T. Hou et. al., IEEE Trans. Device Mater. Rel., vol. 24, no. 2, pp.96-98, February 2003.
[4] A. Gehring et. al., Journal of Computational and Theoretical Nanoscience, Vol.2, 26–44, 2005.
[5] P. Duhan et. al., IEEE Electron Device Lett., vol. 36, no. 8, pp. 739-741, Aug. 2015.
[6] K. Xiong et. al. , Applied Physics Letters, vol.87, 183505 (2005).
[7] BSIM6 Technical Manual [Online]. Available: http://wwwdevice. eecs.berkeley.edu/bsim/
[8] N. Yang et. al., IEEE Trans. on Electron Devices, vol. 46, pp. 292-294, (1999).
[9] Y.C. Yeo et. al., IEEE Trans. Electron Devices, vol. 50, no.4, pp 1027-1035, April 2003.
[10] J. Wu et. al., Proc. Int. Reliability Physics Symp., 1999, pp. 389–395.
[11] C. Svensson et. al., Journal of Applied Physics, Vol. 44, pp. 4657 -4633, 1973.
[12] W. J. Chang et. al., J. Appl. Phys., vol. 89, no. 11, pp. 6285–6293, 2001.
[13] M. Bajaj et. al., IEEE Trans. on Electron Devices, vol. 60, pp. 4152-4158, 2013.
[14] W.R. Harrell et. al.,Thin Solid Films, vol 352, pp. 195-204, 1999.
[15] R. G. Southwick et. al., IEEE Trans. Device Mater. Rel., vol. 10, no. 2, pp.201 -207, 2010.
[16] R. G. Southwick et. al., IEEE Transactions on Device and Materials Reliability, vol. 11, no. 2, pp. 236–243, Jun. 2011.
[17] F. Jimeˊnez-Molinos et. al., Journal of Applied Physics, Vol 90, no. 7, Oct. 2001.
[18] L. Vandelli et. al., IEEE Trans. on Electron Devices vol. 58 no. 9 pp. 2878-2887, Sep 2011.

A New Verilog-A Compact Model of Random Telegraph Noise in Oxide-Based RRAM for Advanced Circuit Design

Francesco Maria Puglisi[°#], Nicolò Zagni[#], Luca Larcher[$], Paolo Pavan[#]

[#] DIEF – Università di Modena e Reggio Emilia – Via P. Vivarelli 10/1, 41125 – Modena (MO) – Italy
[$] DISMI – Università di Modena e Reggio Emilia – Via Amendola 2, 42122 – Reggio Emilia (RE) – Italy
[°]corresponding author email: francescomaria.puglisi@unimore.it phone: +39-059-2056324

Abstract— **In this work, we propose for the first time a Verilog-A physics-based compact model of Random Telegraph Noise (RTN) in Resistive Random Access Memory (RRAM) devices. Starting from the physics of the RTN mechanism in both high (HRS) and low (LRS) resistive states, and combining experimental data with physics-based simulations, we develop and validate a complete compact model of RTN in RRAM devices. The model accounts for the intrinsic randomness in the number of defects contributing to the RTN and their properties. Moreover, it can be readily integrated in existing RRAM device compact models, extending their capabilities. The model is implemented in Verilog-A, and its effectiveness is demonstrated by using it to design the building block of a Truly-Random Number Generator circuit exploiting the RTN randomness as an entropy source.**

Keywords - RRAM; Random Telegraph Noise; HRS; LRS; Compact Model; Verilog-A, Random Number Generator.

I. INTRODUCTION

Resistive Random Access Memory (RRAM) is among the most promising runners in the competition to realize reliable, fast, and high-density embedded and storage class memories [1]. As RRAM technology is entering the industrial phase, Random Telegraph Noise (RTN) mechanisms [2-4] must be included in compact models to design circuits for specific applications e.g., non-volatile memory [5], neuromorphic computing [6], Physical Unclonable Function (PUF) [7], and Random Number Generator (RNG) [8]. Indeed, RTN fluctuations in RRAM can reduce the memory margin [2-4], and can be critical in neuromorphic applications. On the other hand, it has been suggested to exploit the intrinsic randomness of RTN fluctuations as an entropy source to realize RTN-based PUFs and RNGs [9]. From the circuit designer perspective, it is imperative to include RTN effects into RRAM compact models. However, a full compact model of RRAM, including RTN and able to reproduce all the relevant stochastic features of RTN [2-4] in both high (HRS) and low (LRS) resistive states is still missing. An attempt is found in a recent contribution [10], which assumes the presence of a gap in the conductive filament (CF), and is valid only in HRS. Recently, we proposed a compact formula [11] that can be used to estimate the average RTN fluctuation *amplitude* in both resistive states. In this work, we derive and validate a compact model for RTN in RRAM that correctly captures the statistics of the RTN fluctuations properties (i.e., variations in amplitude and transition times) and the variability in the number of defects contributing to the RTN. Though we focus only on HfO_2 MIM structures, very

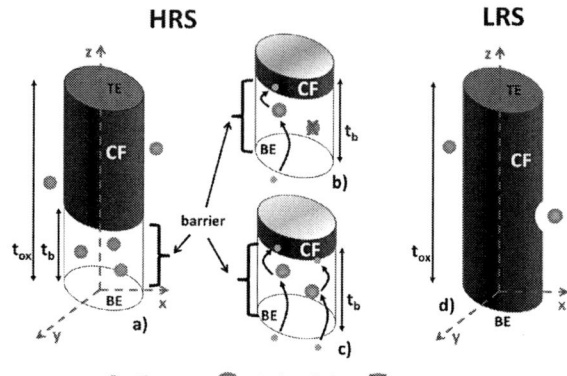

Fig. 1 – (a) Schematic picture of the device and (b-c) of the RTN mechanism in HRS, showing the defects (b) de-activation and (c) activation. The barrier thickness, t_b, and CF thickness, t_{ox}, are reported. (d) Schematic picture of the RTN mechanism in LRS, showing the electrostatic coupling between a charged defect and the conductive filament.

similar RTN trends are found in devices based on different materials [3, 11], which is a significant benefit to obtain a flexible compact model for RTN in RRAMs. The model is implemented in Verilog-A, providing the first SPICE-compatible RTN compact model for RRAMs. Importantly, this model can be easily plugged in existing compact RRAM device models [5, 12], extending their capabilities. We demonstrate the effectiveness of the proposed model by using it in the design of a RNG circuit based on RTN in RRAM. This flexible model provides an invaluable tool for circuit designers to correctly account for the presence of RTN, both in scenarios in which RTN is an undesired effect limiting the design space, and in those in which RTN is purposely exploited.

II. DEVICES AND EXPERIMENTS

We perform extensive RTN measurements on RRAMs with $TiN/Ti/HfO_2/TiN$ structure in many operating conditions (temperature - T, V_{RESET}, and current compliance - I_C). After forming, each device is cycled 100 times; RTN is measured in both HRS and LRS at each cycle in reading conditions (i.e. $V_{READ} = 100$ mV). This results in a wide dataset where device-to-device and cycling variability are considered. RTN data are processed using the Factorial Hidden Markov Model (FHMM) [13], which allows decomposing the multi-level fluctuations into a superposition of two-level signals, each attributed to the activity of individual defects. This allows characterizing every defect contributing to the RTN [13] and to estimate their statistical properties i.e., the amplitude of the RTN fluctuation (i.e. ΔI) and the capture (τ_c) and emission (τ_e) times [2, 13]. All measurements are performed with a Keithley 4200-SCS.

III. THE RTN COMPACT MODEL: PHYSICS, IMPLEMENTATION, AND VERIFICATION

A complete RTN compact model must capture the inherent variability in the number of defects contributing to the RTN and in their statistical properties (ΔI, τ_c, τ_e). These all depend on the operating conditions and on the physical state of the device (i.e. HRS or LRS). In the following, we show how to include these features; the model is then implemented in Verilog-A and verified in a simple test circuit.

A. RTN Amplitude Statistics in LRS

In LRS, a full CF shunts the two electrodes, Fig. 1d. The CF size is controlled by the I_C value used during forming [5, 12, 14] and charge transport shows a ohmic-like behavior. In LRS, RTN is commonly attributed to electron trapping and de-trapping at individual defect sites in the proximity of the CF (i.e. within one Debye length, λ) [3, 11]. The trapped charge perturbs the potential in its surroundings [3, 11], causing a screening effect on the closest portion of the CF, inducing a resistance change, Fig. 1d. In this framework, the relative resistance change ($\Delta R/R$) depends on the CF geometry, i.e. its cross-section, S_{CF}: as the CF gets thinner, the screened portion of the CF gets comparatively wider, with a larger impact on the average relative resistance change [3, 11]. However, while the average $\Delta R/R$ in LRS is CF-size dependent, deviations from the average are CF-size independent, only depending on the distance between the defect and the CF border. This results in the lognormal $\Delta I/I$ (or, equivalently, $\Delta R/R$) distribution in LRS to show the same variance (σ) regardless of the CF size (controlled by I_C) [3, 11]. This is consistent with the experimental observations in Fig. 2, which allows us estimating $\sigma \approx 0.3$. The median $\Delta I/I$ value, M, changes with the CF size:

$$M\left(\frac{\Delta R}{R}_{LRS}\right) = \frac{r_t^3}{2t_{ox} \cdot S_{CF}}; \; \sigma\left(\frac{\Delta R}{R}_{LRS}\right) \cong 0.3 \quad (1)$$

where r_t is a parameter accounting for the effective screening length of the trapped charge in this simplified framework, related to the electron Debye length, λ [3, 11]. This simple formula [11] well captures the experimental trend in Fig. 2.

B. RTN Amplitude Statistics in HRS

The reset operation, driving the device in HRS, leads to the partial re-oxidation of the CF, creating a dielectric barrier [5, 12, 14], Fig. 1a. The thickness of the dielectric barrier is determined by the reset conditions. The HRS current is dominated by the electron Trap-Assisted Tunneling (TAT) at positively charged oxygen vacancy defects (Vo^+) in the dielectric barrier, Fig. 1a-b-c [14-15]. The current understanding of the physics of RTN in HRS is based on the temporary (de-)activation of Vo^+ defects, distributed in the barrier, assisting charge transport [11, 14-15], Fig. 1a-b-c. This (de-)activation mechanism is supposed to be due to charge trapping and de-trapping at additional defects, not involved in current conduction and tentatively identified with oxygen ions (O) [11, 15]. As discussed in previous works [2, 11, 15], the

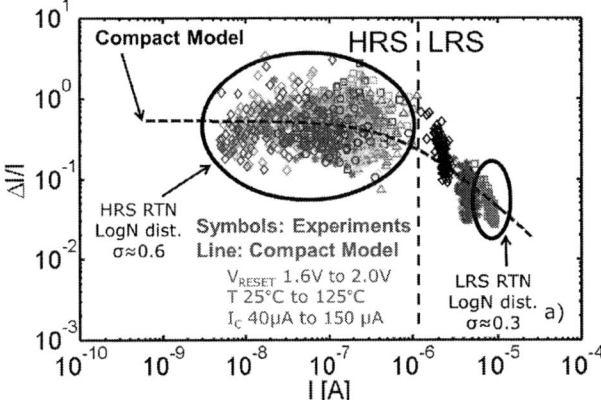

Fig. 2 – Experimental $\Delta I/I$ vs. I (symbols) in HRS and LRS in different operating conditions (I_C, V_{RESET}, T) for five devices (different symbols) with a 5 nm HfO$_2$ layer. Model prediction (dashed line) is in excellent agreement with real data. The extracted σ values of the normal distributions associated with the lognormal $\Delta I/I$ distributions in HRS and LRS agree with the values extracted from simulations and used in the compact model.

complex physics of RTN in HRS points to the average $\Delta I/I$ (or equivalently $\Delta R/R$) to be constant in every operating condition. This picture is fully consistent with the experimental results in Fig. 2, and is corroborated by the results of physics-based kinetic Monte-Carlo simulations, Fig. 3. Simulations include the presence of defects and of the trapped charge, accounting for different charge transport mechanisms, including TAT (details in [2]). We performed simulations considering different applied voltages, barrier thicknesses and temperatures. For each simulation, we consider randomly distributed Vo^+ and O ion defects, and we estimate the $\Delta I/I$ resulting by the Vo^+ activation and de-activation due to trapping/de-trapping at O ions. Results are reported in Fig. 3, where the probability plot of the resulting $\Delta I/I$ in different conditions appears to be log-normally distributed and invariant with operating conditions, in agreement with results in Fig. 2. Notably, both the median and the variance of the simulated distributions agree with the values extracted from experiments, which allows writing:

$$M\left(\frac{\Delta R}{R}_{HRS}\right) \cong \frac{1}{2}; \; \sigma\left(\frac{\Delta R}{R}_{HRS}\right) \cong 0.6 \quad (2)$$

This result confirms that, notwithstanding the intrinsic large variability and the complexity of the RTN mechanism, it is possible to devise a simple statistical description of the RTN fluctuations amplitude in every operating condition.

C. RTN Capture and Emission Times

To reproduce the RTN signals over time, it is also necessary to estimate the capture and emission times, $\tau_{c,e}$, of each defect contributing to the RTN. Since the physical mechanism causing RTN is associated with charge trapping and de-trapping, we can calculate for each defect the associated $\tau_{c,e}$ using the TAT formalism [2, 16], exploiting the compact formulae:

$$\tau_C \propto e^{\left(\frac{d}{\lambda_c}\right)} e^{\left(\frac{E_c(V)}{kT}\right)}; \; \tau_e \propto e^{\left(\frac{d}{\lambda_e}\right)} e^{\left(\frac{E_e}{kT}\right)} \quad (3)$$

These formulae, fully described in [2, 16], require defining the applied voltage, V, the temperature, T, the defect distance to the closest electrode, d, (determined by the defect vertical

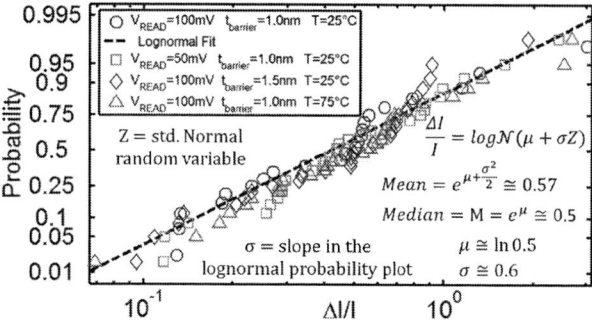

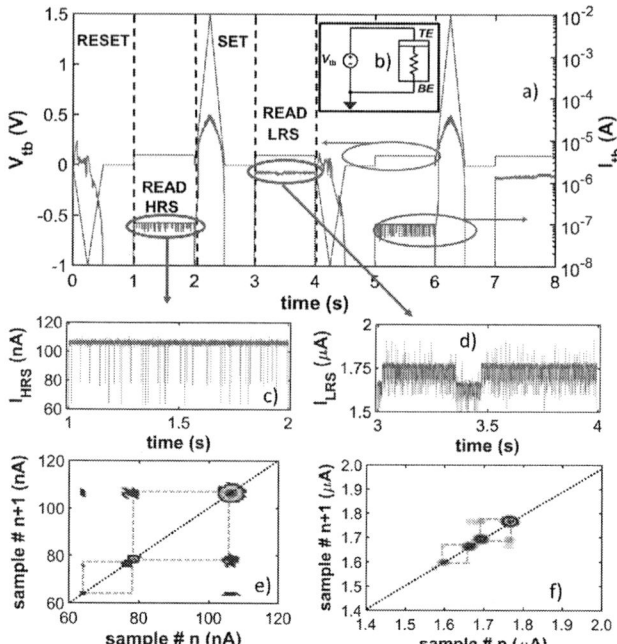

Fig. 3 – Distribution of simulated $\Delta I/I$ in HRS obtained using kinetic Monte-Carlo simulations. The I-time traces related to a 5x5 nm^2 MIM device with TiN electrodes are simulated, including the effect of Vo$^+$ defects activation and de-activation. Vo$^+$ are randomly distributed in the barrier with a density of $N_T = 2 \cdot 10^{21}$ cm^{-3} [2, 11, 15]. Simulations are performed in different conditions (V_{READ}, t_b, and T) and repeated on 10 different device realizations by randomizing defects positions and energies. $\Delta I/I$ distribution is insensitive to operating conditions in agreement with experimental reports.

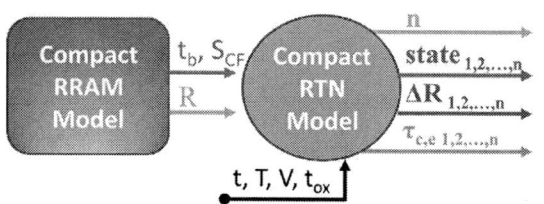

Fig. 4 – Integration scheme of the proposed compact RTN model in compact RRAM device models. The RTN compact model only needs few inputs, easily given by the compact RRAM device model and the simulation framework.

position within the dielectric barrier – in HRS – or alongside the CF – in LRS) and few defect properties (e.g., thermal ionization energy, relaxation energy) embedded in the λ_c, λ_e, E_c, and E_e parameters [2, 14-16] and that only depend on the defect typology (O ions in HRS, Vo$^+$ in LRS).

D. Model Implementation and Verification

The model proposed in this work can be seamlessly integrated into existing RRAM device compact models [5, 12]. Indeed, it requires very few inputs that are easily provided by compact device models, i.e. the oxide thickness, t_{ox}, the device resistance (R), the applied voltage and temperature (V and T), the current simulation time, t, S_{CF}, and t_b – Fig. 4. The algorithm to simulate RTN is composed of the following steps (1 through 4), and has been implemented in Verilog-A. *(1)* At each simulation step, the algorithm checks *if any change in the t_b and/or S_{CF} parameters occurred*: if so, the portion of volume hosting the defects responsible for RTN undergoes structural modifications (e.g., a SET or RESET event occurs). In this case (and at the first simulation step), *the algorithm (re-)instantiates the defects and their properties*. First, it estimates the number of defects contributing to RTN according to the state of the device. This follows the Poisson distribution [17], fully defined by the average number of defects. i.e., the product of the defect density and the volume in which the defects are found. In HRS, the defects average number can be estimated as $t_b \cdot S_{CF} \cdot N_O$, with N_O (in the range $10^{20} - 10^{21}$ cm^{-3} [2-3, 14-15]) being the O ions density in the barrier. Conversely, in LRS the defects responsible for RTN are those distributed around the CF, but not further than r_t from the CF border. Assuming a cylindrical

Fig. 5 – Simulations of resistance switching and RTN readout in HRS and LRS. An RRAM device in LRS with $R_{LRS} \approx 15$ kΩ is instantiated in the circuit (b). The compact device model used in this simulation is found in [5], and has been extended with the RTN module. The voltage waveform in (a), blue line, is applied to perform the reset ($V_{RESET} = -1$ V) and set ($V_{SET} = 1.5$ V) operations, as well as the readout operation (lasting 1 s) at $V_{READ} = 100$ mV both in HRS and LRS. The resulting current, red line in (a), during readout shows RTN signals in both states, zoomed in (c) and (d). The resulting traces are also shown in (e) and (f) using the time-lag plot technique [18], which confirms the presence of the RTN "discrete levels" as spots on the main diagonal (dotted black lines) and "transitions between levels" as spots off the main diagonal (transitions are highlighted by the dashed red boxes).

CF, the average number of defects contributing to RTN in LRS is $t_{ox} \cdot \pi \cdot (r_t^2 + 2r_{CF} \cdot r_t) \cdot N_V$, where N_V is the density of defects (supposedly Vo$^+$) around the CF (in the range $10^{19} - 10^{20}$ cm^{-3} [3]). *(2)* For each instantiated defect, *the algorithm assigns a ΔR value*, randomly extracted from a lognormal distribution, whose parameters are defined by eqn. (1-2). Also, a random vertical position (within the barrier if in HRS or around the CF in LRS) is assigned to each defect, which allows estimating its $\tau_{c,e}$. Finally, each defect is assigned an initial state (i.e. either "filled" or "empty"). *(3)* To properly simulate RTN over time, at each simulation step, besides evaluating if it is necessary to re-instantiate the defect information, *the algorithm also calculates for each defect the probability to make a transition to the other state*, according to the formula in [17]. *(4)* The overall ΔR due to all contributing defect is calculated by summing up the individual ΔR of all defects that are in the "filled" state. The validity of the proposed model is verified using a simple circuit composed of an arbitrary voltage generator and a RRAM device (initially set to the LRS with $R_{LRS} \approx 15$ kΩ) instantiated in a circuit simulator, Fig. 5b. The compact device model used in this simulation is found in [5], and has been extended with the RTN model. The applied voltage waveform (blue line in Fig. 5a) performs the reset and set operations, as well as the readout operations (lasting 1 s) at $V_{READ} = 100$ mV both in HRS and LRS. The resulting current,

978-1-5090-5979-9/17 $31.00 © 2017 IEEE

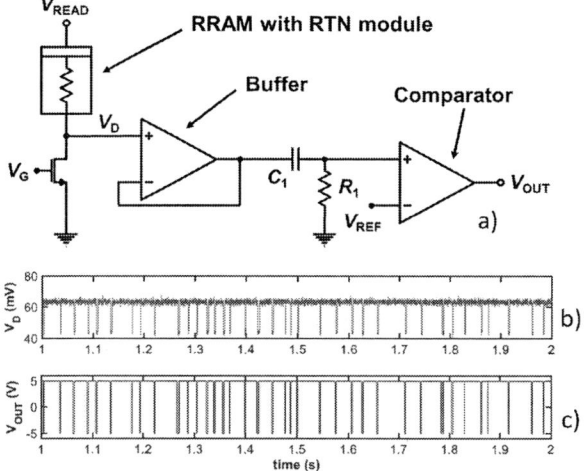

Fig. 6 – (a) Schematic of the Random Number Generator circuit based on the RTN in the RRAM device. A device in HRS with $t_b = 1$ nm is considered. The compact device model used here is found in [5] and has been extended with the proposed RTN module. (b) RTN voltage fluctuations as detected at the transistor drain terminal. (c) The random RTN pattern is successfully reproduced at the circuit output, giving a random bit stream.

red line in Fig. 5a, during readout shows RTN signals in both resistive states, as further confirmed by the zoomed images in Fig. 5c-d. Their time-lag plot representation [18] in Fig. 5e-f further confirms that the RTN properties are correctly simulated.

IV. USING THE MODEL TO DESIGN AN RNG CIRCUIT

The proposed model allows circuit designers easily taking into account the effect of RTN in the circuit design phase. Here we show how the model can be used to design a circuit that exploits the RTN, i.e. an RTN-based RNG. The circuit topology we explore, Fig. 6a, finds implementation in high reliability systems for the generation of truly random numbers [9]. The circuit is composed of the RRAM device in HRS and a series transistor, a buffer with a high-pass filter, and a comparator. It has been implement in Cadence Virtuoso™: we included an RRAM device initialized in HRS, with $t_b = 1$ nm. The compact device model used here [5] has been extended with the proposed compact RTN model. The results of a transient simulation are shown in Fig. 6. The application of a constant voltage, V_{READ}, to the top electrode of the RRAM causes voltage RTN fluctuations to appear at the buffer input, Fig. 6b. The RTN pattern is transferred to the high-pass filter (to get rid of unwanted DC and low frequency components) and compared with a reference voltage. The randomness in the RTN signal produces a random bit stream at the output of the comparator, Fig. 6c. The simulation of this circuit topology shows the potential of the proposed model for advanced circuit design. For instance, the model can be used to estimate the best operating conditions for the RRAM device to optimize the randomness and the uniqueness of the output bit stream.

V. CONCLUSIONS

In this work, we proposed the first Verilog-A physics-based compact model of RTN in RRAMs. The model accounts for the intrinsic randomness in the number of defects contributing

to the RTN and their properties, and is validated against experimental data and physics-based simulations. The model, implemented in Verilog-A, can be easily integrated in RRAM device compact models, and its effectiveness is proved by using it to design an RNG circuit using the RTN randomness as an entropy source.

REFERENCES

[1] Y.-S. Chen et al., "Highly scalable hafnium oxide memory with improvements of resistive distribution and read disturb immunity", Proc. of IEEE Electron Devices Meeting (IEDM) 2009, pp.1-4, 7-9 Dec. 2009.

[2] F. M. Puglisi, L. Larcher, A. Padovani, and P. Pavan, "A Complete Statistical Investigation of RTN in HfO₂-based RRAM in HRS", IEEE Trans. on Electron Devices, vol.62, no.8, pp. 2606-2613, Aug. 2015.

[3] S. Ambrogio et al., "Statistical Fluctuations in HfOx Resistive-Switching Memory: Part II—Random Telegraph Noise", IEEE Trans. on Electron Devices, vol. 61, n. 8, pp. 2920-2927, 2014.

[4] N. Raghavan et al., "RTN insight to filamentary instability and disturb immunity in ultra-low power switching HfOx and AlOx RRAM", Symposium on VLSI Technology, pp. T164-T165, 11-13 Jun. 2013.

[5] F. M. Puglisi, L. Larcher, G. Bersuker, A. Padovani, and P. Pavan, "An Empirical Model for RRAM Resistance in Low- and High-Resistance States", IEEE Electron Device Letters, vol. 34, no. 3, pp. 387-389, Mar. 2013.

[6] J. Woo, K. Moon, J. Song, M. Kwak, J. Park and H. Hwang, "Optimized Programming Scheme Enabling Linear Potentiation in Filamentary HfO2 RRAM Synapse for Neuromorphic Systems", in IEEE Trans. on Electron Devices, vol. 63, no. 12, pp. 5064-5067, Dec. 2016.

[7] A. Chen, "Utilizing the Variability of Resistive Random Access Memory to Implement Reconfigurable Physical Unclonable Functions", IEEE Electron Device Letters, vol. 36, n. 2, pp. 138-140, 2015.

[8] S. Balatti et al., "Physical Unbiased Generation of Random Numbers With Coupled Resistive Switching Devices", in IEEE Trans. on Electron Devices, vol. 63, no. 5, pp. 2029-2035, May 2016.

[9] J. Yang et al., "A low cost and high reliability true random number generator based on resistive random access memory", 2015 IEEE 11th International Conference on ASIC (ASICON), Chengdu, 2015, pp. 1-4.

[10] B. Guan and J. Li, "A compact model for RRAM including random telegraph noise", 2016 IEEE International Reliability Physics Symposium (IRPS), Pasadena, CA, 2016, pp. MY-5-1-MY-5-4.

[11] F. M. Puglisi, P. Pavan, and L. Larcher, "Random telegraph noise in HfOx Resistive Random Access Memory: From physics to compact modeling", 2016 IEEE International Reliability Physics Symposium (IRPS), Pasadena, CA, 2016, pp. MY-8-1-MY-8-5.

[12] Z. Jiang et al., "Verilog-A compact model for oxide-based resistive random access memory (RRAM)", 2014 International Conference on Simulation of Semiconductor Processes and Devices (SISPAD), Yokohama, 2014, pp. 41-44.

[13] F. M. Puglisi, and P. Pavan, "RTN analysis with FHMM as a tool for multi-trap characterization in HfOx RRAM", Proceedings of IEEE International Conference on Electron Devices and Solid-State Circuits (EDSSC) 2013, pp. 1-2, 3-5 Jun. 2013.

[14] G. Bersuker et al., "Metal oxide resistive memory switching mechanism based on conductive filament properties", Journal of Applied Physics vol. 110, no. 12, pp. 124518,124518-12, Dec. 2011.

[15] F. M. Puglisi et al, "A microscopic physical description of RTN current fluctuations in HfOx RRAM", Proceedings of the IEEE International Reliability Physics Symposium (IRPS), pp. 5B.5.1-6, 19-23 April 2015.

[16] L. Vandelli et al., "A Physical Model of the Temperature Dependence of the Current Through SiO₂/HfO₂ Stacks", IEEE Trans. on Electron Devices, vol. 58, no. 9, pp. 2878-2887, Sept. 2011.

[17] F. M. Puglisi, P. Pavan, "Guidelines for a Reliable Analysis of Random Telegraph Noise in Electronic Devices", in IEEE Trans. on Instrum. and Meas., vol. 65, no. 6, pp. 1435-1442, June 2016.

[18] T. Obara et al., "Analyzing correlation between multiple traps in RTN characteristics", 2014 IEEE International Reliability Physics Symposium, Waikoloa, HI, 2014, pp. 4A.6.1-4A.6.7.

978-1-5090-5979-9/17 $31.00 © 2017 IEEE

Ink-jet printed 2D crystal heterostructures

Francesco Bonaccorso

Istituto Italiano di Tecnologia, Graphene Labs,
Via Morego 30, 16163 Genova, Italy
francesco.bonaccorso@iit.it

Abstract— **The availability of graphene and related two-dimensional (2D) crystals in ink form, with on demand controlled lateral size and thickness, represents a boost for the design of printed heterostructures. Here, we provide an overview on the formulation of functional inks and the current development of inkjet printing process enabling the realization of 2D crystal-based heterostructures.**

Keywords—2D crystals; Ink-jet printing; Heterostructures;

I. INTRODUCTION

Heterostructures have already played a central role in technology, for the realization of, *e.g.*, semiconductor lasers, high mobility field effect transistors (FETs) and diodes. Heterostructures based on two-dimensional (2D) crystals (2DHs), *i.e.* isostructural system of 2D crystals assembled by stacking different atomic planes in sandwiched structures, offer the prospect of extending existing technologies to their ultimate limit using monolayer-thick tunnel barriers and quantum wells.[1] In fact, the on-demand assembly of 2D crystals allows the engineering of artificial three-dimensional (3D) crystals, exhibiting tailor-made properties that could be tuned to fit any application. Pioneering research has already demonstrated first proof of principle 2DHs, such as field effect vertical tunneling transistors based on graphene with atomically thin hexagonal boron nitride (*h*-BN) acting as a tunnel barrier,[2] gate-tunable p–n diodes based on a p-type BP/n-type monolayer molybdenum disulphide (MoS$_2$)[3] or proof-of-concept photovoltaic cells.[4]

Although it is not difficult to envision many possible combinations of materials, one stack of many different layers with atomic precision, the practical realization of such vision is much more complicated. The ideal approach would be to directly grow 2DHs where needed, but this target is still far from any practical realization. Currently, three methods have been exploited for the production of 2DHs: (I) layer by layer stacking via mechanical transfer;[5,6] (II) direct growth by chemical vapour deposition (CVD)[7] and molecular beam epitaxy (MBE)[8]; and (III) layer by layer deposition of solution processed 2D crystals. However, at this time all of the aforementioned approaches have limitations. The layer by layer stacking or deterministic placement (I) via mechanical transfer relies on the mechanical exfoliation of layered materials into atomically thin sheets.[9] Moreover, in order to fabricate 2DHs with clean interfaces (*i.e.*, without trapped adsorbates between the stacked layers), which is necessary for long-term device reliability, a dry transfer procedure is preferred. This would avoid the wet conditions with polymer coatings, which suffer from polymer contamination. Even the most developed dry

transfer protocols may not result in perfectly clean interfaces, as some adsorbates may get trapped between the stacked 2D layers.[10] Although this procedure is now optimized to yield sophisticated layered structures, it is limited to vertical structures,[11] it is not suitable for layer registration with the underlying films and, even more critical, it is impractical for high volume manufacturing. The direct growth (II) of different 2D crystals vertically stacked is another approach suitable for the production of 2DHs. First attempts have already shown the feasibility of such processes. Just to highlight some examples, *h*-BN has already been demonstrated to be an effective substrate for the CVD growth of graphene.[12] Vertically stacked 2DHs have been synthesized by the sequential CVD growth of 2D transition metal dicalchogenides (TMDs) on top of pre-existing *h*-BN[13] and graphene[14,15] or by the selenization and sulfurization of elemental metals.[16,17] The co-reaction of Mo and W-containing precursors with chalcogens[18] or the *in-situ* vapor-solid reactions[19] have proven to be other feasible routes for the realization of lateral and vertical 2DHs. Van der Waals epitaxy has also been exploited, using WCl$_6$/S, MoCl$_5$/S and Se as precursors and SnS$_2$ templates.[20] However, these approaches have significant limitations in that monolayer by monolayer growth process conditions have not been established yet. That is, island or 3D growth is observed rather than 2D growth in contrast to what has been observed in CVD graphene growth.[21] Any industrial application will require a scalable approach. To this aim, layer-by-layer deposition (III) from 2D crystal-based inks (Fig. 1)[22,23] could be the right strategy for scalable production of 2DHs.

Here we will present the latest progress on the large-scale placement of 2D crystal-based inks by inkjet printing,[24] which allows printing of layers of different 2DHs on a large scale. We will discuss several issues that need to be optimized, such as the uniformity of large area film stacks, the discontinuity of the individual crystals assembling the 2DHs, which are currently affecting the 2DH (opto)electronic properties.

II. INKJET PRINTING OF 2D CRYSTALS

Inkjet printing is used to print a wide range of (opto)electronic devices.[25,26,27,28,29,30] Many factors influence the printed features. In fact, during an inkjet printing process, a regular jetting from the print-head nozzle is needed to avoid printing instabilities, *i.e.*, formation of satellite drops and jetting deflection.[31] Depending on the ink wettability behaviour at the nozzle, unwanted spray formation may occur instead of a regular jetting.[32] Furthermore, the resolution of the printed feature is influenced by the drop velocity v (*i.e.* 5-10 ms^{-1}) when it impacts onto the substrate.[32]

The formulation of 2D crystals-based (as well as nanomaterials in general) printable inks is rather challenging because the various liquid properties such as density (ρ), surface tension (γ) and viscosity (η) have a strong effect on the printing process itself.[33] These ink physical properties need to be carefully tuned and can be summarized in dimensionless figures of merit (FoM) such as: the Reynolds (N_{Re}) and Weber (N_{We}) numbers,[34,35] and the inverse of the Ohnesorge number, Z ($1/N_{Oh}$), defined as the ratio between the N_{Re} and the square root of the N_{We}.[33] Different Z values for the stable drop formation have been proposed,[36,37] with Z values mostly enclosed in the range $4 \leq Z \leq 14$, although many reports have also demonstrated stable ink-jet printing with Z values of the printing ink outside this range.[38,39,40,41] In particular, nanomaterial-based inks (e.g., polystyrene nanoparticle[38] and graphene-based inks[42]) have also been ink-jet printed with Z values outside the aforementioned range. The morphological properties of the nanomaterials (e.g., lateral size for 2D crystals) dispersed in the ink as well as the formation of aggregates in the ink and their accumulation on the print-head can also contribute to printing instabilities. Dispersed nanomaterials with lateral sizes smaller than $\sim 1/50$ of the nozzle diameter (*i.e.*, $\geq 100\mu m$[43]) can reduce these damaging effects.[44,45] Moreover, wetting and adhesion[31] to the substrate and its distance to the nozzle (*e.g.*, 1-3 mm)[46] are other key requirements for the printing. The behaviour of a droplet which spreads on the substrate under the action of the inertia and surface forces is characterized by the dynamic contact angle θ_c,[47] a parameter linked with the substrate wettability.

Graphene and related 2D crystals are emerging as promising functional materials for ink formulation.[32,42] The first attempts in formulating 2D crystal-based inks for inkjet printing exploited graphene oxide (GO) or reduced graphene oxide (RGO).[48,49,50,51,52,53,54,55,56,57,58,59] Although several processes have been developed to chemically " reduce" the GO flakes in order to re-establish an electrical and thermal conductivity as close as possible to pristine graphene, RGO contains structural defects.[60] Liquid phase exfoliation (LPE) of pristine graphite[61] to obtain un-functionalized graphene flakes is a most promising approach for the formulation of graphene-based inks.[32] Most importantly, LPE allows the formulation of other 2D-crystal-based inks, see Figure 1, starting from the exfoliation of their bulk counterpart.[62]

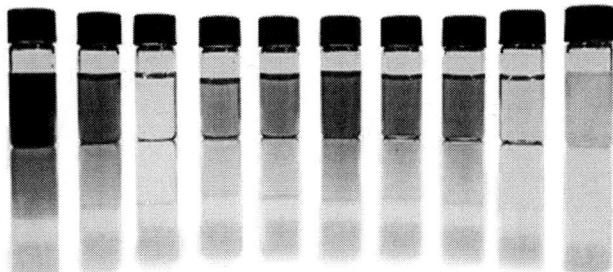

Figure 1: 2D-crystal-based functional inks.

In this case, the process is mostly driven by the choice of solvents able to disperse the flakes.[31,34] The first formulation involved the use of graphene inks prepared in N-Methyl-2-pyrrolidone (NMP)[42] and Dimethylformamide (DMF)[63] to print conductive stripes, achieving sheet resistance (R_S) values of ~ 30 k$\Omega/\square$ on glass. The formation of coffee ring effect when graphene ink is printed on rigid substrates (glass[64] and SiO$_2$[62]) can be overcome by substrate treatments, e.g., hexamethyldisilazane. Alternative routes to avoid coffee ring effects can be either the use of low boiling point solvents, with higher enthalpy of vaporization than water, or substrates that promote adhesion.[65] In the first case, the use of low boiling point solvents for the exfoliation of layered crystals has to take into account the mismatch between the γ of the solvent and the surface energy of the sheets. This issue could be overcome by the exploitation of co-solvents,[66,67] e.g., water/alcohol mixtures, to tune the fluidic properties of the liquid. This allowed for the direct inkjet printing of graphene-based conductive stripes from low boiling point solvents also on flexible substrates achieving $R_S \sim 1$-2 K$\Omega/\square$.[32,68]

The inkjet printing technology has been also recently demonstrated a promising tool to print other 2D crystals (e.g., MoS$_2$, WS$_2$) apart graphene, overcoming several still existing drawbacks for a reliable mass production of high-quality 2D crystal-based films/patterns,[69,70,71] see Figure 2. Despite these progresses, several issues need to be still overcome for the optimization of 2D crystal-based ink-jet printing. The main problem is that the common solvents used in LPE (e.g., DMF and NMP) are toxic and have very low viscosities (< 2 mPa·s), the latter strongly decreasing the jetting performance. In addition, the concentration of 2D crystals in these solvents is low (< 1 g L^{-1}), thus requiring many printing passes to obtain functional films. Another issue to be faced, especially with the use of high boiling point solvents is the required post-processing annealing for solvent removal,[72] which poses severe limitations to the type of substrate to be used for the printing process. Similar issues are also faced in the case of 2D crystals-based inks prepared in aqueous solution, where the surfactants/polymers removal requires thermal and/or chemical treatments,[73] which are often not compatible with the substrate.

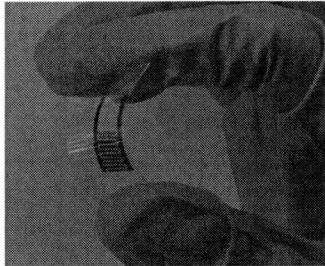

Figure 2: Inkjet-printed interdigitated graphene supercapacitor on PET (100 printing layers).

III. PRINTED HETEROSTRUCTURES

The practical realization of 2DHs, to obtain stacks of many different layers with atomic precision, is a difficult task especially in view of industrial application, which requires a scalable approach. In this context, "layer-by-layer deposition" from 2D crystal-based inks by means of Langmuir–Blodgett,[24] and inkjet printing[69] allows the deposition of layers of different 2D crystal-based heterostructures on a large scale. Although the printing approach is very recent, some proofs of concept devices have already been demonstrated.[33] In this context, the simplest structure is an all-inkjet printed in-plane photodetector based on MoS_2 channel and interdigitated graphene electrodes.[69] Another example is a tunneling transistor, where the tunneling between the top and bottom graphene layer through a TMD layer is back gate controlled.[74]

A key advantage of this approach could be represented by the possibility to integrate/complement other production approaches, for example for the realization of contacts. Very recently, a programmable logic memory device (*i.e.*, graphene/WS_2/graphene) has been realized by inkjet printing technology.[75] An all-printed, vertically stacked transistors with graphene source, drain, and gate electrodes, a TMD channel, and a BN separator has been demonstrated.[78] The proposed printed vertical heterostructure has shown a charge carrier mobility of 0.22 cm^2/Vs,[76] which is however rather low. Thus new insights into the assembly of such printed heterostructures are needed to further improve the performances of such devices.

IV. CONCLUSION

The realization of printed heterostructures based on 2D crystal-inks is now emerging as a possible route for their large scale production. However, this technology is still in its infancy and several issues have to be solved. In fact, apart from the uniformity of large area film stacks, the assembly of such heterostructures suffers from discontinuity of the individual crystals, thus resulting in structures with (opto)electronic properties of lower performance with respect to the one obtained by dry transfer methods such as layer by layer stacking *via* mechanical transfer[5] and direct growth.[7,8] Thus, before the layer-by-layer deposition of 2D crystal-based dispersions and inks can be exploited for the realization of vertical 2D heterostructures with (opto)electronic properties comparable with the ones achieved with the other approaches, a strong experimental effort is needed to fully evaluate the potentiality of this method, overcoming the aforementioned issues.

Another issue to tackle, as in the case of layer-by-layer stacking via mechanical transfer, relies on the fact that the layer-by-layer deposition of 2D crystal-based dispersions and inks can be exploited for the realization of vertical 2DHs but not for lateral ones, which is a main limitation for this approach.

ACKNOWLEDGMENT

We thank A. Ansaldo, N. Curreli, A. E. Del Rio Castillo, E. Petroni for useful discussion. This work was supported by the European Union's Horizon 2020 research and innovation program under grant agreement No. 696656—GrapheneCore1.

REFERENCES

[1] Novoselov, K. S. "Nobel lecture: Graphene: Materials in the flatland." Rev. Mod. Phys. 83, 837, 2011.

[2] Britnell, L., et al. "Field-effect tunneling transistor based on vertical graphene heterostructures." Science 335, 947-950, 2012.

[3] Deng, Yexin, et al. "Black phosphorus–monolayer MoS2 van der Waals heterojunction p–n diode." ACS Nano 8, 8292-8299, 2014.

[4] Britnell, Liam, et al. "Strong light-matter interactions in heterostructures of atomically thin films." Science 340, 1311-1314, 2013.

[5] Mayorov, Alexander S., et al. "Micrometer-scale ballistic transport in encapsulated graphene at room temperature." Nano Lett. 11, 2396-2399, 2011.

[6] Dean, Cory R., et al. "Boron nitride substrates for high-quality graphene electronics." Nat. Nanotech. 5, 722-726, 2010.

[7] Tanaka, T., et al. "Heteroepitaxial film of monolayer graphene/monolayer h-BN on Ni (111)." Surface Review and Letters 10, 721-726, 2003.

[8] Yuan, Xiang, et al. "Arrayed van der Waals vertical heterostructures based on 2D GaSe grown by molecular beam epitaxy." Nano letters 15.5 2015 pp. 3571-3577.

[9] Novoselov, K. S., et al. "Two-dimensional atomic crystals." PNAS, 102, 10451-10453, 2005.

[10] Haigh, S. J., et al. "Cross-sectional imaging of individual layers and buried interfaces of graphene-based heterostructures and superlattices." Nat. Mater. 11, 764-767, 2012.

[11] Geim, Andre K., and Irina V. Grigorieva. "Van der Waals heterostructures." Nature 499, 419-425, 2013.

[12] Liu, Zheng, et al. "Direct growth of graphene/hexagonal boron nitride stacked layers." Nano Lett. 11, 2032-2037, 2011.

[13] Wang, Shanshan, Xiaochen Wang, and Jamie H. Warner. "All chemical vapor deposition growth of MoS2: h-BN vertical van der Waals heterostructures." ACS nano 9, 5246-5254, 2015.

[14] Yu, Woo Jong, et al. "Vertically stacked multi-heterostructures of layered materials for logic transistors and complementary inverters." Nat. Mater. 12, 246-252, 2013.

[15] Lin, Yu-Chuan, et al. "Atomically thin heterostructures based on single-layer tungsten diselenide and graphene." Nano lett. 14, 6936-6941, 2014.

[16] Jung, Yeonwoong, et al. "Chemically synthesized heterostructures of two-dimensional molybdenum/tungsten-based dichalcogenides with vertically aligned layers." ACS Nano 8, 9550-9557, 2014.

[17] Yu, Jung Ho, et al. "Vertical heterostructure of two-dimensional MoS2 and WSe2 with vertically aligned layers." Nano lett. 15, 1031-1035, 2015.

[18] Li, Ming-Yang, et al. "Epitaxial growth of a monolayer WSe2-MoS2 lateral pn junction with an atomically sharp interface." Science 349, 524-528, 2015.

[19] Duan, Xidong, et al. "Lateral epitaxial growth of two-dimensional layered semiconductor heterojunctions." Nat. Nanotech. 9, 1024-1030, 2014.

[20] Zhang, Xingwang, et al. "Vertical heterostructures of layered metal chalcogenides by van der Waals epitaxy." Nano lett. 14, 3047-3054, 2014.

[21] Li, Xuesong, et al. "Large-area synthesis of high-quality and uniform graphene films on copper foils." Science 324, 1312-1314, 2009.

[22] Torrisi, Felice, et al. "Inkjet-printed graphene electronics." ACS Nano 6, 2992-3006, 2012.

[23] Capasso, A., et al. "Ink-jet printing of graphene for flexible electronics: an environmentally-friendly approach." Solid State Comm. 224, 53-63, 2015.

[24] Osada, Minoru, and Takayoshi Sasaki. "Two Dimensional Dielectric Nanosheets: Novel Nanoelectronics From Nanocrystal Building Blocks." Adv. Mater. 24.2, 210-228, 2012.

[25] Withers, Freddie, et al. "Heterostructures produced from nanosheet-based inks." Nano lett. 14, 3987-3992, 2014.

[26] Kang, Boseok, Wi Hyoung Lee, and Kilwon Cho. "Recent advances in organic transistor printing processes." ACS Appl. Mater. Interfaces 5, 2302-2315, 2013.

[27] Kwak, Donghoon, et al. "Self-Organization of Inkjet-Printed Organic Semiconductor Films Prepared in Inkjet-Etched Microwells." Adv. Funct. Mater. 23, 5224-5231, 2013.

[28] Gorter, H., et al. "Toward inkjet printing of small molecule organic light emitting diodes." Thin Solid Films 532, 11-15, 2013.

[29] Eggenhuisen, T. M., et al. "Large area inkjet printing for organic photovoltaics and organic light emitting diodes using non-halogenated ink formulations." J. Imaging Sci. Technol. 58, 40402-1, 2014.

[30] Krebs, Frederik C. "Fabrication and processing of polymer solar cells: a review of printing and coating techniques." Sol. Energy Mater. Sol. Cells 93, 394-412, 2009.

[31] Eom, Seung Hun, et al. "High efficiency polymer solar cells via sequential inkjet-printing of PEDOT: PSS and P3HT: PCBM inks with additives." Org. Electron. 11, 1516-1522, 2010.

[32] Magdassi, Shlomo. "Ink requirements and formulations guidelines." The chemistry of inkjet inks. New Jersey-London-Singapore: World Scientific, 19-41, 2010.

[33] Bonaccorso, Francesco, et al. "2D☐Crystal☐Based Functional Inks." Adv. Mater. 28, 6136-6166, 2016.

[34] Singh, Madhusudan, et al. "Inkjet printing—process and its applications." Adv. Mater. 22, 673-685, 2010.

[35] Batchelor, George Keith. An introduction to fluid dynamics. Cambridge university press, 2000.

[36] Bergeron, Vance, et al. "Controlling droplet deposition with polymer additives." Nature 405, 772-775, 2000.

[37] Fromm, J. E. "Numerical calculation of the fluid dynamics of drop-on-demand jets." IBM J. Res. Dev. 28, 322-333, 1984.

[38] Jang, Daehwan, Dongjo Kim, and Jooho Moon. "Influence of fluid physical properties on ink-jet printability." Langmuir 25, 2629-2635, 2009.

[39] Shin, Pyungho, Jaeyong Sung, and Myeong Ho Lee. "Control of droplet formation for low viscosity fluid by double waveforms applied to a piezoelectric inkjet nozzle." Microelectron. Reliab. 51, 797-804, 2011.

[40] de Gans, Berend-Jan, et al. "Ink-jet Printing Polymers and Polymer Libraries Using Micropipettes." Macromol. Rapid. Comm. 25, 292-296, 2004.

[41] Hsiao, Wen-Kai, et al. "Ink jet printing for direct mask deposition in printed circuit board fabrication." J. Imaging Sci. Technol. 53, 50304-1, 2009.

[42] Jung, Sungjune, and Ian M. Hutchings. "The impact and spreading of a small liquid drop on a non-porous substrate over an extended time scale." Soft Matter 8, 2686-2696, 2012.

[43] Buzio, Renato, et al. "Ultralow friction of ink-jet printed graphene flakes." Nanoscale 2017.

[44] Kim, Dongjo, et al. "Ink-jet printing of silver conductive tracks on flexible substrates." Mol. Cryst. Liq. Cryst. 459, 45-325, 2006.

[45] www.microfab.com/technotes/technote 99-02

[46] Van Osch, Thijs HJ, et al. "Inkjet printing of narrow conductive tracks on untreated polymeric substrates." Adv. Mater. 20, 343-345, 2008.

[47] Saunders, Rachel Elizabeth, and Brian Derby. "Inkjet printing biomaterials for tissue engineering: bioprinting." Int. Mater. Rev. 59, 430-448, 2014.

[48] De Gennes, Pierre-Gilles. "Wetting: statics and dynamics." Rev. Mod. Phys. 57, 827, 1985.

[49] Huang, Lu, et al. "Graphene-based conducting inks for direct inkjet printing of flexible conductive patterns and their applications in electric circuits and chemical sensors." Nano Res. 4, 675-684, 2011.

[50] Jo, Yong Min, et al. "Submillimeter-scale graphene patterning through ink-jet printing of graphene oxide ink." Chem. Lett. 40, 54-55, 2010.

[51] Le, Linh T., et al. "Graphene supercapacitor electrodes fabricated by inkjet printing and thermal reduction of graphene oxide." Electrochem. Commun. 13, 355-358, 2011.

[52] Shin, Keun-Young, Jin-Yong Hong, and Jyongsik Jang. "Micropatterning of Graphene Sheets by Inkjet Printing and Its Wideband Dipole-Antenna Application." Adv. Mater. 23, 2113-2118, 2011.

[53] Shin, Keun-Young, Jin-Yong Hong, and Jyongsik Jang. "Flexible and transparent graphene films as acoustic actuator electrodes using inkjet printing." Chem. Commun. 47, 8527-8529, 2011.

[54] Kong, De, et al. "Temperature-dependent electrical properties of graphene inkjet-printed on flexible materials." Langmuir 28, 13467-13472, 2012.

[55] Tölle, Folke Johannes, Martin Fabritius, and Rolf Mülhaupt. "Emulsifier☐Free Graphene Dispersions with High Graphene Content for Printed Electronics and Freestanding Graphene Films." Adv. Funct. Mater. 22, 1136-1144, 2012.

[56] Zhang, Hui, et al. "Layer-by-layer inkjet printing of fabricating reduced graphene-polyoxometalate composite film for chemical sensors." Phys. Chem. Chem. Phys. 14, 12757-12763, 2012.

[57] Su, Yang, et al. "Reduced graphene oxide with a highly restored π-conjugated structure for inkjet printing and its use in all-carbon transistors." Nano Res. 6, 842-852, 2013.

[58] Y. Su, S. Jia, J. Du, J. Yuan, C. Liu, W. Ren, H. Cheng, Nano Res. 2015, 1-9

[59] Yoon, Yeoheung, et al. "Highly Stretchable and Conductive Silver Nanoparticle Embedded Graphene Flake Electrode Prepared by In situ Dual Reduction Reaction." Sci. Rep. 5, 14177, 2015.

[60] Zhang, Weijun, et al. "Synthesis of Ag/RGO composite as effective conductive ink filler for flexible inkjet printing electronics." Colloid Surface A 490, 232-240, 2016.

[61] Bonaccorso, Francesco, et al. "Production and processing of graphene and 2d crystals." Mater. Today 15, 564-589, 2012.

[62] Hernandez, Yenny, et al. "High-yield production of graphene by liquid-phase exfoliation of graphite." Nat. Nanotech. 3, 563-568, 2008.

[63] Coleman, Jonathan N., et al. "Two-dimensional nanosheets produced by liquid exfoliation of layered materials." Science 331, 568-571, 2011.

[64] Li, Jiantong, et al. "Efficient inkjet printing of graphene." Adv. Mater. 25, 3985-3992, 2013.

[65] Secor, Ethan B., et al. "Inkjet printing of high conductivity, flexible graphene patterns." J. Phys. Chem. Lett. 4, 1347-1351, 2013.

[66] Osthoff, Robert C., and Simon W. Kantor. "Organosilazane compounds." Inorg. Syn. Volume 5, 55-64, 1957.

[67] Bonaccorso, Francesco, and Zhipei Sun. "Solution processing of graphene, topological insulators and other 2d crystals for ultrafast photonics." Opt. Mater. Express 4, 63-78, 2014.

[68] Arapov, Kirill, Robert Abbel, and Heiner Friedrich. "Inkjet printing of graphene." Faraday discuss. 173, 323-336, 2014.

[69] Finn, David J., et al. "Inkjet deposition of liquid-exfoliated graphene and MoS 2 nanosheets for printed device applications." J. Mater. Chem. C 2, 925-932, 2014.

[70] Li, Jiantong, et al. "Inkjet Printing of MoS2." Adv. Funct. Mater. 24, 6524-6531, 2014.

[71] Li, Jiantong, et al. "A simple route towards high-concentration surfactant-free graphene dispersions." Carbon 50, 3113-3116, 2012.

[72] Li, Jiantong, Max C. Lemme, and Mikael Östling. "Inkjet printing of 2D layered materials." Chem. Phys. Chem. 15, 3427-3434, 2014.

[73] Kim, Yong-Hoon, et al. "Controlled Deposition of a High☐Performance Small-Molecule Organic Single-Crystal Transistor Array by Direct Ink-Jet Printing." Adv. Mater. 24, 497-502, 2012.

[74] Georgiou, Thanasis, et al. "Vertical field-effect transistor based on graphene-WS2 heterostructures for flexible and transparent electronics." Nat. Nanotech. 8, 100-103, 2013.

[75] McManus, Daryl, et al. "Water-based and biocompatible 2D crystal inks for all-inkjet-printed heterostructures." Nat. Nanotech. 12, 343-350, 2017.

[76] Kelly, Adam G., et al. "All-printed thin-film transistors from networks of liquid-exfoliated nanosheets." Science 356, 69-73, 2017.

WS₂ transistors on 300 mm wafers with BEOL compatibility

T. Schram, Q. Smets, B. Groven , M.H. Heyne, E. Kunnen, A. Thiam, K. Devriendt, A. Delabie, D. Lin, M. Lux, D. Chiappe, I. Asselberghs, S. Brus, C. Huyghebaert, S. Sayan, A. Juncker[*], M. Caymax, I.P. Radu

IMEC, kapeldreef 75, B-3001 Leuven, Belgium,*COVENTOR, 3, Avenue du Quebec , 91140 Villebon sur Yvette, France

tom.schram@imec.be

Abstract— **For the first time, WS₂-based transistors have been successfully integrated in a 300 mm pilot line using production tools. The 2D material was deposited using either area selective chemical vapor deposition (CVD) or atomic layer deposition (ALD). No material transfer was required. The major integration challenges are the limited adhesion and the fragility of the few-monolayer 2D material. These issues are avoided by using a sacrificial Al₂O₃ capping layer and by encapsulating the edges of the 2D material during wet processing. The WS₂ channel is contacted with Ti/TiN side contacts and an industry-standard back end of line (BEOL) flow. This novel low-temperature flow is promising for integration of back-gated 2D transistors in the BEOL.**

Keywords— WS₂, TMD, 2D , transistor, BEOL, 300 mm

I. INTRODUCTION

2D materials like graphene and transition-metal dichalcogenides (TMD) are being considered as channel materials for future technology nodes [1]. 2D materials promise improved short channel behavior compared to Si, Ge or III/V, due to the intrinsic channel thickness scaling and the absence of surface roughness. The introduction of 2D materials also allows to include transistor functionality in the BEOL (fig.1). The planar nature of 2D materials facilitates the BEOL integration over the front end integration where 3D structures are frequently used.

In this work, we demonstrate back-gated devices that can be used in a configuration where metal interconnect lines are used for gating the 2D materials (fig.1). The back gate configuration also allows gating of the contact regions, since this eliminates the need to dope the 2D materials chemically in the extension region. Only limited chemical doping methods for the 2D materials exist, and most of them are not compatible with a conventional BEOL process [2]. The integration, including the WS₂ deposition, was entirely performed on full-scale 300 mm wafers in the imec pilot line.

II. INTEGRATION

In the proposed BEOL compatible process flow (fig.2), WS₂ is successfully integrated in 300 mm pilot line back-gated transistors. The WS₂ layers are deposited across 300 mm wafers by either plasma-enhanced atomic layer deposition (PE-ALD) or area-selective chemical vapor deposition (CVD). The growths are performed at BEOL compatible deposition temperatures of 300°C and 450°C, respectively. The layers develop a strong, nanocrystalline texture along (0002) despite the low deposition temperatures [3,4,5]. Such low temperature deposition processes avoid the material transfer to target substrate, and they are compatible with temperature sensitive structures, e.g., for the heterogeneous integration with existing Si nanotechnology. The front end integration was not explicitly implemented, but emulated by using a thermally oxidized blanket Si wafer, to allow the evaluation of optional WS₂ recrystallization anneals. The gate bias is applied through the backside of the wafer.

The process flow differs for the approach used for active area formation. The right branch shows the CVD WS₂ route, where a 3 nm Si sacrificial layer is pre-patterned and subsequently selectively converted to WS₂ [3]. In the left branch, WS₂ is deposited by plasma enhanced ALD at a BEOL compatible

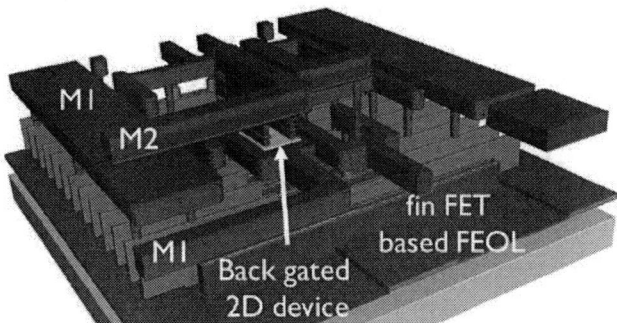

Fig. 1. COVENTOR model of a possible co-integration of fin-FET FEOL device with a planar BEOL 2D transistor (shown with transparent oxide). The 2D transistor uses the M1 as a back gate and M2 for the source and drain connections.

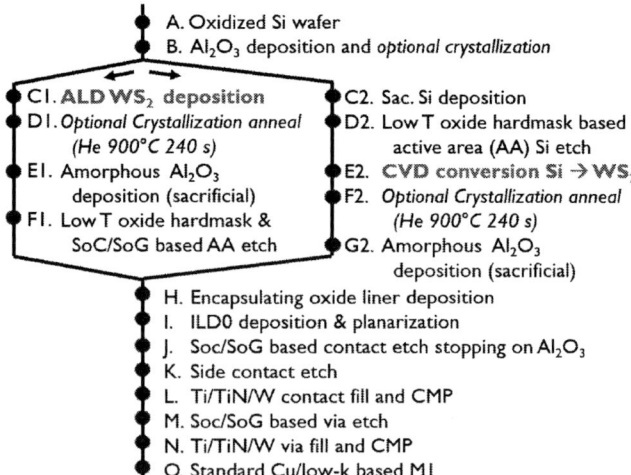

Fig. 2. Schematic process flow used to fabricate the back gated 2D transistors with Ti/TiN side contacts.

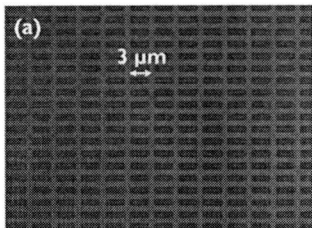 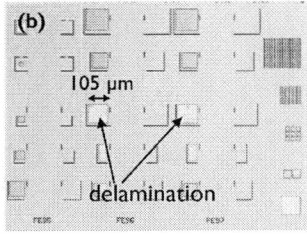

Fig. 3. Optical inspections after selective wet Al_2O_3 removal for active areas with (a) small dimensions (3 µm) and (b) large dimensions (50-100 µm). An enhaced delamination is observed for larger areas.

temperature of 300 °C on the full 300 mm wafer [4,5] and subsequently patterned. Both processes are performed in a hot-wall, showerhead-type ASM PECVD reactor with direct (RF) plasma capability, using tungsten hexafluoride (WF_6), and dihydrogen sulfide (10 % H_2S in He, with 99.9 % pure H_2S) precursors. A H_2 plasma (100 W, 2 Torr) is used as the reducing agent in a ternary PEALD reaction cycle, whereas a sacrificial Si layer is used to reduce the W in WF_6 in the area-selective chemical vapor deposition approach. In an attempt to improve the crystalline quality and grain size of the WS_2, an optional anneal is performed for both approaches [3].

The critical challenge during the 2D material integration is the adhesion loss between the Van der Waals bonded 2D material and the surrounding carrier substrate and capping dielectrics. This is particularly crucial during wet process steps, including immersion lithography and wet post etch strips. The bad adhesion can be explained by the liquid intercalation between the 2D material and the surrounding materials [6]. The intercalation mechanism is often used for the transfer of 2D material flakes. We observe increased delamination for larger areas of WS_2 (fig.3). Therefore, the WS_2 is patterned as soon as possible after deposition and the liquid access to the edges of the 2D material is blocked by adding an oxide cap whenever wet processing is used (figs.4 & 5).

During the contact etch, the delicate 2D material is protected from dry etch induced plasma or sputter damage by using a sacrificial Al_2O_3 etch stop layer. Two approaches were developed to remove the sacrificial Al_2O_3 etch stop layer. In a first approach, the Al_2O_3 layer in the contact region is removed by a selective wet etch while preserving the underlying WS_2 (fig.6a). Indeed, Raman in fig.6b confirms that the full width at half-maximum (FWHM) of the longitudinal acoustic mode (2LA) remains constant after the selective wet etch, compared to the as grown WS_2. This 2LA mode appears as the most prominent vibrational mode for WS_2 at the laser excitation wavelength of 532 nm, and originates from the collective, in-plane vibration of both W and S atoms in the periodic WS_2 lattice [7]. The FWHM of 2LA relates to the disorder in the in-plane atomic lattice of the WS_2 layer. SEM and FIB inspection images during the key critical process steps in the used process flow. Trench contacts are used on the 2D device due to the absence of junctions.

In a different approach, the Al_2O_3 and the WS_2 are dry etched (fig.4K). A conventional Ti/TiN liner and W contact fill is used in the trenches, providing a side contact to the 2D material (fig.4L). This side contact approach is used in this work. The subsequent process steps (fig.2 M to O) are also industry-standard BEOL processes.

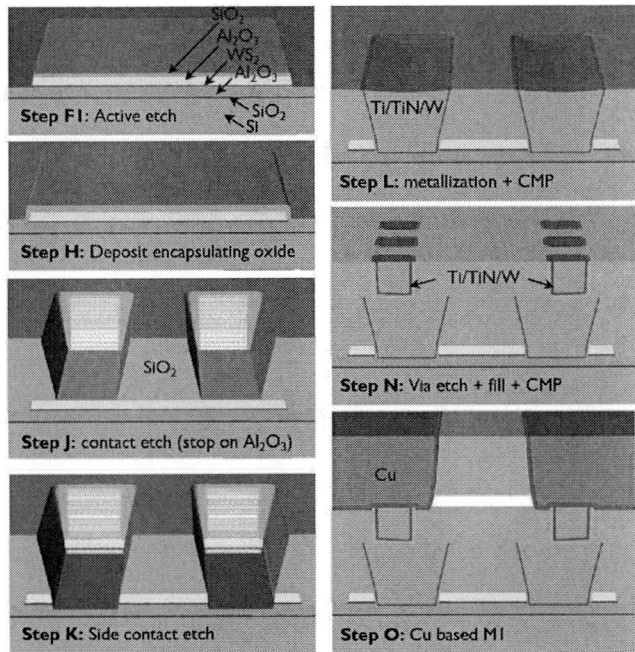

Fig. 4. COVENTOR simulation of the critical steps in the process flow for ALD WS2. (F1) Active etch with sacrificial Al_2O_3 hard mask followed by oxide edge encapsulation. (J) Trench contact etch stopping on the sacrificial Al_2O_3 to protect the 2D material from damage. (K) Side contact etch. (L) Ti/TiN/W contact fill. (N) Via etch fill and polish. (O) Cu based M1 module.

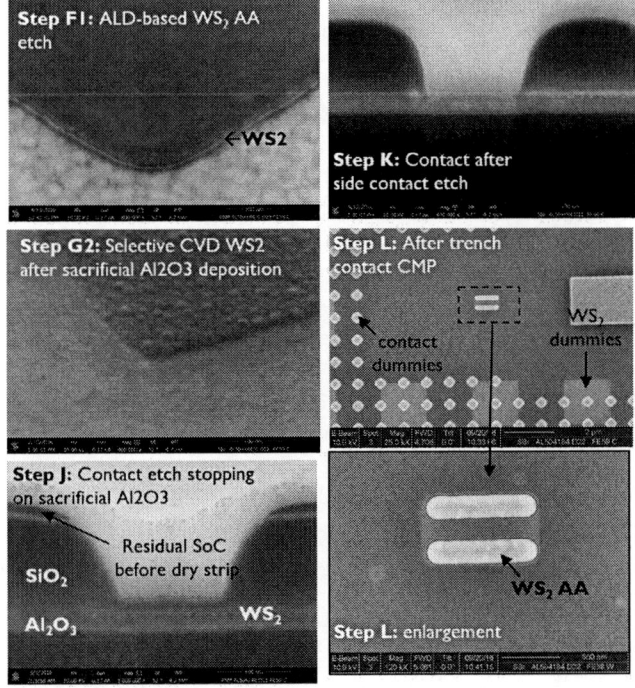

Fig. 5. SEM and FIB inspection images during the key critical process steps in the used process flow. Trench contacts are used on the 2D device due to the absence of junctions.

Using these proposed, sequential process steps, the PEALD WS_2 layers are successfully contacted at the basal plane of the channel by using a side contact integration scheme and using the Si substrate as a back-gate (TEM, fig.7). The direct contact

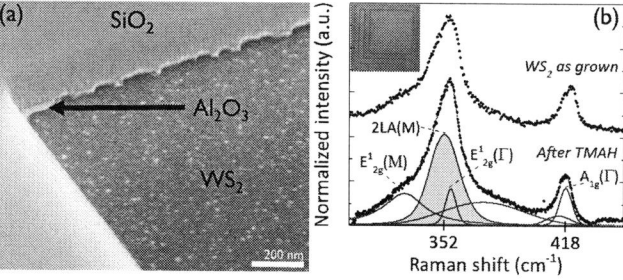

Fig. 6. (a) SEM picture of the WS_2 contact area after selective wet Al_2O_3 removal. (b) Raman spectra of 4 ML WS_2 on 30 nm ALD Al_2O_3 directly after deposition (top) and after selective wet etch in the contact area (bottom), with the measurement location at the contact area as inset. The FWHM of the as grown WS_2 and after selective wet etch remain constant, at 20.5 ± 0.8 cm^{-1} and 20.0 ± 0.6 cm^{-1}, respectively. This confirms that the WS_2 is preserved under the contact region during the selective Al_2O_3 etch stop layer removal. The individual Raman spectra have a vertical off-set for clarity. They are corrected for the background to the Si reference, and normalized to the out-of-plane optical mode A_{1g} for direct comparison.

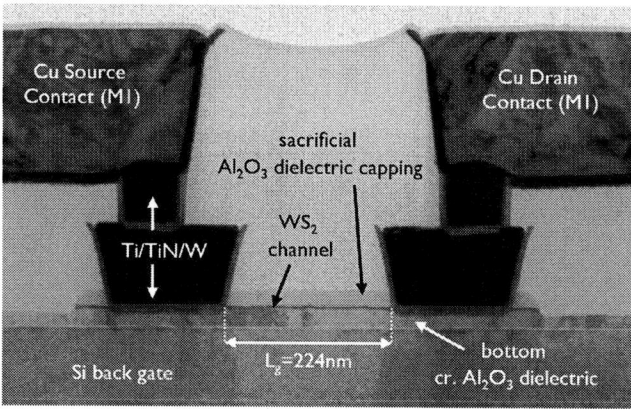

Fig. 10. TEM cross section of the back-gated diffusion-free 2D device with Ti/TiN side contacts. The protective sacrificial top Al_2O_3 layer is still present on top of the active area (AA) and the contact area is deliberately located inside the AA to avoid exposure of the 2D material edges to wet processing.

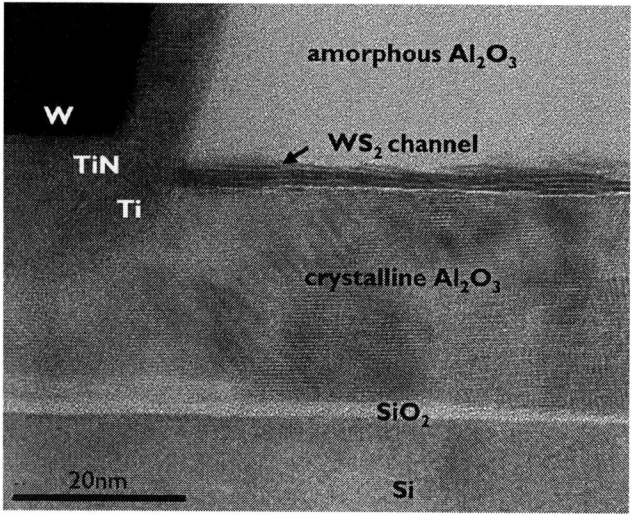

Fig. 9. TEM of the side contact area. The presence of the deposited ALD WS_2 layer is confirmed, as well as the good quality of the Ti/WS_2 side contact.

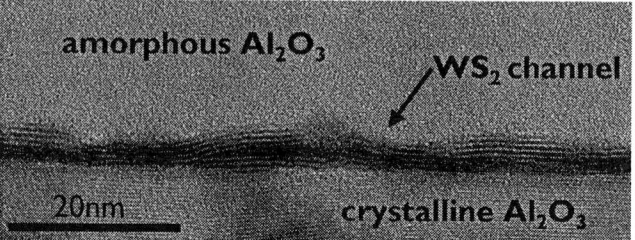

Fig. 8. Detailed TEM of the channel region, confirming the polycrystalline nature of the ALD WS_2 layer of 4-5 monolayers thin.

region between the Ti side contact and the WS_2 channel edge is smooth, and the 2D structure of the WS_2 layers is preserved (TEM, fig.8). The WS_2 channel consists of approximately 4-5 monolayers and is nanocrystalline and strongly textured along the (0002) plane. Moreover, the sequential integration steps have limited impact on the 2D structure and layer thickness of the WS_2, as the number of layers is preserved compared to the as-grown material [5,6].

III. DEVICE ELECTRICAL RESULTS

The I_d/V_g curves for the CVD WS_2 devices (fig.10) show a small ambipolar response with weak gate modulation, attributed to the polycrystalline nature of the WS_2. The PEALD WS_2 devices have a much stronger PMOS modulation (fig.11). Preferential PMOS versus NMOS modulation in 2D devices has also been observed in other work [2,8]. ALD WS_2 devices have I_{max}/I_{min}=8000, but the measurement for higher on-state current at more negative bias is prevented by gate leakage. Reduced leakage is expected with a higher quality oxide.

The stronger modulation for the PEALD WS_2 devices can be explained by the material properties and its growth mechanisms, which depend on the type of precursors, the deposition temperature and the deposition technique. Generally, high temperature CVD processes from metalorganic and -halide precursors (550 °C–900 °C) provide few layered MX_2 films with state-of-the-art semiconductor properties approaching the electronic properties reported for exfoliated flakes of the pristine 2D material [9-10]. However, for the WF_6 and H_2S precursor system, the PEALD WS_2 channel modulates stronger when applying a back-gate bias compared to the CVD grown WS_2. Note that the CVD WS_2 is grown from WF_6 and H_2S through a surface mediated reaction with a sacrificial Si layer. Such reaction maximizes the nucleation density due to a strong,

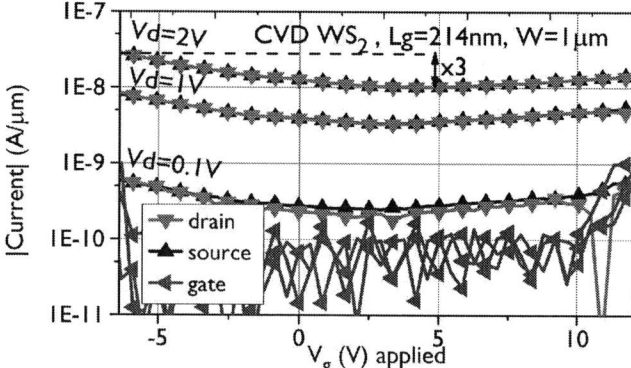

Fig. 7. Transfer characteristics of a CVD WS_2 device. The source, drain and gate contacts are shown in fig.7. A high applied V_g is required due to EOT=15nm.

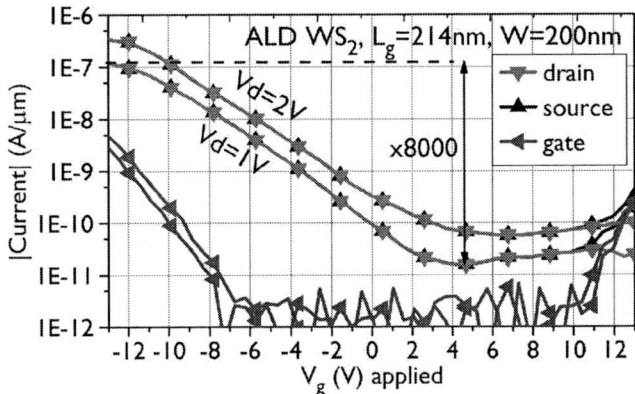

Fig. 13. Transfer characteristics of a ALD WS_2 device have strongly increased PMOS modulation with Imax/Imin=8000.

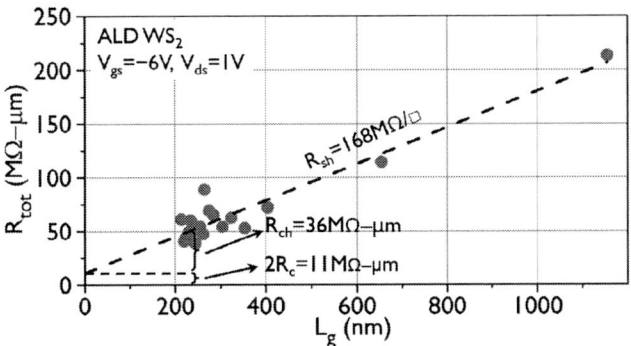

Fig. 12. The TLM method shows Rch is the dominant contribution over Rc, even for devices with smallest Lg=214nm. Rc is extracted at Vgs=−6V for highest accuracy.

preferential reaction between WF_6 and Si layer, and therefore limits the crystal grain size and orientation, as well as the modulation of the channel. On the other hand, it opens opportunities for area-selective deposition as no WS_2 is grown without a Si layer.

Devices with L_g=214nm have a high subthreshold swing SS=4V/dec at room temperature, caused by a high EOT=15nm and high trap density. Assuming dominant oxide-2D interface trapping, we calculate a density D_{it}=8×10^{13}cm^{-2}eV^{-1}. Extrapolating to EOT=1nm, we expect an SS improvement to 300mV/dec. If trapping can be eliminated entirely, we expect SS to improve to 60mV/dec due to the ultra-thin channel.

Dominant transistor channel response is confirmed using an L_g-array (fig.12), from which we extract a side contact resistance of 6 MΩμm and a channel resistance of 36 MΩμm at V_{gs}=-6V. A lower channel resistance and hence higher I_{on} is expected by carefully controlling the crystallinity of the WS_2 layers through fundamental understanding of the growth mechanisms during the PEALD process.

The Si back gate control is confirmed by comparing devices with either low Si doping or degenerate n+ Si doping in the back gate (fig.13). The high back gate doping causes an effective work function decrease resulting in a negative V_g shift.

IV. CONCLUSIONS

We demonstrated for the first time a 2D material integration flow on 300mm wafers, designed for BEOL compatibility, using a few-monolayer WS_2 channel and a Al_2O_3 back gate dielectric.

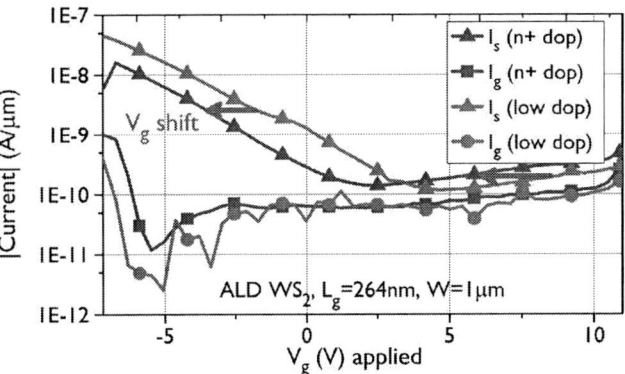

Fig. 11. Adding n+ doping to the Silicon back gate contributes to an effective work function change and negative Vg shift of the input characteristics.

The fragile 2D material was encapsulated under a protective oxide layer to avoid delamination and crystal damage during processing. Further performance improvements are expected by improving the crystallinity of the 2D material and the use of a better back gate dielectric with lower EOT and reduced gate leakage.

Acknowledgments: This work was done under the imec IIAP core CMOS programs. The authors acknowledge support by the EC under the Graphene flagship (contract no. CNECT-ICT-604391).

[1] Sheneve Z. Butler et al. "Progress, Challenges, and Opportunities in Two-Dimensional Materials Beyond Graphene", *ACS Nano* vol. 7 (4), pp. 2898-2926, 2013, DOI: 10.1021/nn400280c

[2] S. J. McDonnell and R. M. Wallace, "Atomically-Thin Layered Films for Device Applications based upon 2D TMDC Materials," *Thin Solid Films*, vol. 616, pp. 482–501, 2016, DOI: 10.1016/j.tsf.2016.08.068.

[3] M. H. Heyne, J.-F. de Marneffe, A. Delabie, M. Caymax, E. C. Neyts, I. Radu, C. Huyghebaert, and S. De Gendt, "Two-dimensional WS 2 nanoribbon deposition by conversion of pre-patterned amorphous silicon," *Nanotechnology*, vol. 28, no. 4, p. 04LT01, 2017, DOI: 10.1088/1361-6528/aa510c

[4] A. Delabie, M. Caymax, B. Groven, M. Heyne, K. Haesevoets, J. Meersschaut, et al. "Low temperature deposition of 2D WS 2 layers from WF 6 and H 2 S precursors: impact of reducing agents," *Chem. Commun.*, vol. 51, no. 86, pp. 15692–15695, 2015. DOI: 10.1039/C5CC05272F

[5] B. Groven, M. Heyne, A. N. Mehta, H. Bender, T. Nuytten, J. Meersschaut, T. Conard, P. Verdonck, S. Van Elshocht, W. Vandervorst, S. De Gendt, M. Heyns, I. Radu, M. Caymax, A. Delabie. "Plasma-Enhanced Atomic Layer Deposition of Two-Dimensional WS_2 from WF_6, H_2 Plasma, and H_2S", *Chem. Mater.*, vol. 29, no. 7, pp. 2927–2938, 2017, DOI: 10.1021/acs.chemmater.6b05214

[6] K. T. He, J. D. Wood, G. P. Doidge, E. Pop, and J. W. Lyding, "Scanning Tunneling Microscopy Study and Nanomanipulation of Graphene-Coated Water on Mica," *Nano Lett.*, vol. 12, no. 6, pp. 2665–2672, 2012, DOI: 10.1021/nl202613t

[7] A. Berkdemir; H. R. Gutiérrez; A. R. Botello-Méndez; N. Perea-López; A. L. Elias; C.-I.Chia; B. Wang; V. H. Crespi; F. López-Urias; J.-C. Charlier; H. Terrones; M. Terrones, "Identification of Individual and Few Layers of WS_2 Using Raman Spectroscopy". *Sci. Rep.*, vol. 3, 1755, 2013, DOI: 10.1038/srep01755

[8] W. Bao, X. Cai, D. Kim, K. Sridhara, and M. S. Fuhrer, "High mobility ambipolar MoS 2 field-effect transistors: Substrate and dielectric effects," *Appl. Phys. Lett.*, vol. 102, no. 4, p. 42104, 2013, DOI: 10.1063/1.4789365

[9] Y. Yu; C. Li; Y. Liu; L. Su; Y. Zhang; L. Cao, "Controlled Scalable Synthesis of Uniform, High-Quality Monolayer and Few-Layer MoS_2 Films". *Sci. Rep.*, vol. 3, 1866, 2013

K. Kang; S. Xie; L. Huang; Y. Han; P. Y. Huang; K. F. Mak; C.-J. Kim; D. Muller; J. Park, "High-Mobility Three-Atom-Thick Semiconducting Films with Wafer-Scale Homogeneity". *Nature*, vol. 520, no. 7549, pp. 656–660, 2015.

200 mm Wafer Level Graphene Transfer by Wafer Bonding Technique

Mesut Inac*[†], Grzegorz Lupina[†], Matthias Wietstruck[†], Marco Lisker[†], Mirko Fraschke[†], Andreas Mai[†], Fabio Coccetti[‡], and Mehmet Kaynak[†§]

*Berlin Technical University, HFT4, Einsteinufer 25, 10587 Berlin, Germany
Email: inac@tu-berlin.de
[†]IHP, Im Technologiepark 25, 15236 Frankfurt (Oder), Germany
[‡]RF Microtech, Via Leone Maccheroni, 06132 Perugia, Italy
[§]Sabanci University, Orta Mahalle, Universite Caddesi 27, Tuzla, 34956 Istanbul, Turkey

Abstract—In this paper, wafer level transfer of graphene on to a dielectric substrate is demonstrated based on SiO_2-SiO_2 fusion bonding and de-bonding processes. The developed technique allows to transfer graphene on 200 mm wafer without any contamination; thus CMOS compatible. The experimental data verifies the successful transfer of the graphene on to another substrate with high quality and a yield value of 98% with average 3.5 $k\Omega/\square$ sheet resistance. To the best of authors knowledge, it is the first time demonstration of the graphene transfer based on SiO_2-SiO_2 fusion bonding and de-bonding process on 200 mm wafer level which would allow a complete integration of graphene material into a CMOS line and opens the way for new devices based on graphene material.

Index Terms—Graphene; Wafer bonding; graphene transfer; CMOS; permanent bonding; wafer de-bonding

I. INTRODUCTION

Graphene is one of the most potent material for the future microelectronic device industry due to its superior electrical and mechanical properties. Since new devices on More-than-Moore concept introduced into literature, graphene is shown as the future material for semiconductor industry [1]. While the technology pushes the devices going to higher frequencies, high frequency devices with graphene have also been introduced [2], [3], as high performance solutions [4]. Due to the fact that graphene is mostly grown on the conductive surfaces, it has to be transferred on another surface to functionalize.

In the literature, there are several methods used to transfer graphene on another substrates [5]. The mostly followed method is by using polymer intermediate layers to transfer graphene. Transfer of graphene on 3- [6] and 4-inch [7] wafers has been demonstrated, but mostly they have small grain sizes [8], small structures [9] or mechanical instability issues [10]. In order to have mass production of devices, graphene has to be transferred without any of these aforementioned issues on larger wafer sizes (i.e. 200 or 300 mm).

Additionally, due to the polymers that are used in the transfer process, surface is already contaminated before the device fabrication. Even though there are optimized methods for cleaning the polymer residues [11], the surface of the wafer is not clean for further fabrication steps. In this perspective, it is required to have cleaner and more practical ways to trans-

fer graphene on to functional surfaces with modern CMOS compatible methods.

In this paper, it is aimed to show CMOS compatible method for graphene transfer on to 200 mm wafer. Utilizing plasma assisted permanent oxide-oxide bonding, donor and target wafers bonded and these bonded wafer de-bonded from an interface which lets graphene layer stay on the target wafer. This method enables to have clean and CMOS compatible graphene transfer on 200 mm wafers. It has been demonstrated that over 98% of high quality graphene successfully transferred onto target wafer. To the best of authors knowledge, this is the first time demonstration of graphene on 200 mm wafer level using wafer bonding / de-bonding technique.

II. GRAPHENE TRANSFER

The transfer of the graphene has been completed in two stages. The first stage is fusion bonding 200 mm donor wafer, which have graphene and intermediate layers on, and 200 mm target wafer which have thermal oxide on the surface, together. The second stage is de-bonding previously bonded wafer stack. After de-bonding layers delaminated from donor wafer and these delaminated layers are transferred on to target wafer. This transfer procedure can be seen in Fig. 1.

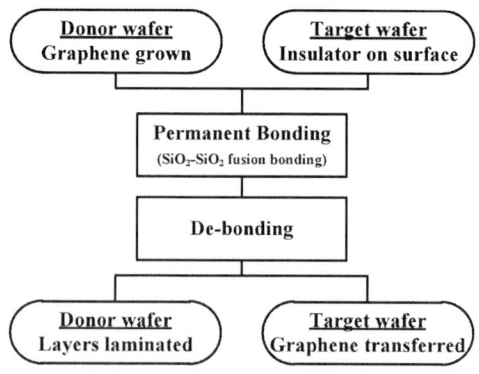

Fig. 1: Flowchart of the transfer process

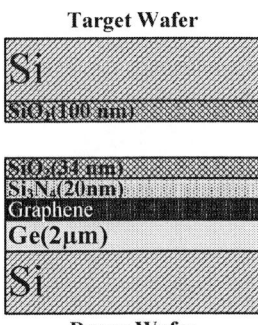

Fig. 2: Graphene on Germanium: Donor wafer and target wafer

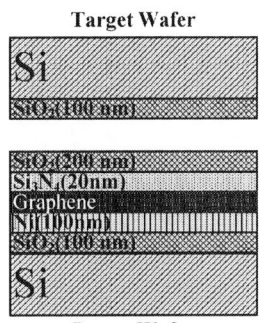

Fig. 3: Graphene on Nickel: Donor wafer and target wafer

Two different donor wafers are prepared for the transfer. On the first donor wafer, graphene is grown on Germanium (Ge) with the method presented in [12]. On the second donor wafer, graphene is grown on top of Nickel (Ni). On both donor wafers, dielectric layers are grown using chemical wafer deposition (CVD) method [13]. The aim of these intermediate layers is bonding donor and target wafer and adhering graphene layer while de-bonding. Therefore, optimized deposition of dielectric layers is required. The target wafer have only 100 nm thermal grown silicon-dioxide layer on top to have high yield and quality SiO_2-SiO_2 fusion bonding.

For the next sub-sections, transfer of the graphene which is on Ge and Ni layer, is detailed. Moreover, analyses of the layers on both donor wafer and target wafer are given.

A. Graphene on Germanium

On the donor wafer, single layer graphene is grown on 2 μm thick, epitaxial-grown Ge layer. On top of graphene Si_3N_4 and SiO_2 layers are deposited 20 nm and 34 nm respectively. On the target wafer, 100 nm thermal oxide is grown. The layers of the donor and target wafers can be seen in Fig. 2.

The donor and the target wafers are bonded after plasma activation, in vacuum chamber under 1.5 kN force and annealed for 1 h at 300 °C. After permanent bonding of silicon-dioxide layers, two wafers are de-bonded. Due to the different interaction energies between the layers on top of Si wafers, the de-bonding occurs at the weakest point of the layer stack in between.

B. Graphene on Nickel

Multilayer graphene is grown on Ni surface which is 100 nm thick laying on 100 nm thermal oxide. On top of graphene layer, there are 20 nm Si_3N_4 and 200 nm SiO_2 layers as intermediate layers. Target wafer is the same as graphene on Ge case: 100 nm thermal silicon-dioxide on top of 200 mm size Si wafer. The layers on the donor and the target wafers that are used in this experiment can be seen in Fig. 3.

Donor wafer and target wafer are bonded together with plasma assisted fusion bonding at the SiO_2-SiO_2 interface and later de-bonded. These wafers are plasma activated and then bonded in vacuum chamber under 1.5 kN force with an

annealing at 300 °C for 1 h. Similarly using the same principle as in the graphene on Ge case, the wafers are de-bonded from the weakest interface between the layers.

III. RESULTS

Wafer stacks are inspected after de-bonding stage. First, visual inspections are performed to get an initial idea about the transferred layers on the de-bonded stack. Then, further measurements are acquired from the de-bonded wafers. In order to better understand the de-bonding interfaces, profile and Raman spectroscopy measurements are performed.

A. Graphene on Germanium

De-bonded wafers are inspected in order to see if there is any transfer in between the wafers. In Fig. 4, the layer transfer can be easily seen. On the target wafer (bottom), light areas are transferred from the donor wafer. Originally 100 nm thermal oxide is grown on Si surface which have dark-blue color. After the transfer of layers, the color becomes metallic and lighter than the original surface. Nearly 60% of the target wafer is lighter, which means the successful transfer yield of 60%.

In order to understand which layers are transferred to the target wafer, profile measurements are performed. The profile

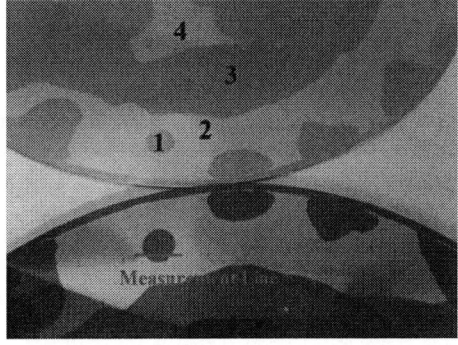

Fig. 4: Graphene on Ge, de-bonded wafer stack. Bottom wafer is target wafer, top wafer is donor wafer. Transferred layers are visible. Profile measurement is taken along the line on target wafer. Raman measurements are taken from sites 1, 2, 3 and 4.

978-1-5090-5979-9/17 $31.00 © 2017 IEEE   217

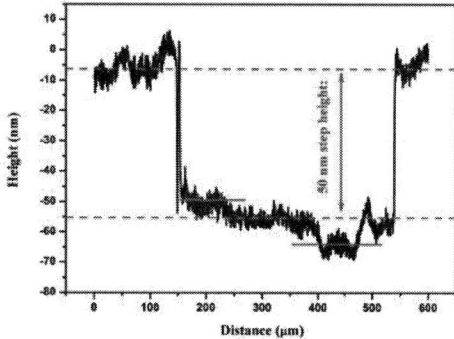

Fig. 5: Graphene on Ge: Profile of de-bonded target wafer along the line in Fig. 4.

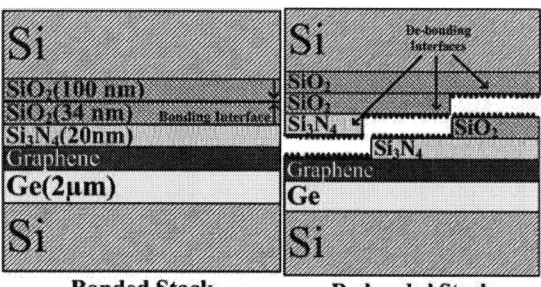

Fig. 6: Bonded graphene on Ge stack (left), de-bonded stack with de-bonding interfaces (right).

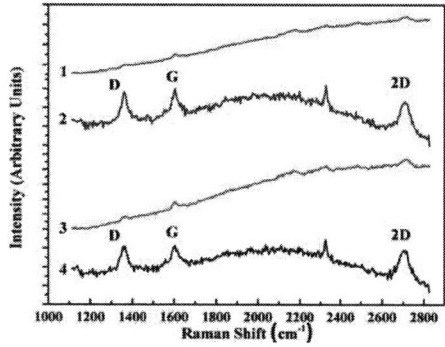

Fig. 7: Raman spectroscopy results of measurement sites shown in Fig 4.

measurement results show at which interface the wafers are de-bonded. The line in Fig. 4 shows the line that profile measurements are taken. According to the profile measurements, there is 50 nm of layer transferred to the target wafer (Fig. 5). In the same figure, it is also visible that there are different steps in the profile, which means that the de-bonding occurred on different interfaces.

In Figure 6, the bonding and de-bonding interfaces can be seen for the graphene on germanium case with respect to the profile measurements. As can be seen from the same figure, de-bonding happens on different interfaces, namely SiO_2-SiO_2 interface and graphene-Si_3N_4 interface, showing the unsuccessful attempts. For the SiO_2-SiO_2 interface de-bonding, the reason can be correlated with the low-quality of the initial bonding process. De-bonding at the graphene-Si_3N_4 interface indicates the real challenge associated with this approach: a relatively strong bonding of the graphene layer to the Ge substrate.

According to the profile measurements, graphene layer is still on the donor wafer, not on the target wafer. In order to verify this, Raman spectroscopy performed on donor wafer. The results are shown in Fig. 7. On sites 2 and 4, it is clearly seen that there is still graphene on the donor wafer by looking at the D and G peaks. These sites are from layer transferred sites, which means that, the layers above graphene is transferred to target wafer. On the sites 1 and 3, because of the non-transferred layers on top of graphene, lower G peak is visible. Therefore, on graphene on Ge case, the transfer under these conditions could not achieved.

B. Graphene on Nickel

After de-bonding of the bonded stack, the visual inspection is performed. The similar color change effect of the layer transfer can be also seen on these wafers: dark blue colored thermal oxide on Si wafer surface change to metallic light color. In this case, around 98% of the area is covered with transferred layer, showing the relatively good yield. The surfaces of the wafers after de-bonding can be seen in Fig. 8.

Profile measurements are done to understand the interface of the de-bonding for the graphene on nickel case. This measurement is taken along the line in Fig. 8, while the results are shown in Fig. 9. It can be seen that nearly 322 nm thick layer is transferred from donor wafer to target wafer, showing the de-bonding occured on Ni layer as seen in Fig. 10.

In order to check the successful graphene transfer to the target wafer, Raman spectroscopy is taken from the target wafer after Ni etch. From the results shown in Fig. 11, it is clearly seen on the sites 1 and 3, graphene is successfully transferred to target wafer. Site 2 and 4 are non-transferred areas, which is also clear with no signals visible on the graphs. The sheet resistance measurements taken from sites 1 and

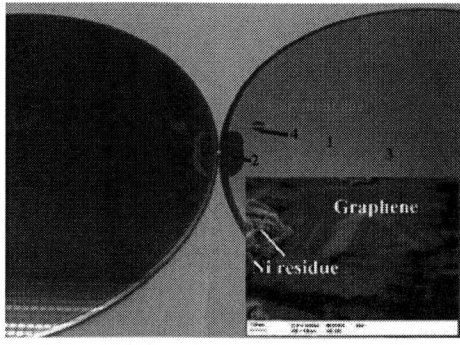

Fig. 8: Graphene on Ni; donor wafer on left, target wafer on right. Profile measurement is taken along the line. 1, 2, 3 and 4 shows Raman measurement sites. SEM image of site 3 after Ni etch is on bottom right.

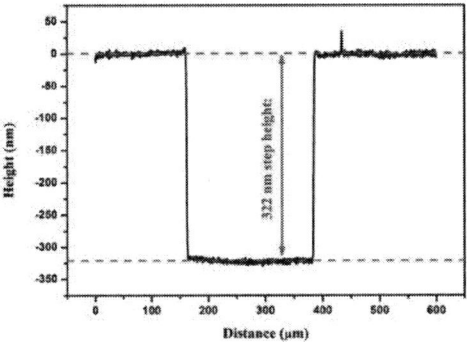

Fig. 9: Step height along the line in Fig. 8.

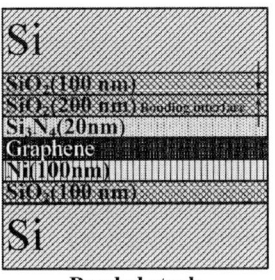 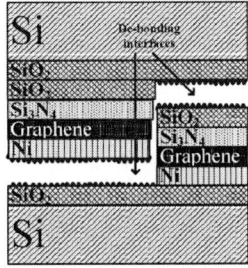

Bonded stack   **De-bonded stack**

Fig. 10: Bonded graphene on Ni stack (left), de-bonded stack with de-bonding interfaces (right).

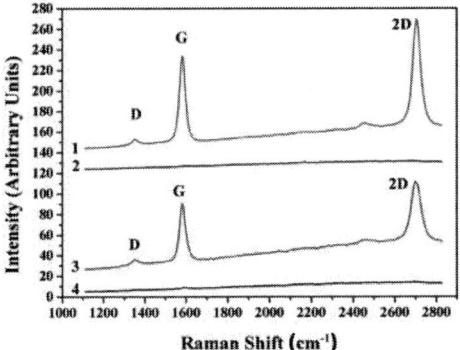

Fig. 11: Raman spectrocopy results taken from the sites in Fig 8.

3 gives an average result of 3.5 $k\Omega/\square$ on the transferred graphene.

In graphene on Ni case, the transfer is very successful with over 98% transfer yield. In order to reach 100% transfer yield, oxide-oxide bonding quality has to be increased. Results show that the weakest point in between the wafers is Ni layer and the wafers will be de-bonded at that interface. Therefore, increase of the SiO_2-SiO_2 bonding quality would yield full wafer transfer.

IV. CONCLUSION

200 mm wafer scale transfer of graphene has been performed by wafer bonding and de-bonding technique for two different cases; graphene on Ge and graphene on Ni. For the case of graphene on Ge, the desired transfer of graphene could not be achieved yet. However, for the case of graphene on Ni, the transfer of graphene was achieved by plasma assisted oxide-oxide fusion wafer bonding and mechanical de-bonding method. The transfer yield of above 98% has been successfully reached for the case of graphene on Ni transfer. The analyses have shown that the yield can be significantly increased by enhancing the SiO_2-SiO_2 bonding process. The developed 200 mm wafer scale technique is fully compatible with the CMOS process and provides the transfer of graphene on to a dielectric layer, thus allowing realizing new type of graphene devices.

ACKNOWLEDGMENT

This work is supported in part within a DFG project of "Integrierte Lab-on-Chip Terahertz-Spektroskopie-Platform in BiCMOS Technologie" (THz-LoC), TI 194/9-1 in Priority Program of Electromagnetic Sensors for Life Sciences (ESSENCE), SPP1857.

REFERENCES

[1] M. C. Lemme, S. Vaziri, A. D. Smith, J. Li, S. Rodriguez, A. Rusu, and M. Ostling, "Graphene for more moore and more than moore applications," in *Silicon Nanoelectronics Workshop (SNW), 2012 IEEE.* IEEE, 2012, pp. 1–3.

[2] A. Bunea, D. Neculoiu, M. Dragoman, G. Konstantinidis, and G. Deligeorgis, "X band tunable slot antenna with graphene patch," in *Microwave Conference (EuMC), 2015 European.* IEEE, 2015, pp. 614–617.

[3] C. Vázquez, A. Hadarig, S. Ver Hoeye, R. Camblor, M. Fernández, G. Hotopan, L. Alonso, and F. Las-Heras, "Millimetre wave transmitter based on a few-layer graphene frequency multiplier," in *Microwave Conference (EuMC), 2015 European.* IEEE, 2015, pp. 510–513.

[4] Y.-M. Lin, C. Dimitrakopoulos, K. A. Jenkins, D. B. Farmer, H.-Y. Chiu, A. Grill, and P. Avouris, "100-ghz transistors from wafer-scale epitaxial graphene," *Science*, vol. 327, no. 5966, pp. 662–662, 2010.

[5] J. Kang, D. Shin, S. Bae, and B. H. Hong, "Graphene transfer: key for applications," *Nanoscale*, vol. 4, no. 18, pp. 5527–5537, 2012.

[6] Y. Lee, S. Bae, H. Jang, S. Jang, S.-E. Zhu, S. H. Sim, Y. I. Song, B. H. Hong, and J.-H. Ahn, "Wafer-scale synthesis and transfer of graphene films," *Nano letters*, vol. 10, no. 2, pp. 490–493, 2010.

[7] A. Smith, S. Vaziri, S. Rodriguez, M. Ostling, and M. Lemme, "Wafer scale graphene transfer for back end of the line device integration," in *Ultimate Integration on Silicon (ULIS), 2014 15th International Conference on.* IEEE, 2014, pp. 29–32.

[8] M. Fujino, K. Abe, and T. Suga, "Large area direct transfer technique for graphene onto substrates using self-assembly monolayer," in *Electronics Packaging (ICEP), 2016 International Conference on.* IEEE, 2016, pp. 619–622.

[9] L. G. De Arco, Y. Zhang, A. Kumar, and C. Zhou, "Synthesis, transfer, and devices of single-and few-layer graphene by chemical vapor deposition," *IEEE Transactions on Nanotechnology*, vol. 8, no. 2, pp. 135–138, 2009.

[10] S. Vaziri, A. D. Smith, G. Lupina, M. C. Lemme, and M. Östling, "Pdms-supported graphene transfer using intermediary polymer layers," in *Solid State Device Research Conference (ESSDERC), 2014 44th European.* IEEE, 2014, pp. 309–312.

[11] M. Her, R. Beams, and L. Novotny, "Graphene transfer with reduced residue," *Physics Letters A*, vol. 377, no. 21, pp. 1455–1458, 2013.

[12] M. Lukosius, J. Dabrowski, J. Kitzmann, O. Fursenko, F. Akhtar, M. Lisker, G. Lippert, S. Schulze, Y. Yamamoto, M. A. Schubert *et al.*, "Metal-free cvd graphene synthesis on 200 mm ge/si (001) substrates," *ACS Applied Materials & Interfaces*, vol. 8, no. 49, pp. 33 786–33 793, 2016.

[13] G. Lupina, J. Kitzmann, M. Lukosius, J. Dabrowski, A. Wolff, and W. Mehr, "Deposition of thin silicon layers on transferred large area graphene," *Applied Physics Letters*, vol. 103, no. 26, p. 263101, 2013.

Epitaxial Growth and Diffusion Characteristics Analysis of Vertical Thin Poly-Si Channel Transfer Gate Structured Pixels for 3D CMOS Image Sensor

Sung-Kun Park[1], Donghyun Woo[1], Min-Ki Na[1], Pyong-Su Kwag[1], Ho-Ryeong Lee[1], Kyoung-Wook Ro[2],
Kyung-Hwan Kim[2], Dong-Kyu Lee[3], Chris Hong[1], In-Wook Cho[1], and Kyung-Dong Yoo[4]

SK hynix, CIS Development Group, [1]Pixel Development Team, [2]Process Development Team, [3]AT Group,
2091 Gyeongchung-daero, Bubal-eub, Icheon-si, Gyeonggi-do, 17336, Korea
[4]Hanyang University, 04763, Wangsimni Road, Seongdong-gu, Seoul, 222, Korea
sungkun.park@skhynix.com; skpark1225@naver.com

Abstract— **This paper reports the epitaxial-Si growth and dopant diffusion characteristics during fabrication of a vertical thin poly-Si channel (VTPC) transfer gate (TG) structured pixel, which is a possible candidate for future three-dimensional (3D) CMOS image sensor (CIS). Due to the increasing demand for higher resolution sensor, major CIS companies have presented various innovative 3D pixel structures of their own design. Recently, by adopting a structural concept similar to that of 3D NAND flash memories, a VTPC-TG structured pixel has been reported. However, grain boundary control and dopant diffusion behaviors in poly-Si have not been identified. The proposed process integration can suppress the dark current caused by grains of poly-Si in the VTPC-TG structured pixel by low temperature solid phase epitaxial growth. In addition, the channel punch-through caused by fast dopant diffusion in poly-Si can be suppressed by a thin poly-Si channel structure and process optimization.**

Keywords—3D pixel; Si epitaxy; thin-film transistor; poly-Si channel; CMOS image sensor; transfer gate

I. INTRODUCTION

Driven by consumer demand for higher-resolution sensors, researches on three-dimensional (3D) CMOS image sensor (CIS) have been started [1]–[7]. We adopted a vertical thin poly-Si channel (VTPC) structure, which is used in 3D NAND flash memories, and modified to fit CIS pixel applications [8]–[10]. In addition, we verified the CIS application of VTPC transfer gate (TG)-structured pixels by capturing the images of a 5 Mpixel back side illumination (BSI) structure test chip [10].

However, unlike 3D NAND flash memories, the grain boundaries of polycrystalline-silicon (poly-Si) which are adjacent to photodiode (PD) act as a dark current source that causes deterioration in the image quality. Until now, an effective grain boundary control method for the VTPC-TG structured pixel had not been identified. Furthermore, different from 3D NAND flash cells, which are connected in series, the TG of CIS is a single transistor having a relatively short channel length, which results in vulnerability to channel punch-through failure. Therefore, dopant diffusion control methods in poly-Si are also critical for fabricating a VTPC-TG pixel.

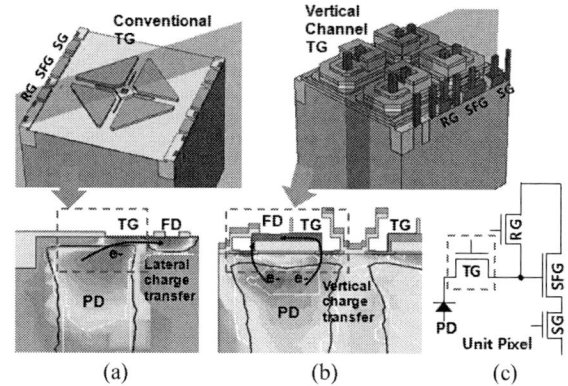

Fig. 1. Comparison of 3D TCAD images of conventional and VTPC-TG pixels. (a) Conventional pixel. (b) Proposed vertical channel TG pixel. (c) TG of four-transistor structure pixel. In this figure, PD, FD, RG, SFG, and SG represent the photodiode, floating diffusion, reset gate, source follower gate, and select gate, respectively [10].

In this paper, for the first time, we report on a grain boundary related defect suppression mechanism for VTPC-TG structured pixels. We also report on the characteristics of dopant diffusion in poly-Si substrate depending on various process methods such as spike annealing and carbon co-implantation.

II. RESULTS AND DISCUSSIONS

A. Fabrication of VTPC-TG Pixels

Fig. 1 shows 3D and cross-section technology computer aided design (TCAD) images of a conventional pixel and the VTPC-TG structured pixel proposed in this study [10]. The proposed VTPC-TG structure uses a vertical charge transfer, in contrast to the lateral charge transfer of conventional TG structures. Shading planes of 3D images indicate cross-section lines. A four-transistor operating four-shared structure pixel was used in both cases. Arrows in the cross-section image indicate the directions of electron movement along the TG channel.

978-1-5090-5979-9/17 $31.00 © 2017 IEEE

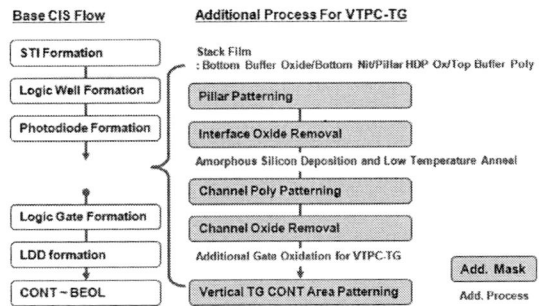

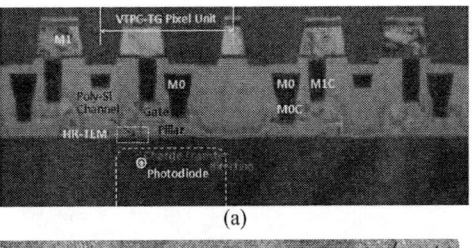

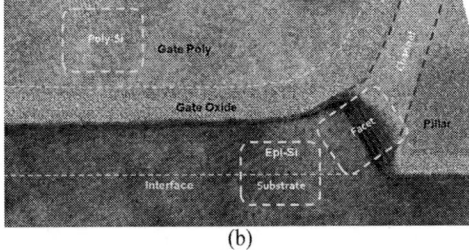

Fig. 2. Concept of VTPC-TG process integration. In addition to base CIS process, five additional mask steps are required for VTPC-TG fabrication [10].

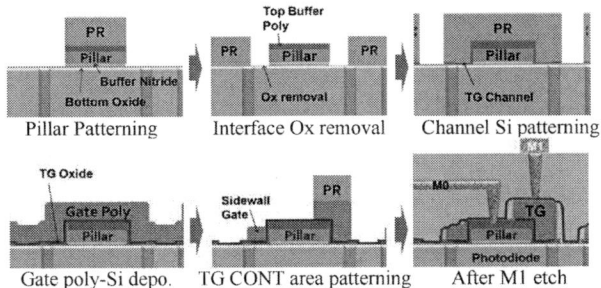

Fig. 3. VTPC-TG fabrication process flow. Five additional mask steps for VTPC-TG and final structure are illustrated.

Fig. 2 shows the process integration concept of the VTPC-TG pixel [10]. In addition to the conventional CIS process, five additional masks are required to fabricate the vertical channel transistor. We used an oxide pillar as a supporting structure for the thin poly-Si channel. In the pillar stack film, the bottom nitride is used as an etch stop layer, and a top buffer of poly-Si is used for the contact (CONT) landing buffer layer. Because of the poly-Si channel thickness of under 100 Å and non-silicide CONT process for dark current suppression, an additional top buffer poly-Si is required on the top of pillar to prevent an increase in the CONT resistance caused by the non-silicide CONT over-etch process.

Fig. 3 shows detailed cross-section illustrations of five additional mask steps for VTPC-TG pixel fabrication. For the formation of TG channel, amorphous-Si was deposited at 500 °C. After that, a long-time low-temperature annealing process was carried out to prevent the PD doping profile changes and large grains growth in the poly-Si channel. Initially, annealing temperature was set sufficiently low to prevent PD profile deformation. However, owing to this low-temperature long-time furnace annealing step which suppresses random direction nucleation of Si in the deposited amorphous-Si, epitaxial-Si was grown in the PD to poly-Si interface area.

B. Epitaxial-Si Growth in the PD-to-Channel Interface

In CIS applications, defect-free PD interfaces is critically important for dark current suppression. Fig. 4 shows cross-section high-resolution transmission electron microscopy (HR-TEM) images of the VTPC-TG at the channel interface to the

Fig. 4. (a) Cross-sectional TEM images of VTPC-TG pixels. (b) HRTEM images of PD-to-channel poly-Si junction area. PD interface area of poly-Si was converted to epitaxial Si after low-temperature annealing process.

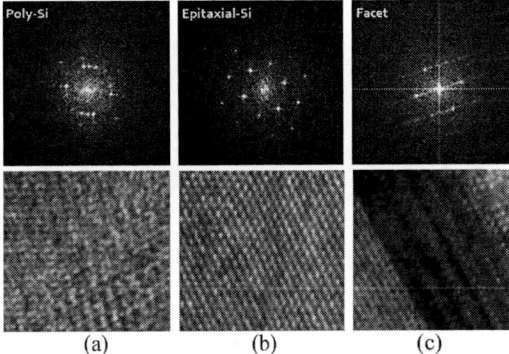

Fig. 5. FFT diffraction patterns showing epitaxially grown Si interface and poly-Si channel depending on location of Fig. 5 (b). (a) Gate area. (b) PD-to-poly-Si channel interface area. (c) TG channel-to-epitaxial-Si facet area.

PD junction area. It was found that as-deposited amorphous Si adjacent to the Si-substrate interfaces was converted to an epitaxial Si layer, thus suppressing possible dark current sources. This phenomenon is known as solid phase epitaxial growth (SPEG) [11, 12]. Unlike conventional Si epitaxy, which requires temperatures above 1000 °C [13], SPEG is conducted at low-temperatures of around 600 °C. Furthermore, if the doping level is below solubility, the doping profile does not change during the SPEG process [11, 12]. Therefore, owing to the SPEG mechanism, we can isolate grain boundaries in the poly-Si channel from the PD region without changing PD profiles that are formed prior to VTPC-TG fabrication. However, the amorphous-Si that was adjacent to the pillar oxide is crystallized to poly-Si with random direction grains.

Fig. 5 shows fast Fourier transform (FFT) diffraction patterns of each part of the cross-section HR-TEM image of Fig. 5 (b). The FFT diffraction patterns and lattice-spacing HR-TEM images clearly show the poly-Si, epitaxial-Si, and facets depending on the specific locations in the VTPC-TG.

978-1-5090-5979-9/17 $31.00 © 2017 IEEE

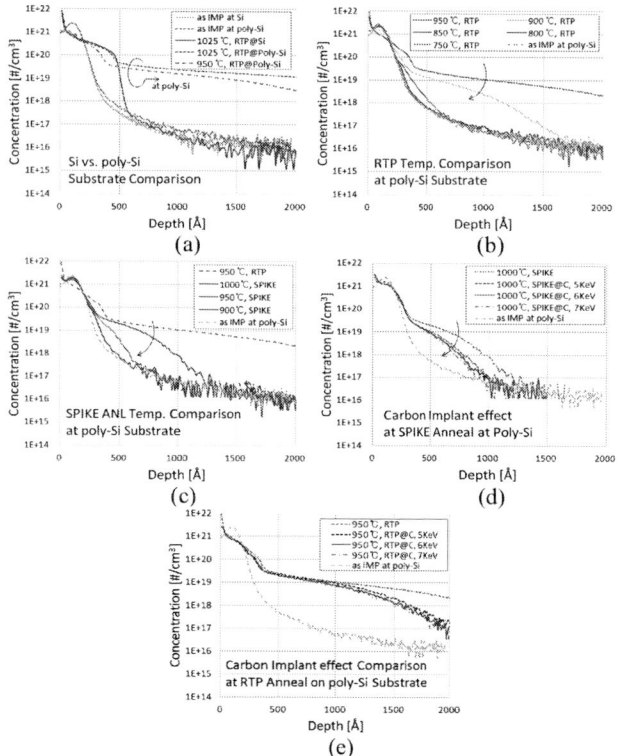

(a)

(b)

(c)

(d)

(e)

Fig. 6. SIMS depth profile comparison results depicting faster dopant diffusion in poly-Si substrate when compared with that in Si substrate, for various source/drain formation conditions including spike annealing and carbon co-implant.

C. Dopant Diffusion in the Poly-Crystalline Si Channel

Until now, poly-Si devices were usually used in thin-film transistors (TFTs), which have a long channel length of over 2 μm for display applications. Since dopant diffusivity for poly-Si is several dozen times faster than for single-crystal Si [14], it is not easy to fabricate a submicron-length poly-Si channel transistor. Furthermore, even though spike annealing and carbon co-implants are usually used in the CMOS process, there are no reports about the diffusion characteristics of spike annealing or carbon diffusion suppression co-implants on poly-Si substrate.

Fig. 6 shows the dopant diffusion characteristics for single-crystal Si and poly-Si depending on various process conditions. In this experiment, 3000 Å poly-Si deposited substrates were used to measure depth profiles using secondary ion mass spectroscopy (SIMS). To minimize dopant diffusion length, arsenic (As) species were used as a dopant of the junction.

Fig. 6 (a) compares the SIMS profiles of as-implanted and after 1025 °C rapid thermal processing (RTP) on single-crystal Si and poly-Si. Before RTP, As concentration in the poly-Si substrate does not show much difference from the single crystal-Si. However, after conventional soak RTP at 1025 °C, the As diffused to 2000 Å in the poly-Si substrate. That is four times longer distance than in single-crystalline Si substrate. The reason for these differences is that the diffusion of As along the grain boundary of poly-Si is much faster than lattice

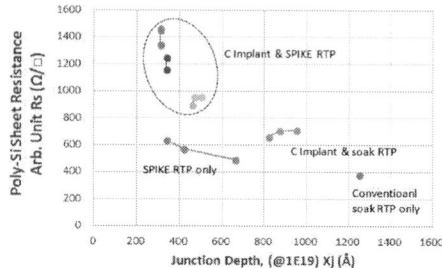

Fig. 7. R_s-X_j characteristics of planar test wafer depending on various RTP conditions in poly-Si. By using spike anneal, we can successfully suppress the fast diffusion of dopant.

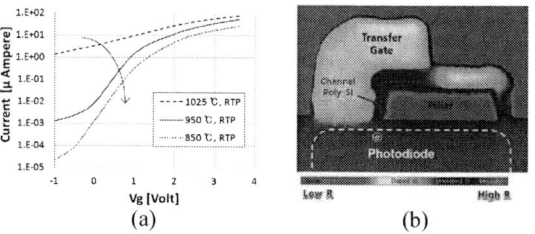

(a)

(b)

Fig. 8. V_g-I_d characteristics of VTPC-TG depending on soak RTP temperatures. Scanning spreading resistance microscopy (SSRM) image of VTPC-TG pixels indicating less dopant diffusion in thin poly-Si channel than bulk poly-Si.

diffusion in single-crystal Si substrate. Fig. 6 (b) compares the diffusion lengths with respect to RTP temperatures. In order to maintain the diffusion length of poly-Si at a level similar to that of a single crystal, a temperature of 850 °C or lower is required. However, for sufficient dopant activation, a higher temperature is preferred. As shown in Fig. 6 (c), the spike RTP indicates less diffusion than conventional soak RTP, even at higher processing temperatures. Fig. 6 (d) shows the carbon co-implant effectively suppressed As diffusion at the spike RTP. However, as shown in Fig. 6 (e), the carbon co-implant effect in poly-Si was not significant when using conventional RTP.

Fig. 7 shows sheet resistance (Rs) measurement results for the previously described process conditions in Fig. 6. When carbon is co-implanted, Rs increases around two times compared with the spike-annealed sample without a carbon co-implantation. The carbon co-implant method is usually used for diffusion suppression in conventional CMOS processes using single-crystal Si substrates. However, because of Rs increases, a carbon co-implantation and spike RTP combination method would not be a good choice as a diffusion suppressor for poly-Si substrate. Considering the junction depth and Rs, spike annealing at 1000 °C is an optimum processing condition for device manufacturing.

TG requires low leakage and sufficient breakdown voltage (BV) for good image quality. As shown in Fig 8 (a), V_g-I_d characteristic of the VTPC-TG does not exhibit the source-drain short phenomenon even though conventional 950 °C RTP is used. This is different from the SIMS result using bulk substrate shown in Fig 6. This discrepancy would be due to the same reason as the effect of macaroni structure in 3D NAND flash [8], because the effect of grain boundary is reduced in

978-1-5090-5979-9/17 $31.00 © 2017 IEEE

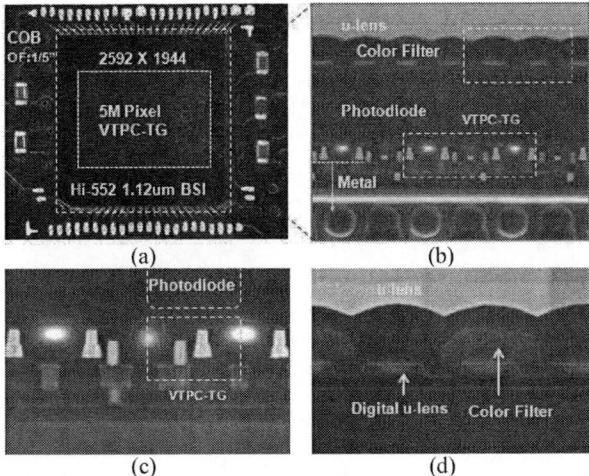

Fig. 9. Chip on board (COB) and focused ion beam (FIB) images of VTPC-TG pixel. (a) COB module image of 1.12 μm, 5 Mpixel BSI structured chip was used for characterization. (b) FIB image of test chip after VTPC-TG pixel adoption. (c) Enlarged images of VTPC-TG pixels. (d) Enlarged image of μ-lens sections, which shows digital μ-lens structure for sensitivity improvement.

Item	Description
Sensor Structure	BSI
Test Chip	Hi-552 (Pixel Replace)
Resolution	5M Pixel (2592X1944)
Pixel Structure	VTPC-TG
CFA Type	Bayer RGB
μ-Lens Type	Dual μ-Lens
Operating Voltage	1.2V/2.8V
Pixel Size	1.12 μm
Fill Factor	48.7 %
Integration Time	33 ms
Light Condition	D65

Fig. 10. An example of 5 Mpixel resolution image [10]. VTPC-TG pixels were adapted on a 5 Mpixel mass-production chip based test chip.

thin poly-Si channel. Furthermore, as shown in the SSRM results of Fig. 8 (b), unlike in the TCAD estimation, the sidewall gate could not act as a gate of the vertical channel transistor.

D. Image Capture Using Product Based Test Chip

Fig. 9 shows the COB module and FIB cross-section images of the mass-production base test chip. Because of the BSI structure, unlike the images of previous section, VTPC-TGs of Fig. 9 (c) are displayed upside down. By replacing the pixel array of an existing 5 Mpixel 1.12 μm BSI product with VTPC-TG pixels, product-level 5 Mpixel images were obtained. The image quality of the epitaxial Si grown test chips improved dramatically than the chips without epitaxial growth. However, even though epitaxial Si growth was applied, the image quality was not as good as that of a production chip with a conventional structure. Fig. 10 shows a real photo image obtained with interpolation and auto white balance, but without any other image optimization techniques [10]. Because the analog chip setting conditions are not optimized, the image quality is not comparable to that of a production chip.

III. CONCLUSIONS

The proposed VTPC-TG integration process effectively suppresses the grain boundary effect by SPEG at the PD junction-to-channel interface. In addition, by optimizing the source/drain junction formation conditions and thin poly-Si channel structure, punch-through caused by fast dopant diffusion in poly-Si can be effectively suppressed.

In poly-Si, when 1025 °C conventional RTP is applied, As diffused four times longer than in single crystal-Si. Unlike an ordinary CMOS process, a carbon co-implant and spike annealing combination is not effective in poly-Si owing to an Rs increase.

5 Mpixel real images were successfully captured using VTPC-TG pixels. Thanks to the PD junction-to-epitaxial-Si channel interface, image quality was improved compared with the PD-to-poly-Si interface structure. Low-temperature and long-time annealing after amorphous-Si deposition is one of the key processes for VTPC-TG fabrication.

REFERENCES

[1] Y. Kitano, "Image sensor, manufacturing apparatus and method, and imaging apparatus," U.S. Patent, 20150029374 A1, Jan. 29, 2015.

[2] X. Fang, "Vertically stacked image sensor," U. S. Patent, 8773562 B1, Jul. 8, 2014.

[3] M-F. Kao, D-N. Yaung, J-C. Liu, C-C. Chuang, F-C. Hung, S-J. Tsai, J-S. Lin, S-T.Tsai, and W-I Hsu, "Structure and method for 3D image sensor," U. S. Patent, 9059061 B2, Jun. 16, 2015.

[4] JC. Ahn, KH. Lee, HG. Jeong, SJ. Choi, and JG. Park, "Image sensors and methods of fabricating the same," U. S. Patent, 9054003 B2, Jun. 9, 2015.

[5] KD. Yoo and KI. Lee, "Image sensor including vertical transfer gate and method for fabricating the same," U.S. patents 9520427 B1, Dec. 13, 2016.

[6] JC. Ahn, KH. Lee, YT. Kim, HG. Jeong, BS. Kim, HK. Kim, et al, "A 1/4-inch 8Mpixel CMOS image sensor with 3D backside-illuminated 1.12 μm pixel with front-side deep-trench isolation and vertical transfer gate," in *Proc. IEEE ISSCC*, 2014, pp.124-125.

[7] T. Shinohara, K. Watanabe, S. Arakawa, H. Kawashima, A. Kawashima, T. Abe, et al, "Three-dimensional structures for high saturation signals and crosstalk suppression in 1.20 μm pixel back-illuminated CMOS image sensor," in *Proc. IEEE IEDM*, 2013, pp. 27.4.1-27.4.4.

[8] Y. Fukuzumi, R. Katsumata, M. Kito, M. Kido, M. Sato, H. Tanaka, et al, "Optimal integration and characteristics of vertical array devices for ultra-high density, bit-cost scalable flash memory," in *Proc. IEEE IEDM*, 2007, pp. 449-452.

[9] KT. Park, DS. Byeon, and DH. Kim, "A world's first product of three-dimensional vertical NAND Flash memory and beyond," in *Proc. IEEE NVMTS*, 2014, pp. 1-5.

[10] SK. Park, YH. Yang, CY. Lee, YJ. Kwon, TS. Shin, JH. Park, et al, "A study of vertical thin poly-Si channel transfer gate structured CIS," *IEEE Electron Device Letters*, vol. 38, no. 2, pp. 232-235, 2017.

[11] J.S. Williams, "Solid phase epitaxial regrowth phenomena in silicon," *Nuclear Instruments and Methods in Physics Research*, vol. 209-201, part-1, pp. 219-228, 1983.

[12] G.L. Olson and J.A. Roth, "Kinetics of solid phase crystallization in amorphous silicon," *Materials Science Reports*, vol. 3, pp. 1-77, 1988.

[13] ML. Hammond, "Si epitaxy by CVD, in Handbook of thin-film deposition processes and techniques" Noyes Publication, 2002.

[14] SW. Jones, "Diffusion in Silicon," IC Knowledge, *IRWS*, 2008. pp. 33

978-1-5090-5979-9/17 $31.00 © 2017 IEEE

Modelling, design and characterization of Schottky diodes in 28nm bulk CMOS for 850/1310/1550nm fully integrated optical receivers

Wouter Diels, Michiel Steyaert and Filip Tavernier
Department of Electrical Engineering, Katholieke Universiteit Leuven
3001 Leuven, Belgium

Abstract—This paper presents N-well Schottky diodes for high speed optical detection in 28nm CMOS technology. These diodes enable fully integrated CMOS optical receivers suited for the 850, 1310 and 1550nm telecommunication bands. The measured 1310 and 1550nm DC responsivity is 0.71mA/W and 0.16mA/W respectively at 1.5V reverse bias when backside illumination is performed while the 850nm responsivity is 0.27mA/W at the same biasing when frontside illumination is done. This is the first reported CMOS photodetector demonstrated at these three wavelengths. The measured capacitance-to-area ratio at zero bias is 1.7mF/m^2.

I. INTRODUCTION

Glass fiber-optic communication links have nearly unrivalled performance concerning data rates and distance. These links generally make use of 1310 and 1550nm light trough a single-mode fiber (SMF) or 850nm light through a multimode fiber (MMF). At the receiving end, the optical receiver typically consists of a high-speed photodetector to convert the optical signal into an electrical current, a transimpedance amplifier (TIA) to convert this current into a voltage and a post-amplifier to amplify this voltage to rail-to-rail levels [1]. As the photodetector is the first block of the receiver, its performance largely determines the overall performance of the full receiver.

For high-end receivers, each of these blocks is typically implemented on a separate chip in a different technology. The photodetectors of these systems are either PiN- or avalanche photodiodes. These diodes are fabricated in dedicated technologies to obtain high responsivity, high bandwidth and low capacitance for low dark currents. While these receivers have excellent performance, this multi-chip implementation and the resulting stringent packaging requirements result in a high cost. On the other hand, Photonic integrated circuits (PICs), keep gaining more attention [2], since these technologies enable integrating optical components, such as lasers, electro-optical modulators and photodetectors, on a single die. However, while CMOS-integrated photonics technologies have been demonstrated [3], most of them make use of relatively old CMOS technologies. Furthermore there is a lack of standardization.

Optical receivers can also be monolithically integrated in (modern) CMOS technologies. However, because there is barely any light absorption in silicon for wavelengths longer than 1.1μm, nearly all research has been aimed at 850nm receivers. As PiN-photodiodes are not available in standard CMOS technologies, the reported optical receivers make use of PN-diodes, some in the linear region [4], [5], others in the avalanche region [6]. P-substrate/N-well junctions provide the largest space charge region (SCR) and smallest capacitance and are thus the preferred choice for photodiodes in the linear region. Although the DC responsivity for 850nm light is relatively high, the low absorption coefficient of silicon results in a low photocurrent bandwidth (1-10MHz) because of carriers generated deep in the substrate [7]. An additional equalizer is necessary to compensate for this intrinsic low bandwidth. Alternatively, spatially modulated light (SML) photodiodes provide bandwidth extension at the cost of reduced responsivity, but ultimately degrade the signal-to-noise ratio (SNR) since half of the light is reflected by the metal coverage. When the reverse bias voltage is increased to high voltages, PN-junctions go into avalanche region. Operation in this region leads to increased responsivity and bandwidth, but also to larger photodiode noise. As junctions with thin avalanche regions result in the lowest excess noise factor [8], P+/N-well(/P-substrate) junctions are the preferred choice for avalanche photodiodes in CMOS.

As mentioned, PN-photodiode-based receivers in CMOS are only suited for short wavelength links (660nm in plastic optical fiber or 850nm in multi-mode fiber). There are thus no CMOS receivers reported for 1310/1550nm SMF links, which are traditionally used for long-haul communication. In [9], N- and P-well Schottky diodes in 40nm CMOS are presented, which can convert 1310nm light through internal photoemission (IPE). These devices are however not optimized for fiber coupled receivers: the large rectangular shape leads to excessive and unnecessary capacitance, while the distributed layout leads to a low diode fill factor, which results in a low responsivity. In this work, two circular plain layout N-well Schottky photodiodes in 28nm CMOS are presented. Their sizes are matched to a SMF and a MMF.

This paper is structured as follows. Section II gives some insight of the mechanisms of Schottky photodiodes, as well as design and layout considerations. In Section III, the I-V, capacitance and responsivity measurement results of the diodes are presented while comparing them to the N-well diode in [9]. Finally, Section IV gives a conclusion.

978-1-5090-5979-9/17 $31.00 © 2017 IEEE

II. CMOS SCHOTTKY DIODES

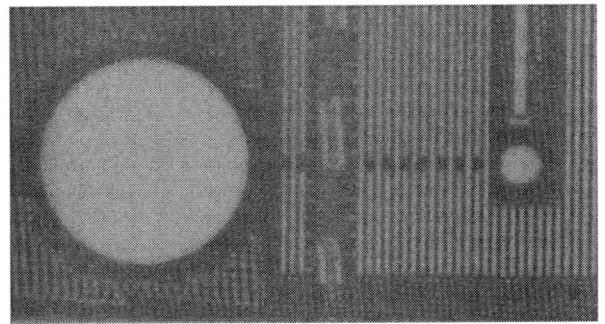

Fig. 1: Zoomed-in photograph of the presented 28nm Schottky diodes. The diode on the left is matched to a MMF, the diode on the right to a SMF.

A photograph of the presented diodes can be seen in Fig. 1. In Fig. 2, the working principle of internal photoemission (IPE) for an N-type Schottky diode can be seen. The light gets absorbed in the metal and if the excited electron has sufficient energy, it can cross the Schottky barrier ϕ_B. The maximum detectable wavelength is thus given by $hc/q\phi_B$, where h is Planck's constant, c is the speed of light and q is the elementary charge. This implies 1550nm and 1310nm can be detected for a Schottky barrier of 0.8eV and 0.95eV respectively. The quantum effiency (QE) is however rather low, since the metal reflects most of the light. The closer the photons are absorbed to the barrier, the higher the probability to cross the barrier. The QE also increases for bigger energy differences between photon energy and ϕ_B [10].

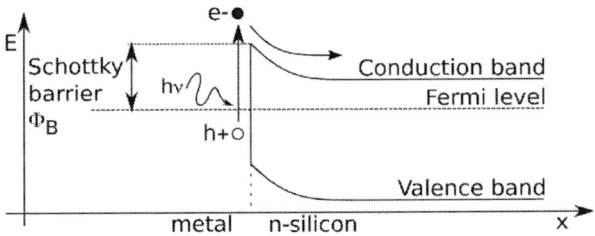

Fig. 2: The principle of internal photoemission: a photon gets absorbed in the metal, exciting an electron to a higher energy level in the process. If the electron has gained sufficient energy, it can cross the Schottky barrier ϕ_B.

A topside and cross section view of the presented diodes can be seen in Fig. 3. Schottky diodes can be realized in CMOS by placing contacts to the relatively lowly doped N-well. A wide junction is formed between the silicon and the silicide. Ohmic contacts to the highly doped N+ layer make up the cathode of the device. Both diodes are circular and have a diameter of 51.3μm and 9μm. These dimensions are chosen as to be matched to MMFs and SMFs, of which the core typically has a diameter of 50-62.5μm and 8-10μm respectively. Light can be impinged from the top (frontside illumination - FSI)

or from the back (backside illumination - BSI). FSI can be performed for the three telecom wavelengths, but only 1310nm and 1550nm can be impinged from the backside due to the non-transparency of silicon for 850nm light. When the chip is illuminated from the top, the photons get absorbed at the top of the M1 layer. The excited electrons must then travel a relatively large distance without losing energy in order to cross the barrier. The probability of these electrons crossing the barrier is much lower than those generated close to the junction, as is the case for BSI. Because of this, BSI results in larger responsivity than FSI. The intrinsic bandwidth of IPE-based photocurrent can easily be in the GHz range: unlike 850nm FSI for CMOS PN-diodes, all carriers are generated very locally and the intrinsic bandwidth is determined by the short transit time to cross the SCR.

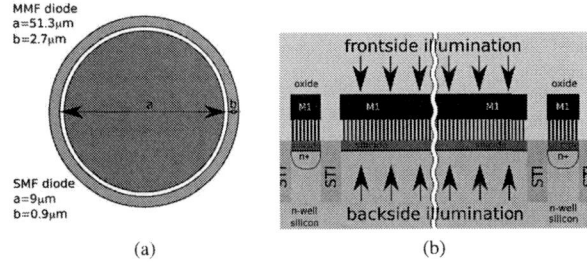

Fig. 3: (a) topside view - red region denotes Schottky regions, green region Ohmic regions (b) cross-section view - silicide and N-well silicon form Schottky region, silicide and N+ region form Ohmic region

The chosen layout maximizes the realizable diode area and responsivity. Whereas the silicide and M1 layers make up the entire area, technology design rules limit the contact width and spacing, resulting in a maximum 10% of the device area covered with contacts. Unlike the diodes in [9], the Schottky photodiodes in this work do not have metal dummies on top of them, which means FSI can be performed on them. This opens up the possibility of 850nm detection.

III. MEASUREMENTS

No standard CMOS process is intended to realize Schottky diodes and thus no model is available. Therefore, the performance of the Schottky diodes is first characterized, after which an analog front-end can be designed. The equivalent schematic of the Schottky photodiode can be seen in Fig. 4. The actual junction can be represented by a voltage dependent resistor rd, a voltage dependent junction capacitance Cd and a photosensitive current source Iph. The series resistance Rs represents the path from the cathode to the junction.

978-1-5090-5979-9/17 $31.00 © 2017 IEEE   225

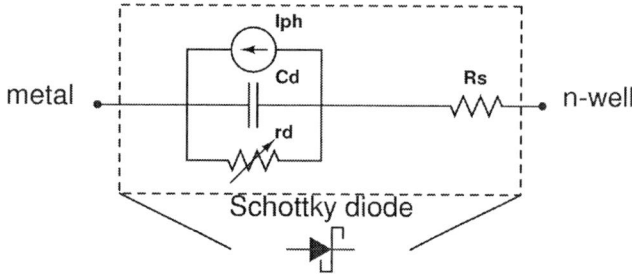

Fig. 4: Schottky photodiode equivalent schematic

A. I-V characteristics

The current-to-voltage characteristics without illumination of the devices, measured using a Keithley 2600 SMU, are shown in Fig. 5. Four important properties can be extracted from these dark current curves: the noise, the junction small-signal resistance rd, the series resistance Rs and the Schottky barrier ϕ_B.

The diode noise is dominated by the shot noise. An expression is given by (1), where I_{ph} is the DC photocurrent, I_{dark} is the DC dark current and BW is the extrinsic bandwidth of the diode.

$$I_n = \sqrt{2q(I_{ph} + I_{dark})BW} \tag{1}$$

As can be seen, the dark current of the two diodes is not larger than a few nA, suggesting a high Schottky barrier. The integrated shot noise current for 1nA dark current, 500nA photocurrent and 10GHz bandwidth is 40nA. As the input-referred noise current of a typical TIA at this bandwidth is much larger [5], this shot noise is of no further concern.

The diode small-signal resistance for reverse bias voltages ranges between 0.1-1$G\Omega$.

The I-V characteristics on a linear scale reveal the series resistance of the MMF and SMF diodes. They are 53Ω and 102Ω respectively. Despite the smaller device size, the series resistance of the SMF diode is larger than that of the MMF diode. This is due to the smaller spacing of the shallow trench isolation (STI), leaving a narrower path for the electrons to traverse. This series resistance due to STI could be minimized by adapting a poly-gate seperated layout as in [11]. Finally, thermionic current theory [10] reveals a zero-bias Schottky barrier ϕ_B of 0.7eV. In comparison with the N-well diode in [9], which has a zero-bias Schottky barrier of 0.55eV, these diodes have a higher Schottky barrier, which results in a lower current. However, as shown in the next subsection, this also results in a lower responsivity.

B. Light responsivity

FSI was performed on a wire bonded chip with 850, 1310 and 1550nm light. The resulting responsivities at different biasing voltages can be seen in Fig. 6. As predicted in section II, the reponsivity increases for shorter wavelengths. The responsivity also increases for increasing reverse bias voltages

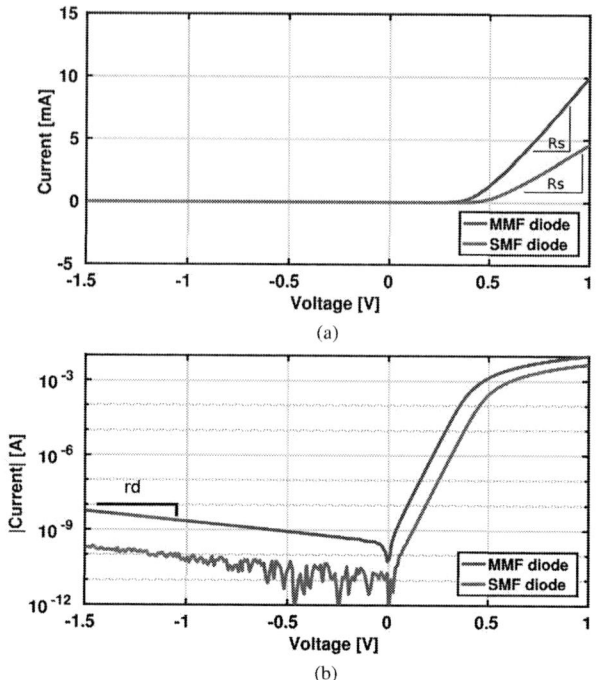

Fig. 5: IV-characteristics plotted (a) linearly (b) logarithmically

due to the barrier lowering effect [10]. At 1.5V reverse bias, the diodes have a responsivity of 0.27mA/W, 5.8μA/W and 1.1μA/W for 850, 1310 and 1550nm light respectively.

For the BSI measurements, a chip was mounted on a PCB by making use of flip-chip bonding. This time, only the responsivities for 1310 and 1550nm light were measured. They are shown in Fig. 7. At 1.5 reverse bias, the diodes have a responsivity of 0.71mA/W and 0.16mA/W for 1310 and 1550nm light respectively. Although the fill factor of these diodes is much bigger (3.5x for diode junctions, 2.7x for contacts) than that of the diode in [9], the BSI 1310nm responsivity is only 1.7x times larger. This can again be explained by the higher Schottky barrier in this technology.

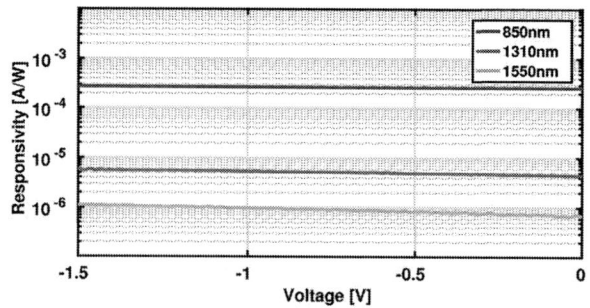

Fig. 6: Schottky photodiode responsivity for FSI with different wavelengths

978-1-5090-5979-9/17 $31.00 © 2017 IEEE   226

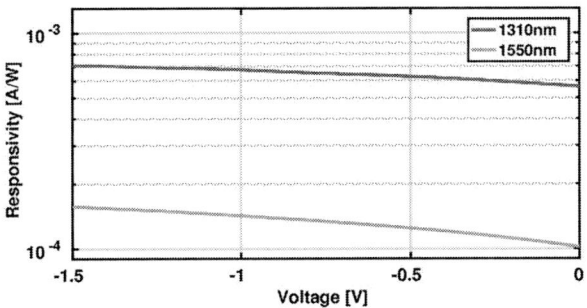

Fig. 7: Schottky photodiode responsivity for BSI with different wavelengths

C. Capacitance extraction

The diode small-signal impedance consists of rd in parallel with Cd. From a certain frequency on, the junction impedance is dominated by the junction capacitance. This small-signal impedance can be obtained through s-parameter measurements. The measurements were performed with a HP 8573V VNA. The resulting impedances and capacitances can be seen in Fig. 8.

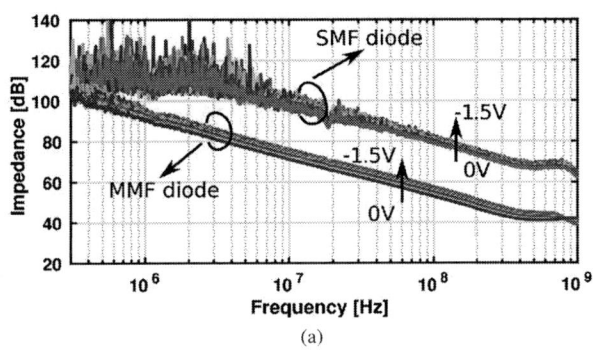

(a)

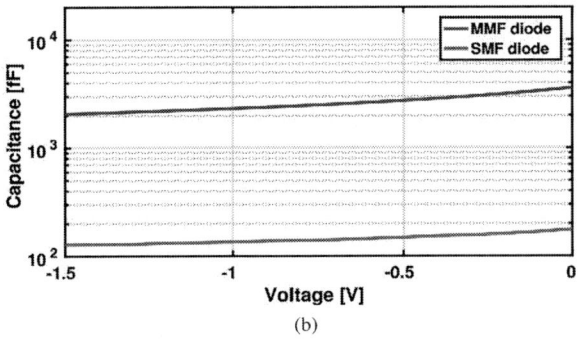

(b)

Fig. 8: (a) Diode impedances extracted from s-parameter measurements for different biasing voltages. (b) Diode junction capacitances for different biasing voltages.

The SMF diode has a capacitance of 175-125fF for reverse biasing voltages of 0-1.5V, while the MMF diode junction capacitance ranges between 3.5-2pF for the same biasing

Spec	40nm N-well diode [9]	28nm N-well diode
Schottky Barrier [eV]	0.55	0.7
Built-in Potential [V]	0.65	0.69
C/A [mF/m^2]	4.8	1.7
850nm Resp. [A/W]	N/A	0.27e-3 (FSI)
1310nm Resp. [A/W]	0.4e-3 (BSI)	5.8e-6 (FSI), 0.71e-3 (BSI)
1550nm Resp. [A/W]	N/A	1.1e-6 (FSI), 0.16e-3 (BSI)

TABLE I: Summary measurements results

voltages. This capacitance-to-voltage relationship reveals that the built-in potential of the Schottky diode is 0.69V. At zero bias, Schottky diodes in this technology have a capacitance-to-area ratio of 1.7mF/m^2, whereas the diodes in [9] processed in 40nm CMOS have a ratio of 4.8mF/m^2. As the built-in potential of that diode is 0.65V, this capacitance-to-area ratio also implies the N-well doping in that technology is higher. While Schottky diodes in this technology achieve a lower responsivity due to the higher Schottky barrier, the lower capacitance in turn relaxes the specs of the analog front-end [4]. The measurement results are summarized in TABLE I.

IV. CONCLUSION

N-well Schottky diodes in standard 28nm CMOS technology have been presented. These diodes can convert a very large spectrum of light to electric current at a large intrinsic bandwidth. Measurement results of I-V characteristics, capacitance and DC responsivity of 850, 1310 and 1550nm light have been shown. This is the first reported 1550nm responsivity of a CMOS photodetector. In comparison to N-well Schottky diodes in 40nm CMOS technology, the diodes in this technology achieve lower responsivity due to a higher Schottky barrier, but a lower capacitance-to-area ratio.

REFERENCES

[1] E. Säckinger, *Broadband Circuits for Optical Fiber Communication.* Wiley, 2005.

[2] A. Rahim *et al.*, "Expanding the silicon photonics portfolio with silicon nitride photonic integrated circuits," *OSA JLT*, vol. 35, no. 4, pp. 639–649, Feb 2017.

[3] Y. A. Vlasov, "Silicon cmos-integrated nano-photonics for computer and data communications beyond 100g," *IEEE Communications Magazine*, vol. 50, no. 2, pp. s67–s72, February 2012.

[4] D. Lee, J. Han, G. Han, and S. M. Park, "An 8.5-gb/s fully integrated cmos optoelectronic receiver using slope-detection adaptive equalizer," *IEEE JSSC*, vol. 45, no. 12, pp. 2861–2873, Dec 2010.

[5] S. H. Huang, W. Z. Chen, Y. W. Chang, and Y. T. Huang, "A 10-gb/s oeic with meshed spatially-modulated photo detector in 0.18-um cmos technology," *IEEE JSSC*, vol. 46, no. 5, pp. 1158–1169, May 2011.

[6] J. S. Youn, M. J. Lee, K. Y. Park, and W. Y. Choi, "10-gb/s 850-nm cmos oeic receiver with a silicon avalanche photodetector," *IEEE QE*, vol. 48, no. 2, pp. 229–236, Feb 2012.

[7] J. Genoe, D. Coppee, J. Stiens, R. A. Vonekx, and M. Kuijk, "Calculation of the current response of the spatially modulated light cmos detector," *IEEE ED*, vol. 48, no. 9, pp. 1892–1902, Sep 2001.

[8] M. M. Hayat, B. E. A. Saleh, and M. C. Teich, "Effect of dead space on gain and noise of double-carrier-multiplication avalanche photodiodes," *IEEE ED*, vol. 39, no. 3, pp. 546–552, Mar 1992.

[9] W. Diels, M. Steyaert, and F. Tavernier, "Schottky diodes in 40nm bulk cmos for 1310nm high-speed optical receivers," in *2017 Optical Fiber Communications Conference and Exhibition*, March 2017, pp. 1–3.

[10] S. M. Sze and K. K. Ng, *Physics of Semiconductor Devices*, 3rd ed. Wiley-Interscience, 2006.

[11] R. Han *et al.*, "A 280-ghz schottky diode detector in 130-nm digital cmos," *IEEE JSSC*, vol. 46, no. 11, pp. 2602–2612, Nov 2011.

Importance of buffer configuration in GaN HEMTs for high microwave performance and robustness

R. Pecheux, R. Kabouche, E. Dogmus, A. Linge, E. Okada, M. Zegaoui, and F. Medjdoub

IEMN - CNRS, Institute of Electronics, Microelectronics and Nanotechnology, UMR8520
Av. Poincaré, 59650 Villeneuve d'Ascq, France
romain.pecheux@etudiant.univ-lille1.fr; farid.medjdoub@iemn.univ-lille1.fr

Abstract— We report on a comparison of the ultrathin (sub-10 nm barrier thickness) AlN/GaN heterostructure using two types of buffer layers: 1) carbon doped GaN high electron mobility transistors (HEMTs) and 2) double heterostructure field effect transistor (DHFET). It is observed that the carbon doped HEMT structure shows better electrical characteristics, with a maximum drain current density I_d of 1.3 A/mm, a transconductance G_m of 500 mS/mm and a maximum oscillation frequency f_{max} of 234 GHz while using a gate length of 220 nm. The low trapping effects together with high frequency performance and excellent electron confinement under high bias enabled to achieve a state-of-the-art combination at 18 GHz of output power density ($P_{OUT} > 6$ W/mm) and power added efficiency (PAE) close to 40% at V_{DS} as high as 30V. At 40 GHz, a PAE above 35% is still observed in spite of the rather large gate length. A key feature is the low gate leakage current of only few tenths of μA/mm that remains stable after many load-pull sweeps at various frequency in the case of carbon doped HEMT, which is attributed to a significant reduction of the self-heating as compared to the DHFET.

Keywords— high electron mobility transistors (HEMTs), double heterostructure field effect transistor (DHFET), GaN, output power density and power added efficiency (PAE).

I. INTRODUCTION

With the development of wireless communication such as 5G or SATCOM, the need and requirements for millimeter-wave high power amplification has significantly increased. GaN-based High electron Mobility Transistors (HEMTs), owing to their outstanding properties, is one of the worthwhile candidate. For high frequency applications requiring short gate lengths, it has been shown that the double heterostructure using a thick AlGaN back barrier with about 8% Aluminum (Al) content enable the combination of high electron confinement, high frequency performance and low trapping effects [1]. It can be noticed that the record frequency performances achieved on devices fabricated by HRL laboratory employed an AlGaN back barrier [2]. However, even though the DHFET structure allows to withstand high electric field and thus benefit from short gate length [3]–[5] the presence of the back barrier has a negative effect on the heat dissipation. Actually, the thermal conductivity decreases with the increase of the Al content into the AlGaN back barrier [6]. This prevents these types of GaN devices to operate at high bias ($V_{DS} > 15$ V) in the millimeter-wave range. As the heat dissipation is a key issue especially for high power high frequency transistors, an alternative buffer

architecture without Al is needed in order to enhance the heat dissipation. The carbon doped GaN buffer (C-doped HEMT) could be an alternative to DHFET structure if this structure could maintain a high electron confinement, a low trapping level and high RF and load-pull performance while delivering higher thermal dissipation.

In this paper, the two types of buffer layer structure, DHFET and C-doped HEMT are compared using a 0.22 μm gate technology.

II. MATERIAL AND DEVICE PROCESSING

The AlN/GaN heterostructures were grown by metal organic chemical vapor deposition (MOCVD) on 4 in. SiC substrates. The HEMT structure consists of transition layers to GaN, a 1 μm-thick C-doped GaN buffer layer followed by an undoped GaN channel, a 4.0 nm ultrathin AlN barrier layer and a 10-nm-thick in situ Si_3N_4 cap layer (Fig. 2.1). The in situ SiN layer is used both as early passivation as well as to prevent strain relaxation. In the second structure called DHFET, the 1.5-μm-thick C-doped GaN buffer layer has been replaced with a 1 μm-thick $Al_{0.08}Ga_{0.92}N$ layer and a 150 nm GaN channel as shown in Fig. 2.1. Room-temperature Hall measurements showed high electron sheet concentrations of 1.8×10^{13} and 1.6×10^{13} cm^{-2} with an electron mobility of 1100 cm^{2}V^{-1}s^{-1} in the HEMT and DHFET heterostructures, respectively.

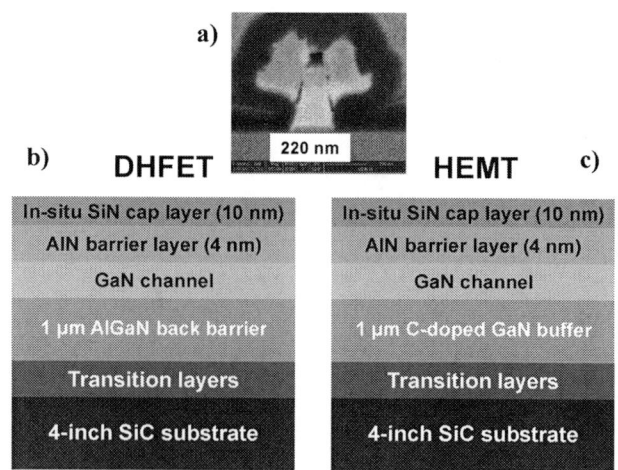

Fig. 2.1. : a) FIB view of the 0.22 μm T-gate and schematic cross section of b) DHFET and c) C-doped HEMT.

A Ti/Al/Ni/Au metal stack followed by a rapid thermal annealed has been used to form the ohmic contacts directly on top of the AlN barrier layer by etching the in situ Si₃N₄ layer. Device isolation was achieved by nitrogen implantation. Ohmic contact resistance (R_c) extracted from linear transmission line model (TLM) structures was as low as 0.25 Ω.mm for both heterostructures. Then, a 0.22 µm Ni/Au T-gate length was defined by e-beam lithography (see Fig. 2.1). The SiN underneath the gate was fully removed by SF₆ plasma etching. The gate-source and gate-drain spacings were 0.3 and 2 µm, respectively, and the device width was 50 µm. Finally, 200 nm Si₃N₄ was deposited as final passivation.

III. DC AND SMALL SIGNAL CHARACTERIZATION

DC measurements have been performed with a Keysight A2902A static modular and source monitor on both structures. The Fig 3.1 shows the typical I-V characteristic of both structures. The gate source voltage was swept from -3V to +2V by step of 1V. For the DHFET, a maximum current drain density (I_{Dmax}) of 1.2 A/mm is observed at V_{DS} = 10V. For the C-doped HEMT, a higher I_{Dmax} of 1.4 A/mm is obtained under the same conditions due to a higher carrier concentration.

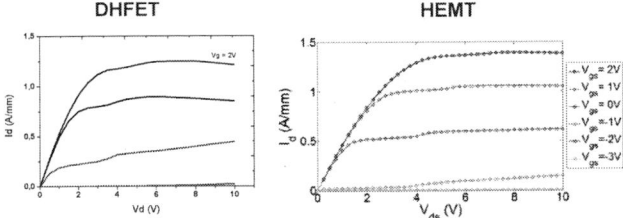

Fig. 3.1. : Output characteristics of DHFET and C-doped HEMT.

The transfer characteristics of both structures at V_{DS} = 6, 8 and 10 V are shown in Fig. 3.2. Excellent device pinch-off are obtained in both cases with a low off-state leakage current below 0.1 mA/mm. This confirms that a good electron confinement can be also obtained without the use of an AlGaN back barrier. On the other hand, a much higher transconductance is observed on the HEMT structure with a G_m around 500 mS/mm at V_{DS}=10V (see Fig. 3.3) against 350 mS/mm at V_{DS} =10V for the DHFET. At these respective biases, the cut-off frequency and maximum frequency oscillation are extracted from the scattering (S) parameters using Rhode and Schwarz ZVA67GHz. The DHFET yields a f_T = 56 GHz and f_{max} = 140 GHz while f_T = 60 GHz and f_{max} = 224 GHz are achieved for the HEMT owing to its higher G_m (see Fig. 3.3). It can be stressed that a high f_{max} / f_T ratio close to 4 is observed in the HEMT structure, which is quite untypical. This is due to the favorable aspect ratio between the gate length and the gate to channel distance (ultrathin barrier), the high carrier concentration, the high G_m and the low parasitic gate capacitance.

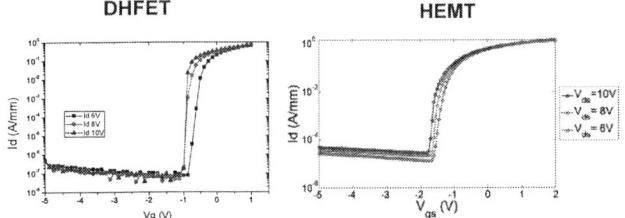

Fig. 3.2. : Transfer characteristics at V_{DS} = 6, 8, 10 V of DHFET and C-doped HEMT.

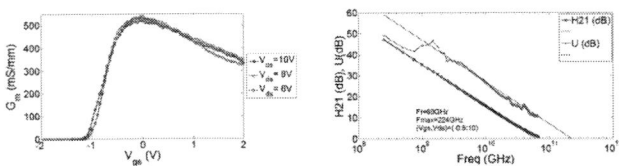

Fig. 3.3. : Transconductance at V_{DS} = 6, 8, 10 V and RF performance of the 0.22 µm C-doped HEMT

IV. LARGE SIGNAL CHARACTERIZATION AT 18 GHZ

Large signal characterizations at 18 GHz have been carried out on both structures in continuous mode (CW) and pulsed mode (1 µs width and 1% duty cycle). For a proper comparison, the same measurement conditions have been used between the HEMT and DHFET. Fig 4.1 shows the CW power performance of a 0.22×50 µm² AlN/GaN carbon doped HEMT and DHFET at 18 GHz with V_{DS} = 10 V. A saturated P_{OUT} (output power density) of 2.2 W/mm was achieved and a peak PAE of 45% associated to a power gain above 15 dB for the carbon doped structure. The DHFET structure shows a P_{OUT} of 1.7 W/mm with an associated PAE of 32%. This gap in output power density and especially in PAE can be explain by the fact that the carbon doped HEMT structure yields a better G_m and a higher maximum drain current density.

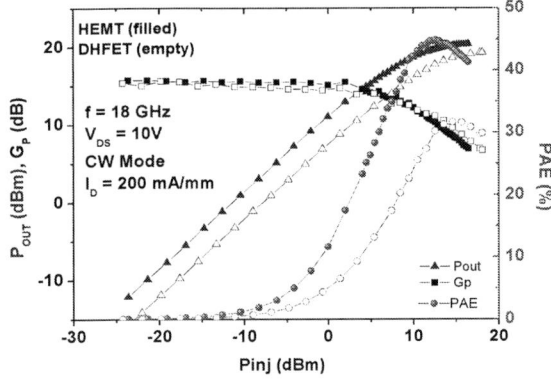

Fig. 4.1. : CW power performance of a 0.22×50 µm² C-doped HEMT (filled) and DHFET (empty) at 18 GHz with V_{DS} =10V.

In can be pointed out, as shown Fig 4.2a, that a peak PAE of 52% at V_{DS} = 10V has been reached with I_d = 80 mA/mm. At V_{DS} = 30V (see Fig 4.2b), a P_{OUT} of 6.3 W/mm has been delivered together with an associated PAE of 38%.

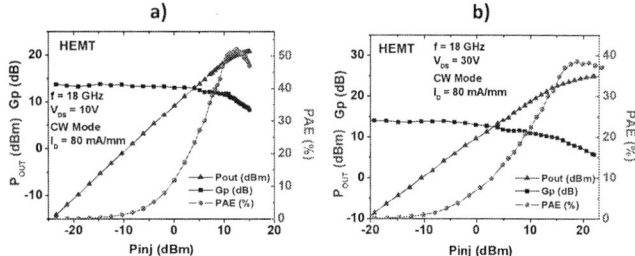

Fig. 4.2. : CW power performance of a 0.22×50 μm² C-doped HEMT at 18 GHz with V_{DS} = 10V (a) and 30 V (b).

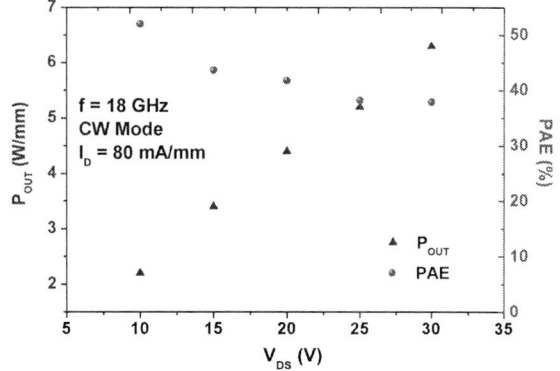

Fig. 4.3. : CW output power density (black) and PAE (red) versus V_{DS} at 18 GHz.

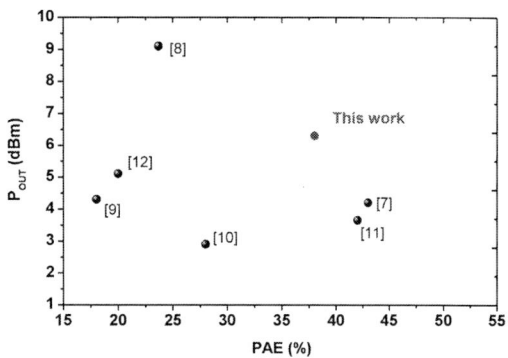

Fig. 4.4. : Benchmark of the output RF power density vs PAE at 18 GHz [7]-[12]

The output power density evolves linearly as a function of the drain bias (see Fig 4.3) with no sign of saturation even at V_{DS} = 30V. This quasi-linear evolution demonstrates the potential of this type of structure as the main limitation results from our nonlinear vector network analyzer (NVNA) system (bias tee limited to 30 V). It is worth noting that the decrease of the PAE is not significant and still remain close to 40% at V_{DS} = 30V. As can be seen from the benchmark in Fig. 4.4, the achieved PAE / P_{OUT} combination at 18 GHz compares favorably to the state-of-the-art.

V. LARGE SIGNAL CHARACTERIZATION AT 40 GHZ

Even though the gate length of 0.22 μm is not considered as suitable for Q band operation, large signal characterizations have been carried out on the same devices at 40 GHz in order

to investigate the potential of this technology for millimeter-wave applications.

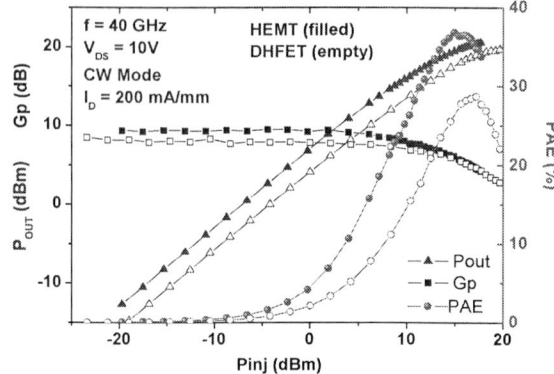

Fig. 5.1. : CW power performance of a 0.22×50 μm² C-doped HEMT (filled) and DHFET (empty) at 40 GHz with V_{DS} = 10V.

Fig 5.1 shows the CW power performance of a 0.22×50 μm² AlN/GaN C-doped HEMT and DHFET at 40 GHz with V_{DS} = 10 V. A P_{OUT} of 2.3 W/mm was achieved, which is in agreement with the expected (calculated) value of 2.5 W/mm at such bias. This confirms the low trapping effects in these devices. The peak PAE is 37% and the linear power Gain is above 9.5 dB. Compared to the C-doped HEMT, the DHFET structure shows a lower P_{OUT} of 1.8 W/mm with a reduced peak PAE of 29% as seen at lower frequency.

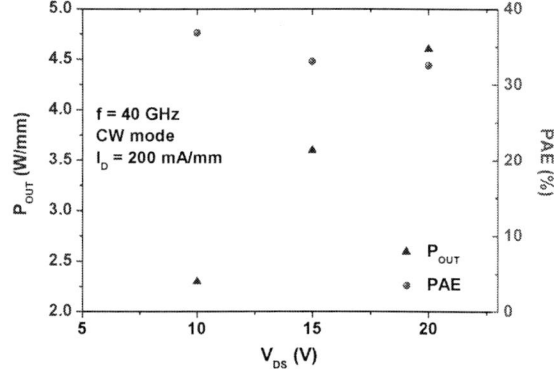

Fig. 5.2. : CW output power density (black) and PAE (red) versus V_{DS} at 40 GHz.

Unlike at 18 GHz, large signal characterizations at 40 GHz were limited to V_{DS} = 20V by our power test bench. Nevertheless, the output RF power density still shows a linear behavior with no saturation up to 20 V drain bias. Again, the PAE decrease as function as V_{DS} is not significant (see Fig 5.2).

VI. ASSESSMENT OF DEVICES ROBUSTNESS

The gate leakage current has been monitored after each load-pull sweep for various V_{DS} (see Fig 6.1) in order to assess the device robustness [13]. The C-doped HEMT structure shows almost no degradation of gate leakage current both in CW and pulsed modes remaining well below 1 mA/mm up to 30V drain bias. However, for the DHFET devices, we systematically observed a strong gate leakage degradation

978-1-5090-5979-9/17 $31.00 © 2017 IEEE   230

resulting in several tenths of mA/mm at $V_{DS} = 20V$. In pulsed mode, the DHFET devices show no gate leakage degradation all the way to $V_{DS} = 30V$. Therefore, this degradation in semi-on state is clearly attributed to the self-heating even though the SiC substrate has a high thermal dissipation. The self-heating resulting from hot electrons induces gate tunneling through the AlN barrier layer. Actually, the poor thermal dissipation of the thick AlGaN back barrier generates a high thermal resistance and thus significantly increase the channel temperature.

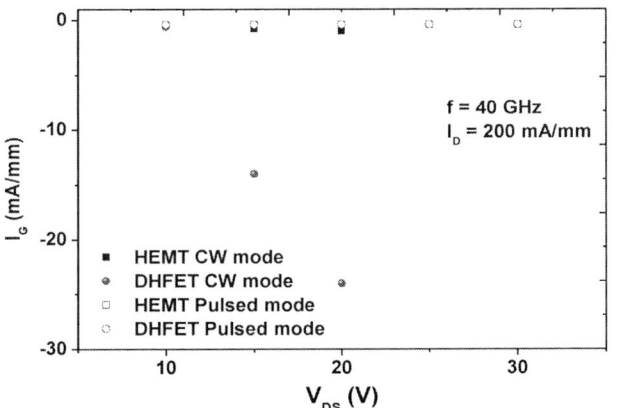

Fig. 6.1. : Evolution of the gate leakage in CW mode (filled) and pulsed mode (empty) of both structures at 40 GHz.

VII. CONCLUSION

This work shows that a careful architecture of buffer layers should be employed to perform not only high performance but also high robustness with GaN devices in the millimeter-wave range. It is shown that the thick AlGaN back barrier (DHFET structure) results in poor device robustness under high electric field as seen from the huge gate leakage increase as a function of the drain bias due to self-heating enhancement. Using the same process fabrication, the C-doped structure delivers higher performances together with much better device robustness with a gate leakage below 1 mA/mm at $V_{DS} = 30V$. In particular, state-of-the-art combination at 18 GHz of RF power density (> 6 W/mm) and associated PAE close to 40% at 30V has been achieved.

ACKNOWLEDGMENT

This work was supported by the French RENATECH network and the French Defense Procurement Agency (DGA) under the project EDA-EuGaNiC and contract FUI-VeGaN. The authors would to thank the company EpiGaN for high quality material delivery.

REFERENCES

[1] F. Medjdoub, M. Zegaoui, B. Grimbert, N. Rolland, and P. A. Rolland, "Effects of AlGaN back barrier on AlN/GaN-on-silicon high electron mobility transistors," Appl. Phys. Express, vol. 4, no. 12, pp. 4–7, 2011.

[2] Y. Tang *et al*, "Ultrahigh-Speed GaN High-Electron-Mobility Transistors With fT/fmax of 454/444 GHz," IEEE Electron Device Lett., vol. 36, no. 6, 2004.

[3] F. Medjdoub, M. Zegaoui, and N. Rolland, "Beyond 100 GHz AlN / GaN HEMTs on silicon substrate," vol. 47, no. 24, pp. 24–25, 2011.

[4] F. Medjdoub, M. Zegaoui, N. Waldhoff, B. Grimbert, and N. Rolland, "Above 600mS/mm Transconductance with 2.3A/mm Drain Current Density AlN/GaN High-Electron-Mobility Transistors Grown on Silicon," vol. 64106, 1882.

[5] F. Medjdoub, B. Grimbert, D. Ducatteau, and N. Rolland, "Record Combination of Power-Gain Cut-Off Frequency and Three-Terminal Breakdown Voltage for GaN-on-Silicon Devices," vol. 44001, pp. 4–7, 1882.

[6] W. Liu and A. A. Balandin, "Temperature dependence of thermal conductivity of AlxGa1−xN thin films measured by the differential 3ω technique," Appl. Phys. Lett., vol. 85, no. 22, p. 5230, 2004.

[7] O. Jardel et al., "Electrical performances of AlInN/GaN HEMTs. A comparison with AlGaN/GaN HEMTs with similar technological process," Int. J. Microw. Wirel. Technol., vol. 3, no. 3, pp. 301–309, 2011.

[8] V. Kumar, G. Chen, S. Guo, B. Peres, and I. Adesida, "Field-plated AlGaN/GaN HEMTs with power density of 9.1 W/mm at 18 GHz," Electronics Lett., vol. 41, 19, 2005.

[9] F. Van Raay, R. Quay, R. Kiefer, M. Schlechtweg, and G. Weimann, "Large Signal Modeling of AlGaN/GaN HEMTs with Psat > 4 W/mm at 30 GHz suitable for Broadband Power Applications," IEEE MTT digest, pp. 30–33, 2003.

[10] F. Lecourt et al., "InAlN/GaN HEMTs on sapphire substrate with 2.9-w/mm output power density at 18 GHz," IEEE Electron Device Lett., vol. 32, no. 11, pp. 1537–1539, 2011.

[11] Z. H. Feng et al., "18-GHz 3.65-W/mm enhancement-mode AlGaN/GaN HFET using fluorine plasma ion implantation," IEEE Electron Device Lett., vol. 31, no. 12, pp. 1386–1388, 2010.

[12] D. Ducatteau et al., "Output power density of 5.1/mm at 18 GHz with an AlGaN/GaN HEMT on Si substrate," IEEE Electron Device Lett., vol. 27, no. 1, pp. 2005–2007, 2006.

[13] M. Dammann et al., "Reliability of AlGaN / GaN HEMTs under DC- and RF-operation," pp. 19–32, 2009.

Shunt capacitive switches based on VO₂ metal insulator transition for RF phase shifter applications

E. A. Casu[1], W. A. Vitale[1], M. Tamagnone[2], M. Maqueda Lopez[1], N. Oliva[1], A. Krammer[3], A. Schüler[3],
M. Fernández-Bolaños[1], A.M. Ionescu[1]

[1]Nanoelectronic devices laboratory (NanoLab)
[2]Laboratory of Electromagnetics and Acoustics (LEMA)
[3]Solar Energy and Building Physics Laboratory (LESO-PB)
École Polytechnique Fédérale de Lausanne (EPFL), CH-1015 Lausanne, Switzerland
emanuele.casu@epfl.ch

Abstract—**This paper presents a wide-band RF shunt capacitive switch reconfigurable by means of electrically triggered Vanadium Oxide (VO₂) phase transition to build a true-time delay (TTD) phase shifter. The concept of VO₂-based reconfigurable shunt switch has been explained and experimentally demonstrated by designing, fabricating and characterizing an 819 µm long unit cell. The effect of bias voltage on losses and phase shift has been studied and explained. By triggering the VO₂ switch insulator to metal transition (IMT) the total capacitance can be reconfigured from the series of two metal-insulator-metal (MIM) capacitors to a single MIM capacitor. Higher bias voltages are more effective in this reconfiguration and give a higher phase shift. The optimal achievable performance has been shown heating the devices above VO₂ IMT temperature. A maximum of 16° per dB loss has been obtained near the design frequency (10 GHz).**

Keywords—vanadium dioxide; phase transition; RF switch; true-time delay; phase shifter; tunable capacitor;

I. INTRODUCTION

Phase shifters are key components for beam-steering implementations, smart adaptive antennas and scanning applications for wideband communications and remote sensing systems. RF distributed MEMS transmission lines (DMTL) have been proven to be an interesting concept to achieve a high phase shift over a wider frequency band compared to traditional solid-state implementations (PIN diodes, GaAs FET, Ferrite materials). Nevertheless, critical issues of MEMS technology, such as reliability, process variability and packaging requirements are still a limiting factor for a widespread implementation.

Strongly correlated functional oxides exhibiting metal to insulator transition have recently emerged in research as promising materials for a large number of applications, including steep transistors [1], RF switches [2,3], reconfigurable filters [4,5] and antennas [6,7]. Vanadium Oxide (VO₂) has proven to be one of the most interesting among these materials thanks to its large change in conductivity between its two states and the possibility of achieving the phase transition by electrical excitation [8].

Compared to MEMS switches, VO₂ switches offer clear advantages such as an easier integration in microelectronic technological processes, smaller footprint and a three order of magnitude faster switching time [9]. A switched line phase shifter with thermally actuated VO₂ switches has been previously demonstrated in microstrip technology [10].

In this paper we present for the first time a shunt capacitive switch reconfigurable by means of electrically triggered VO₂ phase transition to build true-time delay phase shifters by periodically loading a coplanar waveguide (CPW) with the capacitive switches (Fig. 1). We present the concept of the device and we validate it by fabricating, designing and characterizing an 819 µm long unit, able to provide up to 16° phase shift per dB loss at 10 GHz.

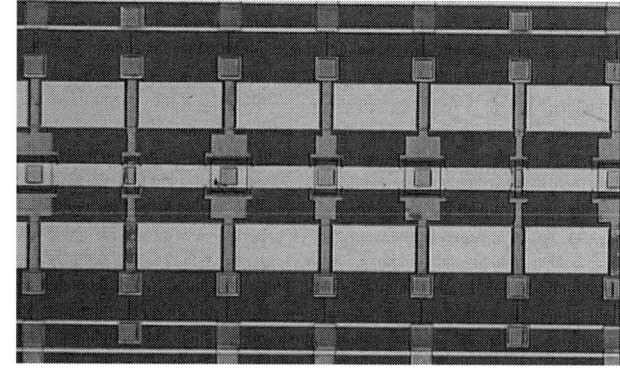

Fig. 1. Optical image of the CPW phase shifter showing the cascaded VO₂-based capacitive shunt switches designed to achieve 3-bits phase states.

II. RECONFIGURABLE CAPACITIVE SHUNT SWITCH

The reconfigurable capacitive shunt switch consists of two fixed MIM capacitors in series, C_S and C_G, where the first can be short-circuited by actuating a VO₂ two-terminal switch (Fig. 2). Below the phase-transition temperature and when no bias is applied, the VO₂ is in its insulating state so that the switch exhibits a high resistance level and can be considered open. The two capacitors are then electrically in series, offering an equivalent capacitance $C_{TOT} = C_G * C_S / (C_G + C_S)$. Whenever a bias larger than the switch actuation voltage is applied, the VO₂ film phase changes to its conductive state and the switch exhibits a low resistance value. In this case the C_S capacitor is short-circuited by the switch and the equivalent capacitance between the signal and the ground line will be simply C_G. In this way the VO₂ switch allows to reconfigure the loading capacitance between C_G and C_{TOT}.

978-1-5090-5979-9/17 $31.00 © 2017 IEEE

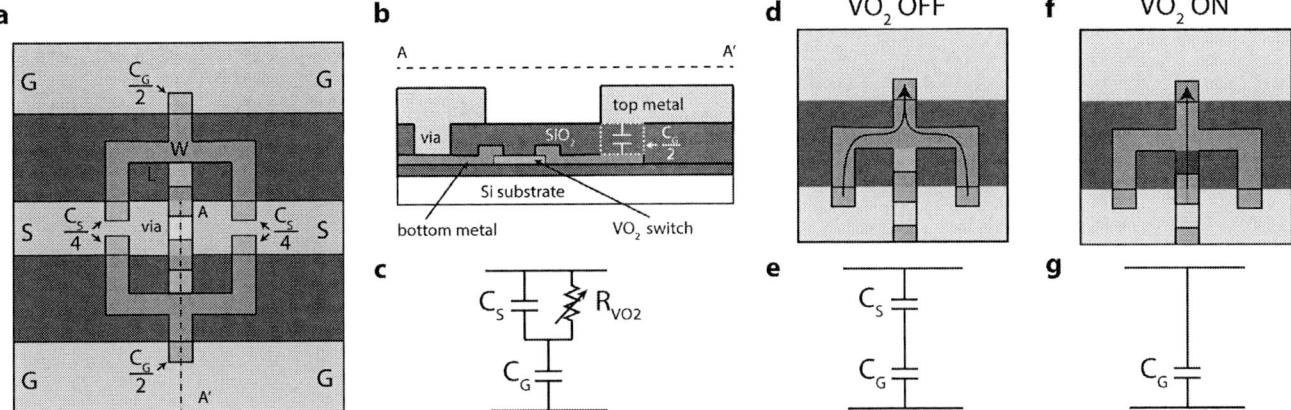

Fig. 2. a) Schematic of the reconfigurable capacitive shunt switch. A via connects the signal top metal line with ametal line underneath in contact with a VO$_2$ switch. Four MIM capacitor C$_S$/4, two per side, lay between the signal line and the underneath metal, whilst other two MIM capacitor C$_G$/2, one per side, lay between the ground planes and the underneath metal. b) Cross-section beween signal and ground plane highlighting the via between the two metals, the VO$_2$ switch and the capacitance C$_G$/2. c) Equivalent circuit of the capacitive divider, with the VO$_2$ switch modeled as a variable resistor R_{VO2}. d) Preferential path seen by the signal when the VO$_2$ is in insulating phase, with an equivalent capacitance given by the series of C_S and C_G. e) Equivalent circuit for the insulating phase, where the switch can be modeled as an open switch ($R_{VO2} > 1$ kΩ). f) When the VO$_2$ film is in its conducting phase the capacitors C_S/4 are short-circuited by the switch ($R_{VO2} \sim 1$ Ω) and g) the equivalent capacitance seen between signal and ground equals C_G.

The VO$_2$ switch can be electrically actuated by means of a bias line decoupled from the RF signal by means of a serpentine resistor realized with a 25 nm-thick Chromium (Cr) film. The switch resistance is in the high state until a critical power is achieved which causes a steep insulator to metal transition (Fig. 3). In order not to affect the RF performance, the length and width of the switch are chosen to be 1 μm and 30 μm respectively, so to obtain a high value of resistance in the off-state (> 1 kΩ) and low value of resistance in the on-state (~1 Ω), while keeping a reasonably low actuation voltage.

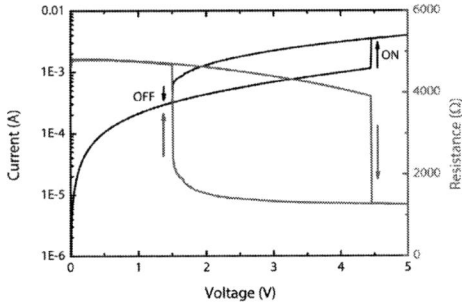

Fig. 3. Current versus voltage electrical characteristic and extracted resistance of the VO$_2$ switch with an integrated serpentine resisistor of 1.2 kΩ in series. The arrows indicates the insulator-metal transition and metal insulator transition.

The phase shifter was fabricated using standard microelectronics processes starting with a high-resistivity 525 μm thick silicon substrate passivated with 500 nm LPCVD-deposited SiO$_2$ (Fig. 4). The VO$_2$ film was prepared by reactive magnetron sputtering deposition starting from a Vanadium target [11]. After the deposition, a resistivity ratio higher than 3 decades was measured with a Van der Pauw measure performed at different temperatures (Fig. 5). The film was then patterned

ing photolithography and wet etching. The bias resistors were realized by lift-off of a 25 nm thick Cr film. A 200 nm thick Al film was subsequently deposited and patterned with lift-off to act as bottom metal. A 300 nm thick SiO$_2$ film was sputtered as insulating layer and as a dielectric for MIM capacitors. Vias were opened by photolithography and dry etching. A final 800 nm thick Al top metal layer was deposited to create the CPW and the contacts on the bottom metal bias lines.

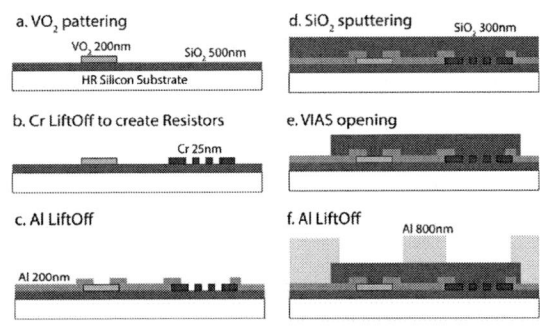

Fig. 4. Fabrication process of the capacitive shunt switch with VO$_2$ switches and integrated bias resistors.

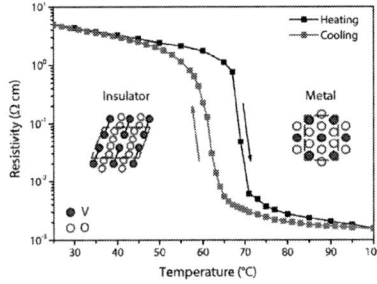

Fig. 5. Dependence of resistivity on temperature for the deposited 200 nm thick VO$_2$ film.

III. Unit Cell Performance

The CPW was designed with a signal width of 100 μm and a ground plane spacing of 150 μm to obtain an unloaded waveguide impedance of 65 Ω. The design of the unit cell for the phase shifter was done following the method described in [12] in order to maximize the phase shift for the minimum insertion loss (IL). Starting from the chosen values of impedances in the ON and OFF state, respectively $Z_{ON} = 42$ Ω and $Z_{OFF} = 58$ Ω, and having chosen the Bragg frequency to be three times the frequency of design for the phase shifter, for a design frequency of 10 GHz the unit cell length was calculated to be 819 μm. The computed capacitances were $C_{ON} = 143$ fF and $C_{OFF} = 26$ fF resulting in a capacitance ratio of 5.5. Thus the MIM capacitances for the reconfigurable capacitive shunt switch are calculated as $C_S = 31.7$ fF and $C_G = 143$ fF.

The device was characterized by using an Anritsu VectorStar MS4647B Vector Network Analyser to measure S-parameters, a HP 4155B Semiconductor Parameter Analyser to provide the bias to operate the VO₂ switches and a Cascade Summit prober with a thermo-chuck to control the substrate temperature.

Fig. 6 shows comparison between Ansys HFSS simulations of S-parameters and measurements with no bias at 20°C (OFF state) and 100°C (ON state), well above the phase transition temperature where the VO₂ film becomes fully conductive (see Fig. 5). When the switches are turned off (20 °C and no bias) the insertion loss is 0.43 dB at 10 GHz, while in the ON state the insertion losses are increased both in simulation and measurements. The difference between measured and simulated values could be attributed to resistive losses along the line and in the VO₂ switch for the ON state.

The devices were measured for different bias values above the actuation voltage of the VO₂ switches (Fig. 7). While the insertion loss seems to increase by increasing the bias, the loss at the design frequency is improved when the switch is at its lowest possible resistance value, obtained measuring at 100 °C. This behavior can be explained looking at the losses not due to reflection. While in the OFF state the insertion losses and no-reflection losses are almost coincident, indicating a good match, in the ON state the behavior varies depending on the bias. At 20 V the IL and total losses are similar, while at 30 V and 40 V a considerable part of the IL is due to the mismatch. At 100 °C the IL are lower than at the considered bias points and the no-reflection losses are minimized, showing better accordance with the FEM simulations.

The measured phase shift with respect to the OFF state increases with the applied bias but tends to saturate around 5 GHz for 40 V bias, while at 100 °C it is linear over the considered frequency band (Fig. 8). The phase shift per dB loss shows as well that the best trade-off is obtained for higher bias and indicates that best performances are obtained at 100 °C, where a maximum of 16° per dB loss is obtained slightly below the design frequency.

The limited performance of the device when electrically actuated suggests that for the used bias voltages, the conduction channel in the VO₂ switch does not extend to the entire film width [13] and its resistance is still not low enough to grant a full capacitance reconfiguration and to prevent significant RF losses. We can assume that by applying a larger bias voltage and thus by injecting a larger current the performances will converge to the one measured at 100 °C.

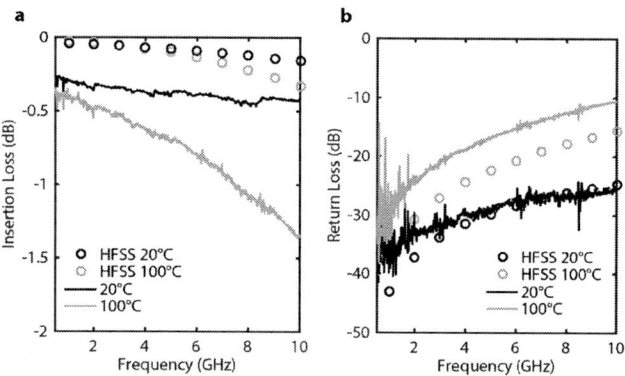

Fig. 6. (a) Insertion Loss and (b) Return Loss of the measured unit cell at 20 °C and 100 °C. Circles correspond to ANSYS HFSS simulations. The simulations have been performed using the VO₂ resistivity measured at 20 °C for the OFF state and 100 °C for the ON state.

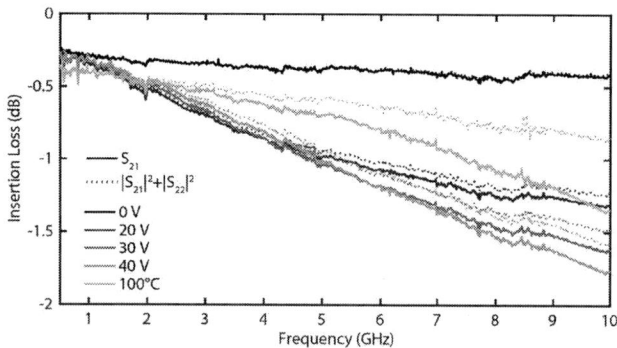

Fig. 7. Insertion loss (continuous lines) and no-reflection losses (dotted lines) versus frequency, measured at 20 °C substrate temperaure for 0 V, 20 V, 30 V and 40 V bias voltage and at 100 °C with no applied bias.

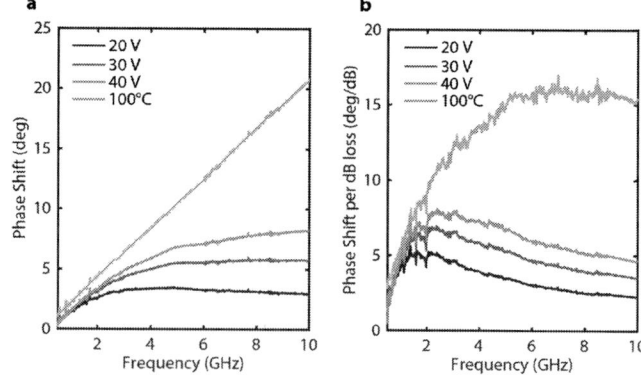

Fig. 8. (a) Phase shift and (b) phase shift per dB loss extracted from S-parameter measurements at 20 °C for 0 V, 20 V, 30 V and 40 V bias voltage at and at 100 °C with no applied bias.

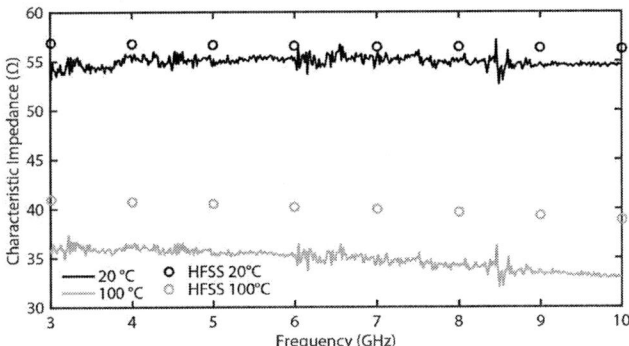

Fig. 9. Equivalent characteristic impedance of the unit cell, extracted from measures at room temperature and at 100 °C (solid lines), and from ANSYS HFSS simulations (symbols).

The equivalent impedance of the loaded line was calculated using the method proposed in [14] and is shown in Fig. 9. In the OFF state the equivalent impedance is about 55 Ω at 10 GHz, not far from the simulated value of 56 Ω. In the ON state at 100 °C the measured impedance is lower than the simulated one, in accordance with the larger measured phase shift and larger insertion loss due to reflection.

IV. PHASE SHIFTER

In order to predict the performance of a phase shifter, the measured S-parameters in OFF and ON states (20 and 100 °C temperature) of the unit cell were mathematically cascaded to consider the presence of more stages in series. The predicted performance of a 6-stages phase shifter targeting 120° shift at 10 GHz is shown in Fig. 10. In order to actuate the different stages it is possible to use multiple bit lines that actuate group of switches to achieve the desired phase shift.

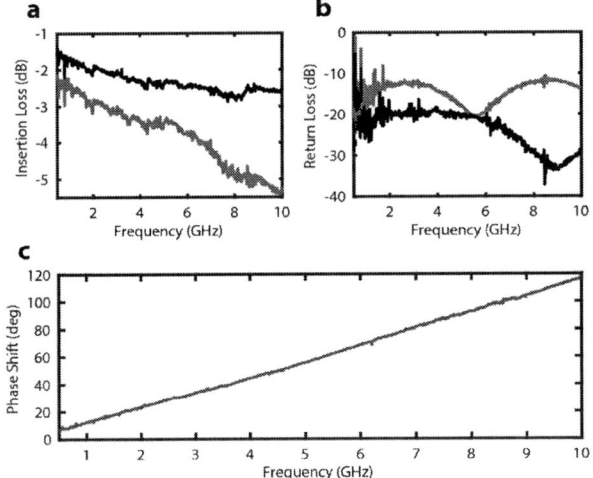

Fig. 10. Predicted (a) insertion loss, (b) return loss and (c) phase shift for a 6-stage phase shifter.

V. CONCLUSION

This paper reports for the first time a VO₂-based capacitive shunt switch as a building block to be cascaded to obtain TTD phase shifters. The working principles as well as the fabrication method have been presented and validated by designing, fabricating and characterizing a unit cell. Insertion losses, largely due to mismatch could be easily reduced by tuning MIM capacitor design. The measurements at different bias voltage and at high temperature have revealed the need of a better optimization of the VO₂ switch in order to have lower resistance values in the ON state to match the good performances at high temperatures. Nevertheless the VO₂-based reconfigurable capacitive switches offer a unique opportunity to build ultrafast and reliable phase shifters.

ACKNOWLEDGMENT

This work has been supported by the ERC Advanced Grant 'Millitech' of the European Commission, the Swiss National Science Foundation (Grant No. 144268) and the Swiss Federal Office of Energy (Grant No. 8100072).

REFERENCES

[1] E. A. Casu et al., "Hybrid phase-change — Tunnel FET (PC-TFET) switch with subthreshold swing < 10mV/decade and sub-0.1 body factor: Digital and analog benchmarking," in *2016 IEEE International Electron Devices Meeting (IEDM)*, 2016, p. 19.3.1-19.3.4. doi:10.1109/IEDM.2016.7838452

[2] S. D. Ha, Y. Zhou, C. J. Fisher, S. Ramanathan, and J. P. Treadway, "Electrical switching dynamics and broadband microwave characteristics of VO2 radio frequency devices," *J. Appl. Phys.*, vol. 113, pp. 1–25, 2013.

[3] W. A. Vitale et al., "Steep slope VO2 switches for wide-band (DC-40 GHz) reconfigurable electronics," in *72nd Device Research Conference*, 2014, pp. 29–30. doi:10.1109/DRC.2014.6872284.

[4] D. Bouyge and A. Crunteanu, "Applications of vanadium dioxide (VO2)-loaded electrically small resonators in the design of tunable filters," in *European Microwave Conference, 2010*, 2010, no. September, pp. 822–825..

[5] W. A. Vitale et al., "Electrothermal actuation of vanadium dioxide for tunable capacitors and microwave filters with integrated microheaters," *Sensors Actuators A Phys.*, vol. 241, pp. 245–253, Apr. 2016.

[6] T. Teeslink, D. Torres, J. Ebel, N. Sepulveda, and D. Anagnostou, "Reconfigurable Bowtie Antenna using Metal-Insulator Transition in Vanadium Dioxide.," *IEEE Antennas Wirel. Propag. Lett.*, vol. 1225, no. c, pp. 1–1, 2015.

[7] W. A. Vitale et al., "Modulated scattering technique in the terahertz domain enabled by current actuated vanadium dioxide switches," *Sci. Rep.*, vol. 7, no. December 2016, p. 41546, 2017.

[8] Z. Yang, C. Ko, and S. Ramanathan, "Oxide Electronics Utilizing Ultrafast Metal-Insulator Transitions," *Annu. Rev. Mater. Res.*, vol. 41, no. 1, pp. 337–367, Aug. 2011.

[9] J. Leroy et al., "High-speed metal-insulator transition in vanadium dioxide films induced by an electrical pulsed voltage over nano-gap electrodes," *Appl. Phys. Lett.*, vol. 100, no. 21, p. 213507, 2012.

[10] C. Hillman and P. Stupar, "An ultra-low loss millimeter-wave solid state switch technology based on the metal-insulator-transition of vanadium dioxide," *Microw. Symp. Dig. IEEE MTT-S Int.*, pp. 0–3, 2014.

[11] W. A. Vitale et al., "Fabrication of CMOS-compatible abrupt electronic switches based on vanadium dioxide," *Microelectron. Eng.*, vol. 145, pp. 117–119, Sep. 2015.

[12] G. M. Rebeiz, "RF MEMS: theory, design, and technology," *John Wiley & Sons, 2004.*

[13] J. Yoon et al., "Investigation of length-dependent characteristics of the voltage-induced metal insulator transition in VO 2 film devices," *Appl. Phys. Lett.*, vol. 105, no. 8, p. 83503, Aug. 2014.

[14] J. Perruisseau-Carrier et al., "Modeling of periodic distributed MEMS - Application to the design of variable true-time delay lines," *IEEE Trans. Microw. Theory Tech.*, vol. 54, no. 1, pp. 383–392, 2006.

Single Event Effects and Total Ionising Dose in 600V Si-on-SiC LDMOS Transistors for Rad-Hard Space Applications

K. Ben Ali[a], P.M. Gammon[b], C.W. Chan[b], F. Li[b], V. Pathirana[c], T. Trajkovic[c], F. Gity[d], D. Flandre[a] and V. Kilchytska[a]

[a] Université Catholique de Louvain, Louvain-la-Neuve, Belgium
[b] School of Engineering, University of Warwick, Coventry, United Kingdom
[c] Cambridge Microelectronics Limited, Cambridge, United Kingdom
[d] Tyndall National Institute at National University of Ireland, Cork, Prospect Row, Ireland
khaled.benali@uclouvain.be

Abstract—**This work presents a novel Si-on-SiC laterally-diffused (LD) MOSFET structure intended to provide high breakdown voltage of 600 V and be resistant for harsh-environment space applications. Single-event effects (SEE) and total ionizing dose (TID) are investigated for the first time in such device. Initially, the considered Si LDMOS structure on SiC suffers from single-event burnout (SEB) at a drain voltage > 175 V, *i.e.* much lower than the target. An optimized LDMOS structure with a heavily doped extended P$^+$ buried region is proposed and shown to be SEB resistant at the target drain voltage of 600 V, even for a highly-energetic ion with a linear energy transfer (LET) of 90 MeV/mg/cm^2. TID simulations indicate that the main concern is the charge build-up in the thick field oxide (FOX). FOX positive charge density beyond 1×10^{11} cm^{-2} causes the breakdown voltage to drop below 200 V. Different oxide types which feature low "net" positive charge build-up have to be considered to allow for higher TID hardness. The proposed Si/SiC structure with a p+ region was shown to be resistant to a combined SEE and TID (in case of limited positive charge build-up in FOX) as well as combined SEE and high-temperature (up to 573 K) environments. In comparison to the equivalent Silicon-on-Insulator (SOI) LDMOS, the Si/SiC LDMOS structure with p+ buried region features similar immunity to SEB but allows for higher TID hardness.**

Keywords — Power devices; LDMOS, Si/SiC, TID, SEE, SEB

I. INTRODUCTION

Among the wide band gap materials, silicon carbide (SiC) has received increased attention because of its potential for high-power and high-temperature applications [1-2]. A novel Si-on-SiC (Si/SiC) LDMOS power device structure proposed in [3] was demonstrated to feature enhanced tolerance to high-temperature effects. The thermal performance of the Si/SiC substrate is almost equivalent to SiC with a much lower temperature difference between the Si/SiC device's internal junction temperature and the local ambient temperature (due to self-heating), compared to bulk Si or SOI [5].

The LDMOS structures (based on surface conduction) are susceptible to permanent damage when exposed to energetic particles. One of the most serious failures in power devices is known to be a single event burnout (SEB) [6]. It is attributed to a feedback mechanism that is due to the activation of a parasitic transistor when a heavy ion strikes the device in the OFF state [7]. Moreover, high voltage LDMOS devices are particularly sensitive to the Total Ionizing Dose (TID) effects, which mainly affect the oxide layers and degrade the oxide/semiconductor interfaces [8].

The Si/SiC substrate is expected to offer higher radiation hardness compared to bulk Si. This is because of a high Si-C bond energy (large bandgap energy of 3.2 eV) and a high atomic displacement threshold in SiC of 21.8 eV [2], which is significantly higher than in Si. Compared to silicon-on-insulator (SOI) substrates, Si/SiC devices are expected to have a higher TID tolerance due to the absence of a thick buried oxide (BOX) layer under the active Si film and thus a lesser effect from radiation-induced oxide charge trapping on the device performance.

This paper investigates SEB, TID and temperature behaviour of the proposed Si/SiC LDMOS structure through 2D simulations aiming to demonstrate its high tolerance to the harsh environmental conditions of space.

II. SI-ON-SIC LDMOS STRUCTURE

The simulations of the LDMOS structure shown in Fig. 1 [3] were performed using TCAD Atlas-SILVACO software [4]. The proposed Si/SiC LDMOS device is implemented in a 1 μm-thick Si film, which lays directly on the high-resistivity semi-insulating SiC substrate. The use of SiC with its high breakdown electric field allows for higher breakdown voltage compared to SOI counterpart [5]. The use of a Si layer instead of SiC to form the device channel allows for a good interface with the gate oxide, high channel mobility and thus high on-current and low on-resistance. Constant doping of 10^{15} cm^{-3}

978-1-5090-5979-9/17 $31.00 © 2017 IEEE

was employed in the 60-μm long drift region [3] to achieve the target breakdown voltage of 600 V. Fig. 2 presents current-voltage (I-V) curves of the studied LDMOS transistor. It can be seen that a breakdown voltage of > 650 V is achievable with the proposed structure, and the threshold voltage, V_{th}, is ~ 1.4 V. It is worth noting that in order to achieve such breakdown voltage in SOI counterpart, thick BOX has to be considered [9]. In our simulations of the 1 μm-thick Si film counterpart SOI structure we employed a linear graded doping with a maximum of 3×10^{16} cm^{-3} and very thick 5 μm-BOX.

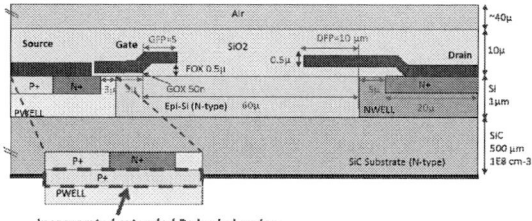

Fig. 1. Simulated Si-on-SiC LDMOS transistors. Inset highlights a modified structure with buried p+.

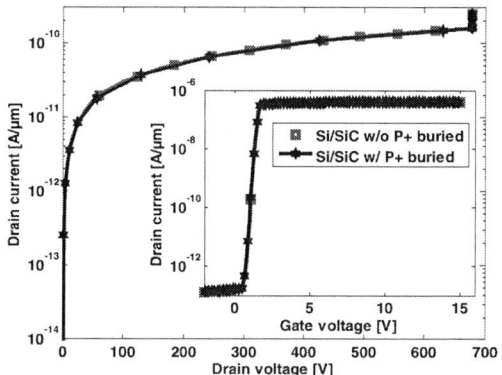

Fig. 2. $I_D(V_D)$ at $V_G = 0$ V of Si/SiC LD-MOS transistor. Inset gives $I_D(V_G)$ curves at $V_D = 1$ V.

III. SINGLE EVENT SIMULATIONS

Single Event Effects (SEE) arise from the interaction of a single particle (*e.g.* alpha or heavy ion) suddenly striking the device, generating electron-hole pairs along its path according to the Linear Energy Transfer (LET) of the particle [10]. Amongst the SEE, SEB is one of the main concerns for LDMOS. It occurs when the parasitic bipolar junction transistor (BJT) is triggered into a self-sustaining forward operating mode as a result of particle-induced current. SEB is sensitive to the particle LET, its trajectory and bias conditions. From the trajectory point of view, the worst-case strike position is at the region where electric field is at its maximum. The most sensitive bias condition for SEB occurrence is OFF state, *i.e.* $V_G = 0$V.

Thus, prior to the SEE simulations, two positions of maximum electric field in the studied device were identified (Fig. 3a). These correspond to the end of gate field plate (GFP) and drain field plate (DFP), x = 35 μm and 78 μm, respectively. Fig. 3b shows the simulated transient current waveforms as a reaction to a heavy-ion strike at one or the other position. It can be seen that even at a drain bias (V_D) of 180 V, the device fails to return to normal/initial operation conditions, self-sustained

current persists and the device burns out. This is confirmed by the time evolution of the holes concentration in response to the ion strike (Fig. 4a). Only when biased at V_D as low as 175 V, the device returns to its normal operation for a LET = 50 MeV/mg/cm^2 after t ~ 40 μs. However, this V_D is much lower than target breakdown voltage of the developed device.

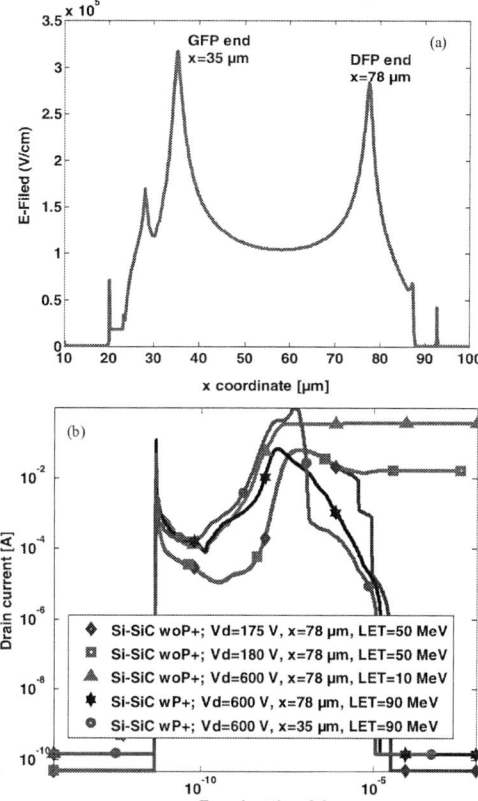

Fig. 3. (a) Electric field as a function of horizontal position (x-coordinate) at depth = 50 nm; $V_G = 0$ V, $V_D = 600$ V. (b) Transient SEE simulations in Si/SiC LDMOS without and with P+ region, at $V_G = 0$ V for different strike conditions. Ion strikes the device at 5ps.

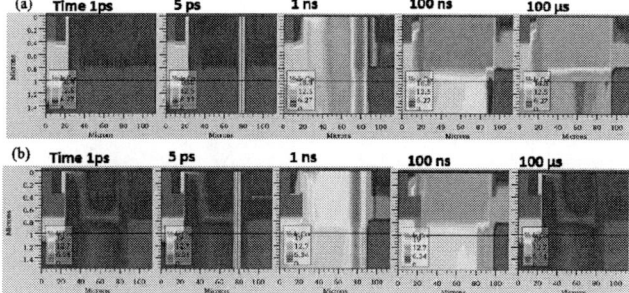

Fig. 4. Simulated holes concentration in Si/SiC LDMOS transistor (a) w/o P+ buried region; $V_D = 180$ V; LET = 50 MeV/mg/cm^2 (SEB) and (b) with P+ buried region; $V_D = 600$ V; LET = 90 MeV/mg/cm^2 (no SEB).

In order to improve SEB tolerance of our LDMOS, the parasitic NPN BJT in the gate-source region has to be suppressed. In this work we propose to incorporate a heavily-doped P+ buried region on the source side as a potential solution (inset in Fig. 1). The introduction of such P+ buried layer was initially discussed in [11]. The highly-doped P+

978-1-5090-5979-9/17 $31.00 © 2017 IEEE

region presents a low resistance path for the collection of holes leading to the reduction of parasitic BJT activation. I-V curves (Fig. 2) confirmed that introducing the P+ buried region does not affect the main device parameters (breakdown voltage, threshold voltage, on-current, etc).

The introduction of the P+ buried region into the Si/SiC LDMOS transistor efficiently suppresses SEB as can be seen in Fig. 3b. Indeed, it resists a heavy-ion strike with very high LET of 90 MeV/mg/cm^2 at the nominal drain voltage of 600 V, suffering no self-sustained current and hence no SEB (Fig. 4b). A similar transient behaviour is observed when the ion hits the device at either position of maximum electric field (Fig. 3b). The maximum current is higher in the case of a strike at x=35 µm (probably as a result of higher E-field at this point), but it drops faster (due to the close proximity of the P+-buried region). Hence, the total collected charge is higher in the case of ion strike at x=35 µm (Q$_{col}$= 4.8x10^{-8} C) than in the case of x=78 µm (Q$_{col}$= 6.8 x10^{-9} C); however, in both cases, the device returns to initial OFF operation without burn-out.

Comparing to simulations of SOI counterpart (not shown here in details due to lack of space) it was revealed that the proposed Si/SiC LDMOS with P+ buried layer features similar immunity to SEB.

IV. TOTAL IONIZING DOSE (TID) SIMULATIONS

Ionization damage is related to electronic bonds disruption, caused by charged particles or gamma-rays, which create free electron-hole pairs in oxide. TID effect results from progressive charge build-up in insulating layers and the creation of interface states. The main consequences of TID are V$_{th}$ shift, mobility (and thus on-current) degradation and leakage current increase. In our simulations, TID effect was emulated through the introduction of fixed positive charge in the 50 nm-thick gate oxide, Qf$_{GOX}$, and 500 nm-thick field oxide, Qf$_{FOX}$. Interface state creation was neglected to the first order. Qf$_{GOX}$ was varied from 0 up to 5x10^{11} cm^{-2}. As the field oxide is almost one order of magnitude thicker than the gate oxide, and the radiation-induced charge density is, to the first order, proportional to the oxide thickness, the tests for various Qf$_{FOX}$ were carried out up to 5x10^{12} cm^{-2}. These values typically correspond to a TID of 10^2-10^3 krad(Si) (depending on the oxide fabrication and bias applied during radiation) [12, 13]. Such dose values cover specifications for space applications (even for long/durable missions).

Fig. 5 presents I$_d$-V$_d$ curves of studied LDMOS transistor at V$_G$ = 0 V for different Qf conditions. Firstly, the breakdown voltage is not affected by Qf$_{GOX}$ up to 5x10^{11} cm^{-2}. The increase of the off-current level for Qf$_{GOX}$ = 5x10^{11} cm^{-2} is due to the V$_{th}$ shift (inset of Fig.5a). Secondly, simulations performed for various Qf$_{FOX}$ revealed that radiation induced charge build-up may result in a very strong breakdown voltage reduction dropping below 200 V (inset in Fig. 5b) for a Qf$_{FOX}$>5x10^{11} cm^{-2}.

From the TID simulations we conclude that the main concern is the radiation-induced positive charge build-up in the FOX. Thus, various solutions for the FOX hardening, as e.g. nitrided oxides or "sandwiched" oxides (SiO$_2$-Si$_3$N$_4$-SiO$_2$ [14], SiO$_2$-BPSG-SiO$_2$ [15]) with low "net" positive charge build-up during radiation have to be considered for the TID hardened device.

It is worth mentioning that the SOI counterpart is more sensitive to TID effects than proposed Si/SiC structure due to charge trapping in the BOX, which is particularly thick (5µm) in our case for LDMOS with a breakdown voltage of 600 V.

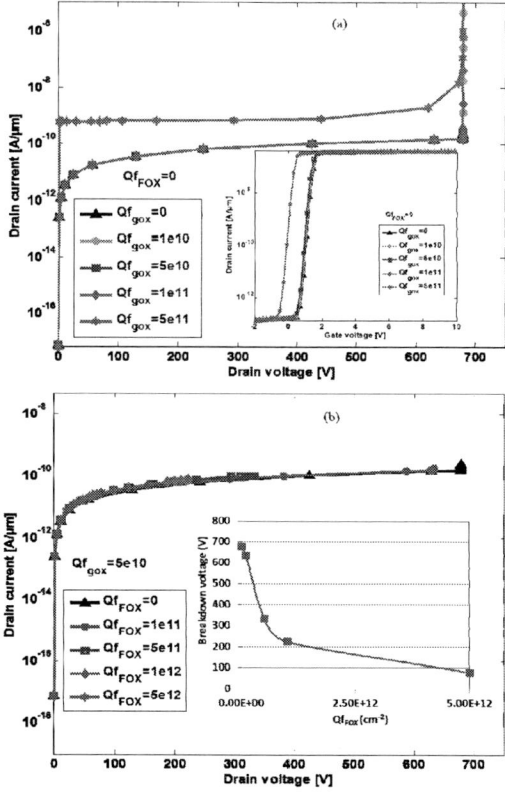

Fig. 5. (a) I$_D$(V$_D$) and I$_D$(V$_G$) of Si/SiC LDMOS for various Qf$_{GOX}$ (Qf$_{FOX}$ = 0). (b) I$_D$(V$_D$) and breakdown voltage for various Qf$_{FOX}$ (Qf$_{GOX}$ = 5x10^{10} cm^{-2}) with P+ buried region.

V. COMBINED TID AND SEE SIMULATIONS ON THE LDMOS STRUCTURE WITH P+ BURIED REGION

The combined effect of TID and SEE was simulated in the device with the P+ buried region. The radiation-induced positive charge in the gate oxide was fixed to 5x10^{10} cm^{-2} and in the field oxide was varied up to 5x10^{11} cm^{-2}. In these simulations, Qf$_{FOX}$ was limited to prevent the breakdown voltage dropping below 300 V (see inset in Fig. 5b). All simulations were done in OFF-regime, V$_G$ = 0 V, and at the V$_D$ close to the breakdown, chosen according to the inset in Fig. 5b. Ion strike was at the end of the drain field plate. Fig. 6 shows SEE current transient for three cases of Qf$_{FOX}$. One can see that self-sustained current does not appear at least up to Qf$_{FOX}$ = 5x10^{11} cm^{-2}.

VI. COMBINED TEMPERATURE AND SEE EFFECTS OF Si/SiC LDMOS TRANSISTORS WITH P+ BURIED REGION

The effect of temperature (T) on the Si/SiC LDMOS with incorporated P+ buried layer was simulated from room-temperature up to 300 °C. Fig. 7 shows I$_D$-V$_G$ curves and inset gives evolution of I$_D$-V$_D$ curves in OFF-state (V$_G$ = 0 V) at different T. As expected: i) leakage current increases with T following intrinsic carrier concentration, n_i; ii) V$_{th}$ reduces

978-1-5090-5979-9/17 $31.00 © 2017 IEEE

with T (~ 1 V over studied temperature range) due to the reduction of Fermi potential and a reduction in depletion width; iii) on-current (and hence on-resistance) degrades at higher T (almost 4x in 27 to 300 °C range) mostly due to mobility reduction. Breakdown voltage increases with T due to reduction of the mean free path of the carriers, thus requiring a higher field to initiate impact ionization. Fig. 8 shows SEE current transient at various T. Self-sustained current (and thus SEB) does not appear up to 300 °C. Maximum transient current decreases and waveform extends slightly to longer time when T increases. This trend can be related to the mobility reduction at higher T as well as to the higher breakdown voltage at high temperature.

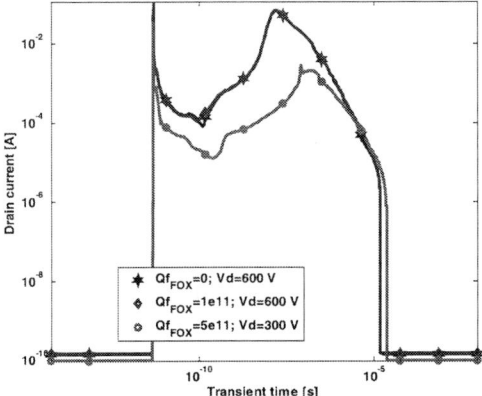

Fig. 6. Transient SEE simulations of Si/SiC LDMOS with P+ buried region with different QfFOX. $V_G = 0$ V, LET = 90 MeV/mg/cm^2.

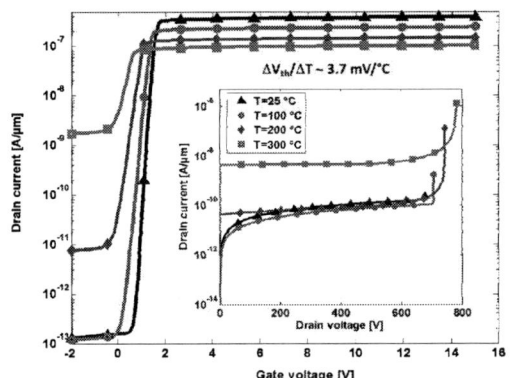

Fig. 7. $I_D(V_G)$ curves of Si/SiC LD-MOS transistor at different temperatures ($V_D = 1$ V). Inset gives $I_D(V_D)$ curves at $V_G = 0$ V.

VII. CONCLUSION

The radiation hardness of a novel Si/SiC LDMOS structure has been studied through physical device simulations. SEE, TID and temperature effects as well as their combinations have been considered. The ways to improve radiation tolerance are discussed. The proposed Si/SiC 600 V LDMOS structure with extended P+ buried region has shown an excellent SEB hardness, as good as SOI counterpart. Moreover, contrary to SOI-based LDMOS, the proposed device does not suffer from TID effects related to the BOX (which is particularly thick for a high breakdown voltage LDMOS). Thus, an optimized 600 V Si/SiC LDMOS is

expected to be very promising for rad-hard space applications at high temperature.

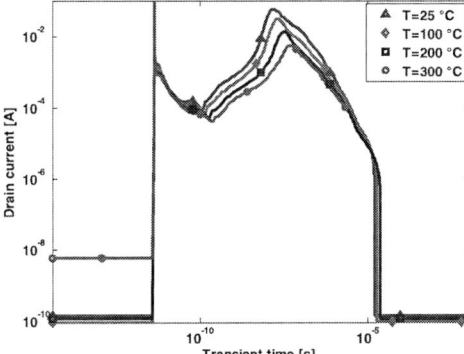

Fig. 8. Transient SEE simulation at various temperatures of Si/SiC LDMOS with P+ buried region. $V_G = 0$ V, LET = 60 MeV/mg/cm^2.

ACKNOWLEDGMENT

The work was funded by EC through SaSHa Project (Si on SiC for the Harsh Environment of Space).

REFERENCES

[1] J.A. Cooper, M.R. Melloch, R. Singh, A. Agarwal, J.W. Palmour, "Status and prospects for SiC power MOSFETs", *IEEE Trans. Electron. Dev.*, Vol. 49, pp. 658-664, 2002.

[2] T. Kimoto, J.A. Cooper, "Fundamentals of Silicon Carbide Technology: Growth, Characterization, Devices and Applications", John Wiley & Sons, 2014.

[3] P.M. Gammon, F. Li, C.W. Chan, et al., "Silicon-on-silicon-carbide power devices for harsh environment applications", *ECSCRM*, Halkidiki, Greece, 2016.

[4] ATLAS user's manual, SILVACO, Inc, www.silvaco.com.

[5] C. Chan, P. A. Mawby, P. M. Gammon, "Analysis of Linear-Doped Si/SiC Power LDMOSFETs Based on Device Simulation", *IEEE Trans. Electron. Dev.*, vol. 63, pp. 2442-2448, 2016.

[6] J. L. Titus, G. H. Johnson, R. D. Schrimpf, K. F. Galloway, "Single event burnout of power bipolar junction transistors", *IEEE Trans. Nucl Sci*, vol. 38, pp.1315–22, 1991.

[7] A. E. Waskiewicz, J. W. Groninger, V. H. Strahan, "Long DM. Burnout of power MOS transistors with heavy ions of 252- Cf", *IEEE Trans Nucl Sci*, vol. 33, pp. 1710–1713, 1986.

[8] P. Fernández-Martínez, F.R. Palomo, S. Díez, S. Hidalgo, M. Ullán, D. Flores, R. Sorge, "Simulation methodology for dose effects in lateral DMOS transistors", *Microelectronics Journal*, Vol. 43, n°1, pp.50-56, 2012.

[9] S. Merchant, E. Arnold, H. Baumgart, S. Mukherjee, H. Pein, and R. Pinker, "Realization of high breakdown voltage (>700 V) in thin SOI devices,"in *Proc.ISPSD*,Apr.1991,pp.31-35.

[10] G.H. Johnson, K.F. Galloway, R.D. Schrimpf, J.M. Palau, C. Dachs, "A review of the techniques used for modeling single-event effects in power MOSFETs", *IEEE Transs Nucl Sci*, vol. 43, pp. 546-560, 1996.

[11] Patrick M.S, "Lateral Power MOSFETs Hardened against Single Event Radiation Effects", *Ph.D Thesis*, University of Central Florida. 2006.

[12] G.C. Messenger, M.S. Ash, "The effects of radiation on electronic systems". 2nd ed. New York: Van Nostrand Reinhold, 1992.

[13] H.J. Barnaby, "Total Ionizing Dose Effects in Modern CMOS Technologies", IEEE Trans. On Nucl. Sci., vol. 53, No. 6, Dec. 2006.

[14] I.P. Barchuk, V.I. Kilchitskaya, V.S. Lysenko, et al. "Electrical Properties and Radiation Hardness of SOI Systems with Multilayer Buried Dielectric", *IEEE Trans. on Nucl. Sci.*, vol.44, No.6, 1997.

[15] R.L. Woodruff, J.T.Chaffee, C. Hafer, "Multi-layered field oxide structure", US Patent 5037718, 1991.

PPAC Scaling Enablement for 5nm Mobile SoC Technology

Mustafa Badaroglu[1*], Jeff Xu[1], John Zhu[1], Da Yang[1], Jerry Bao[1], Seung-Chul (S.C.) Song[1], Peijie Feng[1],
Romain Ritzenthaler[2], Hans Mertens[2], Geert Eneman[2], Naoto Horiguchi[2], Jeffrey Smith[3], Suman Datta[3],
David Kohen[4], Po-Wen Chan[5], Keagan Chen[5], and PR. Chidi Chidambaram[1]

[1]Qualcomm Technology Inc, San Diego, CA USA, [2]IMEC, Leuven, Belgium, [3]University of Notre Dame, Notre Dame, IN, USA,
[4]ASM International, Leuven, Belgium,[5]Applied Materials, Sunnyvale, CA USA
*mustafab@qti.qualcomm.com

Abstract— **We present a 5nm logic technology scaling step-up holistic approach for 5-track standard cell design employing electrically gate-all-around nanowire architecture (EGAA NW) with much reduced parasitic capacitance and increased effective width for better short channel control and stronger drive. We suggest SiGe P-channel by Ge Condensation for intrinsic mobility improvement and substrate strain, conformal wrap-around contact (CWAC) to reduce contact resistance with minimum parasitic capacitance penalty, metal gate (MG) stressor to improve N-channel mobility, EUV single exposure metal patterning with improved tip-to-tip patterning technique for maximum mask count reduction, and Al metallization to reduce metal & via resistances, however still requiring a validation of the proposed electromigration (EM) risk mitigation. We show that finFET can still be extended to 5nm technology to meet Power-Performance-Area-Cost (PPAC) targets. EGAA NW could enable further 50mV less supply voltage to significantly improve 5nm PPAC scaling.**

Keywords—5nm, More Moore scaling, process technology, integration, finFET, gate-all-around, nanowire, PPAC.

I. INTRODUCTION

The grand logic technology scaling challenges beyond 7nm have been: 1) mediocre performance gain at scaled supply voltage (Vdd) because of lack of new device performance elements along with high BEOL wire and via resistances as well as high parasitics at cell level, and 2) ever increasingly high wafer cost and lower yield due to multi patterning and process complexities. There is a need to find out process step-up elements needed to achieve node-to-node scaling targets (every 2-3 years) listed below:

- (P)erformance: >25-35% more frequency at constant power, >10-15% more frequency at scaled supply voltage,

- (P)ower: >40-50% less energy at a given performance,

- (A)rea: >50% less area for the same logic function,

- (C)ost: <30% limited wafer cost increase – 30-40% less die cost for the scaled die with the same function.

While targeting 10-15% frequency improvement at scaled supply voltage and scaled logic area, power density needs to be kept under control because of thermal constraints [1][2]. Power density is given by $C.Vdd_{boost}^2.fmax$ at unit cell footprint where C is the total capacitance, Vdd_{boost} is the boosted (turbo) supply voltage and fmax is the maximum chip frequency at that supply voltage. Power density is controlled by concurrently reducing the capacitance and supply voltage. Ensuring 57% reduction in $C.Vdd^2$ reduction allows a frequency (fmax) increase of 15% at scaled supply voltage while enabling 50% scaled logic area without a penalty in power density. Overview of low-power design methodologies applicable to nanometer-scale technologies is given in [3][4].

Proposed ground rules for the 5nm System-on-Chip (SoC) technology are listed in Table 1. Those rules ensure desired (>50% less) area scaling accompanied with reduced cell height by design-technology-co-optimization (DTCO) constructs [5].

Table 1: Ground rules of 5nm node.

Ground rule	Value
Gate pitch	42nm
Gate length	16nm
Metal1 pitch	28nm
Metalx pitch	28nm
Fin pitch	24nm
Fin Height	36nm

In this paper, we present how to achieve much needed 5nm SoC performance at reduced power and lower cost through novel device, interconnect, patterning, and cell architecture innovations [6]. Finally, PPAC (Power-Performance-Area-Cost) assessment of various 5nm scaling scenarios will be presented to find the optimal scaling path to achieve 5nm mobile SoC PPAC scaling goals.

II. TECHNOLOGY STEP-UPS NEEDED FOR 5NM SCALING

A. High-mobility channel and stress boosters

In order to extend FinFET to 5nm node, SiGe P-channel and compressive N-channel metal gate (MG) stressor are proposed to improve CMOS drive strength. SiGe fin formation by Ge Condensation (Figure 1) to lower wafer cost, eliminate SiGe fin defect, and selectively form SiGe fin for high performance devices. The compressive metal gate (MG) stressor could be formed by creating a compressive stressor above the N-channel fin to achieve > 40% electron mobility

978-1-5090-5979-9/17 $31.00 © 2017 IEEE

gain as shown in Figure 2. Those two knobs could be effectively brought up for full CMOS integration, for instance cyclic-condensation to achieve the desired Ge concentration for PMOS while selective MG stressor for NMOS only to avoid a potential performance degradation for PMOS.

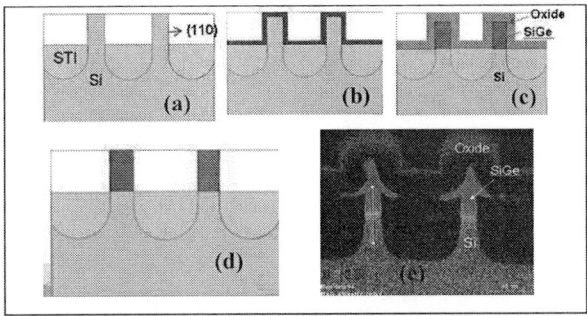

Figure 1: SiGe condensation to form SiGe channel for PMOS mobility improvement. Process flow: (a) Si fin, (b) conformal deposition of polycrystalline or aSiGe, (c) wet and dry oxidation to drive Ge into Si, (d) removal of oxide to form the final SiGe fin (e) Experimental validation at ASM and imec with Ge% around 70%. Residue Ge on STI is due to thick oxide that blocked oxidation. Improvement is in progress with thinner poly SiGe to target 45% Ge fin.

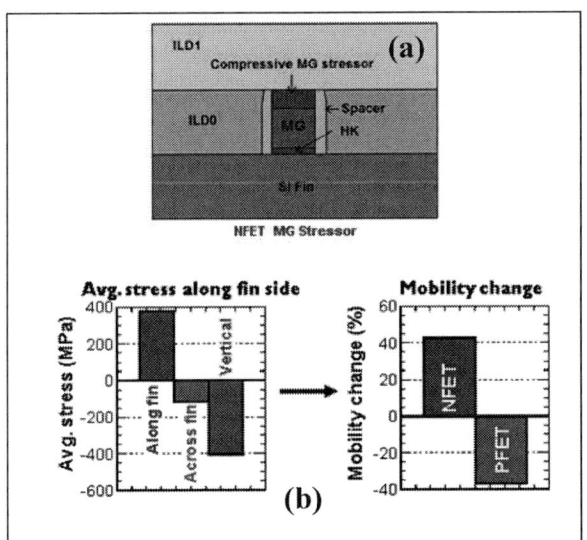

Figure 2: (a) Selective NMOS gate stressor, (b) imec TCAD simulations shows up to 40% mobility boost for NMOS.

B. Self-Aligned Conformal Wrap-Around-Contact

Controlling source/drain series resistance within tolerable limits will become much more difficult. Due to the increase of current density, the demand for lower resistance in smaller dimensions at the same time poses a great challenge [7][8]. One tradeoff of the previously reported Wrap-Around-Contact (WAC) [9] is the uptick of Contact to Gate capacitance due to extending Contact plug down to the bottom of Source/Drain (S/D) shown in Figure 3. An improved self-aligned Conformal Wrap-Around-Contact (CWAC) with 4-6 nm thick Direct

Contact metal or silicide conformably deposited on S/D is proposed to reduce Contact to MG parallel capacitance (Figure 3). CWAC could also relax Contact to S/D enclosure rules and help construct compact standard cells.

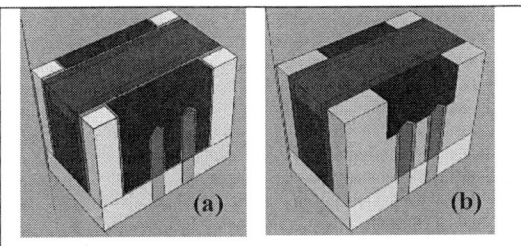

Figure 3: (a) Conventional wrap-around Contact scheme with filling the gap across the full trench, and (b) Conformal wrap-around Contact with capacitance reduction with respect to the conventional scheme.

C. Al dual damascene metallization

Alternative metals such as Co and Ru have been proposed to replace Cu for metallization. However, none of proposed alternative metals outperforms Cu in wire resistance down to wire CD of 14nm. To break wire resistance barrier, Al dual damascene (DD) metallization is explored while solving its electromigration (EM) limitation. Since self-limiting Al_2O_3 will be formed at Al and low-k interface, effective volume of Al at 14nm wire CD will be 80% larger than that of Cu with 2.5nm barrier/liner thickness and 2nm Co EM cap. Al DD fill has been demonstrated using selective CVD Al via fill followed by CVD Al Trench fill with a 380-400C in-situ Al reflow anneal to promote bamboo Al grain structure formation as shown in Figure 4. Figure 5 shows the simulation results that demonstrates the potential 50% wire and via resistance reductions for Al DD metallization because of 80% larger effective volume, larger grain structure, and shorter mean free path compared to conventional Cu DD metallization. Since Al surface is self-passivated, 1X metal airgap integration is possible. Interconnect stand-alone is expected to bring significant AC performance improvement around 10% at sub-7nm technology nodes [10].

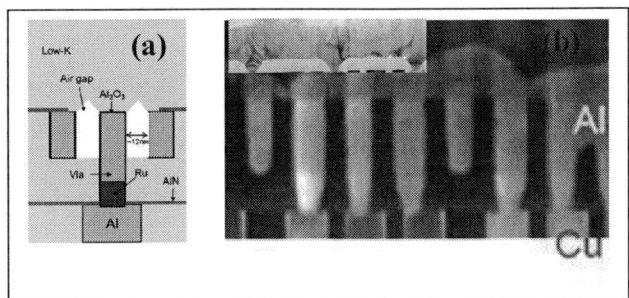

Figure 4: (a) Al interconnect with self-forming barrier allowing airgap integration at tight pitch, and (b) experimental validation of the CVD Al DD fill with bamboo-like grains (inset) at AMAT. Electromigration risk mitigation still needs to be validated.

The biggest EM performance limiter of conventional Al metallization is the so-called "triple points" where three Al grains meet. For wire CD < 50nm with Al reflow anneal, Al is expected to form bamboo-like structure to improve EM life time and match Cu EM activation energy [11].

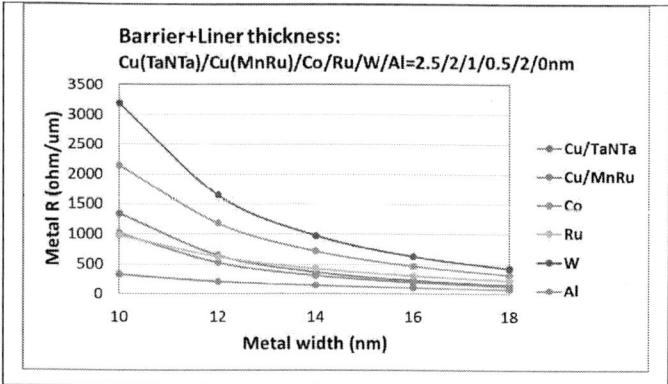

Figure 5: Comparison of unit resistance of Al interconnect versus other metallization schemes including Cu.

D. 5-track standard cell library

Since ground rules scaling alone will not be able to scale logic area by 50%, 5-track standard cell architecture is proposed to achieve 140nm cell height and scale standard cell and logic area over 50%. 5-track cell architecture is shown in Figure 6 along with high aspect ratio power rail (HAR PR) and stapled M1 to M0 power rail for IR drop mitigation. Self-Aligned Gate Contact (SAGC), improved tip-to-tip and extension rule in 1D vertical patterns will be needed at reduced cell height. EUV single exposure for direct patterning of line/space will be needed to reduce the wafer cost and process complexity enabled by reduced mask count.

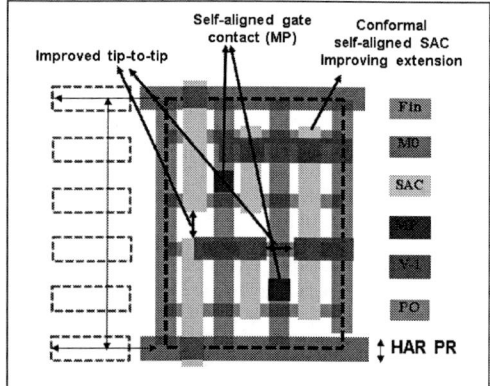

Figure 6: 5-track standard cell architecture template.

III. ELECTRICALLY GATE-ALL-AROUND (EGAA) NANOWIRE

With contacted poly pitch (CPP) scaled to 42nm and physical gate length to 16nm, one can obtain an acceptable DIBL of 80mV/V and subthreshold slope (SS) of 90mV/dec for vertical FinFET per TCAD numerical simulation. To minimize parasitic capacitance penalty of conventional physically GAA NW device architecture, vertically stacked hexagonal NW (Figure 7) is proposed to achieve electrically gate-all-around (EGAA) gate control for superior SCE performance and continuously stacked architecture and thus

provide larger effective channel width for the same structure height than its FinFET counterpart. Much improved SEC performance with remarkably low DIBL of 25mV/dec and much reduced SS of 67mV/V is shown in Figure 8. Cross-section of hexagonal EGAA NW structure is shown in Figure 9 along with Figure 10 showing the first electrical results revealing comparable SEC and better DC performance for the electrically GAA NW compared to the conventional physically GAA NW.

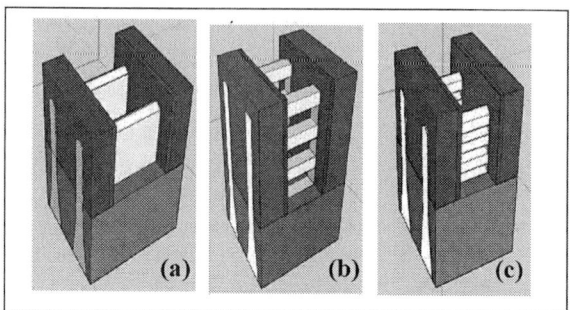

Figure 7: (a) FinFET, (b) Conventional physical GAA with tall structure height to get comparable W_{eff}, and (c) Hexagonal electrically GAA structure with the comparable height and W_{eff} as in FinFET.

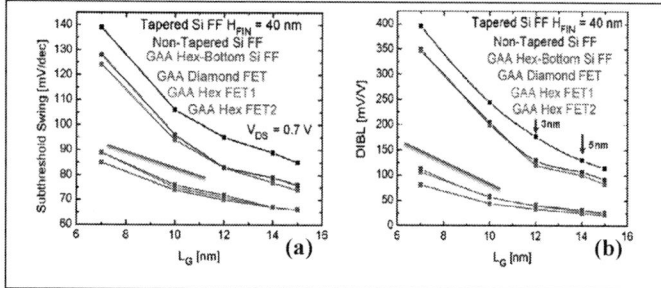

Figure 8: (a) Subthreshold slope, (b) DIBL from FinFET and hexagonal GAA structure as function of Lgate. GAA NW is the only device could support Lg scaling beyond 12nm. TCAD simulations in Notre Dame Univ.

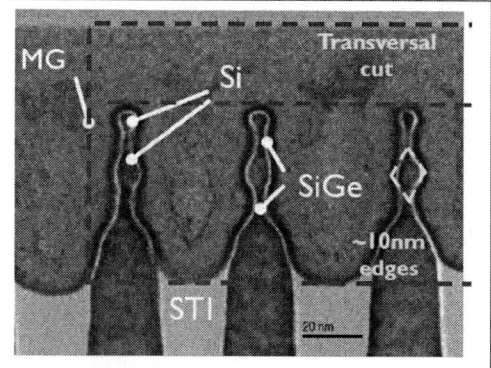

Figure 9: Hexagonal nanowire physically validated and characterized at imec. Inset is a physical structure of HNW.

978-1-5090-5979-9/17 $31.00 © 2017 IEEE

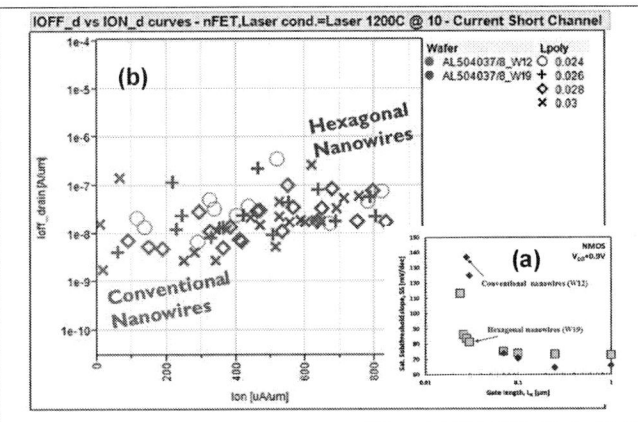

Figure 10: (a) Measured SS as function of gate length and (b) Measured Ion/Ioff . Better performance for electrically GAA Hexagonal Nanowire than conventional physically GAA Nanowire demonstrated at imec.

IV.   5NM PPAC ASSESSMENT

SoC-level PPAC is assessed using a customizable MIT VSM compact model [12]and circuit-level netlist modeling the wire-loaded critical path in SoC. Parasitics coming from the device and standard cell architecture are taken into account. The reference model card and circuit netlist are fit to the 7nm device and SoC data. Corresponding 5nm device model and circuit are scaled to reflect the new ground rules of 5nm. SCE and mobility of new device geometry have been validated in device TCAD. Wireload has been modeled as a star topology in pi2 configuration with its branches driving three D4 cells where each branch takes into account the wirelength and the number of vias as an outcome of physical design.

Simulations show that layout-only scaling FinFET of 7nm node at Vdd of 0.75V to 5nm node at Vdd of 0.7V ends up with 16% degraded performance. With the introduction of all the aforementioned boosters, optimized FinFET improves frequency by 19% with 27% lower power at 0.7V Vdd. Conformal Wrap-Around Contact and Al interconnect bring the highest gains in lifting 5nm FinFET performance to the target levels. Replacing FinFET with EGGA NW and implementing Air Spacer [6] without any layout change achieves remarkable 28% frequency gain and 51% lower power at 0.65V Vdd.

Speed-power improvements for FinFET and EGAA NW are shown in Figure 11. Projected 5nm SoC die size will be scaled by 45% to achieve 35% die cost reduction with contributions from aggressive ground rules scaling, 5-track standard cell architecture, and EUV SE metal patterning.

V.   CONCLUSIONS

Unprecedented innovations are required to break PPAC scaling barrier to provide adequate values and motivate device manufacturers to migrate to 5nm SoC technology. FinFET can be extended to 5nm node with innovations described in this paper. Electrically GAA NW provides an opportunity to further scale 5nm SoC supply voltage Vdd to 0.65V while achieving remarkable 28% frequency gain with 51% power reduction.

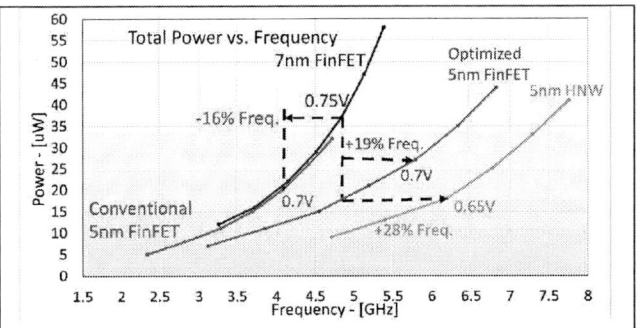

Figure 11: Optimized 5nm FinFET offers 19% frequency gain at 0.7V Vdd while optimized Hexagonal GAA NW gives a remarkable 28% frequency gain at 0.65V Vdd.

REFERENCES

[1]   M.E. Mason, "Design enablement: the challenge of being early, accurate, and complete," VLSI 2012, pp. 145-146, June 2012.

[2]   W. Huang, K. Rajamani, M. Stan, and K. Skadron, "Scaling with design constraints: predicting the future of big chips," IEEE Micro, Vol. 31, No. 4, pp. 16-29, Jul-Aug 2011.

[3]   K. Arabi, "Low power design techniques in mobile processes" ISLPED, pp. 1-2, August 2014.

[4]   S. Kosonocky, T. Burd, K. Kasprak, R. Schultz, and R. Stephany, "Designing in scaled technologies: 32nm and beyond", VLSI, pp. 147-148, June 2012.

[5]   M. Badaroglu, K. Ng, M. Salmani, S.G. Kim, G. Klimeck, C.-P. Chang, C. Cheung, and Y. Fukuzaki, "More Moore landscape for system readiness - ITRS2.0 requirements," ICCD 2014, pp. 147-152, 2014.

[6]   2015-2017 USPO Patents, Filed, Pending and Granted, 2015-2017 – 9257556, 9196583, 9543248, 9306066, 20150262875, 9472453, 9240480, 20160133714, 20160141250, 20160293485, and others.

[7]   S. Datta, R. Pandey, A. Agrawal, S. K. Gupta, and R. Arghavani, "Impact of contact and local Interconnect scaling on logic performance," VLSI 2014, pp. 1-2, 2014.

[8]   Keshavarzi et al, "Architecting advanced technologies for 14nm andbeyond with 3D FinFET transistors for the future SoC applications," IEDM 2011, pp. 4.1.1-4.1.4, December, 2011.

[9]   S.C. Song, J. Xu, N.N. Mojumder, K. Rim, D. Yang, J. Bao, J. Zhu, J. Wang, M. Badaroglu, V. Machkaoutsan, P. Narayanasetti, B. Bucki, J. Fischer, and G. Yeap, "Holistic technology optimization and key enablers for 7nm mobile SoC," Digest IEEE Symp. VLSI Technology, pp. T198-199, Kyoto, Japan, June 16-19, 2015.

[10]   M. Badaroglu and J. Xu, "Interconnect-aware device targeting from PPA perspective," Proc. IEEE/ACM Int. Conf. on Computer-Aided Design (ICCAD), pp 1-6, Austin, TX, USA, November 2016.

[11]   C.V. Thompson and H. Kahn, "Effects of microstructure on interconnect and via reliability: Multimodal failure statistics," J. Electron. Materials, Vol. 22, pp. 581-587, 1993.

[12]   A. Khakifirooz, O. M. Nayfeh, and D. Antoniadis, "A simple semi-empirical short-channel MOSFET current-voltage model continuous across all regions of operation and employing only physical parameters," IEEE TED 2009, vol. 56, No. 8, pp. 1674-1680, 2009.

Hybrid InGaAs/SiGe CMOS Circuits with 2D and 3D Monolithic Integration

V. Deshpande, H. Hahn, V. Djara, E. O'Connor, D. Caimi, M. Sousa, J. Fompeyrine and L. Czornomaz

IBM Zurich Research Laboratory,
Saumerstrasse 4, CH-8803 Rueschlikon,
Switzerland

(Invited Paper)

Abstract—**Advanced CMOS nodes target high-performance at lower supply voltage. High-mobility III-V channel materials have the potential to meet this target. Although III-V materials such as InGaAs are beneficial for nFET channels, SiGe (or Ge) provides better hole mobility and is more suited for pFET channels. Therefore, a InGaAs/SiGe hybrid CMOS technology is being pursued for scaled nodes. There are significant challenges to co-integrate these two materials in a scalable process. In this regard, here, we present some of our recent work in InGaAs/SiGe CMOS integration through a novel direct epitaxy process for co-planar 2D integration. We also present our efforts in 3D monolithic integration of InGaAs-on-SiGe for CMOS and beyond.**

I. Introduction

High-mobility materials such as InGaAs and SiGe alloys are being considered as potential options for channel materials in leading-edge CMOS technology nodes. SiGe is already being introduced as pFET channel material at 7 nm node. InGaAs based alloys have become leading choice in the research for nFET channel material for sub-7 nm node [1]. Therefore, a CMOS technology with hybrid channel materials, namely, SiGe (or Ge) for pFETs and InGaAs for nFETs is being envisaged. So it is imperative to develop a robust and scalable CMOS integration scheme starting from III-V material integration on Silicon to dense circuits. Due to largely different process conditions between InGaAs and SiGe, especially with regards to the thermal budget, developing a CMOS process becomes very challenging. A reliable CMOS integration scheme can only be developed with a proper choice of material integration methods (for III-V on Silicon). Although direct wafer bonding (DWB) based integration led to the first demonstrations of InGaAs/SiGe CMOS [2], it faces challenges in terms of topography. Through a novel direct epitaxy method, we show integration approaches for a conventional co-planar 2D CMOS [3]. Utilizing standard front-end-of-line processes, we demonstrate a dense 2D integration of InGaAs and SiGe channels on Si substrate. Scaled gate length InGaAs nFETs and SiGe pFETs along with inverters and dense SRAM cells with finFET devices are demonstrated.

Besides this, 3D Monolithic (3DM) integration is attracting much attention owing to density scaling benefits and the potential to stack independently optimized multifunctional layers at transistor level [4]. However, due to sequential processing of transistor layers, the thermal budget of top transistor layer can be detrimental to the bottom layer transistors. This puts a limit on the top layer thermal budget, which is typically around 600°C when the bottom layer has Si/SiGe transistors. As conventional high performance Si/SiGe MOSFET process requires high thermal budget, an alternate low temperature top layer Si/SiGe process development is necessary for a 3DM scheme and thus presents challenges to obtain high-performance MOSFETs. On the other hand, the InGaAs MOSFET processing thermal budget is significantly lower and makes it naturally suited to be used as the top layer channel material. Moreover, the higher channel mobility can enable high performance at lower supply voltages. In this context we demonstrate, 3D monolithic integration of InGaAs nFETs on SiGe-on-Insulator pFETs with state-of-the-art device integration on both levels [5]. 3D inverters down to short gate lengths of $L_g \sim 30$ nm are demonstrated. Going beyond CMOS circuits, 3DM integration has also been exploited to demonstrate InGaAs RF devices over SiGe pFETs, paving the way for 'functional scaling'.

II. Co-planar 2D CMOS Integration

A schematic of InGaAs/SiGe co-planar 2D CMOS integration by direct epitaxy is shown in Fig. 1. This method offers a truly co-planar integration along with '-on-Insulator' device architecture which could enable Fully-depleted Silicon-on-Insulator (FDSOI) like technology. It allows a dense co-integration of InGaAs nFET with SiGe pFETs and is scalable to large wafer sizes.

The process flow of the InGaAs/SiGe CMOS integration by direct epitaxy as demonstrated in ref [3] is shown in Fig. 2. Top

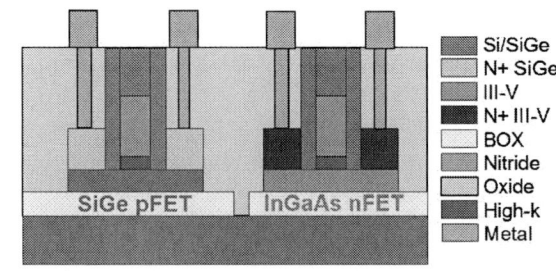

Fig. 1. Schematic showing InGaAs/SiGe co-planar 2D CMOS integration. Adapted from ref [3]

Direct epitaxy

- SiGe-OI wafer
- pFET active region etch
- Empty cavity fabrication
- InGaAs selective epitaxy
- Capping oxide removal
- nFET active region etch
- High-k/metal gate deposition
- Gate patterning
- SiN_x Spacer deposition
- Spacer etch on nFET
- RSD of n+ $In_{0.5}Ga_{0.5}As$ on nFET
- Spacer etch on pFET
- NiSiGe on pFET
- ILD deposition and contact opening
- Metal contacts (M1)

Fig. 2. Process flow of InGaAs/SiGe co-planar 2D CMOS integration by direct epitaxy. Adapted from ref [3].

view SEM images of at various steps of the process flow are shown in Fig. 3. The process starts with a SiGe-OI wafer and firstly, pFET active regions are formed. It is then followed by series of process steps (described in detail in ref [6]) to form SiO2 cavities which have an access to Si substrate through a seed window(Fig. 3b). This location acts as a crystalline seed to start the direct growth of III-V materials as demonstrated in ref [7] [6]. The empty SiO_2 cavities are filled by InGaAs epitaxy step (Fig. 3c). Defect filtering in this growth step occurs with a change in direction of growth from vertical to lateral. The thickness of so grown InGaAs layer is set by the height of the SiO_2 cavity. A 30 nm thick InGaAs (In content = 70 %) layer was grown in ref [3]. After InGaAs growth, the encapsulating SiO_2 which formed the cavities is removed and InGaAs channel is exposed(Fig. 3c). Thus co-planar InGaAs and SiGe active regions are obtained. Now, standard CMOS front-end-of-line processes can be carried out to form InGaAs nFETs and SiGe pFETs. In ref [3], a gate-first flow was carried out. A common high-k dielectric and metal gate stack was deposited and gates were patterned on both n- and p-FET regions. Then spacer layer was deposited. Spacers were first formed only on nFET regions by blocking pFET region during dry etching. Thereafter, highly n-doped InGaAs raised source/drain (RSD) was grown. This was followed by spacer formation on pFET regions and Ni-SiGe alloy formation to form S/D region for SiGe pFETs. Finally encapsulating oxide was deposited, contact holes opened and metallization was carried out. Both planar as well as finFETs were fabricated for InGaAs and SiGe devices. A dense integration of InGaAs and SiGe active areas with 25 nm spacing was demonstrated for the first time on Silicon platform. Although, simplified process

was used for pFET S/D formation, it can readily be extended to state-of-the-art device integration with RSD. Large SRAM arrays with dense cell size were fabricated.

Well behaved device characteristics featuring n- and p-FET gate lengths down to $L_g=35$ nm were demonstrated. For the first time, InGaAs/SiGe SRAM cells were demonstrated down to 0.4 um^2. The electrical characteristics of an inverter, a relaxed cell-size SRAM, and a dense SRAM are shown in Fig. 4. This shows the scalability and robustness of the CMOS integration scheme.

III. 3D MONOLITHIC CMOS INTEGRATION

The process flow of the 3DM integration demonstrated in ref [5] is shown in Fig. 5. First the bottom layer SiGe-OI fin pFETs were fabricated with a gate-first (GF) process as described in [8], [9]. The top layer nFET fabrication was carried out after the silicidation step of SiGe-OI finFET process. It begins with InGaAs layer transfer on top of fabricated pFETs. Firstly, a inter-layer oxide was deposited and chemical-mechanical-polish (CMP) planarization was carried out. The InGaAs layer (53% In content) was transferred on to this oxide with direct wafer bonding from 2 inch InP donor

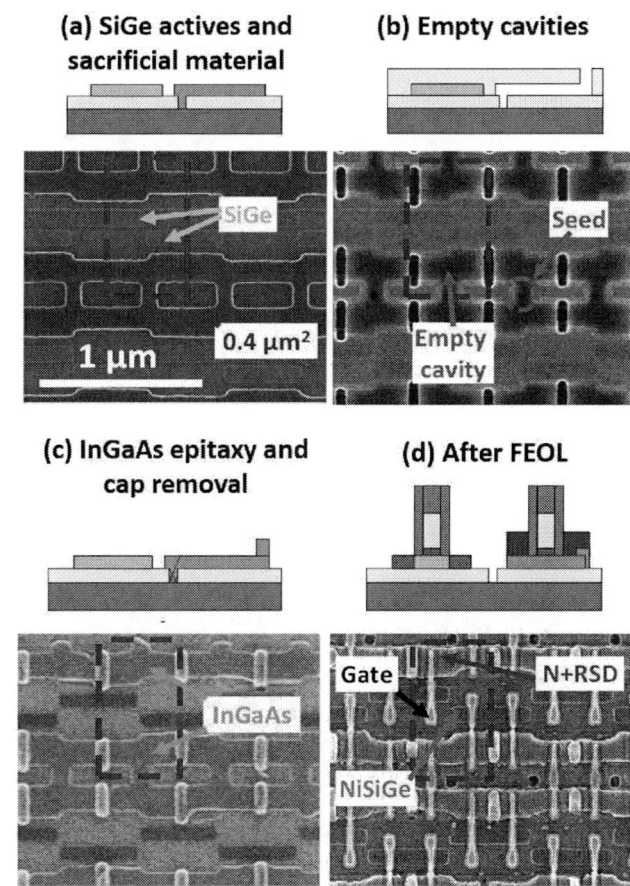

Fig. 3. Top view SEM images of at various steps of the process flow of the InGaAs/SiGe co-planar 2D CMOS integration. Adapted from ref [3].

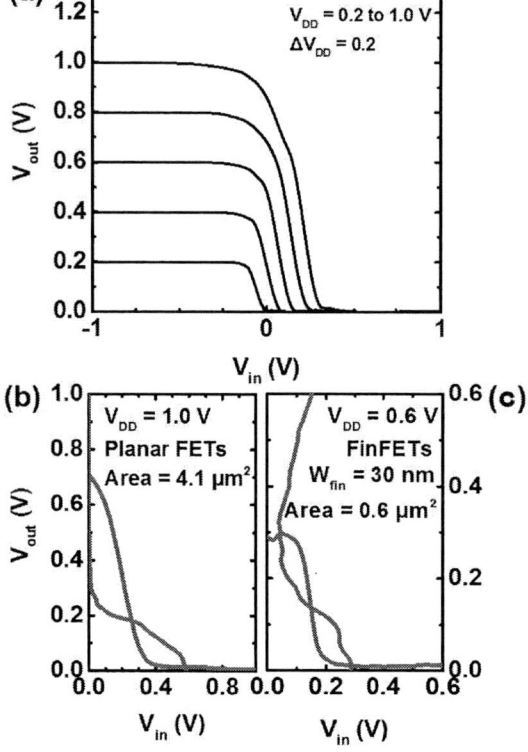

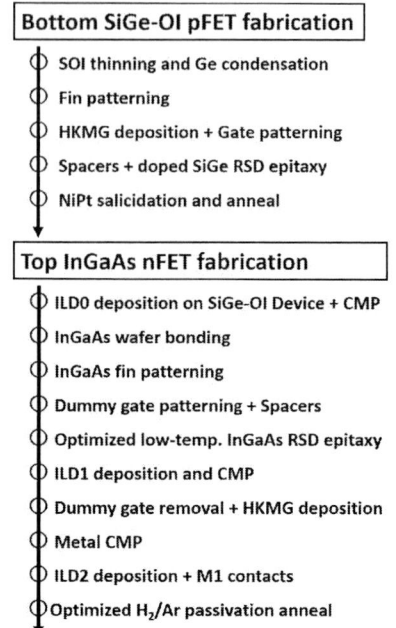

Fig. 5. Process flow of InGaAs/SiGe 3D monolithic CMOS integration. Adapted from ref [5].

Fig. 4. (a) Voltage transfer characteristics a InGaAs/SiGe inverter. (b),(c) SRAM 'butterfly' characteristics. Adapted from ref [3].

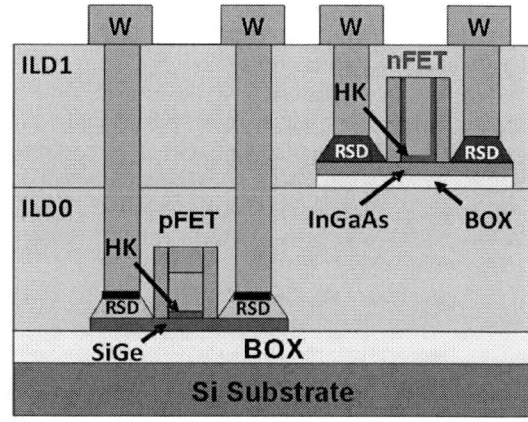

Fig. 6. Schematic showing InGaAs/SiGe 3D monolithic CMOS integration. Adapted from ref [5].

wafers. InGaAs nFET fabrication was then performed with the replacement-metal-gate (RMG) process described in [10]. This involves patterning the active transistor regions followed by a dummy gate stack deposition. After patterning the dummy gate, spacers were formed on either side similar to the bottom pFET process. Then comes the critical step of self-aligned in-situ doped InGaAs epitaxy to form RSD regions. This step has relatively high thermal budget and therefore, was optimized to minimize the process temperature while obtaining high doping in the layer as described in [5]. RMG process steps follow thereafter. An oxide layer was first deposited and planarized to expose the top of dummy gate. Then the dummy gate stack was selectively etched out. An optimized high-k/metal gate stack was deposited followed by metal CMP. Finally, oxide encapsulation was deposited and contact holes were opened to both top and bottom layers. Metallization was completed to create contact pads for both layers. The schematic of the so completed 3D monolithic stack is shown in Fig. 6.

Figure 7 shows the DC $I_d - V_g$ characteristics of top layer InGaAs nFET with $L_g = 70$ nm with a competitive DC performance after the 3DM fabrication. Figure 8 shows bottom SiGe-OI pFET $I_d - V_g$ after top nFET fabrication with excellent electrostatic integrity. Compared to the characteristics of the device before top nFET fabrication [5], there was negligible

impact on the electrostatics, demonstrating the robustness of the process. Thus an optimized performance on both, top-nFET and bottom-pFET layers can be achieved with this 3DM process.

Figure 9 shows the the voltage transfer characteristics (VTC) of scaled 3D inverters with nFET $L_g = 80$ nm (with pFET L_g 30 nm) and L_g 30 nm (for both nFET and pFET). Well-behaved transitions are obtained down to $V_{DD} = 0.25$ V. Therefore the integration scheme shows the potential pathway towards an optimized low-voltage CMOS with with InGaAs-on-SiGe 3DM technology.

Besides CMOS demonstration, the top layer InGaAs nFETs

978-1-5090-5979-9/17 $31.00 © 2017 IEEE   246

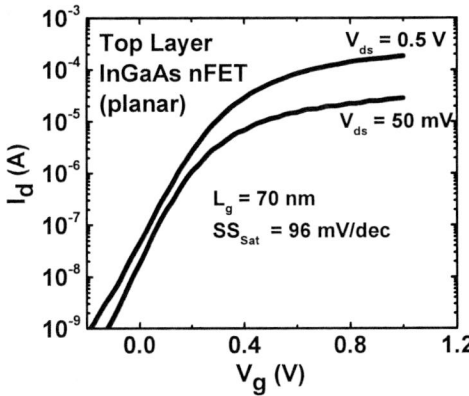

Fig. 7. $I_d - V_g$ characteristics of top layer InGaAs nFET. Adapted from ref [5].

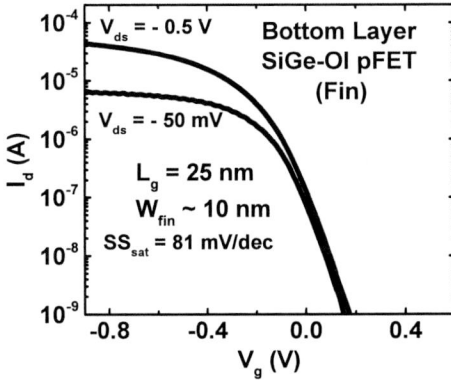

Fig. 8. $I_d - V_g$ characteristics of bottom layer SiGe-OI pFET. Adapted from ref [5].

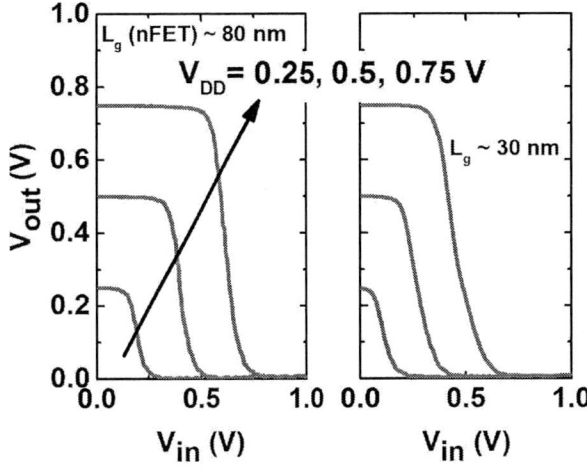

Fig. 9. Voltage transfer characteristics of InGaAs/SiGe 3D inverter. Adapted from ref [5].

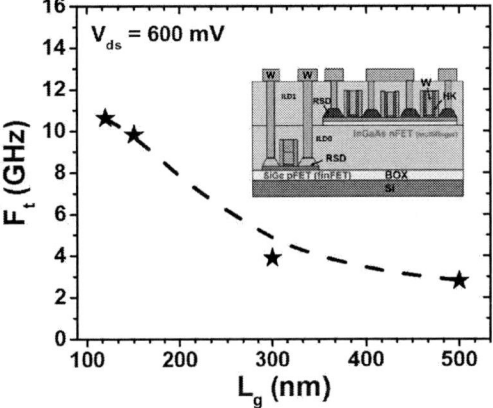

Fig. 10. Cut-off frequency for various gate lengths for top layer InGaAs RFFETs. Dotted line is guide to eye. Inset shows schematic of multifinger-gate structure of RFFETs. Adapted from ref [11].

were also characterized for RF applications [11]. A multi-finger gate structure was used for such devices as shown in the inset of Fig. 10. Cut-off frequency vs. gate length plot for the InGaAs RF-FETs is shown in this figure. Promising first results open up the opportunity for RF-over-CMOS circuits in the 3DM integration.

IV. CONCLUSION

A novel direct epitaxy based co-planar 2D InGaAs/SiGe CMOS integration scheme was demonstrated with potential for scalability to sub-7 nm node technology. Functional CMOS circuits, including dense SRAM cell, obtained through the integration are testimony to the advantages of the scheme. Besides 2D integration, a InGaAs-on-Si(Ge) 3D Monolithic integration scheme was demonstrated that allows low voltage high-performance CMOS circuits as well RF-over-CMOS circuits. Owing to inherent benefits of low thermal budget and high mobility, InGaAs as top layer enables optimized performance for both layers. Thus the building blocks for a high performance, multi-functional 3D Monolithic technology were also demonstrated.

ACKNOWLEDGMENT

Funding from the EU is acknowledged under: ICT-2013-11 COMPOSE3, ICT- 2013-11 IIIVMOS, H2020-ICT-2015-688784-INSIGHT and PEOPLE-2013-IEF FACIT.

REFERENCES

[1] J. A. del Alamo, Nature, 479, pp. 317323 (2011).
[2] L. Czornomaz et al., IEDM , pp. 2.8.1-2.8.4.(2013).
[3] L. Czornomaz et al., VLSIT , pp.1-2.(2016).
[4] M. Vinet et al., S3S Conference, pp. 1-3 (2014).
[5] V. Deshpande et al., IEDM, pp. 8.8.1-8.8.4. (2015).
[6] L. Czornomaz et al., VLSIT , pp. T172-T173.(2015).
[7] H. Schmid, et al., APL, 106, 233101 (2015),
[8] P. Hashemi et al., VLSI, pp. T18-T19 (2013).
[9] P. Hashemi et al., VLSI, pp. 1-2 (2014).
[10] V. Djara et. al, EDL, 37, 2, pp. 169-172 (2016).
[11] V. Deshpande et al.,EuroSOI-ULIS, pp. 127-130 (2016).

Tunable ESD clamp for high-voltage power I/O pins of a Battery Charge Circuit in mobile applications

Mirko Scholz, Geert Hellings, Shih-Hung Chen, Dimitri Linten

imec
kapeldreef 75, 3001 Heverlee, Belgium
email: geert.hellings@imec.be

Abstract—A tunable PNP-based ESD clamp is designed for a 4.5V power IO in a foundry technology. Using Mixed-Mode TCAD simulations, we show that the clamp's trigger and holding voltage can be easily tuned by simple layout modifications. The fabricated clamp was characterized using an on-wafer TLP system, confirming the tunable V_{T1}=13.4-16.8V, with V_{HOLD} slightly above 10V and I_{T2}>1.2A. Finally, the clamp is combined with an off-chip transient voltage suppressors (TVS) to withstand surge stress on system-level.

Keywords—ESD, High-voltage tolerant I/O, SCR, Component-Level ESD, System-Level ESD.

I. INTRODUCTION

Today, many system-on-chip (SoC) and system-in-package (SiP) designs are used in mobile applications. To further reduce the number of components on the mobile application boards several building blocks, which used to be independent ICs, are now integrated in one die or IC package. For example, a low-voltage, low-power RF front-end is integrated together with the high-voltage power management circuit for the charging of the mobile phone battery.

This high level of integration brings new challenges for the ESD protection design. High-voltage tolerant ESD protection solutions need to be implemented in advanced low-voltage CMOS technologies. Thereby, not only component-level targets such as Human Body model (HBM) and Charge Device model (CDM) need to be met. Some pins also need to sustain system-level stress (IEC61000-4-2) and surge stress (IEC61000-4-5). In addition, many designs are developed by fabless design houses who need to rely on a given foundry process. Changing doping profiles is often not a practical way to optimize the ESD protection design.

In this work, we will show the design and development of an ESD protection clamp for a 4.5V power IO using the 1.8 V power domain. Mixed-mode simulations enable to design the ESD clamp and to define the design of experiment (DoE) for on-wafer verification. Transmission Line Pulsing (TLP) is carried out to analyze the clamp operation and robustness in the HBM time domain. Finally the obtained data is compared to TLP data from TVS diodes to judge the robustness of the solution during a surge event applied to the application board.

II. ESD CLAMP DESIGN

A. System-Level and Component-Level ESD co-Design.

Power IO pins usually consist of very wide lateral drain-extended MOS (LDMOS) drivers. During ESD, they fail already at low stress levels during snapback due to non-uniform conduction. For these devices, an ESD designer cannot rely only on their self-protection capability. Instead, a dedicated ESD protection of the driver circuit is required.

Several requirements need to be considered during ESD protection design. Since the ESD clamp protects a power I/O pin, its holding voltage (V_H) should be above the nominal V_{DD}, to prevent Latch-up-like failure. As the pin will connect directly to a "real world" power source, system-level EOS/ESD requirements need to be considered during the protection design. Robustness to both IEC61000-4-2 and the surge standard IEC61000-4-5 are required.

Figure 1 shows an illustration of the component-level design window. Based on latch up specs, the minimum holding voltage is defined as $1.25 \times V_{DD}$. With the nominal I/O voltage of 4.5V, a V_H of 6.75 V is firstly required. However, due to the system-level requirements, the holding voltage of the on-chip ESD protection needs to be considerably higher to avoid that the huge system-level stress would be dumped into the small component-level ESD protection. The V_{T1} (turn-on voltage) for the ESD clamp can be lower or slightly larger than the breakdown voltage of the LDMOS (~15 V). This criterion depends mainly on the behavior of the ESD clamp after turn-on: if a strong snapback can be achieved, the ESD clamp may turn on at a slightly higher voltage than the LDMOS breakdown. If not, V_{T1} of the ESD clamp needs to be significantly lower, to prevent turn on and destructive snapback of the LDMOS. Finally, an I_{T2}>1.3A is required (i.c. 2000V HBM) for the on-chip ESD protection.

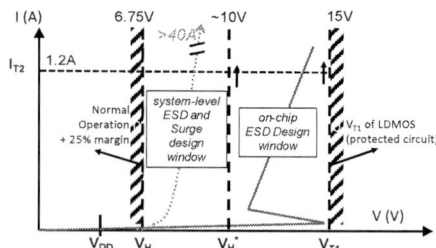

Figure 1: ESD design window for the on-chip ESD protection:
$I_{T2} > 1.2$ A, $V_{DD} = 4.5$ V, $V_H > \sim 10$ V, $V_{T1} < 15$ V

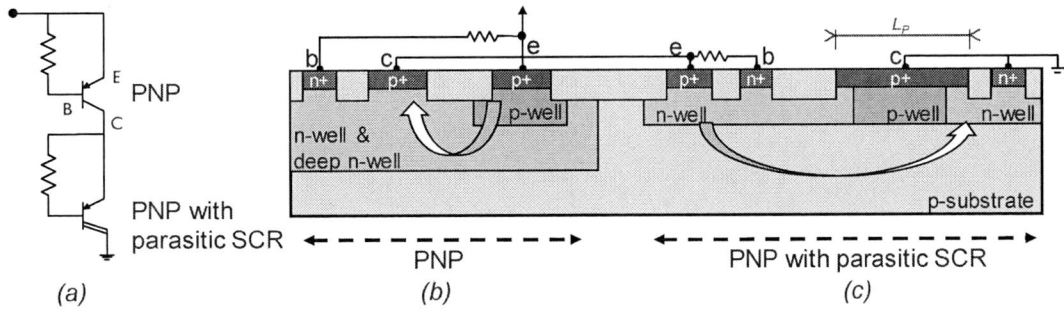

Figure 2: Schematic of the ESD clamp (a) and cross-sections of the PNP (b) and the PNP with parasitic SCR during ESD stress (c); the large arrow shows the main conduction path.

B. Optimized ESD Clamp Design

The proposed clamp design consists of a first PNP in series with a second PNP where a parasitic silicon-controlled rectifier (SCR) is triggered. Figure 2 shows the schematic and cross-section of the ESD clamp and its components. The series-connection of both PNPs is needed to ensure a sufficiently high trigger voltage and holding voltage: a clamp with only a single PNP or SCR would have too low V_H or V_{T1}. Its emitter is placed inside a p-Well which is surrounded by n-Well and deep n-Well for isolation. The collector and base are in the same n-Well. A 10 kΩ resistor between emitter and base provides a bias at the base, keeping the PNP OFF during normal operation.

The second PNP (with parasitic SCR) consists of two n-Wells surrounding a p-Well. The base and emitter are in the same n-Well. The emitter is directly connected to the collector of the first PNP, while the base is connected through a 10 kΩ resistor. The collector is formed by a substrate contact which overlaps partially to the two surrounding n-Wells. The collector and the second n-Well are grounded. During ESD stress, this configuration operates like a SCR. The anode is formed by the PNP emitter and base. The cathode is formed by the substrate contact and the second n-Well.

The authors in [1] proposed a similar clamp for the ESD protection of MOS arrays in a 20V technology. One main disadvantage of that design is the high trigger voltage. The authors proposed changing the doping profiles to lower the trigger voltage. However, this solution is unpractical for our ESD clamp since it is to be made in a foundry process (with fixed doping). Our design therefore relies on moving the location of the triggering in the second PNP is moved to the substrate contact. As shown in the next sections, this allows to tune the turn-on voltage of the clamp and to lower it to a safe level near the breakdown voltage of the internal circuit.

C. Mixed-Mode TCAD simulations

To analyze its operation the cross-section of the new ESD clamp is implemented in the mixed-mode simulator DECIMM [2]. The TCAD cross-section and doping profiles were based on generic 40 nm CMOS platform at the time: without detailed process calibration, such a simulation deck does not allow to predict e.g. exact junction breakdown voltages. Still, relative effect of e.g. layout variations can be predicted (as will be shown). Figure 3(a) shows the simulated electrical field just before device turn-on in the SCR. The breakdown of the clamp is moved clearly to the n-Well/collector junction. Figure 3(b) shows the simulated current density after the clamp has turned on. The highest current density and thus the main conduction path is from the left N-Well to the right N-well through the p-substrate. This is the expected conduction path of an SCR.

The simulation deck allowed to investigate the electrical effect of layout variations on the electrical ESD performance. Changing the length of the p-Well in the SCR is particularly interesting, since it allows to tune the SCR V_{T1}: Increasing the length of the p-Well (L_P) from 1.5 to 3 μm, yields a much higher V_{T1}, while only slightly increasing V_{HOLD} (fig. 4). This can be explained by considering that this p-Well is the base of the NPN transistor inside the SCR. A larger base thus leads to a lower current gain. Consequently, the current required to fire the SCR snap back is also larger, which ultimately leads to a V_{T1} which is a few Volt higher. Once the SCR has turned on, its current conduction is more limited by various series resistances, explaining the small effect of this layout parameter on V_H. Clearly, the demonstrated sensitivity to layout variations allow designers to adapt the clamp's electrical behavior to a targeted application.

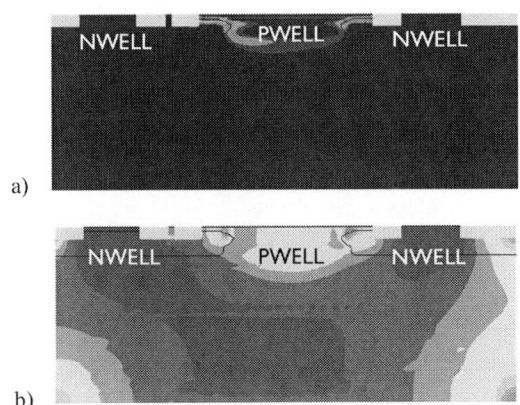

Figure 3: (a) Simulated electrical field before snapback and (b) current density in the SCR during TLP stress.

978-1-5090-5979-9/17 $31.00 © 2017 IEEE   249

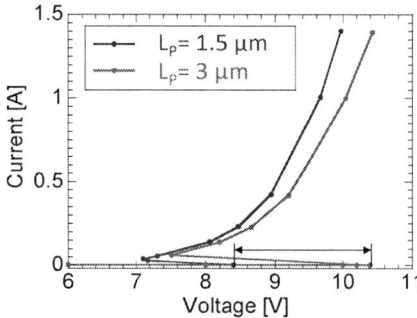

Figure 4: Simulated TLP I-V curves for the ESD clamp, showing the impact of the a larger L_P in the second PNP (parasitic SCR): while the triggering voltage V_{T1} is increased significantly, V_H is only slightly higher.

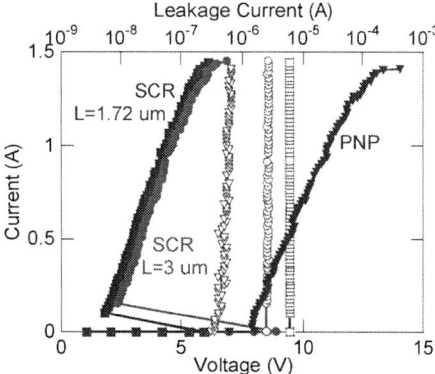

Figure 5: Measured TLP I-V and leakage, as measured for the two components of the designed ESD clamp. Note the large impact of the layout parameter L_P on the SCR turn-on voltage V_{T1}, as predicted by the TCAD.

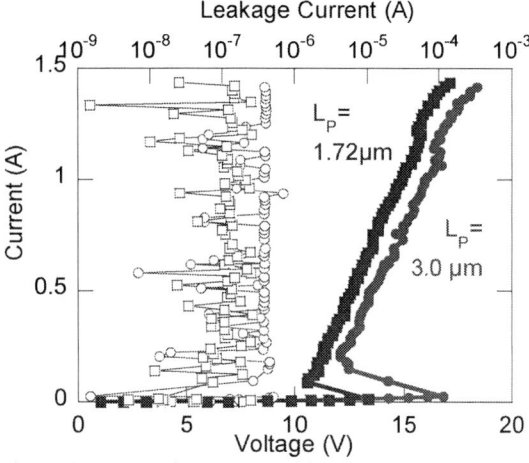

Figure 6: Measured TLP I-V and leakage current, for the ESD clamp, with a layout variation in the SCR ($L_P=1.72 - 3.0\mu m$).

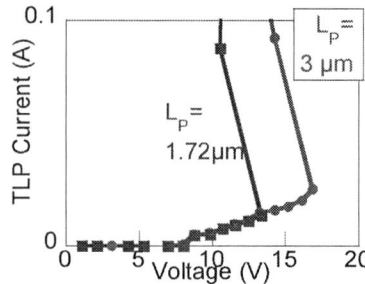

Figure 7: (zoom-in) Measured TLP I-V and leakage current, for the ESD clamp, with a layout variation in the SCR ($L_P=1.72 - 3.0\mu m$): increasing LP makes the SCR snap back occur at a higher voltage V_{T1} by increasing the trigger current needed for snapback.

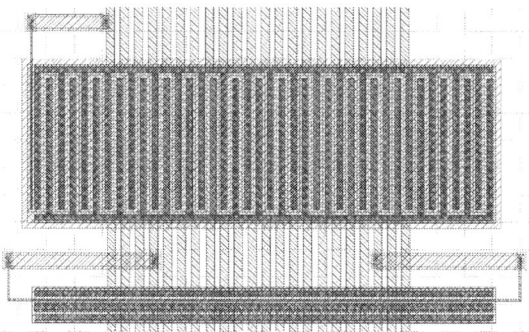

Figure 8: Example layout of the proposed clamp, showing the first PNP (top) and the PNP with parasitic SCR (bottom).

III. CHARACTERIZATION RESULTS

A. Fabrication details

The designed ESD clamp was fabricated in a 40-nm TSMC technology. To this end, the first PNP was implemented in a multi-finger layout. The SCR was implemented in a single finger layout. Its layout width was equal to that of the first PNP. An illustration of this layout is given in figure 8. This efficient layout style is possible because of the SCR, which can conduct much more current than a PNP.

B. TLP Characterization

The clamp and its two components have been analyzed using a Hanwa T5000 Transmission Line Pulsing system with 100 ns pulses, 2ns rise time.

Fig.5 shows the TLP I-V traces for the PNP and the SCR. The PNP has a $V_{T1}=8V$ and shows an R_{ON} of 3.4 Ohm after a minor snapback to $V_H=7.6V$. The failure current $I_{T2}=1.4A$, corresponding to an HBM voltage well above 2kV. The second part of the clamp, the SCR needs a turn-on voltage $V_{T1}=5.9V$, before the typical deep snapback to $V_H=1.6V$. With $R_{ON}=2.8$ Ohm, its ON-resistance is similar to that of the first PNP. Increasing the layout length of the p-substrate (L_P) to 3.0μm allows to modify the turn-on behavior: V_{T1} is increased to 8.9V (+50%), while $V_H=1.9V$ and $R_{ON}=2.7$ Ohm show only small changes. This is indeed a convenient way to tune the turn-on behavior of the ESD clamp, depending on the application.

The TLP characterization of the combined ESD clamp is shown in fig. 6. A trigger voltage of $V_{T1}=13.4$ V is observed for a p-substrate layout width of $L_P =1.72$ μm. A larger $L_P =3$ μm yields a $V_{T1}=16.8$V as expected, based on the TCAD simulations presented in section II.C . Fig.7 allows to look more closely to the mechanism by which L_P determines the SCR trigger voltage V_{T1}: while both versions of the ESD clamp start conducting equally at ~7V, the current required for snap back is considerably larger in the layout with $L_P=3$μm. This proves a different current gain in the SCR. The characterization results show that a L_P of 1.72 um allows to meet the component-level design window for the given application, and that its V_{T1} and V_H are high enough to allow for the system-level protection to operate in the 6-10V range.

IV. SURGE PROTECTION AND SYSTEM-LEVEL ESD DESIGN

Interface circuits protection designs need to operate both in a component-level and a system-level ESD design window. Particularly, the system-level specifications of IEC61000-4-2 and surge protection requirements need to be met. Since the timescale of an overvoltage stress like surge is much longer than that of a typical ESD event (10's of μs instead of e.g. 150ns for an HBM ESD pulse), the contained energy is orders or magnitude larger. Handling these stresses on-chip would require a very large area including an area overhead for the interconnects design. Consequently, an off-chip solution in the form of a TVS (Transient Voltage Suppressor) is preferred.

In addition, for system-level consideration, the entire system then needs to be balanced in such a way that the large current during system-level or surge stresses is fully absorbed by the TVS since it would easily destroy the smaller on-chip protection (i.c. the tunable PNP ESD clamp). Consequently, the TVS should have a lower turn-on voltage than the on-chip protection, in combination with a sufficiently low R_{ON}. A schematic of the protection system design is shown in fig. 9.

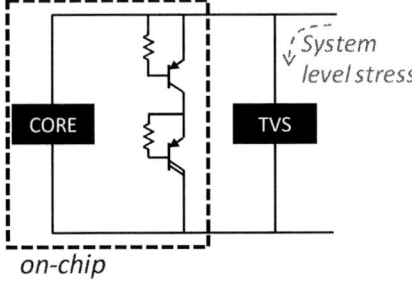

Figure 9: Schematic of the system-level protection. The off-chip Transient Voltage Suppressor should absorb system level/surge stresses.

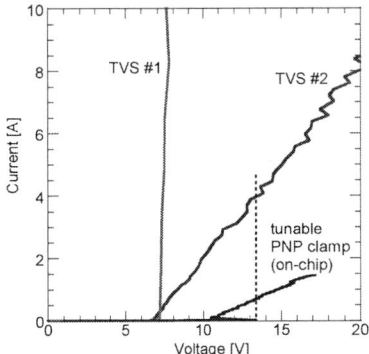

Figure 10: 100 ns TLP I-V curves of two commercially available TVS, compared to the tunable PNP ESD clamp. TVS#1 is more suited for the codesign: the low clamping voltage prevents on-chip protection to turn on.

For this work, two commercial TVS components were selected to meet these requirements (e.g. [3]). As shown in fig.10, both feature a holding voltage well above the $V_{DD}=4.5$V and a sufficiently low R_{ON}. As a result, surge current would fully flow through the TVS #1, rather than entering into the on-chip protection. TVS#2 is less suited, since the higher R_{ON} will cause the on-chip protection to also turn on currents above 4A and absorb part of the surge stress.

V. CONCLUSIONS

A tunable PNP-based ESD clamp is designed for a 4.5V power IO, in a component-level / system-level co-design. The proposed clamp design consists of a first PNP, in series with a second PNP wherein a parasitic SCR is triggered. Using Mixed-Mode TCAD simulations, layout variations of the clamp were found to have a large impact on the clamp's trigger voltage. The clamp was fabricated in a commercial foundry technology and characterized using an on-wafer TLP system, showing tunable $V_{T1}=13.4$-16.8V, with V_{HOLD} slightly above 10V and $I_{T2}>1.2$A. The ability to tune these parameters by layout variations, rather than having to alter the fabrication process, is obviously more cost-efficient. Finally, off-chip transient voltage suppressors (TVS) were selected to absorb the more energetic system level and surge stress without excessively stressing the on-chip protection.

REFERENCES

[1] Da-Wei Lai, Shuang Zhao, Jian Gao and Theo Smedes, "PNP-eSCR ESD Protection Device with Tunable Trigger and Holding Voltage for High Voltage Applications", Proceedings of EOS/ESD Symposium, 2016

[2] Angstrom DECIMM v6.0, Angstrom Design Automation, San Jose, CA, USA.

[3] Semtech uClamp 0571P, datasheet, Semtech Coorporation, CA, USA.

Guidelines for intermediate Back End Of Line (BEOL) for 3D sequential integration

[1]C. Fenouillet-Beranger, [1]S. Beaurepaire, [1]F. Deprat, [1]A. Ayres de Sousa, [1]L. Brunet, [1]P.Batude, [1]O. Rozeau, [1]F. Andrieu, [1]P. Besombes, [2,1]M-P. Samson, [1]B. Previtali, [1]F. Nemouchi, [1]G. Rodriguez, [1]P. Rodriguez, [1]R. Famulok, [1,2]N. Rambal, [1]V. Balan, [1]Z. Saghi, [1]V. Jousseaume, [1]C. Guerin, [2]F. Ibars, [2]F. Proud, [2]D. Nouguier, [2]D. Ney, [1]V. Delaye, [1]H. Dansas, [2]X. Federspiel,[1]M. Vinet

[1] Univ. Grenoble Alpes, F-38000 Grenoble, France
CEA, LETI MINATEC Campus, F-38054 Grenoble, France
[2]STMicroelectronics, 850 rue Jean Monnet, F-38926 Crolles Cedex, France
claire.fenouillet-beranger@cea.fr

Abstract— **For the first time the thermal stability of a new fluorine-free (F-free) W barrier coupled with W interconnections enabling 22% line 1 resistance improvement is evaluated in view of 3D VLSI integration. Integrated with ULK, no resistance nor lateral capacitance degradation is observed up to 550°C 5h while preserving good reliability. For additional thermal stability a TEOS/W stability is demonstrated up to 600°C 2h. Both types of interconnection stacks have been successfully integrated on devices with 28nm design rules and show similar performance for MOSFETs and Ring Oscillators (RO) as compared to the ULK/Cu stack. Finally, iBEOL guidelines are given at the end in view of 3D sequential integration.**

I. INTRODUCTION

An alternative approach to conventional device scaling for future nodes is the 3DVLSI CoolCube[TM] or sequential 3D integration [1]. Compared to TSV-based 3D ICs, sequential process flow offers the possibility to stack devices with a lithographic alignment precision (few nm) enabling via density > 100 million/mm^2 between transistors tiers. To benefit from the full 3D opportunities and avoid global routing congestion [2], there is a need to implement local routing of the bottom tier: inter-tier metal layers need to be incorporated in the technology (Figure 1). As a consequence, intermediate Back End Of Lines (iBEOL) levels need to endure top FET thermal budgets. A reasonable maximum thermal budget for top FET has been determined around 500°C [3-4] implying to find solutions for implementing back end material stable beyond 400°C. Currently the combination of copper with ULK (Ultra Low-K) materials is widely used for standard BEOL (low resistivity & capacitance, and thus speed improvement). However the integration of such materials in the iBEOL of a CoolCube[TM] integration faces a number of challenges. Indeed copper metallization can cause contamination issues in the case of wafer break during the process of the top transistor; where FEOL contamination free environment is required. A solution can be the use of tungsten (W) as it has already been

integrated in the FEOL of several products. Our previous studies [5] highlighted that Cu and W interconnections combined with ULK are stable up to 500°C 2h and 550°C 5h respectively. However, W resistance is still larger as compared to the copper one. Moreover, as presented in the literature, the ULK stability is still questioning beyond 500°C [6]. Indeed, the modification of ULK structure and permittivity during a thermal anneal at temperature higher than 500 °C may increase the leakage and delays of the iBEOL and thus degrade the circuit performances.

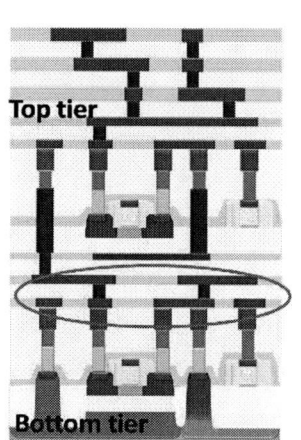

Figure 1: 3D VLSI structure with 2 levels of inter-tiers interconnections.

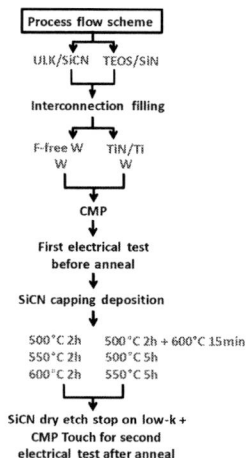

Figure 2: Process flow scheme for BEOL line 1 stack and list of annealings applied on W lines in ULK or TEOS with Ti/TiN or F-free W barrier.

In this paper, for the first time the thermal stability of a new ultra-thin low resistivity F-free W barrier in view of W/ULK line 1 resistance improvement is studied and compared to TiN/

978-1-5090-5979-9/17 $31.00 © 2017 IEEE

W seed through morphological, electrical and reliability characterizations. In addition, this study is completed by investigation of the TEOS/W interconnection stability as an alternative to ULK. Finally, the integration of both ULK/Ti/TiN/W and TEOS/Ti/TiN/W line for a 28nm node design rules technology is validated through NMOS and PMOS transistors characterization and RO measurements. A summary of the suitable iBEOL materials for CoolCube™ integration is given at the end.

II. DEVICE FABRICATION

For this study we have used as reference the standard 28nm design rules damascene BEOL integration of the state of the art FDSOI technology [7]. The Inter Metal Dielectrics (IMD) materials are either ULK material (20 nm SiCNH etch stop layer (k_{as-dep} = 5.6) and 100 nm of porous SiOCH (k_{as-dep} = 2.7) or a 20nm SiN (k_{as-dep} = 7) with 200nm PECVD TEOS (k_{as-dep} =4). Only line 1 level is integrated. After the Ti/TiN/W or F-free W/W filling (even with F-free W barrier an ultra-thin layer of adhesive PVD Ti layer is present in the line 1 as seen on figure 6), the CMP planarization is realized and a SiCN layer of 20nm is deposited as a capping layer during post-annealings to avoid Cu oxidation. The different annealings tested on the line 1 interconnect are summarized in Figure 2. After thermal annealings the SiCN layer is removed by dry etching and a CMP touch is realized in order to enable electrical tests. The different materials splits are highlighted in figure 3.

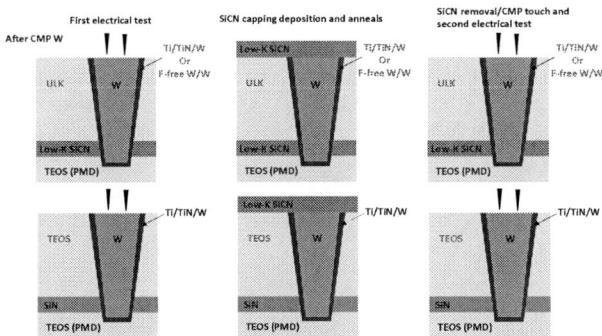

Figure 3: Schematics of process flow scheme for BEOL line 1 stack. ULK/W and TEOS/W stacks are compared.

III. IMPACT OF BARRIER ON W/ULK INTERCONNECTION STABILITY

Figure 4 shows the line 1 capacitance versus resistance measured on a specific multi-fingers/serpentine test structure for W interconnections before and after annealing with the Ti/TiN barrier in ULK. The large spread can be easily explained by the CMP difference between Cu and W and also by the post-processing SiCN removal and the CMP touch steps (penetration of CMP chemicals in the pores of the exposed ULK material). This figure confirms a factor 6 times higher resistance value for the W as compared to the copper one [5] and the W/ULK stability up to 550°C 5h [5]. In addition dies are functional at 600°C during 2h but the measurements dispersion increases as

the temperature increases beyond 550°C. This can be correlated to the ULK instability beyond 550°C as highlighted in [6].

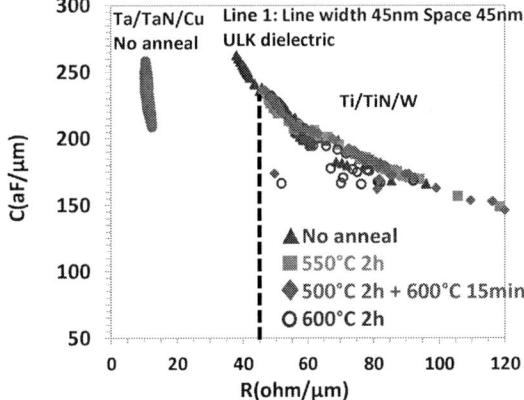

Figure 4: Lateral capacitance versus resistance for line 1 W/ULK interconnection with Ti/TiN barrier before and after annealings (line width and space=45nm). No anneal Cu line is plotted as reference.

Using the new F-free W barrier (Figure 5), allows a 22% reduction in the interconnect resistance. This resistance improvement is correlated to the smaller thickness of the F-free W barrier (4nm instead of 3nm of TiN + 6nm W seed layer mandatory for W filling), that enables to win 27% for W filling volume. Moreover, the stability is demonstrated (as for the Ti/TiN one) up to 550°C 5h. At 600°C drastic and irreversible degradation is pointed out.

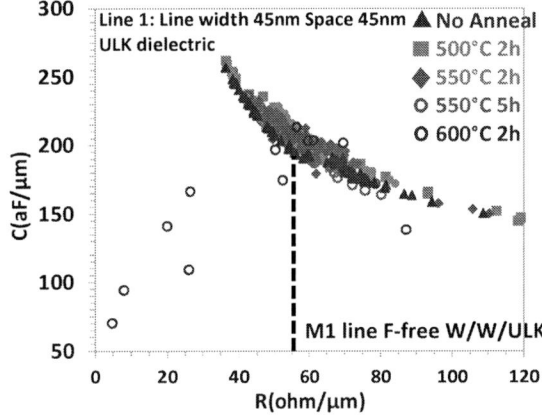

Figure 5: Lateral capacitance versus resistance for line 1 W/ULK interconnection with F-free W barrier before and after annealings (line width and space of 45nm).

The TEM cross section of Figure 6, confirms the accurate morphology of the F-free W interconnections before and after anneal 600°C 2h anneal (no specific defects are observed in both W and ULK materials).

978-1-5090-5979-9/17 $31.00 © 2017 IEEE   253

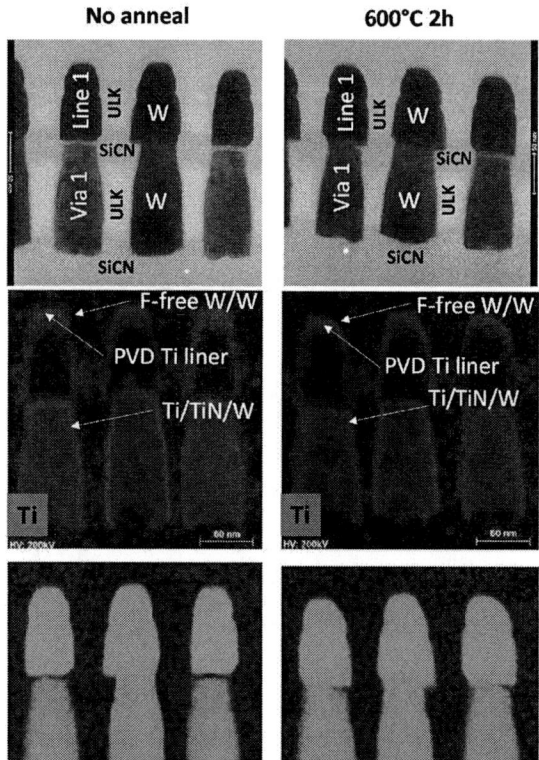

No anneal   **600°C 2h**

Figure 6: Line1 1 and Via 1 TEM cross sections filled with F-free-W/W and Ti/TiN/W respectively before and after 600°C 2h anneal. Line depth ~100nm.

Finally, using the following extrapolation parameters (100 m line, operating at room temperature and failure rate < 1%), the extrapolated lifetime is extracted for W interconnections with the F-free W barrier (figure 7) and compared to our previously published data [8]. Although TTF decreases with increasing thermal budget, a lifetime (line to line) of 10^6 years is calculated for the highest thermal budget, slightly lower compared to Ti/TiN but equivalent to usual Cu/ULK IMD lifetime values. Therefore, W/ULK reliability is not an issue for the 3D sequential integration due to the high initial lifetime since top transistor thermal budget is limited to 500°C for a couple of hours. Moreover F-free W barrier can easily replace the standard Ti/TiN one to win 22% of resistance value.

	No anneal	500°C	550°C	600°C
Ti/TiN/W (years)	4.10^{15} [8]	(5h) 5.10^{11} [8]	(5h) 10^{09} [8]	
F-free W/W (years)	2.10^{12}	(5h) 5.10^{06}	(5h) 2.10^{06}	(2h) 1.10^{06}

Figure 7: Extrapolated lifetime (TTF) of inter metal diel ULK with F-free-W/W or Ti/TiN/W before and after anneals. Years needed to reach the failure rate (<1%) with an operating voltage at 1.115V.

IV. LINE 1 TEOS/TI/TIN/W INTERCONNECTIONS STABILITY

In parallel Ti/TiN/W interconnections in TEOS have been studied in view of offering extended thermal stability. Due to non-optimized stack for the CMP step, the line is deeper (Figure 8) as compared to the previous one (200nm for W/TEOS and 100nm for W/ULK), explaining the lower resistance value (Figure 9).

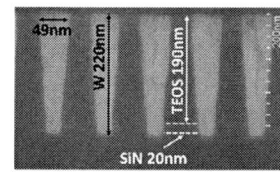 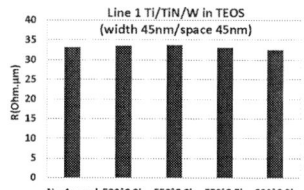

Figure 8: Line 1 SEM cross section filled with Ti/TiN/W in TEOS. Line depth ~220nm

Figure 9: Lateral capacitance versus resistance for W interconnections with Ti/TiN barrier before and after annealings in TEOS.

The median value of the W line resistances have been extracted and highlight no modification up to 600°C 2h annealing confirming the good thermal stability of Ti/TiN/W in TEOS (Figure 9).

The thermal stability of these lines Ti/TiN/W in TEOS have been evaluated after wafer bonding and etch back of the top substrate after a 550°C 2h anneal above the 28nm design rules metal 1 shortloop (Figure 10a). The reference without anneal, presented in Figure 10b, resulted in a high quality layer transfer as observed with acoustic microscopy observation (Only very few bonding defects are observed) [9]. Finally no morphology change are observed after 550°C 2h anneal highlighting the good stability of the interconnections structure after bonding (Figure 10c).

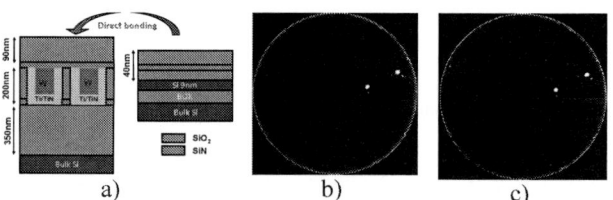

Figure 10: a) Description of the Si layer transfer above W metal 1 level. Acoustic microscopy on the studied TEOS/Ti/TiN/W structure before b) [9] and after c) 550°C 2h anneal. High quality bonding and stability is reached.

V. 28NM DEVICES INTEGRATION WITH LINE 1 TI/TIN W/ULK OR W/TEOS INTERCONNECTIONS

The transistors process flow is based on 28nm state of the art FDSOI technology as described in [7] with a BOX (Buried

978-1-5090-5979-9/17 $31.00 © 2017 IEEE   254

OXide) thickness of 25nm and a high-k (HfSiON)/metal (TiN) gate stack with raised source/drain. Finally a BEOL with either Ti/TiN/W in ULK or TEOS is integrated for the first time and compared to Cu/ULK reference.

Figures 11 highlights the NMOS and PMOS Ion/Ioff characteristics for each variants of interconnections. As expected the Ion/Ioff are matched for all the variants.

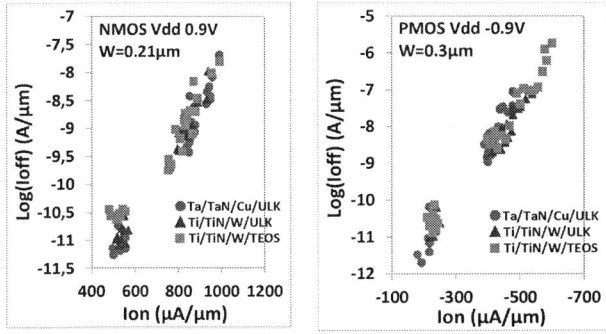

Figure 11: NMOS (left) PMOS (right) Ion/Ioff for transistors with different line 1 material integration.

The TEM cross-sections for the two types of interconnections are presented on Figure 12.

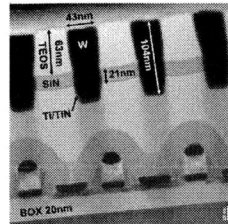 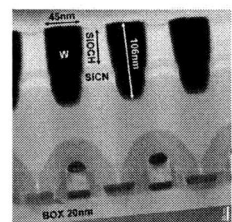

Figure 12: TEM cross sections for W/TEOS (left) and W/ULK (right) 28nm design rules line 1.

In addition, the impact of W/ULK interconnections as compared to Cu/ULK is evaluated through Ring Oscillator (RO) performances measurements. RO at 28nm design rules have been measured and reveal no change of propagation delay with respect to the two types of interconnections (12ps/11ps).

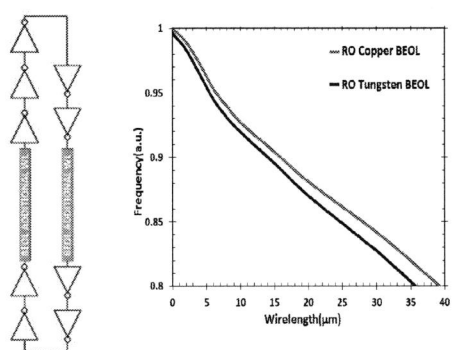

Figure 13: Simulated frequency versus Wire length for a RO FO1 with line1 Cu/ULK or W/ULK. As the wirelength is reduced the frequencies tends to be similar.

It is in line with our simulation results (Figure 13). The evaluation methodology consists on changing the characteristics of intermediary back end of line from copper metal to tungsten, then the parasitic elements are extracted. The simulation focuses in the frequency output. We can observe that as the wire length is reduced the RO frequencies tends to similar values for the two types of interconnections.

VI. CONCLUSIONS

Thanks to morphological and electrical characterization for a 28nm design rules technology, the interest of the new F-free W barrier with W interconnections is highlighted enabling 22% line 1 resistance improvement. No resistance, and lateral capacitance degradation are observed up to 550°C 5h while preserving good reliability. As an alternative to ULK, TEOS/W interconnection highlights a thermal stability up to 600°C 2h, but at the expense of an increase in k value. Finally W/ULK and W/TEOS interconnections have been successfully integrated on devices with 28nm design rules showing similar performances on both MOSFETs and Ring Oscillators (RO) as compared to the Cu/ULK one highlighting the potential of W to replace Cu for the first metal levels without important delay penalty. This is due to the fact that the line are short at this level and enable local routing. Summarizing, F-free W/W/ULK is the best compromise regarding contamination issue and resistance/capacitance compromise and reliability up to 550°C 5h. However, Cu/ULK stays always the best candidates for 3D sequential integration in terms of R,C values, but the reliability versus thermal budget is still to be evaluated.

Metal	Barrier	Permittivity	C, R	Stability	Reliability
W	Ti/TiN	ULK	R x6 /Cu, C ok	Ok up to 550°C 5h	Ok up to 550°C 5h
	F-free W	ULK	R x4 /Cu, C ok	Ok up to 550°C 5h	Ok up to 550°C 5h
	Ti/TiN	TEOS/SiN	R x6 /Cu, C NOK	Ok up to 600°C 2h	N/A
Cu	Ta/TaN	ULK	R, C Ok	Ok up to 500°C 2h	In progress

Figure 14: Summary of the different interconnections stability behavior versus thermal budgets regarding R, C and reliability performances.

ACKNOWLEDGMENT

This work is partly funded by the French Public Authorities through NANO 2017 program and EQUIPEX FDSOI11, ST-LETI Alliance program and by Qualcomm. The authors would like to thank AMAT for F-free W demo.

REFERENCES

[1] P. Batude et al, IEEE VLSI 2015; [2] P. Batude et al, IITC-AMC 2014; [3] C. Fenouillet-Beranger et al , IEEE ESSDERC 2014, [4] C. Fenouillet-Beranger et al, IEEE IEDM 2015, [5] C. Fenouillet-Beranger et al, SSDM 2015, [6] F. Deprat et al, MAM 2016, [7] N. Planes et al, IEEE VLSI Technology, 2012, [8] V. Lu et al, IEEE VLSI 2017, [9] L. Brunet et al, IEEE VLSI Technology, 2016.

978-1-5090-5979-9/17 $31.00 © 2017 IEEE

Device circuit and technology co-optimisation for FinFET based 6T SRAM cells beyond N7

Mohit Kumar Gupta[1,2,*], Pieter Weckx[2], Stefan Cosemans[2], Pieter Schuddinck[2], Rogier Baert[2], Dmitry Yakimets[2], Doyoung Jang[2]
Yasser Sherazi[2], Praveen Raghavan[2], Alessio Spessot[2], Anda Mocuta[2], Wim Dehaene[1,2]
[1]imec, Leuven, Belgium, [2]ESAT-MICAS, KU Leuven, Belgium, email: *Mohit.Gupta@imec.be

Abstract—SRAM paves the way for new technology nodes as it is more prone to failure due to intrinsic devices variability and technology process. To further boost high density SRAM yield and performance we need assist techniques and increased SRAM bit cell size at the expense of area. This paper discusses SRAM design strategies for future technologies nodes like beyond the N7 node, by comparing higher height cells and assist techniques. Although, higher height cells improve variability with scaled nodes but also need assist techniques to lower the operating voltage. Consequently, 122 is shown to meet the yield requirement and the preferred option to use with and without assist circuitry.

Keywords—SRAM, beyond N7, CMOS scaling, memory

I. INTRODUCTION

Mobile application processor energy consumption has been strongly limited by cache energy consumption [1]. Moreover, both energy consumption and cost depends on the SRAM cell area. As a result, 111 configuration (1fin for pull up (PU), 1fin for pass gate (PG) and 1fin for pull down (PD)) is preferred as high density (HD) SRAM cells. However, device variability [2] is shown to reduce the 111 cell noise margin, degrading the minimum operating voltage (Vmin). Furthermore, imbalance between writing and reading stability creates by both the quantization of no. of fins and the reduction in the NMOS/PMOS drive strength ratio for SRAM inverter from 2 to ~1 [3]. Finally, as a result of scaling, improvements in performance are also limited by device capacitance, back end of line (BEOL) capacitance and resistance [4, 5]. The quantization of no. of fin leads to different SRAM cell types (e.g. 112 , 122 and 123), which reduces variability because of increased device size [4,6]. Similarly, to increase the noise margin, several circuit assist techniques can be used, affecting both the performance and the Vmin of the cell [7]. However, assist techniques and higher height cells come at the expense of added area. Due to extra circuitry and complex design, or increase in bit-cell area which in turn results in additional energy consumption. This work presents the SRAM design strategy beyond N7 from device to array, considering the parasitics of the array, the device variability along with fabrication process complexity and cost. A strategy is proposed to design reliable HD SRAM using optimal assist techniques which will improve performance and Vmin. Additionally, the area and energy penalties are presented which provide insights into the gains of SRAM design beyond N7.

II. TECHNOLOGY PROCESS STEPS AND ASSUMPTIONS

This section will first describe the process assumption and pitches used for beyond N7 FinFET and the corresponding SRAM cell layouts. Finally, the assumption on device variability and the benchmarking used to characterize the SRAM cell are discussed.

A. Device

FinFET based devices and their performances are shown to outperform planar devices and used in sub-22nm CMOS technologies [8]. The scaled FinFET technology for beyond N7, used for simulations in this work, has excellent electrostatics with 67mV/dec sub-threshold slope, 49mV/V DIBL and Ioff of 0.02nA/um which is comparable with [9] as shown in Fig 1. A predictive BSIM-CMG based compact model has been used in this paper [5].

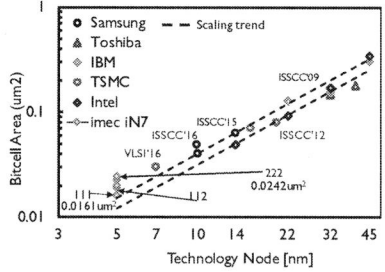

Figure 2. SRAM scaling trend provided by industry with technology node.

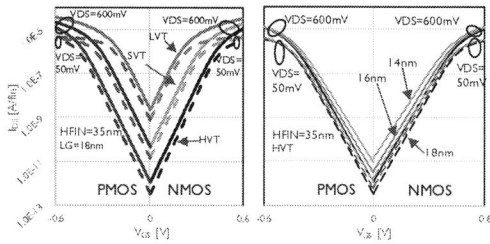

Figure 1. I-V characteristics of FinFET @ iN7 node for different threshold voltage and gate length repectively.

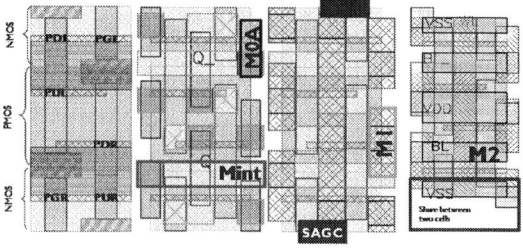

Figure 3. Layout of 111 SRAM with different intergration layers (from Fin to Metal (M2) connection for bitlines and supplies).

978-1-5090-5979-9/17 $31.00 © 2017 IEEE

Cells	111	112	122	123	222	Pitch (nm)
Size (Track=32nm)	6	7	7.5	8.5	9	
Cell Height	192	224	240	272	288	
Via for cross connection	Mint	Mint	Mint	Mint	Mint	
Fin Patterning	SADP	SAQP	SAQP	SAQP	SAQP	24
Fincut Mask	2	4	4	4	2	
Gate	SADP	SADP	SADP	SADP	SADP	42
M0A (Tungsten)	SAQP	SAQP	SAQP	SAQP	SAQP	42
Mint (Tungsten)	SAQP	SAQP	SAQP	SAQP	SAQP	32
M1 (Copper)	SADP	SADP	SADP	SADP	SADP	42
M2 (Copper)	SADP	SADP	SADP	SADP	SADP	42
Via (Copper)	Self aligned contact					42
Technology cost	Low	High	High	High	Medium	
Process	Simple	Complex	Complex	Complex	Complex	

Table I: Table shows the assumed process steps for integration of different layers along with considered pitch and material.

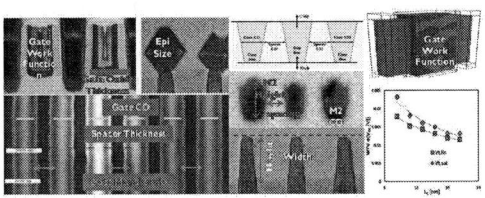

Figure 4. TEM image of FinFET and epi contacts (left). Right figures show the work function trend aith gate length.

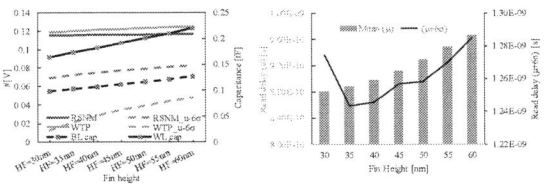

Figure 5. RSNM, WTP, WL and BL cap variation vs fin height (left). Read delay degrades with fin height due to increasing device capacitance (right) however variability reduces. µ+6σ shows improvement with height but upto certain limit.

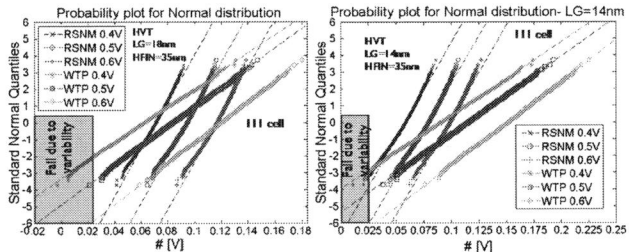

Figure 6. Variability analysis of '111' for different voltages and gate lengths.

B. Layout

Efficient cell layout for SRAM beyond N7 requires expensive processing steps and technology. 193i Lithography enables patterning but, due to resolution limits, fabrication below 36 nm pitch requires self-aligned quadruple patterning (SAQP), which is costly and complex [10]. SRAM bitcell area scaling is proportional to fin pitch and gate pitch which is considered as 24nm and 42nm, respectively. However, for HD SRAM with a 111 cell, SADP requires only 2 FINCUT masks making this cell not only the densest but less complex and expensive to fabricate. Table I shows the assumptions used in this paper for beyond N7 technology. Figure 2 shows the bit cell area trend of SRAM among different industry technologies. 47% reduction in area is shown for 111 cell as compare to recently published high density cell [3] for beyond N7 node. Figure 3 shows the layout of the 111 SRAM cell for different metal layers. The cross

connections between the two inverters is realized using a 32nm pitch horizontal Tungsten as intermediate metal layer Mint (M0) combined with self-aligned contacts, which reduces the impact of overlay by making partial Vias. A vertical Metal 1 is used for word line (WL), while horizontal Metal 2 layer is set for bit lines (BL and complement BL_) and power lines (VDD and VSS).

C. Variability

FinFET threshold voltage variation have improved from planar [8] because of reduced channel doping and higher effective gate area. However, random variations in (V_{th}) such as Random Discrete Dopants (RDD), Metal Grain Granularity (MGG), fin length, fin width and spacer variations due to Line Edge and Width Roughness (LER and LWR) [11-12] still reduce the stability margin for SRAM cell as shown in Fig. 4. Due to the low cannel doping and self-aligned patterning, RDD and LER/LWR have significantly reduced impact compared to MGG. Fig. 4 shows a rapid increase in work function variability due to MGG for shorter gate lengths.

D. Benchmarking of cell

Variability can be mitigated by increasing the Fin height which furthermore increases the drive current within the same footprint. However, it also increases the device's capacitance and reduces the performance benefit [13] as shown in Fig. 5 (a). As leakage increases with drive current along with Fin height scaling, for further analysis, shifting the threshold voltage is done for equal leakage across different fin heights. RSNM and WTP determine the read and write ability of bit cell respectively. Employing higher fin heights results in lower variation [6], but increases capacitance and moreover limits the fin aspect ratio to guarantee fin uniformity [14]. Fig. 5 (b) shows the read delay of 128bit per bitline for different fin heights illustrating the read delay degrades with fin height. A fin height of 35 nm is taken as optimum in all aspects (logic also [13]), having sufficient WTP margin, RSNM, good read delay and a fin aspect ratio of 7. Lower gate length and V_{TH} increases the leakage drastically. Since a cell leakage of 10pA is needed for HD cell [8], for further analysis HVT cell and LG=18nm is considered.

III. SRAM CELLS

The section explains advantages and limitations of higher heights cells as HD SRAM. The impact on both stability and performance depending on PU:PG:PD ratio is discussed.

A. 111 (HD) cell

The 111 cell is favored as it improves power, performance, area and cost, however, it is more vulnerable to variability. Figure 6 shows the distribution of RSNM and WTP for different supply voltages, which depicts that below 600mV the 111 cell is difficult to be used as HD cell for this technology. The WTP and RSNM can be improved by going for higher height cells which reduce variability by the reason of larger device sizes.

B. Higher height cells

1) Parasitics

In this work, we consider four extra SRAM cells topologies: 112, 122, 123 and 222, to improve the performance of the

978-1-5090-5979-9/17 $31.00 © 2017 IEEE

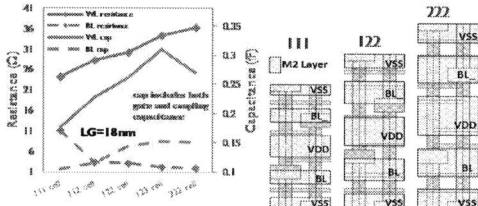

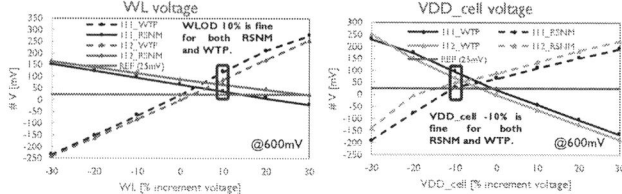

Figure 7. Bitline parasitics vs higher height cells (left). Higher height cells improve resistance but capacitance degrades. Right figure shows the M2 width increment with cell height.

Figure 10. WTP and RSNM variation vs different boosting voltage on WL and VDD cell. With advancement in technology, drive strength of PMOS is improving more as compare to NMOS. WLOD with 10% and VDD collapse of 10% helps in operating 111 SRAM @ 600mV.

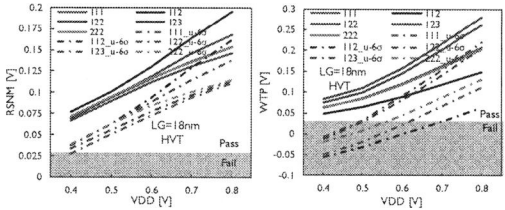

Figure 8. RSNM (left) and WTP (right) variation for different volatges. 122 cell has best WTP while 112 has best RSNM.

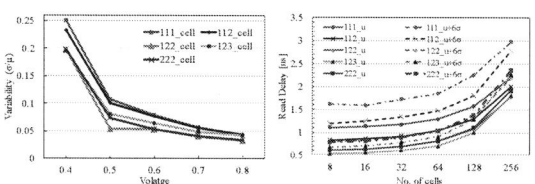

Figure 9. BL swing (σ/μ) showed in left image. With lower supply voltage it becomes difficult to design sense amplifier for 111 and 112 cell. Read delay with 6 sigma margin vs no. of cell shown in right figure.

SRAM. Figure 7.a shows the WL and BL resistance and capacitance per cell. Higher height cells have higher WL resistance and lower BL resistance because of the wider M2 lines which can be used. The WL capacitance consists of the pass gate transistor's capacitance and coupling capacitance which is different among cells (0.1fF/cell and 11Ω/cell for 111 cell, 0.144fF/cell and 3Ω/cell for 122 cell). The capacitance of the pass gate in 123 cell is more than for a 222 cell, due to non-conducting fin underneath the extended WL gate. Of all cells, the 222 cell is the tallest with a height of 288nm which takes 50 % more area as compared. Furthermore, all cells except the 111 cell still require up to 4 FINCUT masks which increases cost even more (Table I).

2) Stability

Figure 8 (a) shows the RSNM for different cells vs. supply voltage along with μ-6σ value, which determines the cell margin [15]. Figure 8 (b) shows the WTP vs supply voltage. If WTP is considered, the 112 cell shows the worst and meets a 10^{-9} failure rate just over 700mV. However, the 122 and 123 cells operates to 500mV and 111 and 222 can operate to 600 and 550mV respectively.

3) Performance

A more accurate sense amplifier (SA) allows for a smaller BL signal swing and hence faster operation, but at the cost of SA area and energy. The read operation is performed by

discharging the bitline when the word line activates. The delay between these two activities gives the read delay. The difference between BL and BL_, also called BL swing, is here set to be 200mV under nominal simulation. Figure 9 then shows the BL swing and read delay distribution under variability. The sense amplifier design for the 111 and 112 cell is challenging as compare to other cells. To cope with this variation, sense amplifiers for these cell types must be large in compared to those of the other cell types to meet the specification, which increases energy consumption (SA energy consumption is approx. 15 % of total read energy consumption [16]).

Finally, Fig. 9 (b) shows the read delay vs. number of cells in n*n array with a supply voltage of 600mV where the read delay includes the WL driver, WL parasitics delay and BL swing delay. *122 meets the yield requirement and is the preferred option to use without assist circuitry. 111 and 112 cannot be used because of poor write-ability as write ability is more crucial than reading due to variability. Cell types 123 and 222 are larger than 122 without bringing significant benefits. However, Vmin is comparable amongst the cells.*

IV. SRAM DESIGN WITH ASSIST TECHNIQUE

SRAM memories operating at low supply voltage draw significant attention as they can enable low power logic, by running on the same supply voltage. To reduce Vmin, several boosting techniques can be used and enhanced the performance [7].

A. Assist techniques

This work explores five assist techniques, a higher cell VDD (VDD_high), a lower cell VDD (VDD_low), word line under drive (WLUD), word line over drive (WLOD) and negative bitline (NEG_BL). VDD_high and WLUD enhance the read stability while VDD_collapse, WLOD, and NEG_BL enhance the write stability. In WLOD, the word line voltage is increased keeping all other voltages at the same level. Similarly, the word line voltage is decreased to enhance RSNM for WLUD. In VDD_high, the supply voltage of the cell is increased which is good for RSNM. Similarly, VDD_low is good for WTP. NEG_BL on the other hand increases the drive strength of the access transistor by increasing both gate and drain voltage with respect to the source.

B. Performance and stability impact

To compare the impact of different assist techniques on 111 and 112 cell stability, we will consider a maximum of 30%

978-1-5090-5979-9/17 $31.00 © 2017 IEEE

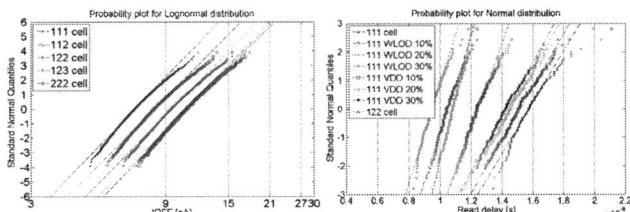

Figure 11. Read delay comparison among assists on 111 and 122 cell (left). Leakage comaprison for higher height cell (right).

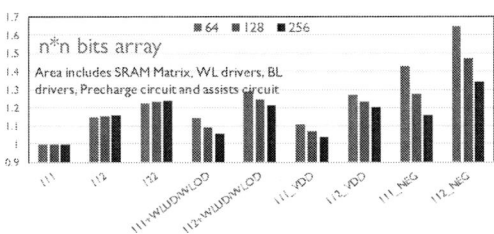

Figure 12. Area comparison among different cells with assist techniques for 20% boosting.

Cell Type	Assist	Read Delay 100mV [ns] #	Read Delay 200mV [ns] #	Write Delay (ns)	Vmin (mV)*
111	Neg_BL	1.25	2.3	0.463	>300mV
	WLOD	0.94	1.63	-----	>300mV
	WLUD	5.89	9.27	0.47	>300mV
	VDD_high	1.16	2.12	1.37	>300mV
	VDD_low	1.39	2.6	0.504	>300mV
112	Neg_BL	1.03	1.83	0.506	>300mV
	WLOD	0.605	1.01	-----	>300mV
	WLUD	3.87	7.49	0.545	>300mV
	VDD_high	1.01	1.79	0.799	>300mV
	VDD_low	1.09	1.97	0.576	>350mV
122	W/O Assist	0.8	1.35	0.45	500mV

*Vmin is calculated according to the importance of assist (stability Vmin considering 30% boosting).
Calculated for 10⁻⁹ error rate (worse case). However, the read analysis is done with 20% boosting @ 600mV.
VDD_high. VDD_low (collapse) and Neg_BL shared along column while WLOD and WLUD shared along row.

Table II. Delay and Vmin analysis for 111, 112 and 122.

boosting for each technique as shown in Fig. 10. It is possible to work at a supply of 600mV with only one assist technique with 10% boosting. Although, going lower than 600mV needs more boosting. Vmin is calculated for all cases to determine the best candidate for CMOS logic co-integration with 30% boosting (Table II). Four techniques perform well for lower voltages up to 300mV, except VDD_low which can operate to 350 mV. VDD_low is limited by the WTP for the 112 cell also results in increased write delay. Of all assist techniques, VDD_high results in higher aging effects [17]. WLOD is a better technique for enhancing read performance but also degrades hold stability and NEG_BL for write performance for both 111 and 112 cells as shown in Table II.

V. DISCUSSION

SADP, FINCUT masks and fin height enables 111 cell type as high density 6T SRAM but it has limited performance and variability resilience. Moreover, as shown in Table II, it is difficult to reach 1GHz frequency [19] at a supply voltage of 600mV is difficult due to BEOL (Fig. 7) and device parasitics with limited read current. Furthermore, as HD memory typically uses 512 bits/bitline [18-19] the bitline for the 111 cell becomes highly resistive and limits the reading and writing

performance. Interestingly, the 122 appears to be the best higher height cell among all considered cell types showing optimized performance, leakage (Fig. 11 (a)) and variability with the lowest area penalty of 25% compared to alternatives 123 and 222. Assist techniques help the 111 cell to achieve higher performance (Fig. 11 (b)) and lower Vmin as compare to 122 (Table II) with almost same area penalty as 122 (Fig. 12) but it makes the design more complex and needs extra power supplies. Moreover, the assist technique boosting voltage is also limited by breakdown voltage of transistor. It has been shown that using 20% boosting for a supply voltage of 600mV significantly accelerates the device aging effect [17]. A way to reduce the boosting voltage is to combine it with the 112 or 122 cell. According to process, the 112 cell is more complicated as compare to 122 [20] cell due to fin cut masks additionally it does not give much performance benefit compared to the 111 cell. Therefore, the 122 configuration is again the viable option for performant high density SRAM using boosting techniques.

VI. CONCLUSION

This paper has discussed SRAM design strategies for technologies beyond the N7 node, by comparing higher height cells and assist techniques. SRAM design governs technology enablement, where HD 111 bitcell is used as the reference design. However, 111 usage is limited due to variability, low performance, back end of line parasitics and demands for lower Vmin. Higher height SRAM cells are alternatives to improve the HD cell stability, but are limited by operating Vmin. Among these higher height cells, the 122 cell is the best cell with performance, leakage and area optimization. Moreover with respect to 122, the SRAM cell improves Vmin around 500mV. To further lower Vmin assist techniques should be used with limited boosting to increase the life span of 122 cell type SRAM.

REFERENCES

[1] T. Song, et al, *ISSCC*, p.17.1-17.3, 2016.
[2] A. Asenov, et al, *IEDM Tech. Dig.* p.1-1, 2008.
[3] S. -Y. Wu, et al, *VLSI Symposium*, p.92-93, 2016.
[4] S. S. Sakhare, et al, *IEEE Trans. On Elec. Dev.*, vol. 62, no. 6, 2015.
[5] M. G. Bardon, et al, *IEDM* Dec.2016, pp.28.2.1, 28.2.4.
[6] M. Pelgrom, et al, *IEEE JSSC*, vol. 24, no. 5, 1989.
[7] E. Karl, et al, *IEDM Tech. Dig.* p.25.1.1-25.1.4, 2012.
[8] S Natarajan, *IEDM*, pp. 3–7, 2014.
[9] C.-H Jan, et al, *IEDM* Dec.2012, pp.3.1.1, 3.1.4,10-13.
[10] A. Malik, et al, *Proc. SPIE* 9422, April 16, 2015.
[11] K. Ohmori et al, *IEDM Tech. Dig.* p.1-4, 2008.
[12] H. F. Dadgour, et al, *IEEE Trans. On Electron Dev.*, vol. 57, no. 10, 2010.
[13] M. G. Bardon, et al, *ICICDT*, p. 1-4, 2015.
[14] R. Jhaveri, et al, *US Patent, US 20140191300 A1*, 2011.
[15] M. K. Gupta, et al, *ICICDT*, May, 2017.
[16] M. E. Sinangil, et al, *IEEE JSSC, 2014*.
[17] M. K. Gupta, et al, *IRPS*, April, 2017.
[18] E. Karl, et al, *IEEE JSSC*, vol. 51, no. 1, 2016.
[19] T. Song, et al, *IEEE, JSSC*, vol. 52, no. 1, 2017.
[20] S. S. Sakhare, et al, *Proc. SPIE* 9427, April 16, 2015.

Microfluidic technology: new opportunities to develop physiologically relevant *in vitro* models

Integrated microfluidic platform for the *in vitro* pre-implantation culture of individual mammalian embryos and their *in situ* characterization

Séverine Le Gac

Applied Microfluidics for BioEngineering Research, MESA+ Institute for Nanotechnology, MIRA Institute for Biomedical Engineering and Technical Medicine, University of Twente, Enschede, The Netherlands

Abstract—Here, we report the development of an integrated sensing platform for the field of assisted reproductive technologies (ART), and more specifically for the pre-implantation culture of mammalian embryos and their *in situ* characterization through evaluation of their metabolic activity. The entire platform consists of a nanoliter-culture chamber, with an integrated oxygen sensor to monitor the respiratory activity of individual embryos, as a marker for their viability and developmental competence. We first discuss the key-advantages of microfluidic technology to realize such an integrated sensing platform. We next present the culture device, and its validation on mouse embryos. This first stage validation reveals that microfluidics supports the full-term development of mouse embryos down to the single embryo level with birth rates comparable to group culture in conventional formats. In a second step, the device is upgraded for the culture of human embryos, and tested on donated frozen-thawed embryos. Finally, we describe an oxygen sensor consisting of an ultra-microelectrode array (UMEA) to be integrated in the culture device. Using this UME-based sensor, we also propose a novel measurement approach at short timescales, which allows reducing drastically the amount of oxygen consumed through the electrochemical measurements. Current work focuses on the integration of the sensor in the culture platform and its validation on biological materials. The integrated platform is currently tested on spheroids, which are used as surrogates of mammalian embryos, before it is applied on mouse embryos.

Keywords—*microfluidics, integrated sensing platform, assisted reproductive technologies, mammalian embryos, ultra-microelectrode arrays, oxygen sensing.*

I. INTRODUCTION

Microfluidics can be defined as a technology allowing the accurate manipulation of small volumes, in the low nanoliter range, using structures having micrometric dimensions [1]. Microfluidics, which is also know as lab-on-a-chip technology (LOC), has reached a mature state and has become very popular in the field of life sciences due to the numerous advantages it offers. Specifically, microfluidic devices enable faster, more sensitive and reproducible analysis using lower amounts of reagents. Furthermore, microfluidics lends itself well to the realization of complex, highly parallelized and integrated platforms.

Originally, microfluidic developments have been driven by the field of bioanalysis. However, applications of LOC have recently diversified and extended to cellular investigations, for which microfluidics presents additional advantages [2, 3]: exquisite control, both spatially and temporally, on a cell microenvironment and the possibility to create *in vivo*–like conditions thanks to the high level of confinement and the laminar character of the flow; dynamic culture conditions; and a unique capability to integrate smart capabilities such as sensors to monitor *in situ*, in real-time and in a non-invasive manner a cell microenvironment and/or a cell behavior. Microfluidic technology is currently utilized for a great variety of applications, ranging from single cell analysis and experimentation [4] to 3D cell culture and investigation [5], and the development of physiologically relevant *in vitro* models, known as organ-on-chip platforms [6].

Here, we focus on one particular field of applications that can greatly benefit from the use of microfluidic technology, for all the aforementioned reasons, which is the field of assisted reproductive technologies (ART) [7, 8]. ART, which include all methods to achieve pregnancy by using artificial means, are increasingly used worldwide, with a total of more than 5 million babies conceived with these techniques. For ART, microfluidics can specifically provide alternative approaches for the various *in vitro* steps of the entire treatment, and thereby remedy currently encountered issues.

In this paper, we particularly focus on two separate steps of the *in vitro* ART protocol, which are namely the pre-implantation culture of the embryos, and their characterization with the purpose of monitoring their growth and identifying embryos with the highest developmental competence for transfer. Specifically, we first report a microfluidic platform developed for the culture of mammalian embryos, and its successful application for the culture of individual mouse embryos. Next, we present the early validation of this microfluidic platform for the culture of donated frozen-thawed human embryos. For the *in situ* characterization of the embryos, we introduce an electrochemical oxygen sensor and a novel sensing principle aiming at minimizing the amount of oxygen consumed during the measurements. Finally, we present the integration of this sensor in a culture microfluidic platform and its early validation and tumor spheroids, used here as surrogates for mammalian embryos.

978-1-5090-5979-9/17 $31.00 © 2017 IEEE

II. MICROFLUIDIC DEVICE FOR THE PRE-IMPLANTATION CULTURE OF MOUSE EMBRYOS

As a first step, a platform is developed for the culture of mammalian embryos, to evaluate if microfluidics can support the pre-implantation growth of individual embryos, down to the single embryo level, leading as well to full-term development. This culture device includes a nanoliter culture chamber, in which embryos are trapped, and from which they can easily be retrieved, for their transfer or further experimentation (Fig. 1) [9]. The microfluidic device is fabricated from PDMS (polydimethylsiloxane) by soft-lithography techniques and it is bonded to a glass substrate.

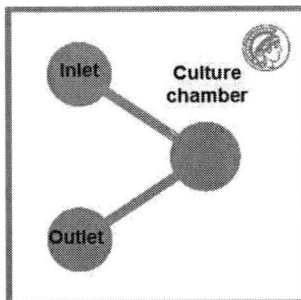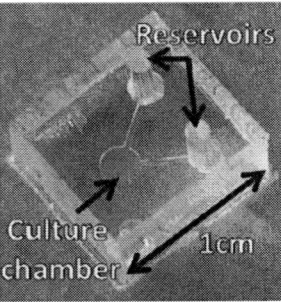

Fig. 1. Design and picture of the microfluidic platform for the culture of individual mouse embryos. The platform includes a culture chamber having a nanoliter volume, and which is equipped with structures at its inlet and outlet to prevent the embryo(s) from escaping during the culture. Interestingly, the structures at the inlet still allow retrieving the embryos for further experimentation or their transfer.

As for any new technology developed for ART, this culture device is first tested on an animal model, which is the mouse here. Naturally fertilized embryos are collected from mice, and these zygotes are distributed in groups of 20 or 5 embryos, or isolated as individual embryos. These embryos are subsequently cultured in microfluidic devices with a culture chamber of 30 or 270 nL, or, as a control, in 5-µL droplets of medium, which are covered with mineral oil to alleviate evaporation issues. Two end-points are considered to evaluate the embryo developmental potential: on one hand, their pre-implantation growth after 4-5 days of *in vitro* culture, which is examined as a blastocyst rate, the blastocyst being the stage the embryos should have reached at that time of their development; and, on the other hand, their full-term growth, or birth rate. For this second end-point, part of the embryos are recovered from the culture platforms after 3.5 days of *in vitro* culture, and transferred into pseudo-pregnant mice to allow them to complete their development.

All microfluidic culture conditions tested here (1, 5, and 20 embryos, and chamber volumes of 30 and 270 nL) give blastocyst rates higher than 90%, while conventional droplet culture for the same embryo group sizes yields significantly lower developmental rates (30-75%), the exact value depending on the embryo group size (Fig. 2) [9]. Regarding single embryo culture, the microfluidic approach proves to be highly promising with a *ca.* 95% blastocyst rate, compared to a low 30% for the droplet format (Fig. 2) [9].

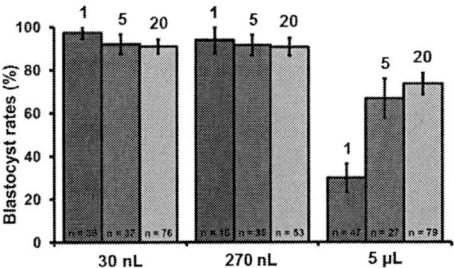

Fig. 2. Pre-implantation development or blastocyst rate for mouse embryos cultured as groups of 5 or 20, or individually, in microfluidic chambers having a volume of 30 or 270 nL or in 5-µL droplets (control).

As a next stage, the embryo full-term development is considered to ensure the used *in vitro* culture conditions do not harm the embryos. Overall birth rates are similar for both culture platforms, microfluidic devices and conventional droplets, and they amount to *ca.* 30% (Fig. 3). However, when examining single embryo culture, significantly superior performance is found in the microfluidic setting (>30% *vs.* 20% for the droplet format), and the smaller the volume for the culture, the higher the birth rate (Fig. 3) [9]. These results collectively suggest that the use of confined culture conditions benefits to embryo growth, down to the single embryo level, and this not only at the pre-implantation stage but also in terms of full-term development. This advantage given by confinement can, for instance, be explained by the fact that embryos secrete growth factors promoting their development [10], and that they can create a niche with a high concentration in these growth factors in a nanoliter chamber. In contrast, in a larger volume, these growth factors are highly diluted, which therefore results in impaired or delayed embryo development, as observed here in droplet culture.

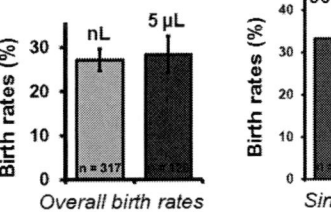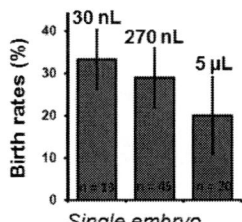

Fig. 3. Full-term development of mouse embryos cultured in microfluidic devices with culture chambers of 30 or 270 nL, or in droplets (control). *Left*: Overall birth rates in droplets *vs.* microfluidics, and *Right*: birth rates obtained for single embryo culture in the three culture volumes tested here – 30 & 270 nL (microfluidics), and 5 µL (droplets).

III. MICROFLUIDIC SYSTEM FOR THE *IN VITRO* CULTURE OF HUMAN EMBRYOS

This single embryo culture approach and microfluidic platform is subsequently tested on human embryos. The design of the microfluidic platform is upgraded to include a larger culture chamber (*ca.* 640 nL) to accommodate human embryos, which are roughly twice larger in size [11], the device being still produced from PDMS and glass using the same fabrication process. Donated human embryos are utilized in this second stage validation (CCMO authorization number NL38300.000.11), which have been kept in liquid nitrogen. These embryos, which are already at day 4 of their development, are first thawed before they are cultured for up to 72 h in the microfluidic devices and, as a control, in 25-µL droplets of medium covered with mineral oil. In total, 120 embryos are included in the study and randomly distributed between the microfluidic systems and the conventional droplets. Embryos are graded at different time points, after 24, 28, 48 and 72 h, in terms of blastocyst rate and stage to evaluate their developmental progression.

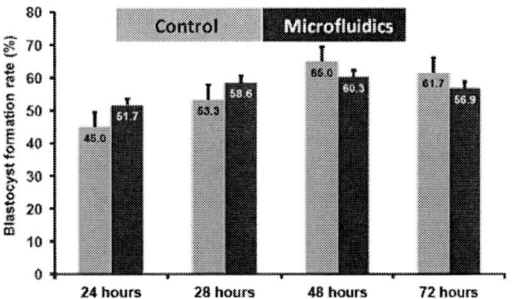

Fig. 4. Developmental rate of single human embryo cultured in PDMS-glass microdevices and conventional droplets, as controls. The developmental rate is evaluated as the blastocyst formation rate, at different time points (24, 28, 48 and 72 h) after introduction of the embryos in the respective culture platforms [11].

This second stage validation confirms the ability of PDMS microfluidic systems to support the growth of single embryos. However, similar growth rates are found for both culture platforms, and no clear advantage of the microfluidic system is seen (Fig. 4) [11]. This difference in outcome, compared to the mouse study, can be explained by the fact that at day 4, human embryos have already undergone essential steps in their development; for instance, genome activation already takes place before the 8-cell stage (day 3 of development).

IV. OXYGEN SENSOR FOR IN SITU MONITORING OF THE EMBRYO GROWTH AND DEVELOPMENTAL COMPETENCE

As a next step towards the development of an integrated platform for single embryo culture and characterization, an oxygen sensor is developed. Oxygen, which is an overall marker for the embryo metabolism, has been acknowledged as an indicator for embryo viability and developmental competence [12]. We have developed an electrochemical sensor consisting of an array of ultra-microelectrodes (UME's) acting as a working electrode. The first prototype for this sensor includes square arrays of 16, 25 or 36 UME's, which have each a diameter of 2 microns and an inter-electrode distance of 20 microns (Fig. 5) [13]. All electrodes (UME array, counter electrode and reference electrode) are fabricated from Pt on glass, and the UME's are patterned in an oxide-nitride-oxide insulating layer.

First, the reducing potential for dissolved oxygen is determined, and a reducing potential of -0.2 V is found in phosphate buffer. To minimize the oxygen consumption during the electrochemical measurements, an innovative

sensing protocol is developed [13]. Specifically, a short (< 5 ms) pulse at -0.2 V is applied and the reducing current continuously measured. At this short time-scale, the diffusion profile around the UME's remains in the linear regime, so that the sensor behavior obeys the Cottrell equation. Subsequently, the dissolved oxygen concentration can be derived from:

$$I\sqrt{t} = mnFA_{UME}\sqrt{\frac{D}{\pi}}C$$

where m is the number of UMEs in the sensors, n the number of electrons involved in the reaction, F the Faraday constant, A_{UME} the surface area of each UME, D the diffusion constant of oxygen, and C the dissolved oxygen concentration.

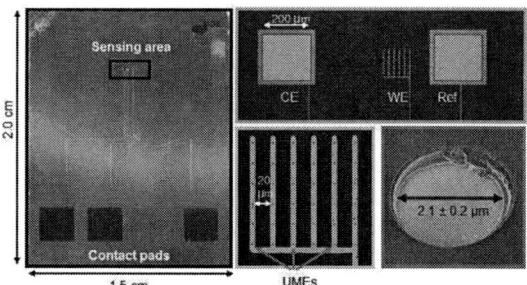

Fig. 5. Design of the electrochemical sensor consisting of an array of ultra-microelectrodes. The sensor includes a working electrode (UMEA), a counter and a reference electrodes, all fabricated from Pt on a glass substrate. Each sensor includes a square array of UMEs with here 36 UMEs patterned in an ONO insulating layer. Each UME is ca. 2 microns in diameter, and the UME's are spaced apart by 20 microns [13].

Using this novel sensing principle, the proposed UME-based sensor is calibrated by varying the concentration in dissolved oxygen through nitrogen bubbling, while monitoring the sensor response as well as the actual dissolved oxygen concentration using a commercial sensor from Unisense. An excellent linear correlation is found between the dissolved oxygen concentration and the sensor response for a 36-UME sensor, with a sensitivity of 0.49 nA.s$^{-0.5}$.L.mg^{-1} (Fig. 6). Finally, the amount of oxygen consumed through the measurements is evaluated; for a 5-ms pulse, ca. 63 fmol of oxygen are consumed, which is significantly lower than when using standard microelectrodes [13].

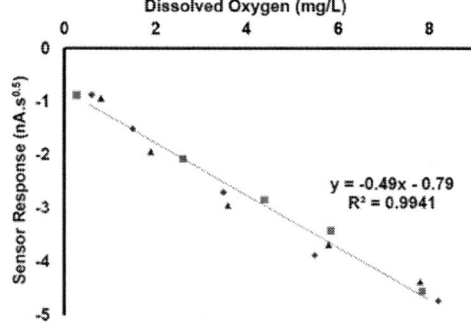

Fig. 6. Calibration of the UMEA sensor in phosphate buffer using the short measurement time protocol. The dissolved oxygen concentration is varied and measured concomitantly with an external sensor and the UMEA-based integrated sensor. Since the UMEA is operated at short measurement times, its response is expressed as nA.s$^{-0.5}$. [13]

Next, the UMEA sensor is integrated in a microfluidic device for combined embryo culture and characterization. To that end, the design of the microfluidic device is altered to include a trapping site for the capture of a single embryo in the close vicinity of the oxygen sensor (Fig. 7). So far, this integrated sensing platform has solely been tested on tumor spheroids, which are utilized as surrogates for embryos for a first stage validation of the integrated sensing platform. After introduction of a spheroid in the device, its oxygen consumption is monitored over time using the integrated sensor and the novel short measurement time approach. Typically, after 2 h monitoring, a decrease of ca. 1 mg/L in dissolved oxygen is measured in the device compared to atmospheric conditions. As a next step, the sensing device is going to be tested on mouse embryos.

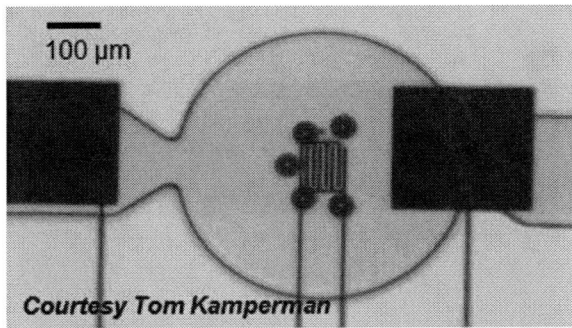

Fig. 7. Picture of the integrated sensing device, which consists of a microfluidic layer with trapping structures for the capture of a single embryo, and a glass substrate which includes the UMEA-based sensor. The sensor is located exactly in the trapping structure.

V. CONCLUSION

The ultimate goal of this project is to develop an integrated platform for combined single embryo culture and characterization based on non-invasive and metabolic parameters. Towards this goal, in a first step, we have realized and successfully validated on mouse embryos a microfluidic platform for single embryo culture. After its adaptation, the same platform has been tested on donated human embryos. In parallel, we have developed an electrochemical sensor consisting of an array of ultra-microelectrodes to monitor *in situ* in the culture chamber the embryo respiratory rate, as an overall marker of its metabolism and viability. We have notably proposed a novel sensing principle to be utilized in combination with this sensor, to minimize the amount of oxygen consumed by the electrochemical measurements. Current work focuses on the integration of the proposed sensor in the culture microfluidic platform, the validation of the entire platform on spheroids which are used as surrogates for embryos, before the integrated sensing platform can safely be evaluated on mouse embryos.

ACKNOWLEDGMENT

The presented work was funded by a Dutch-German bilateral project NWO-DFG (DN 63-258) (microfluidics for mouse embryo and oxygen sensor) and a Grant for Fertility Innovation (GFI) from Merck Serono (microfluidic platform for human embryo culture). S.L.G. would like to thank her collaborators for the presented work: Dr. Telma C. Esteves and Prof. Michele Boiani (MPI, Muenster, Germany), Prof. Verena Nordhoff and Prof. Stefan Schlatt (CeRA, Muenster, Germany), Fleur van Rossem, Johan Bomer, Dr. Zhenxia Hao, Tom Kamperman and Dr. Yawar Abbas (UT, The Netherlands), Dorit Kieslinger, Dr. Carlijn Vergouw, Dr. Elisabeth Kostelijk and Prof. Nils Lambalk (VUMC, Amsterdam, The Netherlands).

REFERENCES

[1] G. M. Whitesides, "The origins and the future of microfluidics", *Nature*, vol. 442, pp. 368-373, 2006.

[2] E. W. K. Young and D. J. Beebe, "Fundamentals of microfluidic cell culture in controlled microenvironments", *Chem. Soc. Rev.*, vol. 39, pp. 1036-1048, 2010.

[3] B. Harink, S. Le Gac, R. Truckenmuller, C. van Blitterswijk, P. Habibovic, "Regeneration-on-a-chip? The perspectives on use of microfluidics in regenerative medicine", *Lab Chip*, vol. 13, pp. 3512–3528, 2013.

[4] S. Le Gac and A. van den Berg, "Single cells as experimentation units in lab-on-a-chip devices", *Tr. Biotech.*, vol. 28, pp. 55, 2010.

[5] N. Picollet-D'hahan, M. E. Dolega, L. Liguori, C. Marquette, Séverine Le Gac, X. Gidrol, D. K. Martin, "A 3D toolbox to enhance physiological relevance of human tissue models", *Tr. Biotech*, vol. 34, pp. 757-769, 2016.

[6] S. N Bhatia and D. E. Ingber, "Microfluidic organs-on-chips", *Nat. Biotech.*, vol. 32, pp. 760-772, 2014

[7] J. E. Swain, D. Lai, S. Takayama and G. D. Smith, "Thinking big by thinking small: application of microfluidic technology to improve ART", Lab Chip, vol. 13, pp. 1213-12224, 2013.

[8] S. Le Gac and V. Nordhoff, "Microfluidics for mammalian embryo culture and selection: where do we stand now?", *Mol. Human Reprod.*, vol. 23, pp. 3-16, 2017.

[9] T. C. Esteves, F. van Rossem, V. Nordhoff, S. Schlatt, M. Boiani, S. Le Gac, "Microfluidic system supports single mouse embryo culture leading to full-term development", *RSC Adv.*, vol. 3, pp. 26451-26458, 2013

[10] B. C. Paria, S. K. Dey SK., "Preimplantation embryo development in vitro cooperative interactions among embryos and role of growth-factors.", *Proc Natl Acad Sci U S A*, vol. 87, pp. 4756-4760, 1990.

[11] D. C. Kieslinger*, Z. Hao*, C. G. Vergouw, E. H. Kostelijk, C. B. Lambalk, S. Le Gac, "In vitro development of donated frozen thawed human embryos in a prototype static microfluidic device: a randomized controlled trial", *Fertil. Steril.*, vol. 103, pp. 680-686, 2015.

[12] F. D. Houghton, J. G. Thompson, C. J. Kennedy, H. J. Leese HJ, "Oxygen consumption and energy metabolism of the early mouse embryo", Mol. Reprod. Dev., vol. 44, pp. 476-485, 1996.

[13] F. van Rossem, J. Bomer, H. L. de Boer, Y. Abbas, E. de Weerd, A. van den Berg, S. Le Gac, "Sensing oxygen at the millisecond time-scale using an ultra-microelectrode array (UMEA)", Sens. Act. A., vol. 238, pp. 1008-1016, 2017.

Development of Ultrasensitive Extended-Gate Ion-Sensitive-Field-Effect-Transistor based on Industrial UTBB FDSOI Transistor

Getenet Tesega Ayele[1, 2, 3, 4], Stephane Monfray[1], Frederic Boeuf[1], Jean-Pierre Cloarec[2], Serge Ecoffey[3, 4], Dominique Drouin[3, 4], Etienne Puyoo[2], Abdelkader Souifi[2]

Email: getenet-tesega.ayele@insa-lyon.fr

[1]STMicroelectronics, 850 Rue Jean Monnet, 38920 Crolles, France. [2]INSA Lyon, 20 Av. Albert Einstein, 69100 Villeurbanne, France. [3]Institut interdisciplinaire d'innovation technologique (3IT), Université de Sherbrooke, QC J1K 2R1, Canada. [4]Laboratoire Nanotechnologies Nanosystemes (LN2) CNRS UMI-3463, Universite de Sherbrooke, QC J1K 2R1, Canada.

Abstract—The proof of concept of a new extended-gate pH sensor, developed on an industrial ultrathin body and buried oxide (UTBB) fully-depleted silicon-on-insulator (FDSOI) transistor, is reported. The strong electrostatic coupling between the front gate and back gate of UTBB FDSOI devices provide a signal amplification opportunity for sensing applications. On the other hand, the biasing capability through a capacitive divider circuit of a floating gate ISFET offers an ample advantage for fabrication of stable and CMOS compatible solid state chemical sensors. In addition, the deep downscaling of the state-of-the-art devices enables it to be sensitive at single-charge-resolution. By integrating aluminum oxide (Al_2O_3) for the pH sensing purpose, we obtained an extended-gate mode ISFET having a sensitivity of 475 mV/pH, which is superior to state-of-the-art low-power ISFETs.

Keywords—Aluminum oxide, capacitive coupling, extended-gate, fully-depleted silicon-on-insulator (FDSOI); ion-sensitive-field-effect-transistor (ISFET), pH sensor

I. INTRODUCTION

Ionic species incorporated in the device during manufacturing of insulated gate field effect transistors (IGFET) introduce variations in the threshold voltage (V_{th}). The principle of ion sensitive field effect transistors (ISFETs), introduced by P. Bergveld in 1970 [1], is to employ this effect in a reproducible and reversible form to measure the activity of an ionic species in an aqueous solution to which the gate insulator is exposed [2]. The structure of ISFET is similar to that of a conventional MISFET except that the metal gate electrode is removed in order to expose the underlying insulator layer to the electrolyte [3]. Starting from the 1980s, the sensitive film could also be integrated on top of a floating metal gate, giving rise to a floating gate (FG) ISFET [4 -9].

Before the introduction of double gate ISFETs, the sensors had a limited maximum sensitivity which is referred to as Nernst limit (59.6 mV/pH at room temperature). However, well enhanced sensitivities surpassing the Nernst limit have been reported utilizing dual gate and SOI devices [10, 11]. Developing our pH sensor on ultrathin body and box (UTBB)

fully-depleted silicon-on-insulator (FDSOI), we obtained a sensitivity amplification arising from the strong electrostatic coupling between the two gates of the device [12]. We designed the ISFET in which no reference electrode is needed due to the incorporation of a floating-gate and a capacitively-coupled control gate. In that case, the front gate bias is applied through a control gate (CG) rather than a reference electrode as in [13]. Therefore, our sensor is highly sensitive and CMOS compatible which is suited for low-power and low-cost ultrasensitive applications expected for the IoT market.

II. SENSING MECHANISM

It is generally considered that the surface charging mechanism for pH-sensing metal oxides is the adsorption of protons or hydroxyl ions by surface hydroxyl groups to form positive or negative sites respectively [14]. Thus the resulting surface charge, which depends on an excess of one type of charged site over the other, is a function of the solution pH value. This phenomenon results in a surface potential which in turn causes a proton concentration gradient between the electrolyte bulk and surface that is mathematically described by Boltzmann equation as in (1).

$$[H^+]_S = [H^+]_B \, exp\left(\frac{-q\varphi}{k_BT}\right) \tag{1}$$

where k_B is the Boltzmann constant, q is the elementary charge, T is temperature, φ is surface potential, $[H^+]_B$ is the proton concentration in the bulk, and $[H^+]_S$ is the proton concentration at the surface (dielectric-electrolyte interface).

Rearranging (1), the surface potential φ is expressed as a function of the proton concentrations at the dielectric-electrolyte interface and in the bulk as given by (2) below, which is also referred to as the Nernst equation [15].

$$\varphi = \frac{k_BT}{q} ln \frac{[H^+]_B}{[H^+]_S} \tag{2}$$

The pH sensitivity of the dielectric film is derived from (2) and is given mathematically as follows in (3) [15]:

$$\frac{d\varphi}{dpH} = -2.3\,\frac{K_B T}{q}\,\alpha \qquad (3)$$

where α is the proton buffer capacity.

The proton buffer capacity (α) is a dimensionless parameter that varies between 0 and 1. At a value $\alpha = 1$, the pH sensing film will have a maximum sensitivity of 59.6 mV/pH at room temperature (Nernstian limit of sensitivity).

III. DEVICE FABRICATION AND CHARACTERIZATION

As a proof of concept, we designed our sensor in the extended-gate mode of operation where the transduction part and the FDSOI were connected externally. In that case, the fabrication of the pH-sensing element was carried out isolated from the transistor, and finally external electrical connections were made for sensitivity characterizations.

A. Fabrication

Our pH sensor was realized utilizing an n-type FDSOI transistor manufactured by STMicroelectronics. This transistor has a channel of thickness $t_{Si} = 6$ nm, buried oxide (BOX) = 25 nm, and equivalent gate oxide thickness = 1.8 nm. The device has a gate width and gate length of 80 nm and 1 µm respectively.

The transduction element was processed on a 750 µm Si-substrate which had a thermally grown 500 nm SiO_2 on top of it. A 10 nm/80 nm Ti/Pt metal layer was then deposited by E-beam evaporation. This metal layer serves as an extension of the FDSOI floating-gate for extended-gate mode operation of the ISFET. 50 nm aluminum oxide (Al_2O_3) was deposited by atomic layer deposition (ALD) on the Ti/Pt metal layer. Finally, we applied silver pastes to complete the external electrical-connections between the Ti/Pt metal layer of the transduction element and the floating metal gate of the FDSOI. The schematic diagram of the electrical connection and system layout for pH-sensing characterization is shown on Fig. 1.

B. Characterization

For our pH-sensitivity characterization, we utilized 3-different pH values from pH 6 to pH 8. When we chose this measurement window, we are aware that for biomedical applications such as pH-measurement and acidity-monitoring of blood, urine, saliva, and intestinal fluids, pH range of 6 to 8 would be sufficient [16-18]. But of course, this doesn't mean that our sensor is limited to this range as it is realized utilizing FDSOI transistor and aluminum oxide sensing film, both of which have wide linear-operating ranges.

At the very beginning, the FDSOI transistor was characterized for its shift in threshold voltage at the back gate as a function of change in voltage at the front gate. Next, the transducer sample was cleaned with acetone/IPA/DI water and dried with nitrogen (N_2), and full electrical connection was made using Keithley 4200 semiconductor characterization system. Images of the setup are indicated on Fig. 2.

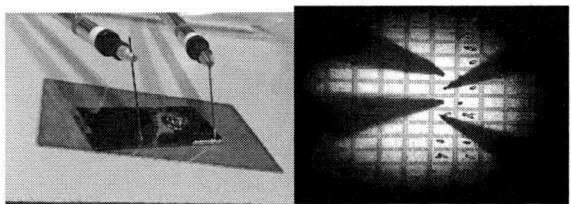

Fig. 2. Images of the transduction part (left) and connection pads of the FDSOI device (right).

At this point, reference I_D-V_{BG} sweep was made before adding any pH solution. Then, 30 µl of pH 8.0 solution was added on the sensing film and I_D-V_{BG} sweeps were made after 1-min, 2-mins, and 3-mins. The pH 8 droplet was removed, and a 30 µl of pH 7.3 droplet was added. Similarly, I_D-V_{BG} sweeps were made after 1-min, and 2-mins. Finally, 30 µl of pH 6.0 droplet was added after removing the pH 7.3 droplet and I_D-V_{BG} sweeps were made after 1-min, 2-mins. From the above experiments, we obtained a big shift in the threshold voltage of the transistor, when the pH value of the solutions was changed

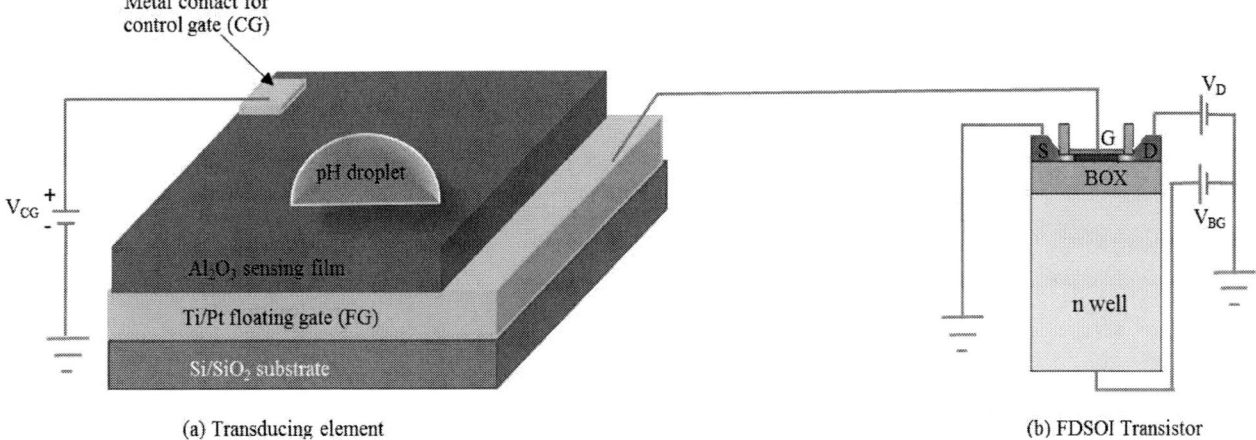

(a) Transducing element

(b) FDSOI Transistor

Fig. 1. Schematic diagram of the extended-gate ISFET.

from one value to the other. We could also observe a stable pH-sensitivity where the drift impact was negligible (compared to the enhanced sensitivity obtained, and considering practical pH measurement durations that last only for a couple of minutes).

The stability of our sensor, with a droplet of pH 8 is presented on figure 6.

IV. RESULT AND DISCUSSION

We characterized the shift in threshold voltage of our FDSOI transistor for different gate voltages, as shown on Fig. 3. It has a coupling factor (γ) of value 13, where γ is the ratio of the shift in threshold voltage at the back gate and the change in applied voltage at the front gate. This means that the change in surface potential occurring at the front gate side appears as a change in threshold voltage at the back gate amplified by 13-times. From this performance of our starting FDSOI transistor, we theoretically expect a maximum pH-sensitivity of 774.8 mV/pH which corresponds to the Nernestian limit of sensitivity of the film.

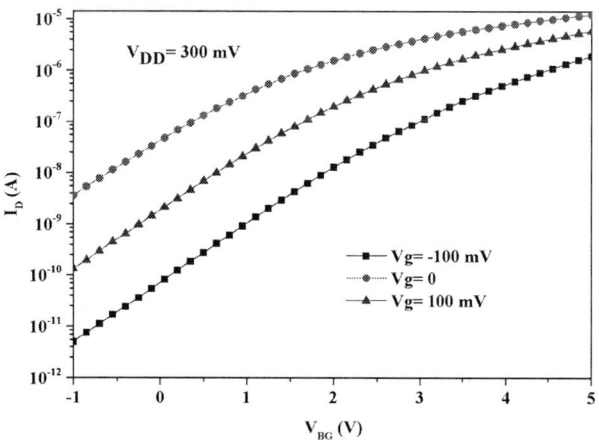

Fig. 3. I_D-V_{BG} characteristics of the FDSOI transistor at different front gate voltages.

Comparing the I_D-V_{BG} characteristics of our extended-gate ISFET, with solutions of different pH values added on the pH sensing film, the corresponding shifts in threshold voltage are presented on Fig. 4.

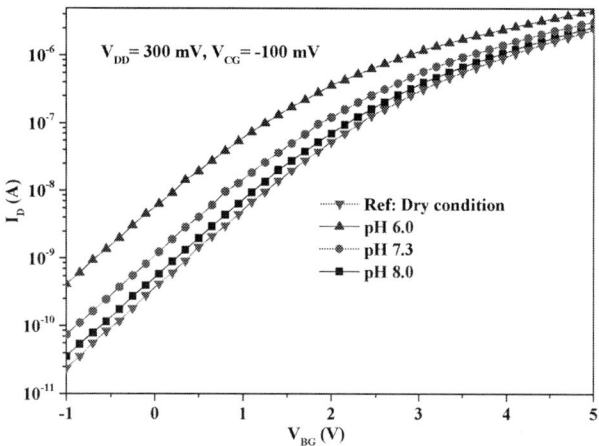

Fig. 4. I_D-V_{BG} characteristics of the extended gate ISFET at different pH values.

Plotting the threshold voltage versus pH value, we extracted the sensitivity of our ISFET, which is presented on Fig. 5. As can be seen on Fig. 4 and Fig. 5, we obtained a pH-sensitivity of 475 mV/pH which is superior performance with respect to the state of the art ISFETs in terms of sensitivity. In addition, our design of biasing through a capacitive divider circuit, makes our sensor CMOS compatible by getting rid of the bulky Ag/AgCl reference electrode. Last but not least, the very high I_{ON}/I_{OFF} ratio of our UTBB FDSOI industrial transistor provided us with a wide linear operating range and high signal to noise ratio (SNR).

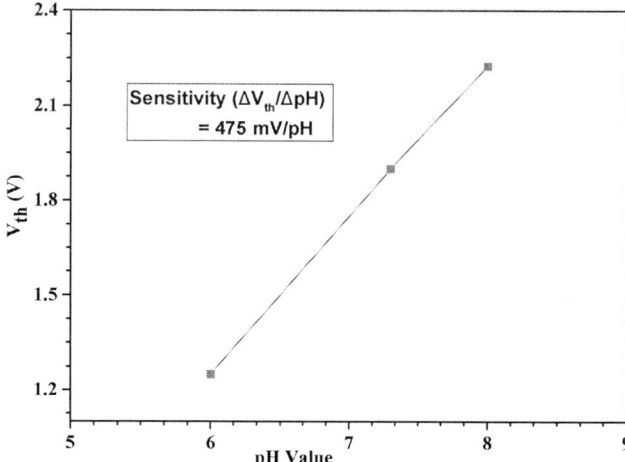

Fig. 5. Extraction of pH sensitivity from threshold voltage shifts of I_D-V_{BG} characteristics at different pH values.

At the different pH values, measurements were repeated to see repeatability and stability. From such measurements, we have observed a reasonable repeatability and stability. From Fig. 6, we can observe a drift free and fast pH-response of our ISFET.

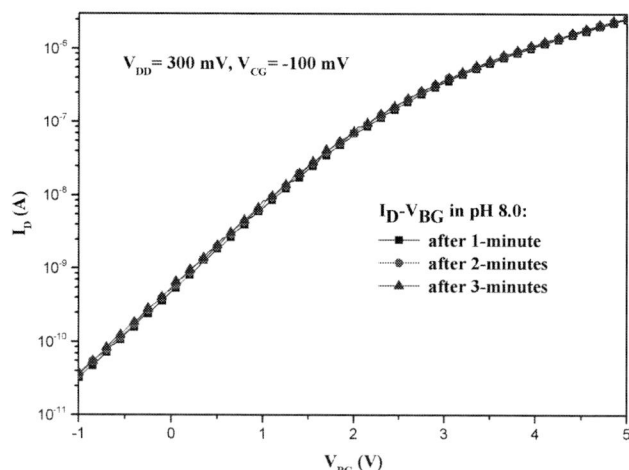

Fig. 6. Repeatability and stability of the I_D-V_{BG} characteristics of the ISFET with a droplet of pH 8 electrolyte

978-1-5090-5979-9/17 $31.00 © 2017 IEEE

V. CONCLUSION AND PERSPECTIVE

Utilizing industrial UTBB FDSOI transistors, we fabricated highly sensitive floating-gate ISFETs. The floating-gate design offers the opportunity to apply a front gate bias through a capacitively-coupled control gate, which makes the ISFET operation very stable and helps to get rid of the bulky reference electrode. This makes such solid state chemical sensors CMOS compatible, and therefore it makes fabrication of deeply downscaled and high density sensor arrays feasible.

Having validated the concept of developing a very sensitive ISFET in the floating-gate mode and utilizing aluminum oxide pH-sensitive layer, we are developing the sensor prototype in back-end-of-line (BEOL) integration. The following process flow will be followed for the back-end-of-line integration.

- Deposition of insulation/passivation layer for electrical isolation
- Patterning by lithography
- Etching for transistor-pad opening
- Metal deposition for source, drain, and back gate metal contacts
- Patterning by lithography
- Etching
- Deposition of pH-sensing film
- Metal deposition for control gate
- Patterning by lithography
- Etching
- Patterning by lithography
- Etching to open metal contacts of source, drain, back gate
- Formation of well for the pH solution

ACKNOWLEDGMENT

Région Auvergne Rhône-Alpes is acknowledged for a Coopera project funding.

REFERENCES

[1] P. Bergveld, "Development of an ion-sensitive solid-state device for neurophysiological measurements," IEEE Transactions on Biomedical Engineering 1 (1970): 70-71.

[2] R. G. Kelly, "Microelectronic approaches to solid state ion selective electrodes," Electrochimica Acta (1977), 22(1), 1-8.

[3] H. Abe, M. Esashi, & T. Matsuo, "ISFET's using inorganic gate thin films," IEEE Transactions on Electron Devices (1979), 26(12), 1939-1944.

[4] L. Lauks, P. Chan, and D. Babic. "The extended gate chemically sensitive field effect transistor as multi-species microprobe," Sensors and Actuators 4 (1983): 291-298.

[5] N. M. Shen, Z. Liu, C. Lee, B. A. Minch, & E. C. Kan, "Charge-based chemical sensors: a neuromorphic approach with chemoreceptive neuron MOS (C/spl nu/MOS) transistors," IEEE Transactions on Electron Devices (2003), 50(10), 2171-2178.

[6] J. Bausells, J. Carrabina, A. Errachid, & A. Merlos, "Ion-sensitive field-effect transistors fabricated in a commercial CMOS technology," Sensors and Actuators B: Chemical (1999), 57(1), 56-62.

[7] K. Levon, A. Rahman, T. Sai, and Z. Ben, "Floating gate field effect transistors for chemical and/or biological sensing." U.S. Patent No. 7,462,512. 9 (Dec. 2008).

[8] L. Bousse, J. Shott, and J. D. Meindl, "A process for the combined fabrication of ion sensors and CMOS circuits," IEEE Electron Device Letters (1988), 9(1), 44-46.

[9] Rothberg, Jonathan M., Wolfgang Hinz, Todd M. Rearick, Jonathan Schultz, William Mileski, Mel Davey, John H. Leamon et al. "An integrated semiconductor device enabling non-optical genome sequencing." Nature (2011), 475(7356), 348-352.

[10] H. J. Jang, and W. J. Cho, "Performance enhancement of capacitive-coupling dual-gate ion-sensitive field-effect transistor in ultra-thin-body," Dep. Electron. Mat. Eng., Kwangwoon Univ., Seoul 139-701, Rep. Korea, Jun. 13, 2014.

[11] Y. J. Huang, C. C. Lin, J. C. Huang, C. H. Hsieh, C. H. Wen, T. T. Chen, L. S. Jeng, C. K. Yang, J. H. Yang, F. Tsui, Y. S. Liu, S. Liu and M. Chen, "High performance dual-gate ISFET with non-ideal effect reduction schemes in a SOI-CMOS bioelectrical SoC," Presented at Electron Devices Meeting (IEDM), IEEE (Dec, 2015).

[12] T. Skotnicki, & S. Monfray, "UTBB FDSOI: Evolution and opportunities," In Solid State Device Research Conference (ESSDERC), (2015) 45th European (pp. 76-79), IEEE.

[13] Qi Zhang, Himadri S. Majumdar, Matti Kaisti, Alok Prabhu, Ari Ivaska, Ronald Österbacka, Arifur Rahman, and Kalle Levon; "Surface Functionalization of Ion-Sensitive Floating-Gate Field-Effect Transistors With Organic Electronics," IEEE, TED (Apr. 2015), 62(4).

[14] Yates, David E., Samuel Levine, and Thomas W. Healy, "Site-binding model of the electrical double layer at the oxide/water interface," Journal of the Chemical Society, Faraday Transactions 1: Physical Chemistry in Condensed Phases 70 (1974): 1807-1818.

[15] Spijkman, M., E. C. P. Smits, J. F. M. Cillessen, F. Biscarini, P. W. M. Blom, and D. M. De Leeuw, "Beyond the Nernst-limit with dual-gate ZnO ion-sensitive field-effect transistors," Applied Physics Letters 98, no. 4 (2011): 043502.

[16] L. Hermansen, & J. B. Osnes, "Blood and muscle pH after maximal exercise in man," Journal of applied physiology (1972), 32(3), 304-308.

[17] D. F. Evans, et al., "Measurement of gastrointestinal pH profiles in normal ambulant human subjects," Gut (1988), 29(8), 1035-1041.

[18] S. Baliga, S. Muglikar, & R. Kale, "Salivary pH: A diagnostic biomarker," Journal of Indian Society of Periodontology (2013), 17(4), 461-465.

Ultrathin Lateral Unidirectional Bipolar-Type Insulated-Gate Transistor as pH sensor

Qinghua Han[1], Anran Gao[1,2], Keyvan Narimani[1], Yuelin Wang[2], Tie Li[2], Siegfried Mantl[1], Qing-Tai Zhao[1]

[1]Peter GrünbergInstitute 9, JARA-FIT, Forschungszentrum Jülich, 52425 Jülich, Germany
[2]Science and Technology on Micro-system Laboratory, Shanghai Institute of Microsystem and
Information Technology, Chinese Academy of Sciences, 200050 Shanghai, China
Email: gaoanran@mail.sim.ac.cn

Abstract—Here, the utilizing of lateral unidirectional bipolar-type insulated-gate transistors (Lubistors) for pH detection was demonstrated for the first time. The high on current and ambipolarity of Lubistors are favorable properties for reliable sensing application. The ultrathin Lubistors were fabricated on 20-nm-thick SOI substrates in planar geometry. The triode-like current–voltage characteristic and excellent gate modulation property for Lubistor were demonstrated. Without any surface modification, this Lubistor based pH sensor was capable of detecting the change of hydrogen ion concentration by the corresponding change in anode current. The anode current revealed nonlinear pH-dependence and complementary change at different working regions. The Lubistor sensor showed stable, reversible and reproducible pH sensing performance. The proposed Lubistor based sensor holds great potential for high reliable biological species detection.

Keywords—Lubistor; pH sensor; hydrogen ion, reliable

I. INTRODUCTION

Ultrasensitive biological and chemical species detection is of great importance for screening and detection of disease, discovery and screening of drugs, as well as biomolecular analysis [1, 2]. In particular, ions detection in liquid field is attractive to a number of fields. The determination of pH values is one of the most important tasks in analytical chemistry. Up to now, different electrochemical and non-electrochemical methods are used in measuring pH values. Semiconductor transistor that exhibits attractive properties, like low cost, small size, low noise, low power and ease to be integrated with electronic circuits [3], is one of the most promising sensor structures for pH sensing.

As we know, the electrical potential change at different pH values is very small. Conventional semiconductor device based biosensors face rising challenges in terms of device stability and poor disturbance resistibility when the device is scaled down. In addition, the signal of conventional semiconductor device is typically very weak and variable particularly in solutions. Thus, possible electrical cross talk and/or false-positive signals may occur and limit its application in real world.

In this study, the lateral unidirectional bipolar type insulated-gate transistor (Lubistor) was proposed for pH sensing. The Lubistor were fabricated on 20-nm-thick SOI substrates in planar geometry with CMOS-compatible

technology. Since the Lubistor works like a forward biased pn junction, high on currents of the device are expected [4], which is favorable for reliable sensing application. Furthermore, the Lubistor can work at both positive and negative gate voltages, revealing ambipolar characteristic. This ambipolarity enables discrimination against false positives by correlating the response versus time from the two types of device behavior, and provides novel sensing strategies that result in a more robust device [5]. Therefore, the proposed Lubistor devices are potential candidates for many biochemical applications.

II. EXPERIMENTAL SECTION

A. Lubistor Fabrication

A Lubistor with a lateral n+-p-p+ structure was fabricated on a (100) SOI wafers with a 145-nm-thick buried oxide layer and a top Si layer of 20-nm-thick. The scheme of the device layout is shown in Fig. 1(a). A total of 6 independent devices are available for use on each chip. The specific fabrication process is as following. The mesa was firstly defined by photolithography and reactive ion etching (RIE). Then the low energy self-aligned implantations for anode and cathode were performed to form effective contact regions. N- type phosphorous implantation for cathode and p-type boron implantation for anode produced fundamental structure of Lubistor. To incorporate the implanted ions in the crystal lattice and cure lattice damage created during implantation while minimizing dopant diffusion, a spike annealing at $1050°$ C was employed. A doping concentration of about $10^{20}/cm^3$ in the doped regions was expected. Then 10nm SiO_2 was thermally oxidized on device. After that, aluminum layer was deposited and patterned by lift-off procedure to form the final contacts. Finally, the whole device was protected with SU-8 while only the device channel and the electrodes were exposed in air.

As indicated in Fig. 1(b), the fabricated Lubistors are grouped in clusters. Such a layout is suitable for multiplexed detection because each device may serve as one unit for sensing a particular solution. The schematic diagram of the completed Lubistor configuration is shown in Fig. 1(c). The anode electrode is positively biased for the device and the gate electrode can be positively/negatively biased to induce inversion layer (electrons/holes). Therefore, in addition

978-1-5090-5979-9/17 $31.00 © 2017 IEEE

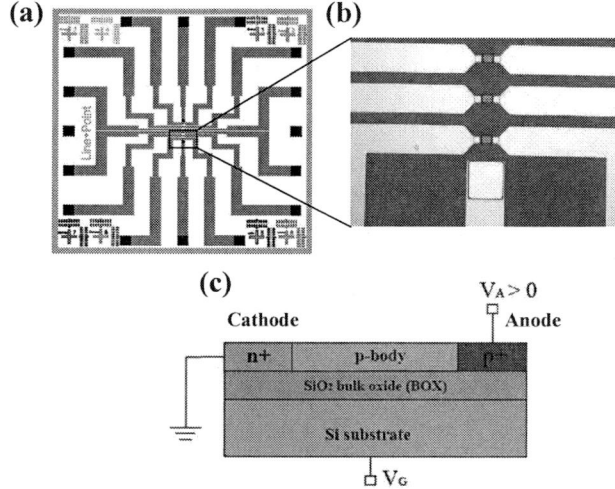

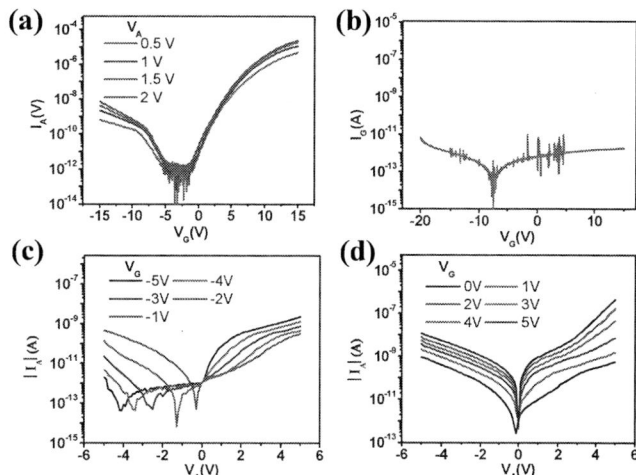

Fig. 1: (a) Scheme of the layout of the device array on the chip. A total of 6 independent devices were are available for use on each chip. (b) The optical micrograph of fabricated device. It was protected by SU-8 as passivation layer except channels and electrodes. (c) Schematic diagram of the completed Lubistor configuration. Lubistor has n+-p-p+ structure with back gate as insulated gate.

Fig. 2: (a) I_A–V_G characteristics of a Lubistor for varying V_A (0.5 to 2 V). (b) The device gate leakage current I_G at V_A=0.5 V. (c) (d) The I_A–V_A characteristics of a device for varying V_G from -5 V to -1 V (c) and from 0 V to 5 V (d).

to the built-in junction between the p-body and n+ cathode, a virtual junction is also formed. When the gate electrode is positively biased, the virtual junction is made between the inversion layer composed of electrons and the p+ anode. When the gate electrode is negatively biased, the junction is made between the inversion layer composed of holes and the n+ cathode.

B. General Sensing Apparatus

For general electrical and pH sensing measurement, the device was characterized by using a Keithley 4200 semiconductor parameter analyzer. Solutions with varying pH values are made from mixtures of 100mM sodium dihydrogen phosphate and disodium hydrogen phosphate. Only high purity water was used as the solvent in this study for the pH sensing.

III. RESULTS AND DISCUSSION

A. Device Characterization

To verify the quality of the Lubistors before pH detection, the electrical characterization of an as-made Lubistor device was carried out. The device properties were measured with varying gate and anode voltages, to evaluate the gate control capability. The device was back gate biased, and I_A-V_G measurements were implemented through the anode and cathode electrodes. Generally, in the measurement of the I_A-V_G curve, the anode-cathode current was measured for varying V_A from 0.5 to 2 V, $\Delta V = 0.5$ V while sweeping the V_G from -15 V to 15 V. As illustrated in Fig. 2(a), the ambipolar characteristic with a strong dependence of I_A on V_G was observed. The current can flow at both negative and positive gate voltages, giving rise to ambipolar conduction. Additional,

the test results demonstrate high on current, indicating that the Lubistor device offers a high on-state performance. The device leakage characteristic was also measured as shown in Fig. 2(b). The I_G-V_G curve reveals small gate leakage current.

Furthermore, the I_A-V_A current-voltage characteristics of Lubistor were measured. In the measurement of the I_A-V_A curve, the current was measured at varying gate voltage from -5 to -1 V (Fig. 2(c)) and from 0 to 5 V (Fig. 2(d)) while sweeping the anode potential (V_A). When the gate voltage is negative, the I_A-V_A curves shift in the positive direction. The anode current increases as the increase of gate voltage when the device is positively biased. The high dependence of anode current on the anode-to-cathode voltage and the gate voltage reveals Lubistor device property.

B. pH Sensing

A Lubistor device can act as a pH sensor with naturally occurring silanol (Si−OH) groups on device surface, which undergo protonation/deprotonation reactions in solutions with different pH values. In this process, surface charges change, which in turn giving rise to conductance variations. A Lubistor device was thus converted into hydrogen ion concentration or a pH sensor.

After device fabrication was finished, the Lubistor was integrated with a mixing cell or solution chamber (Fig. 3(a)). The cell was used for delivering solutions. This cell was placed over the sensor surface and allows the solution to be delivered from the top aperture. In this setup, solutions with different pH values were delivered by replacement methods. For pH sensing experiments, two gate voltages were applied on device as shown in Fig. 3(b). Back-gate acts as a driving gate that controls the current flow, while front-gate acts as a supporting gate that prevents environment noise.

978-1-5090-5979-9/17 $31.00 © 2017 IEEE

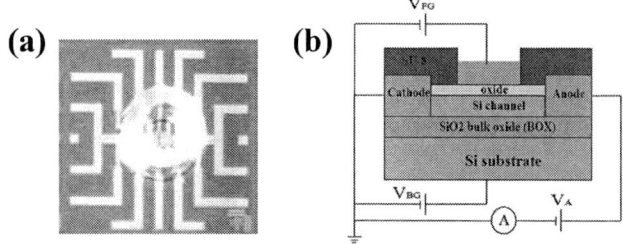

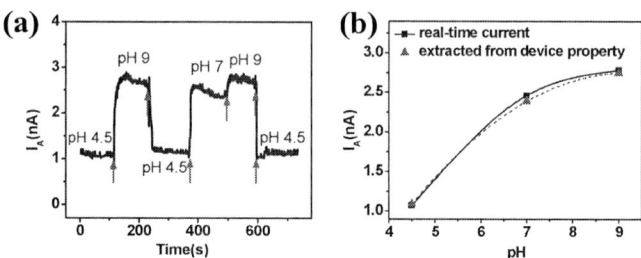

Fig. 3: (a) The picture of the completed device integrated with a sensing chamber. (b) Schematic illustration of the fabricated Lubistor device for pH sensing measurement.

Fig. 4 displays the test results of pH sensor based on Lubistors. All the tests were done at room temperature. The I_A-V_{BG} curves for the device under different pH conditions are shown in Fig. 4(a). An obvious current change at different pH values was observed. The threshold voltage (V_{th}) of the Lubistor device was defined as the gate voltage at anode current of 10^{-9} A based on the constant current method. As the pH value increases, the threshold voltage shifts. The pH sensitivity of 42 mV/pH was obtained, which is a typical value for silicon oxide surfaces [6]. For more explicit illumination, the sensor current versus pH is demonstrated in Fig. 4 (c) and (d) for varying V_{BG}. The p (c) and n (d) -type like device performance were demonstrated at different working regions of Lubistor device. The Lubistor revealed complementary signals at different working regions, providing a simple yet robust means for detection false-positive signals from nonspecific binding onto Lubistor sensor device. By correlating the change of current signals in both regions, the device can also distinguish unambiguously noise.

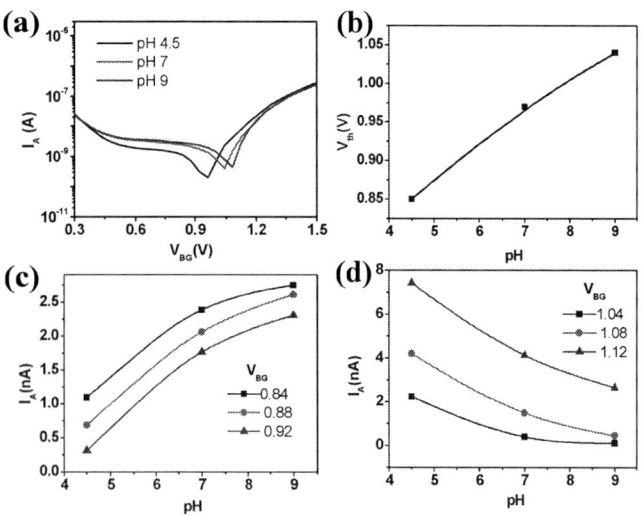

Fig. 4: (a) I_A-V_{BG} curves measured in solution with different pH values for V_{FG}=0 V, V_A=0.5 V. (b) Plots of measured V_{th} versus pH for Lubistor device. (c) (d) The current of sensor versus pH at different gate voltages. P (c) and n (d) -type like device performance were demonstrated.

Fig. 5: (a) Real-time detection of device current at different pHs for V_{FG}=0 V, V_{BG}=0.8 V, V_A=0.5 V. Arrows mark the points when the solutions were changed. (b) The comparation of pH sensor performance between real-time measurement results and the data extracted from device property.

Moreover, the real-time pH measurement was also performed at a constant bias voltage. As illustrated in Fig. 5(a), the real-time current of Lubistor increases with the increasing pH values at V_{FG}=0 V, V_{BG}=0.8 V, V_A=0.5 V. For a given pH value, the current is almost a constant. Furthermore, the changes in current are also reversible for increasing or decreasing pH. A typical plot of current versus pH (Fig. 5(b)) shows that the pH dependence is nonlinear over the pH range from 4.5 to 9.0. The sensing results between real-time measurement and the data extracted from device property are in good consistency, showing that the Lubistor based pH sensor is stable and reliable.

IV. CONCLUSION

In conclusion, the fabrication and operation of a Lubistor device based pH sensor have been demonstrated in this study. The Lubistor with n+-p-p+ structure and back gate has been fabricated with standard semiconductor processing techniques with advantages of low cost for bath production, reproducibility, high controllability and so on. The Lubistor device exhibited excellent gate modulation ambipolar property, laying foundation for great complementary sensing performance. The Lubistor based pH sensor have shown stable, reproducible sensing performance. The nonlinear pH-dependence of the device current is in good agreement with previous reported unmodified device. The Lubistor sensor enables discrimination of possible electrical cross-talk and false-positive singles. Our sensing approach using Lubistors offers a new option for high reliable detection of biological species.

ACKNOWLEDGMENT

We appreciate financial support from Alexander von Humboldt Foundation. This work was partially supported by project of National Natural Science Foundation of China (no. 81402468), Shanghai Youth Science and Technology Talent Sailing project (no. 14YF1407200), Project for Shanghai Outstanding Academic leaders (15XD1504300) and Youth Innovation Promotion Association, CAS (2014206).

REFERENCES

[1] M. Ferrari, "Cancer nanotechnology: opportunities and challenges," Nature Reviews Cancer, vol. 5, pp. 161-171, March 2005.

[2] P. Bergveld, "Development, Operation, and Application of the Ion-Sensitive Field-Effect Transistor as a Tool for Electrophysiology," IEEE Trans. Biomed. Eng. vol. 19, pp. 342-351, 1972.

[3] P. Bergveld and A. Sibbald, "Analytical and biomedical applications of ion-selective field-effect transistors," Amsterdam, Elsevier, 1988.

[4] Y. Omura, "Experimental study of two-dimensional confinement effects on reverse-biased current characteristics of ultrathin silicon-on-insulator lateral, unidirectional, bipolar-type insulated-gate transistors," Japanese journal of applied physics, vol. 46, pp. 2968, 2007.

[5] G. Zheng, F. Patolsky, Y. Cui, W. U. Wang and C. M. Lieber, "Multiplexed electrical detection of cancer markers with nanowire sensor arrays," Nat. Biotechnol. vol. 23, pp. 1294–1301, 2005.

[6] X. T. Vu, J. F. Eschermann, R. Stockmann, R. GhoshMoulick, A. Offenhäusser and S. Ingebrandt, "Top - down processed silicon nanowire transistor arrays for biosensing," physica status solidi (a), vol. 206, pp.426-434, 2009.

A novel approach for scalable sensor arrays using cantilever field-effect transistors

Andreas Hessel, Stefan Scholz, Alexander Pelger, Albert Pfander, Joachim Knoch

Institute of Semiconductor Electronics
RWTH Aachen University
Aachen, Germany
E-mail: Hessel@IHT.RWTH-Aachen.de, Knoch@IHT.RWTH-Aachen.de

Abstract— **Based on a novel approach of reading-out cantilever sensors with an exponential response of the measurement signal on adsorption of e.g. molecules, we study cantilever array sensors that enable the detection and discrimination of various different adsorbates. Different cantilever geometries and different ratios of functionalized and unfunctionalized areas on the cantilevers are used together with a pairwise comparison of the measurement signal, i.e. a shift of resonance frequency of oscillating cantilevers. As a result, a discrimination of different adsorbates becomes possible with a limited number of functionalizations.**

Keywords—cantilever sensor; cantilever array; artificial nose; self-oscillation; cantilever scaling

I. INTRODUCTION

Cantilever sensor arrays are a promising technology [1] [2] for high sensitive molecule detection and thus have been intensively studied in recent years. A mass resolution up to yoctogram (10^{-27} kg) has been reported for a single carbon nanotube resonator [3]. However, the main research focus to-date has been mostly on single devices [4] [5] and their functionalization [6] [7] or arrays consisting of a rather small number of cantilevers, such as the device developed by IBM [8] [9] [10] [11]. Although alternative readout concepts have been reported, e.g. modulation and lock-in amplification [4], Wheatstone bridge circuit [12] or integrated photonics [13], the integration of a highly sensitive and at the same time scalable concept is still challenging. Recently we demonstrated that using a cantilever as a movable gate electrode in a metal-oxide semiconductor field-effect transistor (MOS-FET) deliberately designed to exhibit strong short channel effects (SCE) provides a straightforward, highly sensitive readout signal with an exponential dependence of the drain current on the position of the cantilever with respect to the device. In fact, exploiting the strong current modulation enables a direct connection of two cantilever MOS-FETs to a logic AND gate facilitating digital signal processing without the need for an analog-to-digital converter [14]. Furthermore, the signal variation is also far higher than in other published FET concepts [15] [16] [17] [18] [19]. Here, we employ this readout concept and study its use in an array consisting of a larger number of cantilevers. We show that increasing the array size with suitable geometry of each cantilever drastically improves the performance of the array as a sensor. Exemplarily, we investigate an array consisting of cantilevers with different sizes with just a single type of functionalization.

By changing the ratio of functionalized and unfunctionalized areas the detection and discrimination of different adsorbed molecules is possible. Thus, increasing the array size enables the detection of a large number of different molecules and a reduction of measurement error.

II. CANTILEVER FIELD-EFFECT TRANSISTOR

A. Combination of MOS-FET and cantilever sensors utilising short channel effects

In a cantilever sensor adhesion of a molecule manipulates the cantilever properties, e.g., by deflection or by an increase in mass. In static mode the absolute position of the cantilever tip is monitored so that the deflection can be measured. In our demonstrated cantilever field-effect transistor (C-FET) changing the position of the cantilever results in an exponential change in drain current [14]. However, without deflection the adhesion of molecules cannot be observed. In addition, an exponential dependence of the readout signal on the measured event requires additional circuitry for signal processing and also leads to substantial noise. In the dynamic mode, on the other hand, a change in resonance frequency (1) is measured from which even very small mass changes can be derived according to (2) [2] [20]. Furthermore, the dynamic mode with an oscillation at eigenfrequency is rather stable and suppresses noise. Equation (1) describes how the frequency depends on the cantilever properties height t_i, length l_i, Young's-modulus E and density ρ as well as a factor λ_i that depends on the resonant mode.

$$f_i = \frac{\lambda_i^2}{4\pi} \frac{t_i}{l_i^2} \sqrt{\frac{E}{3\rho}} = c_f \frac{t_i}{l_i^2} \qquad (1)$$

Here, i is an index to refer to a particular cantilever within an array. The frequency shift Δf_i in (2) depends on the additional mass Δm_i on cantilever i, width w_i and length. As our study is restricted to a single material for the cantilevers, i.e. silicon, the formulas can be simplified using a constant c_f.

$$\frac{\Delta f_i}{\Delta m_i} = -\frac{1}{2} c_f \frac{1}{w_i l_i^3} \qquad (2)$$

The drawback is that the dynamic mode requires an actuator with frequency generator and a readout to find the resonance frequency. This often increases circuit and/or measurement complexity.

978-1-5090-5979-9/17 $31.00 © 2017 IEEE

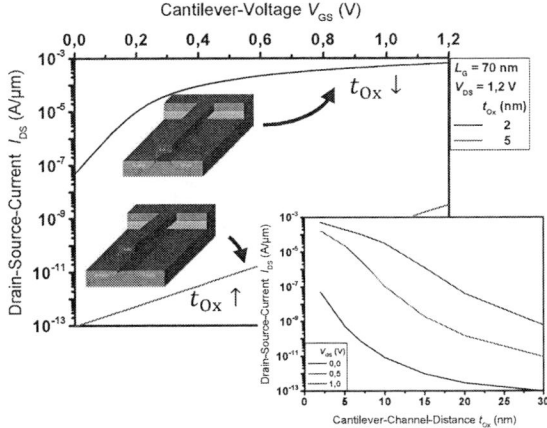

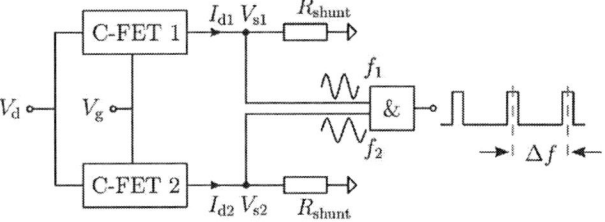

Figure 2: Schematic circuit diagram of a C-FET AND-gate structure.

Figure 1: Schematic and current (simulation) of a C-FET in the on-state. Improved gate-control reduces short-channel effects and increases I_{DS}.

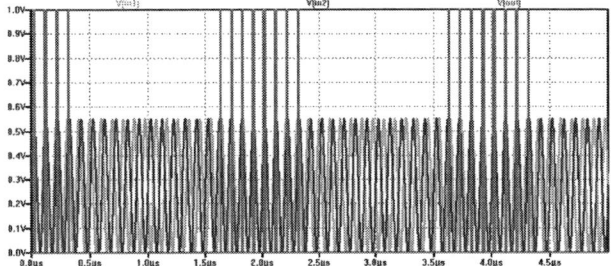

Figure 3: LTSpice simulation of 2 cantilever signals (green, blue) and resulting AND-gate output signal (red).

As mentioned above, we recently demonstrated a novel way of reading-out a cantilever sensor relying on a movable gate FET that shows strong SCE. When the position of the cantilever changes in such a C-FET, SCE are modulated resulting in an exponential change of drain current [14]. Fig. 1 shows electrical simulations carried out with Sentaurus TCAD for devices with 2 different gate-oxide thicknesses. Note, that in order to keep the computational burden as small as possible, we simulated transistors with 70 nm gate length. Results can be transferred to larger gate lengths based on, e.g., constant field scaling. A reduced oxide thickness leads to an improved gate control and higher drain currents. For a constant gate voltage, the difference of the drain currents here is far higher than the modulation obtained due to the change of the gate oxide thickness alone; the reason for this is an appropriate exploitation of SCE and a suitable doping concentration in the channel [14]. The inset of Fig. 1 shows the dependence of the cantilever distance for different gate voltages. Thereof, a feasible operating point with maximum current ratio can be derived. Depending on geometry and operating point the device can be operated in on- or off-state. When the gate dielectric thickness is reduced, the current strongly increases in the on-state due to an improved gate control. In the off-state, on the other hand, the improved gate control reduced SCE and thus reduces leakage, hence the drain current (for details see [14]). Here, the on-state is considered since our experimentally demonstrated C-FET operates best in on-state.

B. AND-gate readout mechanism for C-FETs

To detect the mass change a differential measurement of at least 2 cantilevers can be used which is often easier than the exact measurement of a single frequency and reduces cross-sensitivity e.g. due to temperature influence. An evaluation of different C-FET signals is possible with a simple AND-gate structure as shown in Fig. 2 resulting in a beating of the 2 signals [14]. Thanks to the AND-gate no ac-dc converter is necessary. The frequency of the output signal equals the difference of the 2 cantilever signals. As this frequency is lower than each single signal it is easier to measure. As the sensor signals are relatively close to each other in frequency

and amplitude the dimensioning of the threshold levels for the AND-gate is challenging and results often in several successive peaks as shown Fig. 3. A dead time or an additional counter helps to distinguish between the sequence and the relevant signal.

C. Self-oscillating cantilever sensor

Resonance of a cantilever can be achieved by exploiting self-oscillation [21]. An applied dc-voltage between an additional electrode next to the C-FET and the cantilever is used for attraction. If the cantilever shape and resistance are well designed field emission occurs at the tip at lower distances. This reduces the effective potential and hence electrostatic force so that the spring force becomes dominant and the cantilever moves to the opposite direction as shown in the inset of Fig. 4. As a consequence, no frequency-tunable actuator is necessary which reduces circuit complexity because there is no need of finding the eigenfrequency of the cantilever [14]. Note that the applied voltage slightly reduces the resonant frequency due to electrostatic spring softening which has to be taken into account. Nevertheless this has no influence on the measurement principle as this is equivalent to a change in cantilever geometry and independent of an adsorbed mass.

From the oscillation amplitude an oscillating current can be derived based on the I_{DS} - Δx transfer characteristic of a measured C-FET shown in Fig. 4. As the cantilever moves closer to the channel the current increases from a base level, mainly defined by the drain voltage V_{DS}, to an enhanced level with maximum gate-control at maximum deflection. This increase is several orders of magnitude and can be used as an input for the AND-gate readout. An inset in Fig. 4 demonstrates how such an output signal possibly looks like. For this purpose the amplitude of the measured oscillation has been combined with the transfer characteristic of the C-FET. As shown in experiments, the oscillation is nearly symmetric

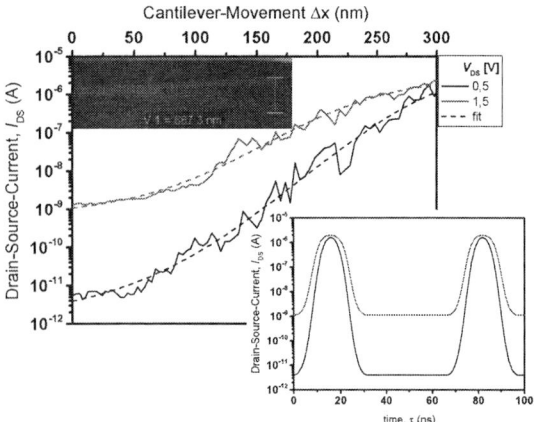

Figure 4: Measured transfer characteristic of a C-FET with cantilever in self-oscillation (left) and extrapolated oscillating current (right).

around the initial position resulting in minimum t_{Ox} of approximately 300 nm as in the case of current measurement. Depending on the amplitude and operating point the shape of the curve differs. As current cannot be suppressed below a certain limit the sinus like function has a cutoff at low current levels determined by V_{DS}.

III. LARGE SCALE CANTILEVER ARRAYS INCORPORATING C-FET BENEFITS

In real life situations the sensor, i.e. the array of cantilevers, is exposed to more than one type of molecule which makes it difficult to extract meaningful sensor signals. To differentiate between different influences a common concept is to functionalize the cantilevers with different materials [9]. This approach is feasible and proven in practice. In this paper we focus for simplicity on 2 different functionalized areas A_i and $\overline{A}_i$ on the same cantilever i, e.g. top part with functionalization and the rest without. For a specific area the increase in mass depends then on the probability for adhesion for each single molecule resulting in a mass area density σ_f for the surface and its area A_i (3). σ_f itself depends on the single masses m_x of all involved molecules M and their specific adhesion probability per unit area μ_x. A change in cantilever geometry, e.g. an increase in length, not only increases the overall surface but also changes the ratio between the different areas. As a consequence the total adhesion probability changes for the specific cantilever which works like an additional functionalization.

$$\Delta m_i = \sigma_f A_i + \overline{\sigma}_f \overline{A}_i$$
$$\sigma_f = \sum_{x=1}^{M} \mu_x m_x$$
$$\overline{\sigma}_f = \sum_{x=1}^{M} \overline{\mu}_x m_x$$
(3)

The frequency of the AND-signal is proportional to the frequency shift of the cantilevers. Thereof the mass change of the cantilevers can be calculated using (4) where Δf_{ij}^0 is the

initial frequency difference between cantilever i and j. This term enables to use cantilevers with dissimilar geometries and can be calibrated with an initial measurement of the unexposed sensor. Moreover, if not all molecules on a cantilever can be removed after the measurement Δf_{ij}^0 compensates this influence.

$$\Delta f_{ij} = \Delta f_{ij}^0 - \frac{1}{2} c_f \left(\sigma_f \left(\frac{A_i}{w_i l_i^3} - \frac{A_j}{w_j l_j^3} \right) + \overline{\sigma}_f \left(\frac{\overline{A}_i}{w_i l_i^3} - \frac{\overline{A}_j}{w_j l_j^3} \right) \right)$$
(4)

For simplification (4) is reorganised in a part depending on the measured signal S_{ij}, the different mass factors σ_f, $\overline{\sigma}_f$ and their related areas A_{ij}, $\overline{A}_{ij}$ (5). A comparison of 2 signals S_{ij} and S_{kl} with 2 non equal sets of cantilevers $\{i, j\}$ and $\{k, l\}$ solves the equation system (6).

$$S_{ij} = \sigma_f A_{ij} + \overline{\sigma}_f \overline{A}_{ij}$$
(5)

$$\sigma_f = \frac{S_{ij} \overline{A}_{kl} - S_{kl} \overline{A}_{ij}}{A_{ij} \overline{A}_{kl} - A_{kl} \overline{A}_{ij}}$$
$$\overline{\sigma}_f = \frac{S_{ij} A_{kl} - S_{kl} A_{ij}}{\overline{A}_{ij} A_{kl} - \overline{A}_{kl} A_{ij}}$$
(6)

With (3) and (6) the masses of the involved molecules can be calculated if at least $3M$ linear independent equations are available. This can be achieved by increasing the number and a sufficient geometrical scaling of cantilevers N. In contrast to other concepts an unfunctionalized reference cantilever is not necessary because it is beneficial to compare all cantilevers pairwise with all others. Due to the simple readout with an integrated MOS-FET and AND-gate this is possible without too much complexity in circuit design. A relation between N and the different masses M as given in (8) can be derived using the number of different coefficients for A_{ij}:

$$\# = \frac{N(N-1)}{2}$$
(7)

$$N \geq \frac{1}{2}\left(1 + \sqrt{5 + 4\sqrt{1 + 13M}}\right) = n(M)$$
(8)

As the number of available equations increases very quickly with N as shown in Fig. 5, it is possible not only to measure many different masses but also to use redundancy to increase the measurement accuracy. Assuming that the individual measurement errors are more or less equal the overall error decreases with the square root of independent solutions (9). If for example 10 instead of 4 cantilevers are used to measure 3 different masses σ decreases by approximately 40%. For 100 cantilever the reduction is already 82%. Additionally the use of more than 2 different functionalized areas on different or similar cantilevers can help to improve measurement accuracy even further.

978-1-5090-5979-9/17 $31.00 © 2017 IEEE

$$\sigma = \sqrt{\frac{n(M)}{N}}\,\sigma_m \leq \sigma_m \qquad (9)$$

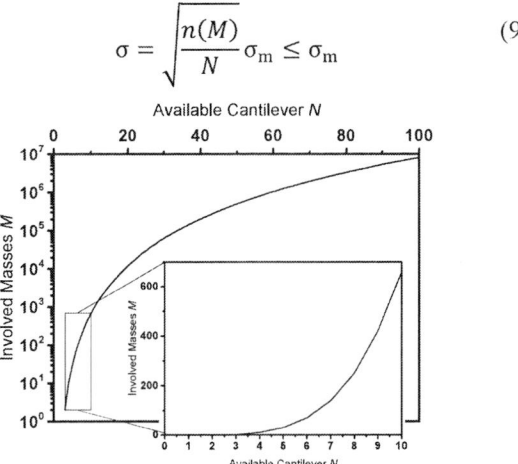

Figure 5: Potential mass resolution depending on the number of cantilevers.

IV. CONCLUSIONS

In this work we showed that an array of cantilevers with different geometrical scaling and an increased number of cantilevers within the array strikingly increases sensor performance. The change in ratio between different functionalized surfaces on a single cantilever works as a different functionalization and enables to measure various types of molecules as long as there is an adhesion probability on at least one of those areas. Cantilever field-effect transistors can be used as readout device for such arrays because cantilever movement is directly transferred into a corresponding current, using a short-channel MOS-FET. Signal variation over several orders of magnitude has been demonstrated with an appropriate C-FET design. This signal is then evaluated by a simple AND-gate structure and results in a digital signal equal to the difference of 2 cantilever frequencies. Hence, a pairwise comparison of each set of cantilever is possible and can be enlarged to a 3 or more signal AND-structure. An additional electrode for self-oscillation enables a simple and feasible device fabrication with no need for tunable actuators or finding the resonant frequency. Furthermore, a low cost production with state-of-the-art MEMS and device technology is possible due to comparably large feature sizes of several hundreds of nanometers.

ACKNOWLEDGMENT

Financial support by the Federal Ministry of Education and Research of Germany (BMBF) under grant no. 16ES0060K and Bosch-Forschungsstiftung is acknowledged.

V. REFERENCES

[1] H.P. Lang, M. Hegner, C. Gerber, Cantilever array sensors, Materials Today 8 (2005) 30–36.

[2] B.N. Johnson, R. Mutharasan, Biosensing using dynamic-mode cantilever sensors: A review, Biosensors and Bioelectronics 32 (2012) 1–18.

[3] J. Chaste, A. Eichler, J. Moser, G. Ceballos, R. Rurali, A. Bachtold, A nanomechanical mass sensor with yoctogram resolution, Nat Nano 7 (2012) 301–304.

[4] Yang, Y. T., C. Callegari, Feng, X. L., Ekinci, K. L., Roukes, M. L., Zeptogram-Scale Nanomechanical Mass Sensing, Nano Lett 6 (2006) 583–586.

[5] E. Gil-Santos, D. Ramos, J. Martinez, M. Fernandez-Regulez, R. Garcia, A. San Paulo, M. Calleja, J. Tamayo, Nanomechanical mass sensing and stiffness spectrometry based on two-dimensional vibrations of resonant nanowires, Nat Nano 5 (2010) 641–645.

[6] M. Okan, E. Sari, M. Duman, Molecularly imprinted polymer based micromechanical cantilever sensor system for the selective determination of ciprofloxacin, Special Issue Selected papers from the 26th Anniversary World Congress on Biosensors (Part I) 88 (2017) 258–264.

[7] Y. Murakami, T. Taniguchi, Z. Zhang, K. Yamashita, M. Noda, M. Sohgawa (Eds.), A highly sensitive Amyloid-β detection by cantilever microsensor immobilized with liposome with incorporated cholesterol and phosphatidylcholine lipid with short hydrophobic acyl chains. 2016 IEEE SENSORS, 2016.

[8] H.P. Lang, R. Berger, F. Battiston, J.-P. Ramseyer, E. Meyer, C. Andreoli, J. Brugger, P. Vettiger, M. Despont, T. Mezzacasa, L. Scandella, H.-J. Güntherodt, C. Gerber, J.K. Gimzewski, A chemical sensor based on a micromechanical cantilever array for the identification of gases and vapors, Applied Physics A 66 (1998) S61.

[9] A. Bietsch, J. Zhang, M. Hegner, H.P. Lang, C. Gerber, Rapid functionalization of cantilever array sensors by inkjet printing, Nanotechnology 15 (2004) 873.

[10] X. Bai, H. Hou, B. Zhang, J. Tang, Label-free detection of kanamycin using aptamer-based cantilever array sensor, Biosensors and Bioelectronics 56 (2014) 112–116.

[11] V. Pini, P. Kosaka, J. Ruz, O. Malvar, M. Encinar, J. Tamayo, M. Calleja, Spatially Multiplexed Micro-Spectrophotometry in Bright Field Mode for Thin Film Characterization, Sensors 16 (2016) 926.

[12] C. Shin, I. Jeon, Z.G. Khim, J.W. Hong, H. Nam, Study of sensitivity and noise in the piezoelectric self-sensing and self-actuating cantilever with an integrated Wheatstone bridge circuit, Review of Scientific Instruments 81 (2010) 35109.

[13] M. Li, Tang H. X., Roukes M. L., Ultra-sensitive NEMS-based cantilevers for sensing, scanned probe and very high-frequency applications, Nat Nano 2 (2007) 114–120.

[14] A. Hessel, S.C. Scholz, A. Pfander, A. Pelger, J. Knoch, Modulated Short Channel MOSFETs as Readout Devices for Cantilever-Based Nano-electromechanical Sensors, Sensors and Actuators A: Physical, unpublished.

[15] G. Shekhawat, S.-H. Tark, V.P. Dravid, MOSFET-Embedded Microcantilevers for Measuring Deflection in Biomolecular Sensors, Science 311 (2006) 1592.

[16] P. Fei, P.-H. Yeh, J. Zhou, S. Xu, Y. Gao, J. Song, Y. Gu, Y. Huang, Z.L. Wang, Piezoelectric Potential Gated Field-Effect Transistor Based on a Free-Standing ZnO Wire, Nano Lett 9 (2009) 3435–3439.

[17] J. Wang, W. Wu, Y. Huang, Y. Hao, <100> n-type metal-oxide-semiconductor field-effect transistor-embedded microcantilever sensor for observing the kinetics of chemical molecules interaction, Appl. Phys. Lett. 95 (2009) 124101.

[18] V. Seena, A. Nigam, P. Pant, S. Mukherji, V. R. Rao, "Organic CantiFET": A Nanomechanical Polymer Cantilever Sensor With Integrated OFET, Journal of Microelectromechanical Systems 21 (2012) 294–301.

[19] M. Barbaro, A. Bonfiglio, L. Raffo, A charge-modulated FET for detection of biomolecular processes: conception, modeling, and simulation, IEEE Transactions on Electron Devices 53 (2006) 158–166.

[20] B. Bhushan (Ed.), Handbook of Nanotechnology, 3rd ed., Springer, Heidelberg, New York, 2010.

[21] A. Ayari, P. Vincent, S. Perisanu, M. Choueib, V. Gouttenoire, M. Bechelany, D. Cornu, S.T. Purcell, Self-Oscillations in Field Emission Nanowire Mechanical Resonators: A Nanometric dc–ac Conversion, Nano Lett. 7 (2007) 2252–2257.

ESD characterisation of a-IGZO TFTs on Si and foil Substrates

Nian Wang[a,b], Shih-Hung Chen[a], Geert Hellings[a], Kris Myny[a], Soeren Steudel[a], Mirko Scholz[a], Roman Boschke[a,b], Dimitri Linten[a], Guido Groeseneken[a,b]

[a] imec, kapeldreef 75, 3001 Leuven, Belgium
[b] KU Leuven, 3001 Leuven, Belgium
E-mail: Shih-hung.Chen@imec.be Geert.Hellings@imec.be

Abstract—**Amorphous Indium-Gallium-Zinc-Oxide (a-IGZO) Thin-Film Transistors (TFTs) integrated with Si based CMOS processes is an emerging technology in ultra-low power applications. ESD characteristics of a-IGZO TFTs with a Si substrate are studied and compared to their characteristics on traditional foil/glass substrate. The ESD performance is shown to be improved, thanks to improved thermal properties of the different buffer material. The layout dependency of ESD behaviors in Si substrate TFTs are investigated in order to meet the 1kV HBM ESD requirement. An on-chip ESD protection design with a gate-coupling TFT is proposed.**

Keywords—*electrostatic discharge (ESD); amorphous Indium-Gallium-Zinc Oxide (a-IGZO); thin-film transitors (TFTs); transmission line pulsing (TLP); Si substrate*

I. INTRODUCTION

Thin-film transistors (TFTs) are widely used as switching elements in various microelectronics and display applications, especially in the large area thin film displays [1,2,3]. Different TFTs such as amorphous Si, polycrystalline Si, oxide-based and even organic materials have been proposed and used for different applications [4]. Particularly, the oxide-based materials draw a lot of attention in the recent years because of the large cost reduction. Among the various oxide-based TFTs, the amorphous Indium-Gallium-Zinc Oxide (a-IGZO) is one of the best materials due to its higher mobility. Recently, the emerging applications of NFC and RFID require the a-IGZO TFTs integrated with Si based CMOS technologies. Previous works have indicated that the Electrostatic Discharge (ESD) damages are a main reliability issue. However, those a-IGZO TFTs has been fabricated on foil or glass substrate [5,6,7,8]. So far, no ESD studies of TFTs on Si substrate have been presented.

In this work, we demonstrate the ESD study of a-IGZO TFTs based on Si substrate and compare the ESD results between the Si and glass substrates. The next section will introduce the TFTs technology with the Si substrate. The ESD behavior analysis with different substrates and layout dependence are shown in Section III.A and III.B. Section IV is a proposed on-chip ESD protection design for this Si based TFT technology.

II. TFTS ON SI SUBSTRATE

The cross-section of the TFT technology is shown in Figure 1. The a-IGZO layer is deposited on the Si substrate with a SiO_2 buffer and a barrier layer. By using the PECVD, SiO_2 is

deposited as the gate dielectric layer. The Molybdenum (Mo) is used as the gate metal (M1 in Figure 1). In between of the gate and source/drain contacts, an intermetal dielectric (SiN_X) is deposited and patterned. The whole a-IGZO layer is N-doped with hydrogen. Heavily doped source and drain regions are defined by the self-align technique in order to reduce the parasitic capacitance and the further RC delay. The obtained a-IGZO TFTs have the channel mobility of 10 cm^2/Vs and their typical operating voltage range is 5~10 V.

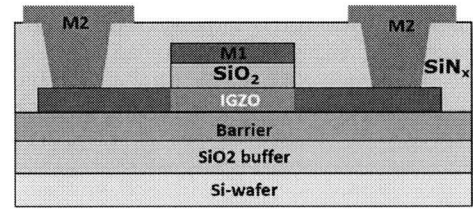

Figure 1. Cross-section of the SA TFT.

III. ESD CHARACTERISATION AND ANALYSIS

The TFTs in diode-mode configuration, which means the gate is shorted to the drain, have the largest current conductance capability under ESD stress [9,10]. TLP stresses with 2 ns rise time and 100 ns pulse width are applied at the gate/drain with source grounded to assess the ESD robustness of the device under test. After each TLP pulse, the off-state leakage at -5V is monitored to judge if the device is fail or not.

A. Impact of Substrate material

Figure 2 shows the TLP I-V curves and corresponding leakage currents of TFTs with Si and foil/glass substrates, with different sizes, respectively. The failure current (It2) of the TFT with Si substrate is ~2× higher than that with foil substrate. For the TFT on Si substrate, the It2 increases from 8.75 mA to 190 mA when the TFT size increased from 15 µm to 250 µm. It increases from 6 mA to 80 mA for the TFT on foil substrate. The higher leakage in Si TFT can be attributed to the patterning issues. Figure 3 shows the normalized It2 and on resistance (R_{ON}) of the TFTs with different device sizes and substrates. With the similar a-IGZO manufacturing technology and the same device size, the normalized It2 and R_{ON} of Si and foil TFT show a significant difference. This implies a difference in thermal dissipation of the two different substrates. Failure are assumed due to thermal failure in the a-IGZO layer. Any gate oxide failure only occurs at higher voltages (e.g. 120 V [7]).

978-1-5090-5979-9/17 $31.00 © 2017 IEEE

Figure 4 shows the schematic of the thermal resistance in these two different substrate TFTs. The total thermal resistance of the two kind of TFTs can be seen as the thermal resistance of each layer in series with each other. In thermal resistance formula, the d is the thickness of the layer (m), A is the area of the layer (m^2), and K is the thermal conductivity of the material (W/(K·m)), which is shown in Figure 4. The total thermal resistance of the Si substrate is ~300× smaller than the polyimide foil substrate TFTs. The number from this simplified calculation result cannot really match to the 2× It2 difference in these two substrates.

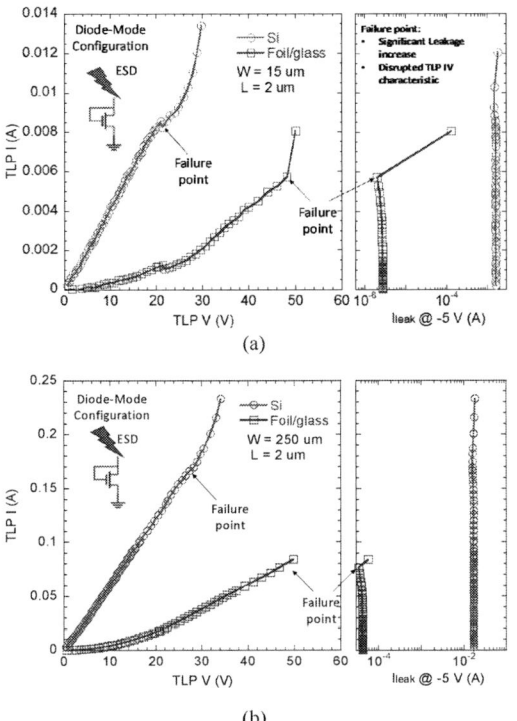

(a)

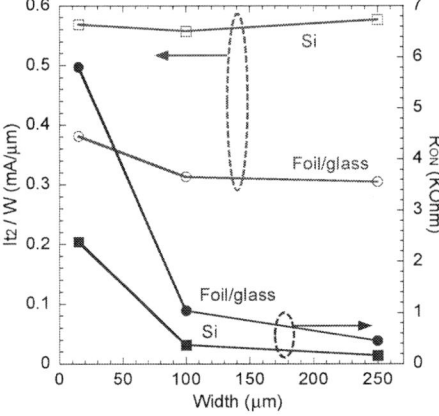

(b)

Figure 2. 100 ns TLP testing in diode-mode for foil/glass and Si substrate TFTs, (a) width of 15 μm and length of 2 μm TFT, (b) width of 250 μm and length of 2 μm TFT.

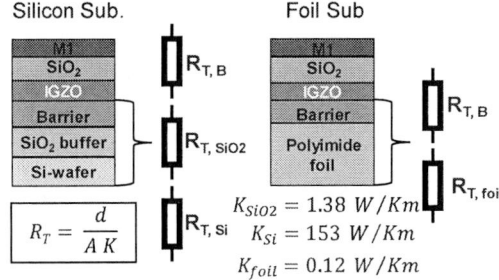

Figure 4. Schematic of the thermal calculation for two types of TFTs.

The total thermal mass (m_T) under 100 ns TLP pulsing stress should be take into account. The thermal mass is determined by the thermal diffusion length (L_T), as illustrated in Figure 5. Note that the thermal diffusion length in the SiO2 is 3× larger. Under 100 ns TLP stress width, the difference of total m_T in these two TFTs only depends on the m_T difference between SiO2 and polyimide foil. In the m_T formula, ρ times c_p is the volumetric heat capacity, shown in Figure 5. The final total m_T is shown in Table 1. The total m_T of Si TFT is ~1.44× larger than total m_T in foil TFT (including the IGZO, barrier layer and the buffer). The maximum thermal TLP energy (E_T, TLP) before the failure is the It2 times failure voltage (Vt2) and times TLP pulse width. The difference of E_T, TLP in these two substrate also approaches to ~1.50, which matches to the difference of the total m_T. The temperature increase generated by 100 ns TLP stress is nearly 300 K in both TFTs. Therefore, the thermal failure temperature of the a-IGZO TFT on both substrates is close to 600 K.

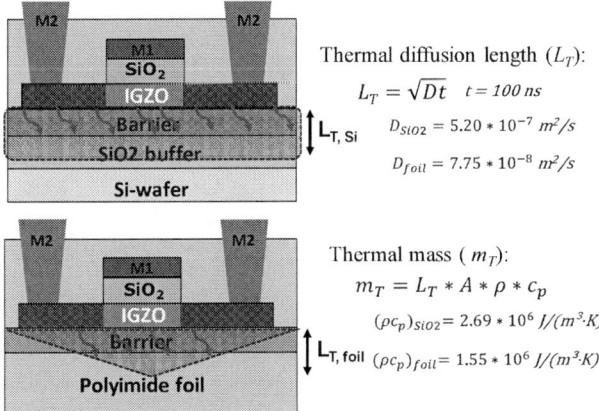

Figure 5. Schematic of the total thermal mass for Si substrate and foil/glass substrate TFT.

TABLE 1 Thermal Calculations of the influence of different buffer layer with 250 um wide and 2 um length TFTs.

Substrate type of TFTs	Si	Foil/glass
Buffer layer	SiO2	Polyimide
Thermal diffusion length (L_T) [m]	2.28E-07	8.80E-08
Total thermal mass (m_T) [J/K]	1.18E-09	8.21E-10
TLP energy in final pulse ($E_{T, TLP}$) [J]	3.60E-07	2.38E-07
Thermal breakdown temperature (above RT)($E_{T, TLP}$/total m_T) [K]	304.29	289.45

Figure 3. Comparsion of intrinsic failure current and on resistance of TFTs in diode-mode with different substrate, different widths and same gate length of 2 μm.

978-1-5090-5979-9/17 $31.00 © 2017 IEEE   277

B. Layout Dependence in Si Substrate TFTs

Figure 6(a) shows the TLP I-V curves with different channel widths of TFTs and the corresponding leakage current. As the channel width increased, the It2 increases and RON decreases. The It2 increases from 48 mA to 122 mA by increasing the width from 100 μm to 250 μm. The RON decreases from 1.8 kΩ to 0.8 kΩ as shown in Figure 6(b). The It2 normalized by channel width is around 0.48 mA/μm which indicates a uniform current distribution along the channel width in Si TFTs under 100 ns TLP stress.

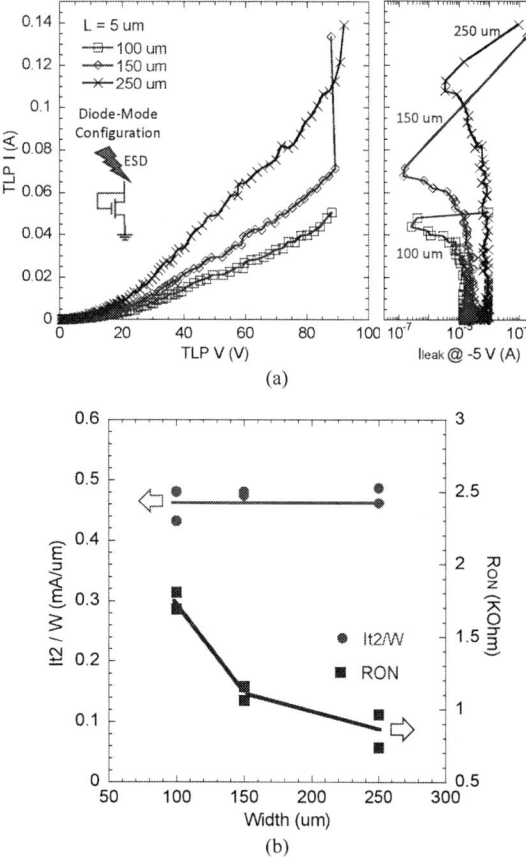

(a)

(b)

Figure 6. 100 ns TLP testing in diode-mode for different widths and same gate length 5 μm, (a) TLP I-V curves and leakage current evolution, (b) failure current and on resistance.

Figure 7(a) shows the TLP I-V curves with different channel lengths of TFTs and the corresponding leakage current. As the channel length increased, the It2 decreases and the RON increases. The It2 decreases from 110 mA to 48 mA by increasing the length from 1.5 μm to 5 μm. RON increases from 0.25 kΩ to 1.75 kΩ as shown in Figure 7(b). The normalized It2 is significantly decreased with increasing channel length. However, the leakage current is much higher in the short channel length TFTs as shown in Figure 7(a). In order to reduce the off-state leakage current of the ESD protection devices in this TFT technology, the longer channel are the preferred design option.

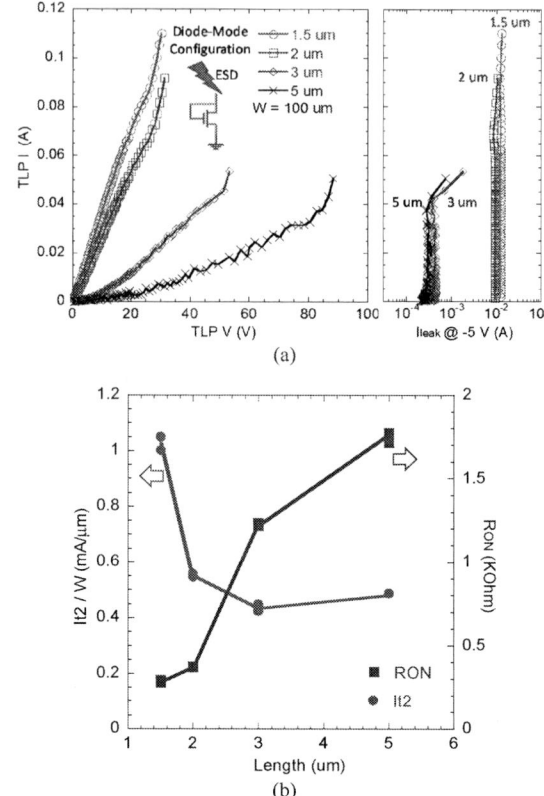

(a)

(b)

Figure 7. 100 ns TLP testing in diode-mode for lengths and same width 100 um, (a) TLP I-V curves and leakage current evolution, (b) failure current and on resistance.

IV. ON-CHIP ESD PROTECTION DESIGN

In order to reach 1 kV HBM ESD robustness, which means the It2 level of 0.67 A, the final ESD protection device is designed with the channel width of 1400 μm and length of 5 μm. Figure 8 shows the TLP I-V curve and corresponding leakage current of the TFT with 1400/5 μm. The It2 is ~0.6 A, which means it will somehow close to or match to the 1 kV HBM target.

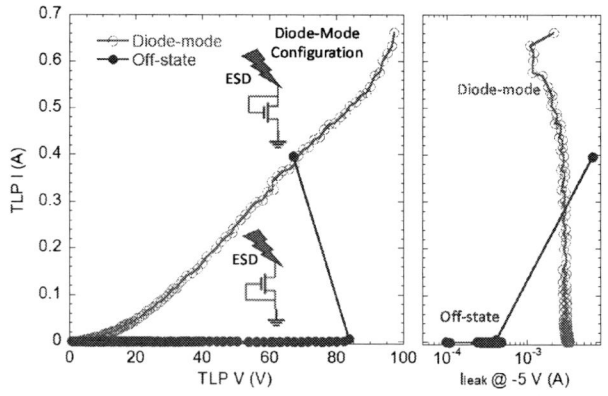

Figure 8. 100 ns TLP testing of 1400 μm wide and 5 μm long diode-mode TFTs.

Figure 9(a) shows the schematic of an ESD protection design with the TFTs under diode-mode configuration. A special gate coupling TFT (M3) with the capacitor-resistor (CR) coupling network as an ESD detection circuit is proposed and placed between VDD and VSS. During the normal circuit operation with a DC voltage signal at VDD, the gate terminal of M3 is connected to the ground through the R. When the positive ESD stress at VDD, the M3 gate terminal will be coupled the drain terminal voltage potential through the C. The equivalent circuits of this gate coupling TFT (M3) under the normal circuit operation and ESD event are shown in the inserted small figures in Figure 9(a). In order to achieve the required function of this gate coupling TFT, the value of the C and the R are designed as 109 pF and 4.3 kΩ respectively. In Figure 9(b), the TLP results of the gate coupling TFT under the positive and negative ESD stress are shown. Under the positive ESD stress, the M3 behaves the same as the TFT with diode-mode configuration, as shown in Figure 8. However, under the negative ESD stress, the It2 of M3 is much lower which cannot be used as an ESD protection device. Therefore, M4 in Figure 9(a) has to be added for the negative ESD stress between the VDD and VSS. In any two-pin combinations, the ESD current can always discharge through the TFT with diode-mode configuration, for example, under PS mode ESD stress, the ESD current discharge through the M1 and M3 with diode-mode configuration.

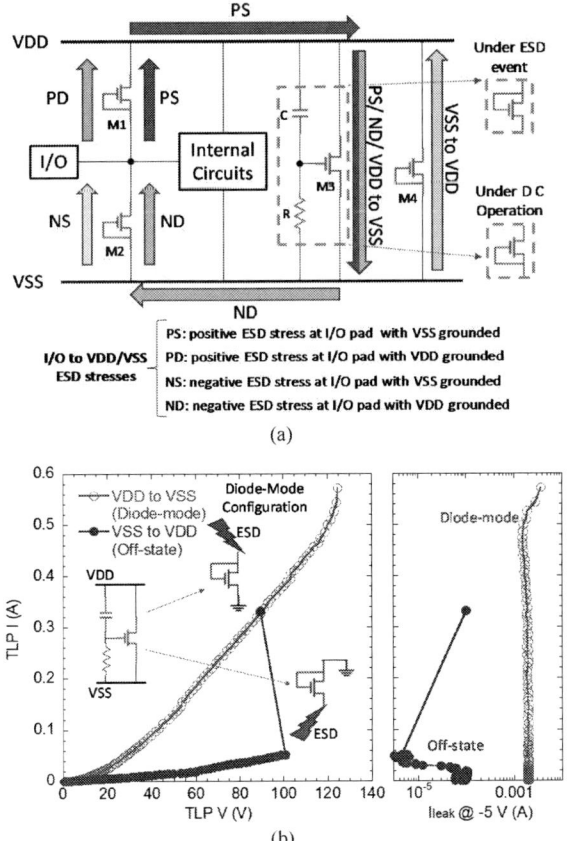

PS: positive ESD stress at I/O pad with VSS grounded
PD: positive ESD stress at I/O pad with VDD grounded
NS: negative ESD stress at I/O pad with VSS grounded
ND: negative ESD stress at I/O pad with VDD grounded

(a)

(b)

Figure 9. (a) Schematic of the protection circuit design, (b) 100 ns TLP testing of 1400 μm wide and 5 μm long power clamp for postive and negative pulsing polarity.

V. CONCLUSION

The ESD behavior of TFTs with two different substrates is presented. A 2× higher It2 (lower V_{T2}) was observed in the TFTs on the Si substrate. This was explained by calculating that the buffer layer used on the Si substrate has a much higher thermal mass than that on the foil substrate: A 1.5× difference of the total thermal mass leads to a capability to dissipate 1.5× more thermal energy during a TLP pulse. The choice of buffer layer is therefore extremely important with respect to thermal behavior and ESD performance. .

Secondly, the It2 of the Si TFTs is increased by increasing the channel width and decreasing channel length. The shorter channel length induces a significant leakage current. The final channel width and length of the ESD protection devices are designed as 1400 μm and 5 μm for 1 kV HBM ESD robustness. An ESD protection design is proposed with a gate-coupling TFT inserted between VDD and VSS. With this gate-coupling TFT, the ESD current can be always discharged through the TFTs with diode-mode configuration.

REFERENCES

[1] K. Myny, S. Steudel, "Flexible Thin-Film NFC Transponder Chip Exhibiting Data Rates Compatible to ISO NFC Standards Using Self-Aligned Metal-Oxide TFTs", ISSCC, 2016.

[2] P. E. Malinowski, "Organic imager on readout backplane based on TFTs with cross-linkable dielectrics", IEEE Photon. Technol. Lett., vol. 26, no. 21, pp. 2197-2200, Aug. 2014.

[3] F. D. Roose, K. Myny, S. Steudel, M. Willigems, S. Smout, T. Piessens, J. Genoe, W. Dehaene, "16.5 A flexible thin-film pixel array with a charge-to-current gain of 59 μA/pC and 0.33% nonlinearity and a cost effective readout circuit for large-area X-ray imaging", ISSCC,2016.

[4] R. A. Street, " Thin-film transistors," Adv. Mater. 21, 2007–2022 (2009).

[5] S. Steudel et al., "Flexible AMOLED Display with Integrated Gate Driver Operating at Operation Speed Compatible with 4k2k", SID Symposium Dig. Tech. Pap., 2015.

[6] S. Steudel et al., "Impact of Buffer Layers on the Self-Aligned Top-Gate a-IGZO TFT Characteristics", SID Symposium Dig. Tech. Pap., 2015.

[7] M. Scholz, S. Steudel; K. Myny, S. Chen, R. Boschke, G. Hellings, D. Linten, "ESD protection design in a-IGZO TFT technologies", Electrical Overstress/Electrostatic Discharge Symposium (EOS/ESD), 2016.

[8] Yuan Liu, Wei-Jing Wu, Zhi-Feng Lei, Lei Wang, Qian Shi, Yu-Rong Liu, Yun-Fei En, Bin Li, "Instability of Indium Zinc Oxide Thin-Film Transistors Under Transmission Line Pulsed Stress", IEEE Electron Device Letters, vol. 35, pp. 1254-1256, 2014, ISSN 0741-3106.

[9] W. Liu et al., "Study of organic thin-film transistors under electrostatic discharge stresses", IEEE Electron Device Letters, 2011.

[10] Y. H. Tai et al., "Test and analysis of the ESD robustness for the diode-connected a-IGZO thin film transistors", IEEE/OSA Journal of Display Technology.

Dopant Diffusion and Segregation, Si-Ge interdiffusion and Defect Engineering in SiGe Devices

Guangrui (Maggie) Xia
Department of Materials Engineering
University of British Columbia
Vancouver, Canada
gxia@mail.ubc.ca

Abstract—Recent research progresses on dopant diffusion and segregation, Si-Ge interdiffusion and defect engineering in SiGe material systems are reviewed, which are relevant to SiGe-based semiconductor devices including SiGe PNP hetero-junction bipolar transistors, metal-oxide-semiconductor field-effect transistors, and Ge-on-Si lasers. Experiment data and continuum modeling are discussed.

Keywords—dopant diffusion and segregation, Si-Ge interdiffusion, defect engineering, SiGe devices

I. INTRODUCTION

In the past two to three decades, due to the compatibility with Si processing and the capability of mobility, strain and energy bandgap engineering, SiGe and Ge have been widely used in electronic and optoelectronic devices such as metal oxide semiconductor field effect transistors (MOSFETs) [1], hetero-junction bipolar transistors (HBTs) [2], Ge photodetectors [3], Ge modulators [4], and Ge-on-Si lasers [5].

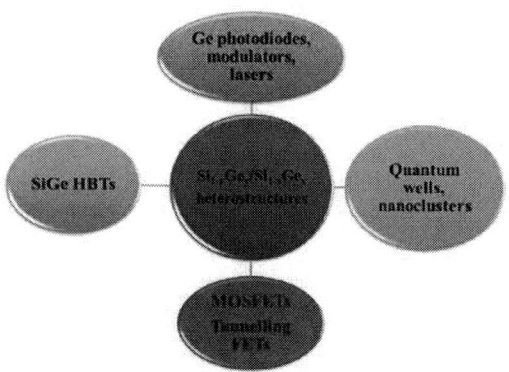

Fig. 1. Important role of SiGe and Ge in semiconductor devices.

Although Si and Ge have lots of common or similar properties, SiGe, SiGe:C and Ge are still different material systems. During the growth and processing, high temperature steps, such as defect annealing, oxidation, deposition, dopant activation annealing are unavoidable. When there is a change in the diffusion media such as a change in Ge and/or carbon (C) concentration or stress or defect concentrations, dopant

diffusion and/or segregation changes. Dopants segregate at $Si_{1-x}Ge_x/$ $Si_{1-y}Ge_y$ interfaces or across a graded Ge concentration slope.

Si-Ge interdiffusion should also be considered, as Ge distribution directly impacts dopant diffusion and segregation, bandgap and stress engineering and many electrical and optical properties associated with the x_{Ge}. Due to the scaling of semiconductor devices, typical diffusion lengths of Si-Ge interdiffusion after high temperature steps in current technologies are comparable to the thickness of SiGe thin films in the devices, which is in 1 to 100 nm range. Si-Ge interdiffusivity increase exponentially with compressive strain. From the Si end to the Ge end, the interdiffusivity increase by five orders of magnitude with the increase of x_{Ge} in typical thermal annealing temperatures. All these effects can make Si-Ge interdiffusivity comparable or faster than dopant diffusivity.

Defect engineering using C incorporation, oxidation and nitridation can be used to change defect concentrations and engineer dopant profiles. Although those effects have been well studied in Si, oxidation and nitridation have not been well studied or applied in SiGe device fabrication.

These three phenomena, namely dopant diffusion and dopant segregation, Si-Ge interdiffusion, and defect engineering, are important topics for SiGe device structure design and processing, as dopant and Ge distribution is one of the most important factors in determining device characteristics and performance. This talk will focus on the three mass transport processes in SiGe epitaxial structures. Experiment data and continuum modelling are discussed.

II. BASE DOPING PROFILE ENGINEEING IN PNP HBTS

A. Background

Complementary SiGe HBTs have many advantages over an NPN-only technology for numerous analog applications requiring high speed, low noise, and large voltage swing [6], [7]. PNP HBTs use P as the base layer dopant and strained SiGe as the base layer material. Our recent studies addressed three problems related to the mass transport phenomena of PNP HBTs.

Texas Instruments, Crosslight Software Inc.and National Science and Engineering Research Council of Canada (NSERC) are acknowledged for funding the studies.

B. Effectiveness of C in P profile control

Although C has been very effective in retarding B diffusion in NPN HBTs, there are limited studies available on the effectiveness of C in PNP SiGe HBTs. Our work quantitatively investigated the C impacts on P diffusion in $Si_{0.82}Ge_{0.18}$:C and Si:C under rapid thermal anneal conditions [6]. The carbon molar fraction is up to 0.32%. We defined that R_C, the carbon impact factor, as the ratio of P diffusivity in $Si_{1-x}Ge_x$:C_y over that in $Si_{1-x}Ge_x$, i.e., $R_C \equiv \frac{D_P^{Si_{1-x}Ge_x:C_y}}{D_P^{Si_{1-x}Ge_x}}$. A smaller R_C is desired for base profile scaling. Fig. 2 shows the relations among the impact factors of dopant diffusion in SiGe:C.

The results showed that the C retardation effect on P diffusion is less effective for $Si_{0.82}Ge_{0.18}$:C than for Si:C. In $Si_{0.82}Ge_{0.18}$:C, there is an optimum carbon content at around 0.05% to 0.1%, beyond which more carbon incorporation does not retard P diffusion any more. This behavior is different from the P diffusion behavior in Si:C and the B in Si:C and low Ge SiGe:C, which can be explained by the decreased interstitial-mediated diffusion fraction $f_I^{P,SiGe}$ to 95% as Ge content increases from 0 to 18%. Empirical models were established to calculate the time-averaged point defect concentrations and effective diffusivities as a function of carbon, and was shown to agree with previous studies on B, P, As and Sb diffusion with C.

C. Coupled diffusion-segregation of P across graded SiGe

Compared to B, P is harder to control, as it segregates towards lower Ge content layers instead of staying inside SiGe base of higher Ge content. The segregation happens simultaneously with P diffusion. Clean experimental data and appropriate modeling were lacking before our work in [7]. In [7], experiments were performed with graded SiGe layers for Ge molar fractions (x_{Ge}) up to 0.18. A coupled diffusion and segregation model was derived, where the contributions from diffusion and segregation to dopant flux are explicitly shown. The model provides a new approach in segregation coefficient extraction, which is especially helpful for heterostructures with lattice mismatch strains. The diffusion-segregation model for P in SiGe alloys was calibrated and $E_{seg} = 0.5\ eV$ is suggested for the temperature range from 800 to 950 □C.

D. Thermal nitridation effect

Thermal nitridation is known to suppress B and P diffusion in Si via exposing bare Si surface in ammonia ambient, where vacancies are injected into Si and retard interstitial diffuser's motion [8]. However, there was no studies available on whether this method also worked in strained-SiGe:C system. If this method is effective, it is a relatively low cost and simple approach to be adopted by industry.

We investigated the thermal nitridation effects on P diffusion in strained $Si_{1-x}Ge_x$ and strained $Si_{1-x}Ge_xC_y$ with up to 18% Ge and 0.09% C [9]. We defined the nitridation impact factor as $R_{NTD} = \frac{D^{nitridation}}{D^{inert}}$. The results show that thermal nitridation can retard P diffusion in SiGe, but the effectiveness of this retardation method decreases with the increasing Ge and C content. When 0.06% and 0.09% C is present in $Si_{0.82}Ge_{0.18}$, thermal nitridation slightly increases P diffusivity compared to the inert condition. The Ge dependence can be explained by the increasing contribution from the vacancy-assisted mechanism for P diffusion in strained SiGe with the increasing x_{Ge}. In terms of interstitial undersaturation ratio $\frac{[I]}{[I]^*}$ in $Si_{0.82}Ge_{0.18}$, thermal nitridation can further decrease $\frac{[I]}{[I]^*}$ by 31 to 53% on top of the C effect. It should be noted that in this study, the Ge profiles were flat in regions of P diffusion. The effectiveness of thermal nitridation for P diffusion in a triangle or trapezoid shaped Ge profile needs to be further investigated for HBT applications.

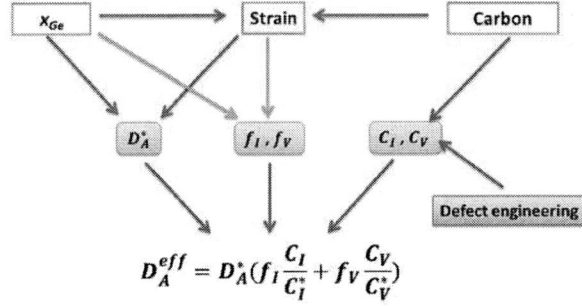

Fig. 2. The relation map among the impact factors of dopant A diffusion in SiGe:C with defect engineering methods. D_A^* is the equilibrium dopant A diffusivity, D_A^{eff} is the effective diffusivity with defect engineering, f_I and f_V are the fraction of diffusion by interstitials and vacancies respectively, C_I^* and C_V^* are the interstitial and vacancy concentration without defect engineering, and C_I and C_V are the interstitial and vacancy concentration with defect engineering. Here the defect engineering methods doesn't include C incorporation.

III. SI-GE INTERDIFFUSION

Topics	Main discoveries
Undoped, low-defect and unstrained SiGe interdiffusivity	$\tilde{D}$ has a near-exponential relationship with x_{Ge}. A unified model from 0 to 100% Ge was built, and verified by data and literature work [10].
Compressive strain impact	$\tilde{D}$ is exponential with compressive stain [11].
Tensile strain impact	Negligible for structures with up to 30% Ge with 1% tensile strain [12].
Strain relaxation effect	Can be modelled with a lower strain for known strain relaxation history [13].
Dislocation impact	Need to add another interdiffusion term, which is significant for low Ge regimes [13].
Doping and impurity impact	P, C and As enhances interdiffusion. B not as much. Need to use Fermi effect to model P effect [14-16].
Oxidation impact	Negligible for structures with Ge 15-60% [17]

Table 1. Summary of factors that impact Si-Ge interdiffusion.

978-1-5090-5979-9/17 $31.00 © 2017 IEEE

Si-Ge interdiffusion happens whenever there is a Ge concentration gradient such as at $Si_{1-x}Ge_x/Si_{1-y}Ge_y$ interfaces and across graded SiGe layers. The interdiffusivity $\tilde{D}$ is influenced by many factors such as temperature, Ge concentration, stress, doping, and defect density. Table 1 summarizes major discoveries on these impacts.

A. Benchmarking Si-Ge interdiffusivity model for undoped, unstrained or tensile SiGe with low defect densities

An interdiffusivity model was established for SiGe interdiffusion under tensile or relaxed strain over $0 \leq x_{Ge} \leq 1$ range, which is based on the correlations between self-diffusivity, intrinsic diffusivity, and interdiffusivity in (1), where D^* is the self-diffusivity and γ is the activity coefficient. It unifies available interdiffusivity models over the full Ge range and applies to a wider temperature range from 840 to 1270 °C at $x_{Ge} = 0$ end and from 580 to 900 °C at $x_{Ge} = 1$ end. The interdiffusivity model of this work has been implemented in major process simulation tools, and the calculation results showed good agreement with experimental and literature data under furnace annealing and soak and spike rapid thermal annealing conditions.

$$\tilde{D} = D_{Si}^*(1 + \frac{\partial \ln \gamma_{Si}}{\partial \ln x_{Si}})x_{Ge} + D_{Ge}^*(1 + \frac{\partial \ln \gamma_{Ge}}{\partial \ln x_{Ge}})(1 - x_{Ge}) \quad (1)$$

B. Compressive strain impact

The role of compressive strain on Si-Ge interdiffusion in epitaxial SiGe heterostructures was systematically investigated both by experiments and by theoretical analysis in [11]. The x_{Ge} studied ranged from 0.36 to 0.75, and the temperature range was 720–880 °C. The epitaxial SiGe structures were kept pseudomorphic during the annealing. Theoretical analysis showed that strain field can add to the interdiffusion driving force from the concentration gradient. This effect can be included in the apparent interdiffusivity $\tilde{D}_{apparent}$ as shown in (2). In (2) and (3), Y_{100} is the biaxial modulus, $f_0(x_{Ge})$ is the Helmholtz free energy per unit volume of the homogeneous solution at the Ge atomic fraction of x_{Ge}, $\tilde{D}^{relax}$ is the interdiffusivity without any strain, ε is the biaxial strain, and q' is the strain derivative of the interdiffusivity. q' was quantitatively extracted from the experimental data as q' = (− 0.081T + 110) eV/unit strain.

$$\tilde{D}_{apparent} = \left(1 + \frac{2 \times 0.0418^2 Y_{100}}{f_0''}\right)\tilde{D}^{relax}e^{\frac{-q'\varepsilon}{kT}} \quad (2)$$

$$f_0'' = \frac{\partial^2 f_0}{\partial x_{Ge}^2} \quad (3)$$

C. Doping and impurity impact

For HBT applications, C, P and B were all shown to enhance Si-Ge interdiffusion [14]. Reference [16] investigated P doping effect on interdiffusion and the modeling of the effect. $Ge/Si_{1-x}Ge_x/Ge$ multi-layer structures with $0.75 < x_{Ge} < 1$, a mid-10^{18} to low-10^{19} cm^{-3} P doping and a dislocation density of 10^8 to 10^9 cm^{-2} range were studied. The P-doped sample shows an accelerated Si-Ge interdiffusivity, which is 2-8 times of that in the undoped sample. The doping dependence of the Si-Ge interdiffusion was modelled by a

Fermi-enhancement factor. The results show that for Si-Ge interdiffusion coefficient is proportional to n^2/n_i^2, which indicates that the interdiffusion in high Ge fraction range with n-type doping is dominated by V^{2-} defects. The Fermi-enhancement factor was shown to have a relatively weak dependence on the temperature and the Ge fraction. The results are relevant to structure and thermal processing condition design of n-type doped Ge/Si and Ge/SiGe based devices such as Ge/Si lasers.

IV. CONCLUSION AND OUTLOOK

Recent research progresses on dopant diffusion and segregation, Si-Ge interdiffusion and defect engineering in SiGe material systems are reviewed, which are relevant to SiGe-based semiconductor devices including SiGe PNP HBTs, MOSFETs, and Ge-on-Si lasers. With the scaling of devices, we expect those phenomena become more of a problem and there are still lots of topics to explore.

Reducing thermal budget is an effective method in controlling undesired dopant movement and interdiffusion. Careful designs of C, Ge and dopant profiles and better epitaxy can also help to control dopant movement. Less compressive strain, lower Ge concentration and defect concentration help to reduce interdiffusion as well. Thermal nitridation as an external defect engineering method is useful to reduce the use of C, but only in a certain Ge window.

Looking forward, we also expect some major changes in the methodology used in studying these phenomena. The major profiling method so far has been secondary ion mass spectrometry (SIMS), which is reliable for profiling concentration profiles thicker than 50 nm or so. For the profiling of thinner layers, SIMS broadening effect can introduce large errors. With scaling, SIMS will become less and less accurate, and a more accurate but much more costly profiling method such as atomic probe tomography (APT) will be needed. APT measures element distribution in 3-dimension (3D) with atomic scale resolution, which will greatly enhance the predictability of current models that have been mainly calibrated against 1D SIMS measurements. So far, stress field has been also difficult to measure in nm scale and in 3D routinely, which may continue to be a problem in the near future.

In terms of modelling, so far, continuum modeling has been sufficient in catching the behaviors of the elements studied in terms of accuracy and fast computation time. With scaling, the dimensions of interest will be decreasing to the extent that continuum modeling will not be as suitable. In that case, other modelling hierarchy such as Kinetic Lattice Monte Carlo (KLMC), Molecular Dynamics, and Density Function Theory (DFT) will be needed. However, in many industry applications, as the processing windows are generally limited, experimental data based continuum modeling may still be important due to its accuracy and fast computation time.

978-1-5090-5979-9/17 $31.00 © 2017 IEEE

REFERENCES

[1] D. Guo et al., "FINFET technology featuring high mobility SiGe channel for 10nm and beyond", VLSI Technology, 2016 IEEE Symposium on VLSI Technology, Honolulu, HI, pp. 1-2, 2016.

[2] D. Yoon et al., "260-GHz differential amplifier in SiGe heterojunction bipolar transistor technology," in *Electronics Letters*, vol. 53, no. 3, pp. 194-196, 2017.

[3] I. G. Kim, K. S. Jang, S. Kim, J. Joo and G. Kim, "High-performance top-illumination type Ge-on-Si photodetectors ready for optical network applications," 10th International Conference on Group IV Photonics, Seoul, pp. 79-80, 2013.

[4] S. A. Srinivasan et al., "56 Gb/s Germanium Waveguide Electro-Absorption Modulator," J. Lightwave Technol. 34, pp.419-424, 2016.

[5] Rodolfo E. Camacho-Aguilera et al, "An electrically pumped germanium laser", Optics Express, vol. 20, pp. 11316-11320, 2012.

[6] Y. Lin, H. Yasuda, M. Schiekofer, B. Benna, R. Wise and G. Xia, "Effects of Carbon on Phosphorus Diffusion in SiGe:C and the Implications on Phosphorus Diffusion Mechanisms", J. Appl. Phys., vol. 116, pp.144904-1 to 144904-9, 2014.

[7] Y. Lin, H. Yasuda, M. Schiekofer and G. Xia, "Coupled Dopant Diffusion and Segregation in Inhomogeneous SiGe Alloys: Experiments and Modeling", J. Appl. Phys., vol. 117, pp. 214901-1 to 214901-7, 2015.

[8] P. Fahey, R. W. Dutton, and M. Moslehi, "Effect of thermal nitridation processes on boron and phosphorus diffusion in ⟨100⟩ silicon," Applied Physics Letters, vol. 43, pp. 683-685, 1983.

[9] Y. Lin, H. Yasuda, M. Schiekofer and G. Xia, "The Effects of Thermal Nitridation on Phosphorus Diffusion in Strained SiGe and SiGe:C", J. Mater. Sci., vol. 51, pp. 1532-1540, 2016.

[10] Y. Dong, Y. Lin, Simon Li, Steve McCoy, and G. Xia, "A unified interdiffusivity model and model verification for tensile and relaxed SiGe interdiffusion over the full germanium content range", J. Appl. Phys.. vol. 111, pp. 044909-1 to 044909-9, 2012.

[11] Y. Dong, W. Chern, P. M. Mooney, J. L Hoyt and G. Xia, "On the role and modeling of compressive strain in Si-Ge interdiffusion for SiGe heterostructures", Semicond. Sci. Technol., vol. 29, p. 015012, 2013.

[12] G. Xia, M. Canonico, O. O. Olubuyide and J. L. Hoyt, "Strain dependence of Si-Ge interdiffusion in epitaxial Si/Si1-yGey/Si heterostructures on relaxed Si1-xGex substrates," Appl. Phys. Lett., vol. 88, pp. 013507-1 to 013507-3, 2006.

[13] Y. Dong et al., "Experiments and Modeling of Si-Ge Interdiffusion with Partial Strain Relaxation in Epitaxial SiGe Heterostructures", ECS Journal of Solid State Science and Technology, vol. 3, pp.302-309, 2014.

[14] Py, M., et al., "Characterization and modeling of structural properties of SiGe/Si superlattices upon annealing", J. Appl. Phys., vol. 110, p. 044510, 2012.

[15] F. Cai,Y. Dong, Y. H. Tan, C. S. Tan and G. Xia,"Enhanced Ge-Si Interdiffusion in High Phosphorus-Doped Germanium on Silicon", Semiconductor Science and Technology, vol. 30, p. 105008, 2015.

[16] F. Cai, D. H. Anjum, X. Zhang and G. Xia, "Study of Si-Ge Interdiffusion with Phosphorus Doping", Journal of Applied Physics, vol. 120, p. 165108, 2016.

[17] G. Xia and Judy L. Hoyt, "Si-Ge Interdiffusion under Oxidizing Conditions in Epitaxial SiGe Heterostructures with High Compressive Stress," Appl. Phys. Lett., vol. 96, pp. 122107-1 to 122107-3, 2010.

Physical Modeling of the Hysteresis in MoS$_2$ Transistors

Theresia Knobloch*, Gerhard Rzepa*, Yury Yu. Illarionov*†, Michael Waltl*, Franz Schanovsky‡,
Markus Jech*, Bernhard Stampfer*, Marco M. Furchi§, Thomas Müller§, and Tibor Grasser*

*Institute for Microelectronics, TU Wien, Gußhausstraße 27–29/E360, 1040 Wien, Austria
Email: knobloch | grasser@iue.tuwien.ac.at
†Ioffe Physical-Technical Institute, Polytechnicheskaya 26, 194021 St-Petersburg, Russia
‡Global TCAD Solutions, Bösendorferstraße 1/12, 1010 Wien, Austria
§Institute for Photonics, TU Wien, Gußhausstraße 27–29/E387, 1040 Wien, Austria

Abstract—The hysteresis in the gate transfer characteristics of transistors made of two-dimensional materials is one of the most obvious problems of this novel technology. Here we attempt for the first time to develop a physical modeling approach for describing this hysteresis in devices based on two-dimensional materials. Our model is based on a drift-diffusion TCAD simulation coupled to a previously established non-radiative multiphonon model for describing charge capture and emission events in the surrounding dielectrics, which are considered the main cause for the observed hysteresis. We validate our model against measurement data on a back-gated single-layer MoS$_2$ transistor with SiO$_2$ as a gate dielectric. Our study provides new insights into the physical reasons for the observed hysteresis, thereby leading the way towards an alleviation of this problem in future devices.

I. Introduction

Molybdenum disulfide (MoS$_2$) is a two-dimensional (2D) material of the large group of transition metal dichalcogenides (TMDs), which has received a lot of attention over the past few years because of its inherent and comparatively large transport band gap ($E_G = 2.48\,\text{eV}$ [1]). This renders it an ideal candidate for applications in digital electronics [2–4], as it enables high current on/off ratios and a large transconductance [5].

However, up to now, MoS$_2$ based FETs have not met the high expectations for example when judging device performance in terms of mobilities. When accurately accounting for the non-negligible contact resistances [6], the mobility extracted for MoS$_2$ layers using multi-terminal measurements at room temperature does not exceed about $100\,\text{cm}^2/\text{Vs}$ [7]. Besides that, while this mobility value lies in a range comparable with standard silicon technology, the large variability observed in the device characteristics and performance issues like the frequently observed hysteresis in the gate transfer ($I_\text{D}(V_\text{G})$) characteristics [8–12] and the typically large drifts of the threshold voltage (V_th) over time [13] are to the present day one of the most critical obstacles inhibiting any industrial applications of MoS$_2$ FETs.

Complementing our previous experimental works [13–15], here we present a detailed study on the main mechanisms governing the hysteresis phenomenon in the $I_\text{D}(V_\text{G})$ characteristics of MoS$_2$ FETs. Our study is based on a drift-diffusion TCAD model [16] coupled to a four-state non-radiative multiphonon (NMP) model, which is necessary to accurately describe charge capture and emission events in the underlying gate dielectric [17]. Here we apply this simulation methodology which was originally developed and established for silicon (Si) technologies [18, 19], to devices based on 2D materials such as MoS$_2$. Our results confirm that the ubiquitous charge trapping at oxide traps is one of the main reasons for the hysteresis in MoS$_2$ FETs [13] and provide new insights into the details of these trapping and detrapping processes.

II. Devices and Measurements

A description of the device fabrication and measurement techniques we use here as a proof-of-concept for our modeling method have been reported in detail elsewhere [13]. For the sake of completeness, we give a short summary of all the details which will be important for understanding the simulation results later on. For demonstration purposes we study a back-gated MoS$_2$ FET using thermal silicon dioxide (SiO$_2$) as a back gate dielectric. The FET is based on a single layer (SL) flake of MoS$_2$ ($d \approx 6.5\,\text{Å}$), obtained via mechanical exfoliation [20]. The titanium/gold (Ti/Au) electrodes for the source and drain contacts were fabricated using electron beam lithography and metal evaporation techniques [8]. As a final fabrication step, the device was annealed in vacuum ($< 5 \times 10^{-6}\,\text{Torr}$, $T = 120\,°\text{C}$) in order to reduce the contact resistances and to remove adsorbed impurities. The $I_\text{D}(V_\text{G})$ characteristics of the MoS$_2$/SiO$_2$ FET were measured using a sweep range of $V_\text{G} \in [-20\,\text{V}, 20\,\text{V}]$ at a temperature of $T = 25\,°\text{C}$ in vacuum ($< 1 \times 10^{-5}\,\text{Torr}$). Several $I_\text{D}(V_\text{G})$s were recorded using a varying sweep rate $S = \Delta V/\Delta t$ (with ΔV as the voltage step and Δt as the time step), which corresponds to a sweep frequency of $f = 1/T$ with T being the total sweep time.

III. Simulation of Initial Devices

Here, the general simulation methodology using drift-diffusion based TCAD [16] is validated against a measured $I_\text{D}(V_\text{G})$ curve. The drift-diffusion equations [21] are computationally very efficient [22] and completely sufficient for describing the charge transport through the channel of these large-area MoS$_2$ FET prototypes. In several recent works [23–25] compact models describing devices based on 2D channel materials with drift-diffusion equations have been developed. The drift-diffusion equations can be used because the lateral dimensions of our devices are in the micrometer range ($W \times L \approx 7.0\,\mu\text{m}^2$ for the device discussed here). As a consequence, the large number of scattering centers in the channel region result in scattering-dominated drift-diffusion charge transport.

We extend an existing drift-diffusion based device simulator [16] to this new device class and use the material parameters summarized in Table I. The list of parameters is divided into two sections. The first section contains material constants, extracted mainly from the thorough DFT study of Rasmussen *et al.* [1, 28] on TMDs. The second section contains material parameters which should be constant but which are strongly influenced either by the defects in the channel region or by the contacts [6, 27]. For these parameters we only give meaningful ranges according to literature, within which the values should be chosen. At the current stage of research these parameters have to be treated as fitting values and have to be adjusted to every device separately. The impact of the most important material

978-1-5090-5979-9/17 $31.00 © 2017 IEEE

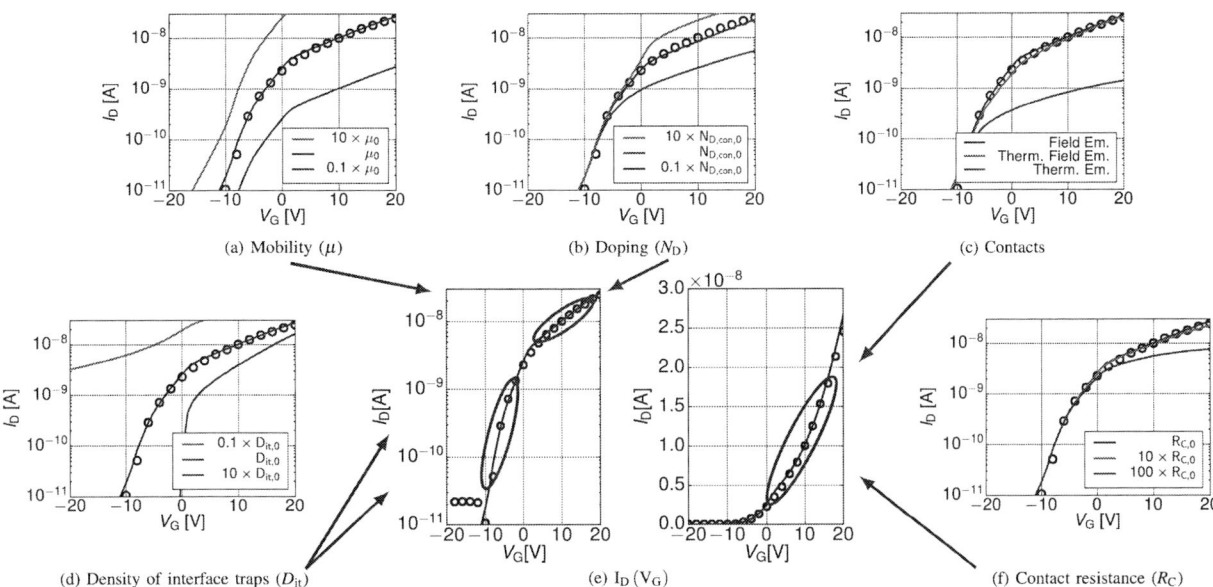

(a) Mobility (μ) (b) Doping (N_D) (c) Contacts

(d) Density of interface traps (D_{it}) (e) I_D (V_G) (f) Contact resistance (R_C)

Fig. 1: I_D (V_G) fit of the drift-diffusion based TCAD model to a measured curve (black circles) and illustration of the impact of the most important physical fit parameters on this characteristic (a)-(d),(f). The material properties on the left hand side (mobility (a), doping (b) and density of interface defects (d)) serve as fit parameters, because in this immature technology there is apparently a huge number of defects in the channel region ($D_{it} \approx 10^{13} \mathrm{cm}^{-2}\mathrm{eV}^{-1}$). The figures on the right hand side (contacts (c), contact resistance (f) and doping (b)) are used to demonstrate the impact of contact-related model parameters on the I_D (V_G). These parameters are especially critical for MoS$_2$ FETs, as these devices are known to be Schottky barrier transistors [26].

parameters on the I_D (V_G) is illustrated in Fig. 1. The central Fig. 1(e) demonstrates the quality of the fit, which can be established with the proposed simulation methodology. Our model is able to capture all aspects of the I_D (V_G) visible on a logarithmic scale as well as on a linear scale.

The doping, the mobility, and the density of interface traps are quantities which are strongly related to the defects in the channel region. Therefore, they are fit parameters for the currently available MoS$_2$ FET prototypes. The mobility (μ) has an impact on the saturation current ($I_{D,sat}$) as well as on V_{th} (Fig. 1(a)). The doping level (N_D) affects only $I_{D,sat}$ (Fig. 1(b)). While these two parameters are inherent parameters of any drift-diffusion model, the impact of interface defects was considered by using the standard Shockley-Read-Hall (SRH) model [31], coupled to the drift-diffusion based TCAD simulator [16]. The density of interface traps (D_{it}) is a very

TABLE I: Simulation parameters used for drift-diffusion based TCAD simulations. The parameters in the first section are material constants, while the parameters in the second section are strongly influenced by defects and contacts [6, 27] and therefore, at the current stage of research, vary from device to device.

Parameter	Value/Range	Reference
Transport band gap (E_G)	2.48 eV	[1]
Electron affinity (χ)	-3.84 eV	[1]
Electron mass (m_n^*)	0.55	[1]
Hole mass (m_p^*)	0.56	[1]
Eff. rel. permittivity ($\varepsilon_r^{\mathrm{eff}}$)	≈ 4	[28]
Contact resistance (R_C)	$[10^4, 10^6]\Omega$	[27]
Work func.diff.(Ti/Au) (E_W)	$[0.05, 0.2]$eV	[27]
Mobility (μ)	$[0.1, 100]$cm^2/Vs	[6, 7, 29]
Doping (N_D)	$< 1 \times 10^{16}$ cm^{-3}	[30]
Den. of interface traps (D_{it})	$[10^{12}, 10^{13}]$cm^{-2}eV^{-1}	[30]

important parameter, affecting at the same time the subthreshold slope and $I_{D,sat}$ through electrostatic doping [25, 32, 33] (Fig. 1(d)). This parameter has been studied in detail by Takenaka et al. [30], who associated the typical density of interface traps observed for MoS$_2$ FETs with sulfur vacancies in the MoS$_2$ layers.

As MoS$_2$ FETs are known to be Schottky barrier transistors [26], the work function differences between the contacts and the MoS$_2$ layer and the contact resistances R_C are very important parameters for an accurate description of the I_D (V_G)s. In Fig. 1(c) the impact of different models for describing the current transport across Schottky barriers is shown. In general one distinguishes between thermionic emission, thermionic-field emission and field emission, depending on whether the thermionic current over the barrier or the tunneling current through the barrier dominates. In the approximation of pure field emission, one usually speaks of an Ohmic contact, while for pure thermionic emission one requires equations describing the transport over Schottky contacts [34, 35]. As stated previously [6, 32], the short tunneling distance in a 2D layer gives rise to large tunneling currents, thereby justifying the approximative modeling of MoS$_2$ FETs with a pure field emission model in the back-gated configuration.

This conclusion renders the work function difference unimportant in the case of back-gated devices [36] while the contact resistance remains an important fit parameter, the impact of which is demonstrated in Fig. 1(f). In relation to the contacts, it has been recently discussed in literature that only the reactions at the interface of the MoS$_2$ layer with the Ti adhesion layer enable a good contact to SL MoS$_2$ through covalent bonding [27, 37]. This coincides nicely with our observation that in our model the doping levels below the contacts ($N_{D,con}$) are by far more important than the intrinsic doping level of the SL MoS$_2$ ($N_{D,L}$) in the channel. Even for intrinsic doping levels of up to half of the effective doping in the contact region due to Ti atoms ($N_{D,L} < 0.5 \times N_{D,con}$) the device behavior remains dominated solely by $N_{D,con}$. From this we conclude that the impact of the Ti

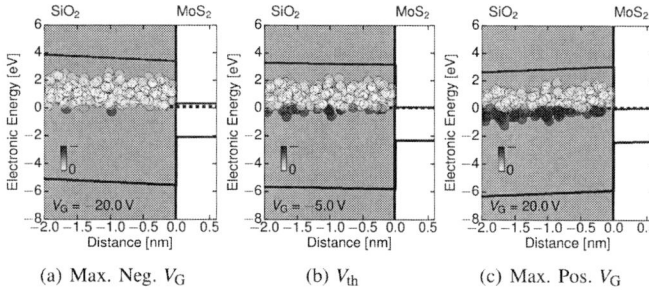

(a) Max. Neg. V_G (b) V_{th} (c) Max. Pos. V_G

Fig. 2: Band diagrams of the MoS$_2$/SiO$_2$ FET at different gate voltages showing the electron trapping band [38, 39] responsible for the hysteresis.

adhesion layer does not lie solely in the establishment of covalent bonds for adequate contacts, but maybe even more importantly in the unintentional but essential introduction of defect states in certain regions of the band gap of SL MoS$_2$, corresponding to an effective doping of the layer, $N_{D,con}$. Therefore, only the impact of $N_{D,con}$ on the $I_D(V_G)$ is demonstrated in Fig. 1(b).

IV. DEFECT MODELING

Having successfully established a good fit of the $I_D(V_G)$ characteristics, we now take the next step towards modeling the hysteresis. In order to see a shift in the threshold voltage between the up sweep and the down sweep of an $I_D(V_G)$, there has to be charge trapping in the vicinity of the channel. While several groups claim that the charge trapping takes place at the interface [11, 40, 41], we argue here in accordance with our previous works [13–15] that the fact that the largest hysteresis is observed for a total sweep time of $T = 200$ s is a strong argument in favor of oxide traps, as they usually have larger time constants than interface traps. What is more, oxide traps are located at a finite distance from the interface, the most important ones for the charge transfer processes lying typically within the first few nanometers. This leads to an increased bias dependence, which is especially important to explain the hysteresis in MoS$_2$ FETs. Interface traps provide trap levels inside the band gap, thus once the Fermi level reaches the conduction band edge (roughly at $V_G \approx V_{th}$) and remains pinned there due to the effective doping of the layer, $N_{D,con}$, there are

to a first approximation no more trapping and detrapping events at interface states. However, exactly these charge capture and emission events for gate voltages above the threshold voltage are the reason for the observed hysteresis.

For the modeling of the hysteresis we use the four-state NMP model, which accurately describes charge transfer reactions in conventional Si/SiO$_2$ devices [17]. It does not only account for the energy balance of the transferred electrons, as it is usually done when using the SRH model [31], but it also considers the energetic relaxation of the structure around the defect, where the electron is captured or emitted [17]. Depending on the microscopic nature of the defect, which has been studied in great detail for SiO$_2$ based on Si/SiO$_2$ FETs [42, 43], one usually speaks either of hole or of electron trapping. As the charge transfer process is exactly the same in both cases, the two processes can only be distinguished by the charge change of the trapping defect in the oxide, which either goes from positive to neutral (hole trap) or from neutral to negative (electron trap).

Thus, in order to explain the hysteresis in MoS$_2$ FETs we use the two known defect bands of SiO$_2$ from silicon technologies [19, 38, 39], with the first being a donor-like hole trapping band located at $E_T^L = 4.6(3)$ eV below the conduction band edge of SiO$_2$ [19], and the second most likely being an acceptor-like electron trapping band at $E_T^U = 2.6(4)$ eV below the conduction band edge of SiO$_2$. The second defect band is less well known, but has already been observed for Si-based devices with dielectric gate stacks [38, 39]. Additionally, it has been used in our previous works for the modeling of the hysteresis and of bias-temperature instabilities in FETs based on MoS$_2$ [13, 15] and black phosphorus [14].

Fig. 2 illustrates how charges are trapped and detrapped in the oxide. At a positive gate voltage, the defect band is bent downwards, leading to more electron trapping, thereby causing a shift in the threshold voltage. However, if the same traps emit their electrons during the down sweep, no hysteresis can be observed. At this point the time constants of the responsible defects, as determined by the four-state NMP model, come into play. A trap can only contribute to the hysteresis if it captures an electron at a high gate voltage and emits this electron not before reaching again the low level of the gate voltage. This means that the electron capture time constant (τ_c) of the respective trap has to be smaller than the electron emission time constant (τ_e) at high gate voltages and vice-versa. For this criterion

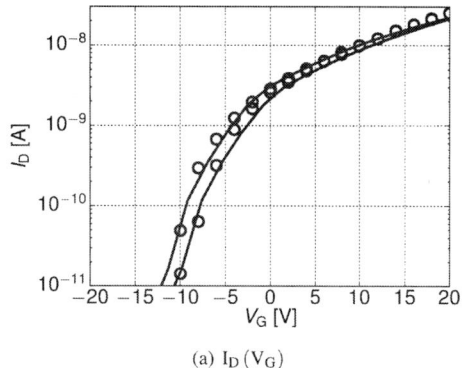

(a) $I_D(V_G)$

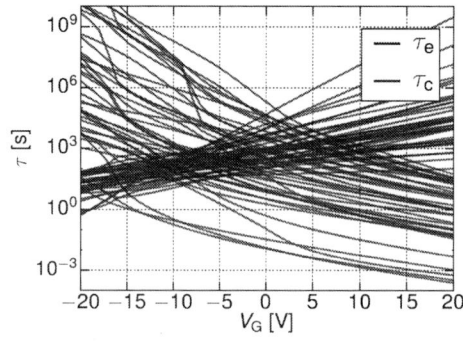

(b) Bias dependence of time constants.

Fig. 3: Established hysteresis fit (red - up-sweep, blue - down-sweep), together with a time constant plot for an exemplary set of defects, selected to display the defects contributing to the hysteresis in our simulations.

the important voltage level is V_{th}, where the hysteresis is extracted, which lies for our devices at around $V_{th} \approx -5\,\text{V}$. If for $V_G < V_{th}$ it holds, that $\tau_e < \tau_c$ and for $V_G > V_{th}$ it holds that $\tau_e > \tau_c$, this trap can in principle contribute to the hysteresis.

The gate bias dependence of the time constants of some selected traps, contributing to the hysteresis in our simulations, are shown in Fig. 3 (b). In Fig. 3 (a) the established fit between the measured $I_D(V_G)$ and the simulated characteristics is presented. Our simulation results clearly corroborate the previously observed [13] PBTI-like hysteresis.

V. CONCLUSIONS

A drift-diffusion based simulation methodology was used to describe the charge capture and emission processes in gate oxide traps resulting in the hysteresis observed in the $I_D(V_G)$ characteristics of back-gated SL MoS_2 FETs. Our results emphasize that the voltage dependence of the time constants of the traps is an essential quantity, which has to be considered when identifying the traps responsible for the hysteresis phenomenon.

ACKNOWLEDGMENTS

The authors gratefully acknowledge financial support through FWF grant n° I2606-N30.

Additionally, we gratefully acknowledge inspiring discussions with Gianluca Fiori (Universtiy of Pisa), Mario Lanza (Suzhou University), Dmitry Polyushkin (TU Wien), and Georg Reider (TU Wien).

REFERENCES

[1] F. A. Rasmussen, et al., "Computational 2D Materials Database: Electronic Structure of Transition-Metal Dichalcogenides and Oxides," *Journal of Physical Chemistry C*, vol. 119, no. 23, pp. 13169–13183, 2015.

[2] B. Radisavljevic, et al., "Single-layer MoS2 transistors," *Nature nanotechnology*, vol. 6, no. 3, pp. 147–50, 2011.

[3] G. Fiori, et al., "Electronics based on two-dimensional materials," *Nature Nanotechnology*, vol. 9, no. 10, pp. 768–779, 2014.

[4] S. Wachter, et al., "A microprocessor based on a two-dimensional semiconductor," *arXiv preprint*, p. 1612.00965, 2016.

[5] Y. Yoon, et al., "How good can monolayer MoS2 transistors be?," *Nano letters*, vol. 11, no. 9, pp. 3768–73, 2011.

[6] S. Das, et al., "High Performance Multi-layer MoS2 Transistors with Scandium Contacts.," *Nano letters*, vol. 13, no. 1, pp. 100–5, 2012.

[7] X. Cui, et al., "Multi-terminal transport measurements of MoS2 using a van der Waals heterostructure device platform," *Nature Nanotechnology*, vol. 10, no. 6, pp. 534–540, 2015.

[8] D. J. Late, et al., "Hysteresis in single-layer MoS2 field effect transistors.," *ACS Nano*, vol. 6, no. 6, pp. 5635–41, 2012.

[9] T. Li, et al., "Scaling behavior of hysteresis in multilayer MoS2 field effect transistors," *Applied Physics Letters*, vol. 105, no. 9, p. 093107, 2014.

[10] A.-J. Cho, et al., "Multi-Layer MoS2 FET with Small Hysteresis by Using Atomic Layer Deposition Al2O3 as Gate Insulator," *ECS Solid State Letters*, vol. 3, no. 10, pp. Q67–Q69, 2014.

[11] Y. Park, et al., "Thermally activated trap charges responsible for hysteresis in multilayer MoS2 field-effect transistors," *Applied Physics Letters*, vol. 108, no. 8, p. 083102, 2016.

[12] J. Shu, et al., "The Intrinsic Origin of the Hysteresis in the MoS2 Field Effect Transistors," *Nanoscale*, vol. 8, pp. 3049–3056, 2016.

[13] Y. Y. Illarionov, et al., "The Role of Charge Trapping in MoS2 / SiO2 and MoS2 / hBN Field-Effect Transistors," *2D Materials*, vol. 3, no. 3, pp. 1–11, 2016.

[14] Y. Y. Illarionov, et al., "Long-Term Stability and Reliability of Black Phosphorus Field-Effect Transistors," *ACS Nano*, vol. 10, pp. 9543–9549, 2016.

[15] Y. Y. Illarionov, et al., "Mapping of Oxide Traps in Double-Gated MoS2 Field-Effect Transistors," *2D Materials*, pp. 1–13, 2017.

[16] Global TCAD Solutions. Minimos-NT Manual, 2017.

[17] T. Grasser, "Stochastic charge trapping in oxides: From random telegraph noise to bias temperature instabilities," *Microelectronics Reliability*, vol. 52, no. 1, pp. 39–70, 2012.

[18] T. Grasser, et al., "The Paradigm Shift in Understanding the Bias Temperature Instability : From Reaction Diffusion to Switching Oxide Traps," *Transactions on Device and Materials Reliability, IEEE*, vol. 58, no. 11, pp. 3652–3666, 2011.

[19] G. Rzepa, et al., "Complete Extraction of Defect Bands Responsible for Instabilities in n and pFinFETs," *2016 Symposium on VLSI Technology Digest of Technical Papers*, pp. 208–209, 2016.

[20] M. M. Furchi, et al., "Photovoltaic effect in an electrically tunable Van der Waals heterojunction," *Nano Letters*, vol. 14, no. 8, pp. 4785–4791, 2014.

[21] W. v. Roosbroeck, "Theory of the flow of electrons and holes in germanium and other semiconductors," *Bell System Technical Journal*, vol. 29, no. 4, pp. 560–607, 1950.

[22] M. Lundstrom, "Drift-diffusion and computational electronics - Still going strong after 40 years!," *International Conference on Simulation of Semiconductor Processes and Devices, SISPAD*, vol. 2015-Octob, pp. 1–3, 2015.

[23] M. G. Ancona, "Electron transport in graphene from a diffusion-drift perspective," *IEEE Transactions on Electron Devices*, vol. 57, no. 3, pp. 681–689, 2010.

[24] D. Jiménez, "Drift-diffusion model for single layer transition metal dichalcogenide field-effect transistors," *Applied Physics Letters*, vol. 101, no. 24, 2012.

[25] S. V. Suryavanshi, et al., "S2DS: Physics-based compact model for circuit simulation of two-dimensional semiconductor devices including non-idealities," *Journal of Applied Physics*, vol. 120, no. 22, p. 224503, 2016.

[26] H. Liu, et al., "Switching mechanism in single-layer molybdenum disulfide transistors: An insight into current flow across Schottky barriers," *ACS Nano*, vol. 8, no. 1, pp. 1031–1038, 2014.

[27] A. Allain, et al., "Electrical contacts to two-dimensional semiconductors," *Nature Materials*, vol. 14, no. 12, pp. 1195–1205, 2015.

[28] F. A. Rasmussen, et al., "Efficient many-body calculations of 2D materials using exact limits for the screened potential: Band gaps of MoS_2, hBN, and phosphorene," *Physical Review B*, vol. 94, no. 155406, pp. 155406–1 –155406–9, 2016.

[29] S. Chuang, et al., "MoS 2P-type Transistors and Diodes Enabled by High Work Function MoO xContacts," *Nano Letters*, vol. 14, no. 3, pp. 1337–1342, 2014.

[30] M. Takenaka, et al., "Quantitative evaluation of energy distribution of interface trap density at MoS 2 MOS interfaces by the Terman method," *IEEE International Electron Devices Meeting*, pp. 139–142, 2016.

[31] W. Shockley, et al., "Statistics of the Recombination of Holes and Electrons," *Physical Review*, vol. 87, no. 46, pp. 835–842, 1952.

[32] J. Appenzeller, et al., "Toward nanowire electronics," *IEEE Transactions on Electron Devices*, vol. 55, no. 11, pp. 2827–2845, 2008.

[33] W. Cao, et al., "A compact current-voltage model for 2D semiconductor based field-effect transistors considering interface traps, mobility degradation, and inefficient doping effect," *IEEE Transactions on Electron Devices*, vol. 61, no. 12, pp. 4282–4290, 2014.

[34] E. H. Rhoderick, et al., *Metal-semiconductor contacts*. Clarendon Press Oxford, 1978.

[35] D. Schroeder, *Modelling of interface carrier transport for device simulation*. Springer Science & Business Media, 2013.

[36] M. S. Fuhrer, et al., "Measurement of mobility in dual-gated MoS2 transistors," *Nature nanotechnology*, vol. 8, no. 3, pp. 146–7, 2013.

[37] S. McDonnell, et al., "MoS2-Titanium Contact Interface Reactions," *ACS Applied Materials and Interfaces*, vol. 8, no. 12, pp. 8289–8294, 2016.

[38] R. Degraeve, et al., "Trap spectroscopy by charge injection and sensing (TSCIS): A quantitative electrical technique for studying defects in dielectric stacks," *Technical Digest - International Electron Devices Meeting, IEDM*, no. 1, pp. 10–13, 2008.

[39] G. Rzepa, et al., "Efficient Physical Defect Model Applied to PBTI in High-k Stacks," *Proceedings of the 2017 IEEE International Reliability Physics Symposium (IRPS)*, pp. XT–11.1 – XT–11.6, 2017.

[40] Y. Guo, et al., "Charge trapping at the MoS2-SiO2 interface and its effects on the characteristics of MoS2 metal-oxide-semiconductor field effect transistors," *Applied Physics Letters*, vol. 106, no. 10, 2015.

[41] K. Choi, et al., "Trap density probing on top-gate MoS2 nanosheet field-effect transistors by photo-excited charge collection spectroscopy," *Nanoscale*, vol. 7, pp. 5617–5623, 2015.

[42] T. Grasser, et al., "On the microscopic structure of hole traps in pMOSFETs," *IEEE International Electron Devices Meeting*, pp. 21.1.1–4, 2014.

[43] Y. Wimmer, et al., "Role of hydrogen in volatile behaviour of defects in SiO2 -based electronic devices," *Proceedings of the Royal Society A: Mathematical, Physical and Engineering Sciences*, vol. 472, p. 20160009, 2016.

Impact of impurities, interface traps and contacts on MoS₂ MOSFETs: modelling and experiments

Gioele Mirabelli, Farzan Gity, Scott Monaghan,
Paul K. Hurley, Ray Duffy

Tyndall National Institute, University College Cork
Lee Maltings Complex, Cork, Ireland

Abstract— **Device modelling is a key enabling capability for the semiconductor industry, especially for process optimisation, and for insight into the physics of novel architectures and materials that are difficult to access experimentally. Despite much innovative experimental work, device modelling capabilities for field effect devices based on Transition Metal Dichalcogenide (TMD) channel materials are at an early stage of development. Properly formulated physics-based models would give a substantial improvement for time- and cost-effective development of TMD devices. In this work, experimental device data was used to develop models and parameter sets in the continuum-based Synopsys Sentaurus Device software. Specifically, few-layer MoS₂ Field-Effect-Transistors (FETs) were systematically electrically characterized, and the modelling of the experimental data focused on the impact of impurities, interface traps, and contact barriers. Furthermore, the experimental MoS₂ FETs device characteristics, combined with the physics based transport models, suggests that the low experimental electron mobility values are a result of a high density of charge impurity defects in the MoS₂ channel. To the best of our knowledge continuum-based TCAD device models did not previously exist for MoS₂, or TMD-semiconductors in general.**

Keywords—MoS₂; Device Modeling; Impurities; Electrical characterization; Physics;

I. Introduction

In recent years the interest in 2D-semiconductors, and Transition-Metal-Dichalcogenides (TMDs) in particular, has increased noticeably. Due to their intrinsic 2D nature the electrostatic control of the gate is higher with respect of 3D-semiconductors, making them more immune to Short Channel Effects [1]. One of the issues associated with 2D-materials is the lack of a suitable large-area growth technique, which limits the systematic study TMD materials grown over large are substrates. Often, TMD films are characterized by defects or grain boundaries, which can limit their electrical performances. Therefore, most of the experimental results in literature have been obtained from flakes mechanically exfoliated from bulk crystals. To date these have shown the best electrical results, with high on/off ratios, low inverse subtreshold slopse and carrier mobilities in the range of tens of $cm^2V^{-1}s^{-1}$ [2].

Nevertheless, a large concentration of impurities was reported in TMD crystals [3], and if in a charged state, these impurities will degrade the carrier mobility and resistivity of the material [4] especially in TMD films comprised of a small (1-10) number of monolayers [5]. In the first part of this work,

we report the electrical characterization of back-gated MoS₂ FETs. Based on the experimental device characteristics a model and parameter set for MoS₂ was implemented in the continuum-based Synopsys TCAD simulator Sentaurus Device, which showed consistent behavior with the experimental data. Charged impurity concentrations, interface states and MoS₂/metal contact barriers were varied to explore the parameters which influence carrier transport in MoS₂ FETs.

II. Experiment

The experimental procedure is reported in Fig. 1. Mechanical exfoliation with scotch tape was used to obtain thin flakes from a MoS₂ bulk crystal. The flakes were transferred on a substrate of 85 nm of dry thermally grown SiO₂ and a highly-boron doped Si handle wafer. The height of the flakes was established by optical color-contrast to be 3-4 layers [6]. Ti/Au metal contact pads and electrodes were defined by electron-beam lithography, followed by metal evaporation and lift off process. 15 kV beam exposures were performed with a Zeiss SUPRA SEM with a Raith Elephy Plus blanker. An adhesion layer of 5 nm of Ti and 45 nm of Au were deposited with e-beam evaporation. For structural analysis, cross-section samples were obtained by FEI's Dual Beam Helios Nanolab 600i system using Ga ion beam. Cross-sectional Transmission Electron Microscopy (XTEM) imaging was carried out using a JEOL 2100 HRTEM operated at 200 kV in Bright Field mode using a Gatan Double Tilt holder. Electrical characterization was performed on a wafer at 25°C in a dry/dark environment using the HP4156C parameter analyzer.

III. Material and Electrical Analysis

Fig. 1 shows the Scanning Electron Microscope (SEM) and cross-sectional Transmission Electron Microscope (TEM)

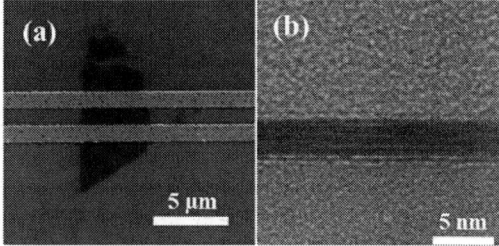

Fig. 1: Top: summary of the experimental process flow and optical image of the contacted MoS₂ flake. Bottom: (a) SEM top-view of the MoS₂ device after contact deposition. (b) Representative high-resolution TEM image showing the layered structure of the sample.

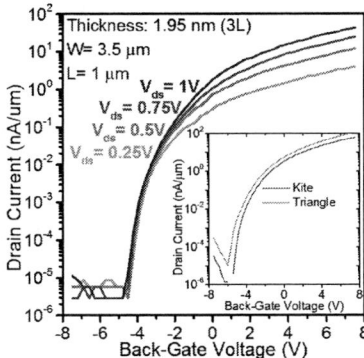

Fig. 2: Transfer characteristics of the Trapezium for different drain-source voltages. Inset: transfer characteristics of the Kite and the Triangle at V_{ds}=1V.

images of the "Trapezium" device, so named because of its shape. The device width is approximately 3.5 μm along the length of the sample and the channel length is confirmed to be 1 μm as determined after the metal deposition. The TEM cross section of the same device, confirms the 3-layers thickness implied by optical contrast.

Fig. 2 shows the transfer characteristics of the device for different V_{ds}. The transfer characteristics at V_{ds}=1V are presented in the inset of Figure 2 for two other back gated MoS$_2$ FETs, termed the "Kite" and the "Triangle" based on the MoS$_2$ flake shapes on which the devices were made. These devices are also 3L thick. The electrical results among the three devices are highly consistent. The drain current at a back gate voltage of 7.5 V and a drain-source voltage of 1V is ~50 nA/μm, limited probably by contact resistance and/or low mobility. The off-current is limited by the sensitivity of the measurement system, hence could potentially be lower. The MoS$_2$ flakes reported here were exfoliated from the same bulk material onto the same SiO$_2$/Si substrate and experienced the same process steps and environment-related exposure. As a consequence, the slight variability among these devices is most probable related to intrinsic factors only, such as defects [7], grain boundaries and/or impurities [8], as opposed to process-related factors such as metal contacting [9] or air exposure [10].

To compare the characteristics of the devices field-effect mobility values were extracted from the derivative of the transfer characteristics. Fig. 3a shows the field effect mobility values plotted against carrier concentration, evaluated considering a linear charge dependence on the gate voltage overdrive. Due the low back-gate voltage (maximum value is 7.5 V) and the relatively thick back gate oxide (85 nm of SiO$_2$) the vertical field contribution to mobility degradation is not expected to be a significant contribution. This is confirmed in Fig. 3, where the field effect mobility increases with the calculated carrier density.

Ma and Jena [11] reported a systematic study of carrier transport in 2D crystals, which are found to be highly dependent on the dielectric environment and ionized impurity density. In particular, the mobility versus electron density for low carrier concentration follows the relationship

$$\mu_{CI} \cong \frac{3500 \cdot 10^{11}}{N_{CI}} \left[A_{CI} + \left(\frac{n_e}{10^{13}} \right)^{1.2} \right] \qquad (1)$$

where μ_{CI} is mobility, N_{CI} is ionized impurity concentration and A_{CI} is a fitting parameter that depends on the dielectric environment. Considering that n_e can be obtained experimentally, impurity concentration can be extracted by fitting the equation with the field-effect mobility of the devices. Fitted data in [11] shows an impurity concentration of 10^{13} cm^{-2}, while in [12], where the same method is used, different levels of impurity concentration (10^{11} and 10^{13} cm^{-2}) have been reported, probably due to the high variability of TMDs and film/device preparation conditions. This is also consistent with the relationship of Hall effect mobility to doping concentration observations in natural and synthetic TMD crystals [4].

Fig. 3b shows the impurity concentration level for each device, extracted from fitting the mobility using eq. (1). The average impurity concentration is 4×10^{13} cm^{-2}. Similar levels of impurity were found before by ICMPS analysis on MoS$_2$ crystals [3]. The target in the ITRS roadmap [13] for unintentional impurity concentration in semiconductors intended for logic device application is for values less than 5×10^{10} cm^{-2}.

IV. TCAD SIMULATIONS

To understand the implications on the electrical performance of such a high impurity concentration the Synopsys TCAD simulator Sentaurus Device [14] was used to develop a model and parameter set for MoS$_2$. To date a clear model for MoS$_2$ is still missing. Due to the lack of a proper growth process, the low yield of the mechanical exfoliation and the high variability of TMDs it is difficult to obtain consistent results and have systematic experiments. The use of a TCAD simulator allows an investigation of what the physical processes are currently limiting device performance.

A schematic cross section of the structure used for simulation studies is shown in Fig. 4. This replicates the actual structure used in the experimental section. The main parameters for MoS$_2$ are obtained from literature. Some of the parameters, such as band-gap [15] and in-plane dielectric constant [16] are dependent on the actual thickness of the MoS$_2$. Then, to match the experimental results of the Trapezium a mobility model dependent on impurities, defective

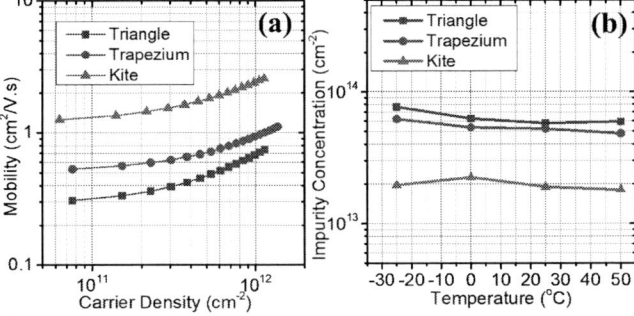

Fig. 3: (a) Field effect mobility versus carrier density extracted from the transfer characteristic at V_{ds}=1V. (b) Impurity concentration extracted from the mobilities in part (a) using eq. (1).

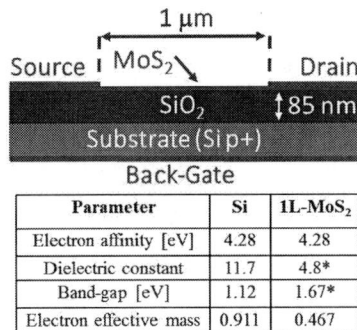

Parameter	Si	1L-MoS$_2$
Electron affinity [eV]	4.28	4.28
Dielectric constant	11.7	4.8*
Band-gap [eV]	1.12	1.67*
Electron effective mass	0.911	0.467

Fig. 4: Schematic of the device structure implemented in Sentaurus device and the main parameters. The symbol * denotes thickness-dependent parameters.

contacts and interface traps were implemented. The Unified Philips mobility model was used to take into account the impurity dependency [17]

$$\mu_{CI} = \mu_{e,N}\left(\frac{N_{e,sc}}{N_{e,sc,eff}}\right)\left(\frac{N_{ref}}{N_{e,sc}}\right)^{\alpha} + \mu_{e,c}\left(\frac{n+p}{N_{e,sc,eff}}\right) \quad (2)$$

where μ_{CI} is the mobility dependent on coulomb impurities, and

$$N_{e,sc} = N_D + N_A + p \quad (3)$$

$$N_{e,sc,eff} = N_D + G(P_e)N_A + f_e\frac{p}{F(P_e)} \quad (4)$$

where N_D and N_A are the ionized donor and acceptor concentrations, p is the hole concentration, $G(P_e)$ and $F(P_e)$ are analytical functions describing minority impurity and electron hole scattering. f_e is a fitting parameter equal to 1. $\mu_{e,N}$ and $\mu_{e,C}$ depend on the maximum and minimum mobility values, set as 1000 and 9 $cm^2V^{-1}s^{-1}$ respectively [10]

$$\mu_{e,N} = \frac{\mu_{e,max}^2}{\mu_{e,max} - \mu_{e,min}} \quad (5)$$

$$\mu_{e,c} = \frac{\mu_{e,max}\mu_{e,min}}{\mu_{e,max} - \mu_{e,min}} \quad (6)$$

The model proposed in [11] was used as reference and the main parameters, i.e., N_{ref} and α, are found to be 4×10^{18} cm^{-3} and 1.0 for monolayer MoS_2. As reported in Fig. 5, these values of N_{ref} and α reproduce the magnitude and trend of the relationship between mobility versus charged impurity density reported for MoS_2 in Ref. [11], especially for high impurity concentration, which is the region of most interest for this study. Impurities are introduced as mid-gap "fixed charges", which in Sentaurus are traps always completely occupied.

Furthermore, barriers were used at the contact to take into account the high Schottky barriers usually found in TMD-FETs. The value used to match the experimental results of the Trapezium is 0.17 eV, which is close to the experimentally reported value [18]. Fig. 6 illustrates the impact of the Schottky barrier height at the metal/MoS_2 interface on the transfer characteristics.

The subthreshold behavior of the device is also dependent on interface traps located at the SiO_2/MoS_2 interface of the back gated device structure. To model the stretch out of the experimental transfer characteristic, acceptor like interface

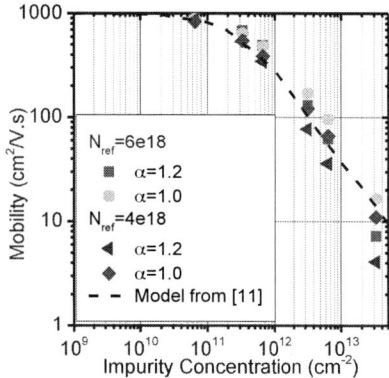

Fig. 5: Mobility versus impurity concentration considering different parameters implemented using the Unified Philips mobility model. The model is based on theoretical calculations for MoS_2.

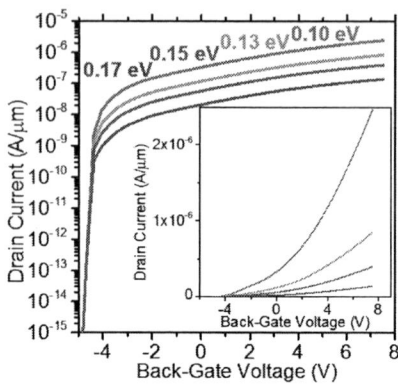

Fig. 6: Simulated transfer characteristics varying the barrier at the source and drain contacts. Inset: same curves in linear scale.

states are introduced close to the MoS_2 conduction band edge [19]. In order to match the subthreshold behavior of the Trapezium 4 trap energies are introduced, with energy levels at $E_{t1}=1.25kT$, $E_{t2}=2.5kT$, $E_{t3}=5kT$ and $Et_4=10kT$ from the conduction band and with concentration of 8×10^{11}, 3×10^{11}, 2.5×10^{11} and 1×10^{11} cm^{-2}, respectively.

Fig. 7 illustrates the difference in the subthreshold behavior of the device without any interface traps, with E_{t1} and E_{t2} only (the traps closest to the conduction band), and E_{t3} and E_{t4} only (the traps closest to the mid-gap). A schematic of the trap energy levels is presented as inset in Fig. 7. As expected the traps closest to the mid-gap affect more the subthreshold behavior, while the ones close to the conduction band affect the on-current. Fig. 8 compares the experimental and the simulated data, with an impurity concentration of 4×10^{13} cm^{-2}.

Since the model and the parameter set are in good agreement with the experimental data, it can be used to gain a deeper understanding of the main physical processes which are limiting the drain to source current of the back gated MoS_2 MOSFETs, focusing on the impact of the impurity concentration. Fig. 9a shows the field-effect mobility of MoS_2 FETs ($V_{ds}=0.1V$) varying impurity concentration for a different number of layers (no interface traps or contact barriers are considered). As can be seen, the mobility is highly dependent on charged impurity concentration and it is probable that the experimental field effect mobility values extracted to date have been significantly influenced (degraded) due to the high impurity concentration.

Experimentally, mobility shows a peak between 5-10 layers [20], but are highly sensitivity to adsorbates on the

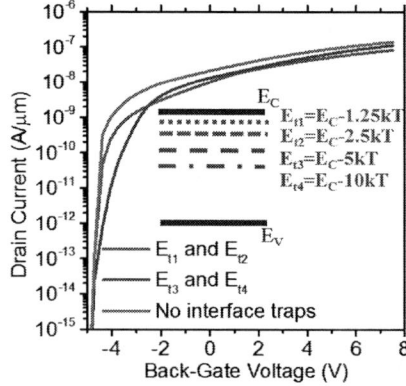

Fig. 7: Simulated transfer characteristics using acceptor traps at different energies. Inset: schematic of the trap energy levels with respect of the MoS_2 bandgap. The transfer characteristics are shifted negatively along the back gate voltage by changing the gate work function. This shift is to match the shift in the experimental characeristics.

978-1-5090-5979-9/17 $31.00 © 2017 IEEE

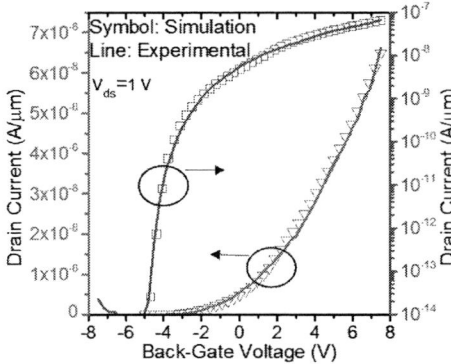

Fig. 8: Transfer characteristics of the experimental and the simulated devices in semi-log and linear scales considering an impurity concentration of 4×10^{13} cm^{-2}, 0.17eV of contact barriers and interface traps at energy levels at E_{t1}=1.25kT, E_{t2}=2.5kT, E_{t3}=5kT and E_{t4}=10kT from the conduction band and with concentration of 8×10^{11}, 3×10^{11}, 2.5×10^{11} and 1×10^{11} cm^{-2}, respectively.

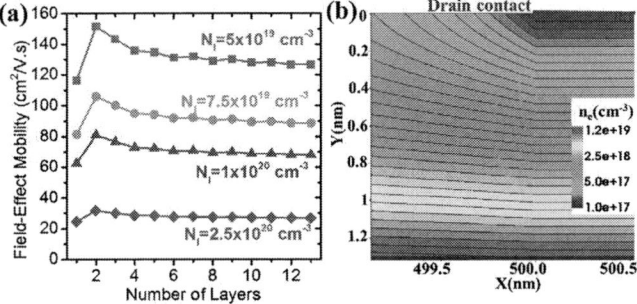

Fig. 9: (a) Field-effect mobility versus number of MoS$_2$ layers considering different impurity concentration. (b) Countour plot of the electron density near the drain contact for 2 layers of MoS$_2$. Both the plots are evaluated at V_{ds}=0.1V and V_{bg}=7.5V.

surface, especially for few layers [21]. Therefore, due to the several imperfections of the material itself, it is possible that these trends might differ when testing purer MoS$_2$. Also, the experimental values of mobility we reported are lower with respect to what is given by the model.

To investigate the field effect mobility further the electron density near the drain contact was considered. Fig. 9b shows the electron contour plot around the drain contact for a 2 layers MoS$_2$ back gated device. Since it is a back-gated FET a change in carrier density from the SiO$_2$/MoS$_2$ interface to the top metal contact exists, resulting in carrier transport perpendicular to the MoS$_2$ planes under the contact region. Thus, the mobility versus thickness trend might be different considering top-gated or back-gated FETs. Also achieving a highly doped MoS$_2$ under the drain and source metal contacts, will significantly reduce contact resistance [22].

V. CONCLUSIONS

In conclusion, MoS$_2$ FETs were electrically characterized from which continuum-based models and parameter sets were developed. It has been show that the experimentally extracted field-effect mobility values are strongly dependent on the impurity concentration present in the material. The trend was confirmed by TCAD simulations, which also showed the impact of interface trap density and contact barriers on the electrical device behavior. Even if promising results are shown to date the true TMD device performance is still masked by

imperfections, which can be tackled by optimizing the growth and device processing conditions.

ACKNOWLEDGMENTS

We acknowledge the support of SFI through the US-Ireland Partnership Programme Grant No. SFI/13/US/I2862 (UNITE) and of the IRC through the Postgraduate Scholarship EPSPG/2015/69. The research was supported in part by the HEA Programme for Research in Third Level Institutions in Ireland under Grant Agreement No. HEA PRTLI5.

REFERENCES

[1] F. Schwierz, J. Pezoldta and R. Granznera, "Two-dimensional materials and their prospects in transistor electronics", Nanoscale, 7, 8261-8283 (2010).

[2] A. Ayari, W. Cobas, O. Ogundadegbe, and M. S. Fuhrer, "Realization and electrical characterization of ultrathin crystals of layered transition-metal dichalcogenides", J. Appl. Phys. 101, 014507 (2007).

[3] R. Addou, et al., "Impurities and Electronic Property Variations of Natural MoS$_2$ Crystal Surfaces", ACS Nano 9 (9), 9124-9133 (2015).

[4] S. Monaghan, et al., "Hall-effect Mobility for a Selection of Natural and Synthetic 2D Semiconductor Crystals", Session 2: More than Moore devices and applications, Heterogeneous Integration, Other, EUROSOI-ULIS 2017, 3-5 April 2017, Athens, Greece.

[5] Z. Y. Ong, and M. V. Fischetti, "Mobility enhancement and temperature dependence in top-gated single-layer MoS$_2$," Phys. Rev. B., vol. 88, (2013).

[6] Hai Li, et al., "Rapid and Reliable Thickness Identification of Two-Dimensional Nanosheets Using Optical Microscopy", ACS Nano 7 (11), 10344-10353 (2013).

[7] S. McDonnell, R. Addou, C. Buie, R. M. Wallace, and C. L. Hinkle, "Defect-Dominated Doping and Contact Resistance in MoS$_2$", ACS Nano 8 (3), 2880-2888 (2014).

[8] S. J. McDonnell, and R. M. Wallace, "Atomically-thin layered films for device applications based upon 2D TMDC materials", Thin Solid Films 616, 482–501 (2016).

[9] A. Allain, J. Kang, K. Banerjee, and A. Kis, "Electrical contacts to two-dimensional semiconductors", Nature Materials 14, 1195–1205 (2015).

[10] G. Mirabelli, et al., "Air sensitivity of MoS$_2$, MoSe$_2$, MoTe$_2$, HfS$_2$, and HfSe$_2$", J. Appl. Phys., 120, 125102 (2016).

[11] N. Ma, and D. Jena, "Charge Scattering and Mobility in Atomically Thin Semiconductors", Phys. Rev.X 4 011043 (2014).

[12] T. Mori, et al., "Characterization of Effective Mobility and Its Degradation Mechanism in MoS$_2$ MOSFETs," in IEEE Transactions on Nanotechnology, vol. 15, no. 4, pp. 651-656, (2016).

[13] International Technology Roadmap for Semiconductors (ITRS, 2015); http://www.itrs.net/

[14] Synopsys Inc., CA, USA, Sentaurus Device User Guide (2015).

[15] H. Zhong, et al., "Interfacial Properties of Monolayer and Bilayer MoS$_2$ Contacts with Metals: Beyond the Energy Band Calculations", Sci. Rep., 6, 21786 (2016).

[16] X. Chen, et al., "Probing the electron states and metal-insulator transition mechanisms in molybdenum disulphide vertical heterostructures", Nat. Commun. 6, 6088 (2015).

[17] D. B. M. Klaassen, "A Unified Mobility Model for Device Simulation—I. Model Equations and Concentration Dependence," Solid-State Electronics, vol. 35, no. 7, pp. 953–959 (1992).

[18] C. Kim, et al., "Fermi Level Pinning at Electrical Metal Contacts of Monolayer Molybdenum Dichalcogenides", ACS Nano 11 (2), 1588-1596 (2017).

[19] W. Cao, J. Kang, W. Liu, and K. Banerjee, "A compact current–voltage model for 2D semiconductor based field-effect transistors considering interface traps, mobility degradation, and inefficient doping effect," IEEE Trans. Electron Devices, vol. 61, no. 12, pp. 4282–4290, (2014).

[20] M. Lin, et al, "Thickness-dependent charge transport in few-layer MoS$_2$ field-effect transistors",Nanotechnology 27 165203 (2016).

[21] D. Lembke, A. Allain, A. Kis, "Thickness-dependent mobility in two-dimensional MoS$_2$ transistors",Nanoscale, 7, 6255–6260 (2015).

[22] G. Mirabelli, et al., "Back-gated Nb-doped MoS$_2$ junctionless field-effect-transistors", AIP Advances 6, 025323 (2016).

978-1-5090-5979-9/17 $31.00 © 2017 IEEE

Electron Mobility in Thin In$_{0.53}$Ga$_{0.47}$As Channel

E. Cartier, A. Majumdar, K.-T. Lee, T. Ando, M. M. Frank, J. Rozen, K. A. Jenkins, C. Liang, C.-W. Cheng, J. Bruley, M. Hopstaken, P. Kerber, J.-B. Yau, X. Sun, R. T. Mo, C.-C. Yeh, E. Leobandung, and V. Narayanan

IBM T. J. Watson Research Center
Yorktown Heights, NY 10598, USA
ecartier@us.ibm.com

Abstract — Channel thickness T_{CH} dependence of electron mobility μ_{EFF} in thin In$_{0.53}$Ga$_{0.47}$As channels was investigated at temperatures T from 35 to 300 K using conventional parametric and pulsed I_D-measurements, including a novel technique with time resolution down to 10 ns. It is show that accurate mobility measurements can be obtained using low T and/or fast pulsed measurements, thus avoiding significant underestimations of μ_{EFF} due to charge trapping with slow/parametric measurements. Furthermore, annealing is demonstrated to strongly suppress charge trapping, which results in μ_{EFF} = 1015 cm^2/Vs at T_{CH} = 7.1 nm, carrier density N_S = 3 × 10^{12} cm^{-2}, and T = 300 K. We demonstrate that room-temperature μ_{EFF} degrades by less than 10% as T_{CH} is scaled from 300 nm down to 7 nm, thus indicating that there is no "mobility bottleneck" down to T_{CH} = 7 nm.

Keywords—III-V FETs; In$_{0.53}$Ga$_{0.47}$As; channel thickness; electron mobility; charge trapping.

I. INTRODUCTION

III-V FETs are being investigated for future technology nodes because many III-V materials have higher electron mobility and higher electron velocity than Si [1]. In order to meet short-channel effect targets, one needs very thin channels [2]. It is well known that electron mobility degrades in thin SOI channels below ~ 5 nm SOI thickness [3]–[5]. Electron mobility degradation has also been observed in In$_{0.53}$Ga$_{0.47}$ channels below ~ 20 nm thickness [6], [7]. This suggests that the control of short-channel effects by thinning the III-V channel thickness might be difficult without substantial mobility loss.

In order to better understand the impact of channel thinning on electron mobility in III-Vs, we performed a systematic study of mobility μ_{EFF} in In$_{0.53}$Ga$_{0.47}$As channels down to channel thickness T_{CH} = 2.1 nm. We demonstrate that conventional parametric measurements can result in a significant underestimation of μ_{EFF} because of charge trapping, which overshadows the impact of channel thinning. We combine conventional parametric mobility and pulsed measurements, including a novel fast-pulsed I_D-V_G technique, with pulse times down to 10 ns, to understand and overcome the measurement limitations caused by fast charge trapping. Finally, we show that annealing leads to significant suppression of charge trapping, thereby yielding μ_{EFF} = 1015 cm^2/Vs at T_{CH} = 7.1 nm, T = 300 K, and carrier density N_S = 3 × 10^{12} cm^{-2} using the parametric method. In the optimized stacks, μ_{EFF} at T = 300 K is shown to degrade by less than 10% as T_{CH} is scaled from 300 nm down to 7 nm, thus indicating that there is no "mobility bottleneck" with channel thickness scaling down to T_{CH} = 7 nm.

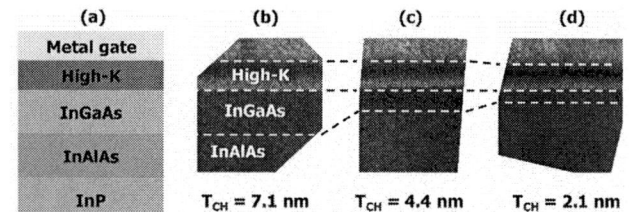

FIG. 1. (a) Layer structure of long-channel In$_{0.53}$Ga$_{0.47}$As/In$_{0.52}$Al$_{0.48}$As FETs on InP substrate. (b)-(d) High-resolution TEM images of fabricated devices with channel thickness T_{CH} = 7.1, 4.4, and 2.1 nm, respectively.

II. FABRICATION OF LONG-CHANNEL DEVICES

The device layers consist of In$_{0.53}$Ga$_{0.47}$As channels of variable thickness and In$_{0.52}$Al$_{0.48}$As barrier layers grown on InP substrates, as sketched in Fig. 1(a). High-resolution TEM cross-section images for three different channel thicknesses in fully processed devices are shown in Figs. 1(b)-1(d). The planar long-channel MOSFETs used in this work contain a high-κ gate dielectric stack with metal gate electrodes. The devices were fabricated using the fabrication scheme described in detail in Ref. [8]. The FET had a channel width, W, of 80 and 100 μm, and gate lengths, L_G, from 10 to 50 μm.

III. RESULTS AND DISCUSSIONS

A. Room-Temperature Data

The measured channel thickness T_{CH} dependence of threshold voltage V_T at T = 300 K is summarized in Fig. 2(a). As expected, V_T increases sharply below T_{CH} ~ 10 nm due to the effects of quantization [3]. Typical room temperature measurements of μ_{EFF} vs. N_S are shown in Fig. 2(b), revealing a strong T_{CH} dependence as summarized in Fig. 3 at N_S = 3 × 10^{12} cm^{-2}. We observe that μ_{EFF} degrades significantly when T_{CH} is scaled to below ~ 15 nm at T = 300 K, a result that is similar to previous reports [6], [7].

The data shown in Fig. 2(b) and Fig. 3 were extracted from parametric drain current vs. gate voltage I_D-V_G characteristics at drain bias V_D = 50 mV and split capacitance vs. gate voltage, C_G-V_G characteristics at 1 MHz [8]. The two DC sweeps were executed sequentially at a similar gate volte ramp rate. To track V_T stability a second I_D-V_G measurement was performed after the C_G-V_G sweep, showing minimal V_T shifts and suggesting gate stack stability. However, as we will show below, this frequently-used stability test using sequential DC I_D-V_G measurements is not sufficient because both charge trapping and charge de-

978-1-5090-5979-9/17 $31.00 © 2017 IEEE

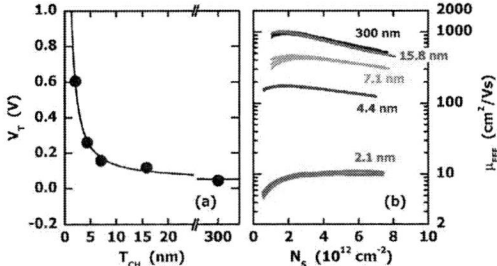

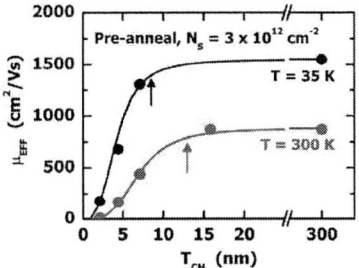

FIG. 2. Room-temperature $In_{0.53}Ga_{0.47}As$ channel data. (a) Threshold voltage V_T vs. channel thickness T_{CH}, and (b) mobility μ_{EFF} vs. carrier density N_S for T_{CH} in the 2.1 to 300 nm range.

FIG. 3. Mobility μ_{EFF} vs. channel thickness T_{CH} of $In_{0.53}Ga_{0.47}As$ channels at carrier density $N_S = 3 \times 10^{12}$ cm^{-2} and temperature $T = 35$ and 300 K. The arrows indicate where μ_{EFF} drops by 10% of its thick-channel value.

trapping/recovery from the gate stack in III-V FETs can be fast compared to the times required for parametric DC sweep sequences. We will also show that even measurement procedures where I_D and C_G are alternatively measured in a single V_G-sweep can be prone to large errors in extracted mobility values due to fast charge trapping.

B. Low-Temperature Data

In order to understand the origin of μ_{EFF} degradation for thinner T_{CH}, we first measured the temperature dependence of μ_{EFF} from $T = 300$ K down to $T = 35$ K. Typical data of μ_{EFF} vs. N_S are shown in Figs. 4(a)-(c) for $T_{CH} = 300$, 7.1, and 2.1 nm, respectively. Mobility μ_{EFF} at $T = 35$ K and $N_S = 3 \times 10^{12}$ cm^{-2} is compared to data at $T = 300$ K in Fig. 3. As can be seen, not only is scattering suppressed at low T but also the onset of μ_{EFF} degradation upon T_{CH} scaling is pushed down to below ~ 8 nm, as indicated by the arrows. To further quantify the change in μ_{EFF} with T, we plot (i) normalized mobility $\mu_{EFF}(T)/\mu_{EFF}(T=35K)$ vs. T in Fig. 4(d), and (ii) $\mu_{EFF}(T=300K)/\mu_{EFF}(T=35K)$ ratio vs. T_{CH} in Fig. 4(e), both at $N_S = 3 \times 10^{12}$ cm^{-2}. These data demonstrate that the temperature dependence of μ_{EFF} sharply increases with decreasing T_{CH}. The larger T-dependence at thinner T_{CH} could be due to: (i) measurement errors caused by charge trapping, (ii) scattering from trapped charge (more prevalent at higher T) that gets amplified at thinner T_{CH} due to larger surface N_S [9], or (iii) the effect of confined phonons [10]. In the following section, we show that (i) is a significant contributor to these trends.

C. Impact of Charge Trapping

We investigated the impact of charge trapping using pulsed I_D-V_G measurements. In Fig. 5(a), we compare pulsed and DC I_D-V_G characteristics for $T_{CH} = 7.1$ nm. The pulsed I_D-V_G data

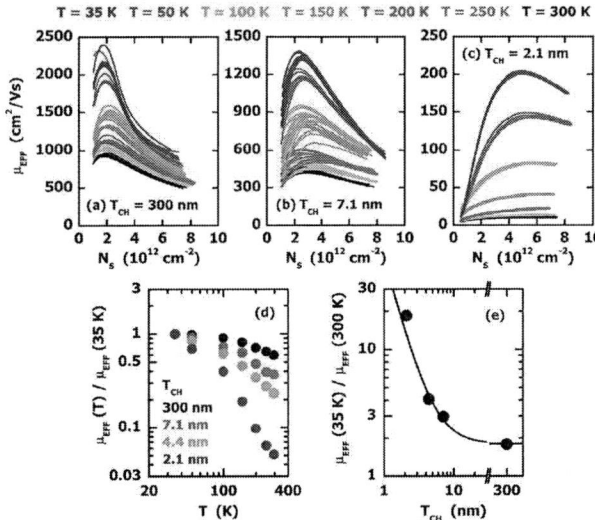

FIG. 4. $In_{0.53}Ga_{0.47}As$ channel data for thickness T_{CH} in the 2.1 to 300 nm range. (a)-(c) Mobility μ_{EFF} vs. carrier density N_S at temperature T in the 35 to 300 K range. (d) Change in mobility with T defined by $\mu_{EFF}(T)/\mu_{EFF}(T = 35$ K) vs. T at $N_S = 4 \times 10^{12}$ cm^{-2}. (e) $\mu_{EFF}(T = 35$ K)/$\mu_{EFF}(T = 300$ K) vs. T_{CH} at $N_S = 4 \times 10^{12}$ cm^{-2}.

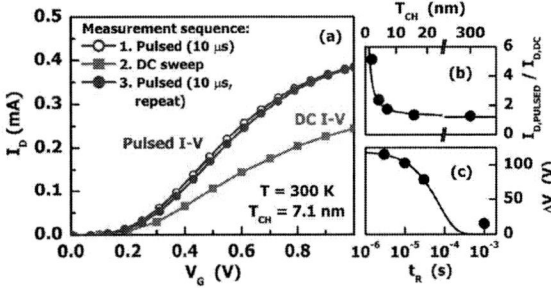

FIG. 5. Room-temperature (a) DC and pulsed I_D-V_G data for $In_{0.53}Ga_{0.47}As$ channel thickness $T_{CH} = 7.1$ nm, (b) $I_{D,pulsed}/I_{D,DC}$ ratio vs. T_{CH}, and (c) threshold voltage shift ΔV_T vs. recovery time t_R after a 10 μs stress at $V_G = 1$ V.

was acquired during the rising edge of a trapezoidal gate pulse with rise time $t_R = 10$ μs and at $V_{DS} = 50$ mV using Agilent B1530A pulsed measurement unit (in the following called pulsed-I_D) while the parametric measurements were performed with Agilent B1511A source-measure units (SMUs) using a step and measure sequence with 30 ms per step (in the following called DC-I_D). It is clear that the pulsed measurements yield significantly larger I_D. Furthermore, both pulsed and DC measurements are found to be reproducible when repeated. Thus, subsequent measurements, as performed here, can yield identical I_D measurements for a given procedure but with vastly different absolute I_D values. This surprising result can be understood when examining the V_T instability of the FET using stress-and-sense procedures that are commonly used to study the positive bias temperature instability (PBTI) of Si-based transistors.

The data in Fig. 5(b) show V_T recovery after a short gate pulse in a FET with $T_{CH} = 7.1$ nm. As can be seen, charge recovery was found to be fast and on a 100 μs time scale, explaining the observations of Fig. 5(a). The observation of fast relaxation is also important in the context of the novel ultrafast,

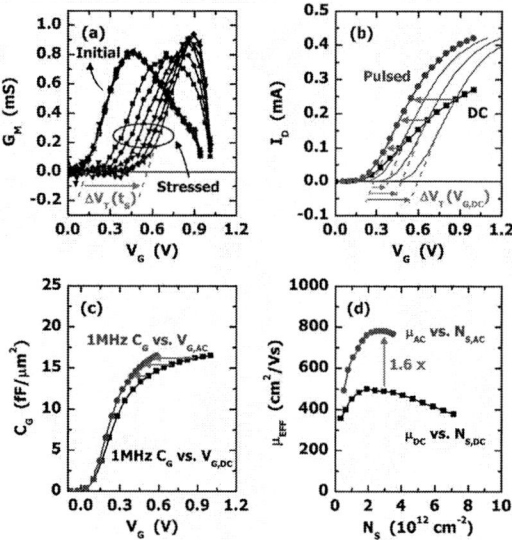

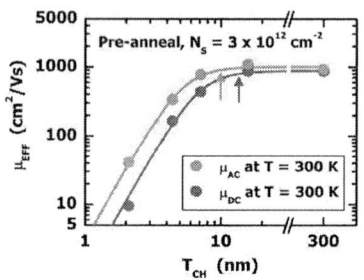

FIG. 7. Mobility μ_{EFF} vs. channel thickness T_{CH} of $In_{0.53}Ga_{0.47}As$ channels at carrier density $N_S = 3 \times 10^{12}$ cm^{-2}. The data sets are (i) μ_{DC} at temperature $T = 300$ K, (ii) μ_{AC} at $T = 300$ K, and (iii) μ_{DC} at $T = 35$ K, where μ_{DC} and μ_{AC} are defined in Sec. III.D and in the caption of Fig 6. The arrows indicate where μ_{EFF} drops by 10% of its thick-channel value.

FIG. 6. Room-temperature data for $In_{0.53}Ga_{0.47}As$ channel with thickness $T_{CH} = 7.1$ nm. (a) Pulsed linear trans-conductance G_M vs. V_G for stress times $t_S = 1$ μs to 100 ms (10 × steps) showing only threshold voltage shift ΔV_T with stress at $V_G = 1$ V. (b) Schematic demonstrating pulsed I_D-V_G curves (1 μs rise time) with V_T shifts and the resultant I_D-V_G curve during DC sweep. (c) Measured C_G-$V_{G,DC}$ curve (same sweep rate as DC I_D-V_G) and corrected C_G-$V_{G,AC}$ curve after accounting for V_T instability. (d) DC mobility μ_{DC} from DC I_D-V_G vs. DC carrier density $N_{S,DC} = \int C_G . dV_{G,DC}$ and AC mobility μ_{AC} from pulsed I_D-V_G vs. AC carrier density $N_{S,AC} = \int C_G . dV_{G,AC}$. Eliminating the artifact of V_T instability leads to 1.6 × higher mobility at $N_S = 3 \times 10^{12}$ cm^{-2}.

10 ns, I_D-measurements discussed later. Furthermore, we found that the $I_{D,pulsed}/I_{D,DC}$ ratio increases at thinner T_{CH}, as shown in Fig. 5(c). This indicates that either more charge trapping occurs in thinner T_{CH} or that thin channel devices are more prone to mobility degradation at the same amount of trapped charge. It is important to note that similar measurements carried out at low temperatures $T < 50$ K yield identical results for DC and pulsed I_D-V_G characteristics, demonstrating that charge trapping is indeed strongly suppressed at low T. This supports the notion that charge trapping causes significant errors in mobility extraction using conventional parametric measurements at room temperature while low-temperature measurements are not impacted by V_T-instability as charge trapping is suppressed.

The impact of charge trapping was investigated in detail by performing charge trapping experiments using the pulsed method. In these measurements, a trapezoidal pulse with rise and fall time $t_R = t_F = 10$ μs was used and a hold time t_H was used to stress the device for stress time $t_S = t_H$ from 1 μs to 100 ms. The pre- and post-stress pulsed I_D-V_G data was acquired during the rising and falling edge of the pulse. From these data, we calculate linear transconductanc $G_{MLIN} = dI_D/dV_G$. Typical data for a stress voltage $V_G = 1$ V at $T = 300$ K are shown in Fig. 6(a). It is clear that V_G stress causes an increasingly larger threshold voltage shift ΔV_T with increasing stress time.

D. Extraction of True Mobility using Pulsed Measurements

The observations in Fig. 6(a) provide a method for extracting the true (trap-fee) mobility from the pulsed I_D-V_G and a corrected C_G-$V_{G,AC}$ characteristics, as follows. The data of Fig. 6(a) implies

that charge trapping effectively leads to a stretch-out of the V_G-axis during DC I_D-V_G sweeps. As sketched in Fig. 6(b), each DC measurement corresponds to I_D measured at increasingly higher V_T, shifted by $\Delta V_T(V_{G,DC}) = Q_{TR}/C_G$, where Q_{TR} is the trapped charge. Since the C_G-V_G characteristics are measured at the same sweep rate as the DC-I_D characteristics (and because full recovery occurs in sequential measurements), it is now possible to correct the C_G-$V_{G,DC}$ as shown in Fig. 6(c) to reconstruct a (trap-free) C_G-$V_{G,AC}$ curve corresponding to the pulsed I_D-V_G characteristics shown in Fig. 6(b). Using the pulsed I_D-V_G data to obtain channel sheet conductance $G_{CH,AC}$ and the corrected C_G-$V_{G,AC}$ data to obtain the corrected carrier density $N_{S,AC} = \int C_G . dV_{G,AC}$, we arrive at the true or AC mobility denoted as μ_{AC} and given by $\mu_{AC} = G_{CH,AC}/qN_{S,AC}$, where q is the charge of an electron. As shown in Fig. 6(d), the true AC mobility μ_{AC} is significantly larger than the mobility obtained by parametric measurements (referred to below as μ_{DC}). The mobility results using the above-mentioned method are summarized in Fig. 7. The data sets are (i) μ_{DC} at $T = 300$ K, (ii) μ_{AC} at $T = 300$ K, and (iii) μ_{DC} at $T = 35$ K, where μ_{DC} and μ_{AC} are defined above and in the caption of Fig 6. The arrows indicate the channel thickness, at which μ_{AC} or μ_{DC} decreases by 10% of its thick-channel value. The μ_{AC}/μ_{DC} ratio increases with decreasing T_{CH}, indicating that mobility underestimation using the conventional DC method is larger at thinner T_{CH}.

E. Effect of Anneals

It has been shown that charge trapping in high-k gate stacks in Si technologies can be reduced by annealing. By employing similar anneals, charge trapping can also by suppressed in the gate stacks used here on $In_{0.53}Ga_{0.47}As$ channels. As shown in Fig. 8, annealing leads to substantial enhancement of μ_{DC} for thin channels as compared to the pre-annealed values. It is interesting to note that post-anneal μ_{DC} values even exceeds pre-anneal μ_{AC} values substantially, as shown in Fig. 9. We conclude from this observation that a true mobility enhancement (not related to V_T shift) is achieved by annealing. This conclusion is confirmed by ultra-fast I_D-measurements using a novel pulse-train method, as depicted in Fig. 10(a). In the pulse-train method, the duty cycle is kept constant at 1 %, allowing for long charge relaxation times after each on-pulse. Furthermore, a low-pass filter is used to integrate the charge over many pulses to effectively convert the pulsed I_D measurement into a DC current

978-1-5090-5979-9/17 $31.00 © 2017 IEEE

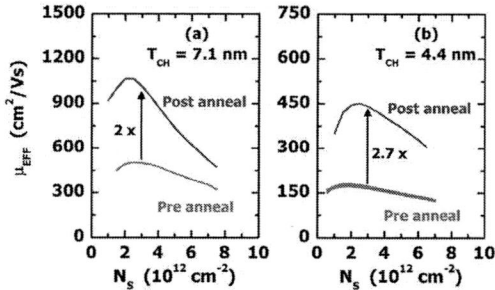

FIG. 8. Pre- and post-anneal room-temperature mobility μ_{EFF} (obtained from DC I_D-V_G) vs. carrier density N_S of InGaAs channels with thickness (a) T_{CH} = 7.1 nm and (b) T_{CH} = 4.4 nm. At $N_S = 3 \times 10^{12}$ cm^{-2}, advanced anneals lead to 2 × and 2.7 × higher mobility at T_{CH} = 7.1 nm and 4.4 nm, respectively.

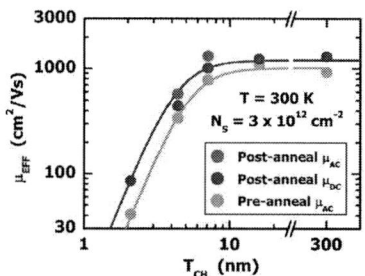

FIG. 9. Mobility μ_{EFF} vs. channel thickness T_{CH} at carrier density $N_S = 3 \times 10^{12}$ cm^{-2} and temperature T = 300 K. The data sets are (i) pre-anneal μ_{AC}, (ii) post-anneal μ_{DC}, and (iii) post-anneal μ_{AC}, where μ_{DC} and μ_{AC} are defined in Sec. III.D and in the caption of Fig 6. For thin T_{CH}, $\mu_{EFF} \propto (T_{CH})^n$ with $n \sim 3.5$.

given by $I_{D,DC} = I_D \times$ duty cycle. As seen in Fig. 10(b), after anneal, I_D is strongly enhanced at short pulse widths and the dependence on pulse width is reduced due to reduced charge trapping.

We performed the same anneals also on MOS capacitors on n-type substrates with identical gate stack. C_G-V_G hysteresis, conductance G_P-V_G, and charge pumping measurements were used to monitor the impact of annealing on slow oxide traps (PBTI-type), interface-states, and "border traps" in the conduction band, respectively. All three types of traps were found to be reduced, contributing to the observed mobility enhancement.

IV. CONCLUSIONS

In conclusion, we have shown that charge trapping leads to significant under-estimation of μ_{EFF} due to V_T-instability. We overcame this limitation by using low T and/or fast pulsed measurements. Both methods are shown to strongly suppress the charge trapping in the gate dielectric, leading to accurate mobility measurements. We also demonstrated that annealing not only suppressed charge trapping but also reduced the interface-state density, thereby leading to a true mobility enhancement. We reported μ_{EFF} = 1015 cm^2/Vs at T_{CH} = 7.1 nm, $N_S = 3 \times 10^{12}$ cm^{-2}, and T = 300 K using conventional parametric measurements in annealed FETs with InGaAs channels. Room-temperature mobility was shown to degrade by less than 10% as the channel thickness was scaled down from 300 nm to 7 nm, thus demonstrating that $T_{CH} \sim 7$ nm can be used to meet short-channel effect targets without severe mobility penalty.

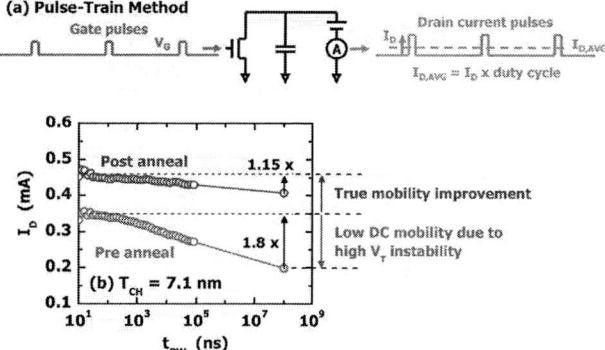

FIG. 10. (a) Illustration of a pulse-train method to measure drain currents in FETs at high speed in large area FETs. The duty cycle is kept constant at 1 %, allowing for long charge relaxation times after each on-pulse. A low-pass filter is used to integrate the drain current over many pulses to boost the measurement sensitivity. (b) Averaged drain current, $I_{D,AVG}$ vs. pulse width, t_{PW}, for InGaAs channels with thickness T_{CH} = 7.1 nm before and after advanced anneal at room temperature.

ACKNOWLEDGMENT

This work was performed by the IBM/Samsung alliance teams. We thank the staff of the IBM MRL for device fabrication. We also thank T. C. Chen and M. Khare for encouragement and management support.

REFERENCES

[1] J. A. del Alamo, "Nanometre-scale electronics with III-V compound semiconductors," *Nature*, vol. 479, pp. 317–323, Nov. 2011.

[2] W. Haensch *et al.*, "Silicon CMOS devices beyond scaling," *IBM J. Res. & Dev.*, vol. 50, pp. 339–361, Jul. 2006.

[3] K. Uchida, J. Koga, R. Ohba, T. Numata, and S. Takagi, "Experimental evidences of quantum-mechanical effects on low-field mobility, gate-channel capacitance, and threshold voltage of ultrathin body SOI MOSFETs," in *Proc. IEDM*, Dec. 2001, pp. 633–636.

[4] K. Uchida *et al.*, "Experimental study on carrier transport mechanism in ultrathin-body SOI n- and p-MOSFETs with SOI thickness less than 5 nm," in *Proc. IEDM*, Dec. 2002, pp. 47–50.

[5] K. Uchida and S. Takagi, "Carrier scattering induced by thickness fluctuation of silicon-on-insulator film in ultrathin-body metal-oxide-semiconductor field-effect transistors," *Appl. Phys. Lett.*, vol. 82, pp. 2916–2918, Apr. 2003.

[6] M. Yokoyama *et al.*, "Extremely-thin body InGaAs-on-insulator MOSFETs on Si fabricatd by direct wafer bonding," in *Proc. IEDM*, Dec. 2010, pp. 46–49.

[7] A. Alian *et al.*, "Impact of the channel thickness on the performance of ultrathin InGaAs channel MOSFET devices," in *Proc. IEDM*, Dec. 2013, pp. 437–440.

[8] A. Majumdar *et al.*, "A four-FET method for extracting mobility in FETs without field oxide," *IEEE Trans. Electron Devices*, vol. 61, pp. 3833–3837, Nov. 2014.

[9] R. Hatcher and C. Bowen, "A comparison of the carrier density at the surface of quantum wells for different crystal orientations of silicon, gallium arsenide, and indium arsenide," *Appl. Phys. Lett.*, vol. 103, art. 162107, Oct. 2013.

[10] R. Kotlyar, B. Obradovic, P. Matagne, M. Stettler, and M. D. Giles, "Assessment of room-temperature phonon-limited mobility in gated silicon nanowires," *Appl. Phys. Lett.*, vol. 84, pp. 5270–5272, Jun. 2004.

Understanding of slow traps generation in plasma oxidation GeO$_x$/Ge MOS interfaces with ALD high-k layers

Mengnan Ke*, Mitsuru Takenaka and Shinichi Takagi

Department of Electrical Engineering and Information Systems, The University of Tokyo, 7-3-1 Hongo, Bunkyo-ku, Tokyo 113-8656, Japan, Phone: & Fax: +81-3-5841-6733,

*E-mail: kiramn@mosfet.t.u-tokyo.ac.jp

Abstract— A small amount of slow trap density in Ge gate stacks is a crucial issue for Ge CMOS, in addition to thin equivalent oxide thickness and low interface state density. In this paper, we study the slow trap position and the generation mechanism in the high-k/GeO$_x$/Ge interfaces fabricated by plasma oxidation. The slow trap density in Al$_2$O$_3$/GeO$_x$/Ge interfaces by plasma pre-oxidation is compared with different GeO$_x$ and Al$_2$O$_3$ thickness, and the interfaces by plasma post oxidation (PPO). ALD Al$_2$O$_3$, HfO$_2$, La$_2$O$_3$ and Y$_2$O$_3$/GeO$_x$/Ge MOS interfaces are compare in terms of the slow trap density. It is found the main slow traps in Al$_2$O$_3$/GeO$_x$/Ge can locate near the GeO$_x$/Ge interfaces for electrons and the Al$_2$O$_3$/GeO$_x$ interface for holes. It is also found that additional slow traps near conduction band side are generated during the PPO process. It is revealed that slow traps are generated by post high-k ALD, while Al$_2$O$_3$ provides the lowest slow trap density in comparison with Y$_2$O$_3$, HfO$_2$ and La$_2$O$_3$ high-k films.

Keywords—Germanium, MOS interface, High-k material, slow trap density

I. INTRODUCTION

Ge is an attracting channel material for next generation MOSFETs because of the higher electron and hole mobility than Si. Here, the GeO$_2$/Ge structure is one of the most promising Ge surface passivation layers because a low interface trap density (D$_{it}$) has been expected. [1-4] Thus, as one of the realistic gate stacks, we have proposed and demonstrated high-k/GeO$_x$/Ge structures realized by plasma post oxidation (PPO), such as Al$_2$O$_3$/GeO$_x$/Ge and HfO$_2$/Al$_2$O$_3$/GeO$_x$/Ge, which have been shown to have 1 nm or thinner equivalent oxidation thickness (EOT) and low D$_{it}$ of ~10^{11} eV^{-1}cm^{-2} [5,6]. However, a large amount of slow traps included in these gate stacks is one of the remaining most critical issue [7-9].

While any defects in Al$_2$O$_3$ have been reported to be responsible for this slow trapping [9], we have recently found that the slow traps could exist inside GeO$_x$ formed by PPO [10]. However, the physical origin of the slow trap generation has not been understood yet. One possible reason for slow trap generation might be any inter-diffusion of Al atoms during PPO. In order to judge the relevance of this mechanism, we compare the interface properties of the Al$_2$O$_3$/GeO$_x$/Ge interfaces prepared by post Al$_2$O$_3$ atomic layer deposition (ALD) [11, 12] on

plasma oxidation GeO$_x$/Ge with the one by PPO, in this study. In addition, the high-k materials deposited on the GeO$_x$/Ge interfaces are varied from Al$_2$O$_3$, HfO$_2$, La$_2$O$_3$ to Y$_2$O$_3$, in order to examine the influence of high-k materials on the generation of interface defects including D$_{it}$ and slow trap density (ΔN$_{st}$). We evaluate the C-V curves, D$_{it}$ and ΔN$_{st}$ of MOS capacitors, where ALD high-k films are deposited on plasma-oxidized GeO$_x$/n- and p-Ge substrates with changing the GeO$_x$ thickness and the high-k materials from Al$_2$O$_3$, HfO$_2$, La$_2$O$_3$ to Y$_2$O$_3$.

II. DEVICE FABRICATION

(100) Ge wafers were cleaned by de-ionized water, acetone and HF. After the pre-cleaning, plasma pre-oxidation was performed by using ECR plasma of Ar (9 sccm) and O$_2$ (3 sccm) at 300 °C under 650 W microwave power for 1-10 s. Subsequently, 1.5 to 2.4-nm-thick Al$_2$O$_3$ or 10-nm-thick Al$_2$O$_3$, HfO$_2$, La$_2$O$_3$ and Y$_2$O$_3$ were deposited at 300°C by ALD. Post deposition annealing (PDA) was performed for 30 min at 400 °C in N$_2$ ambient, followed by formation of 100-nm-thick Au gate electrodes and 100-nm-thick Al back contacts by thermal evaporation. The process flow and the structure of MOS capacitors are shown in Fig. 1. D$_{it}$ is evaluated by the conductance method. Also, ΔN$_{st}$ is evaluated from the hysteresis in the C-V sweep as a function of the effective oxide field (E$_{ox}$). Here, E$_{ox}$ and ΔN$_{st}$ are determined by [5]

$$E_{ox} = (V_G - V_{FB})/CET$$
$$q \, \Delta N_{st} = C_{ox} \, \Delta V_{hys}$$

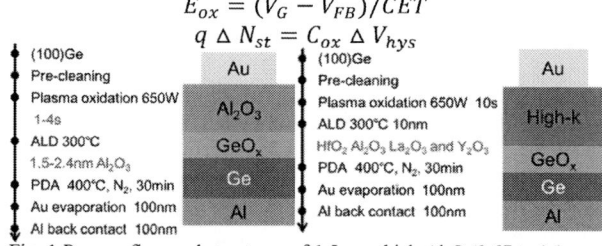

Fig. 1 Process flow and structures of 1.5-nm-thick Al$_2$O$_3$/0.67 to 1.1-nm-thick GeO$_x$/Ge MOS interfaces and 10-nm-thick high-k/1.4-nm-thick GeO$_x$/Ge MOS interfaces.

III. ELECTRICAL PROPERTIES AND DISCUSSIONS

First, we evaluate D$_{it}$ of the Al$_2$O$_3$/GeO$_x$/Ge MOS interfaces with plasma pre-oxidation. The energy distributions of D$_{it}$ are shown in Fig. 2. It is found that D$_{it}$

978-1-5090-5979-9/17 $31.00 © 2017 IEEE

is almost in a same level from 4 to 10×10^{11} eV^{-1}cm^{-2} among the MOS capacitors with the different thickness of Al$_2$O$_3$ and GeO$_x$.

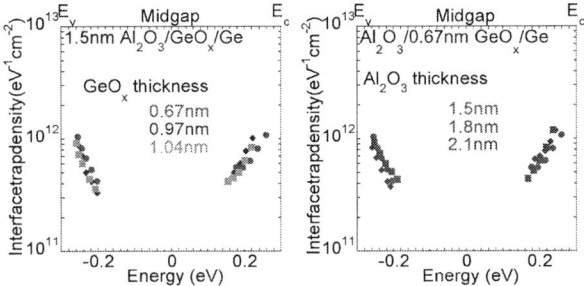

Fig. 2 D$_{it}$ of Al$_2$O$_3$/GeO$_x$/Ge with plasma pre-oxidation.

Fig. 3 shows ΔN_{st} of 1.5 to 2.4-nm-thick Al$_2$O$_3$/0.67-nm-thick GeO$_x$/Ge with plasma pre-oxidation. Here, the GeO$_x$ film thickness is measured by ellipsometry. The variation of ΔN_{st} with changing the Al$_2$O$_3$ thickness is small for both the Al$_2$O$_3$/GeO$_x$/n- and the p-Ge MOS interfaces, meaning that traps inside the Al$_2$O$_3$ layers are not working as the slow traps in the observed hysteresis.

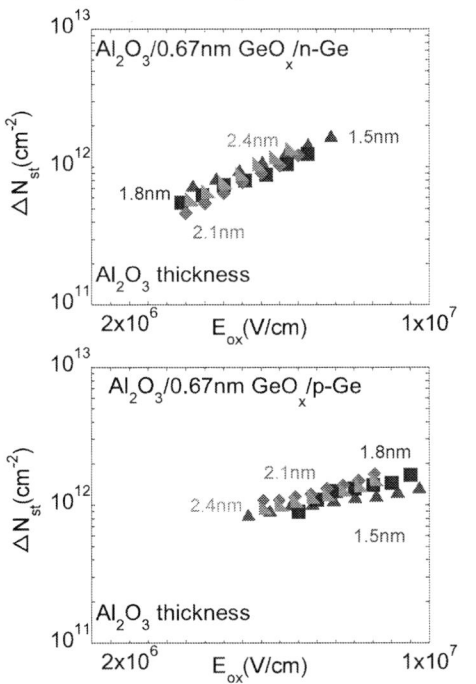

Fig. 3 ΔN_{st} of 1.5 to 2.4-nm-thick Al$_2$O$_3$/0.67-nm-thick GeO$_x$/Ge with plasma pre-oxidation.

Fig. 4 shows ΔN_{st} of 1.5-nm-thick Al$_2$O$_3$/GeO$_x$/Ge with plasma pre-oxidation as a parameter of the GeO$_x$ thickness. It is found that the Al$_2$O$_3$/GeO$_x$/p-Ge MOS interface has the GeO$_x$ thickness variation of ΔN_{st}, while the Al$_2$O$_3$/GeO$_x$/n-Ge MOS interface has no dependence. This phenomenon indicates that slow traps for electrons and holes exist in different positions in the Al$_2$O$_3$/GeO$_x$/Ge MOS interfaces. Fig. 5 shows ΔN_{st} of the 1.5-nm-thick Al$_2$O$_3$/1.04-nm-thick GeO$_x$/n- and p-Ge MOS interfaces by plasma pre-oxidation. Much lower slow trap density for holes than that for electrons is observed.

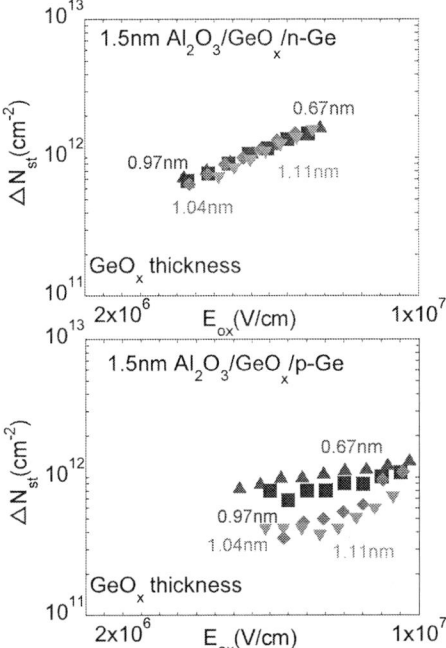

Fig. 4 ΔN_{st} of 1.5-nm-thick Al$_2$O$_3$/GeO$_x$/Ge with plasma pre-oxidation time from 1 to 4s.

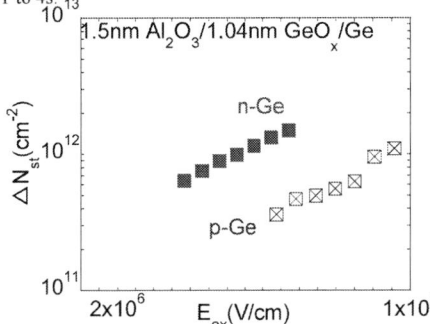

Fig. 5 ΔN_{st} of 1.5-nm-thick Al$_2$O$_3$/1.04-nm-thick GeO$_x$/n- and p-Ge MOS interfaces by plasma pre-oxidation.

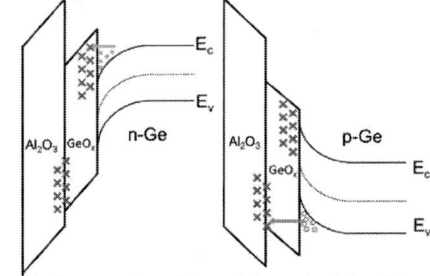

Fig. 6 schematic diagram of possible position and origin of slow traps of 1.5-nm-thick Al$_2$O$_3$/1.04-nm-thick GeO$_x$/n- and p-Ge MOS interfaces by plasma pre-oxidation.

Fig. 6 show a schematic diagram of possible locations of slow traps in the Al$_2$O$_3$/GeO$_x$/Ge MOS interfaces. The meaningful GeO$_x$ thickness dependence and almost no Al$_2$O$_3$ thickness dependence of ΔN_{st} for the p-Ge MOS capacitors suggest that slow traps for holes can locate around the Al$_2$O$_3$/GeO$_x$ interface. Also, the smaller hysteresis and smaller ΔN_{st} for the p-Ge capacitors indicate that slow trap density for holes is smaller than that for electrons. On the other hand, almost no GeO$_x$ and Al$_2$O$_3$ thickness dependencies of ΔN_{st} for the n-Ge

978-1-5090-5979-9/17 $31.00 © 2017 IEEE   297

capacitors suggest that slow traps for electrons can locate near the GeO$_x$/Ge MOS interfaces.

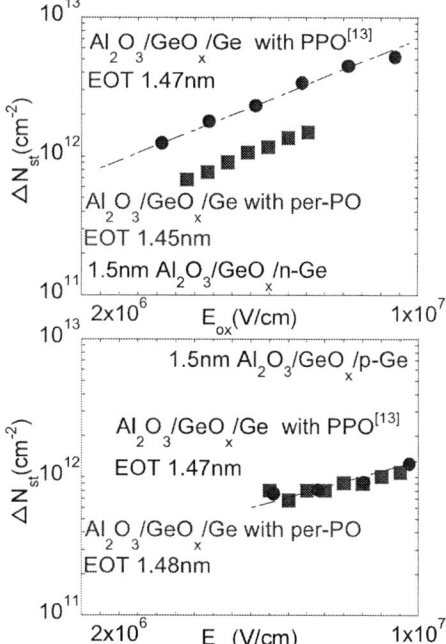

Fig. 7 ΔN_{st} of 1.5-nm-thick Al$_2$O$_3$/GeO$_x$/Ge with plasma post or pre-oxidation with an EOT ~1.5-nm-thick

ΔN_{st} in the Al$_2$O$_3$/GeO$_x$/n- and p-Ge MOS interfaces is compared between plasma pre- and post-oxidation [13] for GeO$_x$ formation under a same EOT. The result is shown in Fig. 7. It is found that ΔN_{st} is significantly higher for plasma post-oxidation than that for plasma pre-oxidation. This result demonstrates that slow traps near the Ge conduction band side are additionally generated during the PPO process, which can be interpreted that the origin of additional slow traps during PPO is attributable to any reaction of Al$_2$O$_3$ and GeO$_x$, and/or inter-diffusion of Al and Ge. It is also found that no additional traps generate near the Ge valence band side.

Next, D$_{it}$ and ΔN_{st} of high-k/GeO$_x$/n- and p-Ge MOS interfaces were evaluated, in order to examine the influence of specific high-k materials on the generation of fast and slow interface traps. Here, Al$_2$O$_3$ was replaced by Y$_2$O$_3$, HfO$_2$ and La$_2$O$_3$, which were also deposited on plasma oxidation GeO$_x$/Ge by ALD.

Fig. 8 shows the C-V curves of 10-nm-thick Al$_2$O$_3$, HfO$_2$, La$_2$O$_3$ and Y$_2$O$_3$/1.4-nm-thick GeO$_x$/n- and p-Ge with plasma oxidation prior to high-k deposition. The C-V curve of Al$_2$O$_3$/GeO$_x$/n-Ge has larger V$_{fb}$ shift than those of the other high-k materials, meaning that more negative charges are included in ALD Al$_2$O$_3$ films. It is observed, however, that the hysteresis of Al$_2$O$_3$/GeO$_x$/n- or p-Ge is significantly smaller than HfO$_2$, La$_2$O$_3$ and Y$_2$O$_3$ on the GeO$_x$/Ge system. Also, La$_2$O$_3$/GeO$_x$/Ge MOS capacitors have the largest capacitance value among all of the capacitors.

Fig. 9 shows the energy distributions of D$_{it}$ of 10-nm-thick Al$_2$O$_3$, HfO$_2$, La$_2$O$_3$ and Y$_2$O$_3$/1.4-nm-thick GeO$_x$/n- and p-Ge MOS capacitors prepared by plasma pre-oxidation. It is found that D$_{it}$ is nearly in a similar level, though HfO$_2$ has slightly higher D$_{it}$. This result indicates

the effectiveness of 1.4-nm-thick GeO$_x$/Ge interfaces formed by plasma pre-oxidation in terms of the D$_{it}$ increase against high-k film ALD.

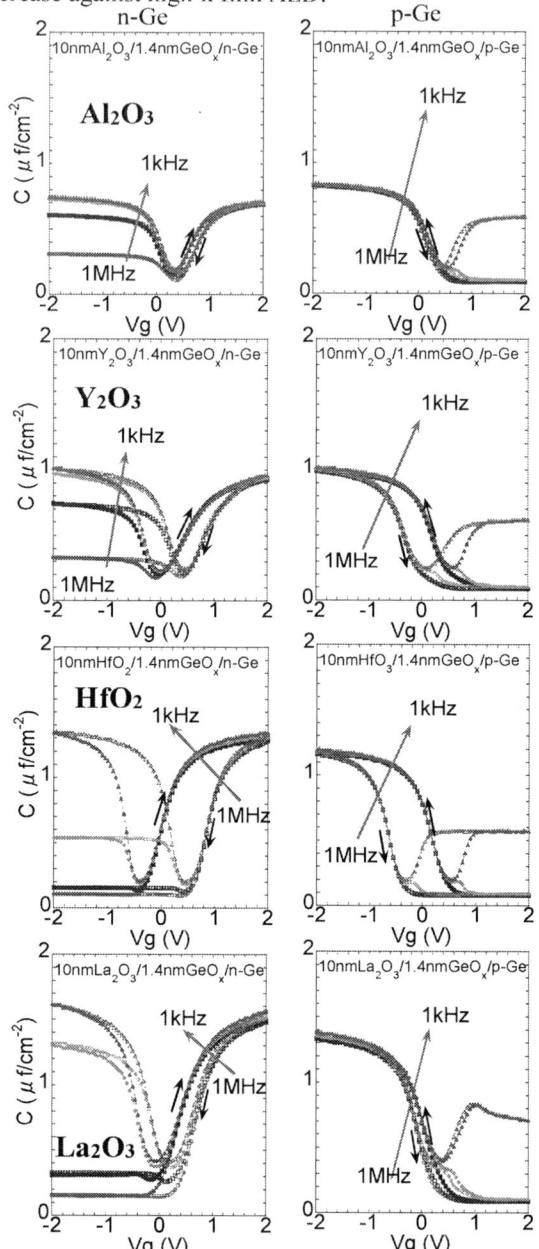

Fig. 8: C-V curses of 10-nm-thick Al$_2$O$_3$, HfO$_2$, La$_2$O$_3$ and Y$_2$O$_3$/1.4-nm-thick GeO$_x$/Ge with plasma pre-oxidation

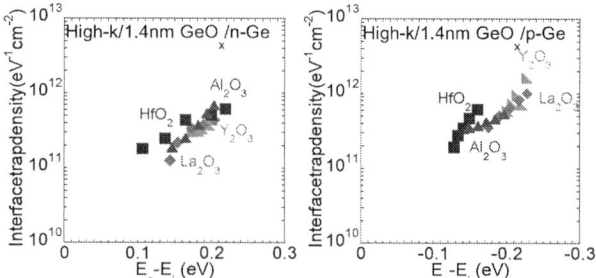

Fig.9: D$_{it}$ of 10-nm-thick Al$_2$O$_3$, HfO$_2$, La$_2$O$_3$ and Y$_2$O$_3$/1.4-nm-thick GeO$_x$/Ge with plasma pre-oxidation

Fig. 10 shows ΔN_{st} of 10-nm-thick Al_2O_3, HfO_2, La_2O_3 and Y_2O_3/1.4-nm-thick GeO_x/Ge with plasma pre-oxidation as a function of E_{ox}. It is found that ΔN_{st} of the Al_2O_3/GeO_x/n- or p-Ge MOS interfaces is significantly lower than those of the other high-k/GeO_x/Ge MOS interfaces. These results indicate that the high-k film ALD can cause the generation of slow traps in GeO_x/Ge or high-k/GeO_x interfaces, which is attributable to intermixing or diffusion of high-k species. Also, the weaker reaction and inter-diffusion is suggested for the Al_2O_3/GeO_x system in comparison with the other high-k materials. On the other hand, La_2O_3/GeO_x/Ge structure provides a different behavior in ΔN_{st}-E_{ox} relation, where ΔN_{st} does not increase or even decreases with increasing E_{ox}, resulting in the smaller hysteresis in higher E_{ox}. The similar properties have been reported in La_2O_3/InGaAs gate stacks [14].

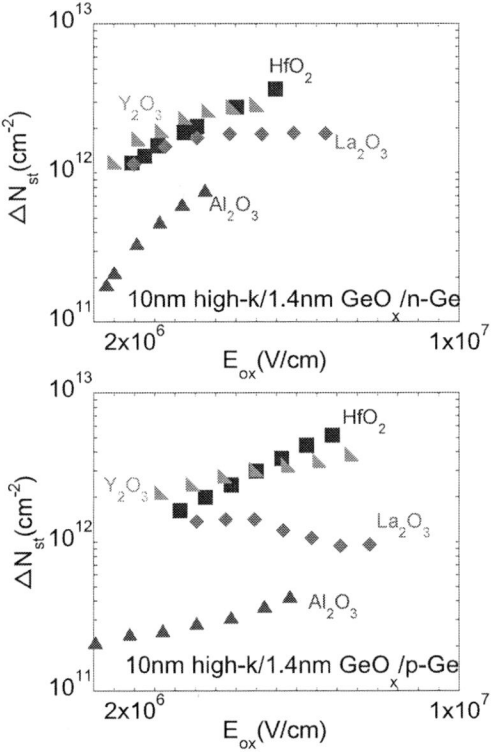

Fig.10: ΔN_{st} of 10-nm-thick Al_2O_3, HfO_2, La_2O_3 and Y_2O_3/1.4-nm-thick GeO_x/Ge with plasma pre-oxidation

IV. CONCLUSION

We studied the location and the possible physical origin of slow trap generation in high-k/GeO_x/Ge MOS interfaces through the comparison between post- and pre-oxidation by ECR plasma. It was found, as a consequence, that the main slow traps in the Al_2O_3/GeO_x/Ge interfaces can locate near the GeO_x/Ge interfaces for electrons and near the Al_2O_3/GeO_x interfaces for holes. We also have revealed that slow traps are additionally generated during the PPO process near conduction band side. It was also found that post high-k ALD creates a large amount of slow traps for the GeO_x/Ge interfaces, while Al_2O_3 provides the lowest slow trap density among the Al_2O_3, Y_2O_3, HfO_2, and La_2O_3 films.

ACKNOWLEDGEMENTS

This work was partly supported by JST CREST Grant Number JPMJCR1332, Japan, and a Grant-in-Aid for Scientific Research (No. 26249038) from MEXT.

REFERENCES

[1] D. Kuzum, T. Krishnamohan, A. J. Pethe, A. K. Okyay, Y. Oshima, Y. Sun, J. P. McVittie, P. A. Pianetta, P. C. McIntyre, and K. C. Saraswat, "Ge interface engineering with ozone oxidation for low interface state density", IEEE Electron Device Lett, vol. 29, no. 4, pp. 328-330, Apr. 2008.

[2] H. Matsubara, T. Sasada, M. Takenaka, and S. Takagi, "Evidence of low interface trap density in GeO₂/Ge metal-oxide-semiconductor structures fabricated by thermal oxidation", Appl Phys Lett, 93 (2008), 032104.

[3] T. Sasada, Y. Nakakita, M. Takenaka, and S. Takagi, "Surface orientation dependence of interface properties of GeO₂/Ge metal-oxide-semiconductor structures fabricated by thermal oxidation", J Appl Phys, 106 (2009), 073716

[4] Y. Fukuda, T. Ueno, and S. Hirono, "Electrical Behavior of Germanium Oxide/Germanium Interface Prepared by Electron-Cyclotron-Resonance Plasma Oxidation in Capacitance and Conductance Measurements", J. J Appl Phys, 44 (2005) 7928.

[5] R. Zhang, T. Iwasaki, N. Taoka, M. Takenaka, and S. Takagi, "Al_2O_3/GeO_x/Ge gate stacks with low interface trap density fabricated by electron cyclotron resonance plasma post oxidation", Appl. Phys. Lett., 98 (2011), 112902.

[6] R. Zhang, P.-C. Huang, J.-C. Lin, N. Taoka, M. Takenaka, and S. Takagi, "High-Mobility Ge p- and n-MOSFETs With 0.7-nm EOT Using HfO_2/Al_2O_3/GeO_x /Ge Gate Stacks Fabricated by Plasma Post oxidation", IEEE Trans. Electron Devices, 60 (2013) 927.

[7] R. Zhang, N. Taoka, P. Huang, M. Takenaka, and S. Takagi, "1-nm-thick EOT High Mobility Ge n- and p-MOSFETs with Ultrathin GeO_x/Ge MOS Interfaces Fabricated by Plasma Post Oxidation", IEDM Tech. Dig. (2011) 642.

[8] J. Franco, B. Kaczer, Ph. Roussel, J. Mitard, S, Sioncke, L. Witters, H. Mertens, T. Grasser, and G. Groeseneken, "Understanding the suppressed charge trapping in relaxed- and strained-Ge/SiO₂/HfO₂ pMOSFETs and implications for the screening of alternative high-mobility substrate/dielectric CMOS gate stacks", IEDM Tech. Dig. (2013) 397.

[9] G. Groeseneken, J. Franco, M. Cho, B. Kaczer, M. Toledano-Luque, Ph. Roussel, T. Kauerauf, A. Alian, J. Mitard, H. Arimura, D. Lin, N. Waldron, S. Sioncke, L. Witters, H. Mertens, L. Ragnarsson, M. Heyns, N. Collaert, A. Thean and A. Steegen, "BTI reliability of advanced gate stacks for Beyond-Silicon devices: challenges and opportunities", IEDM Tech. Dig. (2014) 828.

[10] M. Ke, M. Takenaka and S. Takagi, "Properties of slow traps of ALD Al_2O_3/GeO_x/Ge nMOSFETs with plasma post oxidation" Appl. Phys. Lett., 109 (2016), 032101.

[11] R. Zhang, T. Iwasaki, N. Taoka, M. Takenaka, and S. Takagi, "Suppression of ALD-Induced Degradation of Ge MOS Interface Properties by Low Power Plasma Nitridation of GeO₂", J. Electrochem. Soc. 158(8), G178–G184 (2011).

[12] L. Nyns, D. Lin, G. Brammertz, F. Bellenger, X. Shi, S. Sioncke, S. Van Elshocht, and M. Caymax, "Interface and Border Traps in Ge-based Gate Stacks", ECS Transactions, 35 (3) 465-480 (2011).

[13] M. Ke, X. Yu, R. Zhang, J. Kang, C. Chang, M. Takenaka and S. Takagi, "Fabrication and MOS interface properties of ALD AlYO₃/GeO_x/Ge gate stacks with plasma post oxidation" Microelectron. Eng., 147 (2015)244.

[14] C.-Y. Chang, K. Endo, K. Kato, C. Yokoyama, M. Takenaka, and S. Takagi, "Impact of La_2O_3/InGaAs MOS Interface on InGaAs MOSFET Performance and its Application to InGaAs Negative Capacitance FET", IEDM Tech. Dig. (2016) 322.

Isolation of Nanowires made on bulk wafers by Ground Plane Doping

R. Ritzenthaler, H. Mertens, A. De Keersgieter,
J. Mitard, D. Mocuta, and N. Horiguchi

imec, Kapeldreef 75, 3001 Leuven, Belgium
contact e-mail: romain.ritzenthaler@imec.be

Abstract— **In this work, the electrical isolation of nanowires fabricated on bulk wafers is investigated. It is shown that electrical isolation can be realized with a Ground Plane isolation implant at the beginning of the process flow. For transistors using extensions, it is seen that a relatively high dose of Ground Plane doping is needed in order to avoid punchthrough through a parasitic channel less controlled by the gate (compared to the nanowires electrostatic control). However, the minimum reachable off state leakage current I_{OFF} is also increased for higher Ground plane doping due to junction leakage increase, and may impose a device trade-off between the reachable I_{OFF} target and the short channel effects control. In order to alleviate this issue, it is proposed to skip extensions in order to reduce the dependence of short channel effects to Ground Plane doping dose. Experimental NMOS and PMOS extensionless nanowires are demonstrated without ground plane doping, and feature no I_{ON}/short channel control penalty compared to reference transistors (using extensions).**

Keywords-component; CMOS, Extensions implant, Ground plane doping, Nanowire transistors, Parasitic channel isolation, Scaling.

I. INTRODUCTION

With its better electrostatic control of the gate on the silicon body, nanowires (also called Gate All Around (GAA) transistors) are considered as enablers for future transistors scaling [1]. Horizontal nanowires [2-4] can be considered as a natural evolution of the FinFETs transistors already commercialized by industry.

While devices made on bulk wafers have the advantage of using standard wafers, in that case a planar parasitic channel (**Fig. 1**) is however naturally formed and needs to be isolated. In this work, we investigate the isolation of the parasitic channel by means of Ground Plane implantation. In II, device trade-offs and process windows are highlighted using TCAD (Technology Computer-Aided Design). In III, a strategy to mitigate junction leakage is developed, and an experimental demonstration without loss of drive current/short channel control with regards to reference wafers is done.

II. PARASITIC CHANNEL ISOLATION BY GROUND PLANE DOPING

A. Process

The process flow of the nanowires transistors used in this work is shown in **Fig. 2**. First, SiGe/Si epitaxial layers are grown and patterned. Dummy gate oxide and dummy gate are then deposited and patterned. Next, extensions/Halos if any are implanted, spacers formed, and S/D epitaxial modules grown. After, Interlayer Dielectrics (ILD0) is deposited, and a CMP

(Chemical Mechanical Polishing) is done in order to open the dummy gate. After dummy gate removal, the nanowires are released using a selective SiGe etch [5]. In order to form the final gate stack, gate dielectric and CMOS dual workfunction metal layers are then deposited. The end of processing consists in standard back-end and metal deposition.

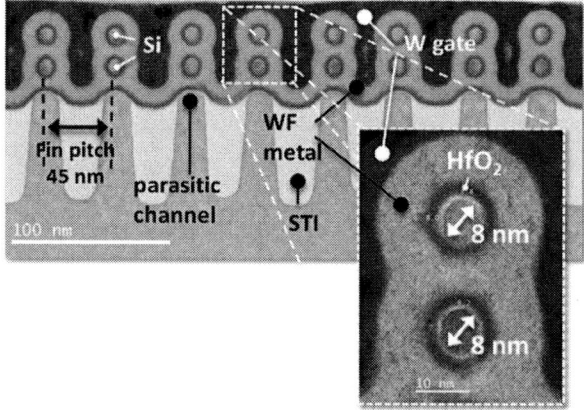

Fig. 1: TEM transversal cross-section of experimental nanowires. The nanowires diameter is 8 nm, and the parasitic channel is clearly visible below the Si nanowires.

On bulk wafers, there is a strong need to isolate the parasitic channel, which is planar by nature (**Fig. 1**). If not performed properly, the short channel control (subthreshold slope, DIBL, I_{OFF}) of the whole structure (nanowires + parasitic channel) may be set by the degraded performance of the parasitic channel and therefore compromise the better electrostatics enabled by nanowires.

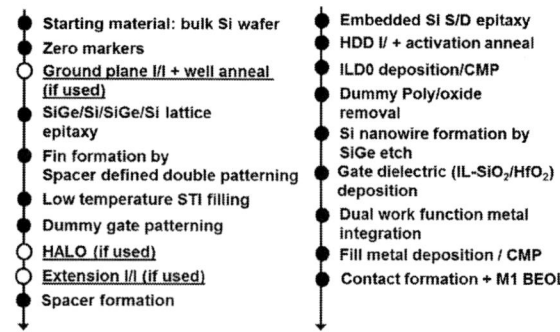

Fig. 2: CMOS process flow for the transistors used in this lot. Ground plane and extensions/Halo doping steps are highlighted.

A solution can be engineered by increasing the parasitic channel threshold voltage in order to make the subthreshold characteristics of the whole structure set by the nanowires electrostatics (**Fig. 3**). This can easily be done with a Ground Plane doping implantation, which is basically a counter doping of the parasitic channel in order to increase its threshold voltage and reduce its subthreshold slope. This solution has the advantage of simplicity, with untitled implants in the wells at the very beginning of the process flow (with implants aligned to the zero markers). A potential drawback is the diffusion of counter dopants in the SiGe/Si epitaxial layers later on, which would results in a dramatic loss of channel mobility (in the Si nanowires) or changes in the SiGe etch rate potentially compromising the selective nanowires release step. It is however shown in [6] that the Ground Plane implant tail is not diffusing into the nanowires and that the SiGe etch rate as a function of Ground Plane doping concentration is kept relatively constant.

B. Device trade-off

While extremely difficult to perform experimentally, the electrical decorrelation of nanowires and parasitic channel can be done easily with TCAD (using Sentaurus Process [7]). It is checked (**Fig. 3**) that the threshold voltage of the parasitic channel is 250 mV higher than the nanowires threshold voltage with the used Ground Plane doping conditions, and that the degraded electrostatics from the parasitic channel is not setting the off state current I_{OFF} of the whole structure.

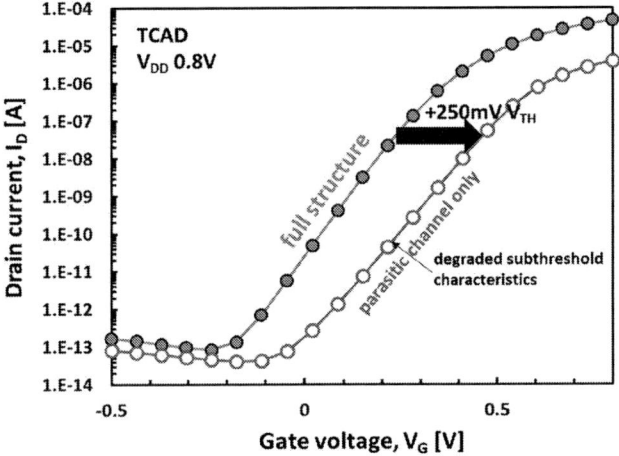

Fig. 3: TCAD simulation showing the drain current I_D vs. gate voltage V_G characteristics of the parasitic channel (open symbols) and of the full structure (closed symbols). The increase of threshold voltage in the parasitic channel after Ground Plane doping is exemplified. Gate length L_G=25 nm.

The deactivation of parasitic channel requires a certain dose of ground plane doping to be effective when extensions are used (extensions implant conditions used here: P 2keV 10^{15} cm^{-2}, Q2, tilt angle 30°, **Fig. 4**). If the ground plane doping dose is too low, the parasitic channel is open at V_G=0V and the whole structure exhibits very degraded subthreshold characteristics (or is even in punchthrough). This has been also observed in experimental transistors (see for example **Figs. 9-11** in [6]). However, the dose amount of Ground Plane doping needed to deactivate the parasitic channel and ensure good short channel control is fairly high (Ground Plane doping dose > 10^{14} cm^{-2}). As a result, the junction in the parasitic channel is built on highly doped regions (Source/Drain, Extensions, Halos, and

channel). The electric field at the PN junction is therefore increasing, resulting in a higher reverse bias diode leakage. This is exemplified in TCAD simulations (**Fig. 5**), where it is seen that the minimal off-state leakage current I_{OFF} that can be reached is significantly increased by an increase of the Ground Plane doping dose. **Fig.6** shows that leakage is occurring not only in the gate vicinity but on the full surface of the junction (therefore discarding a pure GIDL effect).

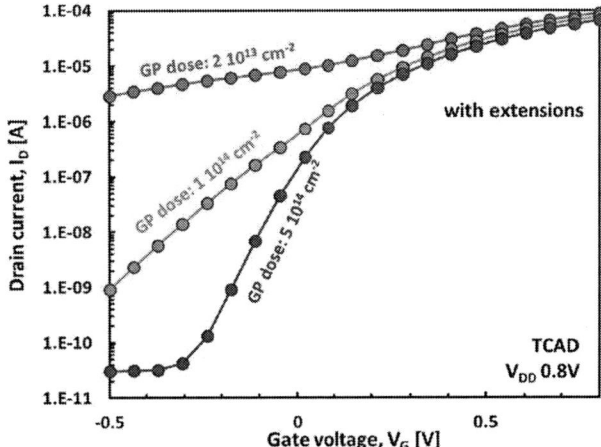

Fig. 4: TCAD simulations of NMOS $I_D(V_G)$ characteristics in the case where extensions are used to dope the nanowires. The need to use high dose Ground Plane (GP) doping in order to avoid the parasitic channel punchthrough is shown. Gate length L_G=25 nm.

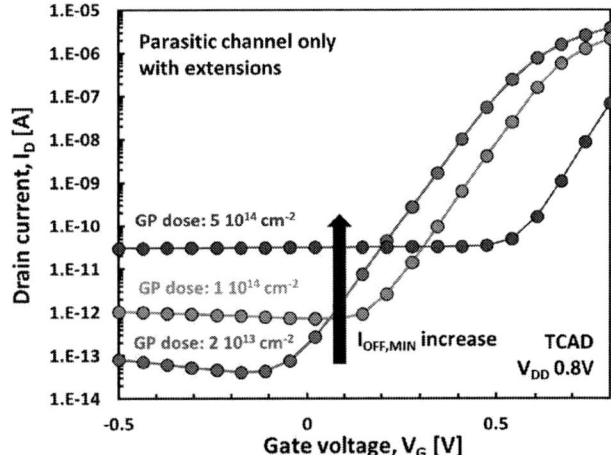

Fig. 5: TCAD simulation of NMOS $I_D(V_G)$ characteristics showing the minimum attainable I_{OFF} leakage increase occurring when the Ground Plane (GP) doping dose is increased. Gate length L_G=25 nm.

As a result, for low power targets the available Ground Plane doping dose is limited by the I_{OFF} increase due to junction leakage. A trade-off between Short channel control and I_{OFF} control must therefore be found, and TCAD simulations show that the Ground Plane doping dose available for a LSTP I_{OFF} target of 10^{-11} A/μm is not sufficient to control SCEs for a target gate length L_G of 24nm (**Fig. 7**). Therefore, alternative solutions must be pursued.

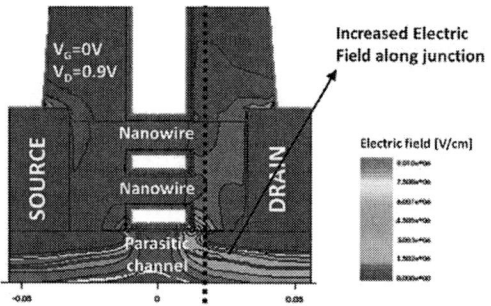

Fig. 6: 2D TCAD cross-section along the Source/Drain axis of a NMOS transistor at V_G=0V and V_D=0.9V for a Ground Plane doping dose of 10^{14} cm^{-2}. The electric field in the parasitic channel is distributed along the whole junction, showing that the I_{OFF} increase in the accumulation regime is related to junction leakage.

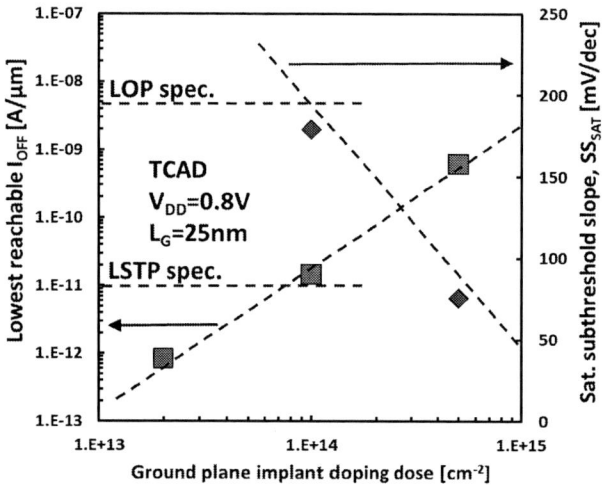

Fig. 7: Minimum attainable off leakage $I_{OFF,MIN}$ and saturation subthreshold slope SS_{SAT} obtained by TCAD simulations vs. Ground plane doping, when extensions are used to dope the nanowires junctions. It is shown that the Ground Plane doping dose necessary to ensure a proper short channel control also leads to junction leakage level incompatible with LSTP I_{OFF} targets. Gate length L_G=25 nm.

III. EXTENSIONLESS DEVICES

A. Junction leakage reduction

In the previous section, it has been seen that the available Ground Plane doping dose is limited by junction leakage for low power I_{OFF} targets. The need for Ground plane doping is coming from extensions implantation and Source/Drain regions lateral diffusion into the parasitic channel. While the Source/Drain height can be controlled (mostly by controlling the recess depth), the case of extensions is thornier since an implant tilt should be used in order to dope the nanowires with the best possible conformity. As a result, it is unavoidable when extensions implants are used to avoid doping the parasitic channel.

One option is therefore to skip extensions, and to let the Source/Drain doping set the junction doping profile. It is noteworthy that skipping extensions may also mitigate the implant knock-in induced Si/SiGe intermixing which might reduce nanowires diameter in the extension regions after nanowires release **[8,9]**. A potential drawback to the

extensionless approach is the need to control very well the dimensions of thinner spacers, resulting in a dramatic drive current loss through external resistance increase if not properly controlled. While allowing to reduce the Ground Plane doping dose and to reach low I_{OFF} levels, the erosion of the threshold voltage margin (defined as the difference of threshold voltage between the nanowires and the parasitic channel) may also lead to a degradation of short channel control in the parasitic channel. TCAD simulations (**Fig. 8**) suggest that short channel control for extensionless devices is set by nanowires and is therefore insensitive to Ground Plane doping. The need to dope the parasitic channel is therefore greatly reduced. Simulated V_{TH} margin (defined as the difference between the parasitic channel and the nanowires threshold voltages) is also maintained with the reduction of ground plane doping (**Fig. 9**), due to an effective gate length L_{EFF} increase of the parasitic channel with junction engineering which reduces the roll-off of the parasitic channel.

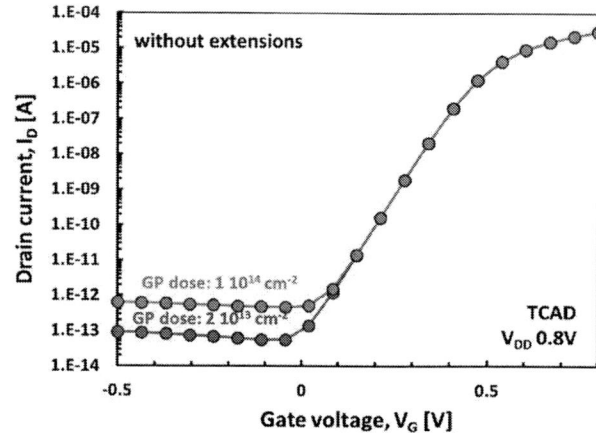

Fig. 8: NMOS TCAD simulations of $I_D(V_G)$ characteristics in the case where extensions implants are _not_ used to dope the structure. The insensitivity of the subthreshold characteristics with the variation of Ground Plane doping dose is shown, as well as the reduction of junction leakages when the Ground Plane doping dose is reduced. Gate length L_G=25 nm.

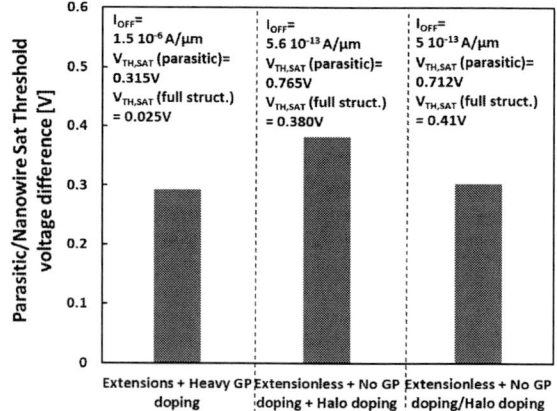

Fig. 9: Simulated threshold voltage margin (defined as the difference of threshold voltage between the nanowires and the parasitic channel) for the NMOS "Extensions+High Ground Plane doping dose" and "Extensionless+No Ground Plane doping" schemes. It is shown that the threshold voltage margin is maintained in extensionless devices even when no ground plane doping is used. Gate length L_G=25 nm.

B. Experimental validation

Extensionless transistors without Ground Plane doping were fabricated using the flow shown in **Fig. 2** and compared to transistors featuring extensions. By carefully controlling the spacers dimensions, it is checked that better drive current can be obtained with extensionless devices for both NMOS and PMOS devices, highlighting the integration of devices without external resistance penalty (**Fig. 10**). Simultaneously, it is checked that the short channel margin of extensionless devices without Ground plane doping is not degraded, as seen in **Fig. 11**. Therefore, it is checked experimentally that the extensionless integration of NMOS and PMOS greatly reduce the need for Ground Plane doping in nanowires made on bulk Si wafer, without drive current or short channel control penalty.

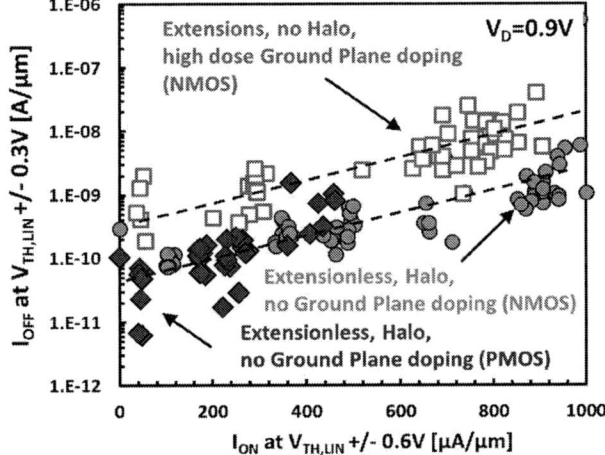

Fig. 10: I_{ON}/I_{OFF} plot for GAA devices with (squares) and without (circles: NMOS; diamonds: PMOS) extension ion implantations (current normalization made by channel perimeter). V_{DD}=0.9V, L_G from 24 nm up to 1 μm. It is shown that the combination "Extensionless+No Ground Plane doping" exhibit better I_{ON}/I_{OFF} performance than the "Extensions+High Ground Plane doping dose" scheme.

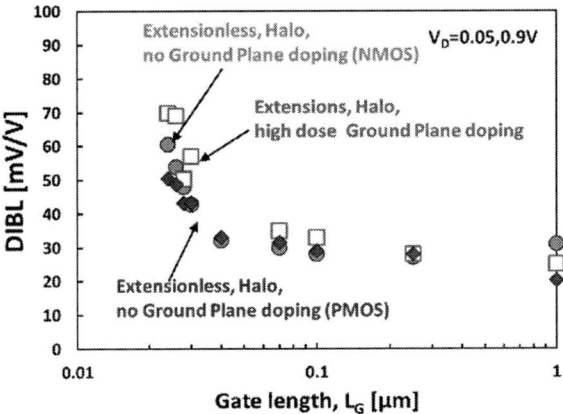

Fig. 11: DIBL vs. L_G for GAA devices with (squares) and without (circles: NMOS; diamonds: PMOS) extension ion implantations (current normalization made by channel perimeter). V_{DD}=0.9V, L_G from 24 nm up to 1 μm. It is shown that "Extensionless+No Ground Plane doping" scheme do not degrade the short channel margin, and is insensitive to the Ground Plane doping dose (highlighting a nanowires driven short channel control).

IV. CONCLUSIONS

In this work, the device trade-offs related to the isolation of parasitic channel for Gate-All-Around devices made on bulk wafers are highlighted. It is shown that an electrical isolation of the parasitic channel is necessary in order to guaranty good short channel performance of nanowires, and that it can be realized relatively easily by using a Ground Plane implantation of the parasitic channel at the beginning of the process flow.

For transistors using extensions, it is seen that a relatively high dose of Ground Plane doping is needed in order to avoid short channel control degradation or punchthrough through a parasitic channel less controlled by the gate (compared to the nanowires electrostatics). However, the minimum reachable off state leakage current I_{OFF} is also increased for higher Ground plane doping due to the junction current increase. It forces a trade-off for low power devices where low I_{OFF} levels are mandatory. In order to mitigate this problem, it is proposed to skip extensions in order to reduce the dependence of short channel effects to Ground Plane doping. Experimental NMOS and PMOS extensionless nanowires without ground plane doping are demonstrated and feature identical I_{ON}/short channel control compared to reference transistors (using extensions).

ACKNOWLEDGEMENTS

The imec sub-22 nm program members, the imec pilot line and amsimec electrical characterization facilities are greatly acknowledged for their support.

REFERENCES

[1] K. J. Kuhn, "Considerations for ultimate CMOS scaling," *IEEE Trans. Electron Devices*, vol. 59, no. 7, pp. 1813-1828, Jul. 2012.

[2] T. Ernst *et al.*, "Novel 3D integration process for highly scalable Nano-Beam stacked-channels GAA (NBG) CMOSFETs with HfO2/TiN gate stack", *IEDM Tech. Dig. 2006*, 2006.

[3] S.-G. Hur *et al.*, "A Practical Si Nanowire Technology with Nanowire-on-Insulator structure for beyond 10nm Logic Technologies", *IEDM Tech. Dig. 2013*, p. 649.

[4] I. Lauer *et al.*, "Si Nanowire CMOS Fabricated with Minimal Deviation from RMG FinFET Technology Showing Record Performance", *2015 Symposium on VLSI Technology Digest of Technical Papers*, p.140.

[5] V. Destefanis et al., "High pressure in situ HCl etching of $Si_{1-x}Ge_x$ versus Si for advanced devices", *Semicond. Sci. Technol.*, vol. 23, p. 105019 (2008).

[6] H. Mertens *et al.*, "Gate-All-Around MOSFETs based on Vertically Stacked Horizontal Si Nanowires in a Replacement Metal Gate Process on Bulk Si Substrates", *2015 Symposium on VLSI Technology Digest of Technical Papers*.

[7] Sentaurus Process User Guide, L-2016.03, Synopsys (2016).

[8] H. Mertens *et al.*, "Vertically Stacked Gate-All-Around Si Nanowire CMOS Transistors with Dual Work Function Metal Gates", *IEDM Tech. Dig. 2013*, p. 649.

[9] W. Vandervorst et al., "A thermal germanium migration in strained silicon layers during junction formation with solid-phase epitaxial regrowth", *Appl. Phys. Lett.*, vol. 86, p. 081915 (2005).

978-1-5090-5979-9/17 $31.00 © 2017 IEEE

In-depth electrical characterization of carrier transport in ambipolar Si-NW Schottky-barrier FETs

Dae-Young Jeon [1,2,3,4]*, Tim Baldauf [5], So Jeong Park [1,2,3,6], Sebastian Pregl [2,7], Larysa Baraban [7], Gianaurelio Cuniberti [7], Thomas Mikolajick [1,2,3], and Walter M. Weber [2,3]

[1]Chair of Nanoelectronic Materials, TU Dresden, Noethnitzer Strasse 64, 01187 Dresden, Germany
[2]Namlab gGmbH, Noethnitzer Strasse 64, 01187 Dresden, Germany
[3]Center for Advancing Electronics Dresden (CfAED), 01062 Dresden, Germany
[4]Institute of Advanced Composite Materials, Korea Institute of Science and Technology, Joellabuk-do 55324, Korea
[5]University of Applied Sciences Dresden, Friedrich-List-Platz 1, 01069 Dresden, Germany
[6]School of Electrical Engineering, Korea University, Seoul 136-701, Korea
[7]Institute for Materials Science and Max Bergmann Center of Biomaterials, 01062 Dresden, Germany
*E-mail: dyjeon@kist.re.kr

Abstract—In this paper the operation mechanism of ambipolar Si-nanowire (Si-NW) Schottky-barrier (SB) FETs is discussed in detail using temperature dependent current-voltage (I-V) contour maps. Thermionic and field emission mechanism limited the overall conduction behavior of ambipolar Si-NW SB-FETs with considerable SB-height. However, Si-channel dominant transports with phonon scattering mechanism occur even in the SB based device at a specific bias condition, where charge carrier injection is saturated with a very thinned SB. Temperature dependent transconductance (g_m) behavior, TCAD simulation and extracted activation energy (E_{ac}) maps also support the explained operation principle of ambipolar Si-NW SB-FETs.

Keywords—*Schottky-barrier transistors; ambipolar behavior; I-V contour map; operation mechanism; transconductance; TCAD; activation energy map.*

I. INTRODUCTION

Nanowire based Schottky-barrier (SB) field-effect transistors (FETs) have been considered for diverse applications due to their inherent electric behavior and simplicity in fabrication. The applications range from printed pH-, bio- sensors [1-3] and as a basis for reconfigurable electronics [4-8] and steep slope devices [6,9]. These applications profit structurally and technologically from a steep junction profile between Si-channel and metallic silicides with an abruptness on the atomic level [10] as well as no need of source/drain (S/D) doping followed by high temperature process [11]. For most metal silicide contacts, ambipolar characteristics of charge carrier transport are observed in SB-FETs due to the injection of both electrons and holes into the semiconductor. Reconfigurable devices make use of the ambipolarity with an additional polarity gate to program electrically unipolar n- and p-type characteristics [4, 12-14]. This leads to new combinatorial circuit topologies where the amount of devices and chip area can be reduced compared to conventional CMOS [8,15]. For the proper operation of the devices, it is important to know what the dominating transport mechanisms are.

In this paper, temperature dependent I-V contour maps ranging from 100 K to 350 K were obtained to get a deeper

insight into the operation mechanism of ambipolar Si-nanowire (Si-NW) SB-FETs and more quantitative information about their performance. SB dominant or Si-channel limited conduction regimes in ambipolar Si-NW SB-FETs are discussed with temperature dependent drain current (I_d), transconductance (g_m) behavior and activation energy (E_{ac}) contour maps. Moreover, TCAD device simulations were carried out to verifying possible conduction mechanisms according to different drain (V_d) and gate (V_g) bias conditions. Figure 1 shows the schematic architecture and typical SEM image of fabricated ambipolar Si-NW SB-FETs with a sharp NiSi$_2$/Si-NW junctions at the S/D sides. Si-NW growth method, detail fabrication process and device structures including a well-defined NiSi$_2$/Si-NW interface of ambipolar Si-NW SB-FETs have been described in previous works [16,17].

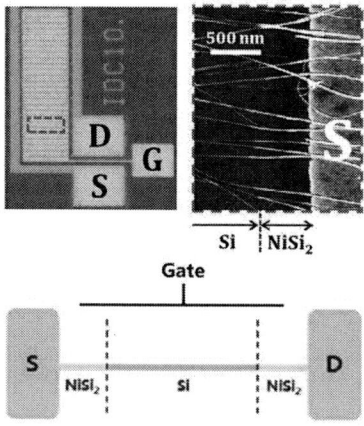

Fig 1. Illustration and SEM image of ambipolar Si-NW SB-FETs with designed channel length of L = 10 μm. In SEM image, bright nanowire segments are NiSi$_2$ and dark ones are Si.

II. TEMPERATURE DEPENDENT I-V CONTOUR MAPS

Temperature dependent I-V contour maps of ambipolar Si-NW SB-FETs, enable a simple distinguishing between electron and hole dominant conduction regimes by varying drain and gate bias, see figure 2. Log(I_d) was plotted as a

978-1-5090-5979-9/17 $31.00 © 2017 IEEE

function of V_d and V_g and diverse color in the I–V map means different amplitude of $\log(I_d)$. Thermionic (TE), thermionic-field (TFE) and field emission (FE) limited conduction regimes can be identified in the I-V map and assigned to different colors such as blue, green and red [16]. Indeed, for decreasing temperature, the TE (blue) regime is significantly widened since thermionic emitted charge carriers overcoming the Schottky or potential barrier can be reduced dramatically at lower temperature. On the other hand, there is only a slight change on FE (red) regime denoting temperature independent field-emission mechanism through a thinned SB.

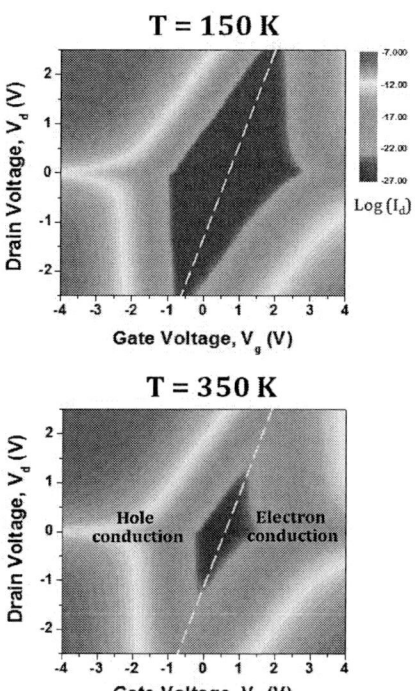

Fig 2. I-V contour maps of ambipolar Si-NW SB-FETs measured at different temperatures.

III.   DOMINANT CONDUCTION MECHANISMS

The different transport regimes are characterized next. Although operation of ambipolar Si-NW SB-FETs is generally governed by the charge carrier transmissibility through the SB, Si-channel limited conduction might also appear as in conventional Si-FETs. This occurs when the resistance caused by the Schottky-junctions becomes smaller than that of the Si-channel, i.e. for SB with a high tunneling probability given by a very thinned barrier-width at a specific bias condition. Figure 3 shows temperature dependent transfer curves with a small drain bias of - 100 mV. With reducing temperature, the drain current is decreasing due to thermionic emission mechanism showing considerable SB effects at the considered bias conditions. However, the trend changes for large positive bias of 2 V, where the drain current is enhanced interestingly with decreasing temperature as shown in figure 4. The band bending at the contacts is strong giving a thinner SB-width and consequently higher barrier transmissibility compared to the smaller biases.   Effective SB-width for hole transport could reach a minimum value. Consequently the contact

resistance induced by the SB is smaller than that of the Si-channel. In addition, hole injection from the drain side saturates with the high drain bias. Phonon scattering dominant conduction in the Si-channel is visible in figure 4 even in the SB based device, since the resistance of the Si-channel is larger than that of the very thinned SB. Furthermore, the zero temperature coefficient (ZTC) point   similar to conventional Si-FETs is also revealed in figure 4 for hole conduction case. This is typically an indication that carrier mobility and threshold voltage increase with reducing temperature [18]. However, the drain current in the electron transport regime is always increasing with increasing temperature without showing the ZTC point within the bias conditions, due to higher SB-height for electrons ($q\varphi B_e \approx 0.66$ eV) compared to the one for holes ($q\varphi B_h \approx 0.46$ eV). The transition border between SB limited conduction and channel limited conduction in SB-FETs is strongly related to the initial SB-height, channel resistance and external bias conditions.

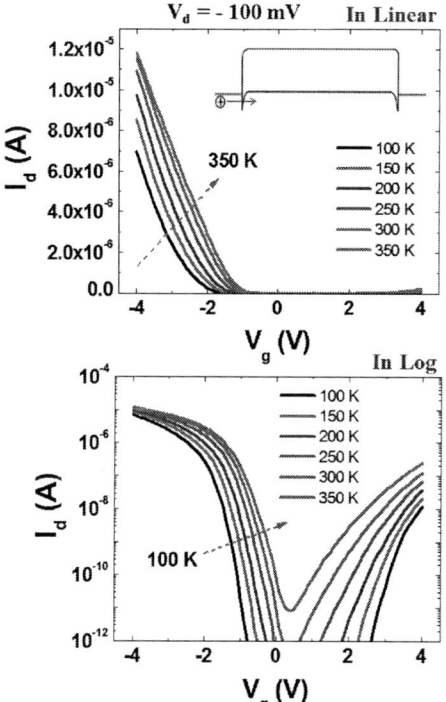

Fig 3. Temperature dependent transfer curves of ambipolar Si-NW SB-FETs with small drain bias of – 100 mV.

IV.   TCAD SIMULATIONS

Figure 5 represents results of TCAD device simulations by considering tunneling through the SB, phonon scattering and surface roughness scattering effects between Si-channel and insulator. A similar trend, such as Si-channel limited conduction behavior with phonon scattering and ZTC point for varying temperature in the p-conduction case, was obtained at high drain bias supporting the experimental observations.

V.   TRANSCONDUCTANCE DEGRADATION

SB based transistors don't generally show a degradation of transconductance (g_m) at large gate overdrive as shown in

978-1-5090-5979-9/17 $31.00 © 2017 IEEE   305

figure 6(a), since carrier injection is enhanced continuously with increasing gate bias [19-20]. However, the g_m degradation caused by surface roughness scattering and series resistance can be observed when carrier injection through the SB is saturated or pinned [16]. Temperature dependence of g_m in figure 6(a) with small drain bias denotes that there are still dominant thermionic emission effects with a considerable SB. On the other hand, in case of high drain bias condition in figure 6(b), a peak and degradation of g_m can be observed and the amplitude of g_m is increased due to reduced phonon scattering with decreasing temperature like in conventional Si-FETs without SB. The high positive drain bias could result in saturation of hole injection with a very thinned SB-width and then, the Si-channel limited conduction behavior appears even in the ambipolar Si-NW SB-FETs.

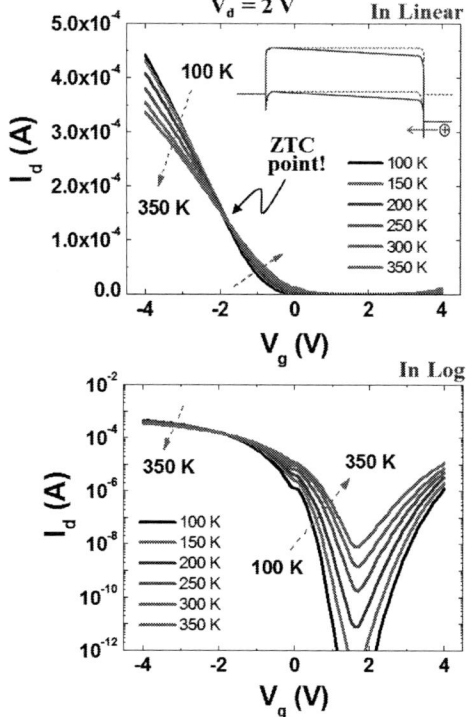

Fig 4. Temperature dependent transfer curves of ambipolar Si-NW SB-FETs with high drain bias of 2 V.

VI.   ACTIVATION ENERGY MAP

For a more detail analysis on the operation mechanism of ambipolar Si-NW SB-FETs, activation energy maps (E_{ac}) were obtained by applying the Arrhenius equation to I-V maps with different temperatures. The result is shown in figure 7. E_{ac} represents the effective thermal energy barrier of the complete system and can thus be correlated in distinct cases to the actual SB-height and to the transmissibility through the barriers. In addition, it shows the regions where potential barriers dominate. The behavior significantly depends on drain bias as well as gate bias. For increasing amplitude of gate bias, both the Schottky and potential barrier decrease. Then, these induce lowering of E_{ac} and reduction of thermionic effects. Interestingly a region of negative E_{ac} was observed at high gate bias regime and is shown in gray color.

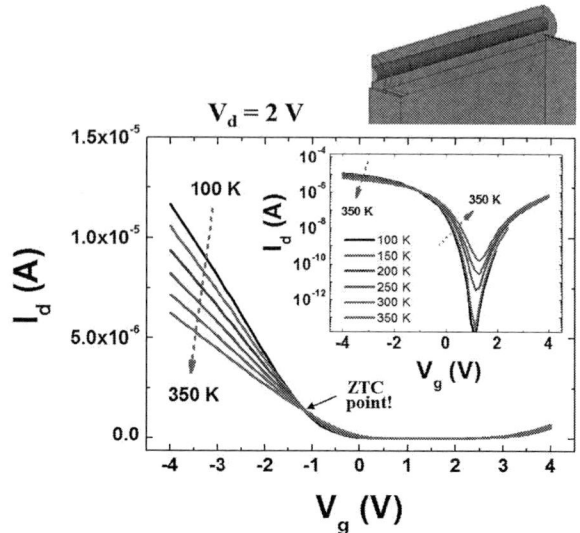

Fig 5. Calculated temperature dependent transfer curves with high drain bias by using TCAD simulation.

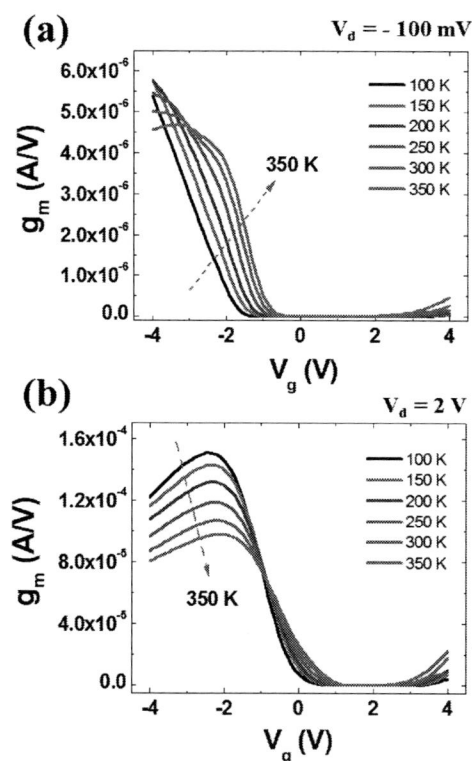

Fig 6. Measured temperature dependence of transconductance in ambipolar Si-NW SB-FETs with different drain bias.

The negative value denotes that thermionic emission is not the dominant conduction mechanism. Possibly Si-channel dominant conduction with saturated hole injection induced by a minimum value of the SB at high gate bias is responsible for this behavior. In addition, the gray regime is more pronounced at high positive drain bias, where phonon scattering dominant conduction behavior has been shown with ZTC point and g_m degradation in figure 4 and 6(b).

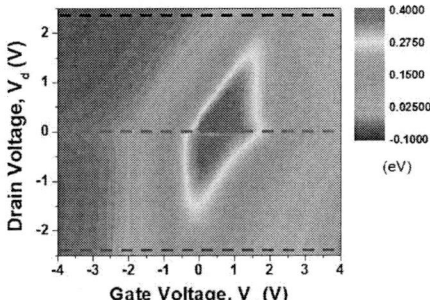

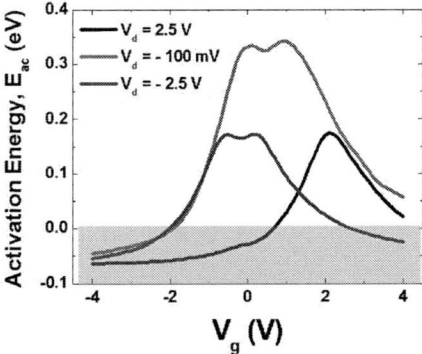

Fig 7. Activation energy (E_{ac}) map obtained from temperature dependent I-V maps by following equation: $\ln(I_d/T^2) \approx -q/kT \times E_{ac}(V)$, where T and k mean absolute temperature and Boltzmann constant, respectively.

VII. CONCLUSIONS

Charge carrier transport mechanism in long ambipolar Si-nanowire (Si-NW) Schottky-barrier (SB) FETs was investigated through an in-depth study of current-voltage (I-V) map with different temperatures ranging from 100 K to 350 K. Different conduction regimes governed by thermionic or field emission were clearly distinguished from temperature dependent behavior of the I-V map. Interestingly, Si-channel dominant conduction behavior, even in the SB based device, was observed with zero temperature coefficient (ZTC) point and transconductance (g_m) degradation at specific bias conditions, since the SB-width reaches a minimum value with saturated charge carrier injection. Those results were verified by TCAD simulation and correspondingly a negative value regime of activation energy (E_{ac}) related to phonon scattering limited operation was also observed. These results are important for the proper design of transducers, sensors and devices for flexible electronics made of Si-NWs with Schottky-junctions.

ACKNOWLEDGMENT

This work was supported by the DFG projects Repronano II (WE 4853/1-2, WE 4853/1-3 and MI 1247/6-2), and the DFG cluster of Excellence Center for Advancing Electronics Dresden (CfAED), Basic Science Research Program through the National Research Foundation of Korea(NRF) funded by the Ministry of Education(2016R1A6A3A11933511) and a Korea University Grant.

REFERENCES

[1] F. M. Zorgiebel, S. Pregl, L. Romhildt, J. Opitz, W. Weber, T. Mikolajick, et al., "Schottky barrier-based silicon nanowire pH sensor with live sensitivity control," Nano Research, vol. 7, pp. 263-271, 2014.

[2] D. Karnaushenko, B. Ibarlucea, S. Lee, G. Lin, L. Baraban, S. Pregl, et al., "Light Weight and Flexible High?Performance Diagnostic Platform," Advanced healthcare materials, vol. 4, pp. 1517-1525, 2015.

[3] J. Schuett, B. Ibarlucea, R. Illing, F. Zoergiebel, S. Pregl, D. Nozaki, et al., "Compact nanowire sensors probe microdroplets," Nano Letters, vol. 16, pp. 4991-5000, 2016.

[4] M. De Marchi, D. Sacchetto, S. Frache, J. Zhang, P.-E. Gaillardon, Y. Leblebici, et al., "Polarity control in double-gate, gate-all-around vertically stacked silicon nanowire FETs," in Electron Devices Meeting (IEDM), 2012 IEEE International, 2012, pp. 8.4. 1-8.4. 4.

[5] A. Heinzig, T. Mikolajick, J. Trommer, D. Grimm, and W. M. Weber, "Dually active silicon nanowire transistors and circuits with equal electron and hole transport," Nano letters, vol. 13, pp. 4176-4181, 2013.

[6] J. Zhang, M. De Marchi, P.-E. Gaillardon, and G. De Micheli, "A Schottky-barrier silicon FinFET with 6.0 mV/dec subthreshold slope over 5 decades of current," in Electron Devices Meeting (IEDM), 2014 IEEE International, 2014, pp. 13.4. 1-13.4. 4.

[7] T. Krauss, F. Wessely, and U. Schwalke, "Electrostatically Doped Planar Field-Effect Transistor for High Temperature Applications," ECS Journal of Solid State Science and Technology, vol. 4, pp. Q46-Q50, 2015.

[8] M. Raitza, A. Kumar, M. Volp, D. Walter, J. Trommer, T. Mikolajick, et al., "Exploiting Transistor-Level Reconfiguration to Optimize Combinational Circuits", Design, Automation & Test in Europe Conference & Exhibition (DATE), 2017.

[9] D.-Y. Jeon, J. Zhang, J. Trommer, S. J. Park, P.-E. Gaillardon, G. De Micheli, et al., "Operation regimes and electrical transport of steep slope Schottky Si-FinFETs," Journal of Applied Physics, vol. 121, p. 064504, 2017.

[10] M. Simon, A. Heinzig, J. Trommer, T. Baldauf, T. Mikolajick, and W. Weber, "Bringing reconfigurable nanowire FETs to a logic circuits compatible process platform," in IEEE Nanotechnology Materials and Devices Conference (NMDC), 2016 IEEE, 2016, pp. 1-3.

[11] S. M. Sze and K. K. Ng, Physics of semiconductor devices: John wiley & sons, 2006.

[12] T. Ernst, "Controlling the polarity of silicon nanowire transistors," Science, vol. 340, pp. 1414-1415, 2013.

[13] T. Mikolajick, A. Heinzig, J. Trommer, T. Baldauf, and W. Weber, "The RFET: a reconfigurable nanowire transistor and its application to novel electronic circuits and systems," Semiconductor Science and Technology, vol. 32, p. 043001, 2017.

[14] A. Heinzig, S. Slesazeck, F. Kreupl, T. Mikolajick, and W. M. Weber, "Reconfigurable silicon nanowire transistors," Nano letters, vol. 12, pp. 119-124, 2011.

[15] W. M. Weber and T. Mikolajick, "Silicon and Germanium Nanowire Electronics: Physics of Conventional and Unconventional Transistors," Reports on Progress in Physics, 2017.

[16] D.-Y. Jeon, S. Pregl, S. J. Park, L. Baraban, G. Cuniberti, T. Mikolajick, et al., "Scaling and Graphical Transport-Map Analysis of Ambipolar Schottky-Barrier Thin-Film Transistors Based on a Parallel Array of Si Nanowires," Nano letters, vol. 15, pp. 4578-4584, 2015.

[17] S. Pregl, W. M. Weber, D. Nozaki, J. Kunstmann, L. Baraban, J. Opitz, et al., "Parallel arrays of Schottky barrier nanowire field effect transistors: Nanoscopic effects for macroscopic current output," Nano Research, vol. 6, pp. 381-388, 2013.

[18] D.-Y. Jeon, S. J. Park, M. Mouis, S. Barraud, G.-T. Kim, and G. Ghibaudo, "Low-temperature electrical characterization of junctionless transistors," Solid-State Electronics, vol. 80, pp. 135-141, 2013.

[19] L. E. Calvet, "Electrical Transport in Schottky Barrier MOSFETs," Yale University, 2001.

[20] S.-J. Choi, C.-J. Choi, J.-Y. Kim, M. Jang, and Y.-K. Choi, "Analysis of Transconductance (g_m) in Schottky-Barrier MOSFETs," IEEE Transactions on Electron Devices, vol. 58, pp. 427-432, 2011.

ESSDERC 2017 Author Index

A

Abdinia, Sahel 98
Adam, Gina Cristina 74
Agarwal, Tarun 54, 106
Alawein, Meshal 94
Aldegunde, Manuel 184
Alomari, Mohammed 196
Alshahed, Muhammad 196

Ando, Takashi 292
Andrieu, François 252
Aritome, Seiichi 10
Asselberghs, Inge 212
Attarimashalkoubeh, Behnoush 152
Avasarala, Naga Sruti 168
Ayele, Getenet Tesega 264
Ayeres de Sousa, Alexandre 252

B

Badaroglu, Mustafa 188, 240

Baert, Rogier 256

Baines, Yannick 126

Bakeroot, Benoit 130

Balaji, Yashwanth 106

Balan, Viorel 252

Baldauf, Tim 304

Bao, Jerry 240

Baraban, Larysa 304

Barraud, Sylvain 66

Baschirotto, Andrea 30, 62

Batude, Perrine 144, 252

Bawedin, Maryline 140

Beaurepaire, Sylvain 252

Beckers, Arnout 62

Bejenari, Igor 90

Bellando, Francesco 78

Ben Ali, Khaled 236

Bernardy, Patrick 42

Besombes, Paul 252

Biscarrat, Jérome 126

Blaise, Philippe 46

Boeuf, Frederic 264

Bonaccorso, Francesco 208

Borghello, Gulio 30

Boschke, Roman 276

Bruley, John 292

Brunet, Laurent 144, 252

Brus, Stephan 212

Bruschini, Claudio 30, 62

Buccella, Pietro 26

Bufler, Fabian 176

Burghartz, Joachim 196

Bylander, Jonas .... No paper

C

Caimi, Daniele 244
Cantatore, Eugenio 98
Cao, Linjun 134
Cartier, Eduard 292
Casse, Mikael 144
Casu, Emanuele Andrea 102, 232
Caymax, Matty 212
Céli, Didier 70
Chan, Chunwa 236
Chan, Po-Wen 240
Charbon, Edoardo 58
Charles, Matthew 126
Chen, Keagan 240
Chen, Shih-Hung 248, 276
Cheng, Cheng-Wei 292

Chiappe, Daniele 106, 212
Chiarella, Thomas 176
Chidambaram, Chidi 240
Chidambaram, Pr Chidi 188
Cho, In-Wook 220
Chombar, Françoise .... No paper
Claus, Martin 90
Clima, Sergiu 168
Cloarec, Jean-Pierre 264
Coccetti, Fabio 216
Coignus, Jean 144
Cosemans, Stefan 256
Cristoloveanu, Sorin 140
Cuniberti, Gianaurelio 304
Czornomaz, Lukas 244

D

Dakran, Mina 196
Dansas, H. 252
Datta, Suman 188, 240
de Keersgieter, An 300
de Souza, Michelly 66
Dehaene, Wim 54, 256
Dehollain, Juan Pablo .... No paper
Delabie, Annelies 212
Delaye, Vincent 252
Deng, Ning 14
Deprat, Fabien 252
Deshpande, Veeresh 244

Detavernier, Christophe 168
Devriendt, Katia 212
Devulder, Wouter 168
Di Piazza, Luca 164
Diels, Wouter 224
DiVincenzo, David .... No paper
Djara, Vladimir 244
Dogmus, Ezgi 228
Donetti, Luca 180
Doria, Rodrigo 66
Dragoni, Alberto 46
Drouin, Dominique 264
Duffy, Ray 288

E

Ecoffey, Serge 264
Elahipanah, Hossein 122
Elmessary, Muhammad 184
Endoh, Tetsuo.................................. No paper
Eneman, Geert.. 240
Eng, Lukas 160
Enz, Christian................................. 30, 62, 78
Esmaeili, Mahsa... 172

F

Famulok, Romain 252

Fariborzi, Hossein 94

Federspiel, Xavier 252

Feng, Peijie 240

Fenouillet-Beranger, Claire 144, 252

Fernandez-Bolaños, Montserrat 232

Fiori, Gianluca 54

Flandre, Denis 148, 236

Florent, Karine 164

Fompeyrine, Jean 244

Frank, Martin M 292

Fraschke, Mirko 216

Furchi, Marco Mercurio 284

G

Gahoi, Amit 110
Gámiz, Francisco 180
Gammon, Peter.................................. 236
Gao, Anran...................................... 268
Gao, Bin 14
García Ruiz, Francisco 180
Garcia-Loureiro, Antonio 184
Garripoli, Carmine 98
Garros, Xavier.................................. 144
Gassilloud, Rémy............................... 144
Gelinck, Gerwin.................................. 98
Ghibaudo, Gérard 144
Ghittorelli, Matteo 98

Ghosh, Ram Krishna 188
Gillot, Charlotte 126
Gity, Farzan................................236, 288
Godoy, Andrés 180
Goux, Ludovic 168
Govoreanu, Bogdan 168
Grap, Thomas 82
Grasser, Tibor 284
Groeseneken, Guido 106, 164, 276
Groven, Benjamin...............................212
Guerin, C...252
Gupta, Charu..................................... 82
Gupta, Mohit.....................................256
Gwoziecki, Romain.............................126

H

Hahn, Herwig 244
Han, Qinghua 268
Haond, M. 148
Hellenbrand, Markus 38
Hellings, Geert 248, 276
Hessel, Andreas 272
Heuken, Lars 196
Heyne, Markus 212

Heyns, Marc 54, 168
Hoffmann, Michael 160
Homulle, Harald 58
Hong, Chris 220
Hopstaken, Marinus 292
Horiguchi, Naoto 176, 240, 300
Hoskins, Brian Douglas 74
Hurley, Paul 288
Huyghebaert, Cedric 106, 212

I

Ibars, F. 252

Ikoma, Ryo 114

Illarionov, Yury Yuryevich 284

Inac, Mesut 216

Incandela, Rosario Marco 58

Inose, Takashi 10

Ionescu, Adrian Mihai 102, 232

Ionescu, Adrian 78

J

Jang, Doyoung 256
Jaoul, Mathieu 70
Jaud, Marie-Anne 126
Jazaeri, Farzan 30, 62, 78
Jech, Markus 284
Jenkins, Keith A 292
Jeon, Dae-Young 304
Jousseaume, Vincent 252
Juncker, A. 212

K

Kabouche, Riad 228

Kalna, Karol 184

Kämpfe, Thomas........................ 160

Kar, Gouri Sankar 168

Kataria, Satender 110

Kavehei, Omid 74

Kawanago, Takamasa 114

Kaynak, Mehmet 216

Kazemi Esfeh, Babak 148

Ke, Mengnan............................... 296

Kerber, Andreas.......................... 134

Kerber, Pranita............................ 292

Kikuchihara, Hideyuki................. 192

Kilchytska, Valeria 148

Kilchytska, Valeriya 236

Kim, Jeeson 74

Kim, Kyung-Hwan........................ 220

Knobloch, Theresia 284

Knoch, Joachim...................... 82, 272

Kohen, David............................... 240

Kovàcs-Vajna, Zsolt 98

Krammer, Anna 232

Kunnen, Eddy.............................. 212

Kwag, Pyong-Su 220

L

Lanza, Mario 86
Larcher, Luca 204
Larrieu, Guilhem 34
Lavizzari, Simone 164
Le Gac, Séverine 260
Leblebici, Yusuf 152
Lee, Dong-Kyu 220
Lee, Heng-Yuan 18
Lee, Ho-Ryeong 220
Lee, Ko-Tao 292
Lee, Kyung Hwa 140
Lemme, Max Christian 110
Leobandung, Effendi 292
Li, Fan 236

Li, Tie 268
Liang, C. 292
Lin, Dennis 106, 212
Lind, Erik 38
Linge, Astrid 228
Linten, Dimitri 248, 276
Lisker, Marco 216
Liu, Rui 18
Lockhart de la Rosa, Cesar J 106
Lorin, Thomas 126
Lu, Anh Khoa Augustin 106
Lu, Cao-Minh Vincent 144
Luong, Gia Vinh 42
Lupina, Grzegorz 216
Lux, Marcel 212

M

Ma, Chenyue 192
Madsen, Jens Kargaard 172
Mai, Andreas 216
Maiti, Tapas K. 192
Majumdar, Amlan 292
Malm, Gunnar 122
Maneux, Cristell 34, 70
Mantl, Siegfried 42, 268
Maqueda Lopez, Mariazel 232
Matsui, Chihiro 6
Mattausch, Hans Juergen 192
Mattiazzo, Serena 30
McAndrew, Colin 22
Medjdoub, Farid 228
Meier, Karlheinz .... No paper
Memišević, Elvedin 38
Meneghini, Matteo 130
Mertens, Hans 240, 300
Micout, Jessy 144

Mikolajick, Thomas 160, 304
Mirabelli, Gioele 288
Miranda, Enrique 86
Mishima, Satoshi 10
Mitard, Jerome 300
Miura-Mattausch, Mitiko 192
Miyaguchi, Kenichi 176
Miyamoto, Hidenori 192
Mo, Renee T 292
Mocuta, Anda 176, 256
Mocuta, Dan 106, 300
Moens, Peter 130
Mohapatra, Nihar Ranjan 200
Monaghan, Scott 288
Monfray, Stephane 264
Moradi, Farshad 172
Mukherjee, Chhandak 34
Müller, Thomas 284
Myny, Kris 276

N

Na, Min-Ki 220
Nagy, Daniel 184
Nail, Cécile 46
Narayanan, Vijay 292
Narimani, Keyvan 268
Navarro, Dondee 192
Nemouchi, Fabrice 252
Ney, D. 252
Ni, Kai 188
Nigam, Tanya 134
Nikonov, Dmitri E. 1
Nili, Hussein 74
Nouguier, D. 252

O

Oba, Tomoaki 114
O'Connor, Eamon 244
Ojha, Apoorva 200
Okada, Etienne 228
Oliva, Nicolo 102
Oliva, Nicolò 232
Opsomer, Karl 168
östling, Mikael 122

P

Palestri, Pierpaolo 42
Pan, Chengbin 86
Parihar, Mukta Singh 140
Park, Hyungjin 140
Park, So Jeong 304
Park, Sung-Kun 220
Parvais, Bertrand 148
Pasini, Luca 144
Pathirana, Vasantha 236
Pavan, Paolo 204
Pavanello, Marcelo 66
Pawlak, Andreas 70
Pecheux, Romain 228

Pelger, Alexander 272
Pesic, Milan 160
Pezard, Julien 34
Pezzotta, Alessandro 30
Pfander, Albert 272
Planes, N. 148
Plissonnier, Marc 126
Popovici, Mihaela 164
Potoms, Goedele 164
Pregl, Sebastian 304
Previtali, B. 252
Proud, F. 252
Puglisi, Francesco Maria 204
Puyoo, Etienne 264

Q

Qian, He.. 14

R

Radu, Iuliana 54, 106, 212
Rafhay, Quentin 144
Raghavan, Praveen .... 54, 256, No paper
Rambal, Nils 144, 252
Ranjan, Rakesh 134
Raskin, Jean-Pierre 148
Raymaekers, Tom 164
Real, Peter .... No paper
Reimbold, Gilles 126
Richter, Claudia 160
Riederer, Felix 82

Ritzenthaler, Romain 240, 300
Ro, Kyoung-Wook 220
Rodriguez, G. 252
Rodriguez, P. 252
Rohbani, Nezam 192
Romana, Giovanii 144
Rossi, Chiara 26
Rozeau, O. 252
Rozen, John 292
Ruffino, Andrea 62
Rupp, Roland 118
Rzepa, Gerhard 284

S

Saeidi, Ali 78
Saghi, Z. 252
Salemi, Arash 122
Sallese, Jean-Michel 26
Sampedro, Carlos 180
Samson, Marie-Pierre 252
Sandrini, Jury 152
Sant, Saurabh 50
Sayan, Safak 212
Schanovsky, Franz 284
Schenk, Andreas 50
Schiffmann, Alexander 192
Scholz, Mirko 248, 276
Scholz, Stefan 272
Schram, Tom 212
Schroeder, Uwe 160
Schröter, Michael 70
Schroter, Michael 90
Schuddinck, Pieter 256
Schüler, Andreas 232
Sebastiano, Fabio 58
Seebacher, Ehrenfried 192
Servalli, Giorgio 156
Shahrabi, Elmira 152
Shen, Tian 134

Sherazi, Yasser 256
Sivaram, Siva .... No paper
Sklénard, Benoît 46
Slesazeck, Stefan 160
Smets, Quentin 106, 212
Smith, Jeffrey 188, 240
Song, Lin 58
Song, Sc. 240
Soree, Bart 54
Souifi, Abdelkader 264
Sousa, Marilyne 244
Spessot, Alessio 256
Stampfer, Bernhard 284
Stefanucci, Camillo 26
Steinmair, Alexander 192
Steudel, Soeren 276
Steyaert, Michiel 224
Stockman, Arno 130
Stolichnov, Igor 78, 102
Stork, Johannes .... No paper
Strangio, Sebastiano 42
Strukov, Dmitri 74
Sun, Xiao 292
Suñe, Jordi 86
Svensson, Johannes 38

T

Tajalli, Alaleh...........................130

Takagi, Hiroyuki 114

Takagi, Shinichi.........................296

Takenaka, Mitsuru 296

Takeuchi, Ken 6, 10

Tamagnone, Michele 232

Tavernier, Filip 224

Thiam, Arame...........................212

Tiedemann, Andreas 42

Torricelli, Fabrizio 98

Trajkovic, Tanya 236

Traoré, Boubacar 46

Trellenkamp, Stefan 42

Trevisoli, Renan 66

U

Uren, Michael.. 130

V

van der Steen, Jan-Laurens 98

van der Veen, Marleen 168

Van Houdt, Jan 164, 168

Vandendaele, William 126

Vianello, Elisa 46

Villena, Marco 86

Vinet, Maud 144, 252

Vitale, Wolfgang Amadeus 102, 232

Vladimirescu, Andrei 58

W

Waltl, Michael 284
Wang, Nian 276
Wang, Shan 14
Wang, Yuelin 268
Weber, Walter M. 304

Weckx, Pieter 256
Wernersson, Lars-Erik 38
Wietstruck, Matthias 216
Woo, Donghyun 220
Wu, Dong 14
Wu, Huaqiang 14

X

Xia, Guangrui 280
Xiao, Na 86
Xu, Jeff 188, 240

Y

Yakimets, Dmitry 256
Yang, Da 240
Yasuhara, Ryutaro 10
Yau, Jeng-Bang 292
Yeh, Chun-Chen 292
Yoo, Kyung-Dong 220
Young, Ian A. 1
Yu, Shimeng 18

Z

Zagni, Nicolò 204

Zegaoui, Malek 228

Zeinali, Behzad 172

Zhang, Chun-Min 30

Zhao, Qing-Tai 42, 268

Zhu, John 240